Microbiology
for the
Health Sciences

FOURTH EDITION

Microbiology for the Health Sciences

MARCUS M. JENSEN
Emeritus Professor of Microbiology
Brigham Young University

DONALD N. WRIGHT
Professor of Microbiology
Brigham Young University

RICHARD A. ROBISON
Associate Professor of Microbiology
Brigham Young University

PRENTICE HALL
Upper Saddle River, New Jersey 07458

Library of Congress Cataloging-in-Publication

Jensen, Marcus M.
    Microbiology for the health sciences / Marcus M. Jensen, Donald N.
Wright, Richard A. Robison.—4th ed.
        p.   cm.
    Rev. ed. of: Introduction to microbiology for the health sciences.
3rd ed. 1993.
    Includes index.
    ISBN 0-13-251464-8
    1. Medical microbiology.   I. Wright, Donald N.   II. Robison,
Richard.   III. Jensen, Marcus M.   Introduction for the health
sciences.   IV. Title.
QR46.J46 1997
616'.01—dc20   96-28137
                                                  CIP

Executive editor: David K. Brake
Development editor: Laura J. Edwards
Editor-in-chief, ESM development: Ray Mullaney
Editorial director: Tim Bozik
Editor-in-chief: Paul Corey
Art director: Heather Scott
Art manager: Gus Vibal
Creative director: Paula Maylahn
Assistant vice-president of production and manufacturing: David W. Riccardi
Executive managing editor: Kathleen Schiaparelli
Assistant managing editor: Shari Toron
Interior design: Circa '86/Carmen DiBartolomeo; Judith A. Matz-Coniglio
Cover design: Heather Scott
Photo research: Mira Schachne
Photo editor: Melinda Reo
Manufacturing manager: Trudy Pisciotti
Production editor: bookworks

Photos in figures 1.5, 28.6, 28.7 and 39.11 are courtesy of Thomas D. Brock.

Human Body Systems Reference insert is from *Fundamentals of Anatomy and Physiology*,
3/E by Martini, © 1995. Adapted by permission of Prentice-Hall, Inc., Upper Saddle River,
NJ.

Color Plate Reference insert photos are courtesy of: Plates 1, 11, 13, 14, 21, 30, and 48:
George Wistreich; Plates 2, 4, 5, 10, 12, 15, 16, 17, 18, 19, 23, 24, 26, 27, 28, 29, 32, 33, 34, 35,
36, 37, 38, 42, 44, 46, and 47 Centers for Disease Control, Atlanta; Plates 6, 8, and 45 J. M.
Matesen, University of Utah; Plates 3, 7, 9, 20, 22, 31, and 43 Donald N. Wright; Plate 25
Laboratory Medicine, Vol. 15:4, April 1984; Plate 39 Howard Hughes Medical Institute;
Plate 40 Burroughs Wellcome, Co., Research Triangel Park, NC; Plate 41 K. F. Bott.

© 1997, 1993, 1989, 1985 by Prentice-Hall, Inc.

Upper Saddle River, New Jersey 07458

Printed in the United States of America
10 9 8 7 6 5 4 3

ISBN: 0-13-251464-8

Prentice-Hall International (UK) Limited, *London*
Prentice-Hall of Australia Pty. Limited, *Sydney*
Prentice-Hall of Canada, Inc., *Toronto*
Prentice-Hall Hispanoamericana, S. A., *Mexico*
Prentice-Hall of India Private Limited, *New Delhi*
Prentice-Hall of Japan, Inc., *Tokyo*
Prentice-Hall Asia Pte. Ltd., *Singapore*
Editora Prentice-Hall do Brasil, Ltda., *Rio de Janeiro*

Brief Contents

Contents

40

Protozoa   493

Preface

Microbiology for the Health Sciences has always been written with the idea that a meaningful introduction to microbiology can be provided for the health-science oriented student in a single semester or quarter of study without demanding earlier exposure to the subject or an extensive background in chemistry and mathematics. Although the text provides the necessary background support for extended study in the discipline, the primary purpose of the text is to provide a useful and basic understanding of the microbe in its role as a disease-producing agent. Thus, the content has been kept at a minimum in an effort to provide only that information essential to an appreciation of microorganisms in the disease process.

Because of its narrower scope, *Microbiology for the Health Science* is intended to be neither an encyclopedic reference of general microbiology nor a detailed analysis of host responses to parasitic microorganisms. Rather, the reader will find that this text provides a succinct, easy-to-read background of the agents of infectious diseases and the human disease processes associated with microorganisms.

In reviewing the fourth edition we have tried to retain the innovative style and student-oriented nature of the first three editions. However, we recognize that advances in the discipline have been rapid and far-reaching, and we've done our best to keep pace in terms of including appropriate updates throughout the text. Our new co-author, Richard Robison, received his doctorate in immunology and has taught the introductory course for many years. His involvement has helped to ensure that the material is up-to-date and accurate in one of the most exciting and rapidly changing fields, immunology.

In a sense, an introductory text like this is in a constant state of revision, because the abundance of scientific discoveries and research milestones of our day demand that we as authors categorize, prioritize, and then integrate new facts and findings into our text—never an easy task, but always a rewarding one.

TAXONOMIC ORGANIZATION

Microbiology for the Health Sciences takes a taxonomic approach to introduce students to the patho-gens of medical significance—that is, the chapters are organized according to taxonomic groups. Within each chapter, important members of the group are discussed in terms of their physical characteristics. This brief introduction to the infectious agent is always followed then by a systematic discussion of the disease(s) it causes: pathogenesis and clinical manifestations; transmission and epidemiology; diagnosis and treatment; and prevention and control. Throughout our many years of teaching, we have found that students are able to assimilate the information more readily in this kind of standardized format.

A **Graphic Sidebar**, which visually depicts the basic characteristics of the group under discussion, is included in the opening page of each organismal chapter. This serves as a quick reference to remind students of the common attributes of members of the group: its basic "cell type"—whether it is a prokaryote, eukaryote, or virus; its Gram morphology (for bacteria); its general shape(s); and its general size or size range.

AN EMPHASIS ON CLINICAL ASPECTS

Designed for students entering nursing or other allied-healthcare professions, this text has always had a clinical focus. However we've introduced several new clinical elements that will help students remember this important aspect of their microbiology course:

Affected Body Systems. This new chapter feature highlights the systems of the body affected by the organism(s) discussed in the chapter. Found near the beginning of each organismal chapter, this provides a quick overview of the clinical discussion ahead.

Disease/Causative Organism Reference and Notifiable Diseases Summary. The printed endpapers of the text feature a quick-reference alphabetical listing of diseases and the organisms that cause them. On the back endpapers, notifiable diseases also summarized by year and numbers of reported cases.

Clinical Notes. One of the most popular chapter features from earlier editions of this text, these real-life accounts—based on the Morbidity and Mortality Weekly Reports (MMWR) issued by the Centers for Disease Control—have been updated and expanded. These valuable boxes focus on pertinent on-going concerns in infectious disease microbiology and provide the student with an opportunity to see the study of microbiology from an applied perspective.

Clinical Summary Tables. Found in the end-of-chapter Material for Review, this new feature provides a succinct review of the clinical content regarding each of the major microbes discussed in the chapter: its virulence mechanisms, the diseases it causes, its mode of transmission, and methods of treatment and prevention.

VOCABULARY DEVELOPMENT

We recognize that building vocabulary is an important task in any scientific discipline. Therefore, we have included the following aids for students:

Key Terms/New Words. The most important words for the chapter are boldfaced where they are defined in the text, while additional terms with which the student might not be familiar (but which are not necessarily "key terms") are italicized. All boldfaced and italicized terms are included in the **Glossary** at the end of the text.

Running Glossary. A running glossary is provided to expand the student's understanding of terms *as they are used.* These bottom-of-the-page entries reduce the time required to read the text and later provide excellent chapter review material for the student

END-OF-CHAPTER REVIEW ELEMENTS

Concept Summary. This numbered list of concepts, written in sentence form, recaps the essential points of the chapter.

Clinical Summary Table. This table provides a quick review of the organisms covered in the chapter and some specifics of the diseases they cause.

Study Questions. These 5–10 questions are designed for students to recall the important facts of the chapter.

Challenge Questions. These 2–3 questions ask the student to go beyond mere memorization of the facts to apply their knowledge to a specific problem.

SUPPLEMENTS

This text is accompanied by a variety of supplements for the student and the instructor.

Combined Instructor's Manual and Test Item File. Each chapter in this resource, developed by Jeffrey Pommerville of Glendale Community College, contains a chapter overview, teaching tips, instructor goals, and student learning objectives—plus answers to the Study and Challenge Questions found in the main text. Additionally, the test item portion contains approximately 30–50 questions for each of the 40 chapters in the book.

Transparency Masters. Transparency masters featuring all of the illustrations from the text is available to qualified adopters of the text.

Student Study Guide. The Study Guide, also developed by Jeffrey Pommerville of Glendale Community College, contains chapter outlines, objectives, vocabulary exercises, and numerous questions and exercises to help students master the material.

Prentice Hall Microbiology Laserdisc. Prentice Hall Microbiology Laser Disc contains more than 2000 images, including full-color micrographs, photographs, illustrations, and animations for use in either a lecture or lab setting. All images are indexed and accessible with or without a bar code scanner. Free to adopters with a minimum of 150 copies.

Prentice Hall/New York Times Themes of the Times. *The New York Times* Themes of the Times consists of selected articles from *The New York Times* dealing with topics related to microbiology. This supplement is updated annually and is available free to adopters, who can order as many copies as the number of new texts purchased.

ACKNOWLEDGMENTS

Our special thanks goes to Jeff Pommerville of Glendale Community College for helping to put final touches on the many new pedagogical features we added in this new edition. We also wish to express our appreciation to the following individuals, who reviewed the manuscript and provided timely insight and suggestions for the text:

Norm Abell
Gateway Community College

Joseph Alls
Northwest Alabama Community College

David Asch
Youngstown State University

Stuart Bradford
Southern Vermont College

Nita Collin
Fort Fange Community College

Richard Crumley
Missouri Western State College

Dan DeBorde
University of Montana

Lawrence Elliott
Western Kentucky University

Helen Foster
Santa Fe Community College

David Gilmore
Arkansas State University

George Heth
St. Louis Community College at Florissant Valley

Afzal Lodhi
St. Louis Community College at Forest Park

Rodney Rogers
Drake University

Don Schnurbush
Independence Community College

Frank V. Veselovksy
South Puget Sound Community College

In addition, we would like to thank the Editorial and Production folks at Prentice Hall—in particular David K. Brake, Executive Editor; Laura J. Edwards, Senior Developmental Editor; Lisa S. Garboski (of bookworks), Production Editor; and Shari Toron, Assistant Managing Editor. The guidance and support of these individuals has been invaluable in the development of this exciting new revision.

Marcus M. Jensen
Donald N. Wright
Richard A. Robison

CHAPTER 15 OUTLINE

15 STREPTOCOCCI

GRAM +

20μm

PROKARYOTE

Graphic Sidebar
Provides visual quick-reference of the common attributes of members of the taxonomic group of the chapter: its basic "cell type"—whether it is a prokaryote, eukaryote, or virus; its Gram morphology (for bacteria); its general shape(s); and its general size or size range.

STREPTOCOCCI: GENERAL CHARACTERISTICS

STREPTOCOCCUS PYOGENES
  PATHOGENESIS AND CLINICAL DISEASES
  TRANSMISSION AND EPIDEMIOLOGY
  DIAGNOSIS
  TREATMENT
  PREVENTION AND CONTROL

STREPTOCOCCUS PNEUMONIAE
  PATHOGENESIS AND CLINICAL DISEASES
  TRANSMISSION AND EPIDEMIOLOGY
  DIAGNOSIS
  TREATMENT
  PREVENTION AND CONTROL

OTHER DISEASE-CAUSING STREPTOCOCCI
  GROUP B
  GROUP C
  GROUP D
  VIRIDANS GROUP

Both pathogenic and nonpathogenic species of streptococci are commonly associated with humans and animals. A wide variety of species are present as normal flora on skin and mucous membranes of all humans. Three species, *Streptococcus pyogenes*, *Streptococcus pneumoniae*, and *Streptococcus agalactiae*, are responsible for most of the streptococcal infections in humans. The pathogenic species produce a wide variety of toxins and cause a wide variety of lesions and diseases. Historically, some streptococcal diseases have been among the most serious diseases of humans. Fortunately, these bacteria are usually easily destroyed by chemotherapeutic agents, and as a result, even though streptococcal infections are still common, their impact on illness and death today is only a small fraction of what it was prior to the 1930s.

Affected Body Systems
Highlights the systems of the body affected by the organism(s) discussed in the chapter. Found near the beginning of each organismal chapter, this provides a quick overview of the clinical discussion ahead.

STREPTOCOCCI: GENERAL CHARACTERISTICS

Streptococci are gram-positive, coccal-shaped bacteria that usually appear in chains of various lengths (Figure 15-1; see Plate 5). These bacteria are moderately resistant to environmental factors; that is, they may remain living for days to weeks after being

expelled from
readily killed
are highly su
therapeutic a
species of the
the skin, nos
humans and
guish clearly

210   Chapter 15 Streptococci

HUMAN BODY SYSTEMS AFFECTED

Streptococci

S. pneumoniae
  ✳ pneumonia
  ✳ pleuritis
S. pyogenes
  ✳ pharyngitis

S. pyogenes
  ✳ glomerulonephritis

S. pneumoniae
  ✳ meningitis
  ✳ otitis media

S. pyogenes
  ✳ impetigo
  ✳ erysipelas
  ✳ necrotizing fascitis

S. pyogenes
  ✳ puerperal fever

SYSTEMIC

S. pyogenes
  ✳ Scarlet fever

Group B streptococci
  ✳ septicemia

Chapter Outline
Lists the topics to be covered in the chapter.

Concept Links
Cue the student that new material is related to or builds on an earlier discussion.

species. This has made a precise classification of these bacteria difficult. Streptococci grow well on blood agar and many species secrete *hemolysins* (enzymes that dissolve red blood cells), which produce patterns of hemolytic zones around the colonies. These hemolytic patterns can be used to make a preliminary identification of streptococcal groups. A clear zone of hemolysis surrounding the colony is called *beta*-hemolysis (∞), a zone with an opaque greenish color is called *alpha*-hemolysis, and some species produce no hemolysis.

The most usable classification system, **the**

Clinical Notes

Provide an applied perspective by focusing on pertinent on-going concerns in infectious disease microbiology. Based on the Morbidity and Mortality Weekly Reports (MMWR) issued by the Centers for Disease Control.

New Artwork

Makes understanding basic concepts of cell biology, genetics and immunology easier. Organismal chapters use art consistently to show pathogenesis of various diseases and graph the incidence of disease occurrence.

CLINICAL NOTE

Outbreak of *Shigella flexneri* 2a Infections on a Cruise Ship

During 29 August–1 September 1994, an outbreak of gastrointestinal illness occurred on the cruise ship *Viking Serenade* (Royal Caribbean Cruises, Ltd.) during its roundtrip voyage from San Pedro, California, to Ensenada, Mexico. A total of 586 (37%) of 1589 passengers and 24 (4%) of 594 crew who completed a survey questionnaire reported having diarrhea or vomiting during the cruise. One death occurred in a 78-year-old man who was hospitalized in Mexico with diarrhea.

Shigella flexneri 2a has been isolated from fecal specimens from at least 12 ill passengers. Antimicrobial susceptibility testing of representative isolates indicated resistance to tetracycline and susceptibility to ampicillin and trimethoprim sulfamethoxazole. The subsequent two cruises of the ship were canceled. Investigation of the mode of transmission is under way (*MMWR* 43:657, 1994).

Shigella species is commonly associated with poor or crowded living conditions. Most of the approximately 20,000 cases occurring annually in the United States (Figure 21-7) are associated with institutionalized individuals, where hygienic conditions may be difficult to maintain because of crowding and lack of individual capabilities. This relationship between an ability to maintain personal hygiene and the frequency of shigella infection is reflected in the age distribution of the disease in the United States, where it is seen most frequently in the pediatric population.

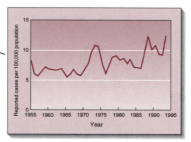

FIGURE 21-7 Reported cases of shigellosis in the United States, 1955–1994. (Courtesy Centers for Disease Control, Atlanta)

Shigellosis is endemic in underdeveloped countries. Historically, dysentery has been a problem in military populations and entire armies have become temporarily disabled when living under unsanitary conditions that commonly exist during wartime. People traveling from countries like the United States often contract bacillary dysentery within a short period after entering a country where dysentery is endemic.

✳ Diagnosis

Diagnosis is made by isolating shigellae from the feces or intestinal tract.

✳ Treatment

In contrast with *Salmonella* gastroenteritis, most cases of shigellosis are improved by chemotherapy. The recent development of *multiresistant* strains of *S. sonnei* (resistant to ampicillin, tetracycline, and trimethoprimsulfamethoxazole) has complicated the approach to therapy, but several available antibiotics remain effective. Oral rehydration and maintaining proper electrolyte balance is an essential component of treatment.

✳ Prevention and Control

Prevention of person-to-person transmission by following good sanitary practices is the most effective means of avoiding shigellosis. Patients with the disease should be isolated.

Multiresistant Bacteria that are resistant to a variety of antibiotics with different mechanisms of antimicrobial action.

---

298 Chapter 21 Enterobacteriaceae

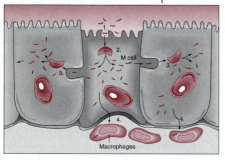

FIGURE 21-6 Invasion of intestinal epithelial cells by *Shigella*. 1. *Shigella* attach to M cells and induce their own phagocytosis. 2. *Shigella* escape the phagocytic vacuole and multiply. 3. Penetration of adjacent epithelial cells with multiplication and cell destruction. 4. *Shigella* released from infected cells are phagocytized by macrophages, thus preventing the spread to deeper tissues.

S. DYSENTERIAE (AND OTHERS): BACILLARY DYSENTERY (SHIGELLOSIS)

✳ Pathogenesis and Clinical Diseases

Following ingestion, the shigellae usually penetrate the large intestine by stimulating the endothelial cells, called M cells, that line the intestine to phagocytize them. However, these "nonprofessional" phagocytes are unable to kill the ingested bacteria and the bacteria then multiply and invade neighboring cells (Figure 21-6). Generally, penetration is not deeper than the submucosal cells. Inflammation, together with sloughing of the epithelial cells, results in ulcerative lesions. After 1 to 3 days of incubation the patient experiences a sudden onset of symptoms—abdominal cramps, fever, and diarrhea. The diarrheal stool frequently contains mucus and blood. Significant loss of water and salts may occur and in young and/or debilitated patients this dehydration and electrolyte imbalance may cause death. In otherwise healthy persons the disease is usually self-limiting and recovery occurs in 3 to 7 days. The death rate from dysentery in young

children is significant in countries with poor sanitation and nutrition.

Infections due to *S. dysenteriae* are always potentially more serious than those due to other species. This organism produces a very powerful exotoxin (*shiga toxin*) that greatly increases its virulence. During a recent epidemic in Central and South America the mortality rate among those infected with this organism was between 8 and 10%. Although not endemic in the United States, this species has recently been introduced by tourists returning from Central America and Mexico. Most residents in areas where dysentery is endemic develop some immunity to the disease either through clinical or subclinical cases. Many such persons, however, remain carriers of the organism and serve as a source of infection for new susceptibles, such as visitors or newborns entering the population.

✳ Transmission and Epidemiology

Transmission is from human to human via the fecal–oral route by "fingers, food, feces, fomites, or flies." Infection can occur with as few as 10 or 10^2 bacteria (this is in contrast to 10^5–10^7 bacteria necessary to cause salmonellosis). Transmission by

Shiga toxin A powerful toxin produced by *S. dysenteriae* that acts on tissues of the central nervous system.

Consistent Chapter Format

Highlights—from a clinical perspective—the *essential* information for each taxonomic group: Pathogenesis and Clinical Disease, Transmission and Epidemiology, Diagnosis, Treatment, Prevention and Control.

Micrographs and Clinical Photos
Show disease organisms and pathological conditions associated with infection.

Pathogenesis Illustrations
Are consistently used in the organismal chapters to depict the means by which organisms cause disease in the body.

FIGURE 31-1 Transmission electron micrograph of herpesviruses showing icosahedral capsids (partially disrupted) surrounded by envelopes (magnified 160,000×). (Courtesy Robley C. Williams, University of California, Berkeley)

of cancer in lower animals are associated with these viruses and several lines of evidence associate herpesviruses, particularly the Epstein–Barr virus, with certain forms of cancer in humans. Women who have cervical herpes infection have a significantly greater incidence of cervical cancer than uninfected women.

HERPES SIMPLEX VIRUSES

Two serotypes of herpes simplex viruses (HSV) have been identified. Type 1 is generally associated with infections of the upper half of the body and type 2 with infections of the genitourinary tract and surrounding tissues. Primary infection usually occurs on the mucosal–epithelial surfaces of the body. Following initial infection, the neurons that innervate the area become infected. The primary site of infection is characterized by a lesion, while infection of the neurons leads to latent infection. Both types may cause disseminated infections in infants and compromised patients.

✳ Pathogenesis and Clinical Diseases

Cold Sores or Fever Blisters. Among the most common of all human infections are *cold sores* or *fever blisters* (herpes labialis), which are usually

caused by the type 1 HSV (Figure 31-2). The recurring lesions on the lips are the clinical manifestation of a complex chronic interaction between the virus and the host. Most newborn infants are not readily infected, possibly as a result of passive immunity that offers some protection against primary infection. Once the passive immunity is gone, the infant is highly susceptible to primary infection. Susceptibility tends to decrease somewhat as the child gets older. However, in conditions of poor sanitation as many as 90% of the population has been infected before adulthood. Persons living under conditions of improved sanitation experience about a 50% infectivity rate. The primary infection is often asymptomatic or is not diagnosed as herpes. Symptoms are seen in 10 to 15% of the cases from 2 to 12 days after being exposed to the virus. The primary lesions may appear as small **vesicles** in the throat, mouth, or nose and go relatively unnoticed. The most noticeable form of primary infection involves the lips, mouth, and gums (*gingivostomatitis*), in which the vesicles rupture and develop into ulcerative lesions. Fever, pain, and irritability usually persist for about 1 week, followed by gradual healing during the second week.

Recovery is associated with a rise in antibodies against the virus. During the primary infection, however, the virus passes along nerve fibers to regional **ganglia**. In the case of gingivostomatitis,

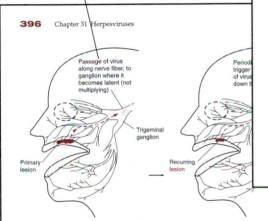

FIGURE 31-2 Herpes simplex fever blister on lower lip 2 days after onset. (Centers for Disease Control, Atlanta)

Vesicle A blister-like structure that contains a clear serous fluid.
Ganglion Major nerve trunks connecting the peripheral nerves to the CNS.

---

396 Chapter 31 Herpesviruses

Passage of virus along nerve fiber, to ganglion where it becomes latent (not multiplying)

Periodi trigger of virus down t

Trigeminal ganglion

Primary lesion

Recurring lesion

FIGURE 31-3 Aspects of the pathogenesis of primary and recurring herpes infections of the lips.

the *trigeminal ganglion* is commonly involved and the virus becomes sequestered in a latent form in this tissue. While in this latent form, the virus cannot be detected by ordinary means. It causes no symptoms and is not affected by antibodies. Periodically, in from 20 to 30% of the general population, these latent viruses become activated and move down the nerve fiber to cause recurring skin lesions at the site of the original infection (Figure 31-3). The frequency of these recurring lesions varies from person to person, ranging from once every few years to about once a month. Various stressful stimuli, such as excessive sunlight, fever, cold winds, emotional stress, and hormonal changes, apparently trigger the reactivation of the virus. As the virus moves down the nerve fibers, it passes directly into the skin cells without becoming exposed to the host's antibodies. The antibodies usually prevent the virus from spreading systemically to other tissues of the body but are unable to prevent the recurring lesions. Recurrent infections are generally less severe than primary infection. Primary and recurring infections may also occur in the eyes, causing a disease known as *herpetic kerato-conjunctivitis*. Lesions on the cornea are most serious, for the accumulating scar tissue may lead to vision impairment.

Occasionally, almost any tissue of the body may become infected with these viruses. Primary and recurring infections may occur on any cutaneous area of the body. Traumatic injury may provide a portal of entry for primary infections that may develop in both children and adults. Such infections have been seen in wrestlers (herpes gladiatorum) due to skin abrasions, or in persons following burns, or on the thumb of a thumb-sucking child or the finger of a dentist (herpetic whitlow). Children with *eczema* may acquire a serious herpes infection over large areas of the body (eczema herpeticum). Herpes simplex viruses may also infect the central nervous system, causing a severe, and often fatal, infection (herpetic encephalitis).

Genital Herpes. A very common sexually transmitted disease is *genital herpes* (see Plate 38). Over 80% of these infections are caused by HSV type 2. In females the vesicles usually occur in the mucosal tissue of the vulva, vagina, or cervix, but any of the genital or surrounding tissues may be involved. These vesicles ulcerate, producing shallow lesions. The symptoms may include malaise, urinary retention, local pain, fever, vaginal discharge, and tender, swollen inguinal lymph nodes. In males the vesicles and subsequent ulcerations

Eczema Inflammation of the skin, often associated with scaling, papules, crusting, and serous discharge.

Boldfaced and Italicized Terms
Highlight new or important words for the chapter. Key terms are boldfaced, while additional new terms are italicized. All boldfaced and italicized terms are included in the **Glossary** at the end of the text.

Color Plate References
Indicate where a color photo is available in the full-color Plate Reference at the back of the text.

Running Glossary
Helps understanding of new terms *as they are used*. These bottom-of-the-page entries reduce the time required to read the text and later provide excellent chapter review material.

has as yet not been associated with any specific clinical disease.

In late 1994, yet another human herpes virus was found, which is associated with a type of cancer called *Kaposi's sarcoma*. Kaposi's sarcoma is found in about 20% of AIDS patients. Some evidence suggests that this herpesvirus, called *Kaposi's sarcoma-associated herpesvirus*, may be the cause of Kaposi's sarcoma in AIDS patients.

MATERIAL FOR REVIEW

CONCEPT SUMMARY

1. Infection due to the DNA herpesvirus group is of considerable attention and concern today. These viruses are responsible for a wide variety of disease conditions in both humans and animals. They cause benign, latent infections, such as cold sores, and extensive life-threatening infections, such as generalized herpes.

2. Herpesviruses are widely known because of current interest in their role as agents of a sexually transmitted disease caused by herpes simplex virus type 2 and because of infectious mononucleosis due to the Epstein–Barr virus.

3. Herpes viruses are among the few viruses for which a specific antiviral therapy has been developed.

4. Although not generally well known, cytomegalovirus infection is extremely common. This agent is responsible for serious, often fatal disease in compromised patients.

Concept Summary
Recaps in numbered-sentence format the essential points of the chapter.

CLINICAL SUMMARY TABLE

Microorganism	Virulence Mechanisms	Diseases	Transmission	Treatment	Prevention
Herpes simplex	Latency	Cold sores Conjunctivitis Genital herpes Encephalitis	Direct and Sexual contact	Acyclovir	Avoid contact
Varicella-zoster	Latency	Chickenpox Shingles	Airborne	Acyclovir	Vaccine
Epstein-Barr	Latency	Infectious mononucleosis	Oral contact	Symptomatic	None

Clinical Summary Table
Provides a quick review of the organisms covered in the chapter and some specifics of the diseases they cause.

STUDY QUESTIONS

1. Briefly describe the host–parasite relationship commonly associated with herpesvirus infections.
2. Why isn't antibody to herpesvirus type I and type II protective against recurrence of the disease?
3. What is the most common site of infection for type II herpes?
4. Why would a viral-component vaccine be particularly useful against herpesviruses?
5. What is the likely source of an outbreak of chickenpox in a community that is apparently free of the virus?
6. What is the biggest risk factor associated with cytomegalovirus infection?

Study Questions
Provide an opportunity for students to quiz themselves about the important facts of the chapter.

CHALLENGE QUESTIONS

1. Why would there be possible opposition to approving a living vaccine against herpes simplex virus infections?
2. Chickenpox in children is usually a relatively harmless disease. Why do health officials say that chickenpox in adults can be very serious?

Challenge Questions
Ask the student to go beyond mere memorization of the facts to apply their knowledge to a specific problem.

Historical Developments in Medical Microbiology

Microbiology as a scientific discipline did not exist 140 years ago. In fact, the men and women who developed the concepts that today are associated with this discipline were largely trained as chemists or physicians. They were concerned with improving the health of the human population and reducing the ravages of infectious diseases, which were major emphases in the development of the science of microbiology. Their discoveries over many years led to an increased understanding and control of infectious diseases such that by the latter half of the nineteenth century, the contagious nature and mode of transmission of many diseases had been demonstrated. This introductory chapter relates many of the historical events and developments that shaped the science of microbiology. As we will see, many discoveries about diseases, coupled with an increased ability to study microorganisms, led to the formation of the germ theory of disease and ushered in the Golden Age of Microbiology (1875–1900). During this time the foundations of the science of microbiology were established, setting the stage for more modern discoveries, including chemotherapy, and modern day advances in molecular biology and genetic engineering.

A BRIEF HISTORY OF INFECTIOUS DISEASES

Infectious diseases have been the greatest pestilences in human history. Only in the past 100 years or so have many of the major infectious diseases been brought under control. In past years, severe outbreaks of infectious diseases periodically swept across nations and through cities, ravaging their inhabitants. Classic examples of these *epidemic*

> **Infectious disease** A disease, such as tuberculosis or chicken pox, caused by a living organism, which in many cases can be transmitted from person to person.
> **Epidemic** The occurrence of a common disease such as influenza in greater than expected numbers; or the occurrence of a rare disease even in small numbers.

scourges include smallpox, typhus, cholera, and plague. Outbreaks of smallpox, which periodically struck cities, would often kill from 10% to 90% of the inhabitants, and the great plague pandemics of the Middle Ages killed an estimated one-fourth of the inhabitants of Europe. Tuberculosis, which develops slowly in the body and rarely causes explosive outbreaks, has always been a major killer disease. Until the present century approximately 30% of the world's inhabitants who reached adulthood died of tuberculosis before reaching old age. Childhood infectious diseases have always been present as well; before 1900 about one of every two children died before reaching 10 years of age, primarily due to infectious diseases.

Today the number of persons dying from infectious diseases in developed countries is only a small portion of what it was in earlier times. Developments that led to the control of many devastating human microbial diseases represent some of the great triumphs of modern technology. Yet many infectious diseases are still not effectively controlled. Moreover, as social and physical living conditions change, new patterns of older diseases, as well as entirely new diseases, such as AIDS, develop. Some of these patterns result from new medical procedures that, while benefiting patients in some ways, render them more susceptible to certain types of infections. A person working in health care must be aware of conditions that cause these changing patterns of infectious diseases and must then be able to adapt procedures to minimize or control such infections. Serious infectious diseases are still widespread in many underdeveloped countries, particularly those diseases caused by parasites and those transmitted because of unsanitary conditions.

DEVELOPMENTS PRECEDING THE GERM THEORY

Developments that preceded and established the basis for the **germ theory** of disease came from three independent branches of research: (1) observations on the contagious nature of disease, (2) experiments with vaccination or immunization, and (3) basic research on the nature of microorganisms. Once the central concept of the germ theory of disease had been established, various subdisciplines arose that specialize in the study of different types of microbial agents as well as the methods used to treat and

control infectious diseases. Some historical highlights of medical microbiology are outlined in Figure 1-1.

✳ Observations of Contagion

The idea that certain diseases could be passed from person to person by contact existed in many ancient cultures. The most notable examples of awareness of **contagion** are biblical references to the disposal of human wastes and regulations to avoid contact with lepers. The ancient Greek civilization was aware of this concept. Aristotle reportedly instructed Alexander the Great to have his armies boil their drinking water and bury their dung. Yet many later cultures seemed unaware of the contagious nature of diseases. One dominant philosophy common before the nineteenth century held that diseases resulted from earthly influences, planetary conjunctions, or supernatural forces. The name "influenza" stems from the Middle Ages when it was thought that certain positions of the stars "influence" the onset of this disease. Despite such philosophies, some early scientists correctly observed the contagious nature of infectious diseases. One such interpretation was recorded in a book entitled *De Contagione et Contagiosis Morbis* (Contagion and Contagious Diseases) written in 1547 by an Italian physician, Girolamo Fracastoro. Fracastoro theorized that tiny imperceptible particles ("seeds of disease") spread from person to person. He postulated three forms of contagion: (1) by direct contact, (2) by *fomites* (a term first introduced by Fracastoro in referring to contaminated inanimate objects), and (3) at a distance—that is, by air or water. Unfortunately, his concepts were several hundred years ahead of their time and generally did not become part of the medical philosophy of his day.

Convincing work on the contagious nature of infectious disease was done in the 1700s by Antonio Micheli, Mathieu Tillet, and Isaac Prevost, who studied the transmission of plant diseases and the prevention of transmission by the use of chemical agents. This work, however, had little impact on those concerned with diseases of humans or animals. In the latter half of the eighteenth century, self-inoculation with the germs of syphilis and gonorrhea by the noted British surgeon John Hunter helped emphasize the concept of contagious diseases. Yet Hunter inadvertently, and somewhat tragically, confused the entire concept. In an attempt

---

Germ theory   The theory that disease is caused by microorganisms.
Contagion   The passing of disease between individuals.

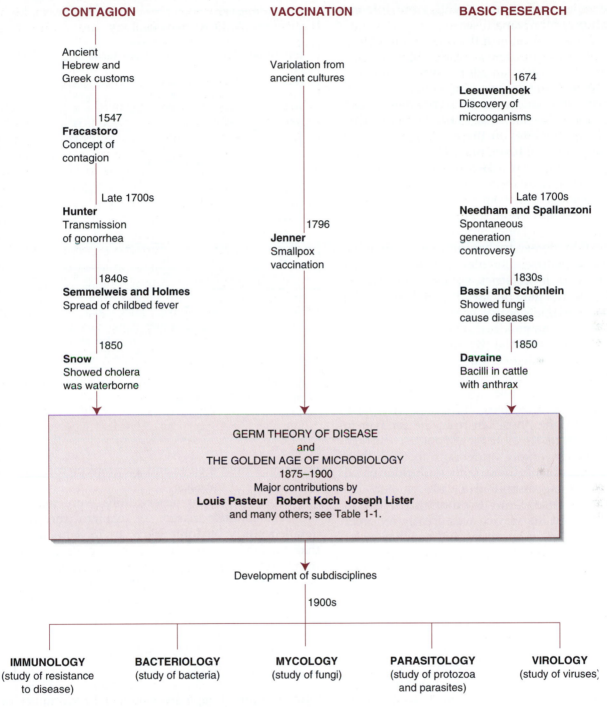

FIGURE 1-1 Historical highlights in medical microbiology, showing three independent branches of research coming together in the latter part of the nineteenth century to formulate the germ theory of disease and usher in the Golden Age of Microbiology. In the twentieth century, subdisciplines of medical microbiology developed.

to prove the contagious nature of gonorrhea, he purposely inoculated the skin of his arm with pus from a person with gonorrhea. His concept was proven when he developed a gonorrheal lesion at the site of inoculation. Unfortunately, the experi- ment was confused because the patient from whom the pus was taken had syphilis as well. This disease was also successfully transmitted to the unfortunate Dr. Hunter and resulted in his untimely death. Because of this error, for years afterward, many persons believed gonorrhea and syphilis were different manifestations of the same disease.

During the first 30 years of the nineteenth century some chemical disinfectants, mostly chlorine compounds used primarily for odor control, came into use in some medical facilities. Other chemicals were used to treat infected wounds, purify water, and disinfect hands. Notable observations were made between 1830 and 1860 on the mode of spread of childbed or puerperal fever, measles, and cholera.

During the 1830s and 1840s independent observations on the transmission of childbed fever were made by the noted American physician–poet Oliver Wendell Holmes and a young Hungarian physician, Ignaz Semmelweis. They proposed methods for controlling this disease by having physicians wash their hands regularly with chloride solutions. Semmelweis, trained in obstetrics, secured a position in that service at the general hospital of Vienna, and while there became greatly concerned about the high death rate from childbed fever in women who gave birth in that hospital. By careful observation (Figure 1-2), he detected certain patterns of this disease. Patients examined shortly after a physician had autopsied a cadaver frequently contracted childbed fever. He theorized that "cadaveric particles" carried by the physician from the cadaver to the patient were responsible for the disease. He then instructed those working under him to wash their hands and instruments thoroughly with chlorinated water after having examined a cadaver or a diseased patient. Consequently, the mortality rate from childbed fever on his service was greatly reduced. Unfortunately, the observations and conclusions of Semmelweis were rejected by many of his contemporaries. Only in subsequent years was the value of his observations recognized and applied as routine procedures in handling patients during child birth.

In 1846 Peter Panum, a 26-year-old Danish physician, was sent by his government to investigate an outbreak of measles in the Faroe Islands. After interviewing thousands of patients, he determined that measles was contracted by contact with a person who already had measles and that the diseases did not arise spontaneously as some authorities thought. He also determined the incubation time for the disease and clearly observed that those who had had the disease were immune when reexposed. His work presented a clear and accurate picture of the *epidemiology* of measles.

I. Abtheilung.
Klinik für Aerzte.

1841	Geburten	3036,	Todte	237,	Percent-Antheil		7.80
1842	»	3287,	»	518,	»	»	15.75
1843	»	3060,	»	274,	»	»	8.95
1844	»	3157,	»	260,	»	»	8.28
1845	»	3492,	»	241,	»	»	6.90
1846	»	4010,	»	459,	»	»	11.44
	»	20042,	»	1989,	»	»	9.92

II. Abtheilung.
Klinik für Hebammen.

1841	Geburten	2442,	Todte	86,	Percent-Antheil		3.52
1842	»	2659,	»	202,	»	»	7.59
1843	»	2739,	»	164,	»	»	5.98
1844	»	2956,	»	68,	»	»	2.30
1845	»	3241,	»	66,	»	»	2.03
1846	»	3754,	»	105,	»	»	2.79
	»	27791,	»	691,	»	»	3.38

FIGURE 1-2 Table from the research of Semmelweis showing a higher death rate (9.92%) from childbed fever in women giving birth in the clinic attended by physicians (Klinik für Aerzte) compared to a death rate of 3.38% in the clinic attended by midwives (Klinik für Hebammen). Translations: *Geburten* = births; *Todte* = deaths; *Abtheilung* = division.

Further important observations were made on the *communicable* nature of cholera through a detailed study by John Snow. In 1854 he showed that persons using water from the Broad Street pump in London were much more likely to contract cholera than those obtaining their water from other pumps (Figure 1-3). He further showed evidence of human fecal contamination at the Broad Street pump and correctly deduced that this contamination was responsible for the outbreaks of cholera.

✷ Immunization

Since antiquity people have observed that individuals surviving one attack of certain diseases were often immune to a second attack. Most notable were observations on the disease smallpox (also called *variola*). It was further noted that both a major and a minor form of smallpox occurred and that recovery

Epidemiology The study of mechanisms and factors involved in the spread of disease within a population.
Communicable Capable of being transmitted from one person to another; a synonym for *contagious*.

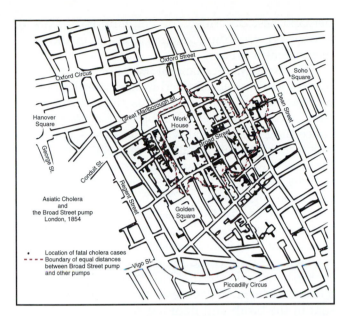

FIGURE 1-3 Map of Broad Street, London, showing the clustering of outbreaks of cholera among persons using water from the Broad Street pump. From the study of John Snow in 1854. (Modified from J. P. Fox, C. E. Hall, L. R. Elveback: *Epidemiology Man and Disease*, Figure 10-10, p. 227. Copyright © 1970 by Macmillan Publishing Company)

✳ Discovery of Microorganisms and Early Basic Research

The discovery of microorganisms in 1674 came not from research by scholars or scientists of that day, but from the astute observations of a layman. Antony van Leeuwenhoek of Delft, Holland, was not well educated in the classical manner of his day, but he had become very skilled in his hobby of grinding glass lenses and making simple one-lens microscopes (Figure 1-5). These microscopes gave magnifications of up to 300 times and through a special method of illumination, which he kept a secret but which was probably a form of dark-field lighting, he was able to observe the fine structure of many materials. While observing pond water in 1674, he was amazed to see many very small creatures, apparently algae or protozoa. He called these creatures *animalcules*. This discovery greatly intrigued Leeuwenhoek, who then spent most of his spare time over the next 50 years making observations of microorganisms that he found in various materials. Encouraged by friends, he communicated his discoveries to the Royal Society of London.

from one conferred immunity against both forms of the disease. Death rates from the minor form of smallpox were considerably less than the risk of dying from the naturally acquired major form. Therefore, in some countries before the nineteenth century, primarily in Asia, Africa, and, to some extent, North America, a procedure called *variolation* was practiced. This procedure entailed purposely exposing persons to the minor form of smallpox. The English physician Edward Jenner, who was aware of the practice of variolation, carried the concept a step further. Jenner observed that milkmaids, who often contracted a mild disease called *cowpox*, rarely came down with smallpox. Therefore, in about 1796 he deliberately inoculated persons with materials taken from cowpox lesions (Figure 1-4). This process came to be known as **vaccination**, a term based on the Latin word for cow (*vacca*). In spite of some early opposition, the practice of vaccination against smallpox eventually became widely accepted and started a process that in time led to the complete eradication of this disease ∞ (Chapter 30).

FIGURE 1-4 Early smallpox vaccination procedure in which cowpox material was introduced into the skin. (Courtesy of Corbis-The Bettmann Archive)

Vaccination A process by which small amounts of infective material, or material similar to that which is infective, is introduced into individuals to increase their resistance to disease; sometimes used in the general sense to refer to the process of immunization.

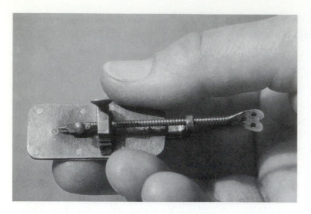

FIGURE 1-5   Replica of the type of single-lens micro-scope used by Leeuwenhoek. The object was placed on the pointed tip and brought into focus by turning the screws. (From Thomas D. Brock, *Biology of Microorganisms*, 7th ed., Figure 1.2, p. 3. Copyright © 1994 by Prentice-Hall, Inc. Reprinted by permission of Prentice Hall, Inc.)

Leeuwenhoek wrote his first letter to the Royal Society somewhat apologetically, for it was in his own simple Dutch dialect and not in the scholarly Latin then customary in scientific writing. Yet using this simple language, he was able to accurately describe his important discovery (Figure 1-6). Impressed with Leeuwenhoek's discovery, the Royal Society encouraged him to continue his observations and correspondence. Between 1674 and the time of his death in 1723, Leeuwenhoek sent over 150 letters to the Royal Society. In his simple, colorful way, Leeuwenhoek accurately described protozoa, fungi, algae, and bacteria. He was honored by being made an honorary member of the Royal Society, the preeminent society for scientific research of his time. Today he is recognized as the "Father of Microbiology." Neither Leeuwenhoek nor his contemporaries made any connection between these recently discovered "animalcules" and diseases, however.

Studies of microorganisms were limited during the first 150 years after their discovery and were considered little more than biological curiosities. Because of technical difficulties, most biologists of that era found it unprofitable to study microorganisms. The Swedish naturalist Carolus Linnaeus, in an attempt to include microorganisms in his scientific classification of plants an animals in 1758, referred to them as the class "chaos." This name perhaps best described the contemporary state of knowledge regarding microorganisms in the eighteenth century.

The systematic laboratory investigation of microorganisms accelerated in the latter half of the eighteenth century. During this time microorganisms became the central subject of a controversy concerning the possibility of *spontaneous generation* (life arising from nonliving matter). Studies by Francesco Redi a century earlier had laid to rest notions such that flies arose spontaneously from decomposing meat. In 1784, however, John Needham, after conducting a series of experiments in which bacterial growth appeared in broth that had previously been boiled, concluded that only spontaneous generation could explain his results. Today it is clearly understood that Needham's findings were a result of the growth of resistant *bacterial spores* present in the broth and poor *aseptic techniques*. In any case, Needham's theory received wide support as well as notoriety. Lazzaro Spallanzani challenged this theory and conducted an extensive series of experiments to show that microorganisms arose only from microorganisms of the same type. The controversy between Spallanzani and Needham continued for many years, increasing the scientific interest in microorganisms. A century later the the-

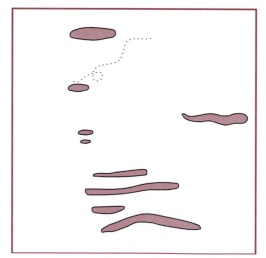

FIGURE 1-6 Reproductions of microorganisms drawn by Leeuwenhoek in one of his letters to the Royal Society. (From C. Dobell, *Anthony and His Little Animals.* New York: Dover Publications, 1960)

Bacterial spore   An environmentally stable and persistent structure formed by some bacteria as protection against adverse conditions.
Aseptic technique   Methods used to minimize the chances of contamination or infection by microorganisms from the environment.

ory of spontaneous generation was finally disproved by Louis Pasteur and John Tyndall.

By the mid-1800s investigators started to recognize the possible role of microorganisms as causative agents of disease. In 1836 Agostino Bassi showed that fungi were the cause of a disease in silkworms; and a few years later Johann Schonlein demonstrated the association of fungi with a human skin disease called *favus*. Gerhard Hansen discovered **bacilli** in cells from leprosy patients as early as 1847. In 1850 Casimir Davaine carried out preliminary studies in which he observed large bacilli in the blood of cattle with anthrax, leading him to suggest that these organisms might be the cause of this disease. Then in 1865 Jean–Antoine Villemin experimentally transmitted tuberculosis to animals. In spite of their potential significance, the full impact of these early studies was not initially recognized; they were, however, a prelude to the formulation of the germ theory of disease and an introduction to the Golden Age of Microbiology that occurred during the latter part of the nineteenth century.

THE GERM THEORY OF DISEASE AND THE GOLDEN AGE OF MICROBIOLOGY

The central figures in establishing microbiology as a science were the French scientist Louis Pasteur and the German physician Robert Koch. Pasteur (Figure 1-7) first gained recognition as a chemist when he successfully separated left- and right-handed crystals of tartaric acid. Then in the 1850s his studies turned to the process of fermentation and he concluded that microorganisms were responsible for this process. In a short series of studies published in 1860 and 1861, Pasteur helped disprove the lingering theory of the spontaneous generation of microorganisms. Because of his knowledge of fermentation, Pasteur was asked to help solve the problem of the so-called "wine disease." In this study, he scientifically determined that undesirable wine was produced by the presence of acid-producing microorganisms. To resolve this problem, he applied mild heat to the grape juice in order to destroy the unwanted microorganisms, a process now called *pasteurization* (Figure 1-8). Pasteur was next requested to cure a silkworm disease that was gradually destroying the French silk industry. After

FIGURE 1-7   A painting of Louis Pasteur by Edelfeld. (Courtesy of Corbis-The Bettmann Archive)

many trials and setbacks and five years of effort, Pasteur demonstrated that a protozoan caused this disease and he introduced methods to limit its spread.

By 1876, Pasteur had turned to the study of the contagious diseases of vertebrates. Carrying out extensive experiments with anthrax and chicken cholera, Pasteur helped establish the causative role of a specific bacterium for each of these diseases. Among his greatest contributions were the development of *vaccines* for their control. This work laid the foundation for an entire scientific discipline known today as *immunology*. In the mid-1880s, Pasteur developed a successful treatment for persons exposed to rabies. A modification of this treatment continues in use even today. More than any other individual, Pasteur helped to move medical science from an empirical to an experimental process.

Bacilli   Bacteria that are elongated or rod-shaped.
Vaccine   A preparation of killed or weakened microorganisms (or their products) that can be used to immunize against disease.

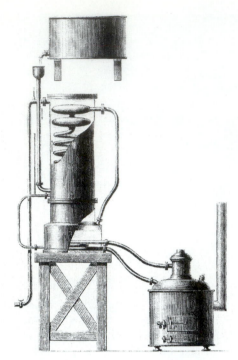

FIGURE 1-8   An early type of apparatus used for the pasteurization of wine. (Institut Pasteur, Paris)

Koch was fascinated with research on microorganisms early in his medical career. Working under makeshift conditions in the 1870s (Figure 1-9), he developed systematic methods for the study of these organisms. His first reports, published in 1876, clearly demonstrated the role of a specific bacterium as the causative agent of anthrax. These efforts enabled him to demonstrate a bacterial cause of disease in large animals and humans. It was while working with anthrax that he formulated a set of postulates, now termed *Koch's postulates*. These postulates (see Chapter 13) are used today as convincing proof of the role of a given microorganism as the cause of a specific disease. By 1880, Koch was receiving increased support for his work and could spend his full efforts in research. He continued to improve the methods used to study microbial diseases, with some of his most important work centering on tuberculosis. By demonstrating the infectious nature of this significant human disease, Koch disproved the theory that tuberculosis was inherited. Before his death in 1910, Koch became a central force for microbiological research and, with his associates, made many notable contributions to medical microbiology.

Great research institutes developed around both Pasteur and Koch and became the major centers for microbiological research during the latter part of the nineteenth century. Through the remarkable scientific skill of these two men and those who worked with them, methods became available to control and prevent many of the common infectious diseases of humans and domestic animals.

While both Pasteur and Koch carried out important work in France and Germany, significant developments were being made in Great Britain by the surgeon Joseph Lister. Lister was familiar with the studies of Pasteur on fermentation and putrefaction and in 1865 he reasoned that microorganisms could be causing the putrefaction or pus formation (*suppuration*) associated with wounds. He further reasoned that dressing the wound with some material capable of killing these germs might help prevent the suppuration. Noting that carbolic acid (phenol) was effective in preventing sewage odors, Lister chose to use this substance as his antiseptic agent. This choice was fortunate and Lister had great success in preventing wound infections. He extended his methods to surgeries by soaking **ligatures** in disinfectants and by performing operations under a spray of phenol (Figure 1-10). Primarily as a result of the work pioneered by Lister, by the end of the nineteenth century aseptic surgery had become a standard procedure in most hospitals.

Numerous discoveries were made by other scientists during this Golden Age of Microbiology (see Table 1-1). By 1900 microbiology was a well-

FIGURE 1-9   A room in Koch's house that was converted into a laboratory and was used during his early studies in microbiology. (Courtesy of Corbis-The Bettmann Archive)

Ligature   Suture; material (often silk) used to close an open wound. Commonly referred to as "stitches."

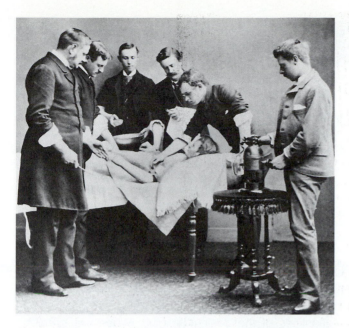

FIGURE 1-10 An operation in Edinburgh, about 1870, using the carbolic spray method of Lister. Masks, gloves, caps, and gowns were not yet being used. (Courtesy of Corbis-The Bettmann Archive)

established science and the relationship between microbes and many diseases had been clearly determined. Even more important, methods for treating and controlling some diseases were developed during this time. The secrets of the great human pestilences of history began to yield to the persistent studies of men such as Emil von Behring, Albert Neisser, Georg Gaffky, Alexandre Yersin, and David Bruce. Together and independently these men unlocked the doors leading to the possibility of human life free from the worry, sorrow, and horror of calamitous epidemic disease. In the early 1900s, numerous contributions broadened our knowledge of microbiology. Out of this early knowledge came the various subdisciplines of medical microbiology shown in Figure 1-1.

CHEMOTHERAPY

Knowing that living biological agents were responsible for disease in humans led naturally to an investigation of the means by which we could come to control infections. One who engaged in this forward-looking research was the German chemist Paul Ehrlich. Ehrlich had studied the action of aniline dyes on many materials and observed that the

TABLE 1-1

Some Important Discoveries Made in Medical Microbiology During the Golden Age of Microbiology, 1875–1900

Year	Discoverer	Discovery
1877	R. Koch	Proved that anthrax is caused by a bacterium
1877	F. Cohn	Demonstrated bacterial spores
1878	J. Lister	First grew bacteria in pure culture
1879	A. Neisser	Discovered the cause of gonorrhea
1880s	L. Pasteur	Developed vaccines
1881	A. Ogston	Discovered staphylococci cause wound infection
1882	R. Koch	Discovered the cause of tuberculosis
1883	T. Klebs	Discovered the cause of diphtheria
1884	A. Nicolaier	Discovered the cause of tetanus
1884	R. Koch	Discovered the cause of cholera
1884	G. Gaffky	Discovered the cause of typhoid fever
1884	E. Metchnikoff	First observed phagocytosis by white blood cells
1887	D. Bruce	Discovered the cause of Malta fever
1890	E. von Behring and S. Kitasato	Discovered bacterial toxins and how to develop antitoxins
1892	W. Welch and G. Nuttal	Discovered the cause of gas gangrene
1892	D. Ivanowski	First demonstrated a virus
1894	A. Yersin	Discovered the cause of plague
1897	E. Van Ermengen	Discovered the cause of botulism food poisoning
1989	K. Shiga	Discovered the cause of dysentery
1900	W. Reed	Showed yellow fever was transmitted by mosquitoes

Molecular Biology and Immunization

From the days of Edward Jenner it has been clear that the process of immunization is not without risk to the patient. Many potential vaccines have either not been developed or are not generally used because of such risks or because of biological limitations. Now, after nearly 200 years of immunization practice, we recognize many of the characteristics of a good vaccine, as well as many of the undesirable effects of vaccine use. Jenner was able to use cowpox as a vaccine for smallpox, but most infectious disease agents have no closely related, relatively harmless microorganisms that can be used for immunization. Thus, scientists have considered numerous alternative approaches in their quest for useful vaccines.

In principle, the development of a microbial vaccine should be relatively simple. It consists of exposing an individual to the causative agent (or part of the agent) of the disease and allowing the individual's body to respond by producing protective *antibodies* against the disease-causing microbe. As with many concepts, theory is easier than practice. To create a vaccine that is maximally useful, scientists must consider questions like these: Which is the best route of vaccine administration; that is, is the vaccine most effective when given by mouth or must it be injected? Is a single dose of the vaccine adequate or must it be repeatedly administered? Can more than one vaccine be administered at the same time, or will multiple components compete with each other and reduce the immunity to all? How old should an individual be in order to receive the vaccine, and what is the optimal age of administration. How can a toxic (poisonous) vaccine be modified such that it can

still be used? These and many more concerns have been addressed by scientists as they have attempted to increase our resistance to serious infectious disease.

With the advent of genetic engineering, a way has been opened to resolve many of the problems associated with the development and use of vaccines. Scientists have been able to identify the specific chemical molecules, known as *epitopes,* of an infecting microorganism that are recognized by individuals who develop immunity to the microorganism. With this information, they have been able to select the microbial gene that is responsible for producing the epitope. The scientists can then engineer the gene into non-disease-producing microorganisms so that it will produce large quantities of the desired epitope. When the specific purified chemical molecules that stimulate immunity to the disease-producing microbe are available in large quantities (without the rest of the organism) it is relatively easy to find answers to the questions raised in the previous paragraph.

There are presently only a few genetically engineered vaccines available, but it can be expected that within very few years these techniques will be used to develop vaccines against viral diseases such as AIDS for which no vaccines are now available. It is also likely that presently available vaccines will be discontinued in favor of vaccines made by these new scientific methodologies. Such vaccines can be expected to be effective even if multiple epitopes are included in the vaccine. They will facilitate immunization of younger children and will reduce the toxic properties that limit the present usefulness of many vaccines.

various parts of cells could be specifically stained with selected dyes. With this idea in mind, he postulated that it should be possible to find a "magic bullet" that would selectively react with an essential part of a microbial cell and not with the human or host cells. He reasoned that such an interaction could be used to selectively destroy invading microorganisms and restore the patient to health. His prodigious work led him through the trial of hundreds of compounds with but little success. But his hypothesis continued to be tested, leading to the firm establishment of the concept of chemotherapy (the use of chemicals to treat disease).

Later developments in chemotherapy in the

1930s revolutionized our ability to treat bacterial diseases. From England in 1928 Alexander Fleming had reported a chance discovery of strong antibacterial properties produced by the secretions of the mold *Penicillium* (Figure 1-11). This substance he called *penicillin*. By 1940 Howard Florey and Ernst Chain had purified this material and demonstrated its usefulness in treating bacterial infections. Stimulated by the advent of World War II, scientists soon found methods for the mass production of penicillin and the first true antibiotic agent became available to the world. The significance of this single event in terms of the reduction in human suffering and misery has been incalculable.

Chemotherapeutic agent   A chemical substance that is harmful to microorganisms and can be used to treat a specific disease.

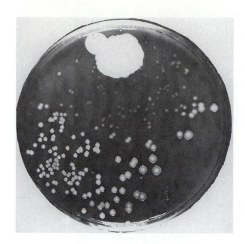

FIGURE 1-11 Photograph of Alexander Fleming's original plate showing the growth of the mold *Penicillium notatum* and its inhibitory action on bacterial growth. (Courtesy of Corbis-The Bettmann Archive)

The development of another important *chemotherapeutic agent* during this era involved several groups of scientists. In the early 1930s Gerhard Domagk, a German pathologist, found that a chemical called *prontosil* had significant antibacterial activity against some bacterial infections. Then in 1935 Jacques and Thérèse Tréfouël, a husband and wife team working in France, discovered that the antibacterial activity of prontosil was due to the sulfanilamide segment of the molecule. This finding rapidly led to the development of numerous "sulfa" compounds that continue to be used as effective antimicrobial agents.

Penicillin and sulfa drugs were the first highly effective compounds developed for treating specific bacterial diseases. Over the past 55 years, untold millions of patients have been treated with these two agents. Moreover, these chemicals have been continually improved over the years, and since 1940 many additional chemotherapeutic agents have been developed for use against many types of infectious disease agents.

MOLECULAR BIOLOGY

Many advances in each of the subdisciplines of microbiology have occurred since the 1940s. The most impressive contributions, however, are those that have led to a preliminary understanding of how molecules work together inside a living cell—a field of study now referred to as *molecular biology*. Because of their relative structural simplicity,

microorganisms and, in particular, viruses have been extensively used in these studies. Using bacterial cells, Oswald Avery, Colin MacLeod, and Maclyn McCarty, working in New York, first reported the genetic role of *deoxyribonucleic acid* (DNA) in 1944. Then in 1953 James Watson and Francis Crick, working in England, resolved the structure of DNA. These early discoveries preceded a rapid advance in understanding as various scientists demonstrated how DNA functions as the genetic storehouse of information and how that information is used to direct the various cellular activities.

Most recently, a subdiscipline of molecular biology referred to as *genetic engineering* has become the central focus of research in efforts to understand and control infectious diseases. By the early 1970s, it had become possible to insert DNA from various sources into the DNA of microorganisms (see Chapter 6). This has allowed the design of microorganisms with new capabilities. Such genetically engineered microorganisms are able to produce

TABLE 1-2

Some Possible Products and Potential Applications of Genetic Engineering

Products	Applications
Enzymes that metabolize petroleum	Clean up oil spills
Insulin	Treatment of diabetes
Human-growth hormones	Treatment of growth disorders
Animal-growth hormones	Stimulate growth for increased meat production
Interferons	Treatment of viral infections and possibly cancer
Pheromones (insect hormones)	Insect control
Endorphins	Pain killers
Antigens	Preparation of vaccines not easily produced by regular methods
Antibiotics	Improved or less expensive antibiotics
Methane or alcohol	More efficient formation of these products from cellulose waste
Nitrogen fixation	Increased soil fertility

Genetic engineering The transfer of genetic material from one organism to another organism.

large quantities of useful products that could not otherwise be easily produced. Such products as human insulin, human growth hormone, various vaccines, and many other products are now produced by these engineered microbes (Table 1-2). The applications of genetic engineering have extended far beyond medical and health concerns. Currently, major advances in genetic engineering are being applied to food production, new energy sources, various industrial processes, the understanding of biological phenomena, and many other applications. The study of molecular biology is still in its infancy, but already constitutes one of the great modern-day technological revolutions; and microorganisms play a key role in this technology.

MATERIAL FOR REVIEW

CONCEPT SUMMARY

1. Microbiology is a rather recent scientific discipline that has grown rapidly during the past 120 years. Historically, the primary concern and interest regarding the microorganism centered around its ability to produce serious, often fatal, disease. Later, microorganisms proved to be a great tool in the study of life and much that we know regarding genetics, biochemistry, cellular physiology, and molecular biology was derived through careful study of these life forms.

2. Microbiology took a great leap forward with general acceptance of the germ theory of disease. This awakening, guided by men such as Pasteur and Koch, led to a period known as the Golden Age of Microbiology during which most infectious disease agents now known were discovered.

3. A parallel science based on a study of resistance to infection developed concurrently with the discovery of infectious agents. This discipline, immunology, ranks today as one of the most challenging and exciting fields of scientific discovery. In time, the development of antimicrobial therapy brought new success to the world of the microbiologist and hope to the whole human family.

4. Recent development and advances in the procedures of molecular biology have stimulated the thinking of scientists throughout the world. With these new technologies, powerful scientific tools have become available by which the secrets of nature are being discovered at an ever-increasing rate.

5. Use of the scientific method evolved slowly into the patterns of research. Application of this process virtually ensures a continuing flow of new understanding about our world.

STUDY QUESTIONS

1. What role did the development of the germ theory of disease play in making microbiology a separate scientific discipline?

2. How has the knowledge that some diseases are communicable led to the methods for control of infections?

3. What could Semmelweis have done to ensure acceptance of ideas regarding the transmission of childbed fever?

4. Review the work of Edward Jenner in terms of the scientific method. Write a probable hypothesis, outline the experiment used to test this hypothesis, and list the results of that experiment.

5. What impact has antimicrobial chemotherapy had on the human lifestyle?

CHALLENGE QUESTIONS

1. Sterilization helped disprove the idea of spontaneous generation. Based on your reading of this chapter, how could you demonstrate that flies do not arise from decomposing meat?

2. From your personal experiences, list three benefits to our standard of living that resulted from the historical developments in medical microbiology described in this chapter.

The Scope of Microbiology

The cell is the basic unit of all living systems and each one contains the unique components of life. Life exists either in multicellular forms or as single cells. The multicellular forms such as plants and animals consist of millions of cells of diverse types that function together and depend on each other. Most single-celled life forms can be seen only with the aid of a microscope and, therefore, are called *microorganisms*—the primary subjects of microbiology. Each is able to carry out the biological functions necessary to perpetuate itself. Microorganisms often have a profound effect on their surroundings and are responsible for many essential biological phenomena which insure the maintenance of balanced life systems. By and large, the overall effect of microorganisms on other life forms is beneficial. Without microorganisms other life forms would ultimately be unable to survive. Although this textbook deals primarily with microorganisms that cause diseases in humans, it is important for readers to be aware of the beneficial and ecological roles of microorganisms in general. This chapter briefly introduces the various groups of microorganisms and some of the nonmedical aspects of microbiology.

BASIC CELL TYPES

Life forms are composed of one of two basic cell types: *eukaryotic* or *prokaryotic* (sometimes spelled eucaryotic and procaryotic). Generally, eukaryotic cells are larger and more complex than the prokaryotic cells and possess a membrane-enclosed nucleus; eukaryotic means "true nucleus" (*eu* = true; *karyon* = nucleus). All plants, animals, fungi, protozoa, and algae are composed of eukaryotic cells. Only the microorganisms classed as bac-teria and the archaea are prokaryotic cells. Prokaryotic cells have no nuclear membrane and therefore lack a true nucleus. The Latin prefix *pro* means "early" and in this instance suggests the somewhat primitive nuclear structure found in bacterial cells.

Eukaryotic microorganisms are sometimes simply referred to as *eukaryotes*, and the prokaryotic

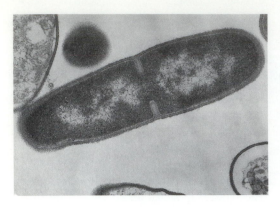 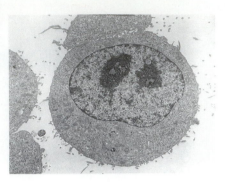

FIGURE 2-1 Electron micrographs of (a) a prokaryotic cell (×20,000) and (b) a eukaryotic cell with a nucleus (×5000). Note absence of a nucleus in prokaryotic cells. (Courtesy of Photo Researchers, Inc.)

microbes as *prokaryotes* (Figure 2-1). The morphology of these cell types is discussed in greater detail in Chapter 4. The differences between eukaryotic and prokaryotic cells are of more than academic interest to the medical microbiologist. Much of the rationale in the treatment of many microbial infections capitalizes on these differences, a concept that is expanded in subsequent chapters.

CLASSIFICATION OF MICROORGANISMS

Classification of microorganisms, and all other forms of life, is an endeavor to arrange the various life forms into related groups. Taxonomy is the term used when referring to the science of classification. All classification schemes start out with major divisions (kingdoms) and move down through a series of progressively smaller and less inclusive categories. The smallest category (species) is one in which all members are alike in all or most major characteristics. The various taxonomic categories used by biologists are defined in Table 2-1. These categories have been very useful in helping biologists organize the myriads of life forms. The most useful system of classification is one that makes use of genetic relationships and places every known living organism into a "family tree." Such a system is called a *phylogenetic* classification and shows evolutionary relatedness. Because of the extremely poor bacterial fossil record, however, the classification of these organisms has largely been based on apparent

relatedness as demonstrated by similarities of function and appearance. This is known as a *phenetic* or *phenotypic* classification. Unfortunately, a phenetic system is not perfect and the classification of microorganisms is frequently readjusted. With the discovery of new microorganisms or new phenotypic properties of an old species, the phenotypic classification of the entire microbial world is subject to change. This problem was evidenced when, in 1980, the International Committee on Systematic Bacteriology agreed to reduce the accepted number of named species of bacteria from more than 30,000 to about 2500 species.

Attempts at a more realistic (genetic) classification of prokaryotes have recently been more successful. Rather than depending on gene expression (phenotype) to characterize these organisms, pre-

TABLE 2-1

The Taxonomic Categories of Living Organisms

Kingdom (major division)
Phylum (groups of related classes)
Class (groups of related orders)
Order (groups of related families)
Family (groups of related genera)
Genus (groups of related species)
Species (living organisms that are alike)

Phylogenetic Referring to the evolutionary relatedness of a group of organisms.
Phenetic Based on the visible features of organisms and their apparent ability to modify their environment.

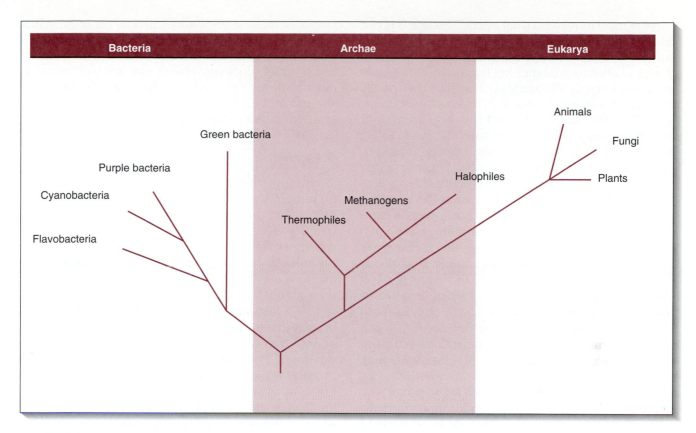

FIGURE 2-2 Universal phylogenetic tree showing relationships among Bacteria, Archaea and Eukarya. Position of the organisms on the tree is based on comparison of ribosomal RNA sequences and distances between organisms shows how closely they are related.

sent day microbial taxonomists can use the tools of the molecular biologist to assist in this difficult problem. A kind of ribonucleic acid (RNA) found in *ribosomes*, called ribosomal RNA (rRNA), has changed very slowly if at all during the evolution of microorganisms. Analysis of rRNA from plants, animals and microorganisms has enabled scientists to construct a phylogenetic tree that includes the prokaryotes as well as the eukaryotes (Figure 2-2). Looking at this tree, it is easy to see that the Archaea branch away from the trunk closer to the main root of the tree than does any other domain of organisms. This suggests that the Archaea are the most primitive forms of life found on earth. This consideration is consistent with the habitat of the Archaea, which often includes high temperatures, high salinity, or low pH. These environmental conditions are thought to reflect those found on earth at the time life first developed.

As can be seen in Figure 2-2, there are three major biologic domains (Bacteria, Archaea, and Eukarya). Two of these (Archaea and Bacteria) contain only prokaryotes, while all eukaryotic organisms, microscopic and macroscopic, are members of the Eukarya. What may be most surprising is the discovery that the prokaryotic Archaea are most closely related to Eukarya and therefore share more common properties with them than they do with bacteria.

Viruses, which are unique subcellular microbiological agents (but not cells), do not fit well into this classification scheme. They are often listed in a separate category.

✳ Nomenclature

Following the recommendations of the great taxonomist Carolus Linnaeus, a system of *binomial nomen-*

Ribosome A ribonucleic acid–protein complex used by cells for the synthesis of new protein.

clature is used in which a dual scientific name is given to each biological species and includes both the *genus* and *species* name of the organism. These names are given in Latin, or are latinized forms, and are usually descriptive or honorary. The species *Pasteurella multocida*, for example, has a genus name to honor Louis Pasteur, who discovered this bacterium, and the species name *multocida* because it is able to infect the cells of many different animal hosts. Likewise, the species *Neisseria gonorrhoeae* has a genus name that honors Albert Neisser, who discovered this bacterium, and a species name that indicates it is the causative agent of gonorrhea. When a species is mentioned repeatedly in the same report, the genus name is written out in full the first time but thereafter is abbreviated to its first letter, for example, *P. multocida* and *N. gonorrhoeae*. By agreement, the genus name is always capitalized and the species name is in lower-case letters. Because of their Latin origins, both names are either underlined or italicized. Often a common name may be used in general references to a given bacterium. For example, *N. gonorrhoeae* may be called the gonococcus.

To achieve uniform naming of microorganisms, an International Code of Nomenclature of Bacteria has been adopted by microbiologists in all parts of the world. Even so, occasional changes in nomenclature are proposed that are not accepted by all microbiologists. To assure some degree of uniformity, all names must be published in the *International Journal of Systematic Bacteriology* to be properly recognized. A periodically updated publication called *Bergey's Manual of Systematic Bacteriology* serves as the standard reference source for the classification and naming of bacteria.

MAJOR GROUPS OF MICROORGANISMS

The environments to which microorganisms have adapted are diverse and, in some cases, extreme. Certain microorganisms grow in hot springs or near deep ocean thermal vents at temperatures above the boiling point, whereas others are able to grow on snow banks and have been found in environments as hostile as that of Antarctica. Microorganisms have adapted such that they are able to grow in a variety of diverse environmental niches (Table 2-2). The purpose of this section is to give a brief overall view of the characteristics of the major groups of microorganisms, with an emphasis on nonmedical areas. The medical aspects will be covered in later chapters. To describe microorganisms it is necessary to use units of measurement that are unfamiliar to most beginning students. As with all scientific measurements, these units are part of the metric system and are shown in Table 2-3.

✳ Fungi

The fungi (singular *fungus*) comprise a large group of nonphotosynthetic eukaryotes, including such diverse organisms as yeast, molds, and mushrooms (Figure 2-3). Yeasts are globular-shaped single-celled organisms about 10–30 μm in diameter that multiply by budding. They are best known for their use as a leavening agent for bakery goods and for their ability to produce alcohol.

Molds are organisms that consist of masses of branch-like filaments called *hyphae* and are most frequently recognized by their fuzzy growth on various foods and other organic matter. Reproduction in fungi usually results from the formation of large numbers of seed-like structures called *conidia*. Fungi are widely distributed in nature and because they are not *photosynthetic* they readily grow in dark, damp places where *organic* matter is found. Only a relatively small number of molds and yeast are able to cause disease in humans (Chapter 39). Many

TABLE 2-2

Common Habitats Where Bacteria Are Present

Habitat	Approximate Number of Bacteria
Garden soil (surface)	9.7×10^6/g
Garden soil (30 cm deep)	5.7×10^5/g
Lake water (shallow)	10^4/mL
Lake water (deep)	10^2/mL
Seawater	1.1×10^3/mL
Human skin	10^6/sq cm
Human mouth	10^7/mL
Human intestine	4×10^{10}/g
Milk	10^3 to 10^6/mL
Cheese	10^8/g
Sunlit surface	Few
Air	Few

Photosynthetic Capable of using light as a source of energy for an organism.
Organic Composed of carbon and hydrogen atoms.

TABLE 2-3

The Units of Measurement Used in Microbiology

Unit of Measurement	Abbreviation	Equivalent
Meter	m	39.37 inches
Centimeter	cm	1/100th of a meter (10^{-2} m)
Millimeter	mm	1/1000th of a meter (10^{-3} m)
Micrometer[a]	μm	1/1000th of a millimeter (10^{-6} m)
Nanometer[b]	nm	1/1000th of a micrometer (10^{-9} m)
Angstrom	Å	1/10th of a nanometer (10^{-10} m)

[a] Formerly called a micron.
[b] Formerly called a millimicron.

fungal species can cause diseases in plants and lower animals, however, and have a significant impact on reducing the world's food supply.

The most notable function of fungi is their ability to secrete powerful enzymes that decompose organic matter. When moisture is present and other environmental factors are not extreme, fungi grow on a wide variety of organic substances. Organic matter in contact with the soil is rapidly decomposed by fungi as part of the natural, essential recycling process in nature. Such products as foods, paper, lumber, fabrics, paint, and rubber can all be decomposed by fungi. This decomposition process may be either beneficial or detrimental, depending on the circumstances; for example, the decomposition of dead plants in the soil is beneficial, whereas the decomposition of lumber stored for building is detrimental. Human societies spend a great deal of time and effort treating and storing materials so that they will not be damaged by fungal growth; even so, large quantities of food and other materials are lost each year due to the action of fungi. On the other hand, some by-products of fungal growth have commercial value and large-scale industrial fermentation processes produce such fungal products as antibiotics, alcohols, cheeses, and solvents. Mushrooms, although not microscopic in size, are actually complex arrangement of single, independently functioning cells that have the appearance of multicellular structures.

✳ Algae

Algae (singular *alga*) are a morphologically and physiologically diverse large group of eukaryotic micro- and macroscopic organisms (Figure 2-4). All contain *chlorophyll*, which allows them to carry out the process of photosynthesis. This process results

in the production of both energy-containing compounds and gaseous oxygen. Many algae occur as single cells, ranging in size from less than 1 μm to upward of 60 μm in diameter. They may be shaped as spheres, rods, or spindles. Other algae occur in multicellular colonies that are often visible to the naked eye and take on a wide variety of shapes. Some, such as seaweed, grow to great sizes and appear much like multicellular plants.

The presence of chlorophyll, carotenoids, and other pigments gives diverse colors to the algal cells and this characteristic is used in their classification. Along with their scientific names, they may also be referred to by color, such as yellow-green, green, red, or brown algae. One major group, the blue-greens, have been discovered to be true prokaryotes and as such are now classed as bacteria with the name *Cyanobacteria*. Thousands of different species of algae exist and are found in most moist environments. Many algae are free living in waters; others grow in soils or on the surfaces of plants and rocks. The wide diversity of algae is reflected by the fact that some grow on ice or snow, whereas others are able to grow in hot springs. The only human health

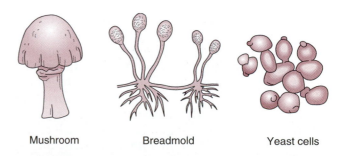

Mushroom   Breadmold   Yeast cells

FIGURE 2-3 Three different types of fungi. The mushroom is close to natural size; the bread mold and yeast are greatly enlarged.

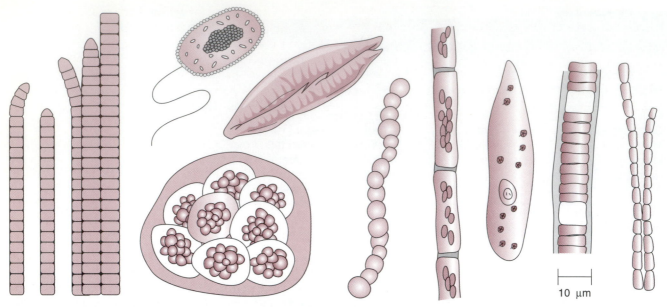

FIGURE 2-4 Different species of algae and cyanobacteria reproduced to scale (see 10-μm bar).

problem commonly associated with algae is the result of the ability of a few algae to produce *toxins* that may be consumed by aquatic animals, such as shellfish. When algae levels are high, during the summer months (Figure 2-5), enough toxins may be retained in the shellfish that they cause illness when eaten by humans. (Plate 1)

Because algae are involved in many biological cycles and are important contributors to the overall balance in nature, they play important roles in the well-being of humans and most other forms of life. They are a primary source of atmospheric oxygen, they aid in soil fertility by adding organic matter, and some species are able to fix atmospheric nitrogen. Some algae are harvested from the sea and used directly as human or animal food. *Agar*, a solidifying agent used in microbial culture media or as a thickening agent in various foods, is extracted from seaweeds. Currently a great deal of interest is being shown in the mass culture of algae as a direct source of food. Large numbers of algae are found in oceans, seas, lakes, ponds, and streams. Small, free-floating algae make up a part of the life forms referred to as *phytoplankton* (Figure 2-6). Plankton are at the beginning of the food chain in aquatic environments and are often consumed by small aquatic animals (*zooplankton*), which, in turn, are eaten by small fish, which are eaten by larger fish, and so on. Some larger aquatic animals, such as the blue whale, eat plankton directly. Essentially all sea and freshwater animals depend on the presence of algae for food. Algae are found in most bodies of water and in depths up to 180 m. The amount of organic matter resulting from the photosynthesis occurring in algae in aquatic environments is estimated to exceed the amount of similar materials produced from all plants on terrestrial surfaces.

✳ Protozoa

The protozoa (singular *protozoan*) are eukaryotic microorganisms that possess many intracellular

FIGURE 2-5 A coastal area showing massive growths of algae (light areas). Photographed from a space satellite during the summer. (NASA Headquarters)

Toxin A poisonous substance produced by some microorganisms.

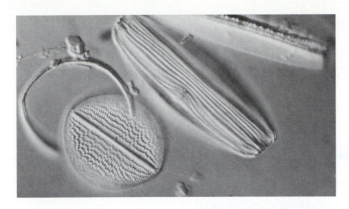

FIGURE 2-6   Freshwater phytoplankton. Photo is by phase-contrast microscopy. (Courtesy Dr. S. Rushforth, Brigham Young University)

components that are characteristic of higher forms of life. Protozoa vary considerably in size and shape and have some form of active locomotion (Figure 2-7). The mechanisms by which they move (flagella, cilia, amoeboid) are major factors in the classification of these organisms.

The smallest protozoa are only a few micrometers in diameter, whereas others may be seen with the unaided eye. Protozoa are able to ingest food particles by folding their outer membrane around the food and then pinching off the membrane to form an intracellular vacuole, a process known as *phagocytosis*. Protozoa thrive in moist environments. They inhabit most bodies of water, are found in soil, and live in the digestive tract of many higher forms of life. Because of their relatively large size and motility, protozoa are easily seen by the micro-

scopic examination of water from such sources as stagnant ponds. Protozoa have a moderate influence on water quality and on the various biological cycles in nature. While the majority of the protozoa do not cause diseases in higher forms of life, some that do are responsible for such serious diseases as malaria and African sleeping sickness (Chapter 40).

✳ **Prokaryotes**

There are two major groups of prokaryotes; *Archaea* (meaning "primitive" bacteria) and *Bacteria*. The Archaea are prokaryotes, but are more closely related to eukaryotic cells than to the Bacteria. They include three groups of organisms: the extreme *halophiles* (salt-loving Archaea), the *thermoacidophiles* (acid-loving Archaea, which grow at high temperatures), and *methanogens* (Archaea that produce methane gas). These organisms share both physiologic and morphologic properties that distinguish them from bacteria. The unique chemical structures associated with the Archaea have led scientists to speculate that eukaryotic cells evolved from this group of organisms. None of the Archaea are currently associated with human disease.

The Bacteria are prokaryotes and are both smaller and less complex than eukaryotic cells. Normally, bacteria have rigid cell walls and are shaped as spheres, rods, or helices (Figure 2-8). Bacteria are found in virtually every environmental habitat and some types have adapted to grow on minimal nutrients or under extreme environmental conditions. Some, for example, are able to grow on simple inorganic compounds, others in hot springs, in cold

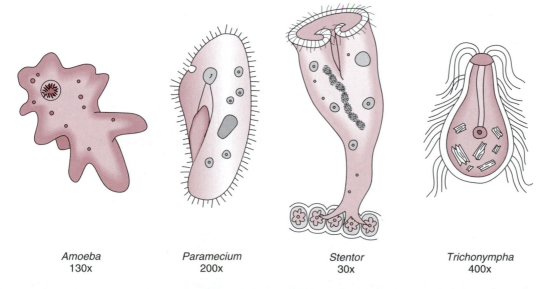

| Amoeba | Paramecium | Stentor | Trichonympha |
| 130x | 200x | 30x | 400x |

FIGURE 2-7   Four different types of protozoa with degrees of enlargement indicated.

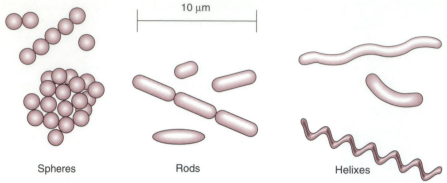

Spheres   Rods   Helixes

FIGURE 2-8   Bacterial cells showing representatives of the three basic cell shapes.

storage food, or on the bottom of the ocean under extreme pressure; still others grow in areas completely devoid of oxygen. Most bacteria multiply through a process known as *binary fission* whereby a cell simply divides into two daughter cells. Under proper conditions, growth may be very rapid with cell division occurring as often as every 12 to 15 minutes. This rapid growth potential may lead to profound changes in the surrounding environment. Although many important infectious diseases of humans are caused by bacteria, most bacteria are not able to cause disease and many bacteria produce beneficial environmental changes, for example, by decompositing of waste products, aiding in soil fertility, or producing useful chemicals.

The cyanobacteria are photosynthetic bacteria formerly referred to as the *blue-green algae*. These organisms are distinguished from other photosynthetic prokaryotes by their ability to produce gaseous oxygen as a product of photosynthesis. They differ from algae in that their chlorophyll is associated with a stack of membranes located just inside the cell membrane, and not in specialized organelles (chloroplasts).

Two groups of small (0.3 to 0.5 μm in diameter) bacteria, called rickettsiae and chlamydiae, are able to multiply only inside living eukaryotic cells. Because of this small size and dependency on living host cells (characteristics that are also shared by viruses), the rickettsiae and chlamydiae have in the past been grouped next to or with the viruses. It is now well established that the rickettsiae and chlamydiae have definite cellular structures and are best considered small, obligately parasitic bacteria. Because of their dependency on living host cells, they have no direct influence on the outside environment. Nevertheless, these microbes can cause some important human diseases (Chapter 27).

✳ Viruses

Viruses are a unique type of biological agent. The microorganisms described previously are complete cells with the capability of carrying out *metabolic* activities and other functions of life. Viruses, however, are not cells and have no independent metabolic activity. Viruses might best be described as independent genes encased in a protein coat. Each virus has either DNA or RNA as its nucleic acid; these molecules contain the genetic information necessary for the production of additional virus particles (Chapter 6 and 28). The virus is covered with a coat called a *capsid* that consists of a geometric arrangement of protein molecules. When a virus enters a living cell, the viral genes are released and the information contained in them may be expressed in the host cell. The information on the viral genes may redirect the cell to make new virus particles and eventually the cell may be altered or destroyed. The only effects produced by viruses are on the host cells; when found outside host cells, viruses are simply a collection of molecules with no apparent life functions. Viruses range in size from about 25 to 300 nm in diameter. Most viruses have definite geometric forms, the most common being spherical; some of the others are rod-, brick-, or bullet-shaped (Figure 2-9). Every form of life, including bacteria, has specific viruses that are able to infect its cells. Their effect on living cells can be destructive and many common diseases of both plants and animals are caused by viruses.

Two subcellular structures, viroids and prions, behave similarly to viruses. *Viroids* are relatively small RNA molecules that have the capacity to produce disease in plants. The exact mechanism by which they carry out this function is unknown. *Prions* are protein molecules that have been studied

Metabolic   Related to the chemical processes occurring within a cell; often associated with energy-producing and energy-using processes.

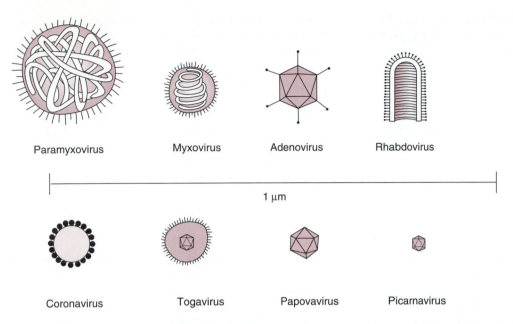

FIGURE 2-9   The shapes and relative sizes of various types of viruses that infect animals.

as the possible cause of a disease in sheep known as scrapie and in cattle as mad cow disease. Little is known concerning these particles but prions may be responsible for human neurologic diseases such as kuru and Creutzfeldt–Jacob disease.

NONMEDICAL ASPECTS OF MICROBIOLOGY

Most of the material in this textbook will deal with medical microbiology; yet microorganisms play an important role in various other areas of human interaction. This section briefly discusses several nonmedical aspects of microbiology.

✳ Soil Microbes

Most microbes occur in the upper 0.5 m of soil, and fertile farm soil may contain as many as 4 to 5×10^9 microorganisms per gram. These large numbers of microorganisms are primarily involved in the decomposition of organic matter. This process is a vital phase in the natural recycling of the elements needed for living systems. The incorporation of nitrogen, sulfur, and carbon into organic matter by plants would quickly deplete the soil of these essential nutrients if they were not recycled by microorganisms. Environmental nitrogen and sulfur cycles have been studied extensively. Microbes have been discovered to play a major role in both of these essential cycles. The photosynthetic algae add organic matter to the

soil by their growth and death. Some bacteria and algae are able to convert (fix) atmospheric nitrogen into a form that can be used by plants. Acids produced by microbes aid in dissolving rocks, an important step in the formation of soil. Both the development and maintenance of fertile soils is to a large extent the product of microbial activity.

✳ Aquatic Microbes

Microorganisms are found in all bodies of both fresh and saltwater. The numbers and types vary greatly, just as the conditions of the water vary. Some microbes are natural inhabitants of water, whereas others are transient, having entered from sewage, land runoff, or other external sources. As in soil, microorganisms are vital links in the recycling of nutrients in the aquatic food chain. Much of the world's food supply is made possible by the activities of these aquatic microbes. Maintaining the proper biological balance of these aquatic systems is one of the great challenges of present-day technology and is essential for the continued survival of most life forms on this planet. An overabundance of algal growth, however, may be detrimental to a body of water; this situation may occur when concentrations of nutrients are high and the water is warm. Such a massive growth of algae is called a *bloom* (see Plate 1). Algal blooms rapidly deplete the available oxygen in the water, resulting in the death of fish and making the water unsuitable for recreational activities. All lakes slowly fill with sediments, and microbial debris contributes significantly to this process called *eutrophication*. Excess nutrients such as phosphate in the water may accelerate the rate of filling, mainly by the increased growth of algae.

In its natural state, water is found either as surface water (lakes, rivers, etc.) or as groundwater that must be obtained from wells or from natural springs. Groundwater is the most common source of water used for household purposes and is always contaminated by microorganisms. To make such water *potable* (safe for drinking), it must be treated. Treatment generally consists of two processes, *filtration* and *chlorination*. Most water is first treated by sand filtration to remove debris and most microorganisms. Water that has been adequately filtered is then rendered completely safe for human use by the addition of only 0.5 parts per million of free chlorine. Chlorine kills those organisms that might have passed through the filtration process and inactivates any that might somehow enter the water system following treatment. Because water can serve as an important vehicle for the transmission of disease-causing microorganisms, proper water treatment is an essential aspect of public health.

A second important aspect of proper municipal water management is associated with adequate treatment of wastewater. This water is normally collected into extensive sewage collection systems and is then carefully processed at a wastewater treatment facility. In these treatment facilities, harmful microorganisms are again filtered from the water, solid waste materials and undesirable molecules are removed, and oxygen may be added back to the water by aeration. Water that has been properly treated is not dangerous for the environment and may be returned to the surface water system. Microorganisms are used to decompose organic matter, one of the significant steps in wastewater treatment. Because water may serve as an important vehicle for the transmission of disease-producing microorganisms, the proper treatment of water for human consumption and the treatment of sewage are of vital importance in maintaining a healthy environment.

✳ Microorganisms in Dairy Products

Milk provides an excellent medium for the rapid growth of many types of microorganisms and great care is needed during the collection and processing of milk to minimize *microbial contamination*. In general, the quality of milk is directly related to the numbers and types of microorganisms it contains, thus the dairy industry spends considerable time and money to control microorganisms.

Bacterial contamination of milk comes from two major sources: *external*, the contamination of milk after it has been collected, and *internal* contamination of milk directly from an infected cow. Either source of contamination may lead to serious human infection and requires the careful handling of fresh milk and proper animal husbandry for dairy cattle. It is almost impossible to prevent some microbial contamination of milk at the time that it is collected. If left unrefrigerated, bacteria quickly multiply and the milk sours. The number of bacteria present in fresh milk is greatly reduced by the process of *pasteurization*. This procedure, first applied by Louis Pasteur to the manufacture of wine ∞ (Chapter 1, p. 7), raises the temperature of the milk to 71.5°C for 15 seconds or 62.9°C for 30 minutes. These temperatures are sufficient to destroy all disease-producing organisms commonly found in fresh milk and render it safe for human use. However, these procedures do not destroy all bacteria that may be present in milk, and even pasteurized milk will quickly spoil if not kept under proper refrigeration. When pasteurization is combined with storage at low temperatures, the storage time of milk is greatly increased.

Certain microorganisms, under controlled conditions, induce desirable changes in milk (Figure 2-10). Dairy items such as yogurt, sour cream, and buttermilk result from the fermentation of milk products through the addition of selected acid-producing bacteria. Cheeses are produced by the controlled growth of selected microorganisms on curdled milk, a process called *ripening*. Different cheeses are produced by the action of different microorganisms.

✳ Microorganisms in Food

Like dairy products, most foods provide excellent media for the growth of many types of microorganisms. Generally, microbial growth in food is undesirable and produces changes in texture and flavor commonly called *spoilage*. Because of their relatively

---

Eutrophication   The process by which a body of water becomes so rich in dissolved nutrients that the resultant algal growth depletes the water of oxygen supplies needed by other organisms.
Microbial contamination   The addition, usually unintended, of viable bacteria to a previously microbial-free environment, or the addition of unwanted organisms to an environment that contains known or desirable microorganisms.

high natural bacterial contamination, certain foods such as fish, poultry, and ground meats spoil easily. Spoilage is not the only concern associated with microbial contamination. Some microorganisms such as *Salmonella* and the virus that causes infectious hepatitis are easily transmitted to humans from some foods. Other bacteria such as staphylococci and clostridia produce toxic waste materials when grown in foods. These toxic substances can lead to severe illness or even death if they are ingested.

Because of such risks, food industries make great efforts to prevent detrimental changes to food by microorganisms. Much of what is done in food processing is directed toward the control of both microbial spoilage and the transmission of harmful microbes. Processes such as drying, smoking, salting, freezing, refrigerating, heat processing (canning), and the use of chemical preservatives like sodium benzoate or calcium propionate are the major methods in preventing or slowing the spoilage of foods. Such food preservation processes are a multibillion dollar industry. Because of the ever-increasing costs of foods and the limited supplies of certain foods in many countries, reducing waste due to spoilage is increasingly important and offers a challenge to food microbiologists. Some estimates indicate that as much as 25% of the world's food supply is lost to spoilage by microorganisms or by infestation by insects or rodents.

To a limited extent, microbes may be used directly as a food source; for example, yeast is used as a food supplement. The potential for using microorganisms as a food source is great and active research projects are currently underway to develop methods of using the rapid-growth capabilities of microorganisms as a means of producing food substances. When compared to other forms of life, microorganisms are much more efficient producers of proteins. A rapidly growing culture of microorganisms under controlled conditions, for instance, is able to produce as much protein in one day as a meat-producing animal can produce in several weeks. Furthermore, microbes do this in a small space, and often by using waste products or inexpensive organic materials as their food source. This area of research offers a possible means of providing a more adequate food supply for the world's increasing population. Currently, the appeal of microorganisms as a basic food for humans is limited. The term *single-cell protein* is used when referring to this type of product to make it sound more appealing. Although most humans are not yet ready to trade their roast beef for "bacterial-protein patties," single-cell protein may become an important food supplement for meat-producing animals that are, in turn, processed for human consumption.

Microbial processes are frequently used to modify natural products for use as food. Beneficial microbial changes in milk have been noted, and microbes are the source of such items as artificial sweeteners and a variety of vitamins. Microbes are even used to modify corn starch to produce fructose, the major sweetener used in the manufacture of soft drink beverages.

✳ Industrial Uses of Microbial By-products

Many by-products of microbial growth are extremely useful. The field of industrial microbiology uses the action of microorganisms to mass-produce many of these products. In some cases, the useful product can be produced only by microorganisms; in other cases, the microbial process is the most economical means of production. To be economically feasible, the raw material should be relatively inexpensive and readily available while the end product must be of greater value. These conditions are fulfilled in many processes, and today, industrial microbial processes are the basis of large commercial industries. Some products produced by microorganisms include solvents, organic acids, alcohols (including alcoholic beverages), enzymes, and antibiotics.

The increasing costs and scarcity of petroleum and natural gas have renewed interest in microbial fermentations that can convert plant materials into methane gas and alcohols. These products can then be used for heating or in internal combustion engines. Because plant materials are renewable, these processes could greatly reduce our dependence on nonreplenishable petroleum reserves.

FIGURE 2-10 Some of the dairy products that are produced by the action of microorganisms on milk components.

✳ Microorganisms as Biological Tools

Microorganisms are widely used as biological models for scientists who wish to study fundamental principles of living systems. When working with bacteria it is possible to start with a single cell that will then replicate so rapidly that within a day trillions of identical cells can be produced. This population of identical cells is much easier to study than a population of mixed cells that might be obtained from animal or plant tissues. Much of what we know about the biological activities of all cells has evolved through studies of bacteria. Some of these basic biological properties of microorganisms are presented in Chapters 4 to 7.

It has now become possible to pass some genes from other forms of life into bacteria and from one type of bacterium to another (Chapter 6). This process, referred to as *genetic engineering* or *recombinant DNA technology* ∞ (Chapter 1, p. 11) allows microorganisms to be used in unique ways. One of the first practical developments of genetic engineering was to place the human insulin gene into a bacterium. As this bacterium divided, the insulin gene divided and was passed into each daughter cell. Within a short time trillions of bacteria, each producing human insulin, were available. Consequently, an unlimited supply of human insulin can be produced for the treatment of diabetes. Many other applications of this new technology now being developed will allow common microbial cells to provide new and beneficial products for humankind (see Table 1-2).

MATERIAL FOR REVIEW

CONCEPT SUMMARY

1. Microorganisms hold an important place in the world of living things. Although unique in many attributes, these single cells exemplify all the characteristics normally associated with biological systems.

2. The microorganisms can be divided into several logical major groups. These groups are classified and named according to the systems used for larger life forms. The smallest and simplest forms of these biological structures are the viruses. More complex and larger are the prokaryotic bacteria and archaea, and three groups of eukaryotic microorganisms (fungi, protozoa, and algae) often included within the designation microorganism.

3. Microorganisms are ubiquitous in their habitat. They are found throughout the world, in, on, and around every conceivable environment. They play a major role in maintaining ecological balance and are often used in procedures that are beneficial to human beings.

STUDY QUESTIONS

1. What limitations are associated with a biological classification system that is entirely phenetic?

2. Could you justify a taxonomic proposal that places both prokaryotes and protozoa in the same kingdom of organisms? Explain.

3. What characteristics of fungi are associated with their ubiquitous presence?

4. Since microscopic algae are rarely involved in human disease, what aspect of these organisms makes them so important?

5. What concerns must be addressed if surface water is to be used for drinking?

6. If all of the bacteria present in milk are not killed during pasteurization, of what value is this process?

CHALLENGE QUESTIONS

1. Suppose you are visiting the Amazon rain forest of Brazil. There, in a sample of pond water, you see with a microscope several small, microscopic organisms. How would you go about determining if the microorganisms are eukaryotic or prokaryotic?

2. Having determined the basic cell type, how would you go about classifying the organisms as Bacteria, Archaea, or Eukarya?

Microscopy

The progress and development of the science of microbiology had to await the availability of tools, techniques, and procedures that could be applied to the study of very small life forms. As you read in Chapter 1, Antony van Leeuwenhoek built the first usable microscopes. With these primitive microscopes, he saw his "animalcules." However, a better understanding of the structure of microorganisms required the development of better microscopes. Today the light and electron microscopes can reveal much about the structure of microorganisms. When coupled with special preparation techniques for microscopy, the astonishing worlds of these prokaryotic and eukaryotic organisms can be clearly seen. This chapter describes the microscopes and optical techniques that have enabled direct visualization of many microbes and indirect observation of others. The chapter also discusses some of the staining techniques used to better examine and understand the composition and function of microorganisms.

MICROSCOPY

Because of the inability of the unaided human eye to perceive objects smaller than 100 µm, a microscope is essential to see microbial cells and to determine their size and shape. Two general types of microscopes are available: the *light microscope* and the *electron microscope*. Light microscopes use visible light waves as the source of illumination and are able to produce meaningful magnifications to about 1000 times (1000×). The electron microscopes use an electron beam as the source of illumination and are able to magnify well in excess of 100,000 times (100,000×). Most light microscopes are relatively easy to operate and are widely used in general and teaching laboratories. Electron microscopes, on the other hand, are large, expensive, complex instruments that are restricted to specialized laboratories with highly trained personnel.

✳ Light Microscopes

The earliest microscopes were little more than sophisticated magnifying glasses. Through these lenses a careful observer, such as Leeuwenhoek, could view objects at a magnification of about 300×. ∞ (Chapter 1, p. 5) A more powerful system of magnification was first developed about 1590 by Zacharias Janssens. His system was an imaginative application of optical principles that resulted in the development of the *compound microscope.* Today, virtually all light microscopes used in microbiology are compound microscopes; in other words, they use a series of lenses to magnify the object. The magnifying lenses are called the *objective* lens and the *ocular* or eyepiece lens. The objective lens is the primary magnifying lens and is positioned close to the material or object to be viewed. The ocular lens is positioned close to the eye of the observer and produces a secondary magnification of the image produced by the objective lens. Light is focused on the object by *condenser* lenses, which are not part of the magnification system. The magnifications of the objective and ocular lenses provide the **total magnification** of the compound microscope. The maximum magnification that can be obtained, under most conditions, from a single glass lens is about 100×, and most compound microscopes have objective lenses that give magnifications of 10×, 40×, or 100×. These lenses are mounted on a rotating base called a *nose piece* so that each lens can be easily moved into position as needed. Most ocular lenses give magnifications of 10×, although lenses of 1.5× to 20× are available. Thus by using the 10× ocular lens with the various objective lenses, total magnifications of 100× (10 × 10), 400× (40 × 10), or 1000× (100 × 10) are obtainable. The lower magnifications are used to observe large areas of the object or relatively large objects, and higher magnifications are used for more detailed observations of selected areas of the object. Magnification of at least 400× is needed for the observation of most microbial cells and 1000× is customarily used. An outline of an optical microscope is shown in Figure 3-1.

It might be asked: Why not use lens combinations such as a 100× objective and a 50× ocular lens to obtain magnification of 5000× or a 100× ocular to obtain a magnification of 10,000× and so on? Theoretically it can be done; however, no added

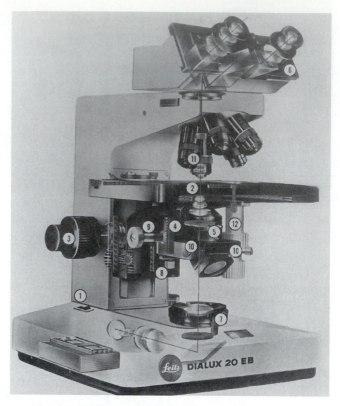

FIGURE 3-1 A light microscope. The light path is shown and the essential parts of the system are indicated: 1, light switch; 2, mechanical stage where the object to be viewed is placed; 3, focus adjustment; 4, condenser lenses; 5, light diaphragm control for condenser; 6, ocular lens or eyepiece; 7, light-focusing lens; 8, condenser adjustment; 9, condenser focus adjustment; 10, condenser light filter; 11, objective lens—this microscope has 5 objectives attached to a rotating base; 12, substage adjustment control. (Courtesy E. Leitz, Inc.)

value is achieved because this greater magnification shows no added detail. It is called "empty" magnification. The reason is that the *resolving power* (the ability to distinguish between two adjacent objects) is primarily a function of the *wavelength* of the type of illumination used. This relationship is defined according to the following formula:

$$\text{Resolving power} = \frac{0.5 \times \text{wavelength of illumination}}{\text{NA}}$$

where NA is the *numerical aperture* of the lens. Obviously the shorter the wavelength of light used

---

Total magnification The magnification of the objective lens multiplied by the magnification of the ocular (eyepiece).
Wavelength The distance between crests of a light wave.
Numerical aperture A measure of the quantity of light entering a lens.

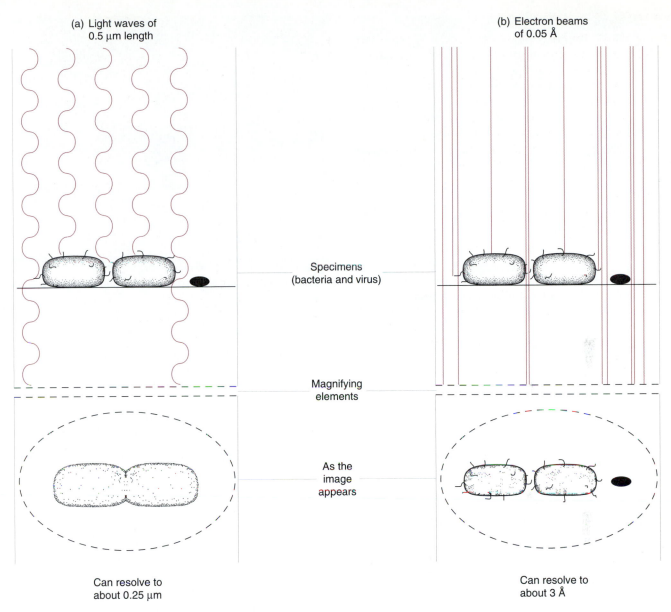

(a) Light waves of
0.5 μm length

(b) Electron beams
of 0.05 Å

Specimens
(bacteria and virus)

Magnifying
elements

As the
image
appears

Can resolve to
about 0.25 μm

Can resolve to
about 3 Å

FIGURE 3-2 The effects of wavelength of radiation source on resolving power of a microscope. (a) Visible light does not detect objects smaller than 0.25 μm and in this example does not show the pili, the virus, or the space between the two bacteria. (b) The electron beams do detect the pili, the virus, and the space between the bacteria.

and the larger the NA of the lenses (up to about 1.4), the greater is the resolving power of the microscope, and the greater the resolving power the greater is the useful magnification that can be obtained. The mid-wavelength of visible light used in optical microscopes is about 0.5 μm. Thus the maximum resolving power obtainable with usual lenses is about one-half the length of the light waves, or about 0.25 μm. The relationship between wavelength and resolving power is shown in Figure 3-2.

The lenses in an expensive microscope must always be corrected for two natural artifacts: *chromatic aberration* and *spherical aberration*. Chromatic aberration is a multicolored fringe that surrounds an object viewed through a single lens, while spherical aberration results from the curvature of the lens itself and makes part of the microscopic field appear to be out of focus. These aberrations can be corrected by adding additional lenses into the light path, each with an aberration that cancels the apparent aberration of the other.

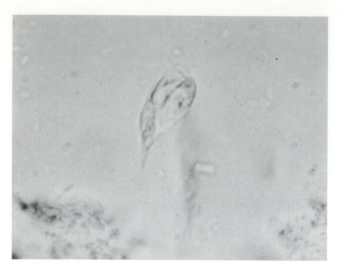 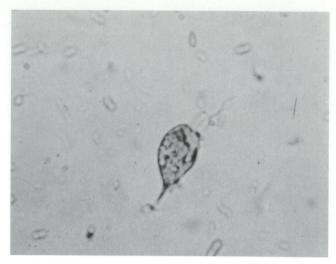

FIGURE 3-3 Photomicrograph of unstained parasite, *Giardia lamblia* (left) and same parasite stained with iodine stain (right). (Courtesy Centers for Disease Control, Atlanta)

✳ **Types of Optical Microscopy**

Several types of optical systems and procedures are used with the light microscope to enhance viewing of microorganisms.

Bright-Field Microscopy. In the most common form of microscopy, the light is focused direct-ly on the object and the resulting image is magnified and observed (Figure 3-3). Thus the image appears as a shaded object in a bright field (background). Most small objects, such as microorganisms, have only slight optical **contrast** from their surroundings and are very difficult to see. To enhance the visual contrast, bacteria are stained with various **dyes** (Figure 3-4).

Dark-Field Microscopy. In the dark-field microscope, light is directed onto the object at an angle so that only the light reflected from the object is magnified and observed (Figure 3-5). The image appears bright in a dark background (Figure 3-6). This form of microscopy is particularly useful for observing the movement of microorganisms, because no staining is needed and living objects can be observed. This procedure is also used to examine organisms that do not stain with simple staining procedures, such as the organism that causes syphilis (Figure 3-6).

Phase-Contrast Microscopy. This type of microscopy uses a special optical system that renders the details of unstained cells much more visible. It is particularly useful in observing living cells. Different components of a cell have slight differences in their refractive indices. These differences cause light rays to be bent or refracted as they pass

FIGURE 3-4 Photomicrograph of stained bacteria.

Contrast The ability to see an object against the background.
Dye A colored chemical compound used to stain microorganisms provides better contrast.

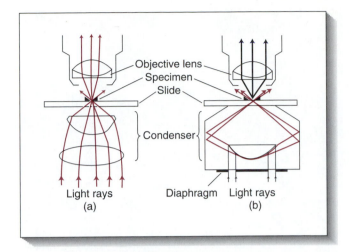

FIGURE 3-5 Bright-field v. dark-field microscopy. In bright-field microscopy (a) light passes directly *through* the specimen, whereas in dark-field microscopy (b) the only light visible is that which has been *reflected* off the specimen. (*Microbiology: Principles and Application, 3/e* by Black, J., © 1996. Adapted by permission of Prentice-Hall, Upper Saddle River, NJ.)

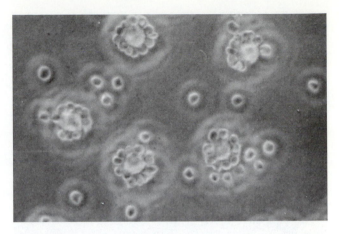

FIGURE 3-7 Phase-contrast micrograph of sheep erythrocytes surrounding mouse peritoneal cells. (Courtesy I. R. Tizard, *Bacteriological Reviews* 35:365)

from one portion of the cell to another. The refracted light is intensified by the optical system of a phase-contrast microscope and forms an image in which much of the fine detail of the cellular components can be seen. Staining is not needed and movement of cellular components can be observed (Figure 3-7).

Fluorescence Microscopy. Fluorescence microscopy allows us to observe materials that *fluoresce*—that is, materials that give off light of one

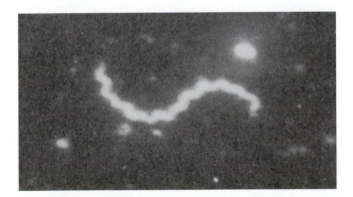

FIGURE 3-6 Photomicrograph of *Treponema pallidum* taken by dark-field microscopy. (Courtesy Centers for Disease Control, Atlanta)

color when subjected to light of another color. Most fluorescence microscopes use ultraviolet light as the primary light source; light given off by the fluorescent material is orange, yellow, or green, depending on the type of fluorescing material. To prevent injury, the ultraviolet light must be filtered before its rays reach the eyes of the observer. Some microorganisms contain naturally fluorescent materials, but in most applications of fluorescence microscopy the microorganisms to be observed must be stained with special fluorescent dyes. The dyes rhodamine, auramine, and fluorescein are most commonly used. They are used directly or attached to specific antibodies (see Chapter 12), allowing attachment to selected microbial cells or only to specific subcomponents of these cells. This procedure has made fluorescence microscopy a useful tool in diagnostic laboratories. Rapid detection and identification of an unknown microbe are possible by adding a known *fluorescent-tagged antibody* to a clinical specimen and then watching for the presence of fluorescence (see Plates 9, 26, 35, and 36). This procedure not only allows a specific identification of microbes but also increases the accuracy and speed of their detection.

✳ **Electron Microscopes**

Electron microscopes can provide much greater magnification than can optical microscopes because they use an electron beam as the source of illumination. The electron beam has a wavelength of only

Fluorescent-tagged antibody An antibody that has a fluorescent molecule attached.

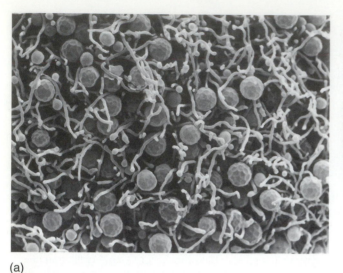

(a)

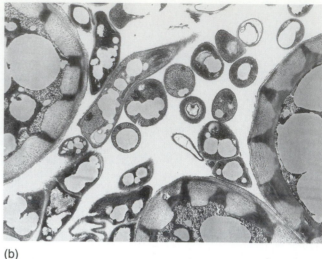

(b)

FIGURE 3-8   Electron micrographs of fungal spores of *Tilletia controversa*: (a) taken with scanning electron microscope; (b) taken with transmission electron microscope. (Courtesy W. M. Hess, Brigham Young University)

0.005 nm, which produces great resolving power. Due to technical problems, however, electron microscopes cannot actually resolve to 0.005 nm. Although several special electron microscopes can resolve to 0.1 nm, most have a resolving power for biological specimens of about 2 nm. This latter figure allows useful magnifications of well over 100,000×. Two general types of electron microscopes exist: *scanning* and *transmission* (Figure 3-8).

Transmission Electron Microscope (TEM).

The transmission electron microscope sends a high-voltage electron beam through the center of a vacuum column (Figure 3-9). The beam strikes the object to be viewed, which then casts a shadow of the object. As the beam continues down the column, electrons are deflected by magnetic fields to produce the desired magnification. The greater the deflection, the greater is the magnification. The electron beam then strikes a fluorescent screen at the bottom of the column, where an outline of the shadow of the object becomes visible. A photographic film is located under the fluorescent screen and a photograph of the object can be made by moving the screen and allowing the electrons to expose the film. The transmission electron microscope has made it possible to see particles as small as viruses as well as the internal structure of cells (see Figures 3-8b, 3-10, 31-1 and 35-1). This information has been

a tremendous aid in increasing our understanding of how biological systems function.

Various methods can be used to coat or stain an object with *electron-dense* materials such as lead or gold. These metals or salts increase the ability of the electron microscope to show contrast and details of minute structures (Figure 3-10). Certain limitations are inherent in the transmission electron microscope. Because the specimen must be observed in a vacuum, it requires dehydration, a factor that, along with the treatment with electron-dense staining materials, frequently causes distortion of the original structure. Moreover, the electrons cannot readily penetrate very dense or thick materials; consequently, the specimen must be spread in a thin film or cut into very thin sections to allow observation of internal components.

Scanning Electron Microscope (SEM).

The scanning electron microscope also uses an electron beam. The electrons are not magnetically amplified within the column, as with the TEM, but are focused into a narrow beam that rapidly scans back and forth over the specimen. The electrons reflected from each scan of the beam are picked up by an electron collector in the same orientation as they are reflected. These electrons produce an electrical current that is sent from the collector to amplifiers where each scan is amplified and the image is

Electron-dense   Contrast provided for electron microscopy using specific metals or salts.

Clinical Microscopy

The word *microbiologist* often evokes a mental image of a scientist looking through a microscope for endless hours. Like many other impressions, however, this one is generally false. Considering the great uniformity of shape and staining properties among bacteria, it is not likely that such intense attention to the microscope would be very productive. There are, however, several circumstances where the microscope is an invaluable tool to the clinical microbiologist. When specimens from infected patients are received by the laboratory, there are generally two questions that need to be answered: Are there any bacteria in the specimen? If there are, what kind are they? The microscope can provide good answers to the first question, and can give considerable help in answering the second.

Specimens from sick patients come either from areas of the body that are normally *sterile* (e.g., blood, or cerebral spinal fluid), or from areas that are the normal habitat of diverse bacteria (e.g., feces, throat, or skin). Proper use of the microscope can help evaluate both kinds of specimens. The microscopic observation of bacteria in a normally sterile body fluid is a strong indicator that these organisms are the cause of a patient's illness. Therefore, with a little practice, a microscopist can provide in a few minutes important diagnostic information that otherwise might take 24–48 hours to obtain. There are limitations to such procedures, however. If, for example, you were to place 10 bacteria into 1 ml of water and then try to detect these bacteria by microscopy, you would probably never be successful. In fact, the concentration of bacteria necessary to see even 1 bacterium per high-power microscope field is on the order of 100,000 organisms per milliliter. Therefore, to improve the probability that smaller numbers of bacteria will not escape observation, specimens are usu-ally concentrated by *centrifugation* before they are placed on a microscope slide. When procedures like this are used, bacteria can be detected in nearly 80% of the specimens from patients with *meningitis* and in over 90% of urine specimens from patients with *cystitis* (see Plate 22).

When specimens are collected from areas of the body where there is a normal **bacterial flora**, the microscopic examination is almost invariably positive for bacteria. The problem is to determine the meaning of such bacteria from a sick patient. Under these conditions, the microscopist looks for other clues to disease, such as the presence of white blood cells (pus) in addition to the bacteria. If such cells are present, then it is a reasonable conclusion that the specimen was obtained from the true site of infection in the patient. With this information, the microscopist then begins the careful examination of the bacterial cells in order to interpret the meaning of their presence: Do they represent normal bacteria or are they responsible for the infection?

Other useful information can be obtained by microscopic examination of clinical specimens. It is usually possible to distinguish bacterial from fungal infections, to identify protozoa, and to give clinical management support to physicians who submit tissue *biopsies* or material from an *abscess* for examination. In some instances it is even possible to make definite diagnosis, for example, of gonorrhea in males or of primary syphilis in any patient. Use of special stains makes possible up to 70% of the diagnoses of tuberculosis and Legionnaires' disease by microscopy (see Plates 20 & 34). The use of special stains and fluorescent antibody procedures continues to increase the number of disease agents that can be rapidly detected and identified by microscopic techniques.

Sterile   The state of being free of microorganisms.
Centrifugation   A technique that spins particles or cells into a highly concentrated mass.
Meningitis   An inflammation of the lining of the brain or spinal cord.
Cystitis   An inflammation of the bladder.
Bacterial flora   Bacteria that normally live on or in the body but do not normally cause disease.
Biopsy   The process of removing tissues from a patient for diagnostic examination.
Abscess   A localized accumulation of pus.

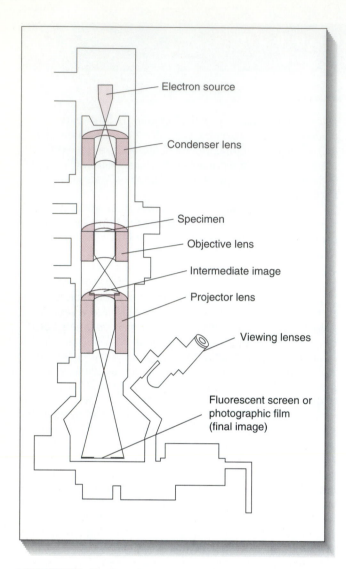

FIGURE 3-9   Operational schematic of an electron microscope. Magnetic fields are used both to focus and to enlarge the electron image of the object. (Adapted by permission of Prentice-Hall, Upper Saddle River, NJ.)

reconstructed line by line on a television screen. The scanning electron microscope is able to produce magnifications up to 100,000×. Nevertheless, much of the work with biological specimens is done at lower magnifications of 20,000×. Important advantages of the scanning electron microscope include its capability to create images in a three-dimensional reproduction with only slight distortion of the specimen (because the specimen need not be cut into thin sections or stained). As a result, materials can be viewed in their natural orientation (see Figures 3-8a, 14-1 and 15-7).

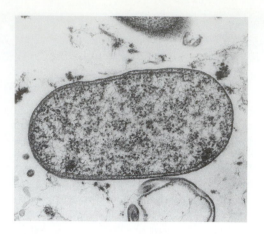

FIGURE 3-10   Electron photomicrograph of bacterial cell (×49,000). (Courtesy W. M. Hess, Brigham Young University)

Today innovations in electron microscopes are appearing rapidly. Instruments are now available that have the capabilities of both the transmission and the scanning microscopes, and some transmission electron microscopes are equipped with x-ray analysis equipment that permits areas of the specimen to be analyzed for the types of chemical elements they possess. New types of microscopes are becoming available, such as scanning tunneling electron microscopes, scanning electrochemical microscopes, atomic absorption microscopes, atomic force microscopes, and confocal laser microscopes. These new instruments provide the means of examining materials at the molecular level.

✳ Preparation of Microorganisms in Microscopy

For an object to be seen, it must contrast with its background. Often, however, objects that we wish to view through a microscope have little contrast from the medium in which they are suspended. To provide the needed contrast when observing with the bright-field microscope the microorganisms are stained. An aqueous or broth suspension of the microorganism is spread in a thin film over an area of a microscope slide and allowed to dry. The slide is then passed two or three times through a flame, which fixes the cells to the surface, and the cells are stained. Simple, negative, or differential staining procedures may be used. *Simple staining* involves adding a single dye to the cells for a time, followed by rinsing of the slide to remove excess stain. The

Simple staining   Adding a single dye to cells, followed by rinsing off the slide to remove excess dye.

TABLE 3-1

Steps in the Gram Stain

Step	Procedure	Results	
		Gram +	Gram −
Initial stain	Crystal violet for 30 seconds	Stains purple	Stains purple
Mordant	Iodine for 30 seconds	Remains purple	Remains purple
Decolorization	95% ethanol for 10–20 seconds	Remains purple	Becomes colorless
Counterstain	Safranin for 20–30 seconds	Remains purple	Stains pink

most commonly used simple stains are methylene blue, crystal violet, and a red dye called safranin. *Negative staining* is a procedure that stains the background but not the cells. For this procedure, either a black dye, called nigrosin, or India ink is mixed with the bacteria and is spread in a thin film on a microscopic slide. The stain will form a dark film over the background but will leave the cells as clear structures. Negative staining is generally used to show the presence of bacterial surface structures such as *capsules*. *Differential staining* procedures use more than one dye and are able to define bacteria or bacterial components, based on their staining characteristics. The most widely used differential staining procedure is the *Gram stain*. This process divides bacteria into two major groups: *gram-positive* and *gram-negative* (see Plates 3, 5, & 10). The steps leading to the reactions of the Gram stain are shown in Table 3-1. These and other special staining procedures help to visualize specific cellular structures, such as bacterial *endospores*.

MATERIAL FOR REVIEW

CONCEPT SUMMARY

1. The microscope is one of the primary tools employed by the microbiologist. Effective use of the microscope requires an understanding of how the properties of light limit our ability to resolve two proximal points as distinct entities.

2. Optical variations on the use of the light microscope to observe microbes in the living state include dark-field, phase-contrast, and fluorescence microscopy.

3. The light and electron microscopes and the procedures used to prepare specimens for microscopy have been useful in providing understanding of cell structure and in helping to define cell composition and function.

Negative staining   Using dye to stain the background but not the cells.
Differential staining   Using more than one dye, which enables identification of bacterial components.
Capsule   A polysaccharide layer surrounding bacterial cells peripheral to the cell wall.
Endospore   A bacterial spore associated with resistance to environmental inactivation.

STUDY QUESTIONS

1. What is the relationship between the resolving power and the useful magnification that may be obtained with the light microscope? What determines the resolving power of the lens system? What is the limit of resolution obtainable with the light microscope?

2. Distinguish between bright-field, dark-field, and phase-contrast microscopy and provide a specific example where each would be the method of choice for observing a culture of bacteria.

3. What advantages does electron microscopy have over light microscopy? What are the disadvantages?

4. Compare the use and the methodology of TEM with SEM. Provide at least one example where each would be the method of choice.

5. Distinguish between and provide an example for each of the following stains: simple, negative, differential.

CHALLENGE QUESTIONS

1. Consider the following situation: You have taken a photograph of a suspension of bacteria with phase-contrast microscopy using a 100× objective lens and a 15× ocular. You then enlarge the photograph three times. What will be the final magnification of the bacteria in your print of the enlargement? (Use the terms *resolution* and *empty magnification* in your answer.)

2. Refer to Table 3-1. In doing a Gram stain, you unintentionally forget to do the decolorization step. When you look at the bacteria with the light microscope, how will the bacteria appear? Could the lack of a decolorization step be important from a clinical microscopy perspective? Explain.

Cellular Anatomy

All cells share some fundamental characteristics in their structural components and biochemical functions. These similarities permit scientists to extrapolate information obtained from the study of one type of cell to other cells, and much of what we know about the biochemical reactions in complex eukaryotic animal cells was first discovered through studies of the simpler prokaryotic cells. Still, there are many differences between cell types, particularly when comparing prokaryotic with eukaryotic cells. A knowledge of these differences is essential in understanding why chemical agents can be selectively used to treat particular diseases. Chemicals, like penicillin, that specifically interfere with the function of components found in prokaryotic cells but are not present in eukaryotic cells have proven to be effective chemotherapeutic agents in treating diseases caused by prokaryotic cells. The following discussion deals with the major cellular components of both prokaryotic and eukaryotic cells.

PROKARYOTIC CELLS

Bacterial anatomy is discussed in this section.

✳ Sizes and Shapes

Bacteria are the smallest independently living cells, with most ranging from 0.25 to 3.0 μm in diameter and 1 to 20 μm in length. Some are slightly larger. Most of the thousands of different species occur in one of only three general shapes: spherical, rod-shaped (cylindrical), and curved (helical) (Figure 4-1).

Spherical bacteria are called *cocci* (singular *coccus*) and each species exhibits one of the characteristic arrangements of cells shown in Figure 4-1. The cylindrical-shaped bacteria are called *bacilli* (singular *bacillus*) or rods; most occur as single cells and not in arrangements like the cocci. However, under some growth conditions, some bacilli also occur in

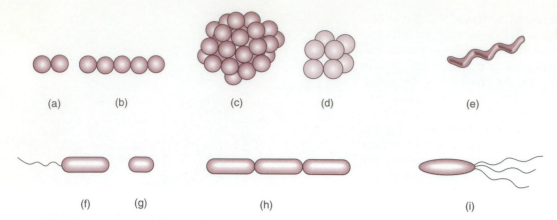

(a) (b) (c) (d) (e)

(f) (g) (h) (i)

FIGURE 4-1 Some variations and arrangements of the three general shapes of bacteria; (a) diplococci, (b) streptococci, (c) staphylococci, (d) sarcinae, (e) spirillum, (f) bacillus with monotrichous flagellum, (g) coccobacillus, (h) streptobacillus, and (i) bacillus with lophotrichous flagella.

pairs or short chains; some characteristically lie side by side in palisade-like arrangements and others may occur in Y or branching-shaped arrangements. Some bacilli are short, stubby rods between 0.5 and 1 μm in length with a diameter only slightly less than the length. Such cells, especially if they have rounded corners, have an elliptical shape and appear to be as much coccal-shaped as rod-shaped; the term *coccobacillary* is sometimes used to describe this shape. Many bacilli have definite rod shapes, with some species being significantly larger than others.

Some helical-shaped bacteria are called *spirilla* (singular *spirillum*) whereas others are grouped into the *spirochetes*. Chains of these organisms occasionally occur. Many variations exist between species as to length and number and amplitude of spirals.

✳ Components of Prokaryotic Cells

Figure 4-2 is a composite drawing of a bacillus-shaped bacterium showing many of the components found in prokaryotic cells. Most species of bacteria can be divided into two different categories based on their reaction to a procedure called the Gram stain. The Gram stain is a cell-staining procedure developed by the Danish microbiologist Christian Gram in the late 1800s. The Gram stain differentiates between two different types of cell walls, called *gram-positive* and *gram-negative*. Briefly, when subjected to the Gram staining procedure ∞ (Chapter 3, p. 33), the gram-positive cell wall retains a crystal violet–iodine dye complex, whereas the gram-negative cell wall does not. The difference between the two types of cell walls are discussed later.

✳ Cell Envelope Structures

Most prokaryotic cells are bound by three structures that are collectively called the *cell envelope*. The innermost structure of the envelope is the *cytoplasmic membrane*, next is the *cell wall*, and external to the cell wall is the *glycocalyx*. The cytoplasmic membrane is much the same in all cells. The composition of the cell wall differs markedly between gram-positive and gram-negative cells, while the nature of the glycocalyx varies from species to species.

Cytoplasmic Membrane. The cytoplasmic membrane, (sometimes called the *cell membrane* or *plasma membrane*) is a thin, fragile membrane located just inside the cell wall; it completely surrounds the internal cellular components (Figure 4-2). This membrane forms a functional barrier between the inside of the cell and the external envi-

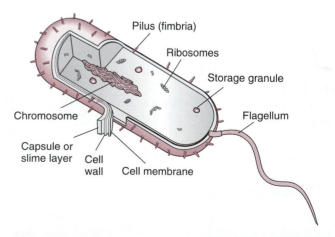

Pilus (fimbria)

Ribosomes

Storage granule

Chromosome

Flagellum

Capsule or slime layer
Cell wall
Cell membrane

FIGURE 4-2 A schematic diagram of a composite prokaryotic bacterial cell.

ronment. Numerous essential biological functions are carried out by the activities of this membrane, including the synthesis of cell wall materials, excretion of enzymes (**exoenzymes**) essential to the nutrition of the cell, determination of selective permeability, and transport of nutrient and waste products into and out of the cell. The cell membrane is the physical site of essential energy-producing chemical reactions such as electron transport (Chapter 5), and is the site of enzymes necessary for DNA replication, cell wall synthesis, and new membrane formation. The structure of the cytoplasmic membrane is similar in both prokaryotic and eukaryotic cells. The functional integrity of this membrane is essential for the survival of the cell, and any process or chemical that disrupts its structure or function causes the death of the cell. Several chemicals that function as disinfectants (Chapter 8) have an effect on this membrane.

The cytoplasmic membrane is only 7 to 10 nm in thickness and consists primarily of **phospholipids** and proteins. The phospholipid molecules have one end that is soluble in water, or **hydrophilic** (Greek *hydro* = water; *philus* = loving), and another end that is insoluble or **hydrophobic** (*phobia* = fear); this situation causes these molecules to spontaneously form a typical double-layer (bilayer) unit membrane when in an aqueous solution, with the hydrophilic ends pointing out and the hydrophobic ends pointing inward (Figure 4-3). Protein islands are imbedded throughout the phospholipid matrix (Figure 4-4). These proteins are essential for transport of *solute* (dissolved materials) across the membrane and into the cell. Two types of transport are possible: *passive*, which does not require the use of energy (e.g., k$^+$ transport out of the cell), and *active*, which requires that the cell use energy to transport the solute, (e.g., transport of sugar). Protein–enzyme systems called *membrane transport proteins*, located in the membrane, are involved in both types of transport. Thus a cell that does not have a transport protein for a specific nutrient is not able to bring that substance across the membrane and into the cell.

Several essential metabolic reactions occur at the

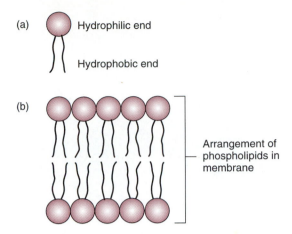

(a) Hydrophilic end

Hydrophobic end

(b) Arrangement of phospholipids in membrane

FIGURE 4-3   (a) The macromolecular phospholipid subunits that are part of biological membranes. (b) Arrangement of phospholipid molecules in membrane.

bacterial cell membrane. The cell membrane is an essential component of the energy-producing system for bacteria and is an essential part of the process by which **adenosine triphosphate (ATP)** is made available to the cell. Another function is the attachment site for the bacterial chromosome (DNA). Replication of DNA may be initiated by growth-related changes in the membrane.

Cell Wall.   The bacterial cell wall is a unique and important structure. No comparable structure is found in any animal cell, and the chemical structure differs from the cell walls found in plants. The cell wall is rigid and gives shape to the cell. Because bacterial cells are directly exposed to the external environment, the cell wall provides a necessary protection for the cell. Bacteria often live in fluids that contain relatively low concentrations of *ions* (atoms with + or − charges), whereas the inside of the cell (cytoplasm) contains high concentrations of ions. Water is drawn to the area of high ionic concentration and thus tends to flow into the cell. This situation creates an **osmotic pressure** inside the cell of up to 20 times atmospheric pressure. Because the internal osmotic pressure of bacterial cells is relatively high, these cells would swell and burst if it were not for the support of the rigid cell wall. Cell

---

Exoenzyme   An enzyme made in the cell and then transported outside where it is functional.
Phospholipid   An important structural molecule in all biological membranes.
Hydrophilic   Capable of dissolving in water (water-loving).
Hydrophobic   Incapable of dissolving in water (water-fearing).
Adenosine triphosphate (ATP)   An important molecule acting as the "energy currency" for all cells.
Osmotic pressure   Hydrostatic pressure on a biological membrane created by unequal concentrations of solute molecules on each side of the membrane.

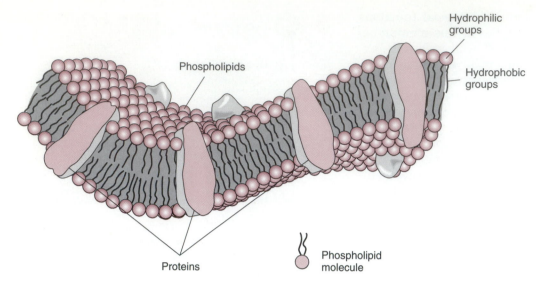

FIGURE 4-4   Cross section of a cytoplasmic membrane composed of two layers of phospholipid units with their hydrophilic ends pointing to the outside and their hydrophobic ends pointing to the center of the membrane. Protein molecules are "floating" in the "fluid-like" phospholipid membrane. (*Biology of Microorganisms, 7/e* by Brock, Madigan, Martinko, Parker, © 1994. Adapted by permission of Prentice-Hall, Inc., Upper Saddle River, NJ.)

walls are porous and allow the free passage of fluids and small molecules. Because the structure of gram-positive and gram-negative cell walls is different they are discussed separately.

Gram-positive cell wall   **Peptidoglycan** is a chemical structure that is unique to bacterial cell walls and makes up a major portion of the gram-positive cell wall. Peptidoglycan forms a coarse, layered, rigid meshwork that surrounds the cytoplasmic membrane and maintains the shape of the cell. A layer of peptidoglycan consists of two types of alternately joined sugar derivatives called *N-acetylglucosamine* and *N-acetylmuramic acid* that form long chains. The chains are cross-connected by short links of amino acids to form a mesh-type structure (much like a chain-link fence). Gram-positive cell walls have many layers of peptidoglycan and the layers are interconnected with other short chains of amino acids (Figure 4-5). This forms a rigid, multilayered shell. The types of amino acids used and the arrangements of cross-linkages may vary slightly among some bacterial species to give subtle differences in the peptidoglycan structure. Gram-positive cell walls may consist of up to 40 layers of peptidoglycan that forms as much as 90% of the cell wall.

In addition to the multilayers of peptidoglycan, gram-positive cell walls may also contain **polysac-**charide molecules called *teichoic acid*. These molecules are attached to the peptidoglycan or to the cell membrane and may extend outward from the cell wall. Although their exact function has not been fully discovered, teichoic acids do provide structural support for the cell wall.

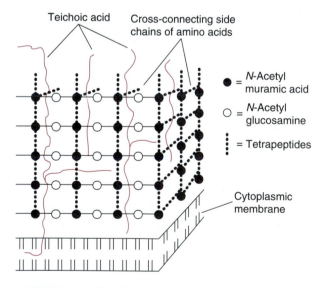

FIGURE 4-5   Schematic diagram of gram-positive cell wall with multiple layers of peptidoglycan.

Peptidoglycan   A large, complex molecule primarily made up of sugar derivatives and amino acids that forms the rigid structural portion of the bacterial cell wall.
Polysaccharide   A very large sugar molecule, composed of smaller sugars units.

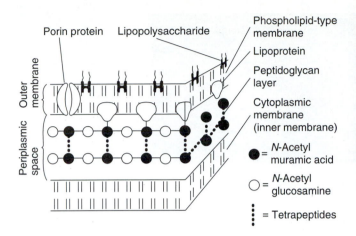

Porin protein   Lipopolysaccharide

Phospholipid-type membrane

Lipoprotein

Peptidoglycan layer

Cytoplasmic membrane (inner membrane)

● = *N*-Acetyl muramic acid

○ = *N*-Acetyl glucosamine

⋮ = Tetrapeptides

Outer membrane

Periplasmic space

FIGURE 4-6   Schematic diagram of gram-negative cell wall.

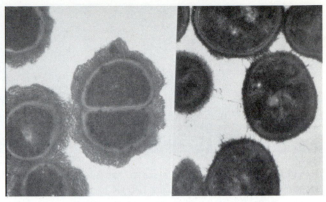

FIGURE 4-7   *Staphylococcus* strain with large glycocalyx (left) and a strain without glycocalyx (right). (Courtesy G. Christensen, *Infection and Immunity* 37:318, 1982: fig. 2, p. 362. Reprinted with permission of ASM.)

Gram-negative cell wall   The gram-negative cell wall is more complex than the gram-positive cell wall. It contains a thin, rigid, inner part often composed of only a single layer of peptidoglycan; this constitutes only from 5% to 20% of the total cell wall. Outside this peptidoglycan layer is an *outer membrane* that is composed of a phospholipid bilayer (similar to the cytoplasmic membrane), *lipoproteins, lipopolysaccharides*, and proteins. Some of these proteins, called *porins*, allow small molecules to readily pass through the phospholipid bilayer. Arrangements of these various components of the gram-negative cell wall are shown in Figure 4-6. The outer membrane forms the major part of the gram-negative cell wall.

The type of cell wall influences certain medical considerations. For example, penicillin is an effective antibiotic because it interferes with the synthesis of peptidoglycan. Thus, penicillin and related antibiotics are generally much more effective against infections caused by gram-positive bacteria that contain large amounts of peptidoglycan. The lipopolysaccharides associated with gram-negative bacteria are also called **endotoxins** and are responsible for some of the symptoms of infections caused by certain gram-negative bacteria. In addition, various components of cell walls may act as attachment molecules for some bacteria, and many of the large molecules function as **antigens** ∞ (Chapter 11, p. 154). Antibodies formed against these antigens may help protect the host against infection or aid in the specific identification of a bacterium (Chapter 12).

Glycocalyx (Capsule and Slime Layer). Most species of bacteria secrete polysaccharides and small proteins that produce a slimy, gel-type material that adheres to the outside of the cell wall (Figure 4-7). When this gelatinous material is dense and present in relatively large amounts it is called a **capsule**. When the layer of this material is thinner, less dense, and less tightly bound to the cell wall than the capsule, it is called the *slime layer*. The term glycocalyx was used in the past when referring to the slime layer. However, the current trend is to use the term glycocalyx when referring to either capsules or slime layers.

A capsule is readily demonstrated by its ability to exclude substances such as India ink (in the standard capsule staining procedure), as shown in Figure 4-8 ∞ (Chapter 3, p. 33). The composition and amount of the capsule varies from species to species (Table 4-1). The capsule may help protect bacteria against the external environment by slowing the rate of dehydration and by holding nutrients next to the cell wall. Of importance in medical microbiology is the ability of a capsule to greatly slow the rate at which white blood cells are able to ingest (*phagocytize*) capsule-possessing bacteria that have invaded the body tissues. Thus, the capsule gives a bacterium a greater opportunity for survival and in turn a better chance to cause disease. The disease-producing capabilities of some bacteria are directly related to the presence of a capsule.

The slime layer also aids in trapping nutrients, slowing dehydration, and binding cells together.

Endotoxin   Part of the outer cell wall membrane of gram-negative bacteria, which can be poisonous.
Antigen   A chemical molecule that can stimulate antibody production in an animal.
Capsule   An amorphous, gel-like material produced by some bacteria and collected around the outside of the cell.

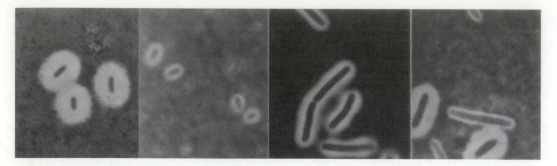

FIGURE 4-8 Various-sized capsules present on different species of the genus *Bacteroides*; magnified 1000 to 1200× (J. L. Babb and C. S. Cummins, *Infection and Immunity* 19:1088–1091, Figure 1, with permission from ASM)

Some of the molecules responsible for microbial attachment are contained in the slime layer. This attachment is an important first step that allows a bacterium to colonize tissues or adhere to a nonliving surface. Various medical problems, such as dental caries, result from this type of specific bacterial attachment to the surface of teeth. Many natural phenomena, as well as medical problems and industrial processes, are either beneficially or adversely affected by the abilities of certain bacteria to attach to given surfaces.

✳ External Appendages

Flagella.   Flagella (singular *flagellum*) are long, rigid hair-like appendages, composed of the protein *flagellin*, that extend out from some bacterial species (Figure 4-9). The purpose of flagella is to provide motility for the cell. Flagella are present on many species of bacilli, on some spirilla, but on very few species of cocci. Flagellin is organized into small aggregate structures that in turn are organized into hollow cylindrical structures. These flagella are attached to the cytoplasmic membrane by a small

hook at the end of the structure (Figure 4-10). Flagella rotate rapidly by using energy derived from an ion **concentration gradient**. The rotation of some flagella has been measured at over 2000 rpm.

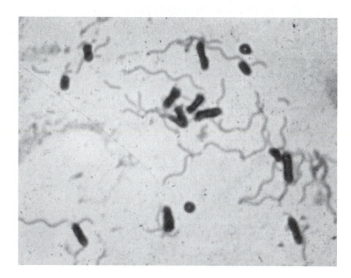

FIGURE 4-9 Bacteria stained with a flagellar stain. Flagella appear as wavy, thread-like structures.

TABLE **4-1**		
Composition of Bacterial Capsules		
Type of Capsule	**Composition**	**Organism**
Polysaccharide	Glucose–glucuronic acid	*Streptococcus*
Dextran	Glucose	*Leuconostoc*
Colanic acid	Various sugars	*Enterobacteriaceae*
Cellulose	Glucose	*Acetobacter*
Polypeptide	Glutamic acid	*Bacillus*

Concentration gradient   The difference in the concentrations of ions or molecules on each side of a biological membrane.

are present on some bacterial species but not on others. Two general types of pili are present. One type is long and hollow and only one or a few are attached to a cell; these are called *conjugation pili* and appear to function as tubes through which DNA may pass either during the process of conjugation ∞ (Chapter 6, p. 80) or when the bacterium is being infected with a virus (bacteriophage, see Chapter 28). Only some bacteria have conjugation pili.

The other type of pili are shorter and are numerous over the entire cell surface. These are called *attachment pili* and function as attachment sites between some bacteria and other surfaces. Some bacteria have attachment pili, others do not. Some microbiologists prefer to use the term *fimbriae* (Latin for "fingers") when referring to attachment pili.

✳ Cytoplasm

All components inside the cytoplasmic membrane are collectively referred to as *cytoplasm*. In all prokaryotic and eukaryotic cells, much of the cytoplasm is made up of proteins, nucleic acids, carbohydrates, and lesser amounts of other substances suspended in fluid. Certain anatomically distinct structures called **organelles** are found in the cytoplasm and are briefly described.

Ribosomes. Ribosomes consist of protein and **ribonucleic acid (RNA)**, and up to 10,000 ribosomes

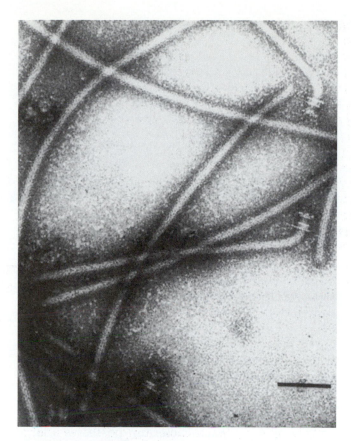

FIGURE 4-10 Electron micrograph of negatively stained flagella removed from *Salmonella typhimurium*. Bar represents 100 nm. (Courtesy T. Suzuki et al., *Journal of Bacteriology* 133:904)

This movement propels the bacterium through fluids. Some bacteria are able to move 10 times the length of their cell in one second; such directional movement of bacteria requires a large amount of available energy.

The term *trichous*, which means hair-like, is used when referring to the arrangements of flagella on bacterial cells (Figure 4-1). Certain bacteria have only a single flagellum and are called *monotrichous;* some have tufts of flagella and are called *lophotrichous;* still others have flagella protruding from all areas of the cell and are called *peritrichous.*

Pili. The structure of pili (singular *pilus;* Latin for "hairs") is similar in some ways to that of flagella, but pili are not associated with motility and are not composed of flagellin. Pili are hair-like structures that project out from the cell wall (Figure 4-11). They

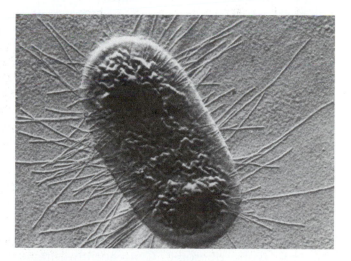

FIGURE 4-11 Electron micrograph of pili associated with *Escherichia coli*. (Courtesy S. Knutton et al., *Infection and Immunity* 44:514)

Organelle An anatomically distinct structure common to eukaryotic cells.
Ribosome A ribonucleic acid–protein complex used by cells for the synthesis of new protein.
Ribonucleic acid (RNA) One of the two kinds of nucleic acid.

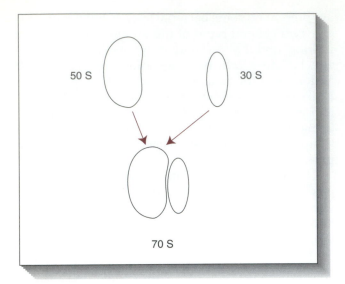

FIGURE 4-12   The two components of ribosomes that combine to form a 70S unit when attached to mRNA during protein synthesis.

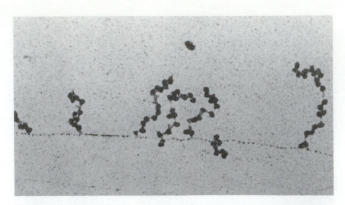

FIGURE 4-13   Polyribosome formation. Ribosomes (black dots) are attached to several RNA molecules and are active in carrying out protein synthesis. (Bar = 0.5 μm.) (From Hamkalo and Miller, "Electron Microscopy of Genetic Material," Figure 6a, p. 379. Reproduced with permission from *Annual Review of Biochemistry*, Volume 42. © 1973 by Annual Reviews, Inc.)

are present in a single cell; they are involved in the important function of protein synthesis. Ribosomes are about 20 nm in diameter and are made up of two unequal-sized lobular subunits, a 30S- and a 50S-sized component (Figure 4-12); (the S refers to Svedberg units, also known as the sedimentation constant, which is determined by the rate that particles can be sedimented in a centrifuge). The two subunits are separated when not involved with protein synthesis. The two subunits are composed of **ribosomal RNA (rRNA)** molecules and many protein molecules. When directly involved in protein synthesis, ribosomes are arranged in chains called *polyribosomes* (Figure 4-13). It is possible to separate particulate cytoplasmic components such as ribosomes from other organelles by the technique of **ultracentrifugation**. Under the high gravitational forces exerted by this procedure, particles separate into layers based on their density. The bacterial ribosomes have a density value of 70S. The 70S size of the prokaryotic ribosome is slightly smaller than the 80S size of ribosomes of eukaryotic cells; and this basic difference is associated with the ability of the antibiotic streptomycin to selectively combine with the prokaryotic ribosome. Consequently, streptomycin can be used as an antibiotic that can interfere

with prokaryotic cells, but will interfere with eukaryotic ribosome activity to a much lesser extent.

Nuclear Region.   The *nuclear region* is that part of the cytoplasm where the DNA molecule is located (Figure 4-14). Each bacterium possesses a large circular DNA molecule that contains the genetic information needed by the bacterium. This "chromosome" appears to be attached to the cytoplasmic membrane at one point. The absence of a membrane surrounding the chromosome is one of the main characteristics used to distinguish prokaryotic cells from eukaryotic cells. ∞ (Chapter 2, p. 14) In addition to the large chromosomal DNA molecule, some bacteria contain smaller circular DNA molecules called **plasmids**.

Cytoplasmic Inclusions.   Granules or globules are observed in many bacteria and are collectively referred to as *inclusions*. Many inclusions are aggregates of lipid, sulfur, carbohydrates, or a form of phosphate called *volutin* that can be stored by the cells as reserve food and energy supplies. The volume of the cell that they occupy depends on the growth rate and nutritional state of the cell. Inclusions are not surrounded by phospholipid bilayer membranes but are often made visible by special stains. A volutin-containing inclusion called

---

Ribosomal RNA (rRNA)   The type of RNA used, along with proteins, to build the structure of ribosomes.
Ultracentrifugation   A technique to separate particles or cell organelles using very high-speed centrifugation.
Plasmid   A small, circular DNA molecule found in some bacteria that is independent of the bacterial chromosome.

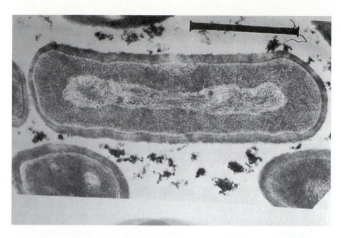

FIGURE 4-14 Electron micrograph of *Bacillus* species showing the central, lighter nuclear region of the cell. Note the absence of a nuclear membrane and the somewhat amorphous nature of the nuclear region. (Courtesy H. Kobayashi et al., *Journal of Bacteriology* 132:262.) Reprinted with permission of ASM.

a metachromatic granule is characteristic of *Corynebacterium diphtheriae* (Figure 4-15).

✳ Endospores

Three genera of gram-positive bacilli, *Bacillus*, *Clostridium* (Chapter 20), and *Sporosarcina*, are able to form a unique structure called an *endospore* or simply a *spore* (Figure 4-16). These spores are formed inside the bacterial cell, hence the prefix

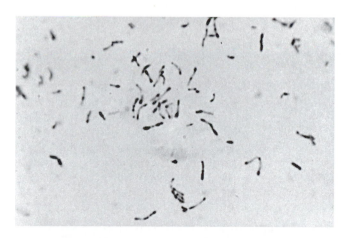

FIGURE 4-15 Gram stain of bacteria with metachromatic granules. These granules give the cells a rough or irregular staining appearance. (Courtesy Centers for Disease Control, Atlanta)

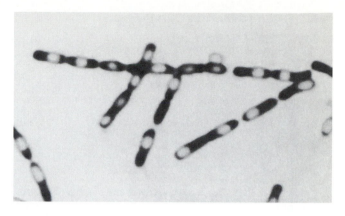

FIGURE 4-16 Photomicrograph of *Bacillus anthracis* showing white endospores (unstained) both within vegetative bacterial cells and free from parental cell. (Reprinted with approval of Abbott Laboratories, all rights reserved by Abbott Laboratories.)

endo. The term *vegetative bacterium* is used to refer to the actively growing, nonspore stage of a bacterial cell. Under optimal conditions of growth, the vegetative bacteria multiply with or without the formation of spores; that is, the spore is not a necessary step in replication. When growth conditions become unfavorable, such as a limitation in nitrogen or an energy source, the formation of endospores is stimulated. One spore develops per cell and forms a dormant, resistant stage for the cell. The endospore is formed by a sequence of changes called *sporulation.* The process of sporulation is very complex and involves the activities of a large number of proteins and enzymes. Following initiation, a chromosome near the terminal end of the cell is surrounded by infolding of the cell membrane. In addition to DNA, this membrane-bounded structure includes ribosomes and enzymes which will be needed for **germination**. This spore "core" is then surrounded successively by a spore wall which contains peptidoglycan (upon germination this material will be the basis for a new cell wall), a relatively thick cortex, and a relatively impermeable proteinaceous layer known as a spore coat. As the spore develops, water is removed to dehydrate and mature the spore. The bacterial spore is the most stable form of life known; consequently, special efforts must be made to destroy spores in order to achieve sterile conditions. These structures pose special concerns for medical and industrial microbiologists.

Germination The process whereby a bacterial endospore develops into a vegetative bacterium.

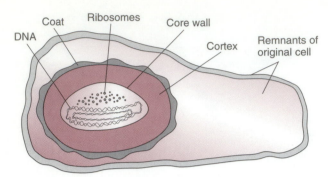

FIGURE 4-17 Major components of a bacterial endospore.

The great environmental stability of the bacterial endospore appears to be related to two major factors: the dehydrated state of the spore and the presence of a unique chemical known as *dipicolinic acid*. Dipicolinic acid is not found in any other living organism. The mature spore is highly resistant to destruction by heat or chemicals and may remain dormant for long periods. Some spores have germinated—that is, developed back into the vegetative stage—after lying dormant for hundreds of years. Germination of bacterial spores requires an activation step and an outgrowth process. Activation is accomplished by external damage to the spore coat by heat or mechanical means; outgrowth then occurs if the cell is in a nutritionally favorable environment.

EUKARYOTIC CELLS

Eukaryotic cells are larger and much more complex than prokaryotic cells (Table 4-2). Many diverse eukaryotic cells exist, ranging from yeast cells (unicellular fungi with some similarities to the prokaryotic bacteria) to highly specialized cells found in multicellular animals. The major organelles found in eukaryotic cells of animals are shown in Figure 4-18.

✳ Components of Eukaryotic Cells

Cell Walls. Of the eukaryotic cells, fungi, algae, and plant cells have cell walls. These walls are much simpler than the cell walls of bacteria. Most eukaryotic cell walls are composed of *cellulose* or other carbohydrates. The functions of eukaryotic cell walls are similar to those performed by prokaryotic cell walls. Animal cells, lacking cell walls, have a much more irregular shape and often are more readily affected by osmotic pressures than are cells enclosed within a wall.

Flagella and Cilia. Some eukaryotes have flagella or cilia for motility. The eukaryotic flagella are more complex than those present on prokaryotic cells but perform much the same functions. Flagella, as well as cilia, are composed of **microtubules** arranged in such a way so they can bend with a whip-like action. *Cilia* are structurally similar to the flagella except that they are much shorter and large numbers are usually arranged over the entire sur-

TABLE 4-2		
Comparison of Prokaryotic and Eukaryotic cells		
Characteristic	**Prokaryotic Cells**	**Eukaryotic Cells**
Nuclear region (nucleus)	Not membrane bound	Membrane bound
Flagella	Submicroscopic	Complex microscopic
Ribosomes	Small (70S)	Large (80S)
Cell wall	Complex peptidoglycan	Simple polysaccharide
Microtubules	Absent	Present
Chromosomes	Singular–circular, no histone	Multiple with histones
Cytoplasmic membrane	Present	Present
Endospores	Present	Absent
Membrane steroids	Absent	Present
Internal membrane-bound organelles	Absent	Numerous; e. g., Golgi mitochondria, chloroplasts, lysosomes

Microtubule A protein tubule that forms part of the structure of eukaryotic flagella and cilia.

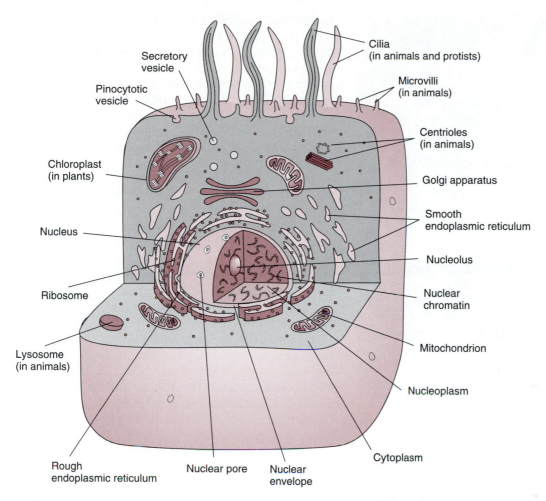

FIGURE 4-18   General eukaryotic cell diagram showing the major cellular components.

face of the cell (see Figure 40-1). Cilia are found only on some protozoa and some specialized animal cells, such as the ciliated epithelial cells of the respiratory tract.

Cytoplasmic Membrane.   This membrane is structurally similar in both prokaryotic and eukaryotic cells. Steroid-type **lipids** are present in eukaryotic cell membranes but not in most prokaryotic membranes. These compounds give animal cell membranes greater strength than membranes without steroid lipids. The cytoplasmic membrane is the outer limiting membrane of most protozoa (and all animal cells).

Nucleus.   The nucleus of eukaryotic cells is a prominent membrane-bound structure that contains most of the genetic material of the cell. A smaller structure inside the nucleus, the *nucleolus,* is

the area involved in the synthesis of ribosomes. Two membranes enclose the nucleus. *Nuclear pores* or holes pass through both membranes (Figure 4-19) and allow the passage of large molecules between the nucleus and the cytoplasm.

Internal Organelles.   Unlike prokaryotes, eukaryotes contain many organelles, many surrounded by a lipid membrane. These organelles provide specific functions to the cell. The *endoplasmic reticulum* is a plate- or tube-like system of membranes used for transport of proteins and lipids. The membrane forming the endoplasmic reticulum often is continuous with the outer of the two membranes of the nucleus.

The function of the eukaryotic ribosomes is the same as the prokaryotic ribosomes—to carry out the manufacture of proteins. Eukaryotic ribosomes are

Lipid   A hydrophobic substance, such as a fat, phospholipid, or steroid.

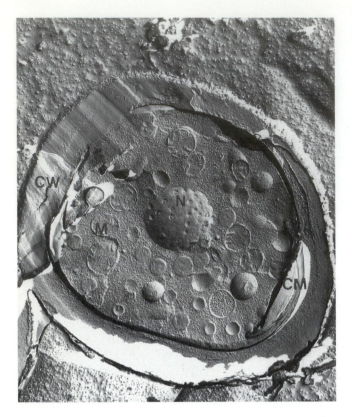

FIGURE 4-19 Electron micrograph of a fungal spore. This freeze–etch replica clearly shows the nuclear pores in the nuclear membrane, cell wall, cell membrane, and cytoplasmic inclusions (×25,000). (Courtesy W. M. Hess, Brigham Young University)

free in the cytoplasm or attached to the endoplasmic reticulum (Figure 4-19). These ribosomes have a size of 80S, which is larger than the prokaryotic ribosomes.

Another membrane-bound structure is the *Golgi apparatus*. This structure is made of a stack of membranes and receive proteins and lipids from the endoplasmic reticulum. A Golgi apparatus then modifies and packages these molecules for transfer to other locations in the cell or for export. Various membrane-bound vacuoles and organelles (mitochondria, chloroplasts, Golgi apparatus) are located throughout the interior of eukaryotic cells.

Lysosomes are membrane-bound organelles that contain digestive enzymes. They are found in animal cells such as the white blood cells that ingest (phagocytize) foreign particles. The lysosomes combine with other foreign particles to produce a *phagolysosome* in which digestion occurs ∞ (Chapter 10, p. 146).

Two other major organelles found in eukaryotic cells are the *mitochondria* and *chloroplasts*. Mitochondria are rod-shaped structures about $1 \times 3\ \mu m$ in size and are associated with ATP energy manufacture. Chloroplasts are prominent chlorophyll-containing structures involved in photosynthesis in algae and plants. The sizes and shapes of chloroplasts vary among the different types of cells. Both mitochondria and chloroplasts contain small amounts of DNA. This genetic material is func-

CLINICAL NOTE

Clinical Application of the Gram Stain

The Gram stain is a useful tool for bacterial identification. In spite of the facts (1) that there are only two reactions (positive and negative), and (2) that all bacteria have only one of three basic shapes (cocci, bacilli, and spirals), these few features can be used to categorize bacteria into fairly specific groups. Under some circumstances, based only on the morphology and Gram-staining characteristics of a bacterial cell, it is possible to designate the genus if not the species of the bacterium under question. Thus, if a specimen contains paired gram-negative cocci, the identification can quickly be narrowed to only two genera of clinically relevant bacteria. Likewise, if the specimen has relatively large gram-positive bacilli that have blunt ends, only two genera are possible. By knowing the morphology and Gram-staining reaction of a bacterium it is often possible for the microbiologist to provide rapid and useful data to the physician.

Even if this were its only use, the Gram stain would be valuable, but this procedure makes another, perhaps even more significant, contribution: The Gram stain can be used as a guide in providing appropriate, early antibiotic therapy for a patient. Bacteria of similar Gram morphologies tend to have similar susceptibility to specific antibiotics. For example, if a patient is reported to have a gram-negative bacillus as a possible cause of a urinary tract infection, the physician can make a likely choice of antibiotic for treatment even before the identity and antibiotic susceptibility of the organism is known. By this means, application of the Gram stain to clinical specimens can have significant impact on therapy. This situation is particularly true for patients suffering from life-threatening diseases, such as meningitis, where the effects of early therapy are especially beneficial.

tional and can determine the amino acid sequences of some proteins. It is thought that both mitochondria and chloroplasts represent highly adapted prokaryotes that live in essential **symbiosis** with eukaryotic cells.

Vacuoles are membrane-bound areas within the cytoplasm of some cells. They are associated with food storage, digestion, osmotic regulation, and excretion of waste products. Their number and size may change with the physiological state of the cell.

MATERIAL FOR REVIEW

CONCEPT SUMMARY

1. Bacteria are prokaryotic cells with limited variation in shape and unique cell wall structures. The composition of the cell wall is a determinant in the staining characteristics of the bacteria, dividing them into gram-positive and gram-negative species.

2. Bacteria possess a number of structures related to cellular function. Among them, flagella, capsules, and endospores are the most noticeable with the light microscope.

3. Eukaryotic cells have many more structures and organelles than do prokaryotic cells. Among them, the nucleus, mitochondria, and chloroplasts are the most prominent with the light microscope. Each type of eukaryotic organelle has a specific function to carry out.

STUDY QUESTIONS

1. Draw and label a cross-section diagram of both a gram-positive and a gram-negative bacterial cell.

2. In what ways can the presence of a bacterial capsule increase the likelihood of bacterial survival in a competitive environment?

3. What are the functional differences between bacterial flagella and pili?

4. A major function of bacterial cytoplasmic membrane is selective transport of materials into or out of the cell. What mechanisms are available to the membrane to carry out this function?

5. List the primary function of each of the following: (a) pili, (b) ribosome, (c) DNA, (d) endospore, and (e) cytoplasm.

CHALLENGE QUESTIONS

1. ATP energy manufacture is required by all cells, prokaryotic and eukaryotic, for survival. Where does this energy manufacture occur in a fungal cell, a protozoan, and a bacterium?

2. Both bacteria and protozoa are single cells that carry out all the reactions required for survival. Explain why the protozoan has so many more organelles than does the bacterium.

3. Based on your knowledge of the bacterial cell, list those structures and characteristics that would be advantageous for a bacterium to cause disease.

Symbiosis　The co-habitation of two different types of organisms.

5 Metabolic Functions

The rapid production of new cellular material requires the breakdown of relatively large amounts of nutrient that must be brought into the cell through the cytoplasmic membrane. Fortunately, the small size of most microbial cells provides for a large surface area relative to cell volume in order to facilitate this rapid growth. In addition, microbial cells must manufacture enough new cellular material to double the cell mass for each generation. These biochemical activities require the presence of many energy-producing and energy-using reactions within microorganisms. To regulate these reactions, the proper enzymes must also be present and active at the right moment.

This chapter briefly outlines the steps involved in the release and transfer of energy from foodstuffs; as well as the production of new building blocks needed for cell growth. The structure and functions of enzymes and other proteins are also discussed. The intention of this chapter is to present a conceptual, generalized view of cellular metabolism, which, by necessity, omits many details of this complex subject.

ENERGY METABOLISM

Any chemical change or reaction that occurs within a cell is called **metabolism**. Reactions involved with the breakdown of food materials for the release of energy for the cell are called *catabolism*. Other reactions involved with the production or building of new cellular components are called *biosynthesis* or *anabolism*. Catabolism and biosynthesis are usually coupled together such that the energy released from catabolic reactions can be directly used for biosynthesis. Figure 5-1 presents a simplified outline of the

> **Metabolism**   The sum of all chemical reactions going on in a cell or microorganism.

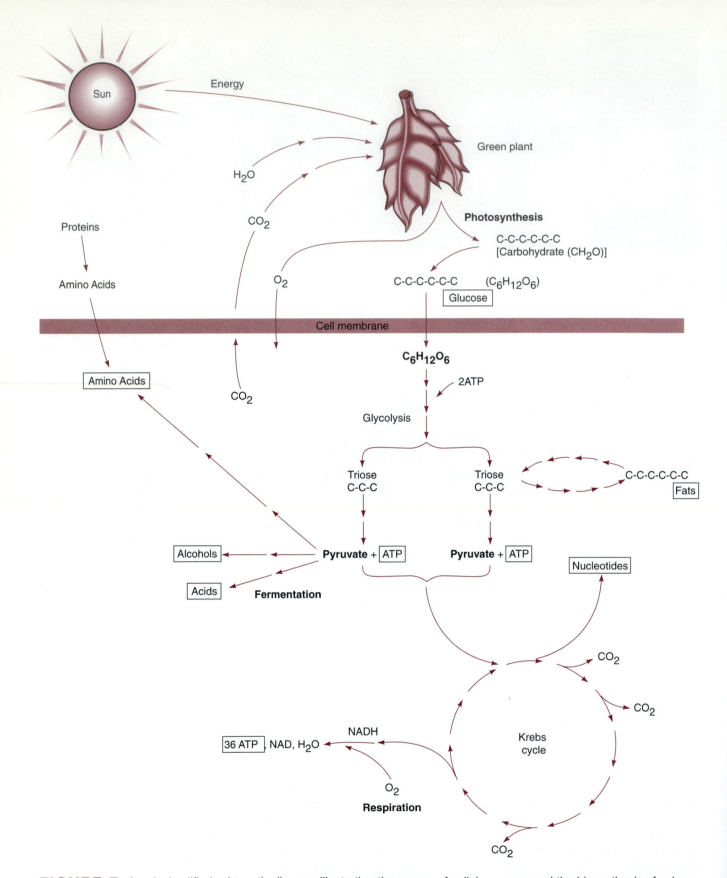

FIGURE 5-1   A simplified schematic diagram illustrating the source of cellular energy and the biosynthesis of cellular materials. The reactions above the cell membrane are representative of photosynthesis (requiring water, carbon dioxide, and light energy for the synthesis of carbohydrates and the release of oxygen). The metabolic reactions occurring in bacterial and animal cells that require chemical energy such as glucose are shown below the cell membrane. These reactions produce the necessary ATP and cellular components for growth while releasing both carbon dioxide and water as end products.

reactions involved in energy metabolism and the biosynthesis of cellular materials.

✷ Photosynthesis

The original source of energy used in the vast majority of biological systems is the sun. *Autotrophic* organisms convert radiant energy from the sun into a form of chemical energy through the process of *photosynthesis*. The available chemical energy can then be used by other living organisms that require organic matter as their source of energy (*heterotrophic* organisms). Green plants growing on land and algae growing in water are responsible for most of the photosynthetic activity occurring on earth.

During the process of photosynthesis, radiant energy from the sun is collected by chlorophyll molecules and is used to rearrange the atoms contained in carbon dioxide (CO_2) and water (H_2O), as follows:

Water is broken into its atomic components, hydrogen and oxygen. The oxygen is then released as a molecule and the hydrogen is combined with CO_2 (Figure 5-2). In this *reduced* state, molecules of CH_2O are combined to form whatever size carbohydrate $(CH_2O)_n$ is required by the cell. Energy is required to form a reduced compound, such as a carbohydrate, and much of this energy is retained in the compound after it is formed. This stored energy can be released for use by a cell at a later time when oxygen is added to the molecule or when hydrogen atoms are removed, returning the molecule to the *oxidized* state (Figure 5-3).

This description of photosynthesis is brief and generalized. Relatively few species of bacteria carry out photosynthesis; in fact, with the exception of the cyanobacteria, bacterial photosynthesis does not include the breakdown of water and proceeds along

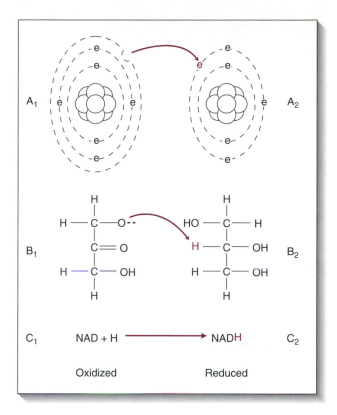

FIGURE 5-3 Oxidation-reduction reactions involve the transfer of electrons. Losing an electron (A_1) results in oxidation, whereas adding an electron results in chemical reduction (A_2). In metabolism, a proton is often transferred with the electron; thus, moving a hydrogen atom (B_1, C_1) results in oxidized molecules, whereas gaining hydrogen produces reduced molecules (B_2, C_2).

a somewhat different pathway from photosynthesis in eukaryotic plants. Even so, these photosynthetic bacteria are able to convert sufficient solar energy to provide for their own metabolic needs.

Various sizes of carbohydrate molecules are formed by photosynthetic plants. The most common molecules are the simple sugars, called *monosaccharides*, usually containing five or six carbon atoms. Two monosaccharides may combine to form a *disaccharide*; many monosaccharides connected together form *polysaccharides*. Starch and cellulose are the major polysaccharides produced by plants (Figure 5-4). These carbohydrates are the major energy-containing food molecules produced by photosynthesis.

$$\text{Sunlight} + 6\,H_2O + 6\,CO_2 \longrightarrow 6\,O_2 + C_6H_{12}O_6$$

FIGURE 5-2 Balanced photosynthetic reaction. Water and carbon dioxide are combined to produce oxygen, carbohydrates, and water.

Reduced   The condition of a molecule to which electrons (hydrogens) have been added.
Oxidized   The condition of a molecule from which electrons (hydrogens) have been removed.

A. Glucose (monosaccharide)

B. Maltose (disaccharide)

C. Starch (polysaccharide)

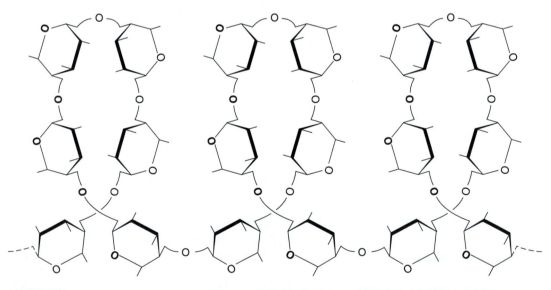

FIGURE 5-4   Carbohydrate monomers (glucose) used to make a large polymer of glucose molecules—a polysaccharide.

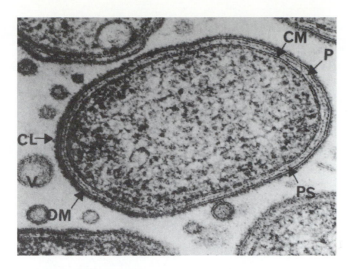

FIGURE 5-5 Electron photomicrograph of a bacillus-shaped bacterium. Note the periplasmic space (PS) between the outer membrane (OM) and the cytoplasmic membrane (CM). (Courtesy T. J. MacAlister. *J. Bacteriol.* 169:3495)

✳ Catabolism

Energy stored in the chemical bonds of carbohydrates and other molecules, such as proteins and lipids, can be made available for biosynthesis through a complex series of catabolic reactions. The main source of energy for most cells is a six-carbon sugar called *glucose*, or a *polymer* of glucose such as starch. Before glucose can be used by the cell, it must be separated from the large polymer. This separation is accomplished in microbial cells by digestive enzymes that are released from the cells into the extracellular environment. In gram-negative bacteria, these extracellular digestive enzymes are often concentrated in the space between the peptidoglycan layer and the outer membrane of the cell wall (*periplasmic space*), where the larger molecules can be broken down more efficiently (Figure 5-5). The glucose molecules contained in starch are relatively easily separated from the larger molecule, whereas the sugars that make up cellulose are more difficult to digest. Some microbes, however, are endowed with a wide array of enzymes that enable them to break down even the most complex and resistant carbohydrates, such as cellulose. Most digestion of the cellulose that is formed by plants is done by microorganisms. Although carnivorous animals are unable to digest plant cellulose carbohydrates, herbivorous animals, such as cattle and sheep, are able to feed on plants and digest cellulose because of the kinds of microorganisms that live in their intestinal tract. These helpful microorganisms (bacteria, protozoa, and fungi) are semipermanent residents of the digestive tract of herbivorous animals, and break down the large polysaccharide molecules into smaller disaccharides and monosaccharides that can then be absorbed by the animal from its intestine.

✳ Metabolic Energy

Glucose provides a cell with a source of readily available energy. However, it is not convenient for a cell to have glucose involved in every chemical reaction that requires energy (*endergonic* reactions). To make energy available for individual reactions, the cell produces one of a number of high-energy compounds, the most common of which is **adenosine triphosphate (ATP)** (Figure 5-6). Energy obtained by metabolic processes is briefly stored in these ATP molecules. ATP contains three phosphate groups, two of which are connected by high-energy bonds that are readily available for various cellular functions. When energy is released, the ATP molecule loses a phosphate group and is changed to an adenosine diphosphate (ADP) molecule. The ADP molecule can be changed back into an ATP molecule when energy and phosphate again become available from the oxidation of other compounds (Figure 5-7).

✳ Glycolysis

Glucose molecules readily pass across the bacterial cytoplasmic membrane; once inside the cell, they enter into a series of reactions that slowly extract the energy stored in these sugar molecules. The first series of reactions in the process of glucose catabolism, called **glycolysis** (splitting of glucose), begins with several changes occurring in the structure of the glucose molecule. As with most chemical reac-

---

Polymer   A relatively large molecule composed of repeating subunit molecules (monomers).
Periplasmic space   The region between the cytoplasmic membrane and the outer membrane where many catabolic reactions occur in gram-negative bacteria.
Adenosine triphosphate (ATP)   A high-energy molecule that is the "energy currency" for all cells.
Glycolysis   A series of reactions involving the catabolism of glucose, resulting in the production of pyruvic acid.

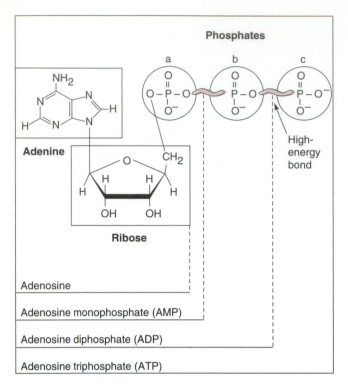

Phosphates

Adenine

Ribose

Adenosine

Adenosine monophosphate (AMP)

Adenosine diphosphate (ADP)

Adenosine triphosphate (ATP)

FIGURE 5-6 Adenosine (a) mono-, (b) di-, and (c) triphosphate showing the position of the high-energy (~) phosphate bonds. (*Fundamentals of Anatomy and Physiology, 3/e* by Martini, © 1995. Adapted by permission of Prentice-Hall, Inc., Upper Saddle River, NJ.)

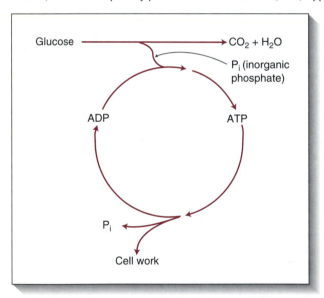

FIGURE 5-7 The ATP-ADP cycle. Energy obtained through the oxidation of glucose is used to form ATP. This energy is then released by ATP to carry out energy-requiring cellular processes. The resultant ADP is then available to be used in the formation of a new ATP molecule.

tions, even those that yield a supply of energy (*exergonic* reactions), it is necessary to initiate the catabolism of glucose by adding energy to the molecule (Figure 5-8). This energy is frequently obtained from the energy available in ATP. In carrying out this reaction, the phosphate group that is removed from ATP is attached to the number six carbon of the glucose, producing glucose-6-phosphate:

In this reaction, the energy from the ATP is transferred to the glucose-6-phosphate molecule, and this molecule becomes less stable and more subject to chemical change than is pure glucose.

Phosphorylation of glucose can occur either directly as the glucose crosses the cell membrane or after it is in the cytoplasm of the cell. After the glucose molecule is phosphorylated it is converted to another six carbon sugar known as *fructose-6-phosphate*, which is again phosphorylated, giving a compound known as fructose-1,6-diphosphate. Thus two units of ATP energy are required to prepare glucose for its final oxidation (Figure 5-8). Further reactions result in the splitting of this six-carbon sugar

Phosphorylation The addition of phosphorus to a molecule. This process is usually accompanied by a transfer of a relatively large amount of energy.

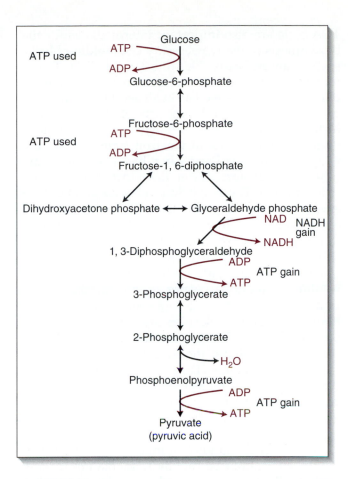

FIGURE 5-8 Simplified schematic of the glycolytic pathway. Note that two ATP molecules are required to activate this pathway. ATP is produced in converting glyceraldehyde phosphate to pyruvic acid. This process results in a net increase of two ATP molecules because the six carbon glucose molecules are converted into two three-carbon glyceraldehyde phosphate molecules.

into two molecules containing three carbon atoms each. These three carbon compounds are further altered to form a compound called *pyruvic acid.* Thus in the reactions that change glucose to pyruvic acid, two ATP molecules are used to prime the reaction; then two ATP units are generated for each of the two pyruvic acid molecules formed. The net gain in energy available to the cell at this point is two ATP molecules, because four were generated and two were used to prime the reaction.

A second significant energy-associated reaction occurs during glycolysis. For each molecule of glucose converted to pyruvic acid, two hydrogen atoms are removed and attached to a carrier mole-

cule known as *nicotinamide adenine dinucleotide,* or simply NAD. This molecule, now reduced to NADH, may be used later by the cell to obtain additional energy (Figure 5-10). In addition to glycolysis (known as the *Embden–Myerhof pathway),* bacteria have other pathways—for example, the *Entner–Doudoroff pathway* and the *pentose phosphate pathway*—by which important energy-containing sugars can be metabolized. In fact, some bacteria have the ability to choose more than one of these pathways as needed.

✱ Fermentation

After pyruvic acid has been produced from glucose, the next step in the metabolic process is determined for the cell by the availability of oxygen. The reactions of glycolysis do not use oxygen and if oxygen is not available (*anaerobic* metabolism), pyruvic acid may subsequently be converted into such products as alcohol or organic acids, a process known as **fermentation** (Figure 5-1). To carry out this process, the NADH that was obtained during glycolysis is now used to reduce pyruvic acid either directly to lactic acid (a process that occurs in human muscle cells) or indirectly to ethyl alcohol (a process that occurs with brewers yeast and some bacteria). The reaction leading to lactic acid

2 pyruvic acid r 2 NADH —R 2 lactic acid r 2 NAD

ensures the cell of a continued supply of NAD for use in glycolysis (Figure 5-8). A net gain of only two ATP molecules is produced in the conversion of one molecule of glucose to two molecules of alcohol or lactic acid. Under these circumstances, most of the energy is still contained in the alcohol or lactic acid molecules. Only certain types of microorganisms are able to produce alcohols or other similar products of fermentation. Many such compounds have commercial value and controlled fermentation is carried out on a mass scale in many industrial processes. Because most of the energy originally present in glucose is still found in the alcohol molecules, and because it is easily released by combustion with oxygen, much effort has been directed toward increasing production of alcohol to be used as a supplemental fuel for internal combustion engines.

✱ Aerobic Respiration

If oxygen is available and the microorganisms can use oxygen (*aerobic* metabolism), pyruvic acid can

---

Fermentation   The conversion of pyruvic acid into such products as alcohol and organic acids.

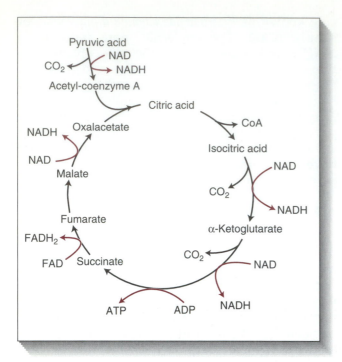

FIGURE 5-9 Schematic of the chemical reactions occurring in the Krebs cycle. Reactions that yield large amounts of energy are shown in color.

TCA cycle are also important intermediates for the biosynthesis of many necessary cell building blocks such as amino acids (Figure 5-1; Table 5-1). Therefore, this cycle is also operational in many anaerobic microorganisms as well as in all aerobes. Significant energy-associated end products of the TCA cycle include NADH and $FADH_2$. These reduced *nucleotides* enter into an **electron transport system** where they are oxidized when hydrogen atoms are ultimately transferred from them to an oxygen atom to form water. These reactions release a great deal of energy as ATP (Figure 5-10). Thus the chemical energy contained in pyruvic acid is slowly released through a stepwise series of reactions (TCA cycle and electron transport) involving the passage of electrons from the carbohydrate to oxygen with the resulting formation of water and the release of CO_2. In all, for every glucose molecule metabolized by the cell, 40 ATP molecules are generated by this process (Table 5-2). Because the cell used 2 ATP molecules to initiate the glycolytic reaction, a net total of 38 available ATP molecules are generated by the oxidation of 1 molecule of glucose. Respiration is a highly efficient energy-releasing system, returning more than 35% of the energy originally available in glucose to the cell.

The energy demands of human cells are such that their needs cannot be met by fermentation for any length of time. This explains our continuing demand for oxygen, which is the terminal *electron acceptor* in the electron transport system. *Anaerobic respiration*, which uses compounds such as sulfur or nitrate instead of oxygen as electron acceptors, is

be oxidized completely to CO_2 and H_2O with the transfer of the energy released from pyruvic acid to ADP molecules. These molecules then become ATP. When reactions use oxygen or other appropriate acceptors the process is called *respiration*. This process proceeds when pyruvic acid enters into a series of reactions forming the *tricarboxylic acid* (TCA) or **Krebs cycle** (Figure 5-9). In the Krebs cycle, the pyruvic acid is further oxidized and hydrogen is transferred to an appropriate carrier molecule, such as NAD.

Pyruvic acid does not directly enter the TCA cycle, but is first modified by the removal of one carbon atom, which is released as CO_2. With each round of the TCA cycle, two additional CO_2 molecules are produced and released. Thus, by the time the TCA cycle goes through two complete revolutions, each of the carbon atoms originally present in glucose have been oxidized to CO_2. The organic compounds (e.g., oxaloacetate) that make up the

TABLE 5-1			
Naturally Occurring Amino Acids			
Alanine	Glutamic	Leucine	Serine
Arginine	acid	Lysine	Threonine
Asparagine	Glutamine	Methionine	Tryptophane
Aspartic	Glycine	Phenylalanine	Tyrosine
acid	Histidine	Proline	Valine
Cysteine	Isoleucine		

Krebs cycle   A series of reactions where pyruvic acid is completely oxidized and hydrogen is transferred to an appropriate carrier molecule, such as NAD.

Nucleotide   A nitrogenous compound found in DNA, RNA and ATP as well as a variety of coenzymes (NAD).

Electron transport system   A series of steps that oxidize NADH and $FADH_2$ by transferring a hydrogen atom from them to an oxygen atom.

Electron acceptor   An atom or molecule (such as oxygen) that is relatively easily reduced by accepting electrons and protons.

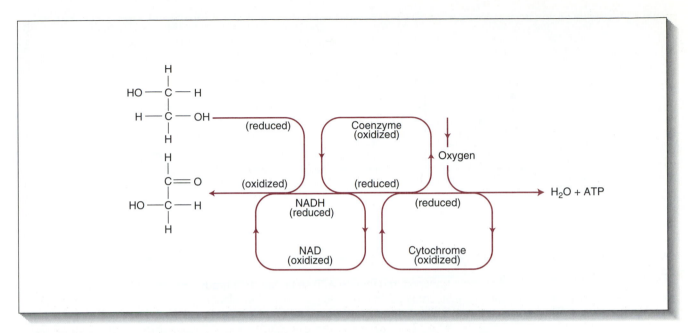

FIGURE 5-10 A schematic diagram representing the transfer of hydrogen from the oxidation of carbohydrate to NAD and from NADH to the electron transport system, which ultimately uses two hydrogen atoms to reduce an atom of oxygen to a molecule of water. These reactions result in a concurrent production of ATP.

known to occur in some bacteria. These reactions make energy available to the bacteria in the absence of oxygen but similar reactions do not occur in animals. The CO_2 that is formed as a by-product of respiration is released from the cell into the atmosphere where it is again able to enter into the reactions of photosynthesis for conversion into an energy-containing carbohydrate. On the other hand, the oxygen given off as a by-product of photosynthesis is used in the respiration reactions of the cell, thus forming one of the important chemical cycles involved in the balance of living systems (Figure 5-1).

Many of the energy-producing reactions occur-

ring within bacterial cells take place at or near the cell membrane ∞ (Chapter 4, p. 36). The membranes serve as physical supports for the enzymes involved. In the electron transport system, the various *cytochromes* that are used to carry the electrons have a fixed orientation in the cytoplasmic membrane. A number of years ago, this observation provided the basis for the present *chemosmotic theory* of active transport in bacteria (Figure 5-11). In this process, electrons are passed from a reduced compound to an oxidized compound by means of the membrane carrier molecules. By means of the orientation of the cytochrome molecules, hydrogen ions *(protons)* are transported to the outside of the mem-

TABLE 5-2

Sources of ATP from Metabolism of 1 Molecule of Glucose

Pathway	Products	ATP Produced
Glycolysis	NADH, pyruvic acid	4
Krebs cycle	NADH, $FADH_2$, CO_2	2
Electron transport	H_2O	34

Cytochrome An electron carrier in the electron transport system.

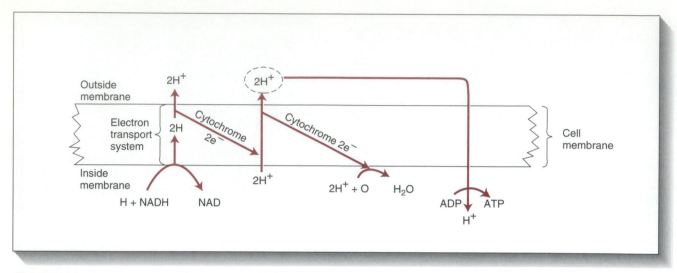

FIGURE 5-11   The use of electron transport to generate ATP. A proton (H$^+$) gradient is established as electron carriers (cytochromes) move electrons down an electrochemical gradient. ATP is generated as the proton returns to the inside of the membrane.

brane, while electrons are transported to the inside; this results in an unequal concentration of protons on the two sides of the membrane. Such unequal distribution of charges produces a force, or pressure, on the membrane to equalize the charge distribution. The cell uses an enzyme, ATPase, to reestablish proton equilibrium by returning protons into the cell, and in the process uses this *proton motive force* to generate ATP. Other forms of work, such as flagellar movement and some membrane transfer protein functions, can also be performed with the proton motive force.

BIOSYNTHESIS

In addition to providing energy through the catabolic reactions just discussed, the metabolic activities of the cell are also engaged in producing building blocks needed for the formation of new cellular components. This process is called *biosynthesis* or *anabolism* and requires energy obtained from the ATP molecules. All living cells consist of a large number of complex organic molecules called *polymers*. Polymers are so called because they are produced by connecting large numbers of smaller subunit molecules called *monomers*. Only about 150 different types of precursor monomers are needed to form the thousands of different polymers found in a living cell. Many monomers are produced through the various metabolic pathways within the cell, whereas others come directly from nutrients

that are carried into the cell through the cytoplasmic membrane (Figure 5-1). Most biological polymers (also referred to as *macromolecules*) are of the following four general types: (1) *polysaccharides*, (2) *proteins*, (3) *nucleic acids*, and (4) *complex lipids*. Polysaccharides are polymers of simple sugars, and were discussed earlier in this chapter. Proteins are polymers that consist of monomers called *amino acids*. The myriads of protein macromolecules found in any living cell are formed from just 20 different types of amino acids (Table 5-1). The structure and functions of proteins are discussed below. Nucleic acids are polymers composed of chains of subunits called *nucleotide bases*. Their structure and functions are more completely discussed in Chapter 6. The lipid complexes vary in composition, with fatty acids, alcohols, sugars, and amino acids as precursor monomers. Some monomer–polymer relationships are shown in Table 5-3.

STRUCTURE AND FUNCTIONS OF PROTEINS

Proteins serve both as important structural components of the cell and as functional molecules, such as enzymes, that regulate the chemical reactions of cells. Controlling the types of proteins produced enables all other characteristics of the cell to be controlled. To help understand this relationship, some knowledge of the structure and function of proteins is needed.

Proton motive force   The state of a membrane resulting from the transport of protons by the action of an electron transport system.

TABLE 5-3	
The Major Categories of Monomers and the Polymer Formed from Them	
Monomer	**Polymer**
Amino acids	Proteins (polypeptides)
Simple sugars	Carbohydrates (polysaccharides)
Fatty acids, monoglyceride	Lipids
Nucleotide bases	Nucleic acids

✳ Molecular Structure of Proteins

The 20 amino acid monomers that constitute the polymeric protein molecules are analogous to letters of the alphabet, whereas the protein molecules are analogous to words of the printed language. The amino acids are connected end to end, forming long chains, and each protein molecule, in order to be formed properly, must contain a specific sequence and number of amino acids, just as correctly spelled words must have a proper sequence and number of letters (Figure 5-12). Yet a protein molecule contains

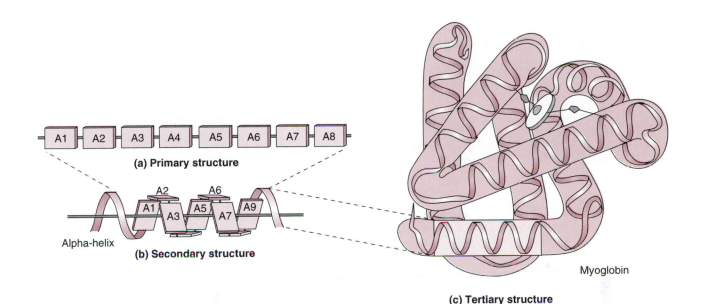

(a) Primary structure

Alpha-helix

(b) Secondary structure

Myoglobin

(c) Tertiary structure

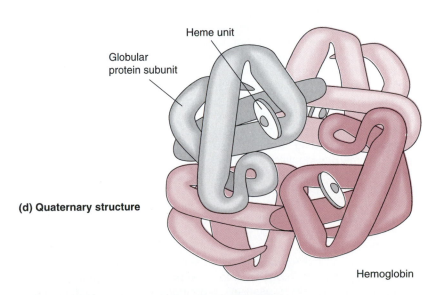

Heme unit

Globular protein subunit

(d) Quaternary structure

Hemoglobin

FIGURE 5-12 Formation of protein molecules from amino acid monomers. Formation and structure of primary, secondary, tertiary, and quaternary structure are shown. (*Fundamentals of Anatomy and Physiology, 3/e* by Martini, © 1995. Adapted by permission of Prentice-Hall, Inc., Upper Saddle River, NJ.)

many more amino acids, usually several hundred, than the number of letters in a word.

Amino acids are so called because they contain an *amine* group (NH_2) at one end and a *carboxylic acid* group (COOH) at the other end of the molecule. The arrangement of atoms in between the amine and the carboxylic acid groups is different for each amino acid. When the amino acids are brought together under specific conditions, a reaction occurs between the amine group of one amino acid and the carboxyl group of the other. In this reaction a water molecule is removed. The bond that forms between amino acids is called a *peptide bond* and a macromolecule containing many amino acids is called a *polypeptide*. A complete protein molecule may consist of one or several polypeptides.

The sequence of amino acids in the protein is called the *primary structure* of the polypeptide. As the polypeptide forms, it coils into a *secondary structure*, an example of which is a spiral or helical arrangement called an *alpha-helix*. Another category of polypeptides is formed by folding amino acids into secondary structures called *beta-pleated sheets*. The secondary structure of proteins is held together by rather weak chemical bonds known as *hydrogen bonds*. Hydrogen bonding interactions occur because of the unequal distribution of positive and negative electrical charges along the molecule. Each complete helical turn in the alpha-helix requires 3.6 amino acid residues. Next, the helix folds on itself and forms chemical bonds between different segments of the molecule. The result of this folding of the alpha-helix is called the *tertiary structure* and gives the protein molecule a specific three-dimen-

CLINICAL NOTE

Bacterial Identification

Although procedures are now available that use bacterial genes, antigens, and growth rates for species identification, the most important technique available to microbiologists of the past relied on the metabolic functions of the organisms to identify and classify bacteria. For example, it seemed self-evident to scientists of earlier years that a bacterium that could metabolize a sugar such as lactose was different from one that could not. Therefore, test after test was made first with one sugar (or with one protein, etc.), and then with a second, and a third, and so forth until the metabolic reactions of bacteria were almost completely cataloged. These metabolic fingerprints are well-known for most bacteria that cause human disease and are still the most commonly used identification procedures in the diagnostic laboratory.

As clinical microbiologists have made considerable use of information regarding bacterial metabolism, they have devised numerous identification schemes to make their task as simple as possible. Because biochemical processes that are common to all species are not useful in discriminating among them, nearly all such schemes begin with a Gram stain, which divides the bacteria into at least two groups. When that information is obtained, the next step is to determine the microbial response to a second biochemical test that will again separate the organism in question into one of two groups, that is, will provide a positive or a negative response. This procedure is repeated until there is only one species of bacteria that possesses the correct response to each of the tests and the identification is

thus secured. For example, if a Gram stain reveals that the organism to be identified is a Gram-positive coccus, the microbiologist can conclude that it is either a *Streptococcus* or a *Staphylococcus*. To separate these genera, the microbiologist might test the isolate for the presence of an enzyme known as *catalase*. If this test is negative, it is necessary to pursue the identification of a *Streptococcus*; if it is positive, the Gram-positive coccus is tested for the presence of an enzyme known as *coagulase*. If this test is positive, the organism is *Staphylococcus aureus*; if it is negative, the organism is one of the other staphylococcal species and additional tests would be used to make a final identification. Using this approach it is possible to obtain a species identification for each microorganism brought into the clinical laboratory. Because there is great need for rapid identification of bacteria that may be causing disease, many of the tests are performed simultaneously.

Knowledge of bacterial metabolism is essential to an understanding of the bacterial identification process. Tests commonly used are able to determine the presence or absence of enzymes; the ability to metabolize various sugars and alcohols, to produce various metabolic end products, to produce various acids during metabolism, to grow on selected energy sources, and to grow in the presence of selected growth inhibitors; and the permeability of the cell to a variety of nutritional compounds. Learning how bacteria respond to this broad variety of metabolic opportunities enables the microbiologist to make accurate decisions leading to bacterial identification.

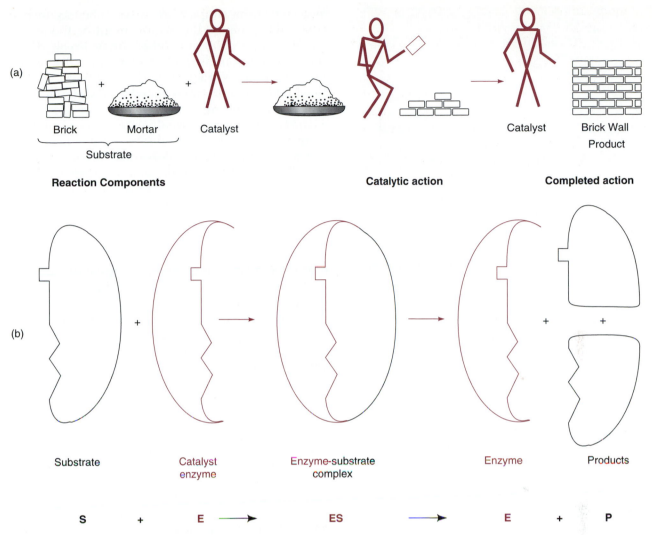

(a)

Brick   Mortar   Catalyst   Catalyst   Brick Wall
Product

Substrate

Reaction Components   **Catalytic action**   **Completed action**

(b)

Substrate   Catalyst   Enzyme-substrate   Enzyme   Products
enzyme   complex

S   +   E   ⟶   ES   ⟶   E   +   P

FIGURE 5-13   A catalyst enters a chemical reaction, facilitates the reaction by increasing the interaction between substrate molecules, and is not changed by the reaction. (a) This can be exemplified by a bricklayer, who uses brick and mortar (substrates) and combines them into a finished product. (b) An enzyme is a biological catalyst. It facilitates a chemical reaction, makes it go, and is left unchanged by the reaction.

sional shape (Figure 5-12). The tertiary structure is largely maintained by strong chemical bonds that form between individual amino acids.

It is apparent then that the primary structure of the polypeptide determines the secondary, which in turn determines the tertiary configuration. In addition, some protein molecules are formed by connecting several different polypeptides. Such an arrangement is called a *quaternary structure.* Any amino acid can be connected to any other amino acid; moreover, because of the large numbers of amino acids in a protein, an almost unlimited number of different types of protein molecules can be formed from the 20 amino acids.

✳ Enzymes and Enzymatic Functions

An important function of some proteins is to serve as **enzymes**. Most enzymes are proteins, but not all proteins are enzymes. Enzymes are the *catalysts* of chemical reactions that occur in living systems. A catalyst is a substance that reduces the energy needed to start a chemical reaction and may increase the rate of the reaction, but the catalyst is not used up in the reaction itself (Figures 5-13 and 5-14). Most chemical reactions that occur in living systems will not proceed without the proper enzymatic catalyst. So without enzymes, life functions, as we know

Enzyme   A catalyst that reduces the energy needed to start a chemical reaction and may increase the rate of the reaction without being used up in the reaction.

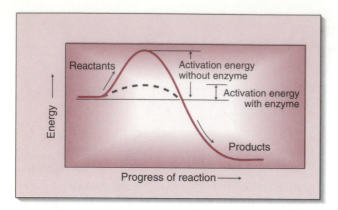

FIGURE 5-14 How a catalyst works. Catalysts such as an enzyme reduce the energy of activation, needed to initiate a chemical reaction.
(*Microbiology: Principles and Applications, 3/e* by Black, J., © 1996. Adapted by permission of Prentice-Hall, Upper Saddle River, NJ.)

them, would stop. Enzymes are very specific and almost every one of the thousands of chemical reactions that occur in a cell requires a distinct and specific enzyme. Thus, a simple bacterial cell must be able to synthesize hundreds of different types of protein molecules just to supply the needed enzymes. The specificity of an enzyme is a result of the three-dimensional configuration (tertiary structure) of the protein molecule. This shape allows the enzyme to react temporarily with, and properly orient, the compounds involved in a chemical reaction so that the energy required to start the reaction is reduced (Figures 5-13 and 5-14). After the reaction occurs, the enzyme disassociates from the products and is free to repeat the process. Most enzymes are able to catalyze thousands of reactions per second.

Some enzymes direct the synthesis of complex molecules from simpler precursor subcomponents; others may change the arrangement of the atoms within a molecule; still others may break down complex compounds into simpler molecules or join two compounds together. The chemical that is acted on by an enzyme is called a *substrate*, and the molecules resulting from the reaction are called *products*. Enzymes are usually named by adding the suffix *-ase* to the name of the substrate or reaction catalyzed. For instance, an enzyme that breaks down proteins would be called a prote*ase*; one that catalyzes the reaction to form a DNA macromolecule would be called a DNA-polymer*ase*. A list of some enzymes and their functions is given in Table 5-4.

What a cell can and cannot do depends on the type of enzymes it possesses. The great diversity of activities of various microorganisms is a function of the types of enzymes they possess. If all the necessary enzymes are formed, the cell will function properly. If an essential enzyme is not properly formed, the cell may die. If an enzyme that catalyzes a minor reaction is missing, the cell may survive but take on different characteristics. Similarly, if a cell acquires the ability to produce a new enzyme, the cell may acquire a new characteristic. Therefore, controlling the synthesis of protein molecules means that all other reactions of the cell are controlled. Fundamentally, the genetic control of a cell or an organism is a function of the control of protein synthesis. The discovery of how a cell can store and transfer to its offspring the information necessary to line up the amino acids into the proper sequences in order to form the proper enzymes has been one of the great scientific achievements of the past 45 years. This topic, known as the central dogma, is discussed in the next chapter.

TABLE 5-4

Enzymes and Enzymatic Functions

Enzyme Class	Examples	Functions
Hydrolase	Amidase, esterase, phosphatase	Hydrolytic cleavage of chemical bonds by the addition of water
Isomerase	Mutase, epimerase	Molecular rearrangement producing one isomer from another
Ligase	DNA synthetase	Linking of simple molecules into complex polymers
Lyase	Deaminase, decarboxylase	Nonhydrolytic cleavage of chemical bonds
Oxidoreductase	Dehydrogenase, peroxidase	Catalysis of oxidation and reduction reactions
Transferase	Transaminase, transmethylase	Transfer of atoms or molecules from one compound to another

MATERIAL FOR REVIEW

CONCEPT SUMMARY

1. Energy for life is primarily a product of the sun. This energy is organized into chemical systems by means of photosynthesis, which occurs in autotrophic life forms. Heterotrophic organisms obtain their needed energy by using autotrophs or the metabolic products of autotrophs for food.

2. Heterotrophic bacteria primarily use carbohydrates as an energy source. The carbohydrate is gradually oxidized through a stepwise rearrangement of the molecule. This energy, obtained in a systematic fashion, is captured in ATP, which is then used by the cell to perform energy-requiring operations, such as cell synthesis.

3. Cellular metabolism is regulated by the availability of enzymes. These proteins ensure the proper, systematic interaction of those molecules necessary to build new cell material. Other enzymes provide orderly catabolic processes that result in a continuous supply of energy to carry out cell processes.

STUDY QUESTIONS

1. Describe the relationship between autotrophs and heterotrophs.
2. How are the processes of photosynthesis and cellular metabolism connected at the molecular level?
3. Why is it important that enzymes have only one or a very limited number of substrates?
4. Draw a diagram showing how the products of glycolysis become the substrates for the Krebs cycle.
5. Explain why it is possible to get a much greater yield of cells when they are grown aerobically rather than anaerobically.
6. How is it possible for single amino acid substitution in a polypeptide to modify the tertiary structure of the protein?

CHALLENGE QUESTIONS

1. Refer to Figure 5-1. Using this diagram, identify those reactions that are anabolic (synthetic) and those that are catabolic.
2. Prokaryotic microbes use the cytoplasmic membrane to synthesize ATP molecules. In eukaryotic microbes, ATP synthesis occurs in the mitochondria. Using Figure 5-11, propose a mechanism for the movement of electrons and protons that would allow for ATP synthesis within the mitochondrion.
3. Refer to Figure 5-14, which shows how a specific substrate binds with the appropriate enzyme to form an enzyme–substrate complex. The enzyme then forms one or more products from the substrate. In Chapter 1, you were introduced to the antimicrobial agents, such as the sulfa drugs. These drugs work by blocking enzyme function. Specifically, they block the conversion of a substrate, called PABA, into the product folic acid. Folic acid is needed to make the nucleotide bases in DNA. Propose a mechanism for the mode of action of sulfa drugs.

CHAPTER 6

Genetic Control of Cellular Functions

Genes are information-carrying molecules that contain all the information needed to direct the proper functions of the cell. When a cell divides, a full complement of the genetic information (genes) must be passed to each daughter cell. Because the individual chemical reactions the of the cell are catalyzed by a specific enzyme, information-carrying molecules of the genes must be able to precisely direct the synthesis of such specific enzymes, when needed. It is now well established that DNA is the information-carrying molecule of the cell. The discovery of this fact was one of the greatest achievements in science. We will follow the chronology of this discovery as we discuss the nature of DNA, how equal complements of DNA are passed to each daughter cell, the means of information storage in the gene, and how this information directs the formation of protein molecules. We also examine how modern scientists are exploiting this knowledge to produce rare biological compounds, create novel plants and animals, and even cure deadly genetic diseases.

DNA AS THE GENETIC MOLECULE

By the 1930s it was recognized that the information-carrying molecules of the cell were in the nucleus. Chemical analysis of the nucleus, however, showed that large amounts of both protein and DNA were present. DNA appeared to be a rather unexciting compound composed of only four monomers known as *nucleotide bases*. These bases varied in concentration among different biologic species, but little was known regarding their function within the cell. On the other hand, proteins, which were better known than DNA, were highly variable molecules comprised of an almost limitless arrangement of 20 unique monomers. The scientists of the 1930s had a choice for the genetic information molecule between the limited 4-letter "alphabet" of DNA, and the 20-letter "alphabet" of proteins. It seemed only logical that protein molecules would somehow contain the genetic information so essential to the function of the cell. However, a series of important experiments conducted in the 1930s and 1940s by Oswald Avery, Colin MacLeod, and Maclyn McCarty demonstrated that DNA was in fact the genetic material.

These investigators extended the work of Frederick Griffith who, in the late 1920s, had studied two forms of the bacterium *Streptococcus pneumoniae*. One of the forms was *virulent* (possessing properties that lead to disease) such that when small numbers were injected into a mouse it resulted in death. The other form of *S. pneumoniae* was *avirulent* (lacking properties that cause disease) and could be injected into a mouse in relatively large numbers without causing any ill effect. When Griffith injected a mouse with a living avirulent *S. pneumoniae* and a killed virulent form at the same time, to his surprise, the mouse died. Further investigation revealed that a living virulent bacterium could be isolated from the dead mouse (Figure 6-1). The only plausible explanation for these results was that some factor relating to virulence must have been transferred from the virulent, but dead, bacterium to the living avirulent form, resulting in a transformation of the avirulent *S. pneumoniae* to virulent form. This was called the *transforming princi-*

ple. Using *S. pneumoniae*, Avery, MacLeod, and McCarty performed experiments that demonstrated the chemical nature of the transforming principle. They showed that the property responsible for virulence in this organism could be passed from one cell to another with pure extracts of DNA (Figure 6-2). Thus, these investigators properly concluded that DNA must be able to carry genetic information that in turn controls cellular function.

✳ Structure of DNA

As is often true with scientific discovery, answers to questions simply lead to new questions. Once DNA was shown to be the genetic material, the question became how could this substance, containing only four monomeric nucleotide bases, supply the massive amount of information needed to control the myriad of cellular functions. Answers to this question, as well as an understanding of the functional properties of DNA, became more apparent in 1953 when James Watson and Francis Crick, working in England, were able to construct a model of the DNA molecule. They showed how the four nucleotide bases fit together. Each nucleotide was known to contain a phosphate group connected to a five-carbon sugar, deoxyribose, which, in turn, was connected to a purine or pyrimidine base. Two pyrimidines, called *cytosine* (C) and *thymine* (T), and

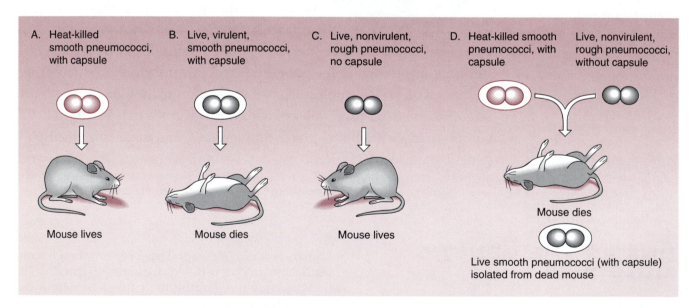

FIGURE 6-1   Griffith's experiments in transformation. (a) Heat-killed smooth pneumococci did not cause disease. (b) Organisms from smooth pneumococci colonies did kill mice. (c) Organisms from live rough colonies were not pathogenic due to absence of the protective capsule. (d) A mixture of live rough and heat-killed smooth pneumococci resulted in a fatal infection due to live type S cells. A factor (later shown to be DNA) from the dead cells transformed the avirulent rough cells to smooth cells.

Later Experiments

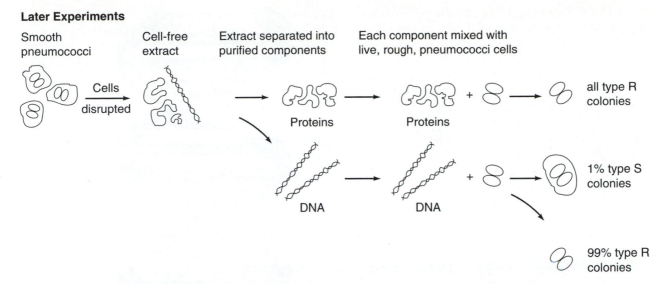

Smooth pneumococci — Cells disrupted → Cell-free extract — Extract separated into purified components — Each component mixed with live, rough, pneumococci cells

Proteins → Proteins + → all type R colonies

DNA → DNA + → 1% type S colonies

99% type R colonies

FIGURE 6-2   Experiments by Avery, MacLeod, and McCarty, which demonstrated the chemical nature of the transforming principle.

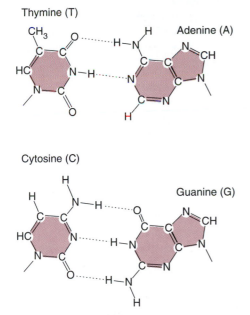

Thymine (T)   Adenine (A)

Cytosine (C)   Guanine (G)

FIGURE 6-3   The structures of the two single-ring pyrimidine bases called cytosine and thymine, and the two double-ring purine bases called adenine and guanine. The highlighted groups participate in the hydrogen bonding that holds the double helix together.

two purines, called *adenine* (A) and *guanine* (G) (Figure 6-3), are associated together in DNA such that the number of adenine-containing nucleotides is exactly the same as the number of thymine-containing nucleotides, and the concentrations of guanine and cytosine nucleotides are also equal to each other.

The Watson–Crick model showed that the DNA molecule was constructed of two long chains of nucleotides intertwined in a *double helix* (spiral). The deoxyribose sugar and the phosphate formed the backbone of each strand. Since one strand was antiparallel (ran in the opposite direction to the other), the purine and pyrimidine bases pointed toward the center and were associated in a specific pairing arrangement to hold the two chains together (Figure 6-4). The specific arrangement was that thymine could pair only with adenine and cytosine could pair only with guanine. This is called *complementary pairing* and is one of the essential features that allows a DNA molecule to replicate into two identical molecules prior to cell division.

✳ Replication

The ability of a cell to provide an exact copy of the genetic information for each daughter cell was essential to the genetic model. Each new cell must be able to perform all of the tasks previously performed by the parent cell, and also provide its own

Complementary pairing   A structural relationship between nucleotide bases that allows adenine to bond with thymine and guanine to bond to cytosine.
Replication   The process by which a cell produces an exact replica of its chromosome.

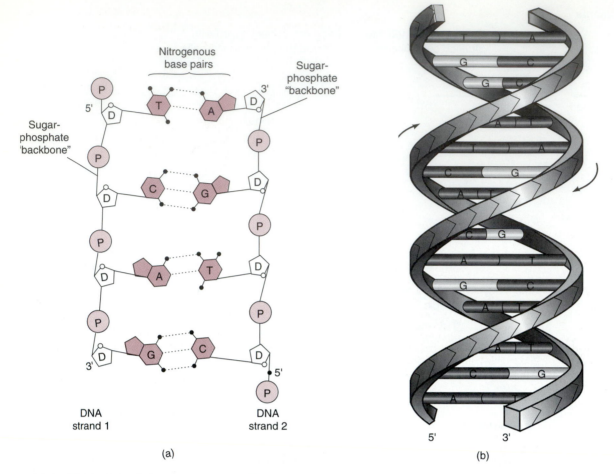

FIGURE 6-4 Structure of a DNA molecule. (a) The molecule is made up of two polynucleotide chains running in opposite directions. The backbone of a chain is composed of alternate sugar (S) and phosphate (P) units. (b) The two chains are coiled in a double helix and connected by hydrogen bonds between the complementary bases (A:T, G:C). (*Microbiology: Principles and Applications, 3/e,* by Black, J. © 1996. Adapted by permission of Prentice-Hall, Upper Saddle River, NJ.)

daughter cells with this capacity. In addition to providing structural information regarding DNA, the Watson–Crick model provided a mechanism to show how the DNA molecule could *replicate*—that is, produce exact copies of itself for each new cell.

Complementary base pairing is the key characteristic that allows the double-stranded DNA molecule to replicate. It is first necessary that each strand of nucleotide bases that makes up the two sides of the double-helix DNA molecule become available as a single-strand *template*. A specific enzyme called *DNA helicase* is able to unwind and briefly separate the two strands of DNA by breaking the hydrogen bonds connecting the complementary bases. Next, other enzymes called *DNA polymerases* bring in new DNA nucleotide bases to their complementary positions; that is, if adenine were the base in the old

strand, only thymine could be paired opposite it, and if guanine were an old base, only a new cytosine base could be paired with it. This "unzipping" of the original double strand and the addition of a new complementary strand to each old strand continues along the entire DNA molecule until two identical molecules are produced. Each of the daughter molecules contains one strand that is from the original molecule and one strand that is made up of new nucleotide bases (see Figure 6-5). This is called *semiconservative replication* because one strand is always conserved.

✳ Genetic Code

Following the work of Watson and Crick, the challenge was to determine how the genetic information

Template A DNA molecule or sequence used as the pattern for making a new nucleotide polymer during replication or transcription.

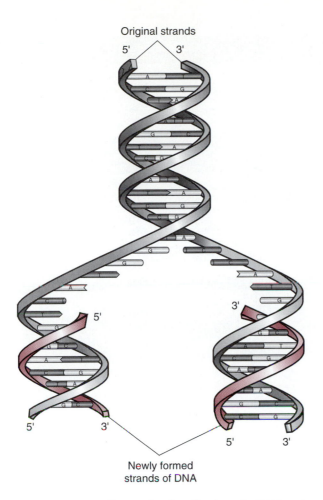

Original strands

5'   3'

5'

3'

3'

5'

5'   3'

Newly formed
strands of DNA

FIGURE 6-5 The mode of replication of a DNA molecule showing how identical sequences of nucleotide bases are maintained in each replicated molecule. (*Microbiology: Principles and Applications, 3/e*, by Black, J. © 1996. Adapted by permission of Prentice-Hall, Upper Saddle River, NJ.)

was encoded in the molecule. The information had to be encoded in the linear sequence of nucleotides along the chains of the DNA molecule. It was therefore necessary to determine how a linear sequence of four nucleotide bases could direct the synthesis of such complex molecules as proteins. That is, how could a 4-letter nucleotide alphabet be used to specify a 20-letter amino acid alphabet? Obviously, a one-to-one nucleotide-to-amino acid relationship would not be sufficient. The next alternative was to consider two neighboring nucleotides as a code for one amino acid, but this makes only 16 (4 × 4) possible combinations. The next alternative was to use a sequence of three nucleotides to code for each amino acid, which would give a total of 64 (4 × 4 ×

4) possible combinations of nucleotide bases. This base arrangement has become known as a *triplet code*. Toward the end of the 1950s, sufficient experimental information had accumulated to show that the triplet code was the one used by DNA. Each sequence of three nucleotide bases in the DNA forms a letter of the **genetic code** (Table 6-1).

The deciphering of the genetic code, that is, discovering which triplet corresponded to which amino acid, was first performed by Nirenberg. By using cell extracts he was able to translate synthetic polymers of selected nucleotides. For example, poly-A (AAAAAA...) messenger RNA (mRNA) resulted in the formation of a peptide constructed entirely of the amino acid lysine. Deciphering the entire code, which is generally universal across species, took about 10 years and was completed in the late 1960s. Because there are 64 possible triplets and only 20 amino acids, some amino acids were found to correspond to more than one triplet. Three triplets, however, did not correspond to any amino acid and were shown to function as punctuation marks in the genetic alphabet. These punctuation marks determine when the messages for specific polypeptides end (Table 6-1). In addition, it was found that nearly every coding sequence began with the same triplet (AUG), signifying the start of the protein.

The DNA of a bacterial cell is a single molecule (about 1 mm in length) that contains approximately 5 million pairs of nucleotides, more than enough to code for the several thousand different proteins produced by the bacterial cell. That is, it takes three nucleotide bases to form one triplet code that corresponds to one amino acid; assuming that an average protein contains 300 amino acids, then 900 nucleotides would be needed to code for that protein. The segment of bases in the DNA molecule that codes for one polypeptide is called a *cistron*.

＊ Protein Synthesis

During the late 1950s important research under the direction of Andre Lwoff, Francois Jacob, and Jacques Monod in France demonstrated how the genetic information of DNA is able to direct the synthesis of proteins. The process of **protein synthesis** frequently seems complex to most beginning students. It is more easily understood if it is remembered what the cell is trying to do. Each chemical reaction carried out by the cell requires the presence

Genetic code The three-base sequences (codons) in the mRNA that specify a specific amino acid.
Protein synthesis The process of transcription and translation by which the information on the DNA is made into proteins.

TABLE 6-1

Genetic Code

First Letter	Second Letter				Third Letter
	U	**C**	**A**	**G**	
U	Phe	Ser	Tyr	Cys	U
	Phe	Ser	Tyr	Cys	C
	Leu	Ser	-STOP-	-STOP-	A
	Leu	Ser	-STOP-	Trp	G
C	Leu	Pro	His	Arg	U
	Leu	Pro	His	Arg	C
	Leu	Pro	Gln	Arg	A
	Leu	Pro	Gln	Arg	G
A	Ile	Thr	Asn	Ser	U
	Ile	Thr	Asn	Ser	C
	Ile	Thr	Lys	Arg	A
	-START-Met	Thr	Lys	Arg	G
G	Val	Ala	Asp	Gly	U
	Val	Ala	Asp	Gly	C
	Val	Ala	Glu	Gly	A
	Val	Ala	Glu	Gly	G

Combining the letters of the codon into triplets composed of 1st, 2nd, and 3rd letters translates directly to a specific amino acid. Thus, UUU translates to phenylalanine, AAA to lysine, and so on.

of a specific catalyst (enzyme) to make the reaction occur. Nearly all enzymes are protein, and therefore proteins control the activities in the cell. To know which enzyme to synthesize and when to produce it, the cell must have control regions contained in the DNA that can be acted upon by environmental stimuli. It should be obvious, then, that protein synthesis is under genetic control of the DNA of the cell and, in turn, cell function is under the control of protein molecules. To get the genetic information contained in the linear sequence of nucleotides along the DNA molecule into a linear sequence of amino acids in a protein molecule, two distinct processes, called *transcription* and *translation*, must take place.

✳ Transcription

Transcription involves the passage of genetic information from the DNA molecule to a second type of nucleic acid called *ribonucleic acid* (RNA). The structure of RNA is similar to that of DNA in as much as it consists of strands of nucleotide bases with a sugar–phosphate backbone. It differs from DNA in that the sugar is ribose instead of deoxyribose, a *uracil* (U) base is used in place of the thymine base in DNA, and RNA is single stranded. In the cell, there are three kinds of RNA, known as *messenger RNA* (mRNA), *ribosomal RNA* (rRNA), and *transfer RNA* (tRNA). Each of these forms of RNA has a unique function related to the synthesis of proteins.

The synthesis of each of the forms of RNA is similar. The four nucleotide bases in RNA—adenine, uracil, guanine and cytosine—share the same complementary relationships, as do the bases in DNA. Furthermore, DNA acts as a template for the synthesis of RNA. Thus, the cell uses an *RNA polymerase* enzyme to synthesize RNA, using one strand of the DNA molecule as a template. The DNA template ensures the proper sequence of bases in the RNA molecule. The process of RNA synthesis is known as **transcription** because the information encoded on the DNA molecule is transcribed onto an RNA molecule using the same nucleotide language, and each base triplet in the DNA molecule is matched by a complementary RNA triplet called a *codon*.

Transcription   RNA synthesis; the process by which information encoded on the DNA is copied onto an RNA molecule.
Codon   Three sequential mRNA bases that match a three-base sequence in the DNA code designating a specific amino acid in a protein.

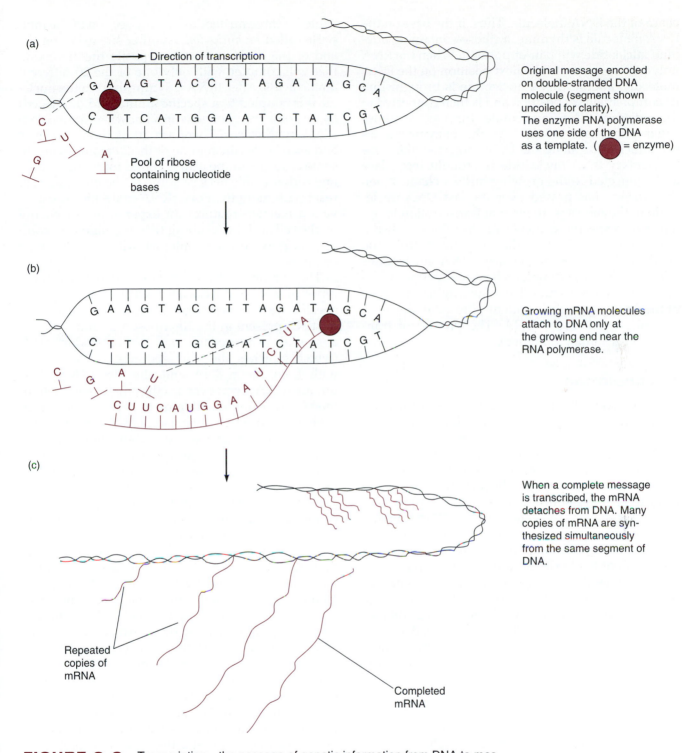

(a) Direction of transcription

G A A G T A C C T T A G A T A G C A
C T T C A T G G A A T C T A T C G T

Pool of ribose containing nucleotide bases

Original message encoded on double-stranded DNA molecule (segment shown uncoiled for clarity). The enzyme RNA polymerase uses one side of the DNA as a template. (= enzyme)

(b)

G A A G T A C C T T A G A T A G C A
C T T C A T G G A A T C T A T C G T

C U U C A U G G A A U C U A

Growing mRNA molecules attach to DNA only at the growing end near the RNA polymerase.

(c)

When a complete message is transcribed, the mRNA detaches from DNA. Many copies of mRNA are synthesized simultaneously from the same segment of DNA.

Repeated copies of mRNA

Completed mRNA

FIGURE 6-6   Transcription—the passage of genetic information from DNA to messenger RNA.

The process of transcription begins at one of several places in the DNA known as a *promoter* site. RNA polymerase binds at this position, and a momentary separation of the strands of the DNA molecule occurs. An RNA nucleotide is brought into the appropriate complementary position on one

Promoter   The sequence of bases on the DNA template that signals the start point for transcription.

chain of the DNA molecule. Thus, if the base on the DNA molecule is thymine, a ribose-containing adenine nucleotide will pair opposite it. When the RNA polymerase moves to the next position on the DNA molecule, another RNA nucleotide is brought into its complementary position and is then connected to the adjacent RNA nucleotide through a ribose sugar–phosphate linkage. As the enzyme passes each nucleotide along the DNA molecule, the complementary RNA nucleotide is brought into place and connected to the growing mRNA chain. When the enzyme has passed over the last DNA nucleotide in the message, the end of transcription is signaled by a specific sequence of DNA bases called a *terminator*. The RNA is then separated from the DNA and the complementary DNA nucleotides rejoin to form the double-helical orientation of the original DNA molecule. Thus the original message of the DNA molecule is both duplicated in the RNA and preserved in the DNA. The process of transcription is shown in Figure 6-6.

✳ Translation

In eukaryotic cells the mRNA passes from the nucleus to the cytoplasm, where translation occurs. In prokaryotic cells translation can begin as soon as the mRNA begins to form, for no nuclear membrane separates the DNA from the protein-synthesizing components (Figure 4-13). The formation of protein from the mRNA message is called translation because it changes the genetic information from the nucleotide language into the amino acid language of proteins. Translation requires the involvement of each form of RNA: rRNA, tRNA, and mRNA.

Ribosomes also play a central role in protein synthesis (Figure 4-12). Each prokaryotic ribosome is composed of two subunits, a large 50S subunit and a smaller 30S subunit ∞ (Chapter 4, p. 42). Both subunits are composed of rRNA and a variety of small proteins. The two ribosomal subunits are not joined together into the 70S ribosome complex unless they are involved in the actual synthesis of a polypeptide.

Transfer RNA is named by its function. Thus, tRNA transfers amino acids from the cell cytoplasm to the ribosome complex in such a way that the correct amino acid can be added on the growing polypeptide chain. Transfer RNAs are relatively short chains of nucleotides that contain a specific

triplet of three nucleotides at one section of the molecule called an *anticodon*. Another site exists on the end of the tRNA molecule opposite the anticodon that will combine with only one of the 20 different amino acids. After tRNA is synthesized in the nucleus, it is coupled to a specific amino acid (activated) by one of a group of enzymes known as aminoacyl-tRNA synthetases. There are 20 of these enzymes and each is specific for one of the 20 amino acids. In some cases an amino acid is coded for by more than one codon (Table 6-1), so some of the enzymes will react with more than one tRNA molecule. The activation reaction requires an expenditure of energy by the cell and the resulting tRNA–amino acid complex is now ready to interact with the ribosome complex.

The synthesis of protein is initiated when the 3 types of RNA are associated with each other in a specific way. A messenger RNA molecule attaches to rRNA present in the 30S ribosome subunit. Next, a special tRNA containing the anticodon UAC (which matches the start codon AUG) associates with both the mRNA and the 30S rRNA. This unique tRNA always carries the amino acid methionine (or N-formylmethionine in bacteria) as seen in Table 6-1. After this *initiation complex* is formed, the 50S ribosome attaches to the mRNA, completing the ribosome complex necessary for protein synthesis.

The process of translation is outlined in Figure 6-7. When the ribosome passes over a codon of the mRNA, a tRNA with the complementary anticodon is connected to the mRNA codon and carries with it its specific amino acid. This amino acid is the one corresponding to the codon on the mRNA and thus, in turn, also corresponds to the complementary triplet code on the DNA molecule. The first amino acid is held in position on the ribosome by its tRNA while the next amino acid is brought into position by its tRNA, and these two amino acids are then attached by means of a peptide bond ∞ (Chapter 5, p. 59). The first amino acid then separates from its tRNA and the "unloaded" tRNA leaves the ribosome. The ribosome continues to move along the mRNA and, as it passes over each codon, the complementary tRNA comes into position and connects the specified amino acid onto the growing polypeptide chain. On reaching a stop codon, translation ceases. The ribosome leaves the complex and the completed polypeptide is released. As each ribosome attaches to the mRNA and then

Terminator   A sequence of bases on the DNA template that signals the end of transcription.
Initiation complex   An arrangement of the 30S ribosome subunit, initiator tRNA, and mRNA in a configuration such that the 50S ribosome can attach, making the synthesis of protein possible.

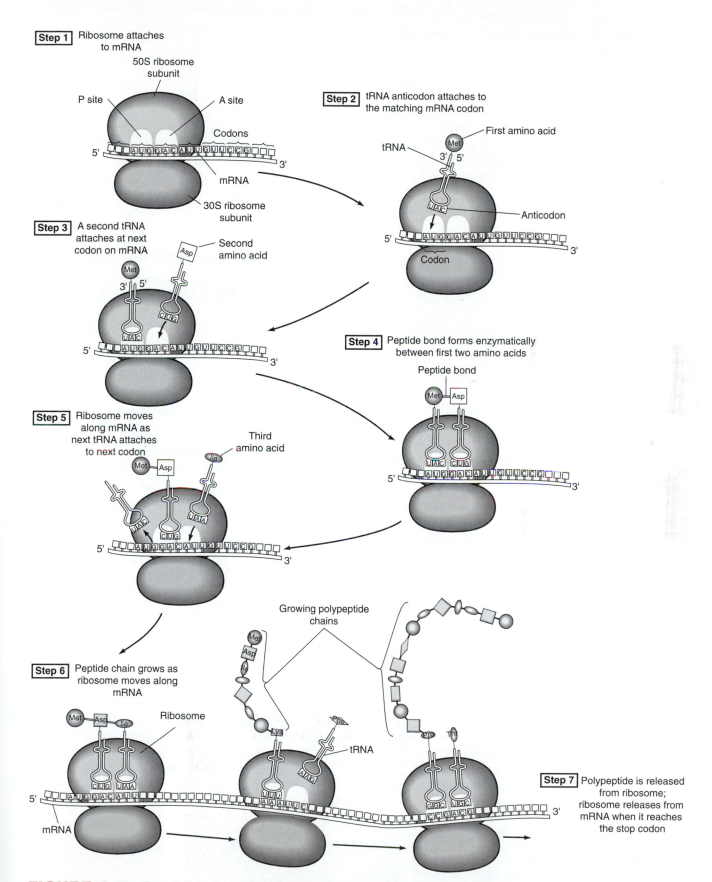

FIGURE 6-7 Translation of genetic information from a nucleotide sequence to an amino acid sequence. (*Microbiology: Principles and Applications, 3/e,* by Black, J. © 1996. Adapted by permission of Prentice-Hall, Upper Saddle River, NJ.)

moves along this molecule, other ribosomes follow in sequence along the mRNA like beads on a string (Figure 4-13). Usually five or six ribosomes will be strung along a mRNA molecule; this complex of ribosomes is called a *polyribosome* ∞ (Chapter 4, p. 42). As ribosomes move along the same mRNA, one after the other, each forms an identical polypeptide. Consequently, thousands of proteins can be rapidly formed from a single mRNA molecule. The mRNA molecule functions only for a limited period and is then broken down by the cell's enzymes. New copies of the mRNA, however, are transcribed from the DNA as needed.

ALTERATIONS IN GENETIC INFORMATION OF THE CELL

The properties expressed by a microorganism in a given environment are a result of the genetic information contained in the DNA and are modified by environmental conditions. The genetic information—that is, the sequence of nucleotide bases in DNA—in a cell is referred to as the **genotype** of the cell. Nevertheless, when observing cellular properties or functions, it is not the genotype (DNA) that is seen but the expression of the genes. This observable property of the cell is called its **phenotype**. The phenotype is always influenced by environmental conditions. When the environment changes, the cell responds by using a new portion of the genome and a different phenotype is expressed. If changes occur in the genotype of a cell, new messages are formed and a new phenotype may be produced.

The genotype of a microbial cell can be altered in three general ways. First, the sequence of nucleotide bases in the existing DNA molecule can be changed; second, the genotype of a cell may be altered by the addition of new DNA; and third, a new genotype results if portions of the DNA are deleted from the cell.

✳ Mutation

Any change in the DNA base sequence is referred to as a **mutation**, and may be the result of either base-pair substitution or the insertion or deletion of a nucleotide. The most common mutational event occurs when a mistake in the replication process

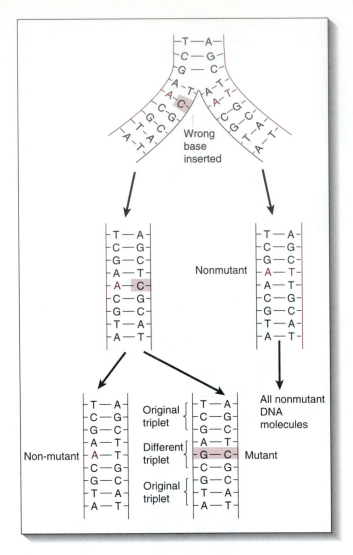

FIGURE 6-8 A point mutation produced by the wrong insertion of the base cytosine in the place of thymine. This results in one different triplet in the mutant DNA molecule and one different amino acid in the resulting polypeptide.

results in a *base pair substitution* where one purine is substituted for the other (e.g., of adenine for guanine) or one pyrimidine for the other (e.g., of cytosine for thymine). These are called *transition mutations*. Such substitutions produce a change (e.g., G-C replaced by A-T pairs) in subsequent DNA replications. If these base pair substitutions, also called *point mutations* (Figure 6-8), remain uncorrected by the DNA polymerase "proofread-

---

Genotype   The total genetic information in the cell. Normally only selected parts of the genotype are used at any one time. Thus, cells may have much greater genetic capability than is observable at any given time.
Phenotype   The observable characteristics of an organism.
Mutation   Any change in the normal DNA base sequence.
Point mutation   A mutation where one base has been substituted by another.

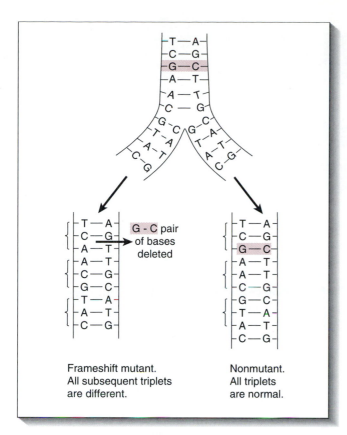

Frameshift mutant.
All subsequent triplets
are different.

Nonmutant.
All triplets
are normal.

FIGURE 6-9   A frameshift mutation in which a pair of bases were deleted. This results in a new sequence of triplets starting at the position of the deletion.

ing" function, they may result in the incorporation of one different amino acid into the corresponding protein; this may or may not cause a change in the phenotypic expression of the cell. Purine to pyrimidine changes and vice versa, termed *transversion mutations*, are also known to occur, but are much less frequent than transition mutations.

A mutational event of more significance occurs when one or two base pairs are inadvertently added to or deleted from the DNA code sequence. These changes result in a change (shift) in the translation of all of the genetic message following the addition or deletion. For example, if the correct DNA message was a sequence reading ACC / TAG / CTA/ TCG/. . . , then a deletion of the second adenine would change the message to read ACC / TGC/ TAT/ CG. . . . Changes of this type are called *frameshift mutations* (Figure 6-9). Frameshift mutations probably occur as a result of a break (nick) in

one of the DNA strands, with the resulting addition or deletion of bases. Frameshift mutation results in an entirely new sequence of amino acids in the corresponding protein; this protein is often nonfunctional, which may cause a pronounced phenotypic change, or death of a cell.

Normally, mutations occur during cell replication at a frequency of between 1:100,000 and 1:10 billion replications for any given phenotypic characteristic. Certain physical or chemical agents, called **mutagens**, may increase the rate at which mutations occur. A rather lengthy list of mutagens can be compiled and includes such physical agents as x-rays and ultraviolet light, and chemical agents such as nitrous acid, acridine dyes, alkylating agents, and *base analogs* (nucleotide-type bases that are slightly different from the normal nucleotides in DNA) (Table 6-2). The agents that increase the mutation rate of bacterial cells usually also increase the occurrence of cancer in humans and animals.

✳ Genetic Exchange

As with higher life forms, bacteria are capable of transferring genetic information (DNA) from one cell to another. However, these small single-celled organisms have evolved not one, but three methods

TABLE 6-2	
Substances That Increase the Frequency of Gene Mutation (Mutagens)	
Mutagen	**Action on Cell**
Chemical	
2-Aminopurine	Base analog; causes transition mutations
Hydroxylamine	Causes G-C to A-T transitions
Nitrous acid	A-T to G-C and G-C to A-T transitions
Nitrosoguanine	Alkylating agent; causes transitions and frameshift mutations
Physical	
X-ray irradiation	Breaks chromosome
Ultraviolet light	Causes thymine-thymine dimers, leading to inaccurate replication

Mutagen   A physical or chemical agent that increases the rate of mutation.
Plasmid   Circular, extrachromosomal DNA that can be transferred between cells. Plasmids replicate independently of chromosomal DNA and provide genetic information that will be expressed in addition to that of the chromosome.

TABLE 6-3

Medically Significant Bacterial Species That Undergo Natural Transformation

Acinetobacter calcoaceticus	*Neisseria gonorrhoeae*
Bacillus subtilus	*Pseudomonas stutzeri*
Bacillus cereus	*Pseudomonas alcaligenes*
Haemophilus influenzae	*Streptococcus pneumoniae*
Haemophilus parainfluenzae	*Streptococcus sanguis*
Moraxella osloensis	

by which gene transfer between two cells can occur: *transformation, transduction,* and *conjugation.*

Transformation. Transformation is the transfer of cell-free fragments of DNA or *plasmids* from one cell to another. This process was referred to earlier in the discussion of Avery, MacLeod, and McCarty's experiments with *Streptococcus pneumoniae* that demonstrated the genetic role of DNA. Transformation of chromosomal DNA depends on a sequence of three distinct events: (1) binding of foreign DNA to a competent cell, (2) uptake of the DNA into the cell, and (3) integration of the foreign (donor) DNA into the chromosome of the recipient cell. These three steps are effectively completed only when the donor and recipient are closely related bacterial species. This process has been demonstrated to occur naturally in only a few species of bacteria (Table 6-3).

In some bacteria, donor DNA is bound to a specific DNA binding protein found on the surface of the recipient cell. After attachment, the DNA fragment enters the recipient cell by passing through the cell wall and cytoplasmic membrane. This process is carried out only with relatively small DNA fragments. Enzymes at the cell surface nick (cut) the donor DNA so that the double-stranded DNA breaks upon entering the recipient, and one of the two donor strands is digested upon entry, leaving a single strand of DNA free to bind with the recipient chromosome. When donor chromosomal DNA enters a cell, it becomes positioned alongside *homologous genes* (genes having a similar base sequence) of the recipient cell. Enzymes cut out (excise) the homologous genes in the DNA of the recipient cell, and the donor chromosomal fragment is integrated into the recipient cell's DNA in place of the excised DNA segment (Figure 6-10). This integration of donor DNA into the recipient DNA is called *recombination.* If the integrated DNA contains genetic information not previously possessed by the recipient cell, the transformation process will cause a genotypic change by adding new information and deleting the information on the excised DNA segment. Plasmids need not recombine with the chromosomal DNA and thus may not delete any information from the recipient cell, but they do add new genetic information.

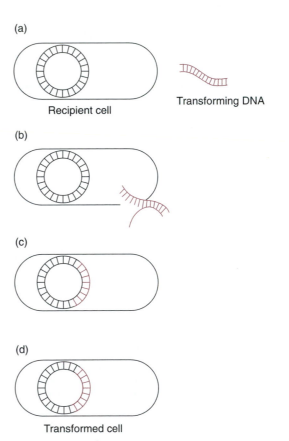

FIGURE 6-10 Transformation of a bacterial cell by integration of DNA. (a) Competent recipient cell and transforming DNA. (b) Transforming DNA attaches to competent cell and is degraded to single strand by cell nuclease. (c) Single-strand DNA pairs with complementary bases of recipient cell chromosome. (d) Integrated transforming DNA is replicated and transformed daughter cells are formed.

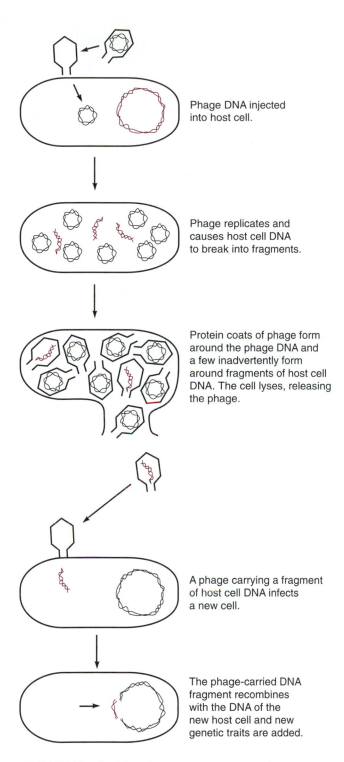

Phage DNA injected into host cell.

Phage replicates and causes host cell DNA to break into fragments.

Protein coats of phage form around the phage DNA and a few inadvertently form around fragments of host cell DNA. The cell lyses, releasing the phage.

A phage carrying a fragment of host cell DNA infects a new cell.

The phage-carried DNA fragment recombines with the DNA of the new host cell and new genetic traits are added.

FIGURE 6-11 Generalized transduction in which bacterial genes are transferred from one cell to another by a phage.

Transduction. Transduction is a process by which a bacterial virus, known as a *bacteriophage* or *phage*, carries a small piece of bacterial DNA from a donor to a recipient cell. Transduction is a more frequent method of bacterial genetic transfer than is transformation, and occurs over a wide range of bacterial species (perhaps all). Two forms of transduction occur: *generalized* and *specialized* or *restricted*. A phage infects a bacterial cell by injecting its nucleic acid directly through the bacterial wall (Chapter 28). Generalized transduction results from a replicative *lytic cycle*. A phage infects a bacterial cell and causes the bacterial chromosome to fragment into many pieces (Figure 6-11). The phage nucleic acid then directs the bacterial cell to synthesize both new phage nucleic acid and protein coats. As new phage particles are assembled, a fragment of bacterial DNA may accidentally become packaged inside the phage protein coat. When several hundred new phages have been assembled, the bacterial cell bursts (*lyses*), releasing the phages into the surrounding medium. When the phage containing a piece of bacterial chromosome later infects a new cell, the bacterial DNA inside the phage is introduced into the new host. If the bacterial DNA fragment recombines with the DNA of the recipient cell it will then be expressed. If a plasmid is transported to the recipient cell by the phage, its genetic messages can be expressed without the need to integrate into the DNA of the recipient cell.

In specialized transduction, the bacteriophage does not initially cause the cell to reproduce new phage particles, but the phage DNA integrates into the DNA of the host cell, a condition called *lysogeny*, and replicates along with the bacterial DNA during cell division (Figure 6-12). In this integrated state, the phage DNA is referred to as a **prophage**. Periodically, lysogenic phages are induced to excise themselves from the bacterial chromosome and go through a replicative *lytic cycle*. In some instances, about 1 in one million cells, these phages do not excise precisely from bacterial DNA, and a piece of host DNA remains attached to the phage chromosome and replicates with it. These DNA molecules are then packaged into the phage coats and are transferred with the phage to new recipients. When these phages infect and integrate their DNA into a new recipient cell, the attached segment of bacterial DNA is also integrated and will induce new genetic

---

Bacteriophage   A virus that infects bacterial cells; also called a phage.
Prophage   The phage DNA integrated into the bacterial chromosome.
Lytic cycle   The series of steps by which a phage replicates and bursts (lyses) the host cell.

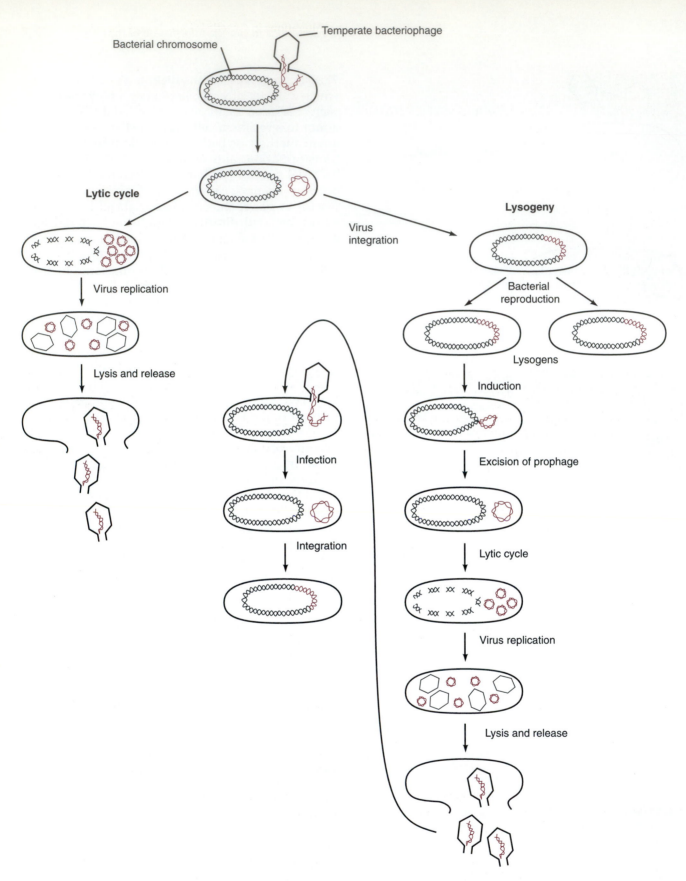

FIGURE 6-12 Life cycle of a temperate bacteriophage and its role in specialized transduction. Note that only genes that flank the integration site on the bacterial chromosome can be transduced in this manner.

TABLE 6-4		
Differences Between Generalized and Specialized Transduction		
Characteristic	**Generalized**	**Specialized**
Genes transferred	Any host gene	Only genes that flank the prophage integration site
Type of bacteriophage	Lytic	Temperate (lysogenic)
Percentage of transducing particles from a single burst	<1%	100%
Requirement of integration into host chromosome	No	Yes
Percentage of transducing particle DNA that is bacterial	100%	<10%

traits in the recipient cells. Some differences between generalized and specialized transduction are listed in Table 6-4.

You might wonder what transduction and lysogeny have to do with clinical microbiology. As you will see in later chapters, the organisms that cause diphtheria, scarlet fever, and botulism are

lysogenized by a *temperate* (*lysogenic*) bacteriophage. In fact, the genes that code for the toxins that are the direct cause of these diseases are carried by the phage. Does this mean that *Corynebacterium diphtheriae* cells that are not lysogenized by a specific phage are completely harmless in an unvaccinated host? The answer is yes. In addition, transduction has

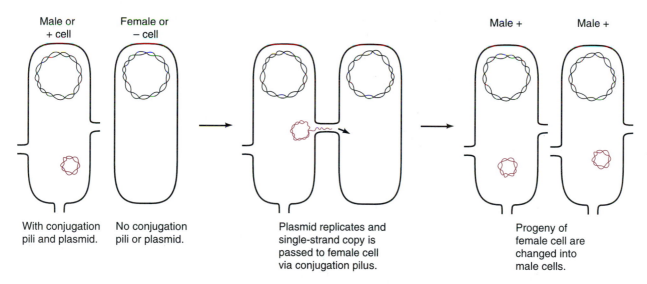

Male or + cell　　Female or − cell　　　　　　　　　　　　　　　Male +　　　Male +

With conjugation pili and plasmid.　　No conjugation pili or plasmid.　　　Plasmid replicates and single-strand copy is passed to female cell via conjugation pilus.　　　Progeny of female cell are changed into male cells.

FIGURE 6-13　A mechanism of conjugation in which genetic traits producing the male characteristics are passed with a plasmid to a female cell through a pilus.

Temperate　Referring to a bacteriophage that does not cause immediate lysis of the infected bacterial cell.
Lysogenic　The ability of a bacteriophage to integrate its DNA into the bacterial chromosome. Under these conditions the cell does not produce new bacterial viruses, but may carry out functions under direction of the virus DNA.

been shown to be involved in the spread of antibiotic-resistance genes in staphylococci.

Conjugation. Bacterial conjugation (Figure 6-13) requires cell-to-cell contact and occurs in a wide variety of both gram-positive and gram-negative bacteria. The process of bacterial conjugation is under the direction of plasmids, which facilitate the transfer of both the plasmid and occasionally chromosomal DNA from a donor to a recipient cell. Conjugation is the only known mechanism for the natural transfer of genetic material between different groups of bacteria. Both transduction and transformation appear to be limited to exchanges between closely related bacterial species.

Plasmid replication is under the control of the plasmid and not the bacterial cell. Many plasmids, particularly those associated with gram-negative bacteria, carry genes necessary to initiate their own transfer between bacterial cells. Such plasmids are termed *conjugative*. The plasmid-transfer genes code for a variety of proteins, the most notable of which is a long thread-like structure the *conjugation pilus* ∞ (Chapter 4, p. 41). The end of the conjugation pilus adheres to any cell with the appropriate receptor site and binds the two cells together. The donor cell (with conjugation pilus) and recipient cell then join with wall-to-wall contact between the two cells. After the cells join, the plasmid replicates and a single-strand copy of the plasmid DNA is transferred to the recipient (female) cell. A complimentary copy of the plasmid DNA is then made in both cells. When the cells break apart, both contain a complete copy of the plasmid and are potential donor (male) cells.

In most cases of conjugation none of the chromosomal DNA is passed. But in a few cases the plasmid integrates into the chromosome of the donor cell, thereby making the chromosome "mobile"—that is, able to be passed to the recipient cell. These cells can transfer large amounts of chromosomal DNA. Under these conditions, the circular chromosome breaks and is transferred as a linear molecule into the recipient cell. Usually before all the donor chromosome can be transferred, the cells separate and the recipient receives only part of the donor DNA. This transferred DNA becomes integrated into the recipient cell DNA and imparts new genetic characteristics to the recipient cell.

In other cases, the plasmid that has integrated into the donor cell chromosome can reverse the process and once again become a free plasmid in the cytoplasm. As in specialized transduction, if the plasmid excises imprecisely, it will carry with it segments of the donor cell chromosome, which then become part of the plasmid. When passed to a recipient cell, this plasmid also carries with it the genetic information picked up from the donor cell chromosome.

✳ The Significance of Mutations and Gene Transfer

Most mutations that occur in microorganisms result in changes that are detrimental to the optimal performance of the *mutant*. These mutants are usually eliminated, because they are less able to survive or compete in the environment. Occasionally, however, a mutation or gene transfer occurs that gives a microbe an advantage in a given environment. This is illustrated by the advantage given to a cell that mutates to antibiotic resistance when grown in an environment containing the antibiotic. This acquired trait allows the organism to out-compete and eventually replace the previous population in that particular environment or else it may be able to grow in an environment where normal microbes are unable to grow. New genotypes that have an advantage, or have adapted to a new environment, may regularly develop in microbial populations where cell division occurs rapidly.

Mutations may be beneficial or detrimental to human needs. The most troublesome changes associated with medical microbiology are strains of microbes that develop resistance to antimicrobial agents. Antibiotic resistance can be acquired through both mutations and gene transfers. Some genes associated with plasmids, or acquired by transduction, impart disease-producing capabilities (virulence factors) to microorganisms (Table 6-5). Those plasmids with a very broad *host range* make an immensely large *gene pool* available to any given bacterial species. Such promiscuous conjugation between unrelated organisms provides a large genetic reservoir for those bacteria which may be subjected to evolutionary pressures.

Mutant A cell or organism containing a mutation.
Host range The kinds of hosts that can be infected by a specific microorganism. Host range is determined by receptors on both the host cells and the microorganism. A parasite with a broad host range can infect many kinds of cells.
Gene pool The total genetic information present in a population at a specific time.

TABLE 6-5

Representative Bacterial Virulence Properties That Are Plasmid-Borne

Property	Bacterial Example
Colonization factor	*Neisseria gonorrhoeae*
	Escherichia coli
Invasiveness	*Shigella flexneri*
	Escherichia coli
Enterotoxin (ST, LT)	*Escherichia coli*
Dermal exfoliative toxin	*Staphylococcus aureus*
Antimicrobial resistance	Various, associated with "R factor" plasmids
Neurotoxin	*Clostridium tetani*
Iron sequestration	Various

Selected mutants and microbes with recombined genes are sometimes of great use. Certain mutants have been selected as highly efficient producers of products of commercial value. For example, when the original *Penicillium* mold (used for producing the antibiotic penicillin) was treated with such mutagens as x-rays and ultraviolet light, mutants that produced a 1000 times greater yield of penicillin were created. These mutants allowed penicillin to be produced much more efficiently. Some mutants of disease-producing bacteria that have lost most of their disease-causing capabilities have been effectively used as live vaccines.

The fact that mutagens of bacteria also tend to cause cancer in humans has prompted the testing of many chemicals each year to determine if they are able to increase the mutation rate in bacteria. These tests take only a few days and are relatively inexpensive. Only those chemicals that are mutagenic for bacteria are then further tested for carcinogenicity by slower and more expensive methods using experimental animals. Thus, by using bacteria in the screening tests, the number of chemicals tested in the animal systems is greatly reduced and overall, many more chemicals can be tested using fewer animals.

GENETIC ENGINEERING

As understanding of DNA function increased throughout the scientific community, it was natural that studies of laboratory-induced changes in gene arrangements would follow. Presently there is great interest in manipulated gene transfer among cells of different species. This technology, called *genetic engineering* or *recombinant DNA technology*, has the potential of developing microorganisms that are able to produce many useful products that are difficult or impossible to produce by other methods ∞ (Chapter 1, p. 11). With this technology it is possible not only to transfer genes from one type of bacterium to another but also to transfer genes from other forms of life into microorganisms and from microorganisms to other types of cells. This technology has grown so rapidly and encompasses such a large body of knowledge that it is beyond the scope of this text.

Perhaps the key discovery that facilitated genetic engineering was that of enzymes that carry out specific DNA modification functions—for example, *restriction endonucleases*, which are able to cleave double-stranded DNA molecules at specific sequences; DNA polymerases, which fill in missing bases in the DNA helix; and *DNA ligase*, which provides a means of linking DNA molecules together. One or more restriction endonucleases occur naturally in most bacterial species and are likely the result of evolutionary selection. These enzymes (hundreds have been studied) may cut DNA molecules at specific base sequences, which occur in *palindromic* order (reading the same in both directions) (Table 6-6). Once cut, foreign DNA is rapidly broken down into nucleotides by nonspecific exonucleases found within the cell. Bacteria are protected against their own restriction endonucleases by a chemical modification of the base sequence at the restriction sites, which are the substrates for their restriction endonucleases.

Many endonuclease enzymes cause staggered cuts in both plasmid and foreign DNA where identical sequences of nucleotides are located. These breaks are such that a short segment of single-stranded DNA is left at the ends of the cut DNA molecules (Figure 6-14). Using this procedure, the nucleotide sequences at the ends of both DNA molecules become complementary. Such complementary single strands are called *"sticky" ends*, and they specifically combine with the single-stranded ends of any other DNA molecule that has been cut with the same endonuclease. The breaks in the phosphate–sugar backbone of these hybrid molecules are then sealed with the enzyme DNA ligase.

Other technical advancements that have helped to facilitate genetic engineering are the ability to determine the sequence of bases in DNA molecules (base sequencing), the capacity to synthesize short segments of DNA molecules (*oligomers*) in the labo-

TABLE 6-6

Representative Restriction Endonuclease Enzymes and Their Cleavage Sites

Enzyme	Source	Site of Action
AatII	*Acinetobacter aceti*	GACGTC CTGCAG
BamHI	*Bacillus amyloliquefaciens*	GGATCC CCTAGG
BglII	*Bacillus globigii*	AGATCT TCTAGA
EcoRI	*Escherichia coli*	GAATTC CTTAAG
HindIII	*Haemophilus influenzae*	AAGCTT TTCGAA
NotI	*Nocardia otitidis-caviarum*	GCGGCCGC CGCCGGCG
PstI	*Providencia stuarti*	CTGCAG GACGTC
SmaI	*Serratia marcescens*	CCCGGG GGGCCC
XhoI	*Xanthomonas holcicula*	CTCGAG GAGCTC
XmaI	*Xanthomonas malvacaerum*	CCCGGG GGGCCC

ratory (synthetic DNA), and the development of ways to use the bacterial enzymes in molecular cloning. Laboratory manipulation of restriction endonucleases provides a method for carrying out molecular cloning, as shown in Figure 6-14. To carry out a cloning experiment, a suitable *vector* (plasmid or bacteriophage) is first chosen. This vector must have the ability to infect the desired host bacterium (*E. coli* is most often used) and to multiply within the host. The DNA to be cloned (foreign DNA) is then selected and both the vector and foreign DNA are treated separately with the same kind of restriction endonuclease. When the endonuclease-treated vector and foreign genes are mixed together in the presence of the DNA ligase, their "sticky" ends join together and the foreign gene becomes integrated into the vector (plasmid). The recombined vector can then be placed back into the host bacterium. The vector now replicates independently of the host, and may produce as many as 2000 copies of both itself and the integrated foreign gene. Recutting the plasmid with the same restriction endonuclease allows the production of a large number of purified cloned sequences. Alternatively, each progeny bac-

terium that contains the foreign gene may be able to produce the protein coded by this gene, provided the proper control regions exist upstream. Because of the rapid growth of microorganisms, large amounts of "foreign" proteins can be produced by such genetically engineered microbes.

Some early applications of genetic engineering include the development of bacteria that can produce human insulin (Figure 6-15) and human-growth hormones. These products have now become available for the treatment of diabetes and children with growth defects. Prior to these developments, insulin was obtained from animals and human-growth hormone was obtained from human cadavers and was in very limited supply.

New vaccines, particularly against viral diseases, that could not be produced economically by conventional methods are now available through genetic engineering. An example is the vaccine for hepatitis B, which consists of the outer surface protein of the virus and is produced by the yeast *Saccharomyces*. In addition, various proteins that regulate and stimulate certain immune functions (Chapter 11) are now being produced by this tech-

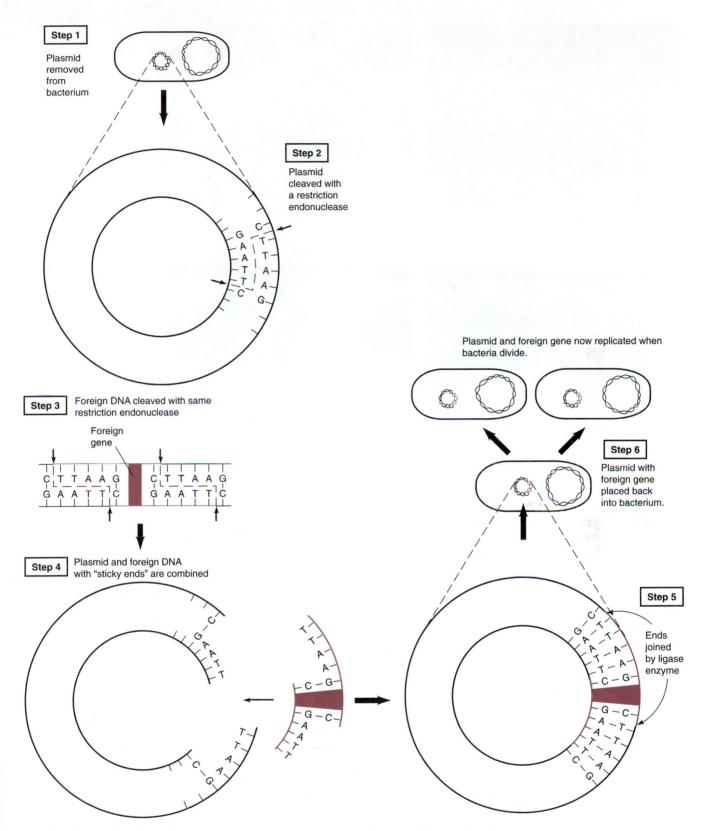

Step 1

Plasmid removed from bacterium

Step 2

Plasmid cleaved with a restriction endonuclease

Step 3 Foreign DNA cleaved with same restriction endonuclease

Foreign gene

Step 4 Plasmid and foreign DNA with "sticky ends" are combined

Plasmid and foreign gene now replicated when bacteria divide.

Step 6

Plasmid with foreign gene placed back into bacterium.

Step 5

Ends joined by ligase enzyme

FIGURE 6-14   A method used to place a foreign gene in a bacterium by genetic engineering.

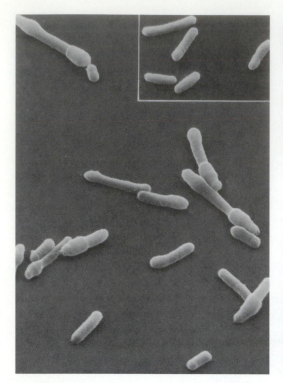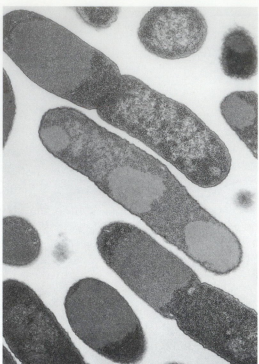

FIGURE 6-15   (a) Scanning electron micrograph of the bacterium *E. coli* that has been genetically engineered to produce components of human insulin. The prominent bulges are caused by the accumulation of insulin inside the bacterial cells; insert shows normal *E. coli* which lack the bulges. (b) Transmission electron micrograph of the insulin-producing *E. coli*, showing prominent inclusion bodies resulting from the accumulation of bacterial-produced human insulin. (Figures 2 and 3 from "Cytoplasmic Inclusion Bodies in *Escherichia coli* Producing Biosynthetic Human Insulin Proteins," D. C. Williams et al., *Science* 215:687–689. Copyright © 1982 by the American Association for the Advancement of Science. Courtesy D. C. Williams, Eli Lilly and Co.)

nology, and with these proteins it may soon be possible to activate specific immune functions as they are needed.

One of the most exciting developments of this technology is the ability to treat genetic diseases by supplying a corrected copy of the dysfunctional gene to the patient. This is called *gene therapy*. It was first done successfully in 1990 by workers at the National Institutes of Health. Two girls born with adenosine deaminase (ADA) deficiency received a corrected copy of this gene that had been placed in their own genetically modified *T lymphocytes*. Although only a single enzyme, ADA is so important that without it the immune system is completely disfunctional. You may remember the dramatic story of David Vetter, the "boy in the bubble," who had this disease and was forced to live in germ-free isolation. Today these girls are alive and well and living a normal life due to this great achievement. Genetic therapy trials are currently underway on other important genetic diseases, such as cystic fibrosis and familial hypercholesterolemia.

To use the rapid growth of microorganisms to increase production and reduce the costs, recombinant DNA technology is being applied in many other areas, such as food and energy production. Genetic engineering is currently creating a major technological revolution.

> **T lymphocytes**   A type of white blood cell involved in an acquired immune response.

MATERIAL FOR REVIEW

CONCEPT SUMMARY

1. The chemical structure of deoxyribonucleic acid and its functional operation form the basis for the principles of inheritance and the control of cell activity. Through the process of replication, genetic continuity is maintained while the processes of transcription to RNA and translation to protein account for the maintenance of cell control. Heritable change in the DNA nucleotide sequence is called mutation and may result in the synthesis of altered proteins.

2. Bacterial genetic exchange occurs through the processes of conjugation, transduction, and transformation.

3. Three structural kinds of RNA are found in living cells. These molecules are essential components of genetic translation.

4. Scientific understanding of the functions of DNA and RNA have made possible transfer and redesign of the bacterial chromosome.

STUDY QUESTIONS

1. Describe the concept of complementary pairing and explain how this principle ensures continuity of genetic information from generation to generation.

2. What role does each of the three RNA molecules play in protein synthesis?

3. In principle, why are frameshift mutations likely to be more damaging to a cell than point mutations?

4. Why are most mutations harmful rather than beneficial?

5. Construct a table comparing the parameters limiting genetic exchange by (a) conjugation, (b) transformation, and (c) transduction.

6. Why was the discovery of restriction endonucleases such an important step in genetic engineering?

CHALLENGE QUESTIONS

1. Suppose you needed to make a synthetic DNA oligomer that would code for the following tetrapeptide: Met-Trp-Phe-His-Asp. How many different oligomers could code for this peptide? What would be the sequence of one of these DNA oligomers? (Remember that DNA must be transcribed to mRNA prior to translation.)

2. One of the early experiments that helped establish DNA as the genetic material involved T2 bacteriophage and specific radioactive labels for DNA and protein. Considering how bacteriophage infect bacteria, design an experiment using this system to prove that DNA and not protein is the genetic material.

3. Why do you suppose that conjugation is not successful between two male bacteria?

Growth and Nutrition of Microorganisms

The materials and procedures necessary for the laboratory cultivation of microorganisms developed over many years and are still developing. Our understanding of bacterial growth and nutrition is critical to the methods used in isolating and identifying disease agents in the clinical laboratory. It is also of vital interest to industries that rely on microorganisms to produce or modify an industrial product. This chapter focuses on a working understanding of the growth of bacteria and helps to define processes routinely used in diagnostic microbiology.

NUTRITIONAL REQUIREMENTS OF BACTERIA

Bacteria were first carefully studied in association with infectious diseases. It was early recognized that most of these microorganisms would not grow on simple substances but required a complex diet, frequently including mammalian body fluids or proteins. Such *nutrients* are needed to supply a source of energy and provide the necessary components for cell growth. All disease-producing bacte-

Nutrients Ingredients used by a living organism to facilitate growth, including carbon and energy sources as well as essential vitamins or minerals.

TABLE 7-1			
Major Elements Essential for Bacterial Growth			
Carbon	Hydrogen	Potassium	Iron
Nitrogen	Phosphorus	Magnesium	Calcium
Oxygen	Sulfur	Sodium	

ria and protozoa, as well as fungi and animal cells, require organic chemical compounds as a source of carbon and energy; such cells are called *heterotrophs*. The methods used by these cells to obtain energy from organic compounds, primarily glucose, were discussed in Chapter 5 ∞ (p. 53).

Certain bacteria, not of direct medical importance, use CO_2 as their source of carbon; such microbes are called *autotrophs*. Some autotrophs obtain their energy from the oxidation of *inorganic* compounds, including nitrates, sulfur, and hydrogen, and are called *chemoautotrophs*. Other autotrophs, such as algae and some bacteria, contain chlorophyll and are able to obtain energy from light through the process of photosynthesis; these microbes are called *phototrophs*, or more precisely, *photoautotrophs* ∞ (Chapter 5, p. 51).

The most common chemical elements needed by bacterial cells are carbon, hydrogen, oxygen, nitrogen, sulfur, phosphorus, potassium, magnesium, calcium, iron, and sodium (Table 7-1). In addition, elements like zinc, molybdenum, copper, and manganese are needed in small amounts and are referred to as *trace elements*. Heterotrophic microbes obtain their carbon from organic compounds, such as sugars, proteins, and lipids. Hydrogen is usually obtained from water, and oxygen from the atmosphere or from water, where it is found in a dissolved state. Nitrogen, sulfur, and phosphorus can be obtained from either organic or inorganic sources. Most of the other needed elements are obtained from soluble inorganic compounds. Some bacteria, especially several of the disease-producing species, require special growth factors, such as vitamins and amino acids, which explains their need for blood or other animal body fluids or proteins.

CULTURE MEDIA

The growth, or *culture*, of a given bacterium requires a culture *medium* (plural *media*) that provides all the essential nutrients, the proper concentration of salts and ions, and the proper pH (relative acidity or alkalinity) for optimum bacterial growth to occur. Moisture is always essential for bacterial growth because the various nutrients must be in a soluble form or in a form that can be solubilized to facilitate diffusion into the cell.

It may be necessary in some studies of bacteria to use a chemically defined medium, called a *synthetic medium*, in which all essential nutrients are supplied as pure chemicals. Such synthetic media are often difficult and expensive to produce for heterotrophic bacteria. Therefore, *complex media* are frequently used in which all necessary ingredients are present but are not precisely defined. Complex media are often mixtures of organic products from plants, animals, or yeasts, along with appropriate salts, and usually contain the nutrients necessary for the growth of a wide range of bacteria. Products like extracts of malted barley, animal tissue, or baker's yeast are frequently used in complex media. Acid or enzyme digests of meats, casein, or soybean protein are also used. These products contain most nutrients, both organic and inorganic, that are needed by even the most *fastidious* microorganisms.

Today almost all culture media formulations are produced by commercial companies and supplied to laboratories as dehydrated products (Figure 7-1). To prepare media at the consumer's laboratory, a specified amount of a dehydrated medium is added to a given volume of distilled water. The medium is then sterilized in an autoclave (see Chapter 8) and dispensed in sterile test tubes or other appropriate containers (Figure 7-2). Currently, many clinical laboratories purchase their culture media already reconstituted and dispensed in appropriate sterile containers.

Culture media are prepared in both liquid (*broth*) and solid forms. The broth media are made simply by dissolving nutrients in water. Solid media are made by adding a solidifying agent to a broth medium. *Agar*, a substance extracted from seaweed, is

---

Heterotroph An organism that requires organic compounds as a source of carbon.
Autotroph An organism that obtains its carbon from CO_2 and energy from light or inorganic compounds.
Inorganic A material lacking carbon, such as, compounds consisting of sulfur, nitrogen, or phosphorus.
Fastidious Requiring special nutrient supplementation in order to grow.

FIGURE 7-1 Some examples of commercially prepared dehydrated culture media. Components and instructions for preparation and use are given on the labels.

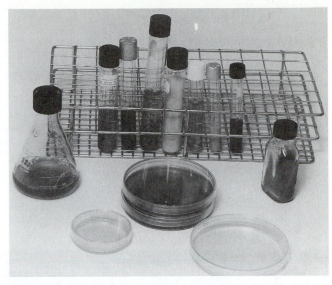

FIGURE 7-2 Sterilized culture media dispensed into various containers. Petri dishes containing solid agar media are in the foreground. Test tubes and flasks may contain either solid or liquid medium.

most often used today as a solidifying agent. We are indebted to the laboratory of Robert Koch for the discovery of agar as an appropriate solidifying agent for growth of bacteria (see Clinical Note). Agar is inert for most bacteria and thus does not change the nutritional qualities of the media. It also has the useful property of melting at a temperature just below the boiling point; yet once melted, it will

remain in the liquid state until it is cooled to about 44°C. Consequently, the solid medium can be incubated at relatively high temperatures if necessary for the growth of bacteria, while the liquid phase

CLINICAL NOTE

The First Use of Agar in Culture Media

Koch recognized that bacteria would effectively reproduce in broth, but if two or more kinds of bacteria were introduced into a broth medium, the resultant growth was a mixed culture. It was extremely difficult to study the properties of any bacterial species as long as it was mixed with other microorganisms. Earlier studies had shown that when placed on a solid medium, such as bread, solidified egg albumen, or the surface of a freshly cut potato, bacteria would grow into a colony consisting of only one kind of bacterial species. The problem for Koch, however, was that such solid surfaces lacked the necessary nutrients to grow disease-producing bacteria, or they produced colonies that could not be easily distinguished from the background surface, or they produced several different kinds of bacterial colonies that were indistinguishable from each other. These problems greatly limited the scientific study of disease-producing microorganisms.

Faced with this problem, Koch tried adding gelatin to clear broth media in order to obtain a solid surface on

which to grow microorganisms from infectious processes. Although this approach enabled Koch to provide the necessary nutritional support for the organisms under study, and growth of various species could be distinguished by their colonial differences, it failed to provide an appropriate solution to the problem because some bacteria digested the gelatin. Further, at optimal bacterial growth temperatures (37°C) gelatin was no longer solid. Therefore, cultures of bacteria in such a system were often equivalent to cultivation in broth.

A solution to the problem came through an observation made by Hesse, one of Koch's assistants. Hesse's wife was aware that in Indonesia a seaweed extract was boiled with fruit juices to provide a jelly-like product. Hesse concluded that if such material could be added to the nutrient broths used to grow bacteria, it might be a satisfactory solidifying agent. The seaweed extract, a complex polysaccharide called *agar*, proved to be an ideal product to solidify culture media.

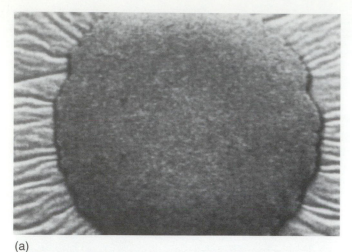
(a)

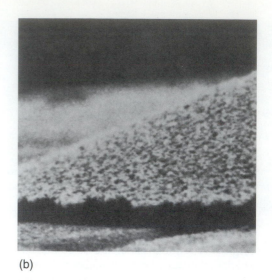
(b)

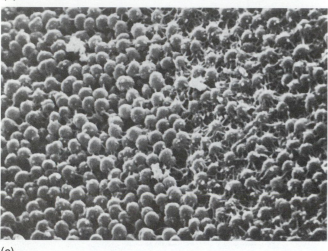

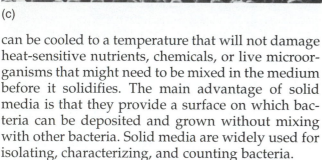
(c)

FIGURE 7-3 Scanning electron micrographs of colonies of *Neisseria gonorrhoeae* shown at increasing magnification: (a) viewed from above at 240× magnification, (b) viewed from the side at 1200× magnification and (c) 6000× magnification showing individual cells. (T. Elmros, P. Horstedt, and B. Winblad, *Infection and Immunity* 12:630–637, with permission from ASM)

can be cooled to a temperature that will not damage heat-sensitive nutrients, chemicals, or live microorganisms that might need to be mixed in the medium before it solidifies. The main advantage of solid media is that they provide a surface on which bacteria can be deposited and grown without mixing with other bacteria. Solid media are widely used for isolating, characterizing, and counting bacteria.

When a bacterial cell is deposited on the solid surface of an agar medium, the cell rapidly divides and its progeny pile up into a mass of identical cells. This mass of cells is called a *colony* (Figure 7-3) and is usually visible to the naked eye by 24 hours. The characteristics of colonies vary among bacterial species and are useful aids in helping to identify a particular species. The colonies may vary in size, texture, contour, margin, and color (Figure 7-4).

✳ Types of Culture Media

Bacteria vary widely in their nutritional requirements and hundreds of different culture media have been formulated to provide optimum nutrition for cultivated microorganisms. Some media are formulated to favor the growth of one type of microorganism over others and are called *selective media*. Selective media often contain ingredients that inhibit the growth of all but a certain group of microorganisms and are used when attempting to isolate these microbes from an environment heavily contaminated with other types of microorganisms. An example of a selective medium is one that contains 7% salt (NaCl). This factor will inhibit or impede the growth of most microorganisms except staphylococci, which, when cultured on this medium, will characteristically outgrow any other bacteria that are present.

Some types of media contain dyes or other chemicals that specifically react with cellular components or growth products produced by a given bacterium. These reactions produce an observable change in the medium or the microbial colony. Such media are called *differential*. They help to differentiate one type of bacterium from others growing on the same surface.

Bacterial species vary in the types of carbohy-

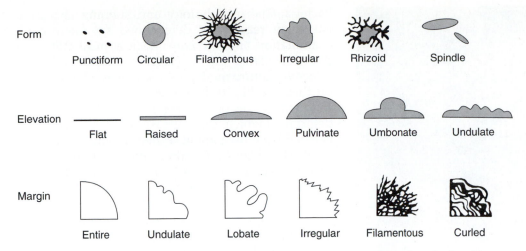

Form | Punctiform | Circular | Filamentous | Irregular | Rhizoid | Spindle

Elevation | Flat | Raised | Convex | Pulvinate | Umbonate | Undulate

Margin | Entire | Undulate | Lobate | Irregular | Filamentous | Curled

FIGURE 7-4   Outline drawings of some of the characteristics of various isolated bacterial colonies.

drates they are able to use for energy. Differential media have been made with many of these specific carbohydrates or other organic compounds as sources of carbon and energy. These media also contain an indicator dye that will change color when the pH of the medium changes. Thus, when a bacterium is able to use the specific carbohydrate and grows sufficiently in the medium to produce enough acid or alkali to change the color of the indicator dye, the color change can be used to assist in identifying the bacterium. Some microbes also produce gases that can be detected by trapping the gas in small inverted vials (*Durham tubes*) that are placed in tubes of broth (Figure 7-5). The production of gas during metabolism is used as an identifying feature for some bacteria.

✻ **Pure Cultures**

Microorganisms of various types are found growing together in natural environments. To study and characterize a particular microorganism, it must first be separated and grown free of other microorganisms; this is called a **pure culture**. The development and maintenance of pure cultures are important basic procedures in microbiology, since it is almost impossible to identify microorganisms unless they can be grown in pure culture. Thus pure cultures are used extensively in the laboratory diagnosis of infectious diseases.

The most common method of isolating pure cultures is called *streaking* (Figure 7-6). Here a wire loop or cotton swab is first placed in contact with

the environmental source to be examined so that the loop or swab picks up a random sample of the microbes present. The loop or swab is then rubbed over one edge of the agar surface contained in a *petri dish*. A petri dish is a small, flat, usually round container with vertical sides and a cover; it permits the liquid agar to harden in a readily available flat surface (Figure 7-2). The loop or swab deposits those bacteria picked up from the source on the surface of the agar. Next, a sterile wire inoculating loop is moved through the deposited bacteria and then streaked over about one-fourth of the untouched agar surface. This step deposits bacteria from the

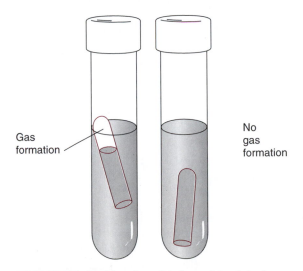

Gas formation

No gas formation

FIGURE 7-5   A method used to detect gas production by a bacterium.

Pure culture   A culture containing only a single species of microorganism.

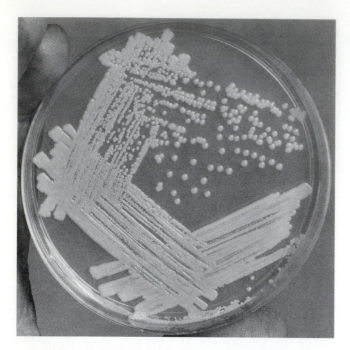

FIGURE 7-6 Streak plate showing thinning out of the bacteria with each additional streaking until well-isolated colonies develop.

area inoculated by the swab along the lines of the streak. The loop is sterilized and moved through the second streaking, followed by streaking over a fresh one-fourth of the plate. This procedure may be repeated one or more times (Figure 7-6). Each streak increasingly dilutes the population of bacteria until single cells are deposited along the streak lines. After incubation for one or several days, individual colonies will develop where the single cells were deposited. Bacteria from each colony can then be transferred to a separate sterile medium to produce a pure culture. To ensure that the culture is pure, a second streak plate may be made from a single colony. This procedure may be repeated a number of times until all the colonies appear identical and the culture is assumed to be pure.

IDENTIFICATION OF BACTERIA

Once a pure culture is obtained, a series of tests can lead to identification of the isolated bacterium. To begin with, a trained microbiologist is able to select those colonies from the primary isolation media that are most likely to represent disease-producing

bacteria. The morphology and staining characteristics can readily be determined by microscopic examination and a colony from a pure culture is then inoculated into a variety of differential and selective culture media. By comparing the reactions on these media with the known characteristics of different species of bacteria, it is usually possible to determine which disease-producing microbe was isolated from the patient. When organisms produce similar biochemical reactions, it is sometimes necessary to use specific antibodies to make a precise identification; this concept is discussed in Chapter 12.

Besides specifically identifying the microbe that is causing the disease, it is important to know which antimicrobials will inhibit its growth and therefore could be used for therapy. Antibiotic susceptibility testing (Chapter 9) is often performed concurrently with the identification tests. Because organism identification and antibiotic susceptibilities are often critical to proper patient care, the time required to obtain these data is an important factor in clinical microbiology. Since traditional procedures usually require several days to complete, time becomes significant in the care of a critically ill patient. Today modern technology is applied in various forms to shorten the time required to obtain the needed information for optimal treatment of the patient.

RECENT LABORATORY INNOVATIONS

Some recent clinical laboratory innovations involve miniaturized units that allow rapid **inoculation** of many types of differential media or enzyme substrates that give rapid identification information (Figure 7-7). Such test kit systems are relatively expensive but they save technician time and reduce the amount of media and space required to run the tests. Instrumentation is now available that can detect the presence of bacteria in normally sterile body fluids, such as blood. Samples of body fluid are introduced into vials of medium containing carbohydrates that have radioactive carbon atoms (Figure 7-8). As the bacteria grow in this medium, radioactive CO_2 is released. This CO_2 can be detected by a sensitive instrument (Figure 7-9), often after as little as 4 to 8 hours of incubation. Other instruments, such as the Autobac (Figure 7-10), can be used to determine antibiotic susceptibility in periods as short as 4 hours.

Inoculation A process whereby microorganisms are placed into or on culture media.

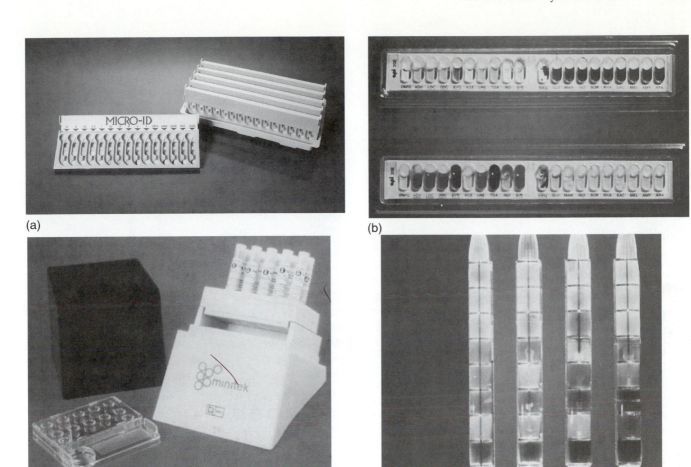

(a)

(b)

(c)

(d)

FIGURE 7-7 A composite of presently available miniaturized, or kit-type, procedures used to identify bacteria in the laboratory. Each of these systems requires isolated organisms in pure culture. The time required for completion of the test varies from 5 to 6 hours for the MICRO-ID (a) and API (b) systems to 24 hours for systems such as Enterotube (d). (a) The MICRO-ID system showing a series of small plastic cuplets that contain identifying chemicals as indicated. (Courtesy Warner–Lambert Company) (b) The API system showing both negative (upper set of reactions) and positive (lower set of reactions) tests for the 20 biochemicals used in the identification scheme. (Courtesy Analytab Products) (c) The Minitek system consisting of small paper disks containing any of many desired biochemicals which are placed into the plastic holder and then inoculated with bacteria (d) The Enterotube was one of the earliest kit-type approaches to bacterial identification; in this system the inoculating needle (seen in the center of the tube) is drawn through a series of small media chambers and the unit is then incubated.

Continued success of research into molecular genetics has provided a tool for rapid, specific microbial detection and identification directly from clinical specimens, such as tissue or sputum. Such specimens can be examined for the presence of individual pathogens by mixing the specimen with a radiolabeled DNA probe ∞ (Chapter 6, p. 81). If microorganisms in the specimen have the same DNA base sequence as the DNA probe, the two will bind together. After any excess probe is washed away, remaining radioactivity is directly proportional to the number of microorganisms present in the specimen (Figure 7-11). Thus by using newer technological advances, it is often possible to provide a physician with vital information within a few hours regarding the nature of the microorganism that may be causing the infection in the patient.

Several advances in automation involve the coupling of microgrowth chambers with sensitive electronic detectors of chemical changes. These systems

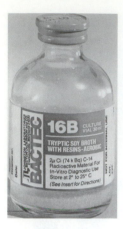

FIGURE 7-8 Radiolabeled medium used in the BACTEC system for the rapid detection of the presence of pathogens in human body fluids.

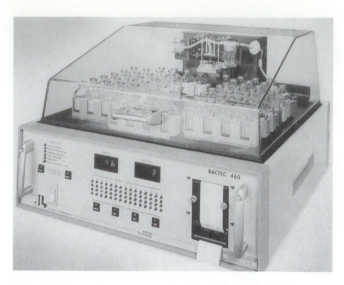

FIGURE 7-9 The BACTEC instrument used to monitor the development of radiolabeled CO_2 from growing bacteria. This instrument continuously monitors each bottle of medium (see Figure 7-8) and prints a report of all positive cultures.

can be interfaced with computers that collect and analyze the data. These instruments, such as the Automicrobic System (Figure 7-12) are able to provide a probable identification of the microbe, as well as information on antimicrobial susceptibility in hours instead of the days required by older traditional methods. New and innovative instruments are being developed each year to aid in the clinical microbiology laboratory, and medical personnel will need continual updating to keep informed of these advances.

MICROBIAL GROWTH

Biological growth may be measured by at least two different criteria. The most frequently observed form of growth is the increase in size of an organism. For bacteria, however, such change in size is a poor criterion of growth; rather, increases in the

FIGURE 7-10 The AUTOBAC microbiology system. The unit on the left is a photometer–computer system that can be used to measure the growth of microorganisms. When properly implemented, this system will both identify bacteria isolated from patients and provide the antibiotic susceptibility profile of such isolates. The units on the right allow instant information retrieval of all cultures processed in the AUTOBAC. (Courtesy Warner–Lambert Company)

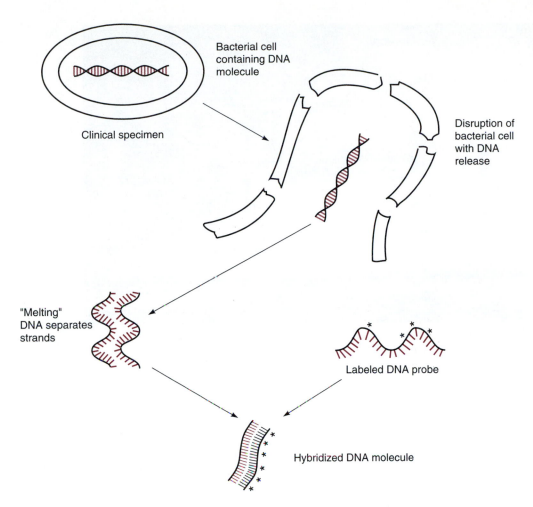

Clinical specimen

Bacterial cell containing DNA molecule

Disruption of bacterial cell with DNA release

"Melting" DNA separates strands

Labeled DNA probe

Hybridized DNA molecule

FIGURE 7-11   Use of DNA hybridization to detect and identify pathogenic bacteria in a clinical specimen.

number of organisms is usually used as a measurement. Therefore, when considering microorganisms, scientists usually count the number of living cells (*viable count*) or all cells (*total count*) in order to determine growth.

The *generation time* of a microbial cell is the time required for one complete cell division. Some microbes are able to divide as rapidly as once every 10 to 15 minutes; others require up to several hours; and a few very slow-growing bacteria may require more than 24 hours per cell division. When proper nutrients are available and other conditions are favorable, the growth of microorganisms can be a dynamic event with profound effects on the surrounding environment. If a bacterial cell were to continue to divide once every 30 minutes, for instance, there would be 64 cells in 3 hours,

17 billion cells in 17 hours, and 280 trillion cells in 24 hours. If this growth rate could continue for 48 hours, the mass of cells produced would weigh many times the weight of the earth. Obviously, such rapid growth cannot continue for very long periods because of the lack of sufficient nutrients. Yet under certain conditions such rapid growth may occur for a short time.

✳ Growth Curve

When microbial cells are placed in fresh nutrient broth under favorable growth conditions but with limited supply of available nutrient, multiplication follows the typical growth pattern (curve) shown in Figure 7-13. A typical *growth curve* exhibits four phases: (1) the *lag phase*, (2) the *log* (logarithmic) or

Generation time   The length of time for a population to double in number; for bacteria, generation time is usually measured in minutes.

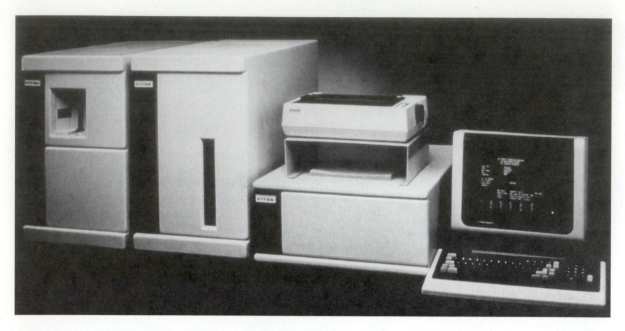

FIGURE 7-12 The Automicrobic, a fully automatic bacterial identification and detection system. (Vitek Corporation, Hazelton, MO)

exponential phase, (3) the *stationary phase*, and (4) the *decline* or *death phase*. This type of growth is referred to as a *batch* or *limited growth* system.

When bacteria are first placed in a fresh medium, a period of adjustment follows during which there is increased metabolic activity preceding cell division. This interval is called the lag phase of growth.

Enzymes and other proteins necessary to make optimal use of substances present in the environment are synthesized at this time. The length of the lag phase is variable among different species of bacteria and is somewhat dependent on the condition of the cells prior to their inoculation into growth medium. Following adjustment during the lag phase, the

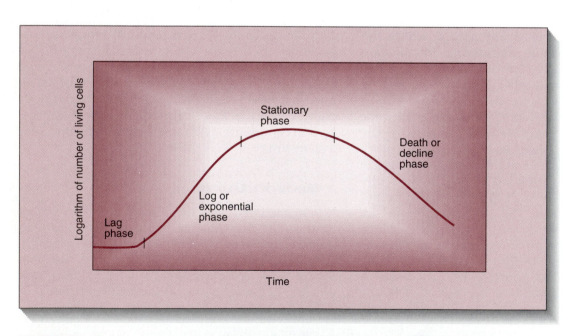

FIGURE 7-13 Phases in the growth curve of a pure bacterial culture in a closed system.

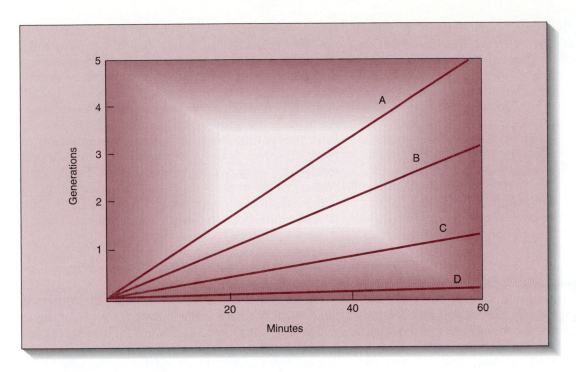

FIGURE 7-14   Growth rate of (a) *Clostridium perfringens*, (b) *Escherichia coli*, (c) *Pseudomonas putida*, and (d) *Mycobacterium tuberculosis*.

cells begin to divide at a constant and maximal rate, with the number and mass of cells doubling every generation; this is the log phase of growth. The number of generations of new cells formed in one hour is known as the **growth rate** of the organism (Figure 7-14). During this phase, the number of new cells increases exponentially and the cells are in a metabolic condition referred to as *balanced growth*. In balanced growth, all measurable components of the cell, such as protein, RNA, DNA or biomass, increase at the same rate. The log phase is usually fairly brief, lasting only 4–10 hours for most rapidly growing bacteria.

Logarithmic growth rapidly depletes the available nutrients, and toxic waste products quickly accumulate. As this happens, growth of cells becomes unbalanced and various cellular components are synthesized at different rates. These factors cause a decrease in, and ultimately cessation of, cellular division. The population of cells may then remain for a period of time in a static condition—that is, no net increase in the number of *viable* cells. This may be the result of cessation of division with

no death, or more likely, a balance between the increase produced by division and the decrease resulting from cell death. This is called the *stationary phase* of the growth curve. Because of unbalanced growth preceding the stationary phase, the cells in this phase are not uniform in composition; they are usually smaller than in the log phase and they are more resistant to environmental changes (e.g., heat, drying, and radiation). Although cells are not actively dividing during this phase, they are not metabolically inactive. Many antibiotics are produced at their highest levels during the stationary phase.

The stationary phase is followed by a period in which the cells gradually die off—the *decline* or *death phase*. Like growth rate, the rate at which cells die is log-linear and a function of both the type of cells and the environment.

Although the characteristic growth curve shown in Figure 7-13 probably occurs only under selected conditions, modifications do happen in nature and in some clinical circumstances. Products like bottled milk, for instance, if not properly refrigerated, could

---

Growth rate   The number of generations of a species within a given length of time.
Viable   Living or capable of reproducing.
Stationary phase   A period during which the number of new cells produced is equal to the number that die. This is not a static condition, and microbes continue to metabolize, multiply, and die during this phase.

support the growth of microorganisms in the logarithmic phase, thus causing rapid souring, while the change in the number of bacteria follows the normal growth curve. In clinical conditions, such as an infection, where an abscess is forming (Chapter 10), a niche may exist that is filled with dead tissue and body fluids that could support the rapid growth of bacteria for a time. In most abscesses the bacteria have reached the stationary phase of the growth curve; in this condition they do not take in many nutrients or other substances from the surrounding environment. Thus antibiotics given to the patient to cure the infection may not effectively penetrate into the abscess and may not be taken up by the bacteria if they do. Consequently, such therapy may fail to reduce the infection. To resolve this problem, it is nearly always necessary to drain abscesses in order to remove the waste products that are inhibiting the growth of the bacteria and preventing penetration of antimicrobial agents. Fresh nutrients then diffuse into the area and the remaining bacteria begin to multiply. If an antibiotic is then given, it will be taken up by the growing bacteria, inhibit their growth, and help cure the infection.

Growth of bacteria in an open environment, such as soil, water, or even the intestine, generally does not follow the curve shown in Figure 7-13. In these circumstances, bacterial growth is most often continuous so that the number of viable microorganisms remains fairly constant over long periods of time.

✳ Environmental Influences on Microbial Growth

Moisture.   Microorganisms grow only when adequate moisture is present. Because microbes exist as single cells, they depend on the continual diffu-

sion of nutrients in solution across their cytoplasmic membrane. Due to the small size of microbial cells, however, the thin film of moisture often present on many substances is enough to support some microbial growth. Keeping materials free of moisture by dehydration is one of the most common methods of controlling the growth of microorganisms and, in turn, preventing the spoilage or decomposition of food or other materials. Frequently, dehydrated foods such as powdered milk contain large numbers of viable organisms. A lack of moisture, however, maintains the microorganisms in a static state so that multiplication cannot occur.

Temperature.   The temperatures at which bacteria will grow are primarily determined by the stability of their membranes and proteins (Table 7-2). Although most bacteria will grow over a range of temperatures, there is an *optimum* temperature at which the growth rate is maximal. There is a minimum temperature below which each microorganism will not grow, but bacteria are not usually killed at low temperatures, and remain in a stationary state.

Each microorganism has adapted to grow within a specific temperature range (Figure 7-15). Some, able to grow at low temperatures, are called *psychrophiles* (cold-lovers). Psychrophiles may grow at temperatures as low as s 10°C; most psychrophilic microbes grow best at about 15°C and grow poorly above 20°C. Because of normal body temperatures, psychrophiles are unable to cause infections in humans. Still, they may cause spoilage of foods or other products stored at low temperatures.

Many microorganisms grow best at temperatures between 20 and 40°C and are called *mesophiles* (middle-lovers). Microorganisms that cause infections in warm-blooded animals are mesophiles and

TABLE 7-2

Optimal Temperature for Microbial Growth

Classification	Temperature Range	Bacterial Examples
Psychrophiles	s 5 to 20°C	*Bacillus globisporus* *Micrococcus cryophilus*
Mesophiles	20 to 40°C	*Escherichia coli* *Haemophilus influenzae* *Staphylococcus aureus* *Neisseria gonorrhoeae*
Thermophiles	40 to 100°C	*Bacillus stearothermophilus* *Bacillus coagulans*
Hyperthermophiles	70 to 110°C	*Pyrodictium brockii*

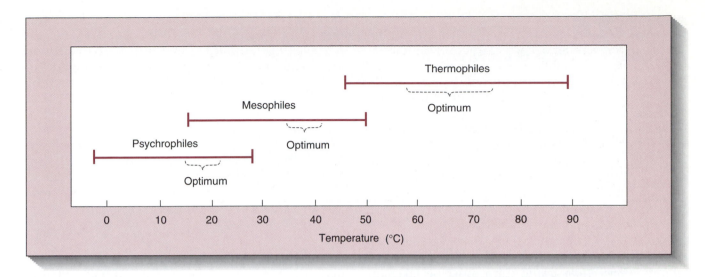

FIGURE 7-15   Categories of microorganisms based on temperature and its effect on their growth rates.

usually have optimum growth temperatures of 35 to 37°C.

Certain microbes have very thermostable proteins and are able to grow at temperatures above 45°C. These bacteria are called *thermophiles* (heat-lovers). Thermophiles are unable to cause infections in humans but are found in such places as natural hot springs. Occasionally thermophilic bacteria cause problems by growing in hot-water systems or in some industrial processes where high temperatures are used. Bacteria whose growth temperature optimum is above 80°C are called *hyperthermophiles*. These organisms are associated with geothermal environments such as deep sea vents. Some hyperthermophiles are able to grow at temperatures as high as 110°C (Table 7-2).

Oxygen.   Many different microorganisms require free oxygen (O_2) for growth and are called *aerobes*. These organisms utilize oxygen as an electron acceptor in the electron transport pathway ∞ (Chapter 5, p. 55), and will grow only in environments where atmospheric or free oxygen is available. These oxygen-associated reactions are essential to the cell and help to make necessary metabolic energy available. These reactions also produce a highly toxic chemical known as *hydrogen peroxide* (H_2O_2). A second very toxic compound known as *superoxide* (O_2^-) is also frequently produced in small concentrations when cells use oxygen as their electron acceptor. Aerobic bacteria depend on the presence of several enzymes, such as *catalase* and *peroxidase* (which remove H_2O_2) and *superoxide dismutase* (SOD) (which removes O_2^-), to protect them from these toxic metabolic products (Table 7-3). Some aerobes, called *microaerophiles*, grow best only when small amounts of oxygen are available.

Certain microorganisms are able to grow in the absence of O_2 and are called *anaerobes*. Several categories of anaerobes exist. Some are able to use O_2 if it is present but can also grow in the absence of O_2; they are termed *facultative anaerobes* (or more recently, *facultative aerobes*). Other bacteria, called *obligate anaerobes*, do not produce protective enzymes such as catalase and SOD and can grow only in the

TABLE **7-3**

Enzymes That Enable Bacteria to Survive in the Presence of Oxygen

Enzyme	Function	Bacterial Example
Catalase	$2H_2O_2 \rightarrow 2H_2O + O_2$	*Escherichia coli*
Peroxidase	$2H_2O_2 \rightarrow 2H_2O$	*Lactobacillus*
Superoxide dismutase	$2O_2^- + 2H^+ \rightarrow O_2 + H_2O_2$	*Streptococcus faecalis*

TABLE 7-4		
Halophilic Properties of Bacteria		
Class	**Allowable Salt Concentration (%)**	**Bacterial Example**
Extreme halophile	5–36	*Halococcus morrhuae*
Halophile	2–20	*Vibrio costicolus*
Marine halophile	0.2–5	*Vibrio parahemolyticus*
Nonhalophile	0–4	*Escherichia coli*

absence of O_2. An example is the *etiologic agent* of tetanus, *Clostridium tetani*. Care must be taken to remove O_2 from the media in which these organisms are cultured. An additional group of microorganisms require CO_2 for growth and are called *capnophiles*. Capnophilic bacteria are found among both anaerobes and aerobes. An example of a capnophile is the organism *Campylobacter jejuni*, one of the leading causes of bacterial diarrhea worldwide. Microorganisms from each of these oxygen-associated categories are able to grow in various habitats of the human body and may cause diseases.

Other Factors.   Bacteria are able to survive and often grow in water with low concentrations of ions because their rigid cell wall protects them against damage due to increased osmotic pressure. This factor is important, for it allows successful waterborne transmission of many diseases. Bacteria vary in their ability to grow in solutions of high *ionic concentration* (Table 7-4). Those that require high salt concentrations for growth are known as *halophiles* (salt-lovers) and can grow in saturated brine solutions. Nonhalophilic bacteria tolerate only moderate salt levels and may be destroyed in high concentrations of salt.

Most bacteria cannot grow in solutions of very high ionic concentration. This makes possible the preservation of certain foods by the addition of high concentrations of salt or sugar. Sugars are used to preserve some foods such as jam and jelly. The addition of high concentrations of sugar increases the osmotic pressure of these foods, making it impossible for most microorganisms to extract enough moisture for their growth. In fact, bacteria are frequently destroyed under these conditions because water is drawn out from the cell into the surround-

ing environment. This loss of cell water (*plasmolysis*) may result in the collapse of the cell membrane in gram-positive bacteria, and collapse of both membrane and wall in gram-negative cells.

The pH of the environment also influences the growth and survival of microorganisms. As is true for temperature, every microorganism has a pH minimum, optimum, and maximum. Those that grow best at a low pH are called *acidophiles*, whereas microbes that prefer a high pH are termed *alkaliphiles*. Most microbes that cause disease in humans grow best at or close to neutrality (pH 7), which is near the pH of most normal body fluids.

COUNTING MICROORGANISMS

It is often necessary to determine the number of microorganisms present in a culture or other material for research or industrial purposes. These determinations help to provide an assessment of the quality of food products and the general sanitary state of a given environment. The more common direct and indirect methods of counting microorganisms are discussed.

✳ Direct Microscopic Counts

Direct microscopic counting is a quick and relatively easy method of determining approximate numbers of microorganisms. Special counting chambers, such as the *Petroff–Hausser chamber* (Figure 7-16), have measured grids marked on the surface and a cover slip held at a precise distance over the slide. The space between the slide and the cover slip is filled with a fluid containing microbial cells; the

---

Etiologic agent   A microorganism that causes a disease.
Ionic concentration   The concentration of ions in a solution.

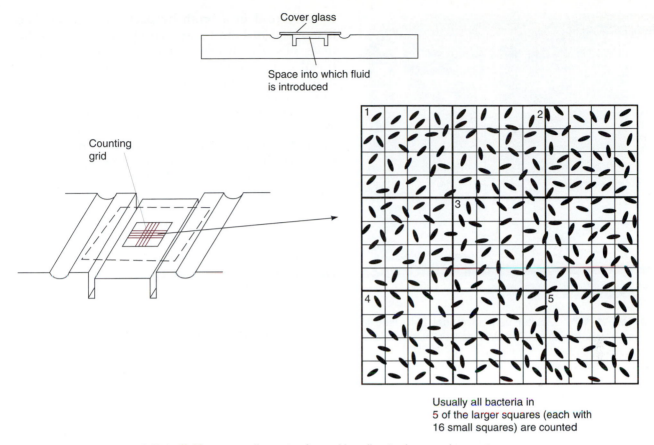

Cover glass

Space into which fluid
is introduced

Counting
grid

Usually all bacteria in
5 of the larger squares (each with
16 small squares) are counted

FIGURE 7-16 A Petroff–Hausser cell counter for making direct microscopic counts of the number of bacteria in a fluid.

number of cells in a specified number of grid squares is counted. Because the volume of fluid over each square is known, the number of microorganisms per unit volume can be determined. Suspended microbial cells are best seen with a phase-contrast microscope ∞ (Chapter 3, p. 28).

Another method of direct microscopic counting is to spread a small known volume of fluid uniformly over a 1-sq-cm^2 area of a microscope slide. The fluid is dried and stained and the number of cells are counted with the high-power (40×) lens of a bright-field microscope. This procedure is often used in counting the number of bacteria in milk.

These direct counting methods have certain disadvantages that may lead to inaccuracies; counting both living and dead cells, difficulty of seeing small cells, and inability to count cells in low concentrations. J.A. Coulter invented an electronic system known as the *Coulter counter* (Figure 7-17), which can be used to directly enumerate bacteria. This counter uses changes in electrical impedance created as the bacteria are drawn through a small opening in an electric field to count the bacteria. Although relatively accurate and extremely fast, the

instrument is expensive and requires that the suspending fluid be absolutely clean and free of all dust or particles, which, like bacteria, would be counted as they pass through the electric field.

✳ Plate Counts

Among the most frequently used methods of counting bacteria are the plate-counting procedures. Such processes are quite accurate in determining the number of living bacteria in a fluid and can measure low concentrations of cells. One method of obtaining plate counts is the *spread-plate method*. For this procedure, a measured volume of the fluid to be examined is spread evenly over the surface of a bacterial growth medium. After incubation for 24 to 48 hours, colonies develop where each viable bacterium was deposited. The number of colonies is then multiplied by a dilution factor to give the original concentration of cells per milliliter. Another process is the *pour-plate method* in which a measured volume, usually 0.1 or 1.0 ml, of test fluid is mixed with a melted agar medium that has been cooled to 48°C. The agar is then poured into a petri plate,

FIGURE 7-17   Coulter counter model ZM. Used to count human or bacterial cells electronically. (Courtesy Coulter Electronics, Hialeah, HI)

where it solidifies. The bacteria are trapped in the agar and each viable organism develops into a colony after incubation for 1 or 2 days.

The number of bacteria is so great in many samples that if undiluted samples were plated directly, the many colonies would grow together to form a solid layer of bacteria and the number could not be determined. Therefore, it is necessary in most cases to dilute the sample before plating. The dilution technique is illustrated in Figure 7-18. The original sample is diluted through a series of tubes; using separate agar plates, 0.1 to 1 ml from each tube is added to an agar plate. The colonies are counted on those plates that produce between 30 and 300 colonies. These numbers have been shown to give the most reliable measure of the correct number of cells. The number of cells in the original undiluted sample can be determined by multiplying the average number of colonies by the reciprocal of the dilution factor and dividing this number by the amount plated.

✳ Indirect Counting Methods

When repeated microbial counts must be made from the same type of liquid, as is often the case in research projects or industrial processes, an indirect method of determining the approximate number of cells can be used. The most frequently used indirect measurement is to determine the degree of turbid-ity produced in a broth by bacterial growth. The amount of turbidity is directly proportional to the mass of cells present and hence directly proportional to the number of cells. The amount of turbidity can be accurately measured by placing a tube of the microbial culture in an instrument called a *spectrophotometer*, which electronically measures the amount of light that is able to pass through the fluid. A turbidity reading can be correlated with a reference number obtained by doing spread-plate counts at different intervals during the growth at the same time that the turbidity is determined (Figure 7-19). Once this curve is obtained, estimates of the number of cells from similar cultures can be quickly made with just the spectrophotometer. In Figure 7-20, it can be seen that an *E. coli* suspension with a 52% transmittance contains approximately 1×10^8 cells/ml.

✳ Environmental Sampling

In many areas, such as a hospital environment, it may be important to know how many bacteria are present in the air, on a surface, or in a fluid. It is necessary to maintain environments with very low numbers of microorganisms in such hospital areas as operating rooms, nurseries, intensive care units, and protective isolation rooms. Various procedures are used to control microbial contamination in these areas, and several air and surface microbial sampling techniques have been developed in order to monitor the effectiveness of these procedures.

Air Sampling.   One of the simplest methods to detect airborne microorganisms is to expose open petri plates containing nutrient agar to the air for specified length of time. This is called the *settling* or *fallout plate method*. A certain percentage of the microbes in the air settles onto the surface, and after incubation, colonies form that can be readily counted. Settling plates give only approximations of the actual number of bacteria present. More precise results are obtained with *impaction air samplers*. Impaction samplers draw a given volume of air through a series of limiting openings so that the velocity of air is increased to a point where airborne particles are impacted onto nutrient agar surfaces (Figure 7-21). The number of colonies on the agar surface gives a fairly reliable count of the number of airborne microorganisms in the sampled air.

Surface Sampling.   The two most common surface sampling methods are *swabbing* and *contact plates*. Swabbing procedures are used for uneven

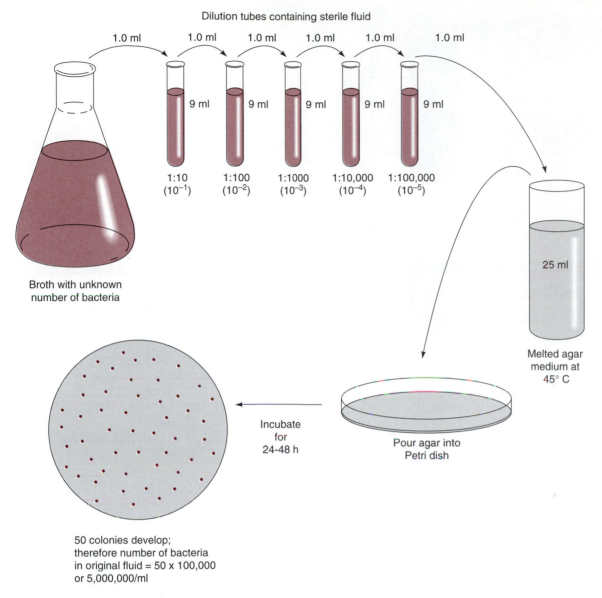

Dilution tubes containing sterile fluid

1.0 ml   1.0 ml   1.0 ml   1.0 ml   1.0 ml   1.0 ml

9 ml   9 ml   9 ml   9 ml   9 ml

1:10
(10^{-1})

1:100
(10^{-2})

1:1000
(10^{-3})

1:10,000
(10^{-4})

1:100,000
(10^{-5})

25 ml

Melted agar
medium at
45° C

Broth with unknown
number of bacteria

Incubate
for
24-48 h

Pour agar into
Petri dish

50 colonies develop;
therefore number of bacteria
in original fluid = 50 x 100,000
or 5,000,000/ml

FIGURE 7-18   Determining the number of bacteria in a broth sample by serial ten-fold dilutions and pour plates.

surfaces, corners, or crevices. The swab is prepared by firmly twisting a material like nonabsorbent cotton over one end of a wood applicator stick and then sterilizing it. To take a sample, the tip of a moistened swab is rubbed slowly and thoroughly over a measured surface area three times. The swab is returned to a known volume of solution and vigorously rinsed. Next, a measured sample of the solution is assayed by the spread-plate or pour-plate method. About 50% of the bacteria on a surface can be picked up via the swabbing method.

Another process that is also about 50% efficient involves the RODAC-type plate (RODAC is an

acronym for replicate organism detection and counting) and is the most widely used method of contact sampling (Figure 7-22). Sampling with these plates, however, is primarily limited to flat surfaces. RODAC-type contact plates are disposable plastic plates that are filled with an agar medium to form a 25-cm^2 convex surface above its sides. A contact sample is taken by pressing the surface of the contact plate on the surface to be sampled for several seconds. The contact plates are then covered and incubated. The number of colonies that develop can be directly counted on the surface of the plate. When sampling surfaces previously treated with

Sample A

Sample B

Sample C

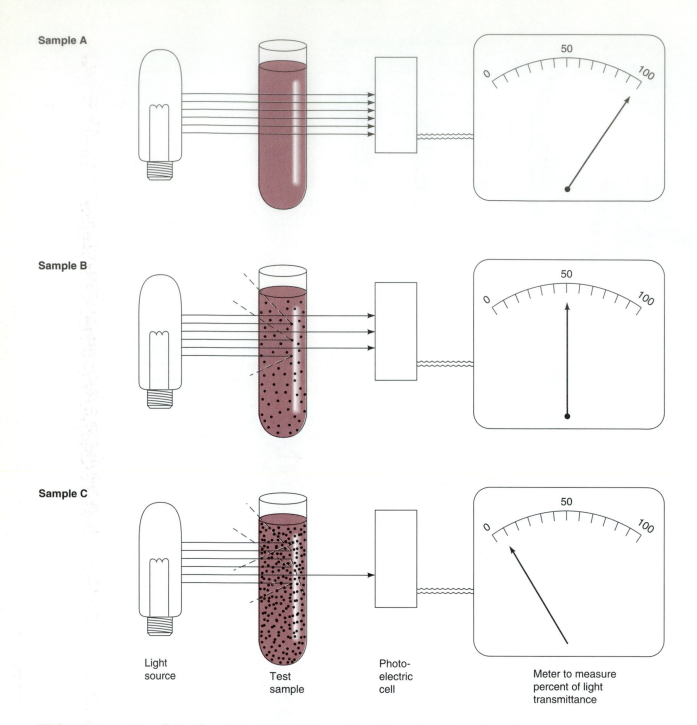

Light
source

Test
sample

Photo-
electric
cell

Meter to measure
percent of light
transmittance

FIGURE 7-19 Estimation of the number of bacterial cells based on the amount of turbidity as measured with a spectrophotometer. Sample A is clear broth and shows 100% light transmittance. Sample B has a moderate cell concentration and sample C has a heavy cell concentration; light transmittance is proportional to the number of cells in the broth.

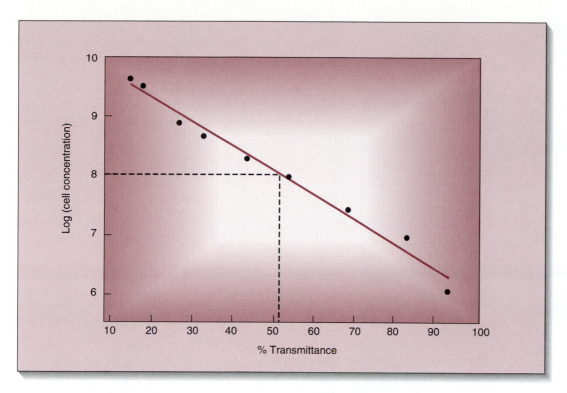

FIGURE 7-20   Linear relationship of % transmittance to the bacterial concentration.

chemical disinfectants (which would inhibit bacterial growth), neutralizers, such as lecithin or Tween 80, must be added to the medium to neutralize any chemical that might be transferred to the agar surface. RODAC plates are also commonly used to directly monitor the bacterial contamination of burn wounds. When used as described, they are less destructive to the injured tissue than swabbing procedures.

Sampling of Fluids.   When fluids may contain moderate-to-large numbers of microorganisms, counts can be made by the spread-plate or pour-plate method, using appropriate dilutions. If very

FIGURE 7-21   The Andersen cascade impaction air sampler. This sampler consists of six metal stages (assembled in background and partially disassembled in foreground) each pierced with 400 holes. The holes in the top stage are larger (far left) and become progressively smaller through to the bottom stage. A petri dish containing nutrient agar is placed under each stage and air is drawn through the unit at the rate of 28 liters per minute. Airborne bacteria are impacted onto the agar surfaces, and as the velocity of air increases as it flows through the progressively smaller holes in the descending stages, the larger particles are impacted onto the upper plate and the progressively smaller particles onto the sequential lower plates. This sampler resembles the respiratory tract, where larger particles are impacted in the upper regions and smaller particles are impacted in the lower region.

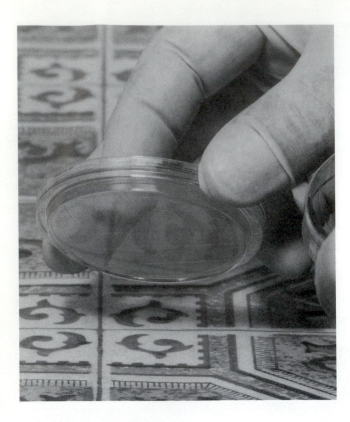

small concentrations of bacteria are present, as is often the case with drinking water, the bacteria may be concentrated by passing a measured volume through a membrane filter that will retain the bacteria. This filter is then placed on the surface of an agar plate and the bacteria trapped on the filter will grow to form colonies that can be counted (Figure 8-5).

FIGURE 7-22 A RODAC-type contact plate being used to collect a surface sample of microorganisms from a floor.

MATERIAL FOR REVIEW

CONCEPT SUMMARY

1. Bacterial nutritional needs in the laboratory are met by adding appropriate nutrients to a solidifying material called agar. By adding specific substances to the agar, media can be made to enrich, select, or inhibit the growth of desired microorganisms. Use of agar media has simplified the process of obtaining bacteria in pure culture. Such pure cultures are essential for the investigation of organisms and their role in disease processes.

2. Species of bacteria are identified on the basis of their reaction to a variety of ingredients that can be added to an agar or broth medium, and to the conditions for growth imposed on the culture.

3. Culture conditions, such as temperature, atmosphere, moisture, and pH, all impact the growth of microorganisms. Whatever the conditions, cultures of microorganisms follow a reproducible pattern of growth known as a growth curve.

STUDY QUESTIONS

1. Construct a table to show a classification of bacteria based on their growth as a function of (a) atmosphere and (b) temperature.

2. What evidence could you suggest to support the statement that most bacteria have a continuous growth pattern in their natural habitat?

3. Use a diagram to show that the change in growth rate from logarithmic to stationary phase is in fact a transition through many growth rate changes.

4. Why was the use of agar as a growth medium component such an important step in the history of microbiology?

5. Write a statement that contrasts differential and selective culture media.

6. If you begin at time X with two bacteria that divide regularly, and each of their progeny divide at the same rate as the parent cells, and after 1 hour you have 32 bacteria, what is the generation time of these organisms?

CHALLENGE QUESTIONS

1. Refer to Figure 7-18. Assuming that the number of bacteria in the broth is 5.0×10^6 as described, how many colonies would you expect if 0.5 ml of the 10^{-4} tube was plated?

2. Most bacteria need to establish a proton gradient across their cytoplasmic membrane in order to survive (see Figure 5-11). Alkaliphiles grow in environments with a high –OH concentration. Since –OH, if present, will rapidly react with H^+ to form water, how do you think alkaliphiles make ATP?

3. Some extreme hyperthermophiles can grow at temperatures exceeding 100°C (Table 7-2). What modifications in cell structure would need to occur to make growth at this temperature possible? What do you suppose happens to those cells when they are cooled to room temperature?

Sterilization and Disinfection

Microbial control in the health care field presents special problems and challenges. Because invasive procedures and more susceptible hosts are common components of a hospital environment, more attention to the reduction or elimination of microorganisms is necessary. The United States Public Health Service estimates that about 10% of all patients in the United States contract infections while in the hospital. These infections add an average of four days to the typical hospital stay, resulting in an additional $2.5 billion in health care costs annually. More importantly, at least 10% of those infected die as a direct result of their acquired infection. Many of these infections, and the corresponding deaths, can be prevented by a better understanding and implementation of proper infection control practices. Thus, it is vital that persons working in the health care services have a clear and correct understanding of microbial control principles and practices. In this chapter we examine a variety of physical and chemical methods used to destroy or to limit the growth of microorganisms on nonliving objects.

DEFINITION OF TERMS

Before beginning a discussion of microbial control methods, it is important to define some of the terms used in conjunction with microbial control. Some terms are absolute, others overlap in meaning, and some are relative, having slightly different mean-

ings in different areas of application. The common use of many of these terms by the lay public has resulted in an increased ambiguity with respect to their proper meaning. For example, one may have heard of surgeons *"sterilizing their hands"* prior to an operation. The term **sterilization**, which refers to a process that destroys all living organisms, is an

---

Sterilization   A process that destroys all living organisms.

absolute term. Any material that has thus been treated is said to be *sterile*, or completely devoid of living organisms. Obviously, living skin cannot be "sterilized." **Disinfection** refers to a process used to destroy harmful microorganisms but usually not including bacterial endospores; a **disinfectant** is a physical or chemical agent that produces this result. Although there are some exceptions, disinfectants are used almost exclusively on *inanimate* objects due to their toxicity for animal tissue. The suffix *-cide* means "to kill." Thus, *bactericides* kill bacteria, *fungicides* kill fungi, *germicides* kill a wide range of microorganisms, and so forth. The suffix *-static* refers to agents that stop the growth of microorganisms; for example, a *bacteriostatic* agent prevents the growth of bacteria. *Sepsis* means the presence of microorganisms in blood or other tissues; thus the term *asepsis* refers to any procedure that prevents microbial access to these areas. **Antiseptics** are agents that can be applied directly to living tissue to reduce the likelihood of infection or sepsis. Therefore, antiseptics are capable of rendering pathogenic microorganisms harmless either by killing them or preventing their growth. Such terms as disinfectant, bacteriostatic, and antiseptic may overlap significantly and all might be applied to the same agent.

The term *contamination* has different meanings in different settings. In the general clinical environment, contamination refers to the presence of disease-producing microorganisms in or on a substance. In more specialized areas, when referring to fluids for intravenous administration or surgical instruments and so on, the presence of *any* microorganism would be considered contamination. *Decontamination*, therefore, means to render a material safe from a microbial perspective. It is similar in meaning to disinfection, but broader in scope to include inactivation or removal of toxic microbial products, such as toxins. The term **sanitation** is often used in public health regulations connected with the food industry. Sanitation refers to any cleaning technique that physically removes microorganisms. This typically involves the use of a detergent only. Sanitation has a very broad meaning and must be defined for each individual application.

PHYSICAL METHODS OF MICROBIAL CONTROL

Physical and chemical methods used to destroy microbes on inanimate objects are generally nonspecific; that is, they destroy a wide variety of different types of living cells. In certain applications, such as the preparation of bandages or instruments to be used in surgery, successful control requires the complete removal or destruction of all microorganisms. In other applications, such as disinfecting a hospital ward, it is practical to remove only the disease-producing microorganisms or reduce their number to such a low level that the chance of infection is remote.

Heat

The use of heat to destroy microorganisms predates their discovery. Today it remains the most widely used method of microbial control. The application of heat has many variations, including temperature, amount of moisture, and transfer rate relating to the use of convection. Microorganisms differ greatly in their susceptibility to heat. Some psychrophiles ∞ (Chapter 7, p. 98) are destroyed by warming to room temperature. Conversely, extreme thermophiles grow and multiply at temperatures exceeding 110°C. A good measure of an organism's degree of heat resistance is the *decimal reduction time* or *D* value. This is defined as the time, at a specified temperature, in which 90% (or one decimal log) of a population is killed. Figure 8-1 depicts theoretical death curves for three different microorganisms at a given temperature. These values are very useful to the food industry in calculating needed processing times. Microbial death (death phase), caused by heat or any of the other methods discussed in this chapter, is a log-linear function similar to growth (see Figure 7-14).

Disinfection   A process used to destroy harmful microorganisms but usually not including bacterial endospores.
Disinfectant   An agent that usually is used to kill microorganisms on inanimate objects.
Inanimate   Not capable of self-movement; usually, but not always, nonliving. Inanimate objects involved in disease transmissions are called *fomites*.
Bacteriostatic   Capable of preventing bacteria from multiplying.
Antiseptic   An agent that usually is applied directly to living tissue to reduce the likelihood of infection or sepsis.
Sanitation   Any cleaning technique that physically removes microorganisms.

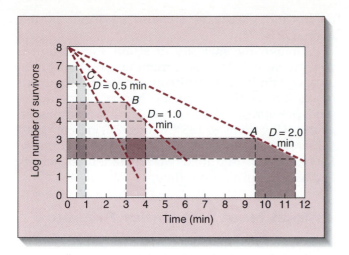

FIGURE 8-1 Theoretical death curves and corresponding *D* values for three different organisms (*A*, *B*, and *C*) exposed to a constant temperature. The time required to kill 90% of each population (a one-log reduction) is reflected in the slope of each line: the steeper the slope, the smaller the *D* value.

✳ Moist Heat Under Pressure

Over 90% of all medical and laboratory products are sterilized by steam under pressure. Because the temperature of water cannot be raised above 100°C at sea level, this procedure must be performed in a pressure chamber called an *autoclave* (Figure 8-2). Steam may be generated within the chamber or introduced from an external source under pressure. As steam enters the chamber, air must be forced out through an escape valve. Once air is expelled, the escape valve is closed and the steam pressure is increased to 1.1 kg/cm^2 (15 lb/in.^{2}). Under these conditions the temperature rises to 121°C. At this temperature, with moisture, cell structures are completely disrupted; proteins and nucleic acids are *denatured* and cell membranes are destroyed. When using an autoclave, it is necessary to allow time for the temperature to penetrate through all the material and then remain at 121°C for 15 to 20 minutes. A large bundle of items such as surgical drapes or

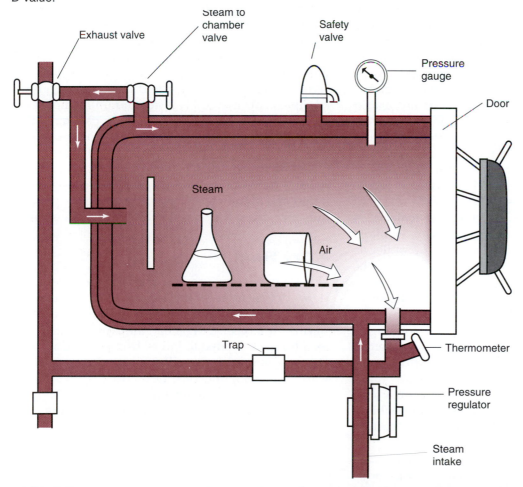

FIGURE 8-2 Schematic drawing of a steam autoclave.

Denature Altering of the tertiary or secondary structure of a protein.

bandages may require exposure times of 30 to 60 minutes or even longer to ensure sterility throughout the package. Items like bandages or surgical instruments that must remain sterile should be wrapped in covers to prevent them from becoming contaminated once removed from the autoclave. Not all materials can be autoclaved. Moisture associated with autoclaving causes such products as dry powders to become soggy, and the heat involved may damage many plastic products or electronic instruments used in hospitals. Fluids that contain *heat-sensitive* components, such as serum, should not be sterilized by autoclaving. Containers of fluids to be autoclaved must not be tightly sealed, for such sealing may prevent the movement of steam to and from their contents, and once sterilization is achieved, autoclave pressure must be released slowly to prevent excessive boiling and evaporation.

Various tests are used to determine if sterility has been attained. Papers impregnated with heat-sensitive chemicals that change color when exposed to a critical temperature are useful but not totally reliable indicators of sterility, since they do not indicate how long sterilizing temperatures have been maintained. The most reliable indicators are *biological monitors*. Typically, these consist of paper strips impregnated with the most heat-resistant bacterial endospores known. Biological monitors are placed in the center of the materials being autoclaved. After the sterilization cycle is completed, the spore strip is placed in contact with a broth medium and incubated. If no growth occurs in the broth, meaning that these spores have been killed, then it can be assumed that all other less heat-resistant forms of life are dead (Figure 8-3). Therefore, the material in that autoclave load can be assumed sterile. Monitors are sometimes placed between wrapped bundles so that they can be removed without unwrapping the sterilized materials. Convenient-to-use, self-contained biological monitors are commercially available and commonly used (Figure 8-4).

Moist heat in the form of pressure cooking is used in home canning and in the commercial canning industry. In these applications it is important to use temperatures and exposure times that will assure the killing of bacterial endospores, and thus protect consumers against the possibility of *botulism* food poisoning. Laboratory culture media and glassware are also sterilized in this manner.

Although the majority of steam sterilization is performed at 121°C for a minimum of 15–20 minutes, certain heat-stable items can be sterilized in a much shorter time by using higher temperatures (and pressures). *Flash autoclaving* is accomplished by exposing items to 135°C (about 2.2 kg/cm^2 pressure) for only 3 minutes. This procedure is common for frequently used items that require both sterility and a quick turnaround time.

✳ Moist Heat Not Under Pressure

Boiling and live steam, not under pressure, destroy most vegetative forms of bacteria within a few minutes. Some viruses, such as certain hepatitis viruses, can tolerate short periods of boiling, and bacterial endospores may survive boiling temperatures for several hours. The moderate heat of **pasteurization** is useful in treating some liquids, such as milk or other beverages. Pasteurization may be accomplished via either of two procedures: the *holding method* or the *flash method*. In the holding method, the liquid is heated in bulk to 62.8°C for 30 minutes. The flash method heats the fluid to 71.7°C for 15 seconds, typically as the liquid flows through heated pipes. This method is faster and much more convenient, making it the most common pasteurization method in use today. Pasteurization does not sterilize, but does kill disease-producing bacteria that might be transmitted by the liquid. It also greatly reduces the numbers of other viable bacteria in the liquid and significantly retards the rate of spoilage, thus increasing the shelf-life of products like milk.

✳ Dry Heat

Hot air ovens are used as dry heat sterilizers, but static dry heat requires higher temperatures for longer periods than moist heat in order to achieve sterilization. Using dry heat, an exposure at 170°C for 2 hours is needed to kill all bacterial endospores. Dry heat is used for sterilizing such items as glassware, powders, and oils. Dry heat requires higher temperatures and longer times because heat transfer is much less efficient in the absence of moisture. To compensate, new high-temperature dry heat

---

Heat-sensitive   Capable of being damaged or destroyed by heat.
Botulism   A disease caused by toxins produced by the bacterium *Clostridium botulinum.*
Pasteurization   The process of using mild heat to kill pathogens.

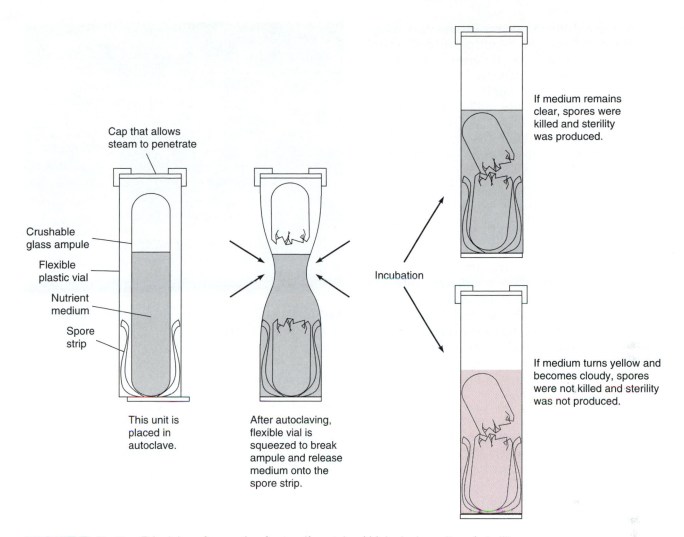

Cap that allows
steam to penetrate

Crushable
glass ampule

Flexible
plastic vial

Nutrient
medium

Spore
strip

This unit is
placed in
autoclave.

After autoclaving,
flexible vial is
squeezed to break
ampule and release
medium onto the
spore strip.

Incubation

If medium remains
clear, spores were
killed and sterility
was produced.

If medium turns yellow and
becomes cloudy, spores
were not killed and sterility
was not produced.

FIGURE 8-3   Principles of operation for a self-contained biological monitor of sterility.

sterilizers have recently become available. These devices operate at 190°C with forced air convection (rapid air movement across the items to be sterilized). This results in rapid heat transfer to the items being sterilized. Although a limited number of items can tolerate such high temperatures, the rapid cycle time (as low as 6 minutes) makes them useful for certain dental instruments, such as orthodontic pliers. Another form of dry heat sterilization is the direct exposure of instruments or inoculating loops to open flames for brief periods, a procedure called *flaming*. Incineration of waste products readily destroys any contaminating microorganism that might be present.

✳ Heat and Chemical Vapor

A sterilization device that combines physical and chemical control methods is the *Harvey Chemiclave*. This device generates an alcohol/formaldehyde vapor at a pressure of 1.1 kg/cm^2 and a temperature of 132°C. The cycle time is similar to that of a steam autoclave. Chemiclaves are used widely in dentistry because their cycle times are relatively short and they lack water vapor, which can dull the fine cutting surfaces of steel instruments.

✳ Radiation

Ultraviolet Light (Nonionizing).   Ultraviolet (UV) light is highly germicidal at wavelengths around 260 nm. This wavelength of light is absorbed by DNA molecules and the increase in energy causes a rearrangement of some chemical bonds. In particular, new chemical bonds are formed between adjacent thymine bases on the same strand of the DNA double helix. This renders the DNA of the UV-irradiated microorganism nonfunctional. Most microorganisms have enzymatic systems that can repair UV-induced DNA damage, as long as it is not too extensive. Extensive damage will overload the system, resulting in cell death. Moderate damage invokes an error-prone DNA repair system, which frequently leads to mutations, many of which will also be fatal.

Sunlight contains UV light; consequently, it has definite germicidal properties. UV light is also produced by mercury vapor lamps. When placed in air ducts or over surfaces, these lamps greatly reduce the number of viable microorganisms in the field of irradiation. UV light does not penetrate solids, including glass and most plastics, a factor that has limited its use to disinfecting surfaces, clear liquids, and air. High-intensity UV lamps located in air supply ducts to such critical areas as operating rooms, nurseries, and intensive care areas can greatly reduce the chance of infections being transmitted to these areas by the airborne route. Also, air leaving contaminated areas, such as isolation rooms, morgues, or laboratories, can be exposed to UV light to prevent the spread of disease-producing microorganisms from such sources. UV light is also damaging to human tissue and eyes and direct exposure should be avoided.

Ionizing Radiation. The shorter the wavelength of electromagnetic radiation, the more energy it possesses. High-energy forms of ionizing radiation, such as x-rays and gamma rays, have respective wavelengths 1000 to 100,000 times shorter than UV light. Radiation with this much energy creates extremely reactive ions, such as hydroxyl radicals, inside the cell (hence the term *ion*izing radiation). These chemicals react quickly with cellular material, resulting in loss of function. For example, the reaction with DNA causes breaks in the strands, which may permanently interfere with DNA replication and therefore any subsequent cell multiplication. They can also react with proteins to cause the irreversible inactivation of essential cell enzymes. Unlike UV light, ionizing radiations can penetrate such products as fabrics, plastics, liquids, and foods to effect sterilization. For these reasons, ionizing radiation is much more efficient at killing microorganisms than UV light. Currently, ionizing radiation is used to sterilize products like surgical sutures and disposable plastic items. Meats, fruits, and some vegetables can be effectively sterilized by ionizing radiation. Such foods have a comparable shelf life to heat-processed canned foods, and are usually much more nutritious and palatable. Radiation-sterilized meats have been used by astronauts during space flights, but as yet this process has had only limited use in the general food industry.

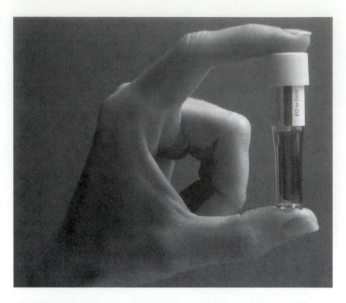

FIGURE 8-4 Example of a commercial self-contained biological monitor.

✳ Filtration

Filtration is an effective means of removing most microorganisms from liquids and gases. Liquids containing heat-sensitive materials can be freed of microbial cells by passing them through filters with pore sizes small enough to retain bacterial cells. Filters made of asbestos, fused glass fragments, or diatomaceous earth have been used for many years. Biologically inert, precisely produced cellulose ester membrane filters are now widely used (Figure 8-5). One with a pore size of 0.22 μm effectively removes all bacteria from a fluid, and use of filters with smaller pores, such as 0.025 μm, will remove most viruses that may be found in solution. Filters of this type are also used to trap and concentrate bacteria that are dispersed in large volumes of liquids. This procedure is useful in water quality testing as well as in some methods that quantitate disinfectant efficacy.

Airborne microorganisms can be effectively removed from air by filtration. Filters made of various fiber media are widely used in air ducts to remove both inert and microbial particles. Varying densities of filter media can remove the desired size and amounts of airborne particles. Special filters, referred to as *absolute* or *HEPA* (*h*igh-*e*fficacy *p*articulate *a*ir) filters, consist of a tightly woven fiberglass medium and effectively remove 99.9% of all airborne particles down to the size of 0.3 μm. HEPA filters remove all types of airborne microorganisms. Even though viruses are as small as 0.02 μm, when airborne they are usually attached to larger particles of water, dust, or dried mucus that are readily trapped by HEPA filters. Air filters are used in air supply systems servicing such hospital areas as operating rooms, nurseries, and intensive care units.

CHEMICAL METHODS OF MICROBIAL CONTROL

Hospitals, clinics, and laboratories rely heavily on chemical antiseptics and disinfectants to reduce or eliminate harmful microorganisms on skin or inanimate objects. Many chemical formulations are commercially available as disinfectants and no single product is suitable for all applications. Several different chemical disinfectants are needed to accommodate the needs of most clinics and hospitals. Persons working in hospitals need to be aware of what disinfectant to use for each type of application. Even though many commercial products are available, most disinfectants belong to one of the categories discussed in the following pages. Figure 8-6 shows the chemical structure of many compounds commonly used as antiseptics and disinfectants.

✳ Factors Affecting Disinfectant Action

Disinfectants destroy or prevent the growth of microorganisms via generalized effects produced on the microbial cells. Some disinfectants are surface-active agents that disrupt the normal functioning of cytoplasmic membranes; others cause denaturation of proteins, such as enzymes, that are essential for cell growth and function. The action and uses of some common groups of disinfectants are shown in Table 8-1. The ability of a chemical disinfectant to act on a microorganism and the extent of that action depend on the following factors.

Time. Not all microbes are killed at the same time after the addition of a disinfectant. Therefore, the disinfectant must remain in contact with the

---

TABLE 8-1

Mode of Action, Uses, and Properties of Major Categories of Chemical Disinfectants

Agents	Major Action	Common Uses	Other Properties	Use Dilution (%)
Alcohols	Lipid solvents Denatures proteins	Skin antiseptics Surface disinfectants	Rapid action Flammable Dries skin	70–80
Mercurials	Inactivates proteins	Skin antiseptics Surface disinfectants	Weak cidal activity Inactivated by organic matter	0.1
Silver nitrate	Denatures proteins	Antiseptic for eyes and burns	Inactivated by organic matter Limited range of microbes affected	1
Phenolic compounds	Disrupts cell membranes Inactivates proteins	In antiseptic skin washes Disinfect inanimate objects	Not inactivated by organic matter Stable Some objectionable odors	0.5–5
Iodine	Inactivates proteins	Skin antiseptic	Soluble in alcohol Rapid action Mixes with soaps	2
Chlorine compounds	Oxidation of enzymes	Water treatment Disinfect inanimate objects	Inactivated by organic matter Flash action Corrosive Irritates skin	0.5 (as bleach)
Quaternary ammonium compounds	Surface active Disrupts cell membranes Denatures proteins	Skin antiseptic Disinfect inanimate objects	Neutralized by soap, organics, and inorganics Nonirritating, odorless	<1
Glutaraldehyde	Inactivates proteins	Cold sterilizing agent for heat-sensitive instruments	Unstable in alkaline form Toxic High activity in alkaline range	2–3

contaminated material long enough to allow for the killing of all microbes present, including the most resistant members. During most short applications, chemical agents do not sterilize; however, if the time is extended to periods of 10 to 12 hours, sterilization with some agents is possible.

Temperature.   The lethal effects of disinfectants increase with increasing temperature. Most disinfecting procedures are carried out at room temperature. The time of exposure must be extended when materials are disinfected at low temperatures.

pH.   The acidity or alkalinity of the environment also influences the interaction of disinfectants with microorganisms and may increase or decrease their action, depending on the agent. Thus, the effects of pH must be considered separately for each disinfectant.

Numbers of Microorganisms.   The time required to eliminate all viable microorganisms is dependent on the initial concentration of microbes. Forty minutes may be required to kill 10^7 cells, but only 15 minutes would be necessary to eliminate 1000 organisms of the same type.

Types of Microorganisms.   Significant variations in susceptibility between species of microbes occur. Microbes are sometimes ordered as to their susceptibility to disinfectants into the following three groups:

> *Group A*, the vegetative forms of most bacteria and *enveloped* viruses that are easily killed by disinfectants
> *Group B*, the more difficult to kill tuberculosis bacillus and nonenveloped viruses
> *Group C*, the highly resistant bacterial endospores

Presence of Extraneous Matter.   Organic materials like soil, blood, and pus may react with some disinfectants and reduce their ability to react with microbes. For this reason, it is strongly recommended that surfaces and materials to be disinfected be thoroughly cleaned before final treatment with the disinfectant.

Proper Exposure.   Care must be taken to ensure proper exposure of all parts of the object to the disinfectant. Tightly packaged material or closed containers may not allow complete penetration or contact with the disinfectant. For example, endoscopic equipment with its many internal sur-

faces and tight crevices has been especially difficult to disinfect, and has been documented as a source of microbial transmission from patient to patient.

Concentration of Disinfectant.   Generally, the more concentrated the disinfectant, the shorter is the killing time. The alcohols are an exception to this general rule. Alcohols operate more efficiently when some water is present; when the alcohol concentration exceeds 90%, a noticeable drop in antimicrobial activity is observed. For most other agents, a direct relationship is seen between concentration and activity. At low concentrations the compound may be only bacteriostatic, whereas at higher concentrations it may be bactericidal. The minimum concentration needed to kill microorganisms varies from microbe to microbe and from disinfectant to disinfectant.

✳ Groups of Chemical Disinfectants

Alcohols.   Two types of alcohols are commonly used to kill microorganisms. *Ethyl alcohol* and *isopropyl alcohol* (Figure 8-6) both kill microbes by disrupting cytoplasmic membranes and denaturing cellular proteins. Alcohols are among the most useful disinfectants and antiseptics. Although they have almost no activity against bacterial endospores, they are highly effective against vegetative bacterial cells, including the tuberculosis bacillus. They also have high activity against fungi and most viruses. Alcohol is very useful as a skin antiseptic because it effectively kills bacteria, has a cleansing effect by removing accumulated lipids, and evaporates without leaving a residue. Although commonly lumped together, ethyl and isopropyl alcohol differ with respect to their antimicrobial spectrum, toxicity, and cost. Isopropyl alcohol is slightly more effective against organisms possessing a membrane, including enveloped viruses. However, it is also more toxic to animals and completely ineffective against some nonenveloped viruses, such as poliovirus. Ethyl alcohol is relatively nontoxic to animals and has good activity against all non-endospore-forming microorganisms. Isopropyl alcohol is much less expensive due to the demand and regulations relating to ethyl alcohol as a beverage. Both alcohols require some hydration for optimal activity, and should be used in concentrations between 70 and 80%.

Many of the active ingredients discussed below

Envelope   A membrane surrounding the protein coat of some viruses.

(a)

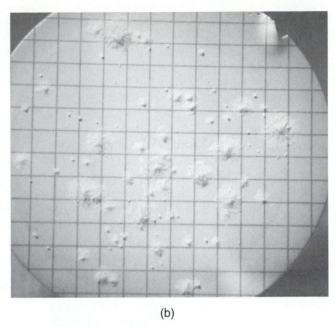

(b)

FIGURE 8-5   Membrane filters used to remove microorganisms from liquids. (a) Position of the membrane filter between a sterile base and funnel. (b) Used membrane filter that has been incubated on growth medium. Each retained colony forming unit has formed a colony on the surface.

are much more effective when formulated in alcoholic solutions. We call these mixtures **tinctures**. Examples are tinctures of iodine, chlorhexidine, or certain quaternary ammonium compounds.

Heavy Metals.   Various mercury-containing compounds, called *mercurials,* were widely used as disinfectants in the past. One such compound, *mercuric chloride,* is quite toxic to humans and is now rarely used. Other mercurial preparations known as *mercurochrome, merthiolate, metaphen,* and *mercresin* are less toxic. Merthiolate is sometimes used as a preservative for vaccines and metaphen may be used to irrigate the *urethra.* The mode of action of mercury is by combining with active chemical (*sulfhydryl*) groups on proteins, including enzymes. The result of such interaction is the inactivation of these enzymes and subsequent cell death. The mercurials, however, are not as effective as many other preparations and are usually corrosive to metallic instruments. Mercurials are not used much today because of their toxicity, tendency to cause allergic reactions, and their ease of neutralization by organic material. In general, other disinfectants are preferred over the mercurials.

Some silver compounds are useful disinfectants, the most common being *silver nitrate.* A 1% silver nitrate solution was once used routinely in the eyes of newborn babies to prevent gonorrheal infections. This procedure was required by law in all 50 states. Ointments containing antibiotics like erythromycin are more effective and less irritating and have largely replaced this use of silver nitrate. Another silver compound is silver sulfadiazine, which is commonly used to prevent infections in burn patients. Colloidal silver compounds that slowly release silver ions are used in some antiseptics and in some filters for water purification. The mode of action of silver compounds is similar to that of mercury.

Copper ions can also inhibit the growth of microorganisms. Dilute solutions of compounds such as *copper sulfate* are sometimes used in aquaria and swimming pools to reduce algal and fungal growth. Copper compounds are also used in some paints as mildew inhibitors.

Tincture   An alcoholic solution; a tincture of iodine is a solution of alcohol that contains iodine.
Urethra   The tube through which urine passes from the bladder to the outside of the body during urination.
Sulfhydryl   Referring to a sulfur bonded to a hydrogen (–SH) in a molecule.

Phenol and Phenol Derivatives. The natural product *phenol*, also called carbolic acid, is fairly toxic to tissues, is corrosive, and has a disagreeable odor; yet historically it has served as an important disinfectant. Pasteur recognized the ability of phenol to prevent decomposition or organic matter. This observation prompted Lister to use the product to prevent infections of surgical wounds. Because of its recognized toxicity, phenol is no longer used as an antiseptic. Various phenol derivatives, called *phenolics*, have been developed that are effective disinfectants and antiseptics, yet do not have many of the objectionable traits of phenol (Figure 8-6). Examples of phenolic disinfectants include *o*-phenylphenol, *p-tert*-amylphenol, and the methylated phenols or *cresols*. Phenolics that have been more extensively modified to reduce toxicity have been useful as antiseptics. These include hexachlorophene, hexylresorcinol, and the phenolic relative chlorhexidine. Phenolics act by damaging cell walls and membranes and by precipitating essential cell proteins. The major advantages of phenolics include their reasonable cost, their compatibility with detergents, and their ability to resist inactivation by organic material. A major disadvantage of phenolics is their lack of activity against endospores and certain nonenveloped viruses such as poliovirus. A few of the phenolics with widespread use are discussed separately below.

Cresols are obtained from coal tars, are less toxic than phenol, and have strong germicidal actions. For many years a mixture of 2% cresol and liquid soap was sold under the trade name of Lysol. This old-type Lysol had a characteristic cresol odor that was familiar to most persons living before the 1940s. In later years mixtures of soaps and improved phenolic compounds with less odor have been marketed under the trade name of Lysol. Lysol spray, one of the most effective surface disinfectants, is a mixture of 79% ethyl alcohol and *o*-phenylphenol.

The phenolic compound hexachlorophene is especially effective against gram-positive microorganisms such as staphylococci and streptococci. Hexachlorophene retains its antimicrobial effectiveness when mixed with soaps or detergents, is nonirritating to skin, and leaves a protective film after application. It is bactericidal at high concentrations and bacteriostatic at lower concentrations. During the 1960s, hexachlorophene had many medical and nonmedical applications. Newborn infants were routinely bathed in mild solutions of hexachlorophene, a procedure that greatly reduced bacterial colonization and subsequent infection by staphylococci. It was also a common ingredient in some underarm deodorants, bar soaps, and even baby powders. However, in the early 1970s, evidence from animal studies suggested that hexachlorophene was absorbed into the blood and caused brain damage. Therefore, restrictions have been placed on its use and it is now a prescription-only product. A 3% concentration is mixed with soaps, detergents, and lotions to form effective antiseptic skin-cleaning products for medical applications. pHisoHex and Hexagerm are two examples.

Chlorhexidine is not a phenolic, but is closely related in structure, as can be seen in Figure 8-6. *Chlorhexidine gluconate* mixed with detergent is a widely used surgical handscrub, obstetrical antiseptic, cleanser for superficial skin wounds, and handwashing agent. It is also relatively nonirritating to mucous membranes and is an effective antibacterial mouth rinse in the treatment of periodontal disease or when normal dental hygiene is not possible. It is highly effective against both gram-positive and gram-negative bacteria and fungi and has largely replaced hexachlorophene in use. It leaves a residual film on skin and is relatively nonirritating, even when used extensively. Unlike hexachlorophene, chlorhexidine is not absorbed into the blood and no similar toxicity has been reported. Hibiclens is an example of a chlorhexidine-containing product.

Halogens. The halogens *iodine* and *chlorine* are among the most useful chemical disinfectants. Chlorine is used widely in water treatment. If chlorine gas (Cl_2), hypochlorite (OCl), or chloramines (NH_2Cl) are added to water, a reaction occurs that liberates hypochlorous acid (HOCl), a very active disinfectant. This uncharged molecule readily penetrates the cytoplasmic membrane of microorganisms and oxidizes sulfhydryl groups of cell proteins. These are the same active areas of protein that combine with heavy metals (see above). Hypochlorites are the common household bleach agents, such as Clorox and Purex. Chlorine compounds are routinely used to sanitize food and dairy-processing equipment and to treat public waters such as swimming pools. Chlorine bleaches are useful household disinfectants and can be used on dishes, utensils, toilets, or other noncorrodible materials. Because of their tissue toxicity, chlorine compounds should not be used on skin or open lesions; they are also corrosive to metals. Chlorine is readily inactivated by organic matter, and dilute solutions easily lose their effectiveness when excessive organic matter is present in solutions or on surfaces. These solutions are also relatively unstable and gradually lose activity with age. You may have

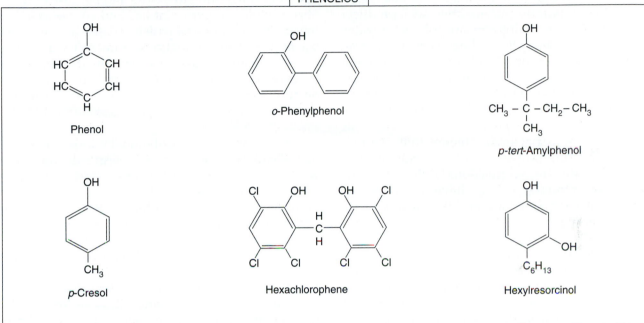

FIGURE 8-6 Chemical structures of common disinfecting compounds.

noticed this if you have ever tried to use bleach that is more than a few years old.

Iodine is among the most effective skin antiseptics. A 2% tincture of iodine is used widely as a skin antiseptic and minor wound cleanser. A 2% solution of iodine in water is also an effective antiseptic. Iodine forms complexes with soaps and detergents without losing its antiseptic qualities; such products are called *iodophors*. Iodophors are soluble in water and gradually release the iodine. They are not as active as tincture of iodine but have the advantage of being less irritating and nonstaining. They also have the cleansing effect of soap or detergent. Iodophors are used as antiseptic soaps, preoperative skin disinfectants, and general disinfectants in medical and industrial environments. Iodines function by iodinating proteins and fatty acids of cells. This inhibits the normal functions of these components. Like chlorines, iodines have a very broad spectrum of activity, but are rapidly inactivated by organic material.

Surface-Active Agents. Various surface-active compounds have *detergent*-like characteristics and cause the destruction of microbial membranes. The *quaternary ammonium compounds*, often referred to as *quats*, are *cationic* detergents and are the only surface-active agents with effective antibacterial activity. These agents are effective against a wide range of vegetative bacteria. They are usually not effective against endospores, the tuberculosis bacillus, nonenveloped viruses, and some other vegetative bacteria. They are also readily inactivated by both organic and inorganic substances. Even material in tap water can destroy the activity of some quats against organisms such as *Pseudomonas aeruginosa*. Quats are used for disinfecting floors, walls, furniture, and other inanimate objects. They have the advantages of being odorless, colorless, tasteless, inexpensive, nontoxic, soluble in water, and active in low concentrations.

Common detergents and soaps are *anionic* detergents and are excellent cleaning agents, but are not antimicrobial. Such surface-active agents can effectively clean large numbers of microorganisms from skin or other surfaces, but are essentially nontoxic to microorganisms. As a general rule, thorough washing with such agents removes in excess of 90% of the microbes present.

Formaldehyde. Formaldehyde is a gas that acts as a fumigant and a gaseous disinfectant. It dissolves in water to make a 37% solution, which is then called *formalin*. Solutions containing 5 to 10% formalin have been widely used for preserving and fixing tissue specimens. Formalin has a disagreeable odor and is irritating to tissues and because of these objectionable properties, it has limited use as a disinfectant in clinics and patient-related activities. Its use in preserving biological specimens used in biology and anatomy classes has been greatly decreased in recent years due to new government regulations on the permissible levels of formaldehyde in the work environment.

Glutaraldehyde. Glutaraldehyde is a pentane dialdehyde (Figure 8-6) and is germicidal against a wide range of microorganisms. Most vegetative microorganisms other than mycobacteria are destroyed within a few minutes. Recent evidence has shown a slower activity against the tuberculosis bacillus due to its waxy cell wall which resists wetting. Glutaraldehyde is used as a cold sterilizing agent for many items that would be damaged by heat. It is widely used for sterilizing dental equipment, items used in inhalation therapy, and equipment with optical lenses such as endoscopes. Glutaraldehyde is most germicidal in the alkaline pH range and so a 0.3% sodium bicarbonate solution is commonly added to a glutaraldehyde solution just before it is to be used. An alkaline pH results, and the solution is then said to be "activated." Although it is more microbicidal, activated glutaraldehyde is also more unstable and slowly forms long polymers with age. An activated solution should not be used longer than 3 to 4 weeks. Materials to be sterilized must be clean and completely immersed in 2-3% activated glutaraldehyde for 10 hours. Glutaraldehyde is irritating to tissues and has a mildly disagreeable odor; thus its use is limited to inanimate objects.

Hydrogen Peroxide. A 3% solution of hydrogen peroxide (H_2O_2) is sometimes used to clean wounds. It is nonirritating to the tissues and has only a brief, mild disinfecting action due to its rapid breakdown to water and oxygen by the enzyme *catalase* ∞ (Chapter 7, p. 99), which is very common in animal tissues and many bacterial cells.

Detergent A compound similar to soap (not a disinfectant) that is used as a cleaning agent because of its ability to emulsify dirt.
Cationic Having positive electrically charged groups.
Anionic Having negative electrically charged groups.

Hydrogen peroxide affects the cells much like the halogens, and inactivates essential protein structures by oxidizing reduced sulfur groups. Some new sterilizing devices have employed hydrogen peroxide at much higher concentrations.

Ethylene Oxide. Ethylene oxide vaporizes readily at room temperatures and is a highly effective sterilizing agent in the gaseous form. It is active against all microorganisms, including bacterial endospores. The major advantage of ethylene oxide gas is its ability to sterilize at moderate temperatures and without high levels of moisture. It is slow acting, however, and 12 hours is required to destroy spores at 70°C. Also, ethylene oxide is explosive when mixed with air; for this reason, it is always diluted with an inert gas, such as carbon dioxide. A 10 to 15% concentration of ethylene oxide is used for sterilization. Special chambers or especially adapted autoclaves are used for this form of gas sterilization. Such items as plastic ware, **catheters**, sutures, electronic instruments, and heart–lung machines that may be damaged by heat are sterilized with ethylene oxide. Over the last few decades, this form of sterilization has become an essential procedure in most hospitals. Items sterilized with ethylene oxide must be well aerated before use to remove any residual toxic gas.

Other Disinfectants. Acids and alkalies have antimicrobial activities due primarily to the free hydrogen or hydroxyl ions. Such acids as *benzoic* or *propionic*, or their salts, are added to foods to help retard spoilage. Some aniline and acridine dyes, such as crystal violet and acriflavine, respectively, have bacteriostatic activities and are used in treating lesions on the skin and mucous membranes.

✳ Evaluation of Chemical Disinfectants

One procedure for evaluating a disinfectant is the *phenol coefficient method*. This method compares the effectiveness of the test disinfectants to that of phenol against the vegetative bacteria *Salmonella choleraesuis*, *Staphylococcus aureus*, and *Pseudomonas aeruginosa*. The time required to kill these bacteria by using dilutions of the test disinfectant, compared to dilutions of phenol, gives a relative strength of the test compound. By comparing all disinfectants to phenol, it is possible to gain a comparison of their relative potency. The major flaw of this method is that some disinfectants tolerate dilution much better than others. The result is that compounds such as quats will have very high phenol coefficients while alcohols will have very low values, even though 80% ethanol is the superior disinfectant. The phenol coefficient method does not give information on the dilution of a given disinfectant that might be suitable for a given object or surface. A second test, called the *use-dilution method*, is now used to determine the concentration of a disinfectant that is needed to kill bacteria effectively. In this test, which has recently been modified, 60 small glass cylinders contaminated with a standardized suspension of one of the test bacteria mentioned above are placed in tubes containing the test dilutions of a disinfectant and left for 10 minutes. If all bacteria are killed on at least 59 of the 60 cylinders, the dilution of disinfectant is considered suitable for use. This is the test that must be passed in order to obtain EPA approval of the disinfectant for hospital use. The entire test must be repeated a total of three times with three different lots of disinfectant, using each of the three organisms mentioned above.

MATERIAL FOR REVIEW

CONCEPT SUMMARY

1. The removal of all living microorganisms from an environment (sterilization) or the removal of most pathogens (disinfection) is usually accomplished by physical or chemical means. Heat, radiation, and filtration are most commonly used to obtain sterile conditions, whereas the bactericidal actions of alcohols, phenols, halogens, and aldehydes are frequently applied for disinfection. Both industrial and

---

Catheter A tube, usually of rubber or plastic, that can be placed into a body cavity (e.g., urethra) or a blood vessel to allow easy drainage of body fluids or the placement of medications into the body.

household uses of these procedures are intended to reduce the number of microorganisms in our environment so that health and safety are maintained.

2. There are a finite number of ways by which bacteria are destroyed by sterilizing and disinfecting agents: membrane disruption, protein denaturation, enzyme inhibition, or nucleic acid (DNA) alteration.

STUDY QUESTIONS

1. Make a table that classifies each of the disinfectants into a category based on their mode of action.
2. What feature of bacteria makes it necessary to heat materials above the boiling temperature to ensure sterility?
3. Which disinfectant would be best suited for: a) disinfection of a countertop; b) disinfection of hospital floors; c) sterilization of an endoscope; d) pre-operative surgical scrub?

CHALLENGE QUESTIONS

1. Refer to Figure 8-1. Assume that you have been given the assignment to manufacture a biological indicator consisting of heat-resistant endospores. If these endospores have a D value of 1.2 minutes, what is the minimum number of endospores you must place on a strip so that some spores will survive 5 minutes in the autoclave?
2. If tinctures of iodine are more effective antimicrobials than aqueous solutions such as iodophors, why are they not routinely used as surgical hand scrubs?

Chemotherapeutic Agents

Disinfectants and antiseptics are useful in destroying microorganisms on inanimate objects and limiting their number on living tissue surfaces like the skin. However, if microorganisms infect living tissues, antiseptics are not the chemical agent of choice. Another group of chemicals, the chemotherapeutic agents, are used to control and hopefully help destroy the invading microorganisms. The continued development of new chemotherapeutic agents over the past 50 years is one of the most important achievements of medical science and has saved millions of lives and alleviated untold suffering. This chapter examines the chemotherapeutic agents, their mode of action, and clinical applications. In addition, microbial resistance and treatment are discussed. Early historical developments of chemotherapeutic agents were covered in Chapter 1.

CHARACTERISTICS OF CHEMOTHERAPEUTIC AGENTS

Antimicrobial **chemotherapeutic agents** are chemicals that can selectively interfere with the growth of microorganisms and yet not interfere significantly with the functions of the cells of the infected animal *host*. This type of activity is known as *selective toxicity*. Generally, diseases caused by bacteria are more effectively controlled by chemotherapeutic agents than are diseases caused by fungi, protozoa, or viruses. A primary reason for this difference is that bacteria are prokaryotic cells and possess some

Chemotherapeutic agent   A chemical agent used to treat diseases.

123

structures and metabolic processes that differ greatly from those of the eukaryotic cells of the animal host. Based on these cellular differences, it is possible to develop chemicals that specifically interfere with prokaryotic cell functions but not with the activity of eukaryotic cells. On the other hand, it has been difficult to find chemicals that selectively inhibit a given type of eukaryotic cell, such as a fungal cell, without interfering with animal cells as well. The development of antiviral agents has been even more difficult because viruses use host cell functions to carry out their replication activities. As yet, few chemical agents have been developed that can interfere selectively with viral activity and not interfere with the functions of the host cells.

An ideal chemotherapeutic agent should possess as many of the following characteristics as possible:

1. Be highly toxic to a large number of pathogens.
2. Have no toxicity for the host.
3. Does not induce the development of *antibiotic resistance* in mutant microbes.
4. Will not induce *hypersensitivities* in the host.
5. Does not interfere with the normal host defense mechanisms.

Unfortunately, the ideal chemotherapeutic agent has not yet been found. Most have some host toxicity and induce varying degrees of hypersensitivity or allow the development of resistant mutant microbes. Thus, some trade-off is given for all applications of chemotherapeutic agents, and benefits to the patient must be weighed against possible adverse side effects.

Only those antimicrobial agents that are the natural products of microorganisms are called **antibiotics**. Antimicrobial compounds that are made in the laboratory but not produced by living organisms are referred to as *synthetic* agents. Because the word *chemotherapy* has been extensively used in connection with cancer therapy, the term *antimicrobial* is often preferred over *chemotherapeutic* when discussing infectious disease.

SYNTHETIC AGENTS

Antimicrobial agents may act by mimicking essential components needed in normal cellular reactions. When present, the antimicrobial agent is taken into a cellular reaction in place of the normal component. Once integrated, the antimicrobial agent prevents the cell from functioning or developing in a normal manner. Extensive efforts have been carried out to develop synthetic chemicals that would interfere with specific microbial functions; yet relatively few useful compounds have been developed. Most useful synthetic antibacterial agents are related to the *sulfonamides*.

✳ Sulfonamides

Since their discovery in the mid-1930s ∞ (Chapter 1, p. 11), the sulfonamides have been important agents in treating a variety of bacterial infections. Sulfonamides, sometimes simply called *sulfa drugs*, are various derivatives of a molecule called *p-aminobenzene sulfonamide* also known as *sulfanilamide* (Figure 9-1).

Mechanism of Action.   The sulfonamides are structurally similar to *p-aminobenzoic acid* (PABA) and function as *competitive inhibitors* of this compound. PABA is an essential component in the synthesis of *folic acid*, which is an essential metabolite for both mammalian and prokaryotic cells. Mammalian cells, however, depend on preformed folic acid obtained in the diet, whereas prokaryotic cells synthesize their own folic acid using PABA. This difference in the source of folic acid allows sulfonamides to function as effective chemotherapeutic agents. In bacteria, the sulfonamide molecule is able to substitute for PABA during the synthesis of

---

Host   An organism that supports the growth of another organism.
Selective toxicity   The quality possessed by a substance that can damage or destroy a living organism in the presence of another organism that remains unaffected.
Antibiotic resistance   A condition in which a microbe is unaffected by the presence of a compound used for antimicrobial therapy.
Hypersensitivity   An immune response that causes an individual to overreact to the presence of an antigen, resulting in an allergic condition.
Antibiotic   Antimicrobial agent that is the natural product of living organisms.
Antimicrobial   Capable of killing or stopping the growth of a microbe.
Competitive inhibitor   A compound that competes with a substrate for position at the active site on an enzyme, but that cannot be changed by the enzyme. These compounds prevent normal, essential enzyme function.

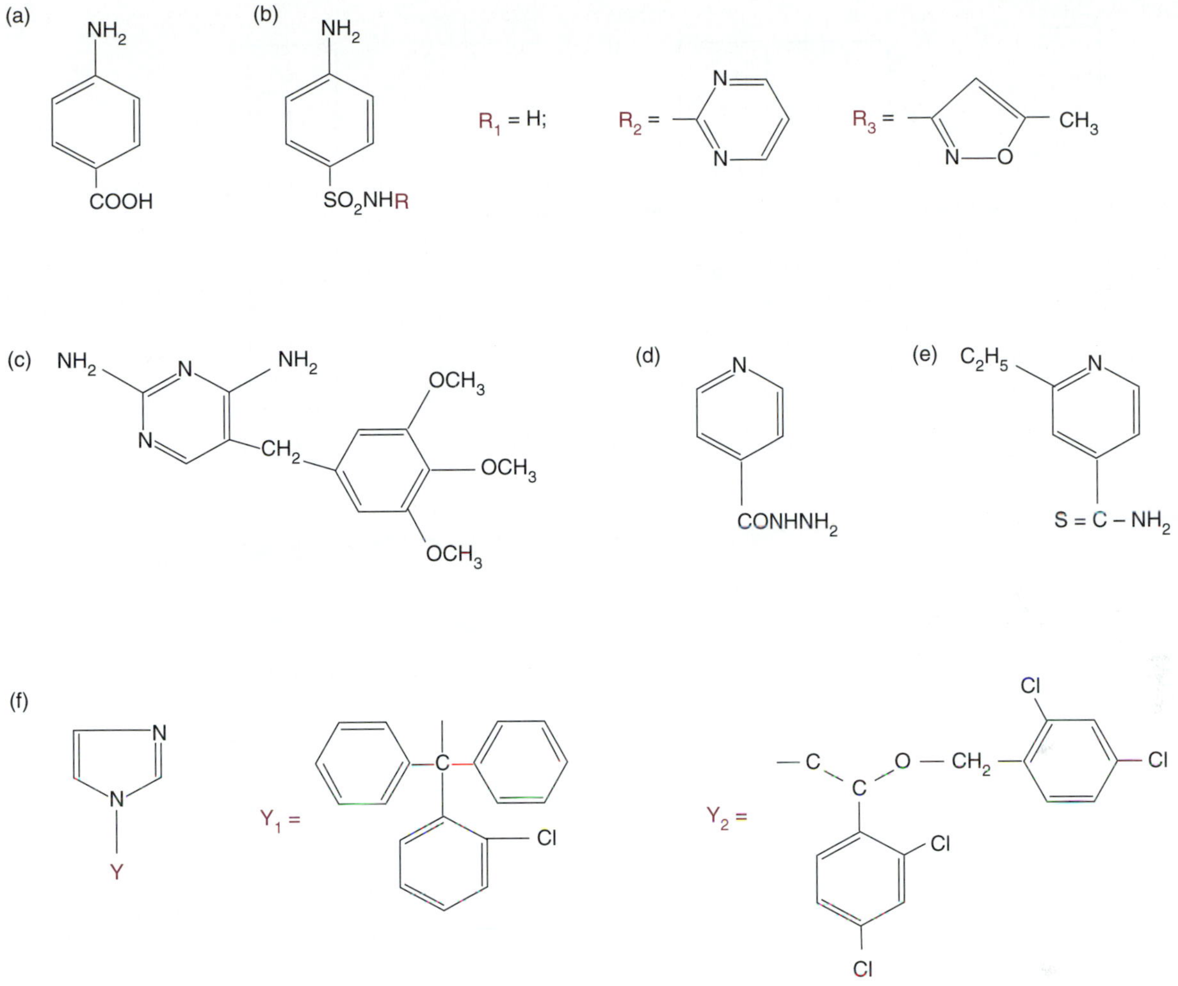

FIGURE 9-1 Structure of some synthetic antimicrobial agents: (a) *p*-aminobenzoic acid, a bacterial metabolite. (b) Sulfonamide, an analog of *p*-aminobenzoic acid. Chemically modified at position "R," the sulfonamide becomes a "sulfa drug," for example, R_1 t sulfonamide, R_2 t sulfadiazine, R_3 t sulfamethoxazole. (c) Trimethoprim, most commonly used in fixed combination with sulfamethoxazole. (d) Isoniazid and (e) ethionamide, which are used almost exclusively as therapy against tuberculosis. (f) The imidazole nucleus, modifications of which produce antifungal agents, for example, Y_1 t clotrimazole, Y_2 t miconazole.

folic acid and this results in the formation of nonfunctional folic acid.

The enzymes that convert PABA to folic acid are unable to distinguish between PABA and sulfonamide. If only a small amount of a sulfa drug is present, most of the enzyme continues to interact with PABA and the cell continues to grow, but at a reduced rate. As the concentration of sulfa increases, there is a greater and greater possibility that the enzyme will find only the sulfa drug to interact with and the cell will stop growing. Such chemical interactions, which depend on the relative concentration of the inhibitor and the normal substrate, are known as *competitive inhibition* reactions (Figure 9-2). That is, there is competition between the two substrates for the *active site* on the enzyme molecule.

In bacteria, just as in humans, the short-term absence of a necessary metabolite does not result in death. However, for a single cell, such a condition results in the cessation of growth. If the needed

Active site That portion of an enzyme where the substrate binds.

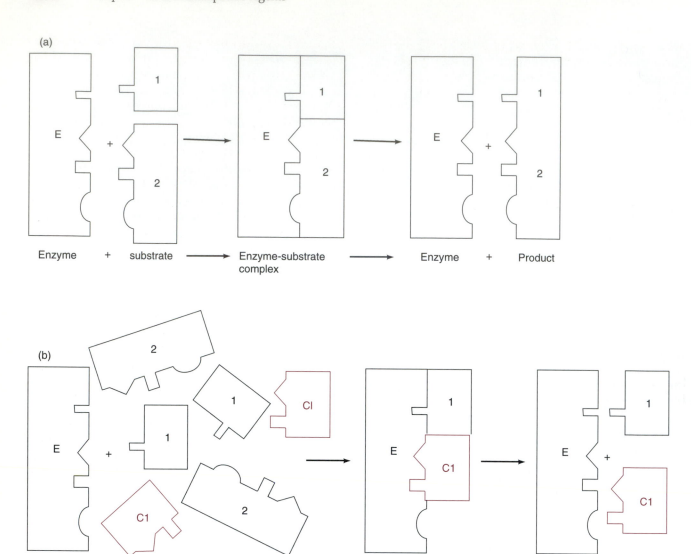

FIGURE 9-2   Representation of competitive inhibition of enzyme function by a substrate analog. (a) Normal enzyme substrate interaction leading to product formation. (b) The addition of a competitive inhibitor to the reaction results in competition for active site on the enzyme. (In this example the competition is between C1 and 2.) The successful formation of product depends on the relative concentrations of C1 and 2.

metabolite is again made available to the cell within a reasonable time, it will once more begin to grow. Such a condition is analogous to placing a culture of bacteria in the refrigerator; although the cells do not immediately die, the cold reduces their rate of metabolism to a point where growth is essentially stopped. When the culture is again placed in the incubator, growth soon returns to normal. This condition of suspended growth is known as *bacteriostasis*. Antimicrobial drugs that lead to such a reversible growth state are called *bacteriostatic* agents. Sulfonamides are an example of a bacteriostatic drug.

Clinical Applications.   Because of the extensive use of sulfonamides during the late 1930s and throughout the intervening years, numerous bacteria are now resistant to these antimicrobial drugs at concentrations normally achieved in patients. For this reason, in spite of the relatively low level of toxicity due to these agents, antibiotics are preferred over sulfa drugs in most clinical treatments. The major exception is in the treatment of urinary tract infections; sulfonamides can reach high levels of concentration in the urine and are generally effective as therapeutic agents. Combinations of antibiotics and sulfonamides have also been used to suppress the number of bacteria in the intestinal tract prior to surgery. The major toxicity problems associated with sulfonamide use are due to some hypersensitivity reactions and the tendency of sulfonamides to crystallize in the kidney with resulting damage to the renal tubules.

One of the newer sulfonamide drugs is actually a combination of a sulfonamide (sulfamethoxazole) and a similar compound (trimethoprim) that also competitively inhibits an enzymatic reaction in the biosynthesis of folic acid. This compound, *trimethoprim–sulfamethoxazole*, is sold by several names, such as *Bactrim* or *Septra*, and has application in a number of serious disease conditions caused by microorganisms of the genera *Shigella*, *Haemophilus*, and *Pseudomonas*. This compound has been particularly effective in reducing the risk of serious infection in kidney transplant patients.

✳ Other Synthetic Agents

Several synthetic agents are widely used in treating tuberculosis and leprosy. The *sulfones* are a group of compounds, related to the sulfonamides, that are effective against infections caused by these acid-fast bacilli. Today the use of sulfones is limited almost entirely to the treatment of leprosy. *p-Aminosalicylic acid* (PAS), an analog of PABA, is an effective bacteriostatic agent for the treatment of tuberculosis but has limited effectiveness against other diseases. *Isoniazid* (INH) is the most widely used agent in the treatment of tuberculosis. The exact mode of action of INH is not known. *Ethambutol* (Embutal or EMB) is another effective antituberculosis agent that is always used in combination with PAS or INH. In fact, therapy for active tuberculous disease should always include multiple antituberculous agents. A number of these agents are listed in Table 20-3, p. 282. The structures of several sulfonamides and related synthetic chemotherapeutic agents as well as *p*-aminobenzoic acid are shown in Figure 9-1.

ANTIBIOTICS

✳ Penicillins

Penicillin is a term applied to a group of closely related compounds produced by various fungi of the genus *Penicillium*. Some penicillins are natural products; others have been chemically altered in the laboratory. Penicillin, the first antibiotic discovered ∞ (Chapter 1, p. 11), has historically been the most useful. During the early 1940s, when penicillin was first used, it became known as a miracle drug

because it could often rapidly cure otherwise fatal diseases. The most dramatic effects of penicillin use were seen on some major killer diseases, such as pneumonia, scarlet fever, staphylococcal diseases, and the sexually transmitted diseases gonorrhea and syphilis. The natural penicillins are primarily effective against gram-positive bacteria, gram-negative cocci, and the syphilis spirochete. Because of its great success, there was a strong tendency during the late 1940s and early 1950s to use penicillin in treating a wide variety of infections. This widespread, and often indiscriminate use, (particularly in the hospital environment) led to the emergence of many penicillin-resistant staphylococci. By the late 1950s the effectiveness of penicillin in treating staphylococcal infections had greatly diminished. At that time, it was discovered that the basic structure of the penicillin molecule could be produced by *Penicillium* molds under special controlled conditions. Various chemical side chains could then be added to this basic structure to modify its action (Figure 9-3). Through extensive, empirical investigations and testing, a series of altered or semisynthetic penicillins were found that have enhanced antimicrobial activities. Several are able to kill bacteria that are resistant to the natural penicillin; others are effective against a wider range of bacteria. A list of some semisynthetic penicillins and their uses is given in Table 9-1. Figure 9-4 is a

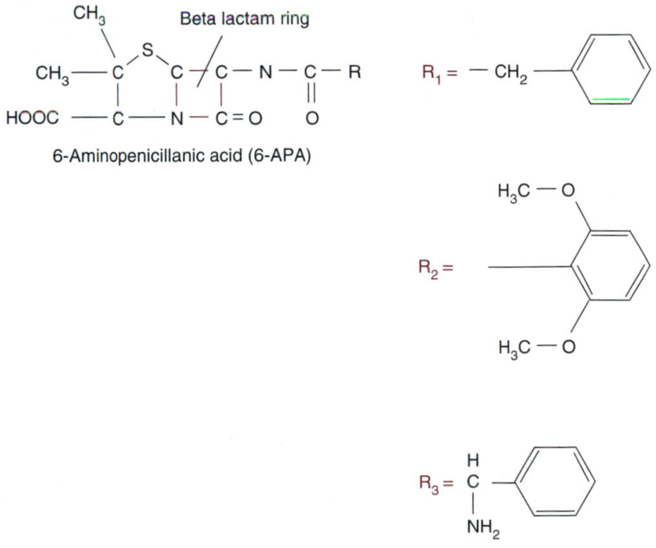

FIGURE 9-3 The structure of penicillin. The common portion of all penicillins is 6-APA. This molecule occurs naturally (R_1 = penicillin G) or may be modified at "R" to produce a variety of "synthetic" penicillins, such as R_2 = methicillin, R_3 = ampicillin.

diagrammatic representation of the development of many available penicillin compounds.

The action of penicillin is to specifically interfere with new bacterial cell wall synthesis during cell division ∞ (Chapter 4, p. 38). In the presence of penicillin, the cross-linkages of the peptidoglycan strands are prevented from forming in the cell wall. Without a complete cell wall, the high internal osmotic pressure of the bacterial cell results in rapid cell lysis. Because mammalian cells have no cell walls, penicillin is generally nontoxic to these cells and can be used in relatively large concentrations. Complications do occasionally result in persons who are allergic to penicillin.

Bacterial resistance to the penicillins results mostly from bacterial production of enzymes called *penicillinases*. Penicillinase splits the beta-lactam ring of the penicillin, producing an inactive molecule (Figure 9-5). Some penicillins, such as penicillin G, are unstable in stomach acid and cannot be taken orally, whereas others, such as penicillin V, are acid stable and can be absorbed intact from the intestinal tract. Penicillin is the treatment of choice when susceptible bacteria cause infection, because it is both less toxic to the host than other antibiotics and, in general, less expensive. It should not, of course, be used in persons who are allergic to penicillin. The appropriate semisynthetic penicillins, such as *methi-cillin, nafcillin,* or *oxacillin*, are the agents of choice in treating penicillin-resistant staphylococcal infections. When a broader spectrum of activity is needed, *ampicillin, carbenicillin,* or *ticarcillin* might be used.

✳ Aminoglycosides

The aminoglycosides include streptomycin and antibiotics with a similar chemical structure. *Streptomycin* (Figure 9-6) is bactericidal against various gram-positive and gram-negative bacteria as well as the tubercle bacillus. It was the first major antibiotic developed after penicillin. When it became available in the mid-1940s, it was effective against many gram-negative bacteria, as well as against the tubercle bacillus that were not susceptible to penicillin. It was the first effective antituberculosis agent to be developed. These antibiotics function by attaching to the small (30S) component of the 70S prokaryotic ribosomes ∞ (Chapter 4, p. 42). Such attachment interferes with the proper initiation of protein synthesis. The 80S ribosomes of eukaryotic cells are not readily affected. Since aminoglycosides cannot be given orally, patients usually must be hospitalized for *parenteral* administration. Resistant mutant bacteria develop when the site of attachment on the ribosome is altered.

TABLE 9-1

Penicillins Commonly Used in Treatment of Bacterial Infection

Name	Source	Common Uses
1. Penicillin G (benzylpenicillin)	Natural	
2. Penicillin V	Natural	Oral penicillin (acid stable).
3. Methicillin	Semisynthetic	Nos. 3 through 6 used in
4. Oxacillin	Semisynthetic	treatment of infections
5. Nafcillin	Semisynthetic	caused by penicillinase-
6. Cloxacillin	Semisynthetic	producing staphylococci. Cloxacillin can be given orally.
7. Ampicillin	Semisynthetic	A broad-spectrum penicillin used against gram-negative bacteria and for patients with endocarditis.
8. Carbenicillin	Semisynthetic	Nos. 8 and 9 are specific-
9. Ticarcillin	Semisynthetic	use penicillins, used to treat *Pseudomonas* infections.
10. Piperacillin	Semisynthetic	A penicillin used in treatment of gram-negative bacterial infections.

Parenteral Given into the body by injection or through a catheter. Many medicines cannot be given orally (by mouth) because they taste bad, are destroyed by stomach acid, or are not absorbed from the intestine.

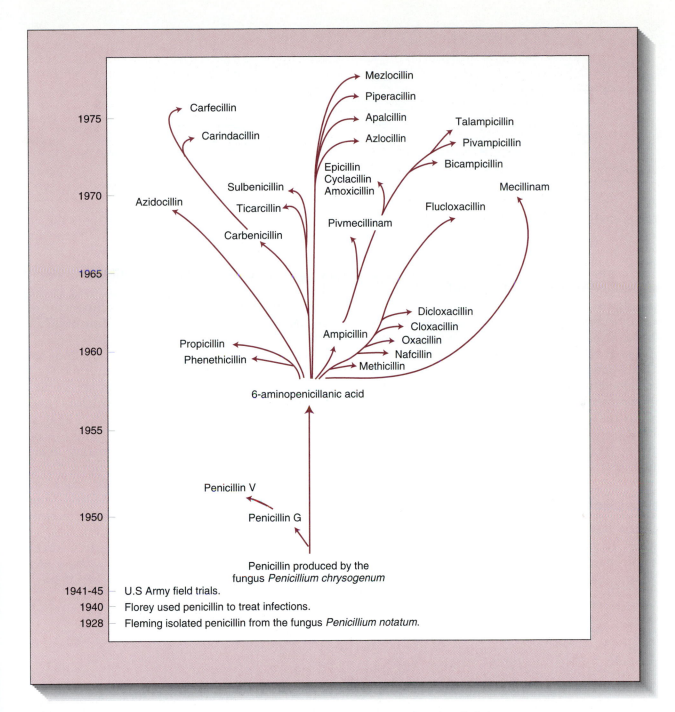

FIGURE 9-4   A schematic representation showing the approximate historical date and genesis of many of today's penicillins. This diagram shows relatedness among penicillins.

Microorganisms such as anaerobes and streptococci are resistant because the cells are impermeable to the aminoglycosides. Some cells become resistant when they are infected with plasmids carrying genes that direct the synthesis of enzymes that modify and inactivate the antibiotics.

Aminoglycosides are among the most toxic of the commonly used antibiotics. Toxicity may lead to kidney damage, hearing loss, or other physical impairments. To prevent the excessive accumulation of antibiotics in patients receiving aminoglycosides, it is necessary to monitor the concentration of the compound present in the body fluids. Such determinations, called *antibiotic assays*, are usually performed on the blood of patients receiving these or other toxic antibiotics. Using this technique, it is

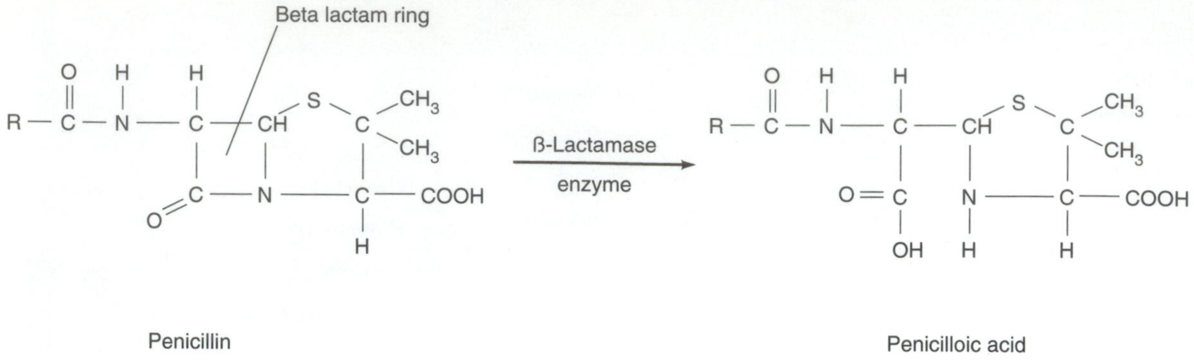

FIGURE 9-5 Inactivation of penicillin by β-lactamase enzyme.

possible to monitor the amount of antibiotic given the patient and to maintain levels below the toxic concentrations.

Several widely used aminoglycoside antibiotics are *amikacin, neomycin, kanamycin, netilmicin, gentamicin,* and *tobramycin.* These antibiotics have a broad antibacterial spectrum and are used to treat serious infections that do not respond well to other chemotherapeutic agents. *Spectinomycin,* which is similar to the other aminoglycosides, has a very specific use in the United States: It is used in the therapy of gonorrhea caused by penicillinase producing *Neisseria gonorrhoeae.*

✳ Tetracyclines

The *tetracyclines* are a small group of closely related antibiotics that are bacteriostatic against a wide range of bacteria including chlamydiae and mycoplasma. Because of this wide range of activity,

they are called *broad-spectrum* antibiotics. Tetracyclines inhibit protein synthesis of growing bacteria by combining with the 30S component of ribosomes and interfering with the binding of tRNA to mRNA.

Some natural and semisynthetic tetracyclines are *chlortetracycline* (Aureomycin), *oxytetracycline* (Terramycin), *doxycycline* (Vibramycin), and *minocycline* (Minocin). Although differing in structure and name, these compounds are mostly effective against the same microorganisms. These antibiotics can be taken orally and are very useful in treating infections that are not responsive to penicillins. Tetracyclines may produce side reactions such as irritation of the intestinal tract, liver damage, and discoloration of teeth when taken before permanent dentition is formed.

The use of tetracycline during the early years of the antibiotic era led scientists to the discovery of a phenomenon generally applicable to broad-spec-

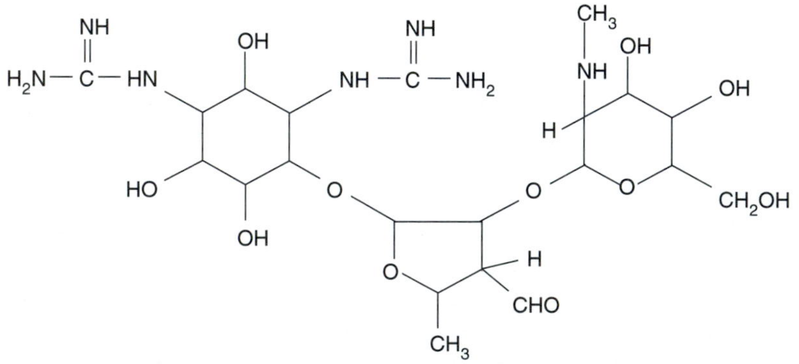

FIGURE 9-6 Chemical structure of streptomycin, first of the aminoglycoside antibiotics to have wide clinical application.

> **Broad-spectrum** Effective against more than one kind of bacterium; usually suggesting antibacterial activity against both gram-positive and gram-negative bacteria.

trum antibiotics. When these compounds are used indiscriminately or for long periods of time, normal bacterial flora, as well as pathogens, are frequently destroyed. The loss of normal flora often allows bacteria that are resistant to the antibiotic to increase rapidly in numbers. The large increase in the number of these organisms frequently results in diseases that will not respond to the antibiotic being used. In some cases, these resistant microorganisms are again reduced to a harmless level when the antibiotic is discontinued and the normal flora returns. Tetracycline is commonly used to reduce the effects of acne in young adult populations. Its use under these circumstances is relatively safe when directed by a physician, but in some cases may result in troublesome yeast infections.

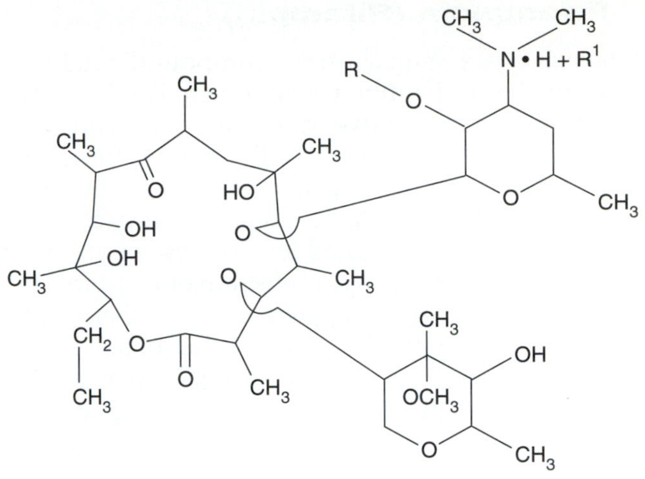

FIGURE 9-7 Structure of erythromycin, most commonly used of the macrolide antibiotics.

✳ Chloramphenicol

Chloramphenicol is a broad-spectrum antibiotic that is effective against a wide range of bacteria, including rickettsize and chlamydiae. It is, however, quite toxic to humans and is generally used only when other antibiotics are not effective. It is the drug of choice under some circumstances, such as in typhoid fever, and is very effective against anaerobic bacteria. It is commonly used in combinations with other antibiotics as empiric therapy in cases of meningitis. Chloramphenicol binds specifically to the 70S ribosomes of prokaryotic cells and blocks protein synthesis by preventing the formation of peptide bonds.

✳ Macrolides

Erythromycin is the most active of a group of related antibiotics called *macrolides* (Figure 9-7). The spectrum of activity of the macrolides is similar to that of penicillin. Erythromycin binds to the 50S ribosome subunit and may be either bacteriostatic or bacteriocidal, depending on the concentration. The mechanism of action is similar to that of chloramphenicol. This antibiotic is most often used as a substitute for penicillin in patients who are allergic to the latter drug. It has been used as therapy in severe cases of acne where tetracycline has not been effective. It is also the agent of choice in treating whooping cough, campylobacter and mycoplasma infections, and Legionnaires' disease. Toxicity is usually not severe, but microbial resistance is common.

Two new macrolides, azithromycin and clarithromycin, have recently become available. These compounds have advantages over the older macrolides in that they have a slightly broader spectrum,

reduced toxicity, and a much longer serum half-life. Both can be taken orally, and azithromycin can be effective when taken only once a day, but must be taken on an empty stomach.

✳ Lincomycin and Clindamycin

These antibiotics have a spectrum of activity similar to penicillin but a basic chemical structure different from that of other antibiotics. They are very effective antistaphylococcal compounds with relatively low toxicity and are useful in treating patients who are allergic to penicillin. The mechanism of action is similar to that of chloramphenicol and erythromycin. *Clindamycin* has been particularly useful in treating disease caused by gram-negative anaerobic bacteria, and patients with toxic shock syndrome.

An increasing use of clindamycin has led to an important observation regarding undesirable reactions during or following antibiotic therapy. Certain patients who had been receiving clindamycin developed a severe, often fatal, intestinal disease known as *pseudomembranous colitis*. It was discovered that this antibiotic caused a severe loss of normal bowel flora except for several resistant bacteria, one of them an anaerobe known as *Clostridium difficile* ∞ (Chapter 18, p. 260). Toxin produced by this organism resulted in the disease. It is now known that this condition can result at any time up to several weeks following the use of almost any antimicrobial agent. Therapy involves both discontinuance of the inciting antibiotic and the reestablishment of normal bowel flora.

✳ **Rifampicin (Rifampin)**

Rifampicin is a semisynthetic compound produced from the natural antibiotic rifamycin. It inhibits the growth of gram-positive, gram-negative, and acid-fast bacteria, can be taken orally, and is effective in low concentrations. It specifically blocks the transcription of mRNA. It has been particularly effective in treating tuberculosis and leprosy. Use of this drug usually results in the rapid selection of resistant bacteria, a feature that suggests using a therapeutic approach that will reduce the number of such resistant organisms. A customary approach when using rifampicin is to give it in combination with another antibiotic and to give them simultaneously to the patient. The rationale for this *combination therapy* lies in an understanding of the genetics of mutation as discussed in Chapter 6. If, for any given microorganism, the chance of developing a mutation that would provide resistance to any given antimicrobial were 10^{-6}, and the chance of developing resistance to a second antibiotic were 10^{-7}, then on the basis of probability, the likelihood that the organism would develop resistance to both antimicrobial agents at the same time would be 10^{-13}, a number so small that such resistance would probably not occur. In spite of the selection of resistant bacteria, this compound is second only to vancomycin as a choice for treatment of infections due to methicillin resistant *Staphylococcus aureus*, a particularly troublesome organism in many hospitals, and is the drug of choice in *prophylaxis* against infection caused by *Neisseria meningitidis*.

✳ **Cephalosporins**

The structure and activity of *cephalosporins* resemble those of penicillin. Cephalosporins are effective against both gram-positive and gram-negative bacteria. Their advantage lies in their broad spectrum of activity as well as their relative resistance to some penicillinases. Research and development in the area of antibiotics are proceeding rapidly with the cephalosporins; perhaps because it is relatively easy to alter the molecule chemically at several points (Figure 9-8), thus producing "new" antimicrobial

agents. These modifications have produced a relatively large number of antibiotics.

Those cephalosporins that have activity primarily against gram-positive bacteria are known as *first-generation* compounds; those active against the *Enterobacteriaceae* (Chapter 21) are generally considered *second-generation* cephalosporins, and those with exceptionally broad spectra that include organisms such as *Pseudomonas aeruginosa* (Chapter 22) are listed as *third-generation* cephalosporins. Although essentially nontoxic, these compounds cannot, with only rare exceptions, be taken orally, and the newer antibiotics are extremely expensive. A number of these newer agents, such as *ceftazidime, cefoperazone, cefotaxime, cefoxitin*, and *cefamandole*, are presently in general clinical use.

✳ **Glycopeptides**

Vancomycin is a sugar-protein glycopeptide antibiotic that inhibits the completion of bacterial cell wall synthesis. This compound has been known for years, but its early use was associated with high levels of toxicity. More recent preparations do not have as many negative side effects, and the compound is now used extensively in treatment of infections due to methicillin-resistant *Staphylococcus aureus* (Chapter 14). It is also effective against some cases of bacterial *endocarditis*. A new glycopeptide, *teicoplanin*, with excellent activity against gram-positive bacteria, is now available and is effective in treating *osteomyelitis* and endocarditis.

✳ **Bacitracin**

Bacitracin is a polypeptide that interferes with the development of the cell wall. It is effective against many gram-positive bacteria. The high toxicity of this agent has limited its use primarily to *topical* ointments.

✳ **Polymyxins**

These antibiotics are simple polypeptides that are quite toxic to humans. Their mode of action is to disrupt the functions of cytoplasmic membranes of

Combination therapy   Treatment with more than one antibiotic at a time; often used when there is doubt as to the cause of the disease, or when there is likelihood that a microorganism may become resistant to one antimicrobial if used alone.
Prophylaxis   Use of an antimicrobial to prevent infection from occurring.
Osteomyelitis   Microbial infection of bone.
Endocarditis   Infection of the tissues lining the inside of the heart or valves in the heart.
Topical   Applied locally, such as to the surface of the skin.

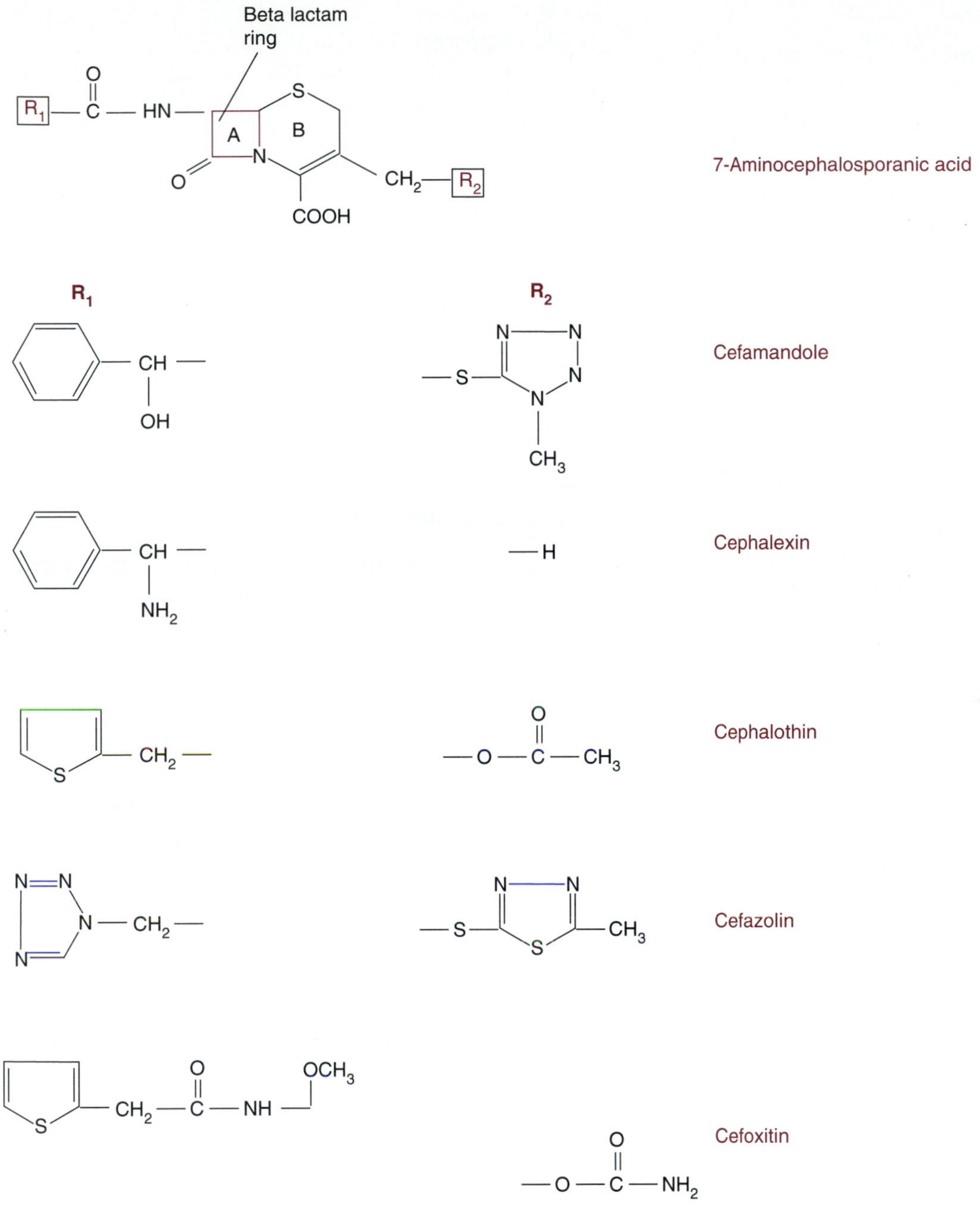

FIGURE 9-8 Chemical structure of some commonly used cephalosporin antibiotics. The 7-aminocephalospornic acid molecule can be modified at R_1 and R_2 (as shown) to produce the indicated antibiotics.

gram-negative bacteria. They are used to treat some of the more resistant gram-negative bacilli, although toxicity severely restricts their use. Several of the more toxic antimicrobials such as *bacitracin, polymyxin,* and *neomycin* can be combined in ointments for topical use only. Neosporin ointment is an example of this combination and is very effective in reducing infection of superficial wounds.

✳ New Classes of Antibiotics

Of several "new" antibacterials, two classes appear to have the potential for extensive clinical use: the *quinolones* and the *penems*.

Quinolones. The quinolones are similar to an older, seldom-used compound known as nalidixic acid. They prevent unwinding of prokaryotic DNA

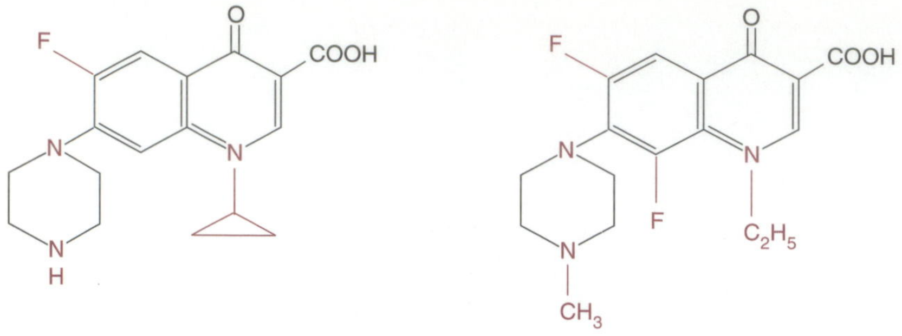

FIGURE 9-9 Two examples of quinolone antibiotics. These compounds interfere with coiling and uncoiling of DNA, resulting in cell death.

so that it cannot be transcribed. These compounds have a very broad spectrum and are used for *systemic* disease, diarrhea, and urinary tract infections. The best known of these agents are *ciprofloxacin* and *ofloxacin*, although many similar compounds are presently being studied (Figure 9-9).

Penems. The penems (Figure 9-10) are remarkable compounds that almost meet the requirements for the "magic bullet" hoped for by Paul Ehrlich many years ago ∞ (Chapter 1, p. 9). *Imipenem* is a commonly used penem. These compounds bind to bacterial penicillin-binding proteins in the periplasmic space. They are essentially nontoxic, very broad spectrum (only a few bacteria are resistant), and very stable. They are effective at low concentration and come close to meeting the conditions of an ideal antibiotic noted earlier in the chapter.

The mode of action of several classes of antimicrobials is shown in Table 9-2.

ANTIFUNGAL AGENTS

✳ Polyenes

Polyenes specifically change the permeability of membranes of fungi and are useful as antifungal agents. *Nystatin* and *amphotericin B* are the major antifungal agents in this category. Nystatin is limited to topical applications because of its toxicity. Amphotericin B is a very toxic compound, but remains one of few effective antifungal agents for serious systemic fungal infections.

✳ Imidazoles

Imidazoles constitute a relatively new approach to fungal therapy. There are several such agents, *miconazole* and *ketoconazole* being perhaps the best known. These agents are a welcome addition to the rather limited antifungal therapeutic options. Their structure (Figure 9-11) is simple and similar compounds have been used to treat diseases due to protozoa and helminths for several years. Their advantages are a limited toxicity and the fact that some can be taken orally. Fungal diseases of the skin, as well as the deep, systemic *mycotic* diseases, respond to these agents.

ANTIVIRAL AGENTS

As noted earlier, development of an antiviral agent is complicated by the fact that viruses do not have their own metabolic machinery. However, viruses induce the formation of several unique viral proteins by infected cells, and their nucleic acids are

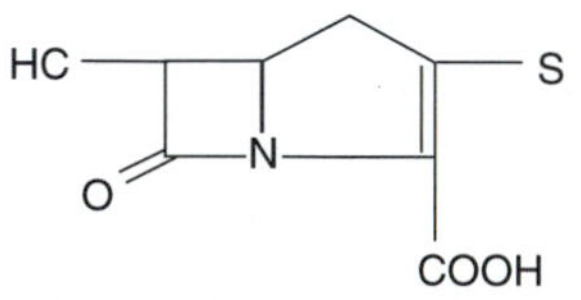

FIGURE 9-10 A new class of antimicrobial, the penem antibiotics.

Systemic Referring to a bodywide infection.
Mycotic Referring to a fungal infection or disease.

TABLE 9-2

Mode of Action of Commonly Used Antimicrobial Agents

Antibiotic Class	Representative	Action
Penicillins	Ampicillin	Inhibits peptidoglycan synthesis; activates autolytic enzymes
Cephalosporins	Cephalothin	Inactivates peptidoglycan synthesis
Tetracycline	Doxycycline	Inhibits protein synthesis by inhibiting aminoacyl-tRNA binding to 30S ribosomal subunit
Chloramphenicol	Chloramphenicol	Blocks protein synthesis by inhibiting peptidyl transferase
Macrolides	Erythromycin	Inhibits protein synthesis by blocking translocation reaction
Glycopeptide	Vancomycin	Blocks early cell wall synthesis
Quinolones	Ciprofloxacin	Inhibits DNA gyrase enzyme; prevents unfolding and refolding of DNA
Polymyxins	Polymyxin B	Prevents membrane transport functions
Aminoglycosides	Amikacin	Prevents protein synthesis by interfering with 30S ribosome function

often somewhat different from those of their mammalian hosts. Scientists have capitalized on these very few differences in viral structure to develop several antiviral compounds. Further research in this area is being vigorously pursued. A list of some presently available antiviral compounds is shown in Table 9-3.

Attempts to provide effective antiviral therapy for HIV-infected individuals exemplifies the problems of developing effective antiviral compounds. In spite of vigorous efforts and millions of dollars in research, there are only about a dozen experimental compounds available. None of these new compounds are completely effective, and all produce considerable toxic side effects.

In 1992, the first clinical tests in humans were conducted on a new class of antiviral compounds called *antisense molecules* (Figure 9-12). An essential step of viral replication is the production of specific viral mRNA molecules (see Chapter 28); the basic concept behind antisense therapy is to develop a molecule that will specifically block this viral mRNA molecule. The antisense antiviral drugs are relatively short chains of about 20 nucleotide bases that are complementary to a specific nucleotide sequence in the mRNA of a given virus. It is hoped

that when these antisense molecules are added to a virus-infected cell they will combine specifically with the viral mRNA and block its translation, and thus stop the replication of the virus. Such a process should have no effect on the normal functions of host cells. While problems exist on how to effectively deliver these antisense molecules into the viral-infected cells, this technology could prove to be the long-sought-after method of developing specific antiviral drugs. Antisense molecules may also prove to be effective against other infectious diseases and perhaps cancer as well.

MICROBIAL RESISTANCE

A major problem associated with chemotherapy is the selection of resistant microorganisms. Microorganisms may spontaneously mutate against a given trait in their environment once in every 10^5 to 10^{10} cell divisions. Because of their rapid multiplication rates, the chance of microbial mutations against a given antimicrobial agent is quite probable. In this situation, the mutant may rapidly multiply in the presence of the antibiotic and produce many resistant progeny. In addition to sponta-

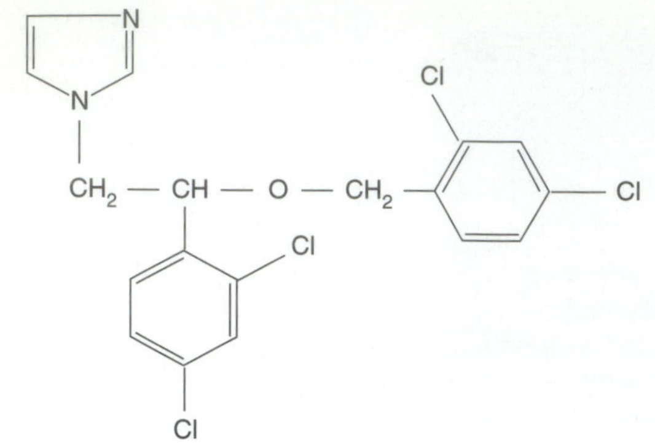

Miconazole

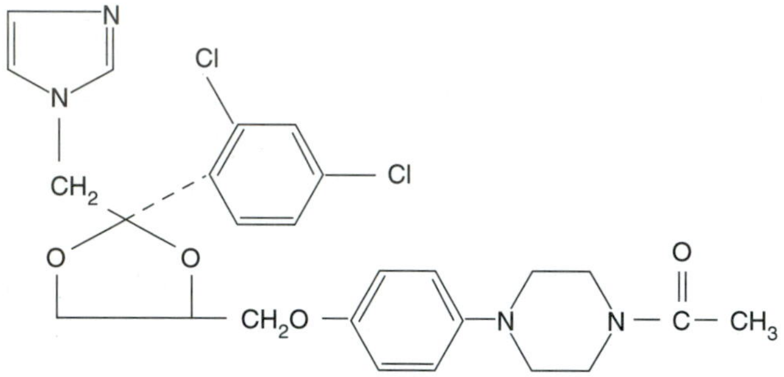

Ketoconazole

FIGURE 9-11   Representative imidazole anti-fungal antibiotics.

neous mutations, genetic resistance may be passed from one bacterium to another by small, circular extrachromosomal DNA fragments called *resistance plasmids* or *R factors* ∞ (Chapter 6, p. 80). A resistance plasmid may contain the genetic information that codes for resistance to one or several antibacterial agents; when the plasmid is passed to a new cell, the trait of antibiotic resistance is also passed.

Certain resistant mutants function by producing enzymes that destroy or alter the chemotherapeutic agent. Others become resistant from changes occurring on the receptor sites to which the chemotherapeutic agent binds. Still others may develop resistance by changes occurring in the permeability of the cell to the chemical agent.

TREATMENT

The procedure used in treating a disease should be designed so that the chance of a cure is maximal and the chance of developing a resistant microbial mutant is minimal. To do this, it is first necessary to determine which chemotherapeutic agents are effective against the disease-causing microbe. This process requires isolating the bacterial cause of disease and testing for its susceptibility to different chemotherapeutic agents. A commonly used method of determining antibiotic sensitivity is to place antibiotic-impregnated disks on a Mueller–

TABLE 9-3

Antiviral Compounds

Compound	Major Viruses Inhibited	Main Clinical Use
Acyclovir	Herpes simplex	Genital and neonatal herpes
Zidovudine	Human Immunodeficiency Virus	HIV infection
Vidarabine	Herpes, Varicella	Herpes encephalitis and keratitis
Amantadine	Myxoviruses	Early treatment or prophylaxis of influenza
Ribavirin	Respiratory syncytial Viruses	RSV pneumonia in infants
Interferon	Hepatitis B	Chronic hepatitis B
Didanosine	Human Immunodeficiency Virus	HIV infection
Trifluridine	Herpes simplex	Herpes keratitis
Idoxuridine	Herpes simplex	Herpes keratitis

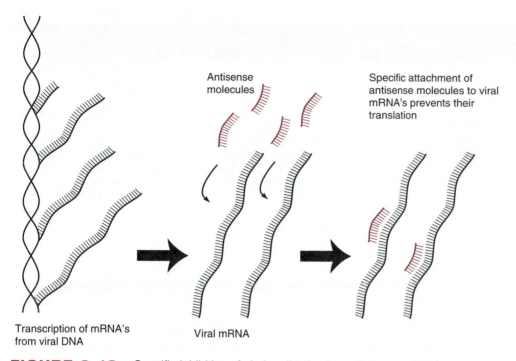

Antisense
molecules

Specific attachment of
antisense molecules to viral
mRNA's prevents their
translation

Transcription of mRNA's
from viral DNA

Viral mRNA

FIGURE 9-12   Specific inhibition of viral replication by antisense molecules.

Hinton agar surface that has been inoculated with a film of the test bacterium (Figure 9-13, Plate 2). A zone of inhibition of bacterial growth is produced around the disks that contain effective antibiotics. This type of testing requires up to 24 hours. With critically ill patients, it may be desirable to initiate treatment before such testing is possible. Such *empirical therapy* is commonly used in clinical situations where time may be critical to the patient, and where it is possible to make a logical choice of antibiotics based on previous experience. Table 9-4 lists those antibiotics that are presently tested against bacteria isolated from human infections. More than one set of antibiotics is necessary for testing, not only because there is considerable variation in the in vitro susceptibility of bacteria, but the concentration of antibiotic obtainable in the many body spaces also varies. Some antibiotics, for instance, can be used to treat a urinary tract infection in which the concentration of antibiotic can reach high levels, whereas those same antibiotics may not reach a high enough concentration in the lung to be useful in treating pneumonia. The several test sets of antibiotic agents also reflect the fact that a finite number of compounds can be conveniently tested at any one time. Rapid susceptibility tests, often

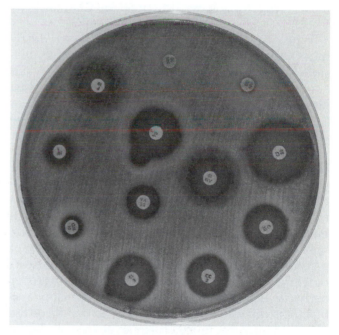

FIGURE 9-13   Antibiotic susceptibility test. The surface of the growth medium is inoculated such that the bacteria will grow as a confluent lawn. Paper disks containing the antibiotic are then placed on the inoculated surface. The susceptibility of the antibiotics is determined by measuring the diameter of the zone of bacterial growth inhibition.

Empirical therapy   Treatment given on the basis of experience, and not as a result of susceptibility testing.

TABLE 9-4

Antibiotics Suggested for Susceptibility Testing Against Different Groups of Bacteria

Gram-Positive Cocci	Enterococcus	Pseudomonas	Enterobacteriaceae
Amikacin	Ampicillin	Amikacin	Amikacin
Cephalothin	Chloramphenicol	Carbenicillin	Ampicillin
Chloramphenicol	Erythromycin	Chloramphenicol	Cefamandole
Clindamycin	Penicillin G	Gentamicin	Cefoxitin
Erythromycin	Tetracycline	Polymyxin B	Cephalothin
Gentamicin		Sulfisoxazole	Chloramphenicol
Kanamycin		Tetracycline	Gentamicin
Penicillin G		Tobramycin	Polymyxin B
Methicillin			Tetracycline
Tetracycline			Tobramycin
Vancomycin			Trimethoprim-sulfamethoxazole

requiring less than one day, are now available through automated systems using sensitive electronic instrumentation for reading results. Some of these rapid tests were described in Chapter 7 ∞ (p. 92).

When an effective antibiotic has been determined, treatment should be initiated as soon as possible or modified to appropriate antibiotics when empirical therapy was initiated. Antibiotics must be given in sufficiently high concentrations to result in a cure, and treatment should continue for some time after the symptoms of the infection have subsided. This procedure gives the best chance of a complete cure and minimizes the chance of selecting resistant mutant microorganisms. If low, rather than adequate, concentrations of antibiotics are given, it is much more probable that resistant mutants will develop. If treatment ends too soon, the infection may reoccur and further increase the probability of resistant mutants developing. Delay in initial treatment may allow the infection to penetrate to deeper body tissues where abscesses may develop and block the diffusion of the antibiotic to the site of the infecting microbe. If bacteria in an abscess are in the stationary-growth phase, antibiotics that are effective only against growing microbes will not work. In these situations, the abscess must be drained before effective treatment can be given.

COMPLICATIONS

Many antibiotics have direct toxic effects on humans and such effects must be considered and weighed against any possible benefits of the antibiotic. Allergies or hypersensitivities may develop against an antibiotic like penicillin, and patients should be questioned about such allergies before treatment is started. Besides eliminating disease-causing microorganisms, antibiotics often destroy many bacteria of the normal flora of the body. A **superinfection** by an antibiotic-resistant, *indigenous* microorganism that is usually held in check by the normal flora may result. Such conditions frequently occur in the intestinal tract of persons on antibiotic therapy and mild intestinal disturbances are sometimes considered a necessary trade-off for the successful treatment of a more serious infection.

Superinfection   A second infection that develops in addition to a previous infection; an infection that occurs as a result of antibiotic treatment.

MATERIAL FOR REVIEW

CONCEPT SUMMARY

1. The concept of selective inhibition is the principle underlying the application of chemotherapy to microbial infections. Careful study of microbial structure, physiology, and metabolism has led to the development of a wide variety of antimicrobial agents useful in combating the infectious diseases of both humans and animals.

2. Both bacteriostatic and bactericidal antimicrobial agents are available. Their action is aimed primarily at the inhibition of essential metabolic activity, interruption of cell wall synthesis, disruption of cell membrane integrity, or suppression of nucleic acid function. Use of these agents has provided a surprising advantage to humans in their quest for good health.

3. A wide variety of antibacterial agents are available for treatment of infectious diseases. However, because viruses depend entirely on host cell metabolic functions, only a few antifungal, antiviral, and antiprotozoal drugs have been effective in clinical practice.

STUDY QUESTIONS

1. What explanation can you give for the very large number of antibacterial compounds available for therapy as compared to the small number of antiviral agents?
2. Describe the concept of selective inhibition.
3. Indicate the mode of action for the following classes of antimicrobial components: (a) sulfonamides, (b) penicillins, (c) tetracyclines, (d) aminoglycosides, (e) quinolones, and (f) cephalosporins.
4. Distinguish among the following terms: antibiotic, chemotherapeutic, antimicrobial, bactericidal, and bacteriostatic.
5. Which antibiotic is frequently used as a substitute for penicillin in patients who are allergic to penicillin?
6. Explain why the penicillins are favored antibiotics under most conditions where they are effective.

CHALLENGE QUESTIONS

1. Explain why streptomycin is not usually the first antibiotic used to treat an infection.
2. Explain why each of the following chemotherapeutic agents will or will not affect cells of the human host: (a) erythromycin, (b) amphotericin B, (c) sulfanilamide

Innate Host Defense Mechanisms

Even in the cleanest of environments, humans are exposed to millions of microorganisms every day. Many are potential disease-causing microbes, such as bacteria, fungi, and viruses. Yet in most cases exposure does not result in an infection or disease because the human body has a number of defense mechanisms to prevent infection. Also, if infection does occur, cellular and biochemical defenses are activated within the body to retard or eliminate the invaders. In this chapter the nonspecific defense mechanisms of the human body are discussed. These include both physical barriers and biochemical agents, as well as cellular and other components of the blood. The next chapter examines the specific defense mechanisms.

HOST DEFENSES

Humans are endowed with a variety of mechanisms that help protect them from the many microorganisms or other foreign agents encountered during their lives. The term **immunity** refers to all of the mechanisms used by the body as protection against microorganisms and other foreign agents (Figure 10-1). Some of these defense mechanisms, called **innate** (natural) **immunity**, are conferred by com-

> **Immunity**   All of the mechanisms used by the body as protection against microorganisms and other foreign agents.
> **Innate immunity**   Inborn or natural immunity; mechanisms of resistance to infection that are not acquired after birth.

IMMUNITY			
LEVEL	**NONSPECIFIC DEFENSE MECHANISMS**		**SPECIFIC DEFENSE MECHANISMS**
First line of defense	**STRUCTURAL** 1. Intact skin 2. Desquamation 3. Mucous membranes	**MECHANICAL/CELLULAR** 1. Mucociliary system 2. Flushing action of fluids (urine, tears, saliva) **BIOCHEMICAL** 1. Acid (skin, stomach, vagina) 2. Enzymes (lysozyme)	1. Secretory antibody
Second line of defense		1. Inflammation 2. Phagocytosis 3. Nonspecific cytotoxic cells (natural killer cells) 1. Complement 2. Interferon	
Third line of defense			1. Specific antibody 2. Cytotoxic T cells

FIGURE 10-1   A functional overview of the body's defenses.

ponents that are part of the body from birth and are always present. These systems function immediately or at short notice to protect against microorganisms or other agents that might land on or invade the body. These mechanisms are also referred to as *nonspecific defense mechanisms* because they respond to all microbes equally and consistently. These innate or nonspecific mechanisms include the external tissues that act as structural barriers to help prevent microorganisms from penetrating the deeper body tissues. Together with mechanical and biochemical mechanisms, these barriers comprise the first line of defense against microbial invaders. If microorganisms do penetrate the first line, however, they encounter secondary lines of defense that include internal innate mechanisms, such as inflammation, phagocytosis by white blood cells, and antimicrobial proteins, such as complement and interferon. Nonspecific cytotoxic cells, such as natural killer (NK) cells are also an important component of the second line of defense.

Antibodies and certain activated white blood cells are important additional defense mechanisms, sometimes called the third line of defense. However, they are acquired only after exposure of the host to an invading microorganism or its *virulence factors*. Antibodies are specific because they react only against the type of microorganisms that originally stimulated their formation. The protection afforded by antibodies is called *acquired immunity* and is discussed in the following two chapters.

NONSPECIFIC EXTERNAL DEFENSE MECHANISMS (FIRST LINE OF DEFENSE)

Nonspecific external defense mechanisms consist of those components of the body that prevent microorganisms from attaching to body tissues or penetrating the deeper, more susceptible tissues. These include mechanisms that kill microbes or inhibit their growth. A functional separation of these mechanisms is shown in Figure 10-1.

✴ Physical Barriers and Chemical Agents

Microorganisms are generally incapable of producing disease if the physical integrity of the tissue barriers remains intact. The intact epithelial membranes, both epidermis and mucous membranes, are the most important defense mechanisms that humans possess against microbial invasion. The parts of the body exposed to the external environment are endowed with a variety of mechanisms to prevent microorganisms from penetrating to the more susceptible internal tissues. The more important mechanisms are described next.

> **Virulence factor**   A structural or physiological character that enables a microbe to cause infection and disease.

Skin. Intact skin provides a barrier that cannot be penetrated by most microorganisms. Most *pathogenic* bacteria are unable to survive on clean healthy skin for any length of time, partly because of the acid pH of the skin, which is inhibitory to most pathogenic bacteria, and partly because of bactericidal acids secreted in the *sebaceous glands* in the skin. An exception is the bacterium *Staphylococcus aureus*, which is able to persist on the skin and is a frequent source of infection when the integrity of the skin is altered.

In addition to being a harsh environment, the skin has another important structural defense. The *epithelium* is constantly growing. Outermost cells are sloughed and replaced by new ones. This process is called *desquamation*, and is an important defense mechanism, since microorganisms attached to these cells are also lost. This effectively prevents the accumulation of large numbers of organisms on the skin's surface.

Mucous Membranes. Body cavities that open to the outside, including those associated with the digestive, genitourinary, and respiratory tracts are lined with mucous membranes that consist of one or more layers of living cells. The cells are bathed in *mucus*, a viscid film that helps trap and remove microbes. These surfaces, however, while offering a valuable protective barrier against most microorganisms, are more easily penetrated by some microorganisms than is the intact outer epidermis.

Eyes. Along with the barrier effect of the intact tissues, the secretion of tears and the movement of eyelids provide a continuous flushing action that disposes of contaminant microorganisms. Tears also contain an enzyme called *lysozyme* that destroys certain gram-positive bacteria by disrupting the integrity of their cell walls.

Outer Ear Canal. The surface of the outer ear canal is lubricated with a waxy deposit that contains effective antibacterial components.

Alimentary Canal. The physical integrity of the mucous epithelium is extremely important in preventing the spread into the blood or deeper tissues of the massive numbers of bacteria found in the mouth and lower intestinal tract. Along with this barrier effect, the flow of saliva and swallowing continually dilute bacteria in the mouth. Saliva also contains secretory antibody and its flushing action has been proven to be important in curtailing microbial overgrowth in the mouth. Stomach acid destroys large numbers of microorganisms that are swallowed. Secretion of mucus along the intestinal tract aids in trapping and removing microorganisms. The mucous secretion contains antibacterial substances and antibodies to help further reduce the numbers of microorganisms.

Genitourinary Tract. The intact epithelium presents a physical barrier and the flushing action of urine keeps most microorganisms restricted to the lower portion of the urethra. In addition, both vaginal secretions and *seminal fluid* contain lysozyme and other antimicrobial substances that prevent microbial growth. The vagina maintains an acid pH throughout the childbearing years, due to the pressure of numerous lactobacilli. This prevents the growth of many pathogens.

Respiratory Tract. The respiratory tract is endowed with a unique series of defense mechanisms against the continuous onslaught it experiences from a wide variety of airborne microorganisms. The positions of various defense mechanisms are shown in Figure 10-2. Nasal hairs are of some value because they induce turbulence of the inhaled air and act as very crude filters. The *nasal turbinates* provide a large exposure surface and cause increased air turbulence. The turbulence results in increased impaction of the larger airborne particles against the surface of the turbinates.

Much of the surface of the nasal cavity is lined with *ciliated epithelium*. Ciliated epithelium contains large numbers of cells covered with cilia ∞ (Chapter 4, p. 45). Each ciliated cell contains several hundred cilia that are rapidly and continuously beating in synchrony. Interspersed between every four to five ciliated cells is a mucus-secreting cell. A film of mucus forms on top of the ciliated epithelium and serves as a sticky surface to trap the airborne particles impacted onto it. Mucus also contains antimicrobial substances. The rhythmic movement of the cilia moves this mucous film, with the entrapped

---

Pathogenic   Referring to the ability to cause disease.
Sebaceous gland   A skin structure that secretes oily substances.
Epithelium   A cellular layer covering internal and external body surfaces that lacks blood vessels.
Mucus   A thick secretion produced by mucous cells that covers mucous membranes. Mucus contains a polysaccharide called *mucin* along with a variety of salts.
Seminal fluid   A secretory fluid that carries sperm.

Defense mechanisms:

1. Nasal hairs
2. Nasal turbinates and mucous secretions
3. Cough reflexes
4. Ciliated epithelium
5. Macrophages and phagocytosis

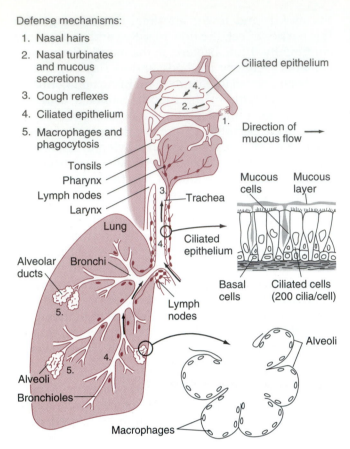

Tonsils
Pharynx
Lymph nodes
Larynx
Lung
Alveolar ducts
Bronchi
Alveoli
Bronchioles

Ciliated epithelium
Direction of mucous flow
Trachea
Mucous cells
Mucous layer
Ciliated epithelium
Basal cells
Ciliated cells (200 cilia/cell)
Lymph nodes
Alveoli
Macrophages

FIGURE 10-2 Defense mechanisms of the respiratory tract.

particles, into the pharyngeal area where the mucus can be disposed of, usually by swallowing. A significant portion of large airborne particles is removed by this mucociliary system.

The smaller airborne particles are carried into the trachea, bronchi, and bronchioles, where many are impacted onto the ciliated epithelium that completely lines these passages. The direction of mucous flow is from the lungs to the opening of the trachea. As mucus accumulates at the top of the trachea, it is removed by "clearing the throat" and disposed of by swallowing. When mucus begins to accumulate or particles become trapped along this ciliated epithelium, the cough reflex is triggered and coughing aids in dislodging and removing these materials from air passageways. Only the smallest microbe-containing airborne particles (those between 5 and 10 μm) are able to penetrate into the air sacs (*alveoli*) of the lungs. Large numbers of phagocytic cells, called *alveolar macrophages*, are located in the air sacs and are able to ingest and destroy most microorganisms deposited in healthy lungs. Overall, the mechanisms of the respiratory

tract, when functioning properly, effectively protect the host against many airborne pathogens.

✳ **Bacterial Interference**

Although not really a defense mechanism mediated by the host, *bacterial interference* is nonetheless important in preventing disease by many potential pathogens. Often the attachment site for a given pathogen is very specific and if the attachment site on a tissue is already occupied by one bacterium, it cannot be readily occupied by a second bacterium. In this regard, we are becoming increasingly aware that many bacteria that make up the normal flora of the host occupy many of these tissue receptor sites. They thus interfere with the attachment of many potentially harmful pathogens and perform a valuable service in helping to protect the host. Also, the normal microbial flora may compete with the invading pathogen for available nutrients and hence suppress their growth. In addition, some bacterial species of the normal flora secrete proteins called *bacteriocins* that specifically inhibit the growth of some invading pathogenic bacteria.

NONSPECIFIC INTERNAL DEFENSE MECHANISMS (SECOND LINE OF DEFENSE)

After a pathogenic microbe has become attached to the tissues of the host, it may be able to pass into a cell because of its own invasive mechanisms or it may be introduced into the deeper, more susceptible tissues through sections of epithelium that have been injured. Once beyond the protective outer barrier of the body, the invading microorganisms encounter a series of internal host defense mechanisms. These mechanisms are closely associated with the activities of the *white blood cells* (WBCs), also referred to as *leukocytes*. The structure and function of these cells, as well as other components of the blood, are discussed next.

✳ **White Blood Cells**

White blood cells can be divided into three categories: *granulocytes, monocytes,* and *lymphocytes*. Specialized types of cells are found within each category. The WBCs are involved in such host defense activities as phagocytosis (the ingesting of foreign particles), inflammation, antibody formation, and cell-mediated immunity. Often different types of WBCs will function together in a coopera-

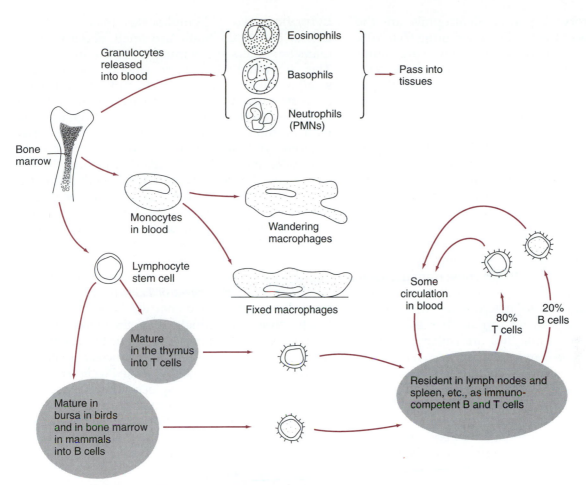

FIGURE 10-3   Development of the different types of blood cells.

tive effort to produce the final result. These interactions are discussed in the following sections and in Chapter 11. The development of the different types of white blood cells is outlined in Figure 10-3.

Granulocytes.   Normal blood contains from 5000 to 9000 WBCs per cubic millimeter and 70 to 75% of them are granulocytes. Granulocytes get their name from the large number of granules present in their cytoplasm. These granules contain chemicals that act as stimulators of certain body responses and chemicals or enzymes that aid in the digestion of microorganisms and other materials. The granulocytes have multilobular nuclei and are 9 to 12 μm in diameter. These cells are formed and mature in the bone marrow and are continually being released into the blood. They circulate for several hours and then pass into the tissue spaces, where they remain for a few days and then die. These cells do not divide after leaving the bone marrow and the body is required to produce about 10^{11} new cells per day.

Many granulocytes are held in reserve in the spleen and bone marrow; when an inflammation occurs, large numbers are released into the blood and the rate of production of new cells is accelerated, increasing the total number of WBCs well above the normal level. This increased WBC count is one of the signs used to diagnose the presence of inflammation in a patient. The different types of granulocytes are *eosinophils*, *basophils*, and *neutrophils*, names derived from their staining characteristics. The full range of functions of the eosinophils is not understood. They seem to be involved with some allergic responses and may function in cell to cell regulation. Although capable of phagocytosis, their major function seems to be in fighting parasitic organisms, such as worms, that are too large to be phagocytized. Eosinophils constitute only 2 to 4% of the total WBCs found in the blood. The basophils are not phagocytic and make up less than 1% of the WBCs. They contain granules of histamine, are involved in allergies, and may be involved in controlling the movement of body fluids involved in the inflammatory response. Neutrophils are the most numerous of the WBCs, constituting 65 to 70% of the total, and are actively phagocytic. Large numbers are involved in the early phases of the *acute inflammatory response*. Because of the varied shapes of their nuclei, they are also called *polymorphonuclear leukocytes* or simply PMNs.

Monocytes. The monocytes are cells that possess a large smooth nucleus and a large area of cytoplasm. These cells are formed in the bone marrow. When they are released into the blood, they are from 14 to 20 μm in diameter. Circulating monocytes make up 1 to 6% of the total WBCs. Monocytes collect in or pass into many different body tissues and further differentiate into **macrophages**. Whether all different types of macrophages come from the same type of monocyte is not known. In the transformation from monocyte to macrophage the cell increases in size up to 50 μm in diameter. Some macrophages are motile and are called *wandering macrophages*. These wandering cells move by *amoeboid* action and are found throughout all tissues and cavities of the body. Other macrophages become attached to the walls of blood capillaries and sinusoids and are known as *fixed macrophages*.

Organ-specific macrophages of both the fixed and wandering type are present at many body locations. Collectively, these comprise a network known as the *mononuclear phagocyte system* (Table 10-1). This was previously called the reticuloendothelial system, or RES, because many of these cells were found in loose connective tissues called *reticulum*, and many were anchored to the endothelial cells that line the sinusoids of organs such as the liver, kidney, and spleen. Macrophages live for months after leaving the bone marrow. Under an appropriate stimulus, macrophages may divide or become "activated" in that their phagocytic and antimicrobial capabilities are considerably increased. All macrophages are active phagocytic cells and play two important roles: (1) removal of particulate antigen, and (2) presentation of processed antigen to lymphocytes to initiate the acquired immune response. The latter role is discussed in Chapter 11.

Lymphocytes. Lymphocytes make up 20 to 25% of the WBCs. Some are small, about 6 μm in diameter; others are as much as 12 μm in diameter. They are round with a large, smooth nucleus and a small amount of cytoplasm. Lymphocytes develop from *stem cells* that are originally produced in the bone marrow. These cells are released from the bone marrow and thereafter differentiate into one of two different types of lymphocytes, *B cells* or *T cells*. The stem cells that become T cells first pass to the *thymus*, where they are influenced by thymic hormones to become T cells. Mature T cells are released from the thymus and enter the circulation. Many T cells circulate throughout the body and constitute about 60 to 80% of the lymphocytes found in the blood. The organ that influences the stem cells to change into B cells has long been recognized in birds to be an organ called the *bursa of Fabricus;* only recently has it been determined that the bone marrow has this function in mammals. From the blood, lymphocytes populate secondary lymphatic tissues, such as the lymph nodes, spleen, and mucosal-associated lymphoid tissue. A majority of the B cells are found in lymph nodes and the spleen. The T cells and B cells are involved in the production of cell-mediated immunity and the production of antibodies (Chapter 11).

✳ **Phagocytosis**

Neutrophils and macrophages are the major cells involved in the phagocytosis and destruction of microorganisms (Figure 10-4). **Phagocytosis** is perhaps the most important secondary defense mechanism of the host once the pathogen has penetrated beyond the epithelial and mucosal barriers. The first step in phagocytosis requires the attachment of the foreign particle to the cell membrane. Some bacteria readily attach to phagocytes and so are readily phagocytized, whereas others will not attach and are thus difficult to phagocytize. Phagocytosis is facilitated if the phagocyte is able to trap the bacterium against a rough surface; in this case, phagocytosis can occur without attachment. Attachment and phagocytosis are much more effective if the

---

Acute inflammatory response   Immediate mechanisms by which host defenses wall off or destroy invading microbes and repair damaged tissue.
Macrophage   A white blood cell found in tissues that phagocytizes foreign material.
Amoeboid   Referring to a crawling movement by certain cells.
Thymus   An endocrine gland located behind the breastbone near the throat. It is large in childhood and decreases in size in adults.
Phagocytosis   A process by which cells ingest particulate matter (such as microbes) from their environment. When this process is carried out by specific host cells it provides a valuable defense against infection.

FIGURE 10-4 Scanning electron micrograph showing a branch-shaped bacterium (*Nocardia asteroides*) being phagocytized by a macrophage. Arrow points to the area where the bacterium is being engulfed. (Courtesy B. L. Beaman, *Infection and Immunity* 15:925–937)

microorganisms are coated with specific antibodies and *complement* components. This is called *opsonization*. This occurs because phagocytes have membrane receptors for these proteinaceous components.

Once attachment of the microbe has occurred, the cell membrane extends around the particle and fuses to form an intracellular membrane-bound vacuole containing the microbe. This vacuole is called a *phagosome*. The phagosome next fuses with the lysosomes ∞ (Chapter 4, p. 46), thereby causing the release of digestive enzymes into the phagosome. This combined structure is called a *phagolysosome*. These enzymes, along with the hydrogen peroxide and other chemicals that are also produced by the

phagocytic cell, readily destroy most microorganisms or break down other particles of organic material that may have been phagocytized. The phagocytosis and intracellular digestion of bacteria are shown in Figure 10-5.

✳ Lymphatic System

The lymphatic system consists of a complex network of thin-walled ducts (*lymphatic vessels*), strategically located filtering bodies called *lymph nodes*, and concentrations of large numbers of lymphocytes in various organs or tissues. The lymphatic vessels are widespread and generally parallel to the blood vessels. They are particularly plentiful in the skin and along the respiratory and intestinal tracts. A major function of the lymphatic vessels is to serve as return ducts for the fluid that is continually diffusing out of the blood capillaries into the body tissues. Before the *lymph* fluid is returned to the blood, it must first pass through several lymph nodes. The major concentrations of lymph nodes that filter fluid coming from the lymphatic vessels of the skin are located in the neck, *inguinal*, and *axillary* regions (Figure 10-6). Deeper lymph nodes are located along the respiratory and intestinal tracts—that is, in areas where microorganisms are most likely to invade the deeper body tissues. The lymph fluid flowing out of the lymph nodes flows into large collecting ducts (e.g., the *thoracic duct*) that eventually empty into the bloodstream. The lymph nodes contain a series of narrow passageways, called

TABLE 10-1

The Mononuclear Phagocyte System

Cell Name	Location	Mobility
Monocyte	Peripheral blood	Circulating
Kupffer cells	Liver	Fixed
Mesangial cells	Kidney (intraglomerular)	Fixed
Alveolar cells	Lung	Wandering
Serosal cells	Peritoneum	Wandering
Microglial cells	Brain	Fixed
Histiocytes	Connective tissue	Fixed

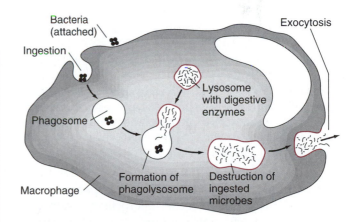

FIGURE 10-5 Mechanisms of phagocytosis and destruction of bacteria by a macrophage.

Lymph A fluid collected from tissues of the body and transported in lymphatic ducts to the venous blood.
Inguinal Referring to the body region near the junction of trunk and thighs.
Axillary Referring to the armpit.

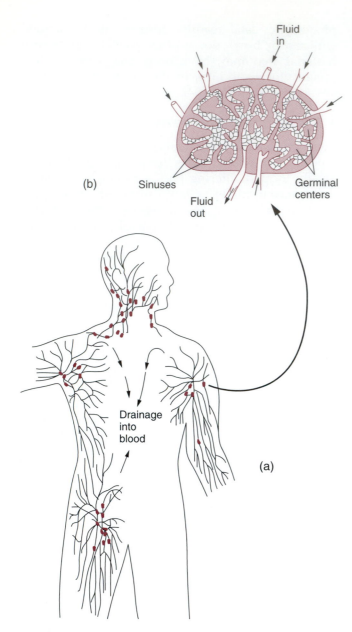

microbes, that may be carried in the lymphatic fluid. Under normal conditions, the lymph nodes successfully prevent invading microorganisms from reaching the bloodstream. When large numbers of bacteria become trapped or multiply in the lymph nodes, inflammation may occur with accompanying swelling and tenderness. The presence of swollen lymph nodes is a useful clinical sign for the diagnosis of an infection in the tissues drained by that node. Lymph nodes also serve as a storage organ for the lymphocytes and are initiating sites for acquired immune responses.

✳ **Other Blood Components**

Red Blood Cells. Red blood cells (RBCs), which constitute 45% of the blood volume, are not discussed, for they have no apparent direct role in fighting invading microorganisms.

Platelets. Blood platelets or thrombocytes are derived from *megakaryocytes*, which, as the name implies, are giant multinucleated cells produced in the bone marrow. Megakaryocytes are continually shedding many small, membrane-enclosed, disk-shaped packets of cytoplasm into the bloodstream, which become the platelets. Platelets are small in size (1-3 μm in diameter) but large in number (2–4.5×10^5/mm^3 blood). Their classic function has been in blood clotting. They rapidly adhere to damaged endothelial cells or fibrin to help stop bleeding of a damaged vessel. More recently, their role in inflammation and the immune response has been confirmed. Platelets contain substances that increase vascular permeability, fix complement, and attract WBCs.

Plasma and Serum. A brief introduction into the fluid components of blood is useful in understanding concepts presented later. When blood is collected and substances are added to prevent clotting, the blood cells can be separated from the fluid by centrifugation. The remaining fluid is called the *blood plasma*. If blood is allowed to clot, then centrifuged, a clear yellowish fluid called *serum* remains. The major difference between plasma and serum is that plasma still contains fibrinogen and other essential components for blood clotting. Both fluids are rich in other proteins. The serum proteins

FIGURE 10-6 (a) The superficial lymphatic ducts and nodes. (b) Cross section of a lymph node. The fluid filters through the sinuses of the node where large numbers of macrophages are located. The germinal centers contain the antibody-producing cells.

sinusoids, that are lined with macrophages. The phagocytic activity of these macrophages removes extraneous particulate matter, including most

---

Sinusoid A small open cavity through which body fluids travel. Sinusoids greatly increase the space available for fluid, causing the fluid to go slowly through these areas.
Megakaryocyte Giant multinucleated cell from which platelets arise.
Globulin A group of proteins found in human serum. The gamma globulins are antibody proteins and are produced by B lymphocytes.

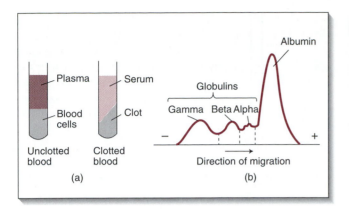

FIGURE 10-7 (a) The source of blood plasma and blood serum. (b) The separation of blood serum proteins based on the rate of migration through an electrical field.

can be separated into two major types called *albumin* and *globulin*. The globulin proteins can be further separated into three types called *alpha, beta*, and *gamma*. Most antibodies that form against microorganisms are found in the gamma globulin fraction of the blood serum and are called *immunoglobulins*. Figure 10-7 shows the formation of blood plasma and serum as well as the separation of serum proteins into their different classes based on their rate of migration through an electrical field (*electrophoresis*).

Complement. The complement system is composed of nearly 30 different serum and membrane proteins, which act in a coordinated manner to protect the body. Approximately 15% of the globulin fraction of serum consists of complement proteins which exist primarily in an inactive form. Once complement activation is initiated, the proteins are sequentially activated through a "domino-like" enzymatic cascade. This results in several protective responses, including (1) the release of *inflammatory mediators*, which increase blood flow, increase vascular permeability, and activate white blood cells, leading to *chemotaxis;* (2) the enhancement of phagocytosis (opsonization), since phagocytes have membrane receptors for certain complement fractions; and (3) damage to microbial cytoplasmic mem-

branes, which frequently results in the lysis of foreign cells.

The complement system can be considered part of both the innate and acquired immune defenses, since it can be activated either nonspecifically or by specific antibody. However, once activated it proceeds in a manner consistent with innate mechanisms, independent of previous antigen exposure. Complement was first discovered to be the major effector mechanism of *humoral immunity*. In fact, these proteins got their name from their ability to "complement" the action of specific antibody. Figure 10-8 shows the two major pathways by which complement is activated. The classical pathway (Figure 10-8a) was discovered first and relies on the prior binding of specific antibody which is discussed in the next chapter. The alternative pathway (Figure 10-8b) was discovered much later, but probably represents a more primitive defense. This pathway is activated by several polymers present on the surfaces of bacteria, fungi, parasites, and even some viruses.

Complement proteins were numbered in the order in which they were discovered. Hence, the sequence in which they are activated does not correspond to their number. Since they were discovered much later, many of the proteins involved in the alternative pathway were given a letter designation. Once a protein is activated it becomes a catalytic enzyme, which activates the next protein in the sequence. This often involves splitting the protein into two parts, both of which are active. By convention, the larger fragment, which usually remains at the site, is given the designation "b". The small fragments that diffuse away and function in chemotaxis and inflammatory mediator release are designated "a". Both pathways of complement activation converge at the level of C5b. From this point they are identical and result in the creation of a large pore-forming molecule called the membrane attack complex. This results in the rapid lysis of foreign cells and the destruction of gram-negative bacteria and some enveloped viruses.

Interferons. The *interferons* are a group of small proteins that constitute part of the body's nonspe-

Electrophoresis A technique used to separate proteins or nucleic acids using an electrical field.
Chemotaxis Movement of a cell toward a chemical influence. Protective phagocytic cells are attracted by chemicals released by damaged body cells.
Humoral immunity Defenses involving antibodies that attack microbes in the body fluids.

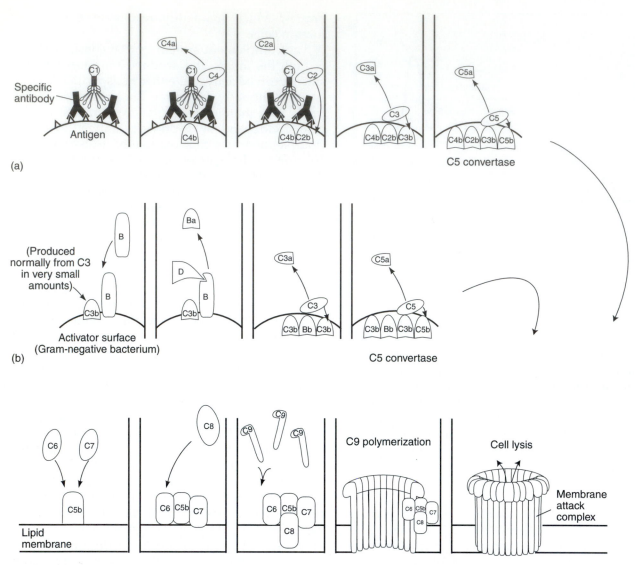

FIGURE 10-8 The activation of complement by the (a) classical and (b) alternative pathways. Results of compliment activation include chemotaxis of white blood cells and inflammatory mediator release ("a" fragments), opsonization ("b" fragments), and lysis of foreign cells (MAC).

cific defense mechanisms. Some interferons are produced by white blood cells as part of the immune response (see Chapter 11) and others are important in protecting host cells against viral infections (see Chapter 29).

✳ Inflammation

When tissue injury occurs, whether by physical trauma or the multiplication of microorganisms, a series of reactions is set in motion to remove or contain the offending agents and repair the damage. This series of events is referred to as the *inflammatory response* and is outlined in Figure 10-9. The inflammatory response is initiated by the release of such chemicals as *histamine*, serotonin, kinins, and prostaglandins. These chemicals are contained in certain tissues and WBCs and are released in response to trauma to the tissues. Histamine is one of the most important chemical mediators and is found in high concentra-

Histamine An important chemical mediator of both the inflammatory response and allergies.

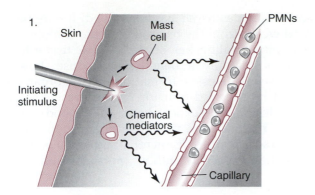

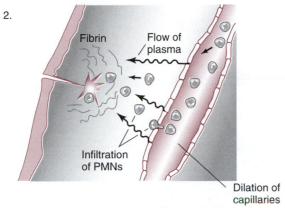

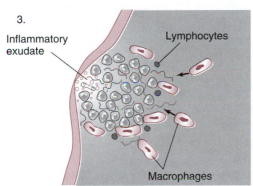

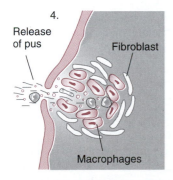
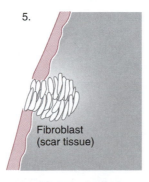

FIGURE 10-9 The progression of an inflammatory response from the initial stimulation to abscess formation, to healing with the formation of scar tissue.

tions in circulating basophils and *mast cells*. Mast cells resemble basophils but are scattered throughout various body tissues, with the largest concentration being in the mucosal tissues. These chemical mediators, often called *vasoactive agents*, can rapidly affect the flow of blood and the permeability of blood vessels in the immediate area. Their release causes the blood vessels in the surrounding area to dilate, which brings an increased flow of blood to the injured tissues. This process occurs within several minutes and is accompanied by an increased permeability of the walls of the blood vessels, which allows an increased amount of plasma and PMNs to move from the vessels into the area of injury. The chemical mediators of inflammation appear to have a chemotactic influence for PMNs that draws these WBCs to the inflammatory site. After the inflammatory response has been initiated, the tissue will be reddened, warm, swollen, and painful due to the increased blood and plasma flow into the area. If the stimulus that initiates the inflammatory response is the growth of microorganisms in the tissues, the PMNs begin collecting in large numbers and will phagocytize and destroy many microbes.

Plasma flowing into the area carries clotting factors, including fibrinogen, which begins to form a network of fibrin strands around the injured area. The fibrin and PMNs continue to build up over several days and are joined by other cells to form a barrier, called the *inflammatory barrier*, around the injured area. The center of the area becomes the battleground in which large numbers of dead bacteria and dead PMNs begin to accumulate in pools of fluids; this fluid is called the *inflammatory exudate* or, more commonly, *pus*. This accumulation of pus circumscribed by the inflammatory barrier is called an *abscess*. As the inflammatory process continues, other types of white blood cells enter the area. Increased numbers of lymphocytes appear and are thought to attract and stimulate macrophages. Macrophages next begin to appear in large numbers to further aid in walling off the injured area, to clear up debris through phagocytosis, and to help in repair. If the abscess has formed in superficial tissue, the inflammatory barrier and skin may break and allow the pus to be expelled. Expelling of the pus, if not accompanied by damage to the inflammatory barrier between the abscess and deeper tissues, may help accelerate healing of the lesion. A pimple or boil is an example of such a skin abscess. If the abscess is in deeper tissues and the inflammatory exudate cannot be expelled, healing is much slower. A standoff may be reached in an abscess in that the bacteria are in an environmental niche of dead tissue that affords them protection from host defense mechanisms and, at the same time, they may be in a metabolically inactive state so that antibiotics may not affect them. Often surgical drainage is needed at this time to clean out the debris or pus.

The last phase of the inflammatory response is the growth of repair cells called *fibroblasts*. The fibroblasts are long, slender cells that form a wall of scar tissue around the injured area. They continue to form scar tissue until the void where the normal cells were destroyed has been filled. It may be several weeks from the beginning of the response until complete healing has occurred. Scar tissue formation is basically a beneficial response; however, excessive scar tissue formation in vital organs may result in impaired function, and excessive scarring is undesirable on exposed areas of the body for cosmetic reasons. The inflammatory response occurs with most injuries; if microbial infection is not present, healing occurs without pus formation.

MATERIAL FOR REVIEW

CONCEPT SUMMARY

1. Each individual is endowed with a number of barriers to infection that function without respect to the kind of possible disease-producing agent. These barriers are called nonspecific host defense mechanisms.

2. Nonspecific host defense mechanisms often function in sequence such that a breach in one barrier leads to the action of the next. These barriers are anatomical, such as the skin; mechanical, such as ciliary movement; chemical, as with lysozyme in tears; and cellular, including both granulocytes and macrophages.

3. The mononuclear phagocyte system has a major responsibility in maintaining body defense against invading microorganisms. Cells and body fluids from this system are primarily responsible for the beneficial inflammatory response associated with many types of infections.

STUDY QUESTIONS

1. Why is an intact epithelium the most significant nonspecific protection against infectious disease?

2. What function do ciliated epithelial cells have in nonspecific disease resistance?

3. What is the fundamental functional difference between granulocytes and monocytes?

4. Would the function of lymphocytes be associated with primary or secondary defense mechanisms?

5. Construct a table or diagram that shows the interrelationships among the granulocyte system, lymphatic system, and the mononuclear phagocyte system.

6. List, in order of occurrence, five functional steps of the inflammatory process.

CHALLENGE QUESTIONS

1. Refer to Figure 10-5 concerning phagocytosis. Patients with chronic granulomatous disease suffer from frequent, serious bacterial and fungal infections. Neutrophils from these patients have a normal chemotactic response and phagocytic ability. However, once phagocytized, the microorganisms are not destroyed. What could be wrong with these neutrophils? List the possibilities.

2. Refer to Figure 10-8 concerning complement fixation. Some patients have a genetic C3 deficiency. Would these patients be healthy? What would be their most common complaint?

11 Acquired Immune Responses

In the previous chapter, we examined innate defense mechanisms, which operate continuously to protect the host. In this chapter, we turn our attention to those protective mechanisms that are inducible by and are specific for a particular invader. Much new information about the immune response and characteristics of the antibodies and cells involved has become available in recent years, and a detailed description is beyond the scope of this textbook. Coverage here is limited to an overall view of the formation and functions of antibodies and the cell-mediated immune response. Also presented are methods used to produce preparations of pure antibodies (monoclonal) in test tubes. We also discuss some detrimental effects that can be mediated by the immune responses.

IgM

ACQUIRED IMMUNITY

Acquired immunity refers to specific host defenses that are initially mobilized during the first encounter of the host with a given microbe or foreign substance. Acquired immune responses improve with each successive exposure to a particular offending agent. Thus, the two key properties of acquired immunity are specificity and memory.

The acquired immune response is a dual system. One part is called the *humoral antibody response* and includes those protective mechanisms mediated by

Acquired immunity Type of specific immunity that develops as a result of exposure to an antigen.

antibody molecules that circulate with the body fluids. The other part is called the *cell-mediated immune response* and involves those mechanisms mediated by specialized cells. The immune response to most foreign substances consists of a mixture of these two types of responses, but one mechanism usually predominates. Both the nature of the substance and the route of exposure to it have been shown to influence the character of this response.

IMMUNOGENS AND ANTIGENS

The basic concept of acquired immunity centers around the ability of the body to react to and eliminate certain types of foreign substances. Any foreign substance that stimulates a host immune response is called an *immunogen* and any substance that reacts with the products of an immune response (antibodies or sensitized lymphocytes) is called an **antigen**. In most applications, the immunogen and antigen are the same substance and in our further discussions only the term *antigen* is used. Antigens are usually proteins, or carbohydrates, or a mixture of the two, with a *molecular weight* over 10,000. Some lipid complexes may also function as antigens. However, lipids and nucleic acids are generally poor immunogens. Once an antibody is formed, it will usually combine only with the type of antigen that stimulated its production. Moreover, the antibody is not formed against the entire antigen but only against certain chemical groups of the antigen called **antigenic determinants**, or *epitopes*. Some low molecular weight compounds, such as certain drugs or antibiotics, are not able to induce an immune response by themselves. However, when coupled with larger molecules, such as proteins, the resultant complex can induce an immune response that includes antibodies directed against the low molecular weight compound. In such cases, the low molecular weight compound is called a *hapten* and the larger molecule is called a *carrier*. Protein antigens and their antigenic determinants relative to parts of a bacterium are shown in Figure 11-1.

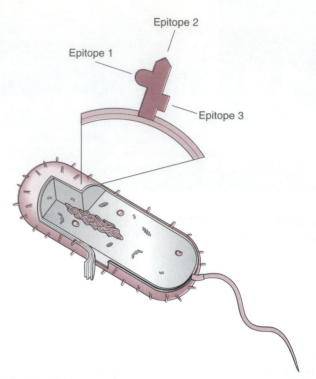

FIGURE 11-1 Antigens associated with bacterial cells. A single cell may contain many different antigens. Each protein may possess a number of different antigenic determinants, as illustrated by the different shapes. Each of these may be recognized by a different specific antibody.

✻ Acquired Immune Responses

Development of the acquired immune response depends on the interaction of several specific types of cells. The cells primarily responsible for an acquired immune response are the *lymphocytes*. However, macrophages and other **antigen-presenting cells (APC)** also play key roles. Each of these cells is derived from stem (precursor) cells, which are most often found in the bone marrow. Once formed, these juvenile stem cells are subject to a variety of maturation processes that lead them to differentiate into specific types of cells ∞ (Chapter 10, p. 144). During this process, some of the juvenile cells mature to become monocytes and then macrophages. Some lymphoid precursors enter the

---

Antigen   Any substance that reacts with products of an immune response.
Molecular weight   The sum of the atomic weights of the atoms in a molecule, or compound.
Antigenic determinant   Those molecular groups to which an antibody reacts; also called an epitope.
Antigen-presenting cell   A cell that presents peptide fragments associated with major histocompatibility class II proteins.

thymus gland and mature as thymus-derived T lymphocytes or *T cells*. Other lymphoid cells enter the bursa of Fabricius (in birds) or remain in the bone marrow (in mammals) and develop into B lymphocytes or *B cells* ∞ (Chapter 10, p. 146).

When the body is exposed to the complex antigens associated with invading microorganisms, both humoral (B cell) and cell-mediated (T cell) immune responses may be stimulated. The acquired immune responses are usually beneficial to the host in that they offer added protection against invading microorganisms. These responses, however, may sometimes react against the host or produce certain types of antibodies that cause undesirable side effects. These undesirable reactions are called *allergies* or *hypersensitivities*. The natural immune responses are discussed under the following general categories: (1) formation of circulating or humoral antibodies, (2) formation of cell-mediated immunity, and (3) allergies and hypersensitivities. The production of monoclonal antibodies, which involves making large amounts of mono-specific antibodies under laboratory conditions, is also discussed.

A GENERAL OVERVIEW OF IMMUNE RESPONSE COMPONENTS

The characteristics of some of the major cell types and components involved in an immune response are briefly discussed in this section. The sequences of events and the roles carried out by components of an acquired response are presented in the sections on humoral and cell-mediated immunity.

✻ B Cells

B cells are derived from a common type of stem cell that matures in the bone marrow of mammals. During this maturation, each B cell is preprogrammed to respond to a unique antigenic determinant. Thus, all B cells are identical except for the specificity of the *protein receptors* that project from their cell membrane. These receptors are, in fact, antibody molecules and are capable of specifically reacting with antigenic determinants that have a complementary configuration. Within the body, B cells are found with antibody receptors having millions of different configurations. However, a single B cell has antibody of only a single specificity. For the B cells to have the genetic information necessary to produce these different protein molecules, it was assumed that millions of different sequences of DNA would have to be present in the genes that direct antibody production. Exactly how this is accomplished by B cells was an enigma for many years. The dilemma was resolved by the work of S. Tonegawa, which demonstrated that the sections of DNA that code for the variable region of antibody are made up of DNA segments from different locations that can be rearranged into millions of different permutations. The rearrangements occur after a stem cell has been committed to become a B cell. Thus, before a person is ever exposed to a foreign antigen, a B cell has already been preprogrammed with antibody against that antigen.

✻ The Major Histocompatibility Complex

There is a region of DNA on human chromosome number 6 that codes for proteins that play a critical role in the immune response. This region is known as the **major histocompatibility complex (MHC)**. The genes in this region are divided into three different classes on the basis of their function. Class I and II genes are extremely *polymorphic*, meaning that many different varieties exist in the human population. It is the unique combination of these proteins on a cell's surface that the immune system uses to identify a cell as either foreign or self. Class III gene products are not expressed on the cell's surface, but rather perform important accessory immune functions.

Class I MHC proteins are expressed on virtually every nucleated cell in the body. Their structure is such that the distal end forms a pocket that accommodates a short *peptide fragment* (most commonly nine amino acids long). These fragments are derived from all of the proteins that that particular cell happens to be making. Thus, T cells that interact with these MHC class I/peptide complexes have an efficient means of not only identifying the cell as

---

Protein receptor   Membrane-bound proteins on the external surface of cells that recognize specific antigenic determinants.
Major histocompatibility complex (MHC)   Cell surface proteins that are essential to self-recognition and immune responses.
Polymorphic   Having many different varieties.
Peptide fragment   A part of a protein.

"self," but also of surveying the types of proteins a particular cell is making. This is especially important in the case of a virus-infected cell, which is clearly "self," but is making foreign viral proteins, and should therefore be destroyed by the surveying T cell (Figure 11-2a).

Class II MHC proteins are similar to class I in that they are membrane-bound *glycoproteins* with a distal peptide binding groove. However, there are several important differences. Class II proteins are expressed only on special APCs, which include several types of macrophages and B cells. Unlike the pocket of class I proteins that is closed at both ends, the groove of class II proteins is open, allowing amino acids of the peptide fragment to extend out both ends. Hence, the peptide fragments that bind class II proteins are longer (usually 13–18 amino acids long). More important is the difference in the origin of the peptide fragment. Class II proteins are "loaded" with a peptide fragment in special *endosomes* that contain degraded protein from outside the cell. This is important because the T cells that interact with MHC class II/peptide complexes are important in initiating and propagating an immune response (Figure 11-2b).

✳ T cells

Pre-T cells that leave the bone marrow are nurtured into mature T cells in the *thymus*. T cells contain sections of DNA (similar to those found in B cells) that can be rearranged into millions of different sequences. These sections of rearrangeable DNA contain the genes that control the configuration of the *T cell receptor* (TCR). Like antibody molecules on the surface of B cells, these TCRs are specific for a unique foreign antigenic determinant. However, unlike B cells, T cells require this foreign antigen to be presented to them by the MHC proteins described above. It is this composite made up of foreign peptide and host MHC protein that the TCR recognizes. This is the reason that the function of T cells is said to be "MHC-restricted." This means that for a T cell to function, the T cell and the cell containing the foreign peptide/MHC complex must be from the same, or a genetically identical, person.

T cells can have many different functions. Since it is not possible to distinguish between different T cells morphologically, scientists have searched for

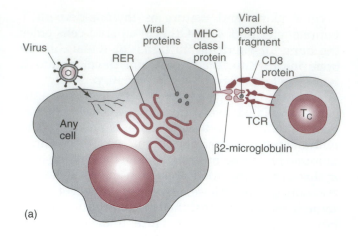

(a)

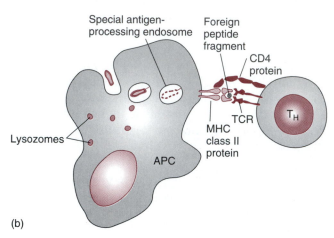

(b)

FIGURE 11-2   (a) Normal presentation of endogenous proteins that are synthesized inside a cell. In this case a virus has invaded, causing viral peptide to be presented in conjunction with MHC class I proteins. This is recognized by a specific cytotoxic T cell (T$_C$), which, when activated, will destroy all cells that bear this complex. (b) Antigen processing and presentation by an antigen-presenting cell (APC) such as a macrophage. In special endosomes, exogenous peptides are degraded and combined with MHC class II proteins. These complexes are presented to helper T cells (T$_H$), which release lymphokines to initiate an immune response.

protein markers on the surface of these cells that correlate with their function. As a result, T cells can be broadly classified into two main groups based on the presence of two key protein markers termed CD4 and CD8. The vast majority of T cells will be either *CD4 positive* or *CD8 positive*. The CD4 positive

---

Glycoprotein   A protein that has simple sugars (monosaccharides) attached at various locations.
Endosome   A membrane compartment in eukaryotic cells that processes proteins.
T cell receptor   A protein receptor on T cells that is specific for a foreign peptide fragment/MHC complex.

T cells include the **T-helper** (T_H) **cells**, which interact with APCs. Once the TCR of a particular T_H cell finds the MHC class II/foreign peptide fragment it was preprogrammed for, it clonally expands and produces helper factors required for the maturation of B cells and **T-cytotoxic** (T_C) **cells** (Figure 11-2b). The CD8 positive cells include the T_C cells, which have a TCR that recognizes a specific foreign peptide fragment in association with an MHC class I protein. Thus, any cell in the body that is manufacturing a foreign protein, due to viral infection or transformation into a cancer cell, can be eliminated by a specific T_C cell (Figure 11-2a).

✳ Natural Killer Cells

Natural killer (NK) **cells** are a third type of lymphocyte that are neither T nor B cells. NK cells are able to recognize and attach to microbe-infected or transformed (cancer) host cells and kill them. These cells are not mobilized by exposure to specific antigens and their numbers cannot be increased by immunization. In this respect, they are part of the innate defense system discussed in Chapter 10.

✳ Macrophages

The phagocytic functions of macrophages were discussed earlier ∞ (Chapter 10, p. 146). Their role in initiating the immune response is to phagocytize and process the foreign antigen, then to present the peptide fragments in combination with MHC class II proteins on their surface membranes to T_H cells. They can also be activated by regulatory proteins produced by T cells. Activated macrophages have increased abilities to kill foreign invaders of many types.

✳ Regulatory Proteins

During the progression of an immune response, soluble proteins are released from various cells and act as signals to regulate certain immune functions. These substances are collectively called *cytokines*. Those cytokines released from lymphocytes (B and T cells) are called *lymphokines*, and those cytokines that send signals between the various leukocytes are called *interleukins*. *Interferons* (IFN) are cytokines that were originally shown to interfere with multiplication of viruses, but some types, such as IFNγ, are now known to have several important immune functions as well.

HUMORAL ANTIBODY RESPONSE

Humoral antibodies are made in response to foreign antigens and are released into the blood and other body fluids. The antigens associated with microorganisms, and foreign proteins in general, are excellent stimulators of the **humoral antibody response.** These antibodies are effective in helping to destroy invading microorganisms and in neutralizing viruses and toxins. When foreign antigens enter the body tissues, a sequence of events occurs that involves interactions between APCs (macrophages), B cells, and T cells (see Figure 11-3). Some of the details of these interactions are still not fully understood; the following presentation is a simplified interpretation of current theories regarding these events.

When a foreign antigen enters the body, much of it is phagocytized by macrophages and some remains circulating free in the blood for some time. The phagocytized antigen is processed in the *phagolysosome* ∞ (Chapter 10, p. 147) of the macrophage and the antigenic determinants are passed to and combine with a MHC class II protein. This MHC class II protein–antigenic determinant complex is transported to the cytoplasmic membrane of the macrophage where it is presented on the cell surface. This antigenic determinant–MHC class II complex will combine only with a T_H cell. Furthermore, the only T_H cell that can combine with this complex is one that possesses a preprogrammed TCR with a specific complementary fit to the presented complex. This combination stimulates the macrophage to produce and secrete a cytokine called *interleukin-1* (IL-1). IL-1 stimulates the helper T cell to mature and begin to secrete a second type of interleukin, called *interleukin-2* (IL-2). IL-2 then stimulates the helper T cell to divide into many identical T cells,

---

T-helper cell   A T-lymphocyte that interacts with macrophages and B cells and to initiate an immune response.
T-cytotoxic cell   A T-lymphocyte that can destroy infected or abnormal host cells.
Natural killer cell   A lymphocytes that can recognize and destroy microbe-infected and cancer cells.
Cytokine   A soluble protein that regulates specific immune functions.
Lymphokine   A cytokine secreted by lymphocytes.
Humoral   Referring to body fluids, such as blood.
Humoral antibody response   The production of antibodies to foreign antigens.

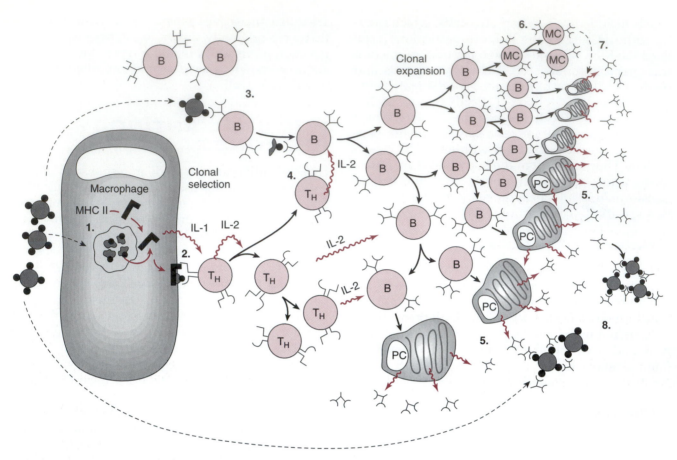

FIGURE 11-3 Theorized sequence of events that are involved in the production of circulating antibodies. (1) Antigens on invading microbes are processed and presented on the surface of a macrophage in combination with class II MHC proteins. (2) A specific helper T cell (T_H) attaches to antigen–MHC complex, interleukin-1 (IL-1) stimulates it to mature and secrete interleukin-2 (IL-2), which stimulates clonal expansion of T cells. (3) The specific B cell (B) combines with its antigen. (4) IL-2 from the helper T cell stimulates the selected B cell to multiply. (5) Most B cells change into plasma cells (PC), which then secrete specific antibodies (✧). (6) Some B cells become memory cells (MC). (7) Memory cells rapidly change into plasma cells when reexposed to the specific antigen. (8) Antibodies combine with antigens on invading microorganisms. The linkage between the T_H cell and the macrophage has been simplified in this figure (for details see Figure 11.2).

each of which secretes additional amounts of IL-2. Meanwhile, some of the same antigen (in its unprocessed native form) is combining with B cells that have antibody receptors for antigenic determinants on the foreign protein. The exact mechanisms of B-cell–antigen interaction are not fully known. Under some conditions, B cells combine with circulating antigen and under other conditions they may also combine with membrane-bound antigen that is only slightly degraded. This process of selecting out from among the repertoire of millions of B and T cells the ones with preprogrammed receptors that specifically fit the antigenic determinants of the foreign antigens is called **clonal selection**.

The B cell–antigen complex needs to be in close contact with the stimulated T_H cell. Whether the selected B cell and selected T cell come into physical contact has not been completely resolved (some theories suggest they do). The IL-2 from the T_H cells now stimulates the selected B cells to divide into many identical B cells, a process called *clonal expansion*. Most of the B cells are further stimulated by

Clonal selection The process whereby exposure to an antigen stimulates a specific lymphocyte clone to divide and differentiate.
Clonal expansion The proliferation of selected lymphocytes.

interleukins to differentiate into **plasma cells** and a few of the B cells are changed into **memory B cells**. The plasma cells contain increased amounts of cytoplasm, endoplasmic reticulum, and ribosomes, which equip them to produce large amounts of proteins. The main proteins produced by the plasma cells are the circulating antibodies, which are called *immunoglobulins* (Ig). It has been estimated that these antibody factories can produce over 2000 antibody molecules per second. As mentioned previously, antibodies contain receptor sites that are produced under the direction of the rearranged genes that were formed when the B cell was originally produced. The memory B cells remain in the body for many years and become involved in the secondary antibody response, which is discussed below.

✳ Structure and Classes of Antibodies

The basic structure of an antibody consists of two *heavy* and two *light* polypeptide *chains* ∞ (Chapter 5, p. 59) connected in a Y-type arrangement (Figure 11-4). The arrangements of amino acids at the top (amino end) of the four chains of the Y are variable and form the specific receptor sites of the antibody. These areas, called the *variable regions*, form two identical unique antigen binding pockets. The remainder of the antibody molecule does not vary; these areas are called the *constant regions*.

Five different classes of antibodies, or immunoglobulins, have been identified and are designated *IgG, IgM, IgA, IgD,* and *IgE*. Each of these classes has a unique heavy chain, the constant regions of which determine the physiological properties of a particular antibody class. The properties of these immunoglobulins are listed in Table 11-1. The IgG antibodies are the major type present in the circulation and consist of a single Y-shaped basic structure. IgG is also present in significant amounts in other body fluids, such as lymph, synovial, and spinal fluids. IgG is an important long-term antibody that helps protect the host against infections and toxins. This antibody is able to cross the placenta in humans and is therefore important in the

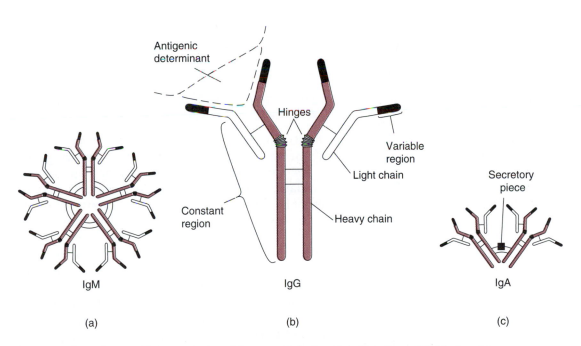

FIGURE 11-4 The structures of immunoglobulins. (a) Structure of IgM showing pentamer arrangement of the basic immunoglobulin Y-shaped units. (b) Details of the IgG structure (scale enlarged). (c) Dimer arrangement of IgA.

Plasma cell   An antibody-secreting cell, derived from B cell, clonal expansion.
Memory B cell   A population of long-lived B cells that act in a secondary antibody response.
Heavy chain   An antibody polypeptide of higher molecular weight.
Light chain   An antibody polypeptide of lower molecular weight.

TABLE 11-1

Properties of Four Classes of Human Immunoglobulins

Property	IgG	IgM	IgA	IgE	IgD
Heavy chain	γ	μ	α	ε	δ
Number of Y-shaped units	1	5	2	1	1
Molecular weight	150,000	900,000	385,000	190,000	180,000
How transferred to offspring	Via placenta	Not transferred	Via milk	Not transferred	Not transferred
Half-life (days)	23	10	6	2	3
Where found in body	45% in blood, 55% in extracellular fluid	80% in blood, 20% in extracellular fluid	Mostly in secretions	Attached to basophils and mast cells	Antibody receptor on mature B cells
Normal serum levels (mg\mL)	13.5	1.5	3.5	0.0003	0.03
Ability to fix complement	++	+++	—	—	—

transfer of passive immunity to the newborn. IgG is also able to fix complement ∞ (Chapter 10, p. 149).

The IgM antibodies consist of clusters of five Y-shaped immunoglobulin basic structures (*pentamers*). These antibodies appear earlier in the infection than the other classes and offer valuable assistance to the host during the critical early stages of an infection. IgM is found mostly in the blood. This class of antibody commonly disappears from the body soon after the onset of the infection and does not cross the placenta. IgM is the most efficient antibody class at fixing complement.

The most common form of IgA antibody is a dimer (two Y-shaped basic structures). These immunoglobulins are readily secreted and, besides being found in the blood, are also found in tears, saliva, mucosal secretions, and breast milk. IgA offers valuable protection against infections of the superficial tissues, such as the mucosal surfaces of the respiratory, intestinal, and genitourinary tracts. Collectively, the body manufactures more IgA per day than any other antibody class.

IgD is found only in low concentrations in the serum. Its main function seems to be as the antigen receptor on B cells. Monomeric IgM can also function in this capacity. IgE antibodies readily attach to *mast cells* and *basophils* as soon as they are formed

and are therefore found only in very low levels in the blood. Recent findings suggest that IgE may play an important role in fighting infections caused by some of the larger parasitic agents. IgE is prominently associated with allergies and this is discussed later.

✳ Primary and Secondary Humoral Antibody Responses

After the first exposure to an antigen in the form of an infecting microorganism, a **primary response** occurs. Small amounts of antibody first appear in a few days, but readily measured amounts usually cannot be detected until about a week after stimulation. A steady rise in antibody concentration occurs over the next 2 weeks until the maximum level is attained at about 3 weeks. The amount of antibody then slowly decreases over ensuing months. A given IgG antibody will remain in the body for only a few months after it is formed; therefore, the prolonged persistence of antibodies must result from the continued activity of specific antibody producing cells. This is usually driven by the presence of the antigen.

If a person has experienced a primary antibody response to a given antigen and is later reexposed to the same antigen, a rapid antibody response is stim-

Pentamer Having five sides or five subunits.
Primary response The response to a first exposure of a host to an antigen.

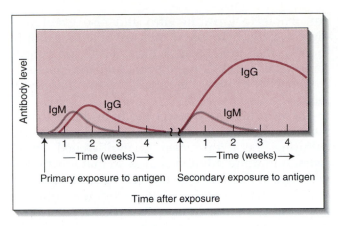

FIGURE 11-5 Primary and secondary responses of IgG and IgM antibodies.

ulated and high levels of antibodies are produced within a few days instead of a few weeks. This is called the *anamnestic*, *booster*, or **secondary response** and is associated with the memory B cells that were produced as part of the primary response. These memory B cells are long-lived and some will remain in the body for years. These cells are preprogrammed such that when they encounter the same type of antigen that stimulated the primary response, they rapidly multiply and change into antibody-producing plasma cells. This mechanism allows the body to replenish its antibody supply before the invading microorganism has had a chance to cause a clinical disease, and in many cases confers life-long immunity to the host. The primary and secondary responses of IgG and IgM are shown in Figure 11-5.

✴ Effect of Humoral Antibodies on Microorganisms

When a microorganism like a bacterium enters the body tissues, antibodies may be formed against such cellular components as different antigenic determinants on the surface, internal proteins, toxins, capsular antigens, or flagella. Thus the antigen–antibody reactions may involve various components of the microorganism with varied effects. Also, antibodies often work in conjunction with other components of the host's defense mechanisms, such as phagocytic cells, natural killer cells, or complement, to bring about the end results. The following categories cover the most common ways

that antibodies protect us against microorganisms.

Promotion of Phagocytosis. Antibodies combined with surface antigens may bind the microbes into clumps by forming antibody bridges between cells. This is called an *agglutination* reaction. These clumps of microbes are much more readily phagocytized than single cells. In addition, some antibodies may first attach to the antigen with their receptor sites and the opposite end of the antibody molecule (called the F_C end) may specifically attach to the phagocytic cell. This step binds the microbe to the phagocytic cell, which greatly facilitates phagocytosis. This process is called *opsonization*.

Interference with Attachment. Antibodies may attach over the surface of the microbes and prevent their attachment to host cells. These are called *neutralizing* antibodies because they neutralize the ability of the microbe to cause infection. This works against many types of pathogens (both viral and bacterial) that depend on attachment to host cells to cause disease.

Neutralization of Toxins. Antibodies may also bind to and neutralize the toxins produced by microorganisms. Such antibodies are called *antitoxins*. Most hospitals maintain a supply of antitoxins for treating snake and spider bite victims (*antivenoms*). However, the most potent poisons known are produced by microorganisms (see the discussion of botulism toxin in Chapter 18). Antitoxins to some of these substances are used by physicians to treat patients with diseases such as botulism and tetanus.

MONOCLONAL ANTIBODIES

Monoclonal antibodies are preparations of large numbers of antibodies in which each antibody molecule is identical, that is, the variable regions of each immunoglobulin molecule are the same. Such pure antibody preparations have many useful applications in various areas of medicine, industry, and research. It is not possible to produce preparations of pure antibodies in intact animals, for animals are invariably exposed to a variety of antigens and most antigens contain a variety of antigenic determinants. Thus their serum will always contain a mixture of antibodies (*polyclonal*). Monoclonal antibodies are produced by cells grown in test-tube (*in*

Secondary response An immune response to second or latter exposure of a host to an antigen.
Monoclonal antibody A population of antibodies with identical variable regions.

vitro) cultures, using a unique procedure developed in the mid-1970s at Britain's Medical Research Council by Cesar Milstein and Georges Kohler. These researchers were able to fuse two types of cells together into a single cell called a *hybridoma*. The hybridoma cells are able to express properties of both parent cells. The two cells used to make monoclonal antibodies are a mouse tumor cell, called a *myeloma cell*, and a specifically programmed B cell from the spleen of an immunized mouse. The myeloma cell has the ability to grow indefinitely in a test tube but is not able to produce specific antibodies. The activated B cell can produce specific antibodies against a single antigenic determinant but cannot grow in a test tube. The myeloma–B cell hybridoma is able both to grow indefinitely in test tubes and to produce specific antibodies. Once a hybridoma that is secreting antibodies against a known antigen is produced, massive cultures of this cell can be propagated and will, in turn, produce large amounts of the pure or monoclonal antibody. Often it is necessary to screen many hybridoma cells to find the one that is producing the desired specific antibody.

The general procedure (Figure 11-6) for producing monoclonal antibodies first entails injecting a mouse with the antigen against which the monoclonal antibodies are to be produced. After several weeks, during which time the specifically programmed B cells have extensively multiplied (clonal expansion), the spleen cells are removed. The spleen cells and large numbers of mouse myeloma cells, which have been grown in vitro, are mixed together in a fusion chamber. The chemical polyethylene glycol has been extensively used as an agent to facilitate the fusion of B cells with myeloma cells. Currently, a process called *electroporation* is also being used to fuse cells, which entails passing a high-voltage, short-duration pulse of electricity through the fusion chamber. Only a few, out of several hundred thousand cells in a culture, fuse to form hybridomas. Therefore, it is necessary to have a method of selecting these few hybridomas from the many unfused B cells, spleen cells, and myeloma cells. This is accomplished by using mutant myeloma cells that will not multiply in a medium that contains hypoxanthine, aminopterin, and thymidine (HAT medium). The cells from the fusion chamber are passed to HAT medium where the unfused myeloma cells are unable to multiply and will die off in a few weeks. The unfused B cells

are unable to grow in vitro and also eventually die out. Only hybridoma cells are able to multiply in HAT medium, because their B-cell component can utilize the HAT ingredients and the myeloma components provide the capabilities to multiply in vitro.

The next task is to select the hybridoma that is secreting the desired antibody. This is done by subculturing the hybridomas in multiwelled *microplates*. A limiting dilution is used so that most wells will contain no more than one cell. After incubation, the fluid from each well is tested for antibody against the specific antigen. A well that tests positive contains the sought-after monoclonal antibody-producing hybridoma. The positive culture is again diluted and passed one or more times to make sure the culture is truly monoclonal. Usually several months of intensive work are required to produce a clone of hybridoma cells that secrete a pure monoclonal antibody.

The next step is to propagate these cells in large numbers. This is often done in an in vitro system where the antibodies are secreted into the medium uncontaminated by other background antibodies. When high purity is not critical, the hybridomas can be propagated in mice where antibodies will be found in high concentrations in the serum and other body fluids.

Because of their specificity for a single antigenic determinant, reactions between monoclonal antibodies and antigens are much more precise than those produced from the antisera of intact animals. Thus monoclonal antibodies are widely used in diagnostic kits to identify isolated microorganisms or detect antigens. Monoclonal antibodies are also useful for specific antiserum therapy against infectious diseases, toxins, and possibly some forms of cancer. In some applications, monoclonal antibodies are attached to specific drugs; when injected into a patient, the antibodies specifically attach to the pathogen or target tissue, which concentrates the effect of the drug against the target and minimizes its effect on normal tissues. Monoclonal antibodies are used in research work in which specific identification of components can be made by immune reactions. They are also used as "handles" to specifically pluck a given substance out of a mixture of chemicals. Today, monoclonal antibody procedures are being applied in many areas of medicine, research, and industry and represent a highly useful technology.

Microplate   A plastic plate containing up to 96 separate wells.

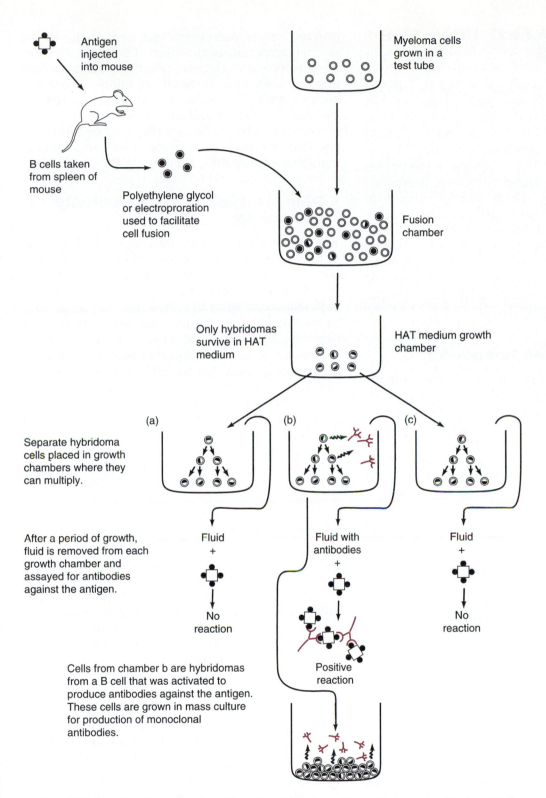

Antigen injected into mouse

Myeloma cells grown in a test tube

B cells taken from spleen of mouse

Polyethylene glycol or electroproration used to facilitate cell fusion

Fusion chamber

Only hybridomas survive in HAT medium

HAT medium growth chamber

(a)   (b)   (c)

Separate hybridoma cells placed in growth chambers where they can multiply.

After a period of growth, fluid is removed from each growth chamber and assayed for antibodies against the antigen.

Fluid + → No reaction

Fluid with antibodies + → Positive reaction

Fluid + → No reaction

Cells from chamber b are hybridomas from a B cell that was activated to produce antibodies against the antigen. These cells are grown in mass culture for production of monoclonal antibodies.

FIGURE 11-6   A schematic outline of a method used to produce monoclonal antibodies.

CELL-MEDIATED IMMUNE RESPONSES

Cell-mediated immune responses can be antigen-specific or nonspecific. Specific mechanisms are mediated by various types of T lymphocytes. Nonspecific responses involve mostly NK cells and macrophages, but can also include neutrophils and eosinophils. Both specific and nonspecific responses can be mediated by localized cytotoxins. In contrast to humoral responses, which function mainly to eliminate extracellular pathogens and their products, cell-mediated immune responses cause destruction of host cells that have been invaded by microorganisms, or have become altered, or are foreign to the host. The major types of antigen-specific responses involve *cytotoxic T cells* (T_C) and *delayed-type hypersensitivity T cells* (T_{DTH}).

✳ Cytotoxic T-Cell Response

A simplified overview of the events that occur in the cytotoxic T-cell (T_C) response is shown in Figure 11-7. The initial stages of this response are the same as those outlined above for the humoral antibody response. That is, some of the invading antigen is first processed by an APC and then presented on the surface of this cell in combination with MHC class II proteins. A CD4 positive T_H cell that has a complementary TCR to the antigen–MHC class II complex attaches to the APC. The APC secretes IL-1, which stimulates the T_H cell to secrete IL-2, which in turn stimulates T_H cell expansion.

In the meantime, the invading microorganism (e.g., a virus) has multiplied inside of the host APC and antigenic determinants have been presented on the cell membrane in combination with the MHC class I proteins. A CD8 positive T_C cell, which also has preformed complementary TCRs, specifically attaches to this antigen–MHC class I complex. This allows the IL-2 produced by the helper T cell to stimulate this selected T_C cell to multiply into many identical progeny cells (clonal expansion). These progeny cells circulate throughout the body and when they encounter this same foreign antigen–MHC class I complex on the surface of other host cells, they specifically attach to this complex. Once attached, the T_C cell firmly binds to the host cell and releases various compounds that cause lysis of the targeted cell. A major protein released is called *perforin*, which inserts into and forms holes in the plasma membrane of the infected host cell. This immune response works very effectively against virus-infected cells and bacterial infections where the bacteria multiply inside of host cells. T_C cells are also able to attach to and destroy some types of cancer cells. In addition, T_C cells play a major role in the rejection of transplanted organs (such as a kidney) that have incompatible MHC antigens.

✳ Delayed-type Hypersensitivity Response

This response requires an initial sensitization phase of 1–2 weeks following primary exposure to an antigen. The reason it is called a *"delayed-type hypersensitivity"* is that a maximum response requires 2–3 days after exposed to an antigen and can sometimes be pathological in nature (see the discussion of hypersensitivities below). However, in most cases this response is both beneficial and necessary for protection against intracellular parasitic bacteria and fungi.

An outline of the events of this response is shown in Figure 11-8 . This response uses the same T_H cells described above. The major effector cell is a CD4-positive T cell called the delayed-type hypersensitivity T cell (T_{DTH}). It is believed that this is a specialized T_H cell capable of producing a potent array of cytokines, whose main effect is recruitment and activation of local macrophages. By clonal selection, a specific T_{DTH} cell attaches to the antigenic determinant–MHC class II protein complex that is presented on the surface of a macrophage. Under stimulation of the IL-2 from the T_H cell, this selected T_{DTH} cell multiplies into many identical cells (clonal expansion) that circulate throughout the body. When these T_{DTH} cells encounter the specific antigenic determinant to which they are programmed they are stimulated to produce and release various lymphokines. These lymphokines attract and then activate macrophages that produce additional cytokines. The final result is the destruction of host cells that contain the foreign antigen. Following are some of the major cytokines and their functions:

Macrophage-chemotactic factor, causes macrophages and other leukocytes to migrate to the site.

Migration-inhibition factor, inhibits the migra-

Cell-mediated immune response   A response in which immune cells (T-helper, T-cytotoxic) react against specific foreign antigens.

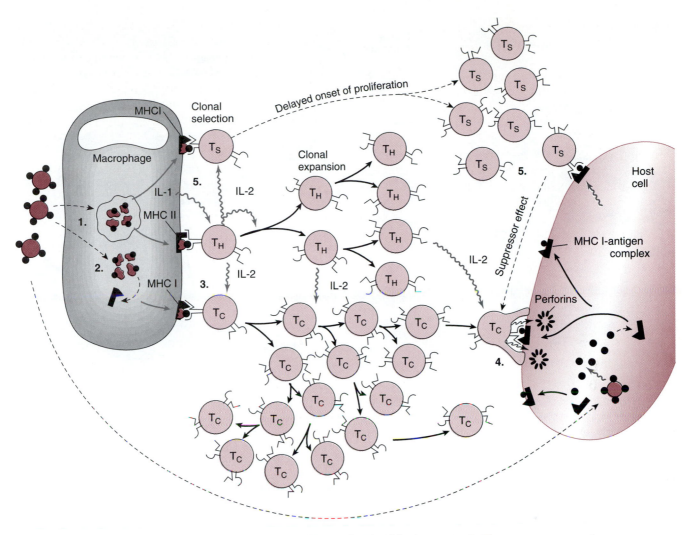

FIGURE 11-7 Theorized sequence of events that are involved in the cytotoxic T-cell response. (1) Exogenous antigen is processed and presented to a helper T (T$_H$) cell. (2) Endogenous antigen is processed by a different mechanism and is presented in combination with class I MHC protein. (3) A specific cytotoxic T cell (T$_C$) attaches to the antigen–class I MHC protein complex and is stimulated to multiply by interleukin-2 (IL-2) from helper T cells. (4) The circulating T$_C$ cells specifically attach to host cells that are infected with the invading microbe and release perforins and other substances that destroy the infected cells. (5) Some theories suggest suppressor T cells (T$_S$) are selected and eventually suppress the immune response. The linkages between the T$_H$ and T$_C$ cells and the macrophage have been simplified in this figure (for details see Figure 11.2).

tion of the macrophages away from the site. *Macrophage-activating factor*, stimulates the macrophages to increase their ability to ingest and destroy foreign antigens.

Gamma interferon, is able to do the three functions listed above, plus induce the amount of MHC class II protein to increase on cells.

Lymphotoxin and tumor necrosis factor, have the ability to kill certain tumor cells.

Interleukin-2, causes expansion of the T$_{DTH}$

clones and also attracts and activates macrophages and NK cells.

The combined effects of these cytokine-induced responses usually cause the destruction of the antigen-containing cells and any altered host cells. An inflammatory response ∞ (Chapter 10, p. 150) is often part of this reaction with the associated *erythema* (reddening) and *edema* (swelling), and scar tissue may eventually form. This reaction is also the

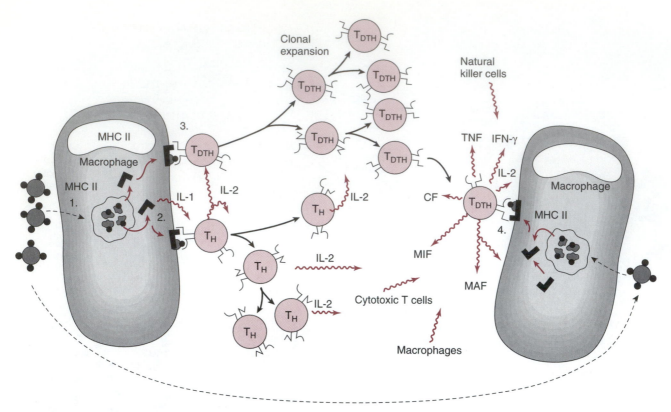

FIGURE 11-8 Theorized sequence of events that are involved in the delayed-type hypersensitivity immune response. (1) The antigen is taken up and processed by a macrophage. (2) The antigen is presented to a specific helper T cell (T_H). (3) The antigen is presented to a specific delayed-type hypersensitivity T cell (T_{DTH}) that is stimulated to multiply by interleukin-2 (IL-2) from helper T cells. (4) A specific T_{DTH} cell attaches to other macrophages that have processed and presented the same antigen. This stimulates the T_{DTH} cells to release among others the following lymphokines: macrophage-chemotactic factor (CF), macrophage-inhibitory factor (MIF), macrophage-activating factor (MAF), tumor necrosis factor (TNF), gamma interferon (IFN), and IL-2. These lymphokines attract macrophages to the area and cause them to become activated. Activated macrophages destroy the infected cells. The linkages between the T_H and T_{DTH} cells and the macrophage have been simplified in this figure (for details see Figure 11.2).

basis of skin tests that are used in the diagnosis of such diseases as tuberculosis ∞ (Chapter 20, p. 275).

An important chronic extension of a delayed hypersensitivity response occurs when the inducing antigen cannot be cleared. Some intracellular pathogens such as *M. tuberculosis* are able to continue to grow within macrophages and contain large amounts of waxy lipid in their cell walls. When this type of antigen persists, the delayed hypersensitivity reaction intensifies and the foci of infection are walled off by the dense packing of numerous macrophages. This is called a "granulomatous reaction" and can result in extensive cell death due to

the many destructive substances liberated by the macrophages.

✳ Natural Killer Cells

Natural killer cells have no specific antigen receptors such as those found on B and T cells, and they are not antigen-specific or MHC-restricted. Yet, they are able to recognize and attach to cancer cells and other abnormal cells (the mechanisms for this recognition are not yet known). Once attached, they destroy the targeted cell in much the same manner as target cells are destroyed by T_C cells (Figure 11-

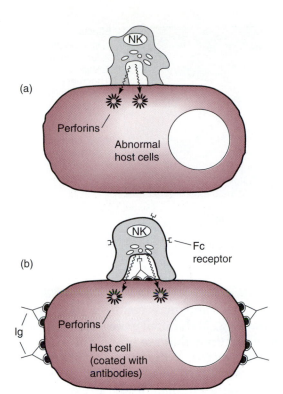

(a)

Perforins

Abnormal
host cells

(b)

NK

Fc
receiver

Fc
receptor

Perforins

Ig

Host cell
(coated with
antibodies)

FIGURE 11-9   Theorized mechanisms used by natural killer (NK) cells to destroy abnormal host cells. (a) Attachment to host cells is not dependent on the presence of specific antigens or MHC proteins. The mechanism of killing is similar to that used by cytotoxic T cells. (b) Alternatively, NK cells are targeted to cells coated with antibody and bind to them via specific Fc receptors. Once attached, the mechanism of killing is similar to that used by cytotoxic T cells.

9a). These cells may also play a role in certain acquired immune responses described above.

Some NK cells, which were previously called just "killer cells," have receptor molecules on their surface that attach to the F_C end of antibody. Thus, these cells work in concert with the humoral immune response. When antibodies have attached to specific antigens on the surface of cells, NK cells are able to attach to and kill these antibody targeted cells. This process is called antibody-dependent cell-mediated cytotoxicity (ADCC) (Figure 11-9b). The mechanisms that cause cell death are similar to those used by T_C cells.

Suppressor and Memory Functions

It has been fairly well documented that factors are present in the body that are able to dampen or suppress the specific immune responses to prevent them from overreacting. It has been postulated for some years that a class of T cells called *suppressor cells* carry out this function (Figure 11-7) but as yet, the identity of these cells and their modes of action have not been clearly determined.

Some of the selected lymphocytes are long-lived and continue to circulate in the host as memory cells. Should the host be reexposed to the same antigen at a later time, these memory cells would recognize that antigen and rapidly remobilize the full immune response. This is especially true for T lymphocytes and the cell-mediated immune response.

HARMFUL EFFECTS OF THE IMMUNE RESPONSE (HYPERSENSITIVITY)

Sometimes immune responses are harmful to the host because they react against host tissues or stimulate the release of excessive amounts of substances that cause tissue destruction. Such reactions are referred to as **hypersensitivities** or *allergies*. These harmful reactions depend on various interacting factors, such as the nature and dose of antigen, route of exposure, and the physiological conditions of the host. They may involve either the humoral or the cell-mediated immune systems. These hypersensitivities have been categorized into four types. Types I, II, and III are mediated by specific antibody, whereas type IV is caused by a cell-mediated immune response.

Type I—Immediate Hypersensitivity

A type I reaction is the typical allergic response mediated by antibodies of the IgE class. Antigens that induce allergies are called *allergens*. Many different substances are capable of functioning as allergens. Some of the more common allergens are plant pollens, animal hairs, house dust, and food. The mechanism of stimulation and formation of IgE is

---

Hypersensitivity   Referring to the immune system over-reacting to a specific antigen; also called *allergy*.
Allergen   A normal substance that triggers an allergic response in a sensitized person.

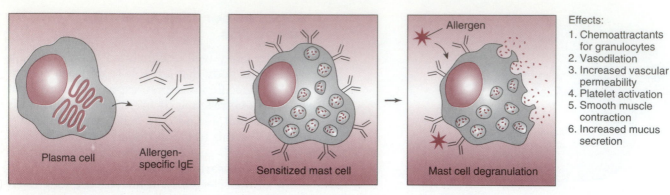

Effects:
1. Chemoattractants for granulocytes
2. Vasodilation
3. Increased vascular permeability
4. Platelet activation
5. Smooth muscle contraction
6. Increased mucus secretion

Plasma cell

Allergen-specific IgE

Sensitized mast cell

Allergen

Mast cell degranulation

FIGURE 11-10   Type I hypersensitivity; an immediate-type allergic response. IgE is produced by allergen-specific plasma cells and attaches to mast cells by the Fc region. Reaction is seen after the allergen binds to cell-bound IgE and stimulates the mast cell to release mediators, including histamine, which induce the listed effects.

similar to the mechanisms used to produce the other immunoglobulins. The amount of IgE that is produced in response to an allergen is under T-cell control. Once formed, IgE molecules do not remain free in the circulation, but rapidly attach to either mast cells (tissues) or basophils (blood). This attachment involves the *Fc region* of IgE and is such that the antibody reactive sites are pointing out from the cells and are still free to react with the antigen (Figure 11-10). Once this sequence of events occurs, the host is sensitized to the specific allergen. When the host is subsequently exposed to that same allergen, it specifically attaches to the reactive sites of the IgE antibodies that are already attached to the mast cells. This antigen–antibody reaction triggers the release of *histamine* and other related substances from the mast cells (Figure 11-10).

Histamine has immediate effects on the surrounding tissues, such as dilation and increased permeability of blood vessels, contraction of smooth muscles, and increased mucous secretions. The types of symptoms are in part a function of the route of exposure to the allergen. If this reaction occurs along the mucosal lining of the upper respiratory tract following inhalation of the allergen, the symptoms are the familiar runny nose, watery eyes, itching, and sneezing, characteristic of hay fever. Reaction in the lungs may cause asthma. If the reaction occurs through contact of allergens with mast cells located in the skin, hives (*urticaria*) may result. When allergens are ingested, reactions occur along

the intestinal tract and may result in such symptoms as vomiting, abdominal pain, diarrhea, and hives. Those reactions that are limited in scope are called *local anaphylaxis*. If the reaction between the allergen and sensitized mast cells or sensitized basophils is of sufficient magnitude, as would occur with the injection of the allergen into the blood stream, *systemic anaphylaxis* may occur. This situation involves a massive dilation of blood vessels causing life-threatening hypotension and shock. Contraction of smooth muscles may result in asphyxiation and death of the patient within a few minutes after the injection. Systemic anaphylaxis may occur in allergic persons following injection with penicillin or the sting of an insect. An injection of epinephrine may counter the effects of systemic anaphylactic shock. Syringes loaded with epinephrine are common in "bee-sting" kits.

Allergy Testing.   *Skin testing* can be used to determine the type of substance to which a person is allergic. Extracts of suspected allergens are injected intradermally (between the layers of skin) or a drop of the extract is placed on an area of skin that has been lightly scratched. If the person is hypersensitive, an immediate (within minutes) reaction of reddening and swelling will occur at the site of contact with the allergen. Often a person will be tested simultaneously for sensitivity to a number of different allergens by injecting extracts of each into different areas of skin.

Fc region   The base portion (end opposite the antigen-binding site) of the Y-shaped antibody.
Skin testing   A procedure characterized by multiple skin injections, used to determine to what substances a person is hypersensitive (allergic).

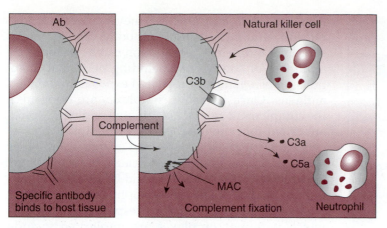

Effects:
1. Complement-mediated cell lysis (MAC)
2. Opsonization for phagocytosis (C3b and Ab)
3. Chemoattractants for inflammatory cells (C3a, C5a)
4. Damage from released lysosomal contents (Neutrophils)
5. Antibody-dependent cell-mediated cytotoxicity (ADCC) Natural killer cells

FIGURE 11-11   Type II hypersensitivity; a cytotoxic response. Cell-specific antibody binds tissues. Complement is activated with its subsequent effects: chemotaxis, opsonization, inflammation, and cell lysis through formation of the membrane-attack complex (MAC). Neutrophils may cause further damage as they attempt phagocytosis and release the contents of their granules.

Control of Allergies.   Once the specific aller-gen is identified, a person can take the necessary precautions to avoid the substance. Food allergies are most easily avoided, but people can also move from a geographic area where an allergen of plant origin is present or can avoid a certain species of animal and so on.

A person may become *desensitized* by receiving injections of extracts of an allergen. When intro-duced into the deeper tissues by injection, the aller-gen is thought to stimulate the formation of IgG antibodies. These antibodies are then able to react with the allergen on subsequent natural exposures and neutralize its ability to react with the IgE anti-bodies attached to the mast cells. Thus, no allergic reaction occurs.

Perhaps the most popular method of relieving allergy symptoms is the use of various medications. *Antihistamines* are widely used and have the direct effect of neutralizing the action of histamine. Other substances that act as decongestants, expectorants, or suppressants of pain and so on may be used in conjunction with antihistamines to bring added symptomatic relief. *Corticosteroids* may also be effec-tive in treating some allergies but should be used only with proper medical direction, for they have a suppressive effect on the entire immune system and may render the recipient more susceptible to infec-tious diseases in general.

✳ Type II—Cytotoxic Hypersensitivity

Type II reactions are invoked when a specific anti-body causes the destruction of host cells (Figure 11-11). The antibodies involved are typically IgG or IgM. These antibodies bind to specific proteins on cells and target them for destruction. The mecha-nisms whereby these cells are destroyed include complement-mediated lysis, phagocytosis or the destructive substances they release, and ADCC mediated by NK cells, as discussed above. A classic example of type II hypersensitivity is **hemolytic disease of the newborn** (*erythroblastosis*

Desensitized   Referring to a person who is no longer sensitive to an allergen.
Corticosteroid   A steroid derivative used to suppress some allergic reactions.
Hemolytic disease of the newborn   A type II hypersensitivity where maternal anti-Rh antibodies cross the placenta and bind to Rh antigens of fetal red blood cells; also called erythroblastosis fetalis.

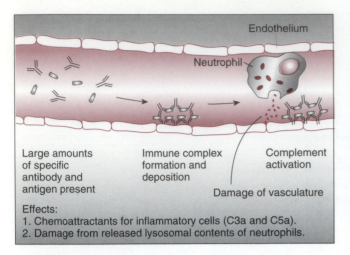

Effects:
1. Chemoattractants for inflammatory cells (C3a and C5a).
2. Damage from released lysosomal contents of neutrophils.

FIGURE 11-12 Type III hypersensitivity; immune-complex-mediated damage. When large amounts of specific antibody and antigen are present, insoluble immune complexes form and are deposited at various body locations, such as the kidney. Complement fixation is followed by the same effects listed in Figure 11-11. The difference between type II and type III hypersensitivity is that in type III, the site damaged may have nothing to do with the specificity of the inducing antibody, only with where the immune complexes are ultimately deposited.

fetalis), wherein maternal anti-Rh IgG crosses the placenta and binds the Rh antigen on fetal red blood cells. This mediates their destruction through the mechanisms described above, resulting in a life-threatening anemia and toxicity for the fetus.

⁕ Type III—Immune Complex Hypersensitivity

Type III hypersensitivity is caused by the deposition of **immune complexes** (*antigen-antibody complexes*) at various body locations (Figure 11-12). Some of the most common sites affected are the blood vessels, kidneys, joints, lungs, and skin. Immune complexes form at these locations whenever there are significant amounts of both antibody and its corresponding antigen. This situation can be induced by a number of conditions, including persistent infections, *autoimmunity*, or repeated exposure to inhaled antigens. Once immune complexes are deposited, the complement cascade is initiated and neutrophils are called into the area. As they attempt to phagocytize the complexes, lysosomal enzymes and other toxic substances spill out and cause tissue damage. Complement fragments C3a and C5a cause degranulation of mast cells and basophils, which further intensifies the inflammatory response. Diseases in which immune complexes are implicated include *rheumatoid arthritis, systemic lupus erythematosus,* and *fibrosing alveolitis.*

⁕ Type IV—Delayed Hypersensitivity

Type IV hypersensitivity is the only one mediated by cells rather than antibody. These allergies consist of an immune reaction that develops beginning 18 to 24 hours following contact with the antigen and peaks in 2–3 days; hence, the designation of *delayed hypersensitivity*. Various substances can elicit this type of response, including metals, cosmetics, microbes, and plant products. Contact with the skin is the most common route of exposure. These allergies involve the delayed-type hypersensitivity T-cell response described earlier (Figure 11-8). No visible reaction is seen following the first exposure, during which time the initial sensitization occurs. However, on subsequent exposures the sensitized T_{DTH} cells react with the antigen and induce the gradual localized reaction of reddening and swelling that is mediated by the activity of large numbers of activated macrophages. Familiar examples of type IV hypersensitivity include skin reactions to poison ivy, nickel-containing jewelry, and tuberculin used in tuberculosis skin tests.

Immune complex   An antibody–antigen complex that is usually eliminated from the body by phagocytic cells.
Autoimmunity   A hypersensitivity whereby individuals are sensitive to antigens on cells of their own body.

MATERIAL FOR REVIEW

CONCEPT SUMMARY

1. The acquired immune response results from antigenic stimulation of lymphocytic cells in the body. These lymphocytes are of two general types: B cells and T cells.

2. The humoral immune response is characterized by the production of specific immunoglobulins, called antibodies, produced by B lymphocytes in response to antigenic stimuli.

These antibodies are grouped into five classes of immunoglobulins. Each type of immunoglobulin acts uniquely in disease limitation by serving as agglutinins, anti-toxins, opsonins, and inhibitors of pathogen attachment.

3. Acquired cell-mediated immunity is largely a function of T cells. These cells are activated by the presence of antigen so

that they fulfill numerous functions in the process of disease control.

4. Some individuals overreact or react in an abnormal way to the presence of antigen in their environment. Such overactivity is known as hypersensitivity. Hypersensitivity reactions are often damaging to the host and may require considerable effort and expense to control.

STUDY QUESTIONS

1. Describe the role in the immune response of each of the following cell types: (a) macrophage, (b) T-helper and T-cytotoxic cells, (c) B cells, and (d) plasma cells.

2. How does each of the following participate in the immune response? (a) interleukins, (b) immunoglobulins, (c) complement, (d) lymphokines, and (e) interferon.

3. Draw a diagram that can be used to explain the concept of antigenic memory.

4. How did the work of Tonegawa help to explain the clonal selection theory?

5. Describe the functional differences between the variable and the constant regions of the immune globulins.

6. Diagram the cellular processes that lead to immediate hypersensitivity.

CHALLENGE QUESTIONS

1. Hemolytic disease of the newborn is caused by maternal anti-Rh antibody crossing the placenta and binding to and destroying fetal red cells. You know that people of blood type O have anti-A and anti-B antibodies in their serum. Why is erythroblastosis fetalis not a major problem in the case of a mother with blood type O carrying a fetus with blood type A?

2. Assume that scientists, using genetic engineering technology, succeed in creating a novel protein

(with a molecular weight of 100,000) that is completely different from any protein on this planet. If you were injected with this protein, would you form antibody against it? Explain why or why not.

3. Some viruses, when they infect human cells, cause the cells to stop manufacturing MHC class I proteins. Explain if and how this would interfere with the immune response. What defense mechanisms would you rely on to stop this type of viral infection?

Applications of the Immune Response

Acquired immunity resulting from the production of circulating antibodies or specific T cells may offer protection for prolonged periods or even for the lifetime of the host. The host actively produces its own antibodies, stimulated either by a primary exposure to an infecting microorganism (primary antibody response) or an ensuing exposure to the same microbe (secondary antibody response). Long-lived immunity in humans and other animals can also be acquired without exposure to the virulent microbe by receiving a vaccination (immunization). In this case, immunity is acquired or promoted without the person or animal suffering the signs and symptoms of the disease. It is also possible to confer a short-lived immunity by the transfer of antibodies from one host to another. This chapter examines these applications of the immune response and describes how antibodies in serum or antigens can be quantitated, and how this information can be used to aid in the diagnosis of diseases.

TYPES OF IMMUNITY

The active production of antibodies within the host is called **active immunity**. It is stimulated either by a natural infection (*natural active immunity*) or by artificial exposure to antigens in a vaccine ∞ (Chapter 1, p. 4). The transfer of antibodies from one host to another is called **passive immunity**. It can be termed *natural passive immunity* when antibodies are passed from mother to fetus or offspring.

Active immunity   A condition in which antibodies are actively produced within the host.
Passive immunity   A condition in which antibodies are passed from one host to another.

Artificial passive immunity occurs when antibodies are received by injection. Thus, traditionally the types of acquired immunity are divided into four categories:

1. Natural active immunity
2. Artificial active immunity
3. Natural passive immunity
4. Artificial passive immunity

✳ Natural Active Immunity

This type of immunity develops after recovery from a naturally acquired infectious disease. During the process of the infection, large amounts of microbial antigens stimulate the production of large amounts of antibodies. Usually both humoral antibodies and cell-mediated immunity are stimulated. These immune responses remain active in the body for years. During this time any reexposure to the same pathogen would result in the rapid destruction of the invading microbe. Even if the time between the first exposure and reexposure has been so long that most of the original antibodies have disappeared, the recall or anamnestic response rapidly stimulates the production of new antibodies before the invading microbe is able to cause a serious or even a clinically evident disease. Immunity of this type is more effective against diseases in which the microorganisms must pass through the blood or into deeper tissues, compared to infections of such superficial tissues as the respiratory or intestinal mucosa. However, even when complete immunity is not maintained, the reinfections are generally much less severe than the original infection. This basic concept of an anamnestic response was shown in Figure 11-5.

✳ Artificial Active Immunity

The immune mechanism of the body can be stimulated to produce antibodies when antigens are introduced by artificial means—that is, in the form of a vaccine. Four general categories of vaccines are used: *killed, subunit, toxoid*, and *attenuated*.

Killed Vaccines. Killed vaccines are made by culturing large numbers of a pathogenic microorganism and then subjecting them to treatment by heat or chemicals. When just the right amount of treatment has been given, the components of the microbe that are necessary for multiplication are destroyed, but the antigens are not altered. If these killed microbes are injected into the tissues of the host, they will be phagocytized by macrophages and passed to the lymphatic tissues, where specific antibodies will be produced. With one injection it is usually not possible to get enough antigen into the tissues to produce a maximum stimulation of antibodies. To obtain the maximum antibody stimulation, the immunization must be repeated about every other week for a total of three or four injections. The effectiveness of this type of vaccine may be enhanced by mixing the killed microbes with such materials as mineral oils or alum. These materials are called *adjuvants* and slow down the clearance of the vaccine from the tissues, thereby providing a longer exposure to the antigen. Generally, adjuvants are not used in vaccines for humans, for they may cause considerable discomfort at the injection site.

The immunity induced by a killed vaccine does not persist as long as the immunity induced by a natural infection. Often the immunity has disappeared or is greatly reduced after several years. Then if a single dose of the vaccine, called a *booster* dose, is given, the anamnestic reaction is stimulated and high levels of antibodies are produced in just a few days. If booster doses are given every few years, an immune state can be maintained continuously.

Subunit Vaccines. Some vaccines use purified fractions of the microbe rather than whole-cell preparations. Several newer vaccines, for example, use only the capsule of the bacteria as the immunizing agent. Such fractioned vaccines help reduce undesirable side reactions caused by endotoxins and other materials found in an intact microbial cell. A new generation of purified antigen vaccines produced by genetic engineering is being developed ∞ (Chapter 6, p. 82). Such vaccines contain only the antigens that stimulate protective antibodies, thereby reducing the potential for adverse side reactions. This technology is particularly useful in providing purified viral antigens that cannot be produced economically by conventional methods. An example is the current hepatitis B vaccine that is composed of viral surface antigen made by yeast cells through recombinant DNA technology.

Toxoid Vaccines. Diseases like diphtheria and tetanus are caused by exotoxins, and the antibodies formed against the exotoxins are called *antitoxins*. When these antitoxins combine with toxins

---

Booster A second or subsequent dose of antigen given to increase the level of immunity.
Toxoid A modified toxin that has lost its toxic properties but retains its antigenic properties.

they inhibit the ability of the toxins to bind to host cells and cause disease. Vaccines against such toxin-mediated diseases contain inactivated toxins, called *toxoids*, rather than the killed microbial cells. Toxins are converted into toxoids by treatment with chemicals or by exposure to heat. The antibody response to toxoids is similar to that produced by killed vaccines in that a series of primary doses and periodic booster doses are needed for maximum and continued protection.

Living Attenuated Vaccines. *Attenuated* means "weakened" or "reduced in force." These vaccines contain microorganisms with a low virulence ∞ (Chapter 13, p. 188). Under most conditions they produce only a mild or subclinical infection. Attenuated microbes proliferate in the tissue to yield a mass of antigens similar to the amount produced by the natural disease; thus a single dose

stimulates a high antibody level that lasts for a prolonged period. In some cases, the living vaccine can be given by a natural route of infection—for example, the oral administration of the live polio vaccine. Such vaccines are easier to administer than those that must be given by hypodermic injection.

A disadvantage of living vaccines is that occasionally, mild or sometimes even active disease may result from the vaccine itself. This factor is of particular concern in persons with defective or suppressed immune mechanisms and such persons should not receive living vaccines. Similarly, women who are pregnant should avoid receiving living vaccines because the attenuated microbes may cause infection and disease in the developing fetus. The most commonly used vaccines and their recommended administration schedules are listed in Table 12-1.

TABLE 12-1

Currently Available Vaccines

Vaccine[a]	Type of Immunogen		Recommended Schedules[b]
Commonly Recommended Immunizations			
DPT	Diphtheria and tetanus toxoid with killed *B. pertussis*	P	4 doses, 2, 4, 6, 15 months
		B	4–6 years and every 10 years
Polio	Live attenuated oral, trivalent (types 1, 2, and 3)	P	2 doses, 2, 4, months
		B	14 months, 6 years
MMR	Live attenuated measles, mumps, and rubella	P	1 dose, 15–18 months
Influenza	Formalin inactivated virus	P	Persons at high risk
		B	Annual
HbCV	Type b polysaccharide–protein conjugate	P	4 doses 2, 4, 6, 15 months
Hepatitis B	Recombinant HBsAg	P	3 doses, newborn, 1–2, 18 months
		B	At-risk adults
Immunizations Recommended for Specified Circumstances			
Pneumococcus	Capsular polysaccharide of 23 most common types	P	Persons over 2 years at high risk
		B	Unknown
Cholera	Phenol inactivated *Vibrio cholera*	P	Persons traveling to cholera area
		B	Every 6 months
BCG	Attenuated *Mycobacterium bovis*	P	Persons at increased risk
		B	Not recommended
Typhoid	Live, attenuated *Salmonella typhi*	P	4 doses, alternate days
		B	5 years
Rabies	Killed attenuated virus	P	Preexposure, persons at high risk Postexposure, all persons
Smallpox	Live attenuated vaccinia virus	P	Selected laboratory workers
Yellow fever	Live attenuated virus	P	Persons traveling to endemic areas
		B	10 years
Meningococcal	Polysaccharide (types A, C, Y, W-135)	P	Persons at increased risk
Varicella	Live, attenuated virus	P	Immunocompromised children

[a]DPT, diphtheria, pertussis, tetanus; MMR, measles, mumps, rubella; HbCV, Haemophilus conjugate vaccine; BCG, bacille Calmette–Guerin (tuberculosis).

[b]P, primary immunization schedule; B, booster immunizations

✴ Natural Passive Immunity

The newborns of many species receive antibodies from their mothers that protect them during the critical early phase of life. Some mammals, such as calves and pigs, obtain passive antibodies in a substance called *colostrum*, which is present in their mother's milk during the first week after delivery. With humans and some other mammals, much of the passive transfer of antibodies occurs across the placental barrier; these are IgG antibodies and at birth, the infant has the same IgG profile as that found in its mother. The role of colostrum and the passage of antibodies to the infant in human milk are not clearly understood. Some IgA antibodies appear to be passed by this means and may offer protection to the mucosal surfaces of the respiratory or intestinal tract. Natural passive immunity lasts from 3 to 6 months in humans. Birds effectively pass antibodies to their offspring via egg yolk.

✴ Artificial Passive Immunity

Under certain circumstances a person may receive, by injection, a supply of preformed antibodies taken from another host or produced as *monoclonal antibodies*. This artificially passed immunity lasts only for several weeks. Before the development of antibiotics, it was a common practice to produce high levels of antibodies in horses or cows, for instance, and then collect their serum and use it as "antiserum" to treat various infectious diseases of humans. Although this procedure had some value, serious side reactions often occurred due to type III hypersensitivities that developed against the animal serum ∞ (Chapter 11, p. 170). Antiserum therapy is still used against some infectious diseases, but it is more effective in treating diseases caused by toxins. Whenever possible, monoclonal antibodies or human antiserum, particularly the *gamma globulin* fraction, are used because they consist of concentrated human antibodies and therefore have a much reduced chance of inducing hypersensitivity. Antisera can effectively treat such diseases as diphtheria, tetanus, and botulism poisoning that are caused by toxins. In addition, antisera can be produced against various venoms and then used to treat victims of bites by venomous animals and insects. Passive immunity may also be used as a prophylactic (preventive) measure for high-risk persons who may have recently been or may likely become exposed to an infectious agent.

MEASUREMENT OF ANTIBODIES AS A DIAGNOSTIC TOOL

The presence of humoral antibodies in someone's serum indicates that the person has either had a specific disease or has been immunized against that disease. Furthermore, the level of antibodies may give some indication as to how recently this person was exposed to the disease.

An analysis of the levels of specific antibodies present in the serum of a patient is an important diagnostic tool. In diagnosing an infectious disease, it is important to obtain a serum sample from the patient as early as possible, preferably during the first few days of illness. Called the *acute phase* serum, this sample should contain no or only low levels of antibodies against the microorganism causing the disease. Several weeks later, a second serum sample should be taken. It is called the *convalescent phase* serum and should contain higher levels of antibodies against the disease-producing microorganisms (Figure 12-1). If the increase in antibody levels between the acute and the convalescent phases is significant, usually fourfold or greater, it is assumed that the disease was caused by the microorganism against which the antibodies were formed. Before this diagnostic test can be run, a pathogenic microorganism is needed to serve as the test antigen. Ideally this microorganism would have been isolated from the patient during the illness. If the signs and symptoms of the disease are compatible with the type of disease caused by the isolated

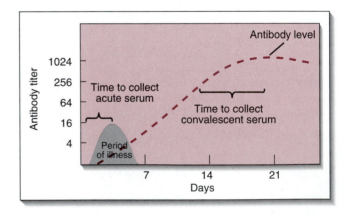

FIGURE 12-1   The relationship between the period of illness and the time to collect acute and convalescent phase serum samples.

Gamma globulin   That portion of blood serum containing antibody.

pathogen, the diagnosis of the disease based on this evidence alone is fairly certain. If a specific rise in antibodies occurred against this pathogen, the diagnosis is confirmed. If no specific pathogen was isolated from the patient, however, the signs and symptoms of the disease might be consistent with a disease caused by any one of several different pathogens. Because diagnostic laboratories generally have commercially prepared antigens obtained from the more commonly encountered pathogens, a laboratory could be asked to determine the levels of antibodies in the acute and convalescent phase sera against the suspected pathogens. The pathogen against which a significant rise in antibody level is seen is assumed to be the cause of the disease.

In other situations, it may be desirable to know if a person had previously had a specific disease. A woman in the early stages of pregnancy, for example, may have been exposed to rubella, a disease known to pose a serious threat to the developing fetus should the expectant mother become infected. It would be useful to know if the expectant mother had antibody immunity to rubella. In this case, the serum of the expectant mother could be tested for antibodies against rubella and the results of the test would then determine the course of action to follow.

The analysis of amounts and types of antibodies has many applications in diagnostic medicine, epidemiology, and research work. The study of antigen and antibody reactions in vitro (in test tubes) is called *serology*. Some of the more commonly used methods of detecting and measuring antibodies and antigens are discussed in the following pages.

SEROLOGIC TESTS

Either a known antigen or a known antibody is needed for most serologic tests. Many routinely used antigens or antibodies (antisera) are available from commercial sources. If a known antigen is used, the serum samples can be tested to determine if they contain specific antibodies against this antigen. On other occasions it may be necessary to identify an unknown antigen. Generally, the antigen to be identified is a microorganism that cannot be completely identified by its physiologic or morphologic characteristics. This unknown microorganism is tested against known selected antisera; when specific reactions occur, a specific identification can be made.

The relative amount of antibodies in a serum sample is measured by determining the extent to which the serum can be diluted and still produce an observable antigen–antibody reaction. The term *titer* refers to this relative amount of antibodies. It is common to use either 2-, 4-, 8-, or 10-fold serial dilutions of serum in determining the antibody titer. If the highest dilution in which a detectable antigen–antibody reaction occurred was a dilution of 1:256, for instance, then the antibody titer would be reported as 256.

✳ Agglutination Tests

Agglutination tests use whole cells or insoluble particles about the size of cells as antigens (Figure 12-2). When antibodies attach to the antigenic determinants on these cells or particles, they form bridges that result in aggregates of interconnected particles. Such aggregates or agglutinations are visible to the naked eye (Figure 12-2b). Agglutination tests are used to detect antibodies against bacterial cells, to type unknown bacteria, or to determine blood types. Since they are easily visualized, agglutination reactions are also used in various artificially constructed tests. In the artificial tests, known antibodies or soluble antigens are attached to small particles of latex or to tanned red blood cells. Such treated particles may then be used in agglutination tests to detect the corresponding antigen or antibody in test specimens.

✳ Precipitation Tests

Precipitation tests use soluble antigens. The interaction of proper ratios of these antigens with specific antibodies results in the formation of insoluble lattices. Precipitation reactions can best be visualized at the interface between a solution of an antibody and a solution of antigen. These tests are carried out in small-diameter tubes or *capillary tubes* in which the antigen is layered over the antiserum. A ring of precipitation will be seen at the interface of the two solutions if specific antibodies and antigens are present. These tests are called *ring tests* and *capillary tube tests* (Figure 12-3).

A convenient and widely used method of demonstrating precipitation reactions is by *immunodiffusion*, a procedure that is also called the *Ouchterlony* method (named after its discoverer). Solutions of

---

Titer The concentration of specific antibody present in the serum of an individual.
Capillary tube A glass tube with a very small opening running through the tube. The opening is usually less than 1 mm in diameter.

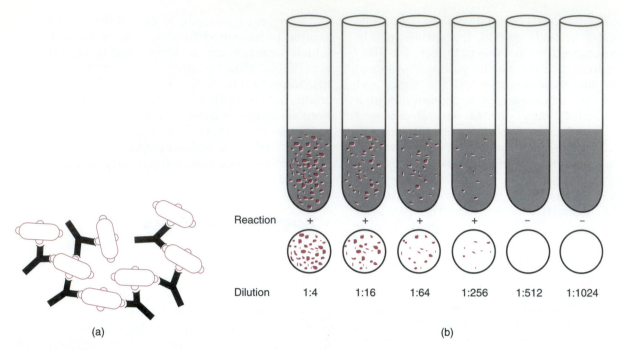

(a)

Reaction   +   +   +   +   −   −

Dilution   1:4   1:16   1:64   1:256   1:512   1:1024

(b)

FIGURE 12-2 Agglutination reactions. (a) Agglutination of bacterial cells by IgG antibodies. (b) Tube agglutination reactions in fourfold dilutions of serum. The reaction is positive through a dilution of 1:256, therefore the titer is 256.

specific antigens and antibodies are placed in adjacent wells that are formed in an agar gel. A line of precipitation occurs where the diffusing antigen and specific antibodies come together. This precipitated antigen–antibody complex becomes fixed in the agar and is readily visualized (Figure 12-3b, c).

Precipitation reactions can be used to quantitate an antigen or antibody similar to the titer obtained in the tube agglutination test described above. A variation of immunodiffusion called *single radial immunodiffusion* is used for this purpose (Figure 12-4). In this test, specific antibody or antigen, at a

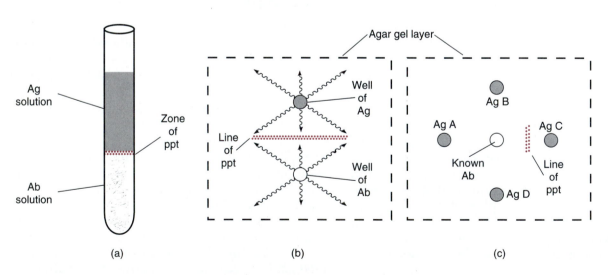

(a)   (b)   (c)

FIGURE 12-3 Precipitation (ppt) reactions between antigens (Ag) and antibodies (Ab). (a) A precipitation reaction in a small diameter test tube. (b) Precipitation reaction in an agar gel. (c) Use of the agar gel method to identify unknown antigens that will react with a known antibody; in this case, antigen C is identified as the specific antigen.

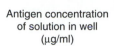
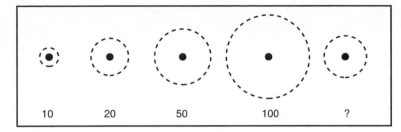

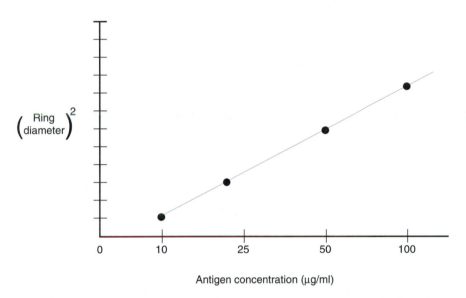

FIGURE 12-4 Single radial immunodiffusion. In this case, the concentration of an antigen in solution is determined by the diameter of the precipitation ring formed in antibody-containing agar. The ring diameter is translated to antigen concentration by comparison with standard curves.

known concentration, is added to the agar gel prior to solidification. A specific amount of the corresponding antigen- or antibody- containing solution is then added to a well cut in the agar. The substance diffuses in the agar until the concentration is reduced enough to allow precipitation to occur. The higher concentration of the substance, the larger the diameter of the ring will be. Standard curves can be constructed to allow an accurate quantitation of the substance in question. These tests are commonly used to measure the amount of different antibody classes in serum. Commercial kits containing the necessary components are available. For example, a gel containing anti-α heavy-chain antibody is used to quantitate the amount of IgA in serum.

✳ Neutralization Tests

Neutralization tests use living microorganisms or their toxins and expose them to antibodies in test tubes. After the antibodies have had a chance to react with the microbe or toxin—usually for a period of an hour or two—the mixture is inoculated into a suitable experimental host. If the specific antibodies have reacted with the substance in question, it will be unable to cause disease in the host; that is, it is neutralized. This test is often used to identify viruses and bacterial toxins.

✳ Complement Fixation Tests

Complement is activated by certain antigen–antibody reactions ∞ (Chapter 10, p. 149). Once complement has been fixed this way, components are hydrolyzed and are not free to enter into subsequent reactions. By finding out if the complement has become fixed, it is possible to determine if an antigen–antibody reaction has taken place. Therefore, the amount of complement that has been fixed is a sensitive measure of the number of antigen–antibody reactions that have formed. The steps

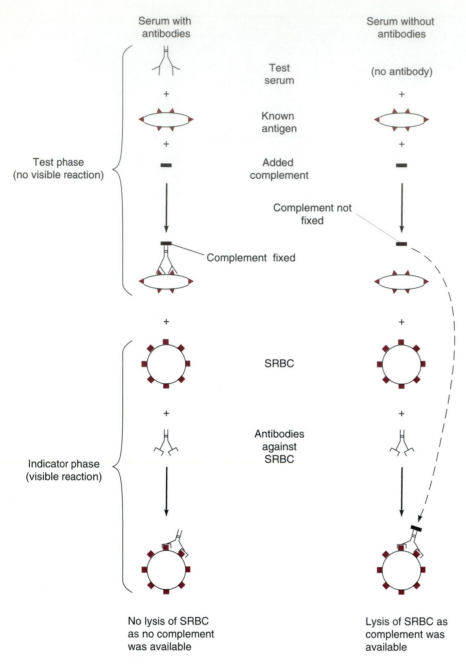

FIGURE 12-5 Complement [-] fixation test using a known antigen [◁▷] to test serum for the presence of antibodies [⅄] against this antigen. This test shows the response of serum samples with and without specific antibodies. The indicator phase of this test uses antibody-coated sheep red blood cells (SRBC). When complement becomes fixed in the test phase, it is then not available to form a complex with and cause lysis of the SRBC in the indicator phase of the test. Thus no lysis means the test is positive for the presence of antibodies in the original serum.

of a complement fixation test are outlined in Figure 12-5.

If a serum sample is being analyzed by the complement fixation test for antibodies against a known antigen, it is first heated to 56°C for 30 minutes to destroy any residual complement that might be present. It is then serially diluted. Next, the known antigen and a standard amount of complement are added to the tubes of diluted serum. If the serum contains specific antibodies to the antigen, a complex of complement–antigen–antibody will form and the complement will be fixed. This reaction is not visible. Therefore, an indicator system is used to tell if the reaction has occurred. The indicator system consists of "sensitized" sheep red blood cells (SRBC), meaning they are coated with specific antibody. The sensitized SRBC are added to the original test system. If a specific reaction did not occur in part one, the complement is still available to complex with the indicator system, which results in the lysis of the red blood cells. This reaction is readily visible because the solution turns from cloudy to clear as the red cells lyse. If the SRBC do not lyse, it is assumed that the original reaction fixed the complement and so enough specific antibodies must have been present in the original serum. This is an indirect test for antibodies and adequate controls are needed to show that all components are working properly before any assumption can be made.

✳ Fluorescent Antibody Tests

It is possible to attach (*conjugate*) certain fluorescent dyes to known antibodies. The antibodies can then be mixed with test microorganisms. If a specific reaction occurs, the antibodies, along with the fluorescent dye, will attach to the microbial antigens. This reaction is carried out on a microscope slide. When the slide is placed under a fluorescence microscope ∞ (Chapter 3, p. 29), the areas where specific antigen–antibody reactions have occurred will fluoresce. Using this method (Figure 12-6), it is possible to identify a specific type of microbe in a mixed microbial population.

✳ Hemagglutination-Inhibition Test

Hemagglutination-inhibition tests, which are relatively simple to conduct, are useful in detecting antibodies against certain types of viral infections. Some viruses, such as influenza virus, specifically attach to red blood cells and cause the cells to agglutinate. This process is called *hemagglutination* and is a useful method for detecting the presence of such viruses. If coated with antibodies, the virus cannot cause hemagglutination. Thus, in the hemagglutination-inhibition test the serum is reacted with a known virus and then the red blood cells are added. If specific antibodies are present in the serum, they will coat the virus and inhibit the occurrence of hemagglutination. If hemagglutination does not occur, the presence of specific antibodies is indicated.

✳ Radioimmunoassay (RIA)

The RIA is a highly sensitive test that uses radioactive materials to detect small amounts of antigens that may be present in tissues, body fluids, or blood. Its primary uses include detecting hepatitis B antigens in blood collected for transfusion and the early detection of small amounts of human chorionic gonadotropin in the serum of pregnant females (Chapter 31). Although a very useful test, it does suffer from certain disadvantages: It is complex, requires expensive equipment, and uses radioactive materials. Therefore, this test can be run only in specially equipped laboratories.

✳ Enzyme-Linked Immunosorbent Assay (ELISA)

The ELISA test is nearly as sensitive as the RIA test, but it does not require radioactive materials and can be done with less elaborate equipment. It can be used to measure the amounts of either antigens or antibodies in a test sample. The procedure for detecting antibodies is shown in Figure 12-7. Here, a known antigen is adsorbed onto the wall of wells in special plastic plates. Then serum to be tested for antibodies is added to the wells. If specific antibodies are present, they will attach to the adsorbed antigen; the wells are then rinsed to remove all components of the serum except the attached antibodies. If the test serum is from humans, the next step would be to add specially prepared antibodies against human gamma globulin; these antibodies are produced in an animal such as a goat or rabbit. The antihuman antibodies are conjugated to an enzyme. These conjugated antibodies will react with the human antibodies that are attached to the adsorbed antigen. The wells are again rinsed to remove any unattached enzyme-conjugated antibodies. Next, a substrate is added that will be acted on by the enzyme in such a way that a color is produced. The intensity of the color can be measured

Conjugate Attach; proteins to which a specific substance or molecule has been attached are said to be conjugated to the added molecule or substance.

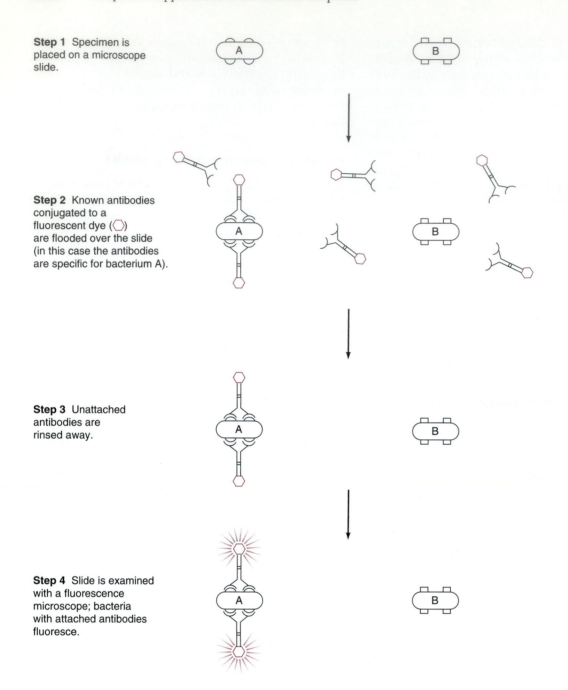

Step 1 Specimen is placed on a microscope slide.

Step 2 Known antibodies conjugated to a fluorescent dye (◯) are flooded over the slide (in this case the antibodies are specific for bacterium A).

Step 3 Unattached antibodies are rinsed away.

Step 4 Slide is examined with a fluorescence microscope; bacteria with attached antibodies fluoresce.

FIGURE 12-6 Fluorescent antibody procedure for identification of a bacterium in a clinical specimen.

by a spectrophotometer or, less accurately, by the eye and is proportional to the amount of the antibody that attached to the antigen. When all reagents are available, the ELISA test can be carried out in less than one hour. This test is very useful in applications where a rapid detection of antigens or antibodies is needed and it is adaptable to most routine serologic tests. It is used as the primary screen-

ing test for the diagnosis of antibodies to HIV, the AIDS virus.

✱ Immunoblot (Western Blot) Procedures

Immunoblotting is a very sensitive procedure that can specifically detect small amounts of either anti-

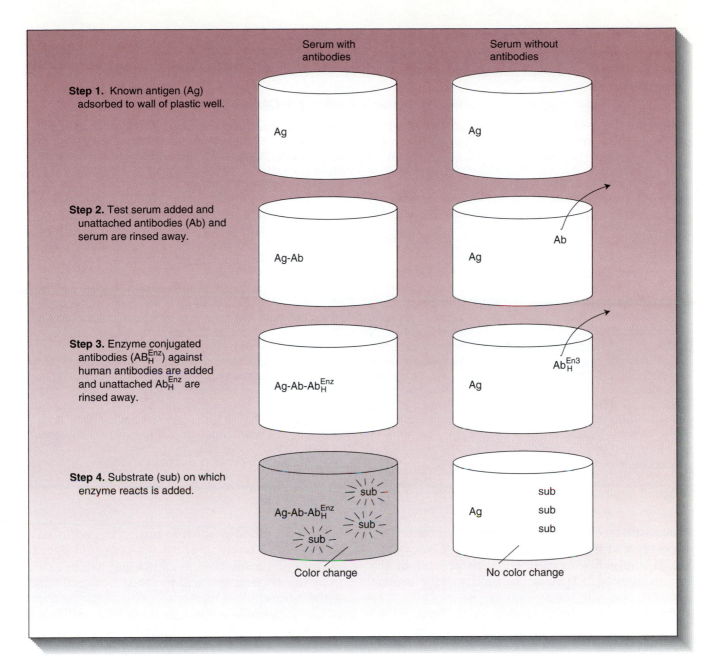

Serum with antibodies | Serum without antibodies

Step 1. Known antigen (Ag) adsorbed to wall of plastic well.

Ag | Ag

Step 2. Test serum added and unattached antibodies (Ab) and serum are rinsed away.

Ag-Ab | Ag   Ab

Step 3. Enzyme conjugated antibodies (AB_H^{Enz}) against human antibodies are added and unattached Ab_H^{Enz} are rinsed away.

Ag-Ab-Ab_H^{Enz} | Ag Ab_H^{En3}

Step 4. Substrate (sub) on which enzyme reacts is added.

Ag-Ab-Ab_H^{Enz} sub   sub   sub | Ag   sub   sub   sub

Color change | No color change

FIGURE 12-7   Indirect ELISA test to detect antibodies in human serum.

gens or antibodies in a clinical specimen. This procedure is widely used to detect antibodies to HIV. The first step in the HIV antibody test is to start with a known purified preparation of HIV that is treated with a detergent, called sodium dodecyl sulfate, that solubilizes the viral proteins. The proteins are then passed through a gel by electrophoresis and the different proteins are separated into bands based on *molecular weight*. Next the protein bands are blotted from the gel onto a nitrocellulose membrane (this is aided by an electrical current). The above steps are often done by commercial companies and the prepared nitrocellulose membranes are supplied to diagnostic laboratories. In the

Molecular weight compound.   The combined atomic weight of atoms in an ion, molecule, or

CLINICAL NOTE

FDA Approves, Pediatricians Endorse Chicken Pox Vaccine

The U. S. Food and Drug Administration (FDA), ushering in an era in which virtually all the major childhood diseases are preventable, recently licensed a chicken pox vaccine. The American Academy of Pediatrics subsequently endorsed this vaccine, recommending that it be administered routinely to children between the ages of 12 and 18 months as well as to older children who have not yet had chicken pox.

Welcome though this approval may be, it came after more than a decade of study, leading many people to ask why the process took so long, particularly since a similar vaccine has been in use for many years in Japan, Korea, and several European countries.

The new chicken pox vaccine, Varivax, is being manufactured and marketed by Merck & Co., Inc., of Whitehouse Station, New Jersey. Until the vaccine's approval, chicken pox had become the principal childhood ailment for which a vaccine was not generally available. Vaccines already are in widespread use in pediatric medicine for preventing disease caused by a range of pathogens, including influenza, diphtheria, tetanus, pertussis (whooping cough), hepatitis B, measles, mumps, rubella, and polio. The chicken pox vaccine is expected to cost pediatricians $39 per dose,

according to company officials. A recent study (estimates were based on a slightly lower per-dose cost of $35) projects annual savings of $384 million in overall health-care costs from routine use of the chicken pox vaccine in the United States.

"This vaccine has been studied in approximately 11,000 individuals, and we expect it to be 70 to 90% effective in preventing chicken pox," says FDA Commissioner David Kessler. As most parents and pediatricians already know, chicken pox is a highly contagious disease—usually a nuisance and sometimes life threatening—that afflicts nearly 4 million people annually in the United States, the majority of them young children. Typically, it is benign in children, but it can be quite severe in teenagers and adults or when complications arise.

Despite what some critics claim, the explanation for the chicken pox vaccine's long road toward FDA approval does not rest entirely with the agency. The manufacturer of the chicken pox vaccine, Merck, and the nature of the chicken pox virus also have contributed to the drawn-out approval process (*ASM News* 61:291, 1995).

diagnostic laboratory, the nitrocellulose membrane is covered with the serum sample and then incubated. If antibodies against HIV are present in the serum, they will specifically bind to HIV protein antigens on the nitrocellulose membrane. Any unbound human immunoglobulin is removed by washing. To determine if antibodies have bound to HIV antigens, an antihuman antibody (detecting antibody) that is conjugated to an enzyme is added to the membrane. This detecting antibody will bind with HIV antibody, which in turn is bound to HIV antigen. After washing away any unbound detecting antibody, a substrate is added that will be acted on by the enzyme and will produce a colored band on the nitrocellulose membrane where HIV protein–antibody complex is located. If no HIV antibodies were present in the original serum sample, no colored band would be seen on the nitrocellulose membrane.

MATERIAL FOR REVIEW

CONCEPT SUMMARY

1. Immunity to infectious disease may result from natural events, such as recovery from the disease or exposure to the agent without the development of recognizable symptoms, both of which could lead to long-term active immunity. Natural events may also lead to transitory protection, called passive immunity, such as the immunity of the newborn obtained from the mother prior to birth.

2. Immunity may also result from artificial events such as the pur-

poseful administration of antigenic material (a vaccine) from the pathogen to induce artificial active immunity. Protective antibody produced in another host and administered to an individual who may have been exposed to the infectious agent leads to temporary protection termed artificial passive immunity.

3. Artificial active immunity rep-

resents the most effective means for the control of many infectious diseases. The vaccine used for immunization may consist of killed pathogens, toxoids (inactivated toxins), or attenuated living cells or viruses. Although the latter often provides easy and effective immunization, some element of risk may exist since the vaccine contains living material.

4. The laboratory determination of the presence and amount of antibody present in blood serum is collectively called serology. A variety of methods are employed that depend on the visualization of the antigen–antibody reaction in dilutions of serum by physical or chemical means.

STUDY QUESTIONS

1. The quality and length of immunity is usually much greater for natural active immunity than for any of the other categories. What feature of natural active immunity accounts for this observation?
2. Compare the use of killed vaccines with the use of living attenuated vaccines. Which often provides superior immunization? What are some of the hazards associated with the use of living attenuated bacteria and viruses? [A note on the history of using attenuated viruses

for the treatment of rabies: In 1884 Pasteur developed the idea of attenuating a virus by passage through an unusual host, a rabbit, and using it to treat what was then invariably a fatal disease (see Chapter 36).]
3. Why is it recommended that the DPT vaccine series be completed by 6 months of age?
4. Although the use of artificial immunizing methods has decreased in modern times, it is still a very effective treatment for persons exposed to toxins. Why?

5. Define each of the following terms: (a) titer, (b) complement, (c) radioimmunoassay (RIA), (d) enzyme-linked immunoabsorbent assay (ELISA).
6. Why is the hemagglutination–inhibition assay a useful test for certain viral infections?
7. Explain how the complement fixation test is run. Why is it essential to destroy the complement in the patient's serum? Why is the primary antigen–antibody reaction coupled to a lytic, sheep red blood cell system?

CHALLENGE QUESTIONS

1. The DPT vaccine consists of diphtheria and tetanus toxoids and whole cells of *Bordetella pertussis*. When hosts vaccinated with DPT were compared with those vaccinated with the DT vaccine (two toxoids only), it was found that the antibody titers to the toxoids of those that received the DPT vaccine were much higher. Explain the

immunological reasons for these observations.
2. Suppose you work for the state Health Department and they send you to investigate a charge that a local restaurant is using "strange" meat in their burritos. You have specific anti-cow, anti-horse, anti-rat, and anti-dog antibodies available to you. Outline a quick test you could

perform to determine if any of these meats are being used.
3. Suppose you are performing a Western blot as described in the text, but you are specifically interested in which HIV proteins are bound only by IgM antibodies. How would you modify your procedures to detect this?

This chapter focuses on the microorganism as a cause of infectious diseases, whereas Chapters 10, 11, and 12 focused on the response (immunologic processes) of the host to challenge from microorganisms. Whether a microorganism is able to infect or cause disease in a host depends on a complex interaction of the host's defense mechanisms and the virulence factors of the microbe. If this interaction favors the host, infection may not occur; if the interaction favors the microorganism, disease will usually result. This chapter looks at the interactions between human hosts and microorganisms. It examines the microbes that normally live on or in the body, the mechanisms they use to exit and enter the body, and the ways microbes maximize their chances of evading the host defenses.

HOST–PARASITE RELATIONSHIPS

Over the years, various terms have been used to describe and categorize microbes into different groups relative to their interaction with human or other hosts. The beginning student should bear in mind that it is difficult to categorize biological phenomena neatly without some overlapping, particularly when attempting to categorize the complex interactions between microbes and human hosts. The concern here is not with the many indirect beneficial or detrimental effects of microorganisms on humans in the general biological cycles in nature, such as nitrogen fixation or spoilage of food, but with those situations in which the microbes are present in or on the tissues of humans or are involved in human diseases.

The term **pathogen** refers to a microorganism that is able to produce disease; *pathogenic* is used as an adjective when referring to such microorganisms. *Nonpathogens* are microbes that are not able to cause disease. The dividing line between pathogens and nonpathogens is not distinct. A large gray area exists in which microbes may be either pathogenic or nonpathogenic depending on the interactions of many traits of both the microbe and the host. This group of microorganisms is often referred to as opportunistic pathogens. The term *virulence* refers to the disease-producing powers or potency of a pathogen; It is sometimes convenient to refer to pathogens as having high, moderate, low, or no virulence. The word **pathogenicity** is frequently used interchangeably with the term virulence. The term *pathogenesis* means the development of a disease and is used when referring to the mechanisms, sequence of changes, and processes that occur in the development of a disease.

The terms *infection* and *disease* are often used interchangeably. By technical definition an **infection** occurs when a microbe is able to overcome the defense barriers and live inside the host; tissue damage may or may not result. **Disease** refers to those conditions where the host's tissues are damaged or their function is altered by the microorganisms. When referring to infections, it is often convenient to use the term *clinical* or *apparent* infection when signs and symptoms of the disease are apparent, and *subclinical* or *inapparent* infection when no apparent signs or symptoms of disease are produced. The term *colonization* is used when referring to the ability of a microorganism to establish itself on a body surface (Figure 13-1). It follows then that colonization is an expression of infection and does not necessarily lead to disease.

Another term that helps describe interactions between a host and microorganisms is *parasitism*. A **parasite** is an organism that lives in or on the host and derives its sustenance from the host. If this parasitic relationship is beneficial to both the host and the parasite, the relationship is called *mutualism*. If the parasite causes no damage to the host, it is referred to as a *commensal*. Of course, a parasite that causes disease is a pathogen. It is possible for a parasitic microbe to change from the commensal to the

FIGURE 13-1 A strain of the bacterium *E. coli* specifically adhering to and colonizing the intestinal epithelium, at the tip of a villus in the ileum, of a pig. Scanning electron micrograph; bar equals 10 μm. (Courtesy B. Nagy, *Infection and Immunity* 13:1214–1220)

pathogenic relationship and vice versa. When the host's defense mechanisms are weakened, a commensal organism of moderate or low virulence that is normally unable to cause disease in a healthy person may then be able to multiply at a faster rate or gain access to deeper tissues and cause a clinical disease; under such conditions the microbe is referred to as an *opportunist*. Under other conditions, the balance between the host and a potentially virulent microbe may be such that the microbe is able to multiply and persist in the host as a commensal; yet when this microbe is transmitted to a new host a clinical disease may result. In such cases, the first host is called a *carrier*. Most microorganisms are not parasites and are able to use nonliving materials as nutrients; these microbes are called *saprophytes*. Most saprophytes are nonvirulent, but a few are able to cause diseases in humans.

The outcome of an infection is a result of an interaction among the integrity of the host's defense mechanisms (discussed in previous chapters), the number of microorganisms present in the host, and the virulence of the invading microorganisms. Perhaps of greatest concern in clinical medicine today is the patient whose natural defense mecha-

Pathogen   A microorganism capable of causing disease in a host.
Pathogenicity   The ability to cause disease.
Infection   The growth of a microorganism on or in a host.
Disease   A change in the state of health of the body resulting in inability to carry out all normal functions.
Parasite   A microorganism that derives nutrients from its host.

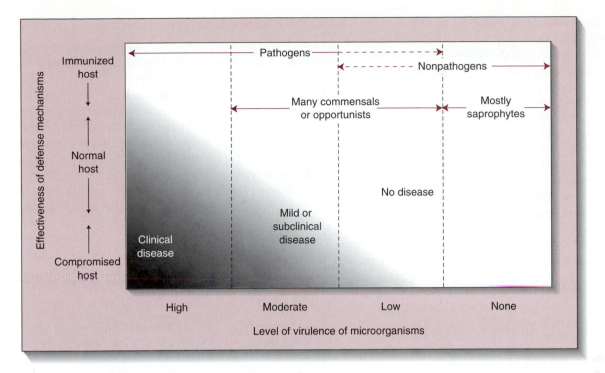

FIGURE 13-2 The interaction of microorganisms of varying levels of virulence with hosts of varying levels of resistance, and the resultant clinical outcome (the density of the shading represents the approximate severity of the disease).

nisms have been impaired or suppressed by other illnesses, genetic defects, or medical treatments; such a person is referred to as a **compromised** host. In most cases, for a microorganism to be a successful pathogen, it must possess those qualities that will allow it to enter the host tissues, resist the host's defense mechanisms, multiply, and cause damage to or malfunction of those tissues.

The relationships between microorganisms and the host are presented in Figure 13-2. The varying levels of the host's defense mechanisms are shown at the left, with the immunized host, resistant to highly virulent microorganisms, shown at one extreme and the compromised host, highly susceptible to low virulent microorganisms, shown at the opposite extreme. The varying levels of microbial virulence are shown at the bottom of the figure. The high degree of resistance of an immunized host is normally specific only against the type of microorganism against which it was immunized. A normal, unimmunized host may have varied responses to microorganisms of high and moderate virulence, and a high degree of resistance toward infections by microbes of low virulence.

Along with the virulence of the microbe and lev-els of resistance of the host, the number of microbes present in the host is an important factor in host–parasite outcome. This added dimension is shown in the following conceptual formula:

$$ID = \frac{N \times V}{R}$$

where ID means infectious disease, V is the virulence of the microorganism, N is the number of microorganisms, and R (resistance) stands for the level of effectiveness of the host's defense mechanism. Based on the interrelationships presented in this formula, the following examples could be postulated. Moderate to small numbers of low-virulence microorganisms would not produce disease in a normal host, but they may be able to produce disease in a compromised host with impaired defense mechanisms; however, a large number of such organisms may be able to produce disease in a normal host. In the case of compromised patients, the resistance component of the formula is reduced, causing these individuals to become more susceptible to infections.

The problem of the compromised patient is of increasing concern today. Many newer medical

Compromised   Reduced resistance with more susceptibility to infection.

treatments and procedures, while providing significant improvements in many medical conditions, interfere with the normal defenses of the patient against microbial infections. An example is the treatment of a cancer patient with drugs that slow down or stop the growth of the cancer cells, but at the same time interfere with the ability of white blood cells to fight infections ∞ (Chapter 10, p. 144). Although the compromised patient may be given many meaningful years of added life by these new products, the patient's balance with microbes in his or her environment has been altered, and this factor must be considered in the overall management of the case. Many infections encountered in medical practices today result from low-virulence microorganisms that are normally found on the tissues of the host (*normal flora*) but are able to cause human infections due to the compromised condition of a person's defense mechanisms.

NORMAL MICROBIAL FLORA OF HUMANS

The term **normal microbial flora** refers to those microorganisms that are the usual inhabitants of healthy individuals. Most would be considered harmless commensal microorganisms; yet some are potential pathogens—that is, opportunists—and do cause disease when the balance between them and the host is altered. Under normal conditions, however, humans are able to live in healthy balance with the microorganisms that constitute their normal flora. In fact, some microbes that make up the normal flora may be beneficial to the host by interfering with the growth or attachment of pathogenic microbes ∞ (Chapter 10, p. 144), development of a normal immune response, and synthesis of vitamin K.

The common types of microorganisms normally found on different parts of the body are mentioned in this section. The names of many microorganisms referred to here will be unfamiliar to most beginning students, but are being introduced at this time so that this section may serve as a reference source. Many microbes that are part of the normal flora, but that are also able to cause infectious diseases, are discussed in greater detail in subsequent chapters.

✳ Location of Normal Flora

Skin.   The type and number of bacteria vary from one area of skin to another. The superficial layers of the skin (squamous epithelium) contain many dead cells that are colonized with bacteria of low virulence called *Staphylococcus epidermidis*. This part of the skin may also contain small numbers of the more virulent *Staphylococcus aureus*. Low numbers of *Streptococcus* and some gram-negative bacilli may also be present. Gram-positive bacilli, referred to as *diphtheroids*, are of very low virulence and are widespread as a normal inhabitant of the skin. An anaerobic bacterium called *Propionibacterium acnes* inhabits the *sebaceous glands* and is not destroyed by most methods used to disinfect the skin. Fortunately, it is a very low-grade pathogen, but it may work *synergistically* with other bacteria in the development of acne.

Respiratory Tract.   The *nares* (nose) may contain large numbers of such bacteria as *Staphylococcus epidermidis*, *S. aureus*, and diphtheroids. A gram-negative coccus called *Branhamella catarrhalis* and an opportunistic pathogen called *Haemophilus influenzae* are periodically present. Large numbers of viridans streptococci, diphtheroids and *B. catarrhalis* also inhabit the pharynx. The pathogenic bacteria *Streptococcus pyogenes*, *Streptococcus pneumoniae*, *Neisseria meningitidis*, and *H. influenzae* may periodically be carried in the pharynx. As a result of effective defense mechanisms that rapidly remove contaminating microorganisms, the normal trachea, bronchi, bronchioles and alveoli are usually free of microorganisms ∞ (Chapter 10, p. 143).

Mouth.   The mouth contains large numbers of diverse types of bacteria, and marked variations, kinds, and numbers are associated with the presence or absence of teeth or of tooth decay. Some bacteria are of low virulence but may cause troublesome infections if introduced into deeper tissues—for example, by a bite. Other bacteria are able to cause tooth decay, and many others are nonvirulent. Several types of microorganisms commonly found in the mouth are listed in Table 13-1. When clinical specimens are being collected from the throat or lower respiratory tract, care should be taken to minimize contamination of the swab or aspirate with oral secretions, because the large number of bacteria

---

Normal microbial flora   Microorganisms normally found on or in the body that do not cause disease in the normal host.
Synergistically   Capable of working together. Two organisms are synergistic when they are able to produce a host response greater than the sum of the effects they produce when acting alone.

TABLE 13-1

Genera of Bacteria Most Commonly Found in the Human Oral Cavity

Anaerobic	Facultative Anaerobic
Actinomyces	Haemophilus
Bacteroides	Lactobacillus
Peptostreptococcus	Neisseria
Veillonella	Staphylococcus
	Streptococcus

in the mouth are able to obscure the pathogen being sought.

Stomach.   The high acid content of the human stomach keeps the number of viable microorganisms in this organ low. Recent research has shown that one bacterial species, called *Helicobacter pylori* colonizes the crypts of the stomach where they are protected from the stomach acid; these bacteria cause stomach ulcers and are technically not normal flora. Various microbes, however, are able to survive passage through the stomach.

Intestines.   The microbial content of the small intestines changes drastically from relatively few transient bacteria in the upper portion to massive numbers in the lower section. The contents of the colon provide an ideal environment for the growth of many species of microorganisms. It has been estimated that over 100 different species of bacteria may be common inhabitants of the colon. Their concentration may reach 10^{11} cells per gram of fecal material. This represents about a third of fecal weight. Intestinal flora become established early in life and generally humans live in healthy balance with these microorganisms. Such microorganisms may offer several benefits to humans by producing certain vitamins, aiding in food digestion, and stimulating the immune responses of the host. These masses of common bacteria may also help suppress the growth or block the attachment of pathogens that might enter the intestinal tract.

The common bacteria of the colon can be categorized either as *facultative* or *obligate* anaerobes ∞ (Chapter 7, p. 99). The facultative anaerobes

include the gram-negative bacilli commonly called the enteric bacilli, which include such genera as *Escherichia*, *Klebsiella*, *Enterobacter*, and *Proteus*. Also present are staphylococci and streptococci. The yeast *Candida albicans* is frequently present. For many years it was assumed that the facultative anaerobes were the major inhabitants of the colon. However, the use of improved anaerobic culture methods has shown that over 99% of the colon bacteria are obligate anaerobes. They include various species of the genera *Bacteroides*, *Fusobacterium*, *Clostridium*, *Peptostreptococcus*, *Eubacterium*, and *Streptococcus*.

Genitourinary Tract.   Some bacteria are found in the lower portion of both the male and female urethra. Normal bladders, ureters, and kidneys are free of microorganisms. The female genital tract has a complex and varying microbial flora. With *menarche* the vaginal and cervical tissues become populated with lactobacilli that produce lactic acid and maintain the pH of these tissues at 4.4 to 4.6. This acid environment inhibits the growth of gram-negative enteric bacteria but allows growth of such microorganisms as bacteroides, diphtheroids, staphylococci, enterococci, and *Candida albicans*. The microflora of the vaginal canal undergo some cyclic fluctuations with hormonal variation.

Internal organs, cavities, and fluids—for example, the heart, liver, spleen, brain, peritoneal cavity, blood, lymph, and spinal fluid—of healthy individuals are normally free of microorganisms.

TRANSMISSION OF MICROORGANISMS

Most microbial diseases are categorized as **communicable** (infectious), which means that the microbe itself must be transmitted to the host. The source of the microbe may be another host or some nonliving *disease reservoir*. To be a successful pathogen of a communicable disease, a microorganism must be able to

1. Survive passage from one host to another or from the reservoir to the host
2. Attach to or penetrate the host's tissues
3. Withstand (for a period of time) the host's defense mechanisms

---

Communicable   Able to be transmitted between hosts.
Disease reservoir   A natural source of disease agent. Such reservoirs may be sick patients, asymptomatic carriers, animals, recovered patients, or environmental sources.

4. Induce damage to or malfunction of the host's tissues

✳ Exit of Microorganisms from the Host

Microorganisms found in the mouth and respiratory tract are expelled to some extent during normal speech and breathing. Singing and shouting expel larger numbers and coughing and sneezing expel massive numbers. Many microorganisms are dispersed (*aerosolized*) on small airborne bits of mucus and saliva. The moisture evaporates within a very brief period and the remaining particle is called a *droplet nucleus*. These microbe-laden particles may remain suspended in air currents to be carried to new hosts. In addition to being aerosolized, saliva may serve as a *vehicle* for microbial transmissions by kissing or *expectoration*. Bacteria attached to the skin are continually being shed from the body. Massive numbers of microorganisms are present in feces and are readily spread to new hosts living under conditions of poor sanitation. Urine and other secretions of the urinogenital tract may contain some microorganisms, but host-to-host transmission from these tissues usually results only from direct contact. Blood from a healthy person is free of microorganisms; however, blood from persons with certain diseases contains pathogenic microorganisms that may be taken up and transmitted by bloodsucking *arthropods*, by blood transfusions, by accidental contact with blood or blood products, or by sharing blood contaminated needles or instruments. Milk may act as a vehicle for the microorganisms shed from a lactating female with infection of the mammary gland. The routes of exit of microorganisms from the body are shown in Figure 13-3.

✳ Routes of Transmission and Entry of Microorganisms Into the Host

Most microorganisms (normal flora) shed from a healthy person will be the same as those found in persons about them; thus the exchange of these microorganisms is of little consequence. If a person is a carrier of a pathogen or has an active infection, however, the spread of that person's microorganisms may cause infection in a new host. Some pathogens are transmitted to humans from animals or birds and, in the case of some fungal infections, the infectious spores are carried from the soil to humans by airborne route. Various modes of transmission are shown in Figure 13-4 and possible routes of microorganism entry into the body are shown in Figure 13-5.

Airborne Transmission. The majority of the infectious diseases of humans in developed countries are of the respiratory tract and most are transmitted through the air. Microorganisms are vigorously aerosolized due to the increased coughing, sneezing, and secretion of mucus associated with a respiratory infection. Many aerosolized microbes become associated with droplet nuclei, whereas others are deposited on such surfaces as floors, clothing, and bedding. The microorganisms present on various objects may become aerosolized on dust particles due to physical movement of air currents and may be inhaled by persons in the area. This form of transmission is most efficient in enclosed, crowded buildings and is an important source of infections in hospitals. Airborne bacteria may also cause infections by settling onto open wounds during surgery or when bandages are changed.

Mouth. Many microorganisms enter the body by the ingestion of contaminated foods or water, by kissing, or by placing contaminated objects in the mouth.

Bites. A significant number of infectious agents are transmitted from host to host by biting insects. Some insects transmit the pathogen mechanically by first feeding on an infected host and then feeding on a new host with its contaminated mouthparts. However, in most instances, after the insect has ingested contaminated fluids from an infected host, the pathogen proliferates in the insect. When such an insect bites a new host, large numbers of the pathogen may be inoculated into the body tissue.

Some infections, such as rabies, are transmitted by the bites of mammals. Because of the large and varied numbers of microorganisms in the mouth, troublesome infection may result from bites, particularly human bites.

Aerosolized   Dispersed particles suspended in air.
Vehicle   Any object or substance that can carry microorganisms from one host to another.
Expectoration   Saliva and other fluids in the mouth that are expelled by spitting.
Arthropods   A large group of invertebrate animals, many of which have biting or sucking mouthparts, such as the mosquito and other insects.

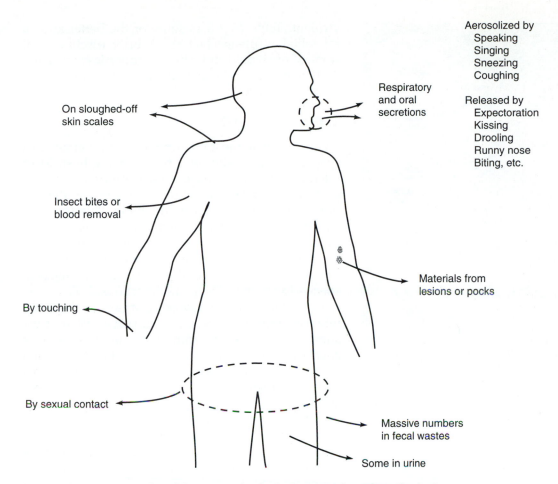

FIGURE 13-3 Possible routes of exit of microorganisms from the body.

Contact. Transmission by contact merits further explanation because of its significance in routine medical care. *Direct contact* entails the touching of infected tissues with uninfected tissues. Some pathogens require intimate contact for their successful transmission and are most effectively passed from person to person by sexual contact or kissing. A person with an infected skin lesion may readily transmit the infection by direct contact. This mode of transmission is of special concern to medical personnel, for they may inadvertently transmit such an infection to patients; conversely, a patient may transmit an infection to attending personnel.

Indirect contact requires intermediate objects, called *fomites*, to transmit the pathogen. The fomites become contaminated by coming in contact with pathogens from an infected patient. Then at some later time the pathogen may be transferred to a second person who comes in contact with the contaminated fomite. In medical practice, such items as catheters, clothing, instruments, toys, and books may serve as fomites in transferring infections between patients. Use of disposable items, proper handwashing, and gloving, gowning, and similar procedures used to prevent transmission of infections in hospitals or clinics should receive continual emphasis. Water and food are the most commonly involved fomites in our everyday environment.

Endogenous Spread. The preceding examples describe the *exogenous* spread of microbes—that is, microbes coming from a source outside the host. It is also possible for microbes to spread from one part of a host to another, a process called *endogenous* spread. Many opportunistic pathogens found in the upper respiratory tract, mouth, or intestinal tract may be spread to open lesions or other tissues and cause serious infections. Some common examples are the spreading of intestinal bacteria into the urethra to cause a urinary tract infection, the "leakage" of intestinal bacteria across the intestinal mucosa into the blood and lymphatic systems, and the spread of oral microbes to eyes by moistening contact lenses with saliva or to wounds by licking.

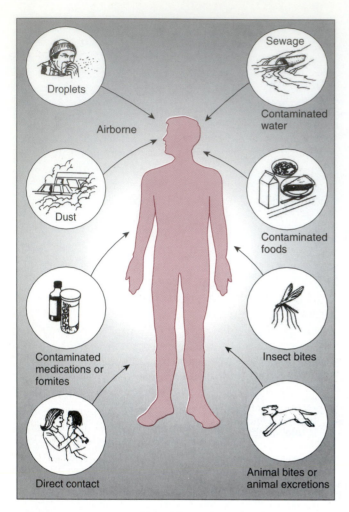

FIGURE 13-4 Possible routes of transmission of microorganisms to humans.

Possible routes of endogenous spread of microorganisms are shown in Figure 13-6.

MICROBIAL PATHOGENICITY AND VIRULENCE

Microbes possess a variety of virulence mechanisms that enable them to produce disease. In some cases, the mechanisms of pathogenicity and virulence are well characterized; at other times the processes by which a microbe causes disease are not well understood. Most bacteria use one or more of a group of virulence factors to produce disease. Those that have many of these properties are usually highly

virulent. Table 13-2 lists some of the better-known microbial virulence factors. A brief discussion of some of the better-understood properties associated with microbial virulence are presented in the following paragraphs.

✳ Attachment

To infect a host, most microbes must first attach to a specific receptor site on a tissue of the host. Most microbes that lack the chemical groups that take part in this specific attachment process are flushed or otherwise expelled from the body. This specificity, which is also called tissue *trophism*, is determined by traits of both the host and the pathogen. The specificity of receptor sites may vary from organ to organ within the body of the host and from one host species to another. Thus, one infectious disease may involve only specific tissues or organs of the body, whereas another disease may involve different tissues. Similarly, some animal species that possess *attachment sites* are highly susceptible to a given pathogen whereas another species that lacks such a site is completely refractory to that same pathogen. The significance of specific attachment sites for the initiation of viral infections has been recognized for many years. Today it is known that specific tissue attachment is also important for the initiation of bacterial infections. Microbial attachment mechanisms include pili and specialized surface proteins ∞ (Chapter 4, p. 41).

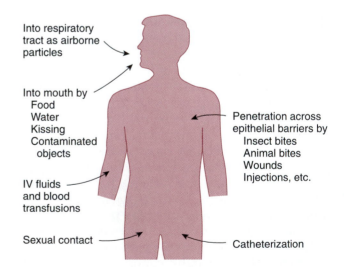

FIGURE 13-5 Possible routes of entry of microorganisms into the body.

Attachment site The location at which an organism attaches itself to host tissues. This is usually determined by specific receptor sites on the host cell and pathogen.

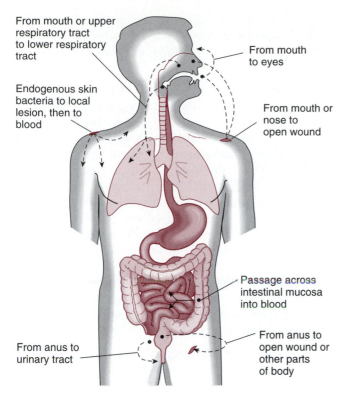

FIGURE 13-6 Possible routes of endogenous spread of microorganisms.

✳ Circumvention of Defense Mechanisms

Various mechanisms present in some bacteria increase their survival time in the host. One such mechanism is the ability to decrease the rate at which pathogens are phagocytized. This is commonly due to the presence of a capsular substance on the surface of the bacterium. Many bacteria possess *capsules* ∞ (Chapter 4, p. 39) that block the attachment of the bacterial cell to phagocytic cells and thus interfere with phagocytosis. Another method used by some bacteria to decrease phagocytosis by host cells is destruction of phagocytes. Some pathogenic strains of staphylococci and streptococci secrete a substance called *leukocidin* that kills white blood cells before they can phagocytize the bacteria. Certain microbes, known as intracellular parasites, are readily phagocytized but are able to withstand the destructive mechanisms functioning in nonstimulated macrophages. Intracellular parasites not only survive and continue to multiply while inside the phagocyte, they may also be transported by the phagocyte to other tissues while being protected from antibodies and other antimicrobial substances in the body fluids. Bacteria may use one or more virulence mechanisms to insure intracellular survival. They may escape from the phagosome before it fuses with a lysosome. By so doing, they avoid the destructive components of the lysosome. Bacteria may also force phagocytosis by nonprofessional phagocytes. Once inside these cells they are free from destruction by professional phagocytes such as white blood cells or macrophages. Nonprofessional phagocytes usually lack lysosomes and are unable to kill the bacteria that have forced their way into the cell.

An enzyme called *hyaluronidase* that breaks down hyaluronic acid is secreted by some pathogens. Hyaluronic acid is a substance that binds host cells

TABLE 13-2		
Some Bacterial Virulence Factors		
Virulence Factor	**Characteristics**	**Bacteria or Example**
Toxin	Endotoxin, lipopolysaccharide gram-negative cell wall, heat-stable, pyrogenic	Gram-negative bacteria
	Exotoxin, polypeptide, antigenic, highly toxic, toxoids formed	Diphtheria toxin Staphylococcal enterotoxin Tetanus neurotoxin
Enzyme	Protein, specific activity	Collagenase: *Clostridium* Coagulase: *Staphylococcus* Hyaluronidase: *Streptococcus*
Capsule	Polysaccharide or polypeptide, antiphagocytic	*Haemophilus influenzae* *Streptococcus pneumoniae* *Bacillus anthracis*
Pili/fimbriae	Attachment to host cells	*Neisseria gonorrhoeae*

together, therefore the production of hyaluronidase may aid the spread of the bacteria between cells of a tissue. Some bacteria possess enzymes called *fibrinolysins* that dissolve the fibrin clots that form part of the barrier of the inflammatory response ∞ (Chapter 10, p. 150). Such bacteria are not readily contained by the inflammatory reaction.

Some bacteria circumvent host defenses by variation of their antigenic appearance. They may produce a sequence of antigens, which makes it difficult for the host to produce effective antibody for protection. Others, such as *Staphylococcus aureus*, may coat themselves with host proteins that make the bacteria invisible to the host antigen–antibody–complement protection system. *Staphylococcus aureus* possess a surface protein called *protein A*. Protein A interacts with the constant region of IgG molecules, thereby pointing the reactive (variable) ends of the antibody molecules away from the bacterium, which minimizes the effects of antibody immunity against these staphylococci.

✳ Tissue Damage

Stimulation of Inflammatory Responses.
Certain bacteria stimulate vigorous inflammatory responses that, in turn, alter the function of the normal host tissues. *Streptococcus pneumoniae*, for example, a major cause of bacterial pneumonia, induces an acute inflammatory response in the lungs that results in a rapid accumulation of fluid *exudate*. These fluids interfere with the normal functions of the lungs. The tubercle bacillus induces a strong cell-mediated immune response that results in the formation of nodules of scar tissue and eventually causes the destruction of normal lung tissue.

Some bacterial toxins, such as the toxin responsible for a *syndrome* called toxic *shock*, stimulate the host immune system to such an extent that nearly all the symptoms of the disease are a direct result of the overactive immune response. These toxins, known as *superantigens*, cause large numbers of T cells to respond with the production of cytokines such as interleukin-2 and tissue necrosis factor alpha. These compounds, when produced in excess, cause nausea, vomiting, malaise, fever, and shock in the host.

Secretion of Enzymes.
Various enzymes secreted by certain pathogenic bacteria are thought to contribute to the disease process by destroying tissues. *Collagenase* and *lecithinase* are enzymes produced by bacteria that cause gas *gangrene*. These enzymes break down tissue fibers and cell membranes and probably contribute to the cell destruction associated with gangrene. *Hemolysins* are enzymes that destroy red blood cells by making pores or channels in the cell membrane that permit the contents of the cell to leak out. They are produced by several different pathogenic bacteria. The direct contribution of hemolysin, lipase, or nuclease, if any, to pathogenicity is not known. *Lipases* and *nucleases* are bacterial enzymes that break down fats and nucleic acids.

Production of Toxins.
Some microbes produce toxins that are important mechanisms in the production of diseases. There are two general categories of toxins, **exotoxins** and **endotoxins**. Exotoxins are usually proteins that are secreted from the bacterial cell into the surrounding medium. Endotoxins are part of the gram-negative cell wall and only small amounts may escape into surrounding fluids from living bacteria. Greater amounts of endotoxin are released when the bacteria die and their cell walls disintegrate. The two types of toxins differ markedly in potency and functions, as noted in Table 13-3.

Exotoxins are extremely powerful biological poisons. The botulism food-poisoning toxin is the most powerful chemical toxin known, and a similar toxin produced by *Clostridium tetani* has a lethal dose for a human of less than 1 μg of purified tetanus toxin. Put in other terms, 1 gram of toxin would be sufficient to kill more than a million people. The action of exotoxins on the host is highly specific, often involving a single essential chemical reaction in the

---

Exudate   A secretion from vessels that collects in body tissues and spaces, or from the tissue that is discharged outside the body.

Syndrome   A number of symptoms occurring together that characterize a specific disease.

Shock   A marked, sudden depression of physiologic function, such as blood pressure, that is often life-threatening.

Superantigen   An antigen that stimulates an extreme immune response.

Gangrene   Tissue death due to the loss of a blood supply.

Exotoxin   A soluble toxin often secreted into the environment, including host tissue.

Endotoxin   A toxin derived from the cell wall of gram-negative bacteria.

TABLE 13-3

Comparison of Characteristics of Exotoxins and Endotoxins

Characteristics	Exotoxins	Endotoxins
Potency	High	Low
Effects on Cells	Specific	Nonspecific
Stability to heat	Labile (inactivated at 60 to 80°C)	Stable (resists 120°C for 1h)
Forms toxoids	Yes	No
Composition	Proteins	lipopolysaccharide

host's tissues. The botulism and tetanus toxins specifically interfere with the transmission of nerve impulses, and diphtheria toxin specifically inhibits protein synthesis by preventing the elongation of the polypeptide chain during the process of translation ∞ (Chapter 6, p. 72). The actions of the exotoxins are covered in greater detail later in this text when the specific diseases are discussed. Exotoxins can be converted into toxoids that are used as vaccines to stimulate specific antitoxin immunity against the toxin. These antitoxins readily neutralize the toxins and play an important role in recovery, treatment, and immunity against diseases caused by exotoxins. Fortunately, most exotoxins are readily destroyed by heat, an important factor in reducing the number of outbreaks of botulism food poisoning.

Endotoxins possess characteristics that differ considerably from those of the exotoxins. Endotoxins are less potent and larger amounts are needed to induce disease symptoms. Their effects on the host are general, producing such clinical signs and symptoms as fever, diarrhea, and circulatory disturbances, including shock. Endotoxins are composed of complexes of polysaccharides and phospholipids. They cannot be converted into toxoids and are highly resistant to inactivation by heat. The endotoxins are found primarily in the cell walls of gram-negative bacteria, whereas exotoxins are produced by both gram-positive and gram-negative bacteria.

Exotoxins can cause disease even when the producing bacterium is restricted to a relatively superficial tissue, such as mucosal epithelium. Or, as in the case of food poisoning, exotoxin production may occur remote from the host. In contrast, endotoxins generally affect the host when they are released following the death and destruction of large numbers of gram-negative bacilli that may be infecting the tissues. Endotoxins undoubtedly con-

tribute significantly to the pathogenesis of many diseases caused by gram-negative bacteria. It is possible to reproduce some disease symptoms in a host by injecting purified endotoxins. Occasionally, endotoxin-contaminated intravenous fluids have inadvertently been infused into a patient, resulting in serious reactions, including shock. This situation can even occur in fluids that have been sterilized in an autoclave, for endotoxins are not readily destroyed by heat.

DETERMINING ETIOLOGY

The mere presence of a microorganism in a lesion or diseased tissue is not sufficient to prove that the microbe is the cause (the *etiologic* agent) of the disease. This problem was recognized early in the study of infectious diseases by Robert Koch, who developed criteria to be used to incriminate a suspected microbe as the etiologic agent of a given disease ∞ (Chapter 1, p. 8). These criteria, known as *Koch's postulates*, are as follows:

1. The suspected microorganism must be found routinely in hosts with the disease.
2. The microorganism must be isolated from the host and grown in pure culture.
3. When microbes from the pure culture are inoculated into a healthy, susceptible host, they must be able to cause the same disease.
4. The microorganism must next be recovered from the experimentally infected host.

Koch's postulates have served as useful criteria in establishing the etiology of many distinct infectious diseases. Nevertheless, the etiologic role of certain microbes in more subtle diseases—particularly those caused by the synergistic reaction of several microbes and those diseases occurring in compromised hosts caused by opportunist microbes—has not been as readily demonstrated by Koch's postulates.

HOST FACTORS

The advances of modern medicine have had profound effects on host–parasite relationships. The great plagues of the past, such as cholera, diphtheria, dysentery, and tuberculosis, have largely yielded to the efforts of medical science. Improvements in diagnosis, therapy, and prevention have relegated many former leading causes of death, such as smallpox, yellow fever, and plague, to historical interest. Not that some of these diseases do not still occur; they do, and they are serious infections, but the frequency of occurrence is rare and the outcome is often favorable. The question might be asked: If science has had such great success in preventing and curing microbial disease, what need is there for a continuing interest in medical microbiology?

The answer to this question is not so much in terms of the microbe, as in terms of the human host. Today, by far, the majority of serious infectious diseases are **nosocomial** (hospital-associated). These diseases often occur in patients who have, by past medical standards, undergone heroic medical procedures. While such procedures have preserved the lives of many patients, they have also left the patients highly vulnerable to infections by the microflora of the hospital environment. Such infections (pneumonia, wound and urinary tract infection, and *bacteremia*) would likely not have occurred without attendant compromise of host defenses. In addition to nosocomial infections, a variety of community-acquired infectious diseases have developed that are based on changes in the normal host immune state; these include diseases such as pneumocystis pneumonia, Legionnaires' disease, AIDS, *Mycobacterium avium* infections, and diarrhea due to *Cryptosporidium*. These appear to be the plagues of the present and future, filling the void left by their departed predecessors. We again turn to the science of microbiology in hopes of finding a cure.

STUDY OF INFECTIOUS DISEASES

The following chapters present a general view of many diseases of humans that are caused by microorganisms. The various diseases are discussed according to the taxonomic categories of the causative microorganism. This approach provides a concise coverage of the subject. Diseases caused by bacteria are described first, followed by viral, fungal, and protozoal diseases. The section on bacterial diseases starts with diseases caused by the gram-positive cocci, followed by the diseases caused by gram-negative cocci, spirochetes, acid-fast bacilli, and gram-positive and gram-negative bacilli. DNA-containing viruses appear first in the section on viral diseases and are followed by the RNA-containing viruses. In the discussions of infectious diseases, emphasis is given to the mechanisms of disease production (pathogenesis), clinical characteristics, and modes of transmission, treatment, and prevention—that is, those concepts of greatest concern to paramedical personnel who are directly involved in patient care. Less emphasis is given to the detailed methods used by microbiologists in the diagnostic laboratory.

Most of the following chapters cover the diseases caused by a single genus or by a group of related microorganisms. The coverage of most diseases includes the following general sections:

1. *Causative microbe.* Contains a brief description of some general morphological and physiological properties of the microorganisms, along with information on classification when appropriate.
2. *Pathogenesis and clinical diseases.* Includes a discussion of the mechanisms whereby the microbe is able to cause diseases and the types of clinical diseases resulting from the infection.
3. *Transmission and epidemiology.* The mode of transmission and the relationship of various factors in influencing the distribution and frequency of diseases in various human populations are discussed.
4. *Diagnosis.* The general procedures used to identify a given disease in a patient are described.
5. *Treatment.* The responses of the infection to the various chemotherapeutic agents, antitoxins, or other medications are discussed.
6. *Prevention and control.* The various methods used to prevent or reduce the amount of contact with infectious agents are described. The use of vaccines and preventive (prophylactic) treatments where applicable is also discussed.
7. *Clinical notes.* Following the discussion of many topics throughout the text, a report on an actu-

---

Nosocomial   Relating to a hospital.
Bacteremia   Presence of dividing bacteria in the blood.

al occurrence of a specific medical problem is given. These notes are largely taken from the U.S. Department of Health and Human Services publication called the *Morbidity and Mortality Weekly Report* (*MMWR*). This document is published weekly by the Centers for Disease Control (CDC) in Atlanta, Georgia. Along with the selected case reports, *MMWR* gives a weekly update on the rate of various diseases in the different states. The clinical notes presented in this textbook were selected to demonstrate current problems encountered

with infectious diseases. They are also selected to help amplify and show applications of some concepts related to the epidemiology, treatment, diagnosis, or control of these diseases. In some cases, the reports have been modified from the original in order to make them more understandable to the introductory student or simply to reduce their length. Some of the clinical notes in the early chapters were written by the authors and relate to clinical application of microbial concepts and practice.

MATERIAL FOR REVIEW

CONCEPT SUMMARY

1. The interaction between a host such as a human and the microbial world can best be viewed as a seesaw where the fulcrum is moved in favor of one or the other as the balance of normality is altered. Thus, classic nonpathogens can become pathogens to a host suffering from environmental pressures, whereas a virulent microorganism may at best coexist with a host that is well adapted to its environment.

2. All life-forms, humans included, are endowed with a contingent of microorganisms collectively referred to as normal flora. These organisms, although capable of producing disease, frequently provide a beneficial component to life and participate in maintaining a normal environment for their hosts.

3. Microorganisms are transferred from one host to another through a variety of mechanisms, including food and water, human or animal bites, contact, and aerosols.

4. Characteristics that endow a microorganism with virulence and facilitate its survival in the host, leading to pathogenicity, are those characteristics that

permit the organism to withstand the host defenses, maintain residence in the host, and produce factors damaging to host tissue.

5. Bacterial virulence factors include the production of exotoxins, endotoxins, and enzymes. These substances circumvent the host defense mechanisms and produce tissue damage, resulting in disease.

6. The application of Koch's postulates has resulted in the determination of the etiology of infectious diseases.

STUDY QUESTIONS

1. Distinguish between the following terms: (a) pathogen and nonpathogen, (b) virulent and avirulent, (c) infection and disease, (d) parasite and saprophyte, and (e) carrier and case.

2. What roles (positive or negative) are played by normal microbial flora?

3. Why are the human mucosal surfaces inhabited by such large numbers of "normal flora"?

4. The use of public health and personal sanitation methods have had their greatest effect on reduction of infectious agents that are transmitted by what means?

5. Because of proximity requirements, agents of disease that are transmitted by direct contact are somewhat unique. Control of these infections seems to depend less on public health procedures and more on _____?

6. Prepare a list of characteristics that are known to be responsi-

ble for microbial virulence.

7. Why are antitoxins effective in protecting against exotoxic diseases, but have little role in pro-

tection against endotoxin?

8. Write a paragraph that explains the rationale for acceptance of Koch's postulates.

9. Describe the relationships between a potential host and its microbial environment.

CHALLENGE QUESTIONS

1. Discuss how a pathogen is transmitted by each of the following: (a) air, (b) arthropods, (c) fomites, and (d) expectoration.

2. Explain how you would go about showing that a species of *Streptococcus* causes strep throat.

Human Body Systems Reference

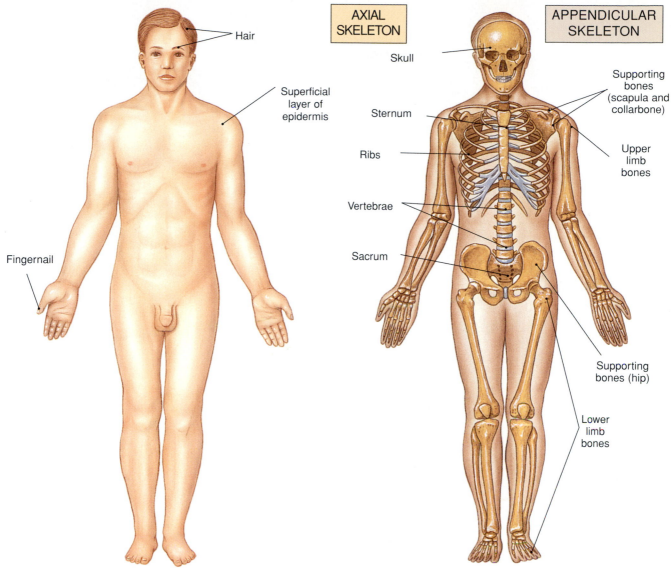

AXIAL SKELETON

APPENDICULAR SKELETON

Hair

Superficial layer of epidermis

Fingernail

Skull

Sternum

Ribs

Vertebrae

Sacrum

Supporting bones (scapula and collarbone)

Upper limb bones

Supporting bones (hip)

Lower limb bones

(a) The Integumentary System

(b) The Skeletal System

Organ	Primary Functions
EPIDERMIS	Covers surface, protects underlying tissues
DERMIS	Nourishes epidermis, provides strength, contains glands
HAIR FOLLICLES	Produce hair
Hairs	Provide sensation, provide some protection for head
Sebaceous Glands	Secrete lipid coating that lubricates hair shaft
SWEAT GLANDS	Produce perspiration for evaporative cooling
NAILS	Protect and stiffen distal tips of digits
SENSORY RECEPTORS	Provide sensations of touch, pressure, temperature, pain
SUBCUTANEOUS LAYER	Stores lipids, attaches skin to deeper structures

Organ	Primary Functions
BONES (206), CARTILAGES, AND LIGAMENTS	Support, protect soft tissues; store minerals
Axial skeleton (Skull, vertebrae, sacrum, ribs, sternum)	Protects brain, spinal cord, sense organs, and soft tissues of chest cavity; supports the body weight over the lower limbs
Appendicular skeleton (Limbs and supporting bones)	Provides internal support and positioning of limbs; supports and moves axial skeleton
BONE MARROW	Primary site of blood cell production

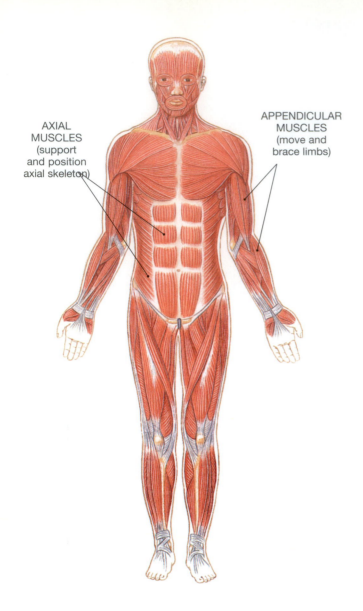

AXIAL
MUSCLES
(support
and position
axial skeleton)

APPENDICULAR
MUSCLES
(move and
brace limbs)

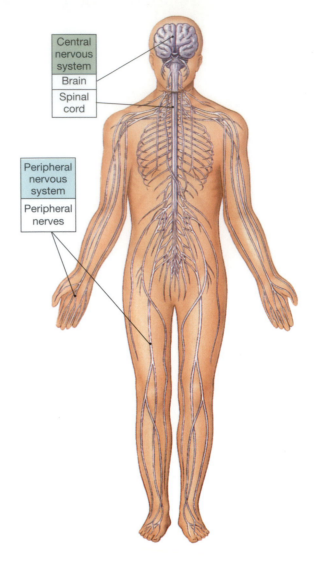

Central
nervous
system

Brain

Spinal
cord

Peripheral
nervous
system

Peripheral
nerves

(c)   The Muscular System

Organ	Primary Functions
SKELETAL MUSCLES (700)	Provide skeletal movement, control entrances and exits of digestive tract, produce heat, support skeletal position, protect soft tissues
TENDONS, APONEUROSES	Harness forces of contraction to perform specific tasks

(d)   The Nervous System

Organ	Primary Functions
CENTRAL NERVOUS SYSTEM (CNS)	Control center for nervous system: processes information, provides short-term control over activities of other systems
Brain	Performs complex integrative functions, controls voluntary activities
Spinal cord	Relays information to and from the brain, performs less complex integrative functions; directs many simple involuntary activities
PERIPHERAL NERVOUS SYSTEM (PNS)	Links CNS with other systems and with sense organs

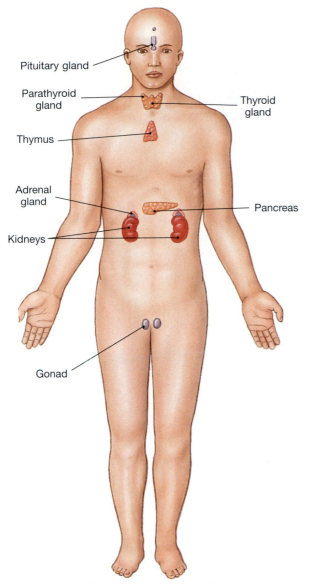

Pituitary gland

Parathyroid gland

Thyroid gland

Thymus

Adrenal gland

Pancreas

Kidneys

Gonad

(e) The Endocrine System

Organ	Primary Functions
PITUITARY GLAND	Controls other glands, regulates growth and fluid balance
THYROID GLAND	Controls tissue metabolic rate and regulates calcium levels
PARATHYROID GLAND	Regulates calcium levels (with thyroid)
THYMUS	Controls lymphocyte maturation
ADRENAL GLANDS	Adjust water balance, tissue metabolism, cardiovascular and respiratory activity
KIDNEYS	Control red blood cell production and elevate blood pressure
PANCREAS	Regulates blood glucose levels
GONADS	
Testes	Support male sexual characteristics and reproductive functions *Figure k*
Ovaries	Support female sexual characteristics and reproductive functions *Figure k*

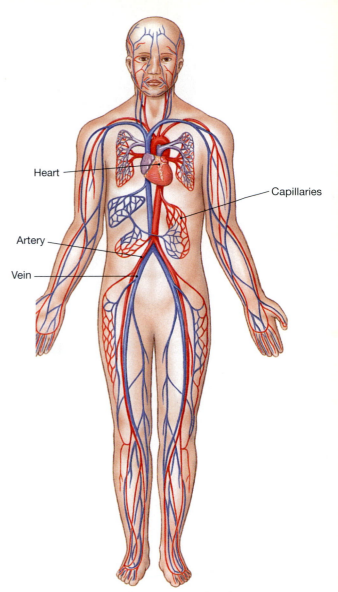

Heart

Capillaries

Artery

Vein

(f) The Cardiovascular System

Organ	Primary Functions
HEART	Propels blood, maintains blood pressure
BLOOD VESSELS	Distribute blood around the body
Arteries	Carry blood from heart to capillaries
Capillaries	Site of diffusion between blood and interstitial fluids
Veins	Return blood from capillaries to the heart
BLOOD	Transports oxygen and carbon dioxide, delivers nutrients, removes waste products, assists in defense against disease

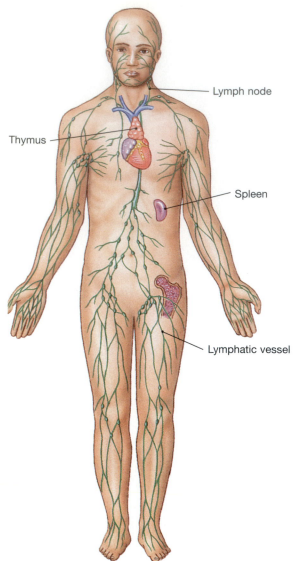

(g) **The Lymphatic System**

Organ	Primary Functions
LYMPHATIC VESSELS	Carry lymph (water and proteins) from peripheral tissues to the veins of the cardiovascular system
LYMPH NODES	Monitor the composition of lymph, engulf pathogens, stimulate immune response
SPLEEN	Monitors circulating blood, engulfs pathogens, stimulates immune response
THYMUS	Controls development and maintenance of one class of lymphocytes (T cells); immature lymphocytes and other blood cells are produced in the bone marrow

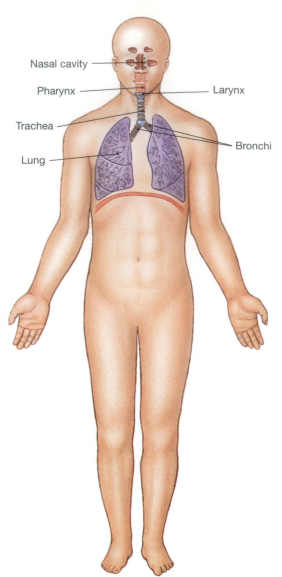

(h) **The Respiratory System**

Organ	Primary Functions
NASAL CAVITIES	Filter, warm, humidify air; detect smells
PHARYNX	Chamber shared with digestive tract; conducts air to larynx
LARYNX	Protects opening to trachea and contains vocal cords
TRACHEA	Traps particles in mucus; cartilages keep airway open
BRONCHI	Same as trachea
LUNGS	Include airways and alveoli; volume changes responsible for air movement
ALVEOLI	Sites of gas exchange between air and blood

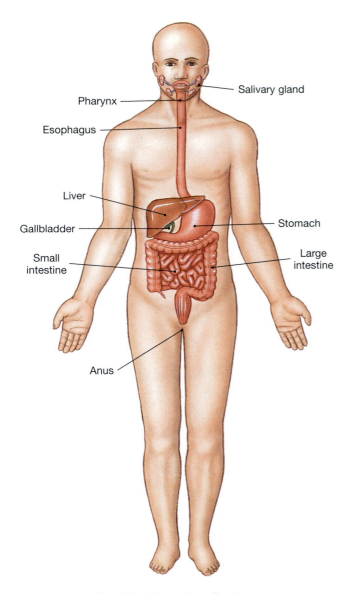

Pharynx
Salivary gland
Esophagus
Liver
Gallbladder
Stomach
Small intestine
Large intestine
Anus

(i) The Digestive System

Organ	Primary Functions
SALIVARY GLANDS	Provide lubrication, produce buffers and enzymes that begin digestion
PHARYNX	Passageway connected to esophagus
ESOPHAGUS	Delivers food to stomach
STOMACH	Secretes acids and enzymes
SMALL INTESTINE	Secretes digestive enzymes, absorbs nutrients
LIVER	Secretes bile, regulates blood composition of nutrients
GALL BLADDER	Stores bile for release into small intestine
PANCREAS	Secretes digestive enzymes and buffers; contains endocrine cells *Figure e*
LARGE INTESTINE	Removes water from fecal material, stores wastes

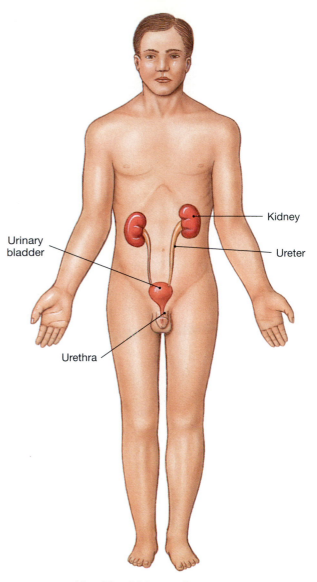

Kidney
Urinary bladder
Ureter
Urethra

(j) The Urinary System

Organ	Primary Functions
KIDNEYS	Form and concentrate urine, regulate blood pH and ion concentrations; endocrine functions noted in *Figure e*
URETERS	Conduct urine from kidneys to urinary bladder
URINARY BLADDER	Stores urine for eventual elimination
URETHRA	Conducts urine to exterior

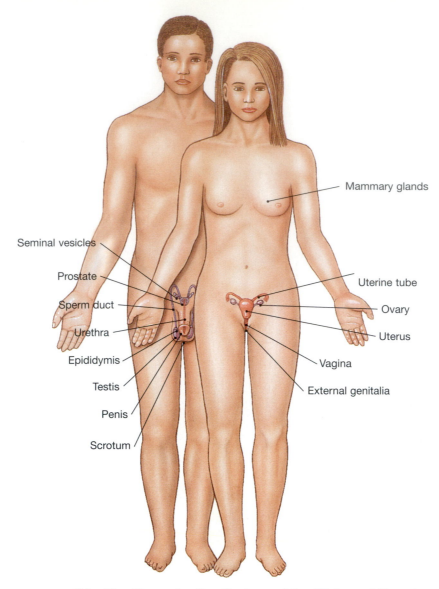

Mammary glands

Seminal vesicles

Prostate

Sperm duct

Urethra

Epididymis

Testis

Penis

Scrotum

Uterine tube

Ovary

Uterus

Vagina

External genitalia

(k)   The Reproductive System of the Male and Female

Organ	Primary Functions
TESTES	Produce sperm and hormones *Figure e*
ACCESSORY ORGANS	
Epididymis	Site of sperm maturation
Ductus deferens (sperm duct)	Conducts sperm between epididymis and prostate
Seminal vesicles	Secrete fluid that makes up much of the volume of semen
Prostate	Secretes buffers and fluid
Urethra	Conducts semen to exterior
EXTERNAL GENITALIA	
Penis	Erectile organ used to deposit sperm in the vagina of a female; produces pleasurable sensations during sexual act
Scrotum	Surrounds and controls temperature of the testes

Organ	Primary Functions
OVARIES	Produce ova (eggs) and hormones *Figure e*
UTERINE TUBES	Deliver ova or embryo to uterus; normal site of fertilization
UTERUS	Site of embryonic development and diffusion between maternal and embryonic bloodstreams
VAGINA	Site of sperm deposition; birth canal at delivery; provides passage of fluids during menstruation
EXTERNAL GENITALIA	
Clitoris	Erectile organ, produces pleasurable sensations during sexual act
Labia	Contain glands that lubricate entrance to vagina
MAMMARY GLANDS	Produce milk that nourishes newborn infant

CHAPTER

14 Staphylococci

OUTLINE

Staphylococci are perhaps the best examples of bacteria that have great pathogenic potential, yet are able to live in symbiotic balance with their hosts. In spite of their ability to produce serious, life-threatening diseases, pathogenic staphylococci are present on the skin or mucous membranes of most humans. Generally, they act as opportunists, causing infections only in damaged tissues. These infections may be serious, and hospital-acquired staphylococcal diseases are a major problem. The prevention and control of these infections depend on the combined efforts of all hospital personnel.

BACTERIA

The spherical-shaped bacterium called *Staphylococcus aureus* is the causative agent of a wide variety of human infections. Many strains, with varying degrees of virulence, exist and are frequently carried on the skin, in the nose, and around the rectum of healthy persons. These bacterial cells are gram-positive, about 1 μm in diameter, and usually occur in grape-like clusters (Figure 14-1), a characteristic that provided the basis for their name (Greek *staphyle* = bunch of grapes). This bacterium is facultatively anaerobic, non-spore forming, and some strains have notable capsules. **Blood agar** is the

Blood agar   Agar to which blood was added before the agar solidified.

HUMAN BODY SYSTEMS AFFECTED

Staphylococci

S. aureus
* neurological infections
* meningitis

S. aureus
* osteomyelitis
* septic arthritis

S. aureus
* pneumonia

S. aureus
* endocarditis

S. aureus
* pyelonephritis

S. aureus
* food poisoning
* enterocolitis

S. aureus
* cellulitis
* boils and other pus-producing lesions
* impetigo
* scalded skin syndrome

S. aureus
* toxic shock syndrome

medium generally used for its isolation from infected tissues, but most common media will support its growth. *S. aureus* produces round, raised, opaque colonies that usually have a golden-yellow (aureus) color (Plate 4). The golden color is the result of the production of a lipid pigment contained in the organism. The morphology of staphylococci is shown in Figure 14-1. *S. aureus* is differentiated from other staphylococci by its ability to clot plasma; this species secretes an enzyme called *coagulase*, which activates the clotting mechanism of normal plasma. Staphylococci that do not secrete coagulase are of no or low virulence and are currently classified into about 26 species. The most ubiquitous coagulase-negative species is *S. epidermidis*, a common inhabitant on human skin. Two of

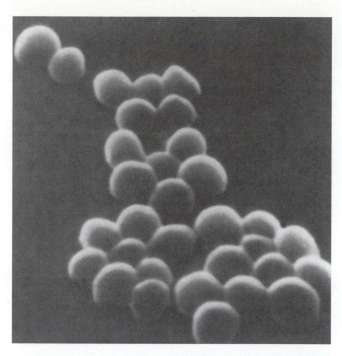

FIGURE 14-1 Scanning electron micrograph of a staphylococcus showing the characteristic cluster arrangement (magnification 8000×). (A. S. Klainer and I. Gies, *Agents of Bacterial Disease*. Harper & Row: Hagerstown, MD, 1973. Figure 12a, p. 18)

the coagulase-negative staphylococci, *S. epidermidis* and *S. saprophyticus*, are occasionally responsible for human disease.

STAPHYLOCOCCAL DISEASES

✱ Pathogenesis and Clinical Diseases

Several features contribute to the difficulty often encountered in controlling and treating staphylococcal infections. *S. aureus* is widespread, being found on the tissues of many healthy individuals; from 30 to 50% of the population at any given time (Figure 14-2). It is very stable, surviving for months when dried in pus or other body fluids, and it is more resistant to common disinfectants than most other *vegetative* bacteria. Genetic traits are readily

transferred between strains of *S. aureus* by plasmids and phages ∞ (Chapter 6, p. 76). This has led to the emergence of many strains that are resistant to commonly used antibiotics.

Although most strains of *S. aureus* are of relatively low virulence and usually harmless when restricted to the superficial layers of intact skin, they are often able to cause infection once they gain entry into damaged skin or deeper body tissues. The staphylococci are prime examples of *pyogenic* (pus-producing) bacteria. Infections due to these organisms are characterized by **suppuration** and localized inflammation. *Staphylococcus* is the first of four such groups of pyogenic cocci; *Streptococcus pyogenes, S. pneumoniae,* and *Neisseria* are discussed later. The pathogenicity of *S. aureus* appears to be associated with the production of various enzymes and toxins and includes such substances as hemolysins, coagulase, lipases, leukocidin, hyaluronidase, and a fibrinolysin. Some *S. aureus* secrete a toxin called *exfoliatin* that causes the peeling of superficial skin layers of infected persons. About 50% of the strains of *S. aureus* may secrete any of several **enterotoxins** that cause acute intestinal symptoms (food poisoning) when ingested in cont-

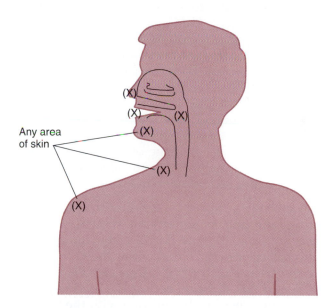

FIGURE 14-2 The sites (X) of the upper body where *S. aureus* are often found but do not necessarily cause disease.

Vegetative Referring to the nonspore stage of a bacterium.
Suppuration An accumulation of white blood cells resulting in the formation of pus.
Enterotoxin An exotoxin produced by a variety of bacteria that is absorbed through the intestinal mucosa. Most such toxins cause nausea, vomiting, and/or diarrhea.

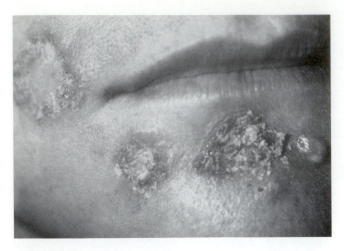

FIGURE 14-3 Impetigo caused by a staphylococcus. This is an infection of the skin most often occurring in children. (*Microbiology: Principles and Applications, 3/e* by Black, J., © 1996. Reprinted with permission of Prentice Hall, Inc., Upper Saddle River, NJ.)

aminated foods. Some *S. aureus* secrete a toxin that causes toxic shock syndrome.

Superficial Infections. Even though staphylococci are able to infect all tissues of the body, the most common infections are of the superficial tissues. *S. aureus* is a common cause of boils, carbuncles, impetigo (Figure 14-3), and infections of surgical or accidental wounds and burns. Characteristically, these infections form an *abscess*—a localized lesion with a cavity of destroyed (necrotic) tissue filled with pus (suppuration). Scar tissue forms on healing. These infections are usually treated with topical antibiotics, or if abscesses are present, by surgical drainage of the pus. In otherwise healthy individuals, these infections tend to resolve without serious consequences.

Systemic Infections. Various forms of trauma resulting around a superficial abscess, such as squeezing a boil, may force large numbers of staphylococci into the blood where they are carried to many tissues of the body, thus causing *systemic* infection. If this situation occurs, the body's natural defenses may not be able to cope with the bacteria and abscesses may develop in various deep organs, such as the liver, lung, or brain tissue. Such infections are always serious and are often life-threatening. The presence of bacteria in the blood is called *bacteremia* or, more commonly, blood poisoning, and is a condition associated with a high mortality. *S.*

aureus may settle in the lungs to cause **pneumonia** or in the pelvis of the kidneys to cause pyelonephritis. In young children the staphylococci have a tendency to infect the bone, causing *osteomyelitis*. Osteomyelitis is frequently a long-term chronic disease. If untreated it may lead to loss of bone function or even death. When present in bone joint spaces the disease is septic *arthritis*. When staphylococci infect the heart chambers or valves, the disease is called *endocarditis*. When they infect the meningeal tissues that line the central nervous system, it is called *meningitis*. Both endocarditis and meningitis are serious life-threatening infections.

Toxic Shock Syndrome and Scalded Skin Syndrome. *Toxic shock syndrome* usually begins suddenly with high fever, vomiting, and diarrhea. In some cases, sore throat, headache, muscle aches, shock, kidney failure, and a red skin rash may be seen. Patients frequently shed the skin from their hands and feet after recovery. Without prompt supportive treatment, death may result. This syndrome was first recognized when a sharp increase in the number of cases was reported in the late 1970s and 1980. Intensive epidemiologic studies demonstrated that most cases occurred in females during or immediately after their menstrual period, and that most cases (99%) were associated with the use of superabsorbent tampons. Staphylococci that produced *toxic shock syndrome* toxin were found in high concentrations in the vaginal canal of affected females. It is now theorized that the introduction of superabsorbent tampons in the late 1970s created a condition in a small percentage of users that permitted the excessive growth of toxin-producing staphylococci. Once this association was recognized, the use of such tampons was discouraged and the number of cases of toxic shock syndrome greatly decreased (Figure 14-4). Some nonvaginal cases occur in both men and women.

Infections of infants or young children with exfoliatin-producing strains of *S. aureus* may cause a condition termed *scalded skin syndrome*. Wide areas of skin become denuded and have the appearance of scalded skin. This syndrome may be fatal, but is usually benign. Most patients are only moderately ill and recover uneventfully.

Food Poisoning and Colitis. Food poisoning is not an infection but an *intoxication* resulting from the ingestion of food containing preformed staphy-

Pneumonia A condition in which a fluid exudate collects in the air spaces of the lung.
Intoxication The ingestion of a microbial toxin that causes a disease.

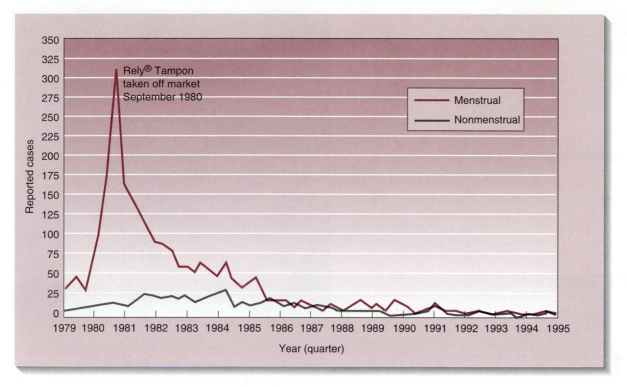

FIGURE 14-4 Reported cases of toxic shock syndrome in the United States, 1979 to 1994 (from CDC annual summaries).

lococcal enterotoxins. During the preparation of foods, it is very easy for the food handler to contaminate the food with staphylococci from the nose or from a skin lesion. If this food is not properly cooled and refrigerated at 4°C, the staphylococci may multiply and release enterotoxins. Enough enterotoxin may be produced in 2 to 6 hours to cause severe symptoms. Foods that are not cooked after preparation—for example, potato salads and cream pies—are common sources of this type of food poisoning. Staphylococci also reproduce rapidly in cooked meats. Their ability to grow at relatively high (9%) salt concentrations makes cured and salted meats such as ham special concerns. Enterotoxins are fairly stable to heat and are not destroyed in foods that are cooked at moderate temperatures after the toxins have been formed. Such symptoms as severe nausea, vomiting, abdominal pain, diarrhea, and prostration may begin to occur as early as 2 hours after ingesting the toxin. There is an absence of fever and complete recovery generally occurs within a day or two. A number of enterotoxins are produced by *S. aureus*. These are designated A through E and produce similar symptoms.

In persons treated with oral broad-spectrum antibiotics, the normal microflora of the bowel can be greatly reduced. This may allow a selective overgrowth by *S. aureus* that is antibiotic-resistant and also produces an enterotoxin B. Enterotoxin B causes direct damage to the intestinal mucosa and results in a condition called **staphylococcal colitis**, which consists of diarrhea, abdominal cramps, and fever.

Other Staphylococcal Infections. The coagulase-negative staphylococci are not as virulent as *S. aureus* and are responsible for fewer infections. However, two of these organisms, *S. epidermidis* and *S. saprophyticus*, are becoming in-creasingly common causes of human infection. *S. epidermidis* is the most common cause of infections in patients with implanted **prosthetic** (artificial) devices, such as

---

Staphylococcal colitis An inflammation of the intestine (colon) caused by *Staphylococcus aureus*.
Prosthetic An artificial or manufactured substitute for a failed body structure, such as an artificial knee joint or heart valve. Other long-term artificial materials that may be implanted for a variety of reasons are also referred to as prosthetic devices.

heart valves and artificial joints. It is also commonly found as a cause of infection of intravenous catheters and feeding lines used in seriously ill patients. *S. saprophyticus* looks very much like *S. epidermidis* but is resistant to the antibiotic novobiocin. This organism has become a relatively frequent cause of urinary tract infection in sexually active young women.

✳ Transmission and Epidemiology

Because all humans are intermittent carriers of *S. aureus*, sources of infections are often difficult to determine. In many cases, the source may be endogenous; that is, it originates with the patient. A burn on the hand, for example, could become infected from bacteria carried on the patient's own skin or in the nose. Spreading from person to person may occur by direct or indirect contact. Medical personnel who work with patients with staphylococcal infections must use good aseptic procedures to avoid transmitting staphylococci to other patients. Nurses or physicians who are carriers or who have skin lesions may unwittingly infect patients. Care must be taken that the hands of medical personnel or contaminated instruments do not spread *S. aureus* from one patient to another.

Because many strains of staphylococci are widespread, specific identification methods are needed to distinguish between strains when tracing the route of transmission of an infection. A laboratory procedure called *phage typing* is used to make this distinction (Figure 14-5). *Phages* are viruses that attack and destroy only specific host strains of bacteria ∞ (Chapter 28, p. 367). By using a series of staphylococcal phages against the staphylococci isolated from a given environment, it is possible to determine which staphylococci are identical to or different from the strain isolated from a lesion or from foods. Different staphylococcal strains have different patterns of resistance and susceptibility to the phages used. Some isolates of *S. aureus* cannot be typed by this procedure, and in recent years DNA fingerprinting has been used to type *S. aureus* in epidemiological studies.

✳ Diagnosis

Diagnosis of staphylococcal infection is generally based on isolation of the organism from a lesion or site of infection. *S. aureus* on blood agar plates will

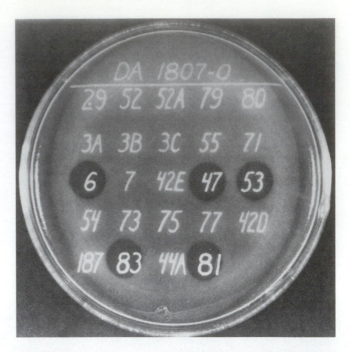

FIGURE 14-5 Phage typing of *Staphylococcus aureus*. The agar plate was first inoculated with a pure culture of *S. aureus* to form a complete layer of bacteria. A suspension of each test phage was dropped on its designated location (indicated by number). Clearing resulted where the phage destroyed the bacterium. In this example, the test bacterium was susceptible to phages 6, 47, 53, 81, and 83. (Courtesy of Centers for Disease Control, Atlanta)

produce a zone of clear **hemolysis** around its typically golden-yellow colony (see Plate 4); *S. epidermidis* and *S. saprophyticus* are rarely hemolytic and produce white colonies. The pathogenic strains of *S. aureus* are coagulase positive. These characteristics of *S. aureus*, along with its ability to ferment mannitol, are used to differentiate it from the ever-present coagulase negative staphylococci. Some *S. aureus* produce white colonies; thus colony pigmentation is not a dependable characteristic for identifying this bacterium.

✳ Treatment

One of the cardinal features of *S. aureus* is its ability to develop resistance against chemotherapeutic agents. During the 1950s, most hospitals and hospital personnel (up to 90%) became colonized with strains that had developed resistance to penicillin. New antibiotics have since been developed and are

---

Hemolysis   The breaking (lysis) of blood cells. Staphylococci often cause hemolysis of the cells in blood agar.
Methicillin-resistant   Resistant to methicillin, a penicillinase-resistant antibiotic. Methicillin-resistant organisms create a difficult treatment problem because of the limited number of antibiotics that can be selected for use.

CLINICAL NOTE

Nasal *S. aureus* Poses Risk for Postoperative Sternal Infection

Sternal wound infection is a serious and common complication of thoracic surgery. More than half these infections are caused by *Staphylococcus aureus*. Investigators at a tertiary hospital in Rotterdam, the Netherlands, showed that nasal carriage of *S. aureus* was a major risk factor.

Over a three-year period, 1980 patients underwent *sternotomy* for cardiac surgery. All had preoperative nasal cultures and started systemic cephalosporin prophylaxis an hour before surgery. Forty patients (2%) developed a sternal *S. aureus* wound infection and were compared with 120 controls. Preoperative nasal carriage of *S. aureus* was a significant predictor of wound infection, as were insulin-dependent diabetes and younger age. Patients with wound infection had a crude odds ratio for nasal *S. aureus* of 9.6, or 8.8 after excluding diabetic patients. Nasal and sternal *S. aureus* strains were identical in all 10 patients for whom typing was done.

Editorial Note: *S. aureus* wound infection is clearly something to avoid: Case patients in this series spent one month longer in the hospital, incurred substantially greater morbidity, and had much higher mortality than controls: 10 versus 0.8%. Although systemic antibiotics have failed to reduce nasal *S. aureus* carriage, topical mupirocin has been quite effective and should be evaluated as a simple strategy for *perioperative* prophylaxis. (*Journal of Infectious Disease* 171:216, 1995).

---

still relatively effective. Emergence of resistant strains against these new antibiotics has been slowed by avoiding abuses in antibiotic therapy ∞ (Chapter 9, p. 136). When treating staphylococcal infections, the most effective antimicrobial agents should first be determined by sensitivity testing and then vigorously administered until the infection is cured.

Of recent concern has been the development of staphylococcal resistance to penicillinase-resistant antibiotics. These organisms are termed **methicillin-resistant**, and they pose a serious treatment problem. The number of staphylococcal infections due to methicillin-resistant strains has risen from a few isolated cases in the 1970s to more than 18% of cases by 1992.

In addition to antibiotic therapy, surgical treatment is frequently necessary to effect a cure from staphylococcal infection. Abscesses should be drained, when possible, to remove inflammatory debris that may block the diffusion of antimicrobial agents to the site of the bacteria; surgical *debredment* for osteomyelitis and removal of infected prosthetic devices may be essential components of treatment in these diseases. Surgical drainage may remove an environmental niche where the staphylococci are in a metabolically inactive stage and thus not affected by the chemotherapeutic agent. Also, such accumu-

lations of pus or debris may protect the infecting organism from normal host defenses.

* Prevention and Control

Most individuals seem to develop antibodies against staphylococci early in life. These antibodies may contribute to the resistance that most healthy individuals have against this bacterium; however, in injured tissues or in compromised patients this immunity is often not sufficient to prevent infection. No successful vaccines are available against *S. aureus*.

Prevention of staphylococcal infection is primarily a function of good *aseptic* techniques in hospitals and clinics where both infectious and susceptible patients congregate. Methods must be used to minimize the spread of microorganisms in critical areas. First, patients with open infected wounds should be isolated. Second, personnel who work in critical areas, such as operating rooms or newborn nurseries, should be screened to determine if they are carriers of drug-resistant strains. Those who are carriers should be restricted from these areas until their condition is cleared up. Any type of infected lesion on hospital personnel should be promptly treated and precaution should be taken to protect patients from infection from this source.

Sternotomy Incision into or through the sternum.
Perioperative Near the time of an operation
Debredment Removal of dead tissue and foreign matter from a wound by scraping.

The newborn nursery presents special problems in that about 90% of infants become carriers of *S. aureus* during the first 10 days of life. It is important that they do not become colonized with the virulent drug-resistant strains prevalent in hospitals. Skin disinfectants, especially chlorhexidine, are useful in decreasing the staphylococcal carrier rate in infants ∞ (Chapter 8, p. 118). The practice of immersing (bathing) newborns in hexachlorophene solutions has been discontinued to reduce the chance of any toxic reaction, but discrete washing of the skin with

this disinfectant is appropriate and effective. In some cases, it has been beneficial to deliberately colonize infants with a nonvirulent strain of *S. aureus*. This step seems to interfere with subsequent colonization by virulent strains.

Air-handling systems should be designed to carry patient-generated airborne bacteria away from other patients or hospital personnel and to prevent any recirculation of these microorganisms to other areas of the hospital.

MATERIAL FOR REVIEW

CONCEPT SUMMARY

1. Staphylococci are ubiquitous gram-positive cocci not infrequently associated with infections of both humans and animals.
2. Of the several species of *Staphylococcus*, *S. aureus* is the most virulent and most com-

monly isolated from human infections. This organism is typical of other pyogenic cocci in that infections are characterized by the production of pus.
3. Numerous disease conditions may occur due to the staphylococci, ranging from food poi-

soning through osteomyelitis and pneumonia to the relatively recently described toxic shock syndrome. Treatment of many of these infections requires surgical intervention along with aggressive antibiotic therapy.

CLINICAL SUMMARY TABLE

Microorganism	Virulence Mechanisms	Diseases	Transmission	Treatment	Prevention
Staphylococcus aureus	Hemolysins Leukocidins Hyaluronidase Fibrinolysin Enterotoxins Toxic shock toxin Exfoliatin	Skin and wound infections Internal infections Toxic shock syndrome Scalded skin syndrome Food poisoning	Person to person Opportunists	Some antibiotics (many resistant strains)	No vaccines Good nursing practices Isolation Aseptic procedures

STUDY QUESTIONS

1. List five characteristics of staphylococci that are related to virulence of these organisms.
2. What features of toxic shock syndrome and scalded skin syndrome are shared in common?
3. With what kinds of infection are coagulase-negative staphylococci commonly associated?
4. What feature of *S. aureus* has made therapy of infections due to this organism particularly difficult during the past few years?
5. What characteristic of coagulase-negative staphylococci has made control of infection very difficult?

CHALLENGE QUESTIONS

1. Refer to Figure 15-4. The tampons that were associated with toxic shock syndrome in the early 1980s are no longer on the market. Explain why there still are a few cases of the syndrome.
2. Refer to Figure 15-5. Explain how a phage infection could produce a clearing of the staphylococci.

CHAPTER

15 STREPTOCOCCI

OUTLINE

Both pathogenic and nonpathogenic species of streptococci are commonly associated with humans and animals. A wide variety of species are present as normal flora on skin and mucous membranes of all humans. Three species, *Streptococcus pyogenes*, *Streptococcus pneumoniae*, and *Streptococcus agalactiae*, are responsible for most of the streptococcal infections in humans. The pathogenic species produce a wide variety of toxins and cause a wide variety of lesions and diseases. Historically, some streptococcal diseases have been among the most serious diseases of humans. Fortunately, these bacteria are usually easily destroyed by chemotherapeutic agents, and as a result, even though streptococcal infections are still common, their impact on illness and death today is only a small fraction of what it was prior to the 1930s.

STREPTOCOCCI: GENERAL CHARACTERISTICS

Streptococci are gram-positive, coccal-shaped bacteria that usually appear in chains of various lengths (Figure 15-1; see Plate 5). These bacteria are moderately resistant to environmental factors; that is, they may remain living for days to weeks after being expelled from the body. However, streptococci are readily killed by common disinfectants and most are highly susceptible to a wide range of chemotherapeutic agents, including penicillin. Bacterial species of the genus *Streptococcus* are widespread in nature and are part of the normal bacterial flora of the skin, nose, mouth, and mucosal surfaces of humans and animals. It is often difficult to distinguish clearly among many of the streptococcal

HUMAN BODY SYSTEMS AFFECTED

Streptococci

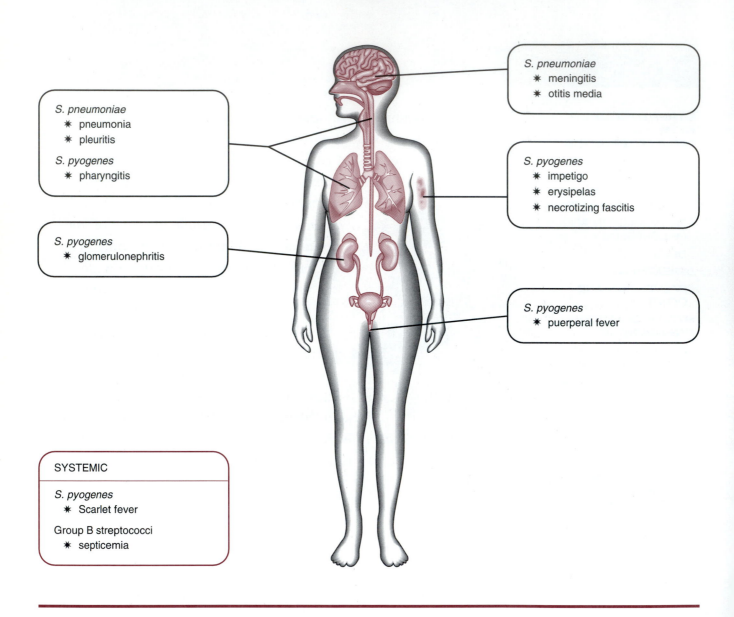

S. pneumoniae
* pneumonia
* pleuritis

S. pyogenes
* pharyngitis

S. pyogenes
* glomerulonephritis

SYSTEMIC

S. pyogenes
* Scarlet fever

Group B streptococci
* septicemia

S. pneumoniae
* meningitis
* otitis media

S. pyogenes
* impetigo
* erysipelas
* necrotizing fascitis

S. pyogenes
* puerperal fever

species. This has made a precise classification of these bacteria difficult. Streptococci grow well on blood agar and many species secrete *hemolysins* (enzymes that dissolve red blood cells), which produce patterns of hemolytic zones around the colonies. These hemolytic patterns can be used to make a preliminary identification of streptococcal groups. A clear zone of hemolysis surrounding the colony is called *beta*-hemolysis (see Plate 6), a zone with an opaque greenish color is called *alpha*-hemolysis, and some species produce no hemolysis.

The most usable classification system, **the**

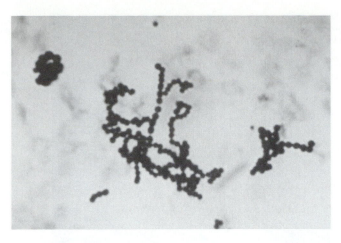

FIGURE 15-1 Photomicrograph of streptococcal cells.

Lancefield system, uses differences in carbohydrate antigens located in the cell wall of streptococci. Under this system, streptococci have been divided into 20 major groups, designated A through V (omitting I and J). Some of the most important streptococcal infections of humans are caused by the species *Streptococcus pyogenes*, which belongs to Lancefield group A. Group A can be further subdivided into some 80 types based on differences in an antigen called the *M-protein*. The M-protein is important because it is responsible for virulence; antibodies formed against it give protection to the host. The other groups or categories of streptococci also include some species that are associated with human infections and are listed in Table 15-1.

STREPTOCOCCUS PYOGENES

Various extracellular and intracellular substances produced by *S. pyogenes* contribute to their ability to withstand the body's defense mechanisms and cause tissue damage. Several more prominent substances and their effects are as follows:

1. *Capsule*, helps retard phagocytosis.
2. *M-protein*, is antiphagocytic.

3. *Erythrogenic toxins*, produce the fever and rash associated with scarlet fever.
4. *Streptolysin O* and *streptolysin S*, two separate hemolysins that lyse red blood cells and damage various host cells. Streptolysin O is oxygen labile; streptolysin S is stable in oxygen.
5. *Streptokinase*, a fibrinolysin that digests fibrin in the inflammatory barrier ∞ (Chapter 10, p. 150).
6. *Hyaluronidase*, the spreading factor. Hyaluronidase is an enzyme that breaks down hyaluronic acid, which acts as an intracellular glue.
7. *Protein F*, promotes adherence to pharyngeal cells.

In addition, various other cellular substances also contribute to the pathogenicity and/or virulence of *S. pyogenes*. The major types of clinical conditions associated with group A streptococci are discussed next.

✳ Pathogenesis and Clinical Diseases

Sore Throat (Pharyngitis). Streptococcal **pharyngitis**, referred to as "strep throat," is very common. Group A streptococci are the major cause of bacterial pharyngitis. Other groups of strepto-

TABLE 15-1	
Non-Group A Streptococci That Cause Disease in Humans	
Organisms	**Disease**
Group B streptococci	Neonatal meningitis, sepsis
Streptococcus groups C, G	Occasional wound infection, pharyngitis
Group D—enterococci	Urinary tract infection, bacteremia, endocarditis
Viridans streptococci	Endocarditis, rare meningitis or cystitis
Streptococcus pneumoniae	Pneumonia, wound infections, sepsis, meningitis

Lancefield system   A classification of streptococci based on the carbohydrate antigens present in the wall of the cell.
M-protein   A major virulence factor of streptococci. Located on the cell surface, this antigen can be used to separate group A streptococci into over 80 serotypes.
Pharyngitis   Infection of the pharynx. The pharynx is the upper portion of the throat. This disease is usually accompanied by enlarged lymph nodes, erythema, and pain.

cocci (C and G) are occasionally involved. Children up to 15 years of age average about one infection per year. Each infection is caused by a different M type, because antibody immunity will generally offer protection for some time against repeated infections with the same M type. This infection is not life-threatening, but strains vary in virulence, and infections with highly virulent bacteria may cause a severe sore throat with fever, headache, swollen cervical lymph nodes, and **purulent** exudate in the throat. Infections in older children and adults tend to be milder and less frequent, due in part to the antibody immunity that has developed against many strains encountered in earlier childhood. The infection may also spread from the throat directly into the lungs to cause pneumonia, or may penetrate into the pleural cavity, causing severe inflammation (pleuritis). It is important to seek therapy for streptococcal pharyngitis in order to prevent subsequent **sequelae** such as rheumatic fever.

Scarlet Fever.

Scarlet fever results from the production of erythrogenic toxin, called streptococcal pyrogenic exotoxin (SPE), by the streptococci that are causing pharyngitis or other relatively benign streptococcal infections. Three types (SPE A, SPE B, and SPE C) of toxin may be produced by *S. pyogenes*. The toxin diffuses into the blood and is carried to the skin, where it causes a diffuse reddish rash within one to two days after the initial symptoms of pharyngitis. The more virulent erythrogenic toxin-producing strains are quite invasive and are able to spread through the lymphatic system and into the blood. They may then be carried throughout the body, resulting in infections of the joints, bones, endocardium, skin, and so on. This virulent form of scarlet fever is quite severe and was a common cause of death in the era before penicillin. The outer layers of affected skin are frequently sloughed during recovery from scarlet fever (Figure 15-2).

Streptococcal Toxic Shock-like Syndrome.

Group A beta-hemolytic streptococci recently have been found to be the cause of a syndrome that resembles staphylococcal toxic shock. Patients with this illness have high fever, **erythema**, and often a purpuric (purplish) skin rash that may become gangrenous. The death of adjacent host tissue cells (*necrotizing fasciitis*) caused by this infection has

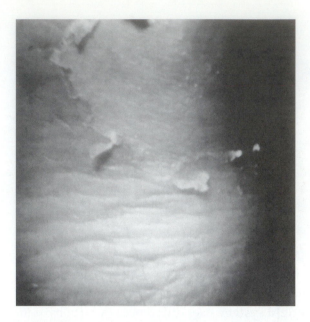

FIGURE 15-2 Skin peeling from scarlet fever patient. (Courtesy J. M. Matsen, University of Utah)

prompted the news media to improperly call these "flesh-eating" streptococci (see Clinical Note). Patients may also experience renal failure, pneumonia, and cardiac failure. This serious disease often begins as a cellulitis (inflammation of subcutaneous tissue) or pharyngitis and rapidly progresses to the more serious generalized disease within 24 hours. Positive bacterial cultures may be obtained from a variety of body sources.

In the years before the availability of penicillin, a streptococcal disease referred to as toxic scarlet fever was well-known. Streptococcal toxic shock-like syndrome presents a clinical picture similar to that reported for toxic scarlet fever and it may be that streptococcal toxic shock-like syndrome represents a resurgence of this old disease. The streptococci isolated from streptococcal toxic shock-like syndrome produce an abundance of a toxin called *streptococcal pyrogenic exotoxin A*, and earlier isolates from toxic scarlet fever also produced large amounts of this toxin. Mortality rates from streptococcal toxic shock-like syndrome are near 30%, and vigorous therapy with intravenous penicillin is necessary to control this infection. There have been several recent outbreaks of streptococcal pneumonia, possibly due to the same severe strains of streptococci as those causing streptococcal toxic shock-like

---

Purulent   Associated with the production of pus.
Sequela   A condition or illness directly relating to an earlier condition but developing some time after the first illness.
Erythema   Redness of the skin resulting from dilation of the capillaries.

CLINICAL NOTE

Invasive Group A Streptococcal Infections: United Kingdom, 1994

On 27 May, 1994, the Communicable Disease Surveillance Center in England reported that six persons in Gloucestershire had disease characteristic of invasive group A streptococcal infection (GAS) with necrotizing fasciitis. Three patients died. Patients ranged in age from 46 to 68 years. Group A streptococcal isolates from blood or joint fluid from five patients were typed by the Public Health Laboratory Service *Streptococcus* Reference Laboratory. Four different types were identified (M1, M3, M5, and M-nontypeable).

Since 1992, the total number of laboratory reports of systemic GAS in England and Wales has remained stable; during the first 16 weeks of 1994, a total of 200 blood isolates were reported, compared with 212 and 200 during the first 16 weeks of 1993 and 1992, respectively.

Editorial Note: The report from the United Kingdom underscores the potential for severe disease associated with GAS. GAS is associated with a broad spectrum of complications in humans, the most common being streptococcal pharyngitis. Serious invasive disease, which occurs less commonly, is defined by isolation of the bacteria from usually sterile sites and is associated with case-fatality rates of 10–20%. One form of invasive GAS, necrotizing fasciitis, is characterized by destruction of muscle and fat tissue.

Based on extrapolation of incidence rates determined by active surveillance in four states during 1989–1991, 10,000–15,000 cases of invasive GAS occurred annually in the United States; necrotizing fasciitis occurred in 5–10% of patients (case-fatality rate: 28%) (CDC, unpublished data, 1992). These findings were consistent with a retrospective review of all invasive GAS in Pima County, Arizona, during 1986–1990; in this review, necrotizing fasciitis was identified in 6.5% of infections. Interest in necrotizing fasciitis as a serious manifestation of invasive GAS increased in 1989 following a report of 20 patients with group A streptococcal toxic-shock syndrome, of whom 11 had necrotizing fasciitis; a subsequent case definition for this syndrome included necrotizing fasciitis as one component. Since 1991, there has been no active surveillance for invasive GAS in the United States; although passive surveillance exists, this disease is not reportable in most states.

Rapid treatment is necessary to reduce the risk for death, and penicillin remains the treatment of choice for GAS. Although penicillin resistance has never been identified in group A *Streptococcus*, some strains are resistant to erythromycin (which is recommended as therapy in penicillin-allergic patients). In addition to antibiotics, surgical intervention is usually needed in cases of necrotizing fasciitis. The occurrence of the cluster of necrotizing fasciitis in England and the recent recognition of a streptococcal toxic-shock syndrome underscore the potential for group A streptococci to cause severe illness and new clinical syndromes and the need to monitor clinical manifestations and changes in the epidemiology of these infections (*MMWR* 43:401, 1994).

---

syndrome. In these cases of pneumonia, even with therapy, the death rate has exceeded 50%. Most of the streptococcal strains isolated from toxic shock-like syndrome belong to M-protein groups 1, 3, or 18.

Puerperal Fever. *Puerperal fever*, also known as *childbed fever*, is a group A streptococcal infection of the uterus that may occur in women shortly after childbirth (postpartum). The streptococci often spread rapidly from the inflamed uterus to the blood, and the resulting widespread infection may cause death. This disease captured the interest of Semmelweis ∞ (Chapter 1, p. 4), and was one of the first examples of human-to-human transmission of infection.

Infections of the Skin. *S. pyogenes*, like *Staphylococcus aureus*, is able to cause lesions on the skin where prior injuries, such as insect bites, wounds, and burns, have occurred. *Impetigo* is commonly caused by streptococci, and *erysipelas* is a specific type of severe streptococcal skin lesion. Impetigo is a relatively common, usually benign, superficial infection of the skin. Lesions may occur any place on the body and are characterized as blisters that break and dry with crusty scabs. Children are the most common victims of this disease, which is relatively easily treated with topical antibiotics (Figure 15-3). Strains of streptococci that cause impetigo are different from those that cause pharyngitis and are often associated with subsequent development of glomerulonephritis.

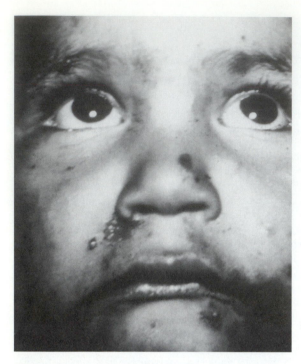

FIGURE 15-3 Streptococcal impetigo.
(Courtesy J. M. Matsen, University of Utah)

Erysipelas often occurs on the face and probably starts from streptococci coming from the throat or nose and entering a skin abrasion. The lesion spreads outward, causing marked erythema and swelling (edema) of the skin with a sharply defined advancing edge. Recovery usually takes a week or longer, and if no treatment is given, this condition may be fatal in some patients.

Poststreptococcal Diseases.

Disease that occurs as a direct result of streptococcal infection but occurs sometime after the infection is known as poststreptococcal disease. The poststreptococcal diseases are *glomerulonephritis* and *rheumatic fever*, and have their onset from 1 to 4 weeks after an *acute* streptococcal infection. These two conditions are not infections, but occur as a direct result of previously untreated streptococcal pharyngitis or **pyoderma**. In spite of much research, the mechanisms and pathogenesis of these diseases are not well understood. Evidence indicates that they are immune complex diseases ∞ (Chapter 11, p. 170), which means that there is an interaction between antibodies and tissue-associated antigens, which, in turn, induces an inflammatory response. The inflammatory response results in the formation of scar tissue, which replaces some of the normal body tissue. The antibodies involved are those produced in response to the streptococcal infection.

Glomerulonephritis involves the basement (filtering) membrane of the glomeruli of the kidneys. The inflammation and scarring of these membranes may result in severe kidney malfunctions and, in some cases, death. This disease is induced by several serotypes of group A streptococci, and may result following infection of the skin, most often pyoderma. Primary infections with these *nephritogenic* streptococci are infrequent, and only about 0.5% of individuals with streptococcal disease go on to develop poststreptococcal glomerulonephritis. About 80 to 90% of the cases undergo slow spontaneous healing, whereas the others develop a chronic form of the disease. Recurrent attacks are rare.

Rheumatic fever develops as a sequela in 0.1 to 3% of untreated patients following acute pharyngitis and may be caused by almost any one of the group A strains, although M3, M5, and M18 strains are most often involved. The disease usually develops in children 5 to 15 years of age. Some signs and symptoms are fever, malaise, inflamed and aching joints (arthritis), subcutaneous lesions, and heart lesions. These complications may occur alone or in various combinations. However, about half the cases may be mild and go undiagnosed. Injury to the heart (*rheumatic heart disease*) is the most serious effect of rheumatic fever and currently is the most common cause of permanent heart valve damage in children. Subsequent infections with group A streptococci may aggravate and cause recurrences of this disease. Because many different strains of streptococci cause rheumatic fever, persons who have had one attack must guard against reinfection and often use penicillin as prophylaxis. During the past few years there has been an increase in the number of cases of rheumatic fever, but the number is still much lower than was reported 30 years ago.

✳ Transmission and Epidemiology

The various strains of *S. pyogenes* are widespread in humans, many of whom are asymptomatic carriers. These bacteria are found in the respiratory tract, mouth, intestines, or on the skin of about 5% of the general population in the summer and up to 15% in the winter. The carrier rate is generally higher in

Acute   A current, rapid, short manifestation of disease symptoms.
Pyoderma   Infection of the skin by a pus-producing (pyogenic) bacterium.

children between 1 and 15 years of age. Because of this high carrier rate, streptococci are readily transmitted when large numbers of persons share common environments. Persons with acute infections are an obvious source of infection, and transmission can readily occur from these individuals via respiratory droplets or by direct or indirect contact. When highly virulent strains are carried into elementary schools, sharp outbreaks of pharyngitis and scarlet fever may occur. Due to the accumulative buildup of antibodies to many different types, outbreaks among adults are less likely.

∗ Diagnosis

Diagnosis of a streptococcal infection is based on both clinical and laboratory findings. The basic laboratory procedure is to swab the site of infection and either culture the material from the swab on a blood agar plate or use a rapid streptococcal antigen detection test. Cultures are examined after 24 hours of incubation for the presence of beta-hemolytic colonies, which, if present, are further studied to determine if they are group A streptococci. The antigen detection tests are commonly employed in physicians' offices and medical clinics. These procedures afford rapid (less than 2 hours) results and are a fairly sensitive method for detection of group A streptococci from cases of pharyngitis. A negative antigen detection test should be confirmed by culture.

On the basis of clinical examination only, streptococcal infections may be difficult to distinguish from some staphylococcal infections. To differentiate between the two types of bacteria following culture, a *catalase test* is used. This test is done by placing a small amount of growth from a colony in hydrogen peroxide; if the catalase enzyme is present, bubbles of oxygen are released. Staphylococci are positive for catalase and streptococci are negative. *S. pyogenes* can be distinguished from other beta-hemolytic streptococci by placing a paper disk containing a low concentration of the antibiotic bacitracin onto a freshly swabbed agar plate; growth of *S. pyogenes* will be inhibited by the bacitracin, whereas that of the other streptococci will not.

Glomerulonephritis and rheumatic fever are diagnosed primarily on the basis of clinical and serologic findings. These are postinfection sequelae and attempts to culture streptococci are not successful. However, because they follow streptococcal infection, the patient should be expected to have a high antibody titer to streptococcal antigens. The most useful tests measure antibody to either streptolysin-O or DNAse B antigens. A high titer to either

of these antigens suggests a recent group A streptococcal infection.

∗ Treatment

The beta-hemolytic streptococci are highly susceptible to most antimicrobial agents. Penicillin is effective in most cases and is the antibiotic of choice. Other antibiotics can be used in patients who are allergic to penicillin. Erythromycin is generally recommended. Because streptococci tend to develop resistance to tetracyclines and sulfa drugs, these agents should not be used; aminoglycosides are not effective.

Penicillin or cephalosporin therapy will usually produce a rapid cure of most acute infections. Symptoms usually resolve within 24 hours of treatment, and destruction of the bacteria is usually complete within 10 days. Treatment should be started as soon as practical to reduce the chance of the subsequent development of rheumatic fever and to reduce spread of the organism. Prompt diagnosis and treatment are particularly important in children, for they have the greatest predisposition to rheumatic fever. Early treatment is also important in helping to reduce the chance of transmitting the infection. Moreover, therapy can be used to try to clear streptococci from known carriers; this factor is particularly important for medical personnel who work with highly susceptible patients.

∗ Prevention and Control

Vaccines are not available for the above streptococcal disease; and because of the large numbers of serotypes, the development of vaccines is considered unlikely. The best control measure is to use common sense procedures to prevent the transmission of infection. Individuals with a known infection should be isolated; for instance, a child with untreated pharyngitis should not attend school, and an infected patient in a hospital should be isolated from other patients. Medical personnel have a special responsibility to follow standard aseptic procedures to avoid transmitting these organisms between patients. Care should be used in delivery rooms and in working with new mothers to prevent transmission to their highly susceptible uterine tissues.

The best means to prevent rheumatic fever and glomerulonephritis involves early diagnosis and treatment of children who have acute streptococcal infections. The recurrence of rheumatic fever is prevented by prophylactic (preventive) penicillin treatment—that is, daily doses of penicillin. Often persons who received severe heart damage from the first attack of rheumatic fever must be given penicillin therapy for years or for their lifetime. In some of these cases, it is difficult to resolve the balance between benefits and abuses of such long-term penicillin therapy.

STREPTOCOCCUS PNEUMONIAE

S. pneumoniae is a gram-positive ovoid shaped coccus that characteristically occurs in pairs or short chains (Figure 15-4). Paired cocci are so common that these organisms are frequently referred to as diplococci. The virulent strains possess a prominent capsule that plays an important role in the pathogenesis of pneumococcal diseases by retarding the rate of phagocytosis. The capsular material is composed of polysaccharide and is *antigenic*. These antigens have been used to subdivide the pneumococci into more than 80 **serotypes**. Protection against infection is type-specific and depends on antibody to the capsular polysaccharide. The organism can be easily cultivated on enriched media supplemented with 5% blood. *S. pneumoniae* produces alpha-type hemolysis, which may cause it to be mistakenly identified as a viridans streptococcus. Pneumococci are traditionally highly susceptible to penicillin as well as other antibiotics; however, some penicillin-resistant strains are now creating treatment problems for seriously ill patients. These organisms do not survive long outside the body and are readily destroyed by disinfectants.

✳ Pathogenesis and Clinical Diseases

Infections due to *Streptococcus pneumoniae* (pneumococcus) rank among the important causes of human illness. Before the era of antibiotics, pneumococcal pneumonia was the leading cause of human death. Even with the use of antibiotics, it is still responsible for significant mortality and is ranked number 6 among the top 10 causes of death in the United States. Pneumococcal pneumonia is a potential threat to every hospital patient or person with a compromised host defense mechanism. Medical personnel must be continually alerted to the possi-

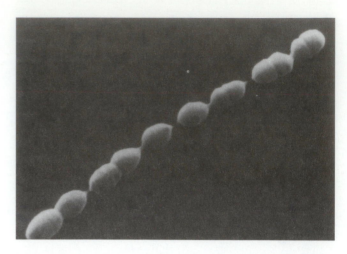

FIGURE 15-4 Scanning electron micrograph of *Streptococcus pneumoniae* showing the chain arrangement and the tendency to form pairs of cocci (diplococci) (magnification 10,000×). (A. S. Klainer and I. Gies, *Agents of Bacterial Disease.* Harper & Row: Hagerstow, MD, 1973. Figure 5-3, p. 69)

ble occurrence of this disease when caring for all patients.

Pneumonia.   *S. pneumoniae* is widespread in the general population. From less than 10% to over 40% of healthy persons may be carrying this bacterium in their respiratory tract. The higher **carrier rates** are most common in children during the winter months, and correlate with increased occurrence of *pneumococcal pneumonia*. The major mechanism of virulence is the prominent capsule that impedes phagocytosis ∞ (Chapter 4, p. 39). Strains that lack a capsule are uniformly avirulent, and individuals with specific antibodies against the capsular material are resistant to infection. A cytoplasmic protein, called *pneumolysin*, is released when pneumococci lyse. Pneumolysin binds to cholesterol in host cell membranes and disrupts them by forming pores. This toxin along with several other bacterial products, such as hydrogen peroxide, may play a role in bacterial virulence. In a healthy active person the macrophages are usually able to control any pneumococci that may enter the lower respiratory tract. However, when the defense mechanisms of a host become suppressed, the pneumococci are able to multiply and establish a focus of infection. As the pneumococci begin to multiply, they induce an

---

Antigenic   Capable of stimulating antibody synthesis.
Serotype   A microorganism with a unique antigen such that it induces antibody specific for that organism.
Carrier rate   The percent of individuals infected with a pathogen.

CLINICAL NOTE

Emergence of Penicillin-Resistant *Streptococcus pneumoniae*: Southern Ontario, Canada, 1993–1994

Streptococcus pneumoniae is a leading cause of infectious disease-related illness and death in the United States, accounting for an estimated 3000 cases of meningitis, 50,000 cases of bacteremia, 500,000 cases of pneumonia, and 7 million cases of acute otitis media each year. Penicillin has been the antibiotic of choice for the treatment of infections caused by *S. pneumoniae*; since the mid-1980s, the prevalence of penicillin-resistant *S. pneumoniae* has increased substantially worldwide. In Canada, a strain of pneumococcus with reduced susceptibility to penicillin was first reported in 1974; based on surveys during 1977–1990, rates of resistance to penicillin were 2.4, 1.5, and 1.3% in the provinces of Alberta, Ontario, and Quebec, respectively. To determine whether the prevalence of penicillin resistance had increased among pneumococcal isolates, investigators from the University of Toronto tested the susceptibility of strains collected from a Toronto hospital and from a surrounding region in southern Ontario during June–December 1993 and March–June 1994. This report summarizes the results of this investigation.

A total of 202 isolates (196 from noninvasive sites [i.e., sputum]) of *S. pneumoniae* were tested, including 122 isolates obtained from the private laboratory and 80 from the hospital emergency department. Of the 202 isolates, 16 (7.9%) were penicillin-resistant, including four with high-level resistance; 11 had been obtained from eye, ear, or sputum samples from children (eight of 68 aged <5 years) in outpatient settings and 5 from sputum, blood, cerebrospinal fluid, and eye samples from adults in the emergency department.

Penicillin-susceptible strains generally were susceptible to other antimicrobial agents. However, high proportions of penicillin-resistant *S. pneumoniae* isolates were resistant to tetracycline (63%), trimethoprim/sulfamethoxazole (56%), erythromycin (50%), and cefuroxime (38%). High-level resistance to ceftriaxone occurred in four (25%) of 16 penicillin-resistant isolates; high-level resistance to penicillin was present in three of the four isolates resistant to ceftriaxone. All isolates were susceptible to vancomycin and imipenem. Serotypes of the penicillin-resistant pneumococci tested in the Canadian Streptococcal Reference Laboratory (Edmonton, Alberta) were 19F (five isolates), 9V (two), 23F (two), and one each of 6A, 6B, and 19A; four were nontypeable.

Editorial Note: The findings in this report suggest an increased prevalence of penicillin-resistant *S. pneumoniae* in metropolitan Toronto compared with that in a similar study in Toronto in 1988 (1.5%). By selecting all pneumococcal isolates from a large outpatient reference laboratory and hospital emergency department in metropolitan Toronto (97% of which were obtained from noninvasive sites), the study provided an indication of the antimicrobial resistance patterns among pneumococci circulating in the community and reflects a trend of emerging pneumococcal drug resistance in North America and other countries. For example, in the United States during 1987–1992, the prevalence of high-level resistance to penicillin increased more than 60-fold, from 0.02 to 1.3% in pneumococcal isolates collected from sentinel sites. The proportion of pneumococcal isolates resistant to penicillin has ranged from 2 to 26% in selected communities in the United States, indicating substantial geographic variability in prevalence of penicillin resistance (*MMWR* 44:207, 1995).

acute inflammatory reaction with a rapid infiltration of *edematous* fluid. This fluid nourishes the bacteria, which, in turn, increases their rate of multiplication and stimulates a greater inflow of fluids. The initial accumulation of fluids also impedes phagocytosis and may impede the infiltration of antimicrobial agents into the lesion. The infiltrated fluids, which contain many polymorphonuclear cells (PMNs) and red blood cells (RBCs), may rapidly fill the infected lung lobe causing areas of *consolidation*; this is called *lobar pneumonia*. This fluid takes on the characteristics of typical inflammatory exudate. In less serious cases, the increased number of PMNs, along with the macrophages, are able to contain the pneumococci; the infiltrated fluid then disperses and the lung tissue is restored to its original condition without permanent damage. When given early, antibiotics greatly accelerate the recovery process.

In more serious pneumonias, the accumulation

Edematous Referring to an accumulation of body fluids.

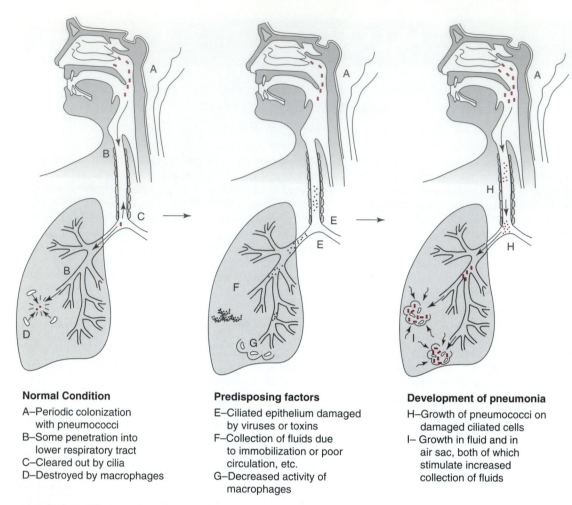

Normal Condition

A–Periodic colonization
with pneumococci
B–Some penetration into
lower respiratory tract
C–Cleared out by cilia
D–Destroyed by macrophages

Predisposing factors

E–Ciliated epithelium damaged
by viruses or toxins
F–Collection of fluids due
to immobilization or poor
circulation, etc.
G–Decreased activity of
macrophages

Development of pneumonia

H–Growth of pneumococci on
damaged ciliated cells
I– Growth in fluid and in
air sac, both of which
stimulate increased
collection of fluids

FIGURE 15-5   Predisposition to and the development of pneumococcal
pneumonia.

of fluid continues and spreads from one lobe to an-other and areas of consolidation develop in which the pneumococci are less susceptible to phagocyto-sis and antibiotics. The pneumococci may spread from the lungs into the pleural cavity or pericardi-um and cause abscesses in these areas. The infection in the pleural space is called *pleurisy*. Such infec-tions may continue to expand, resulting in the death of the patient. On the other hand, if specific anti-bodies develop in time, a crisis may be reached, fol-lowed by recovery.

The signs of classical pneumonia begin with the rapid onset of shaking chills and fever between 38.8 and 41.4°C. Severe chest pains are often pre-sent. A cough develops and the **sputum** is purulent and rust-colored (contains RBCs). In many un-treated cases, the crisis is normally reached in about 5 to 10 days, followed by recovery. Overall, the

death rate in untreated cases is 30% compared to 5% in treated cases. The outcome is greatly influenced by the age and underlying predisposing conditions of the patient. The stages in the develop-ment of pneumococcal pneumonia are outlined in Figure 15-5.

Other S. pneumoniae Infections.   Other infectious processes associated with *S. pneumoniae* include *otitis media* (middle ear infection), septi-cemia, wound infections, and meningitis. Mortality from pneumococcal disease is highest for meningi-tis (up to 55%) and septicemia (40%), in spite of appropriate antibiotic therapy. Otitis media, while painful, is relatively benign and most commonly occurs in young children. Infection may also spread via the blood or by direct extension to other body tissues. Septicemia and meningitis are common out-

Sputum   Expectorated matter, such as mucus, derived from the lower respiratory tract.

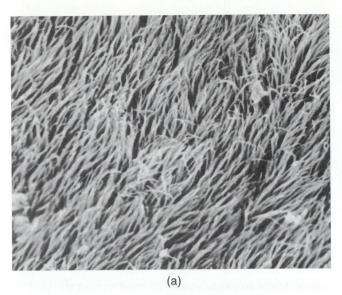
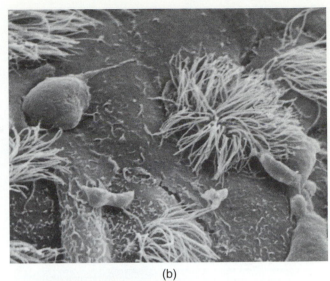

(a)   (b)

FIGURE 15-6 (a) Scanning electron micrograph of normal ciliated epithelium showing a continuous protective blanket of cilia. (b) Ciliated epithelium damaged by a viral infection which renders the respiratory tract much more susceptible to secondary bacterial infections. (S. E. Reed and A. Boyd, *Infection and Immunity* 6:68–76, Figures 1 and 4, with permission from ASM)

comes when the pneumococci spread beyond the respiratory tract; currently pneumococci are one of the most frequent causes of these infections in young children.

✳ Transmission and Epidemiology

Streptococcus pneumoniae is an exclusively human pathogen and, in most cases, the source of human disease is endogenous—that is, from the pneumococci already present in the respiratory tract. Person-to-person aerosol transmission ∞ (Chapter 13, p. 192) of pneumococci readily occurs, but the disease occurs only in those with predisposing conditions. The most common *predisposing factors* are viral infections of the respiratory tract (Figure 15-6); physical injury to the respiratory tract from inhaling toxic or irritating substances, including anesthetic gases; prolonged immobilization in bed, which may result in accumulation of fluids in the lungs; alcoholism; increasing age; diabetes; and immunodeficiency diseases, such as Hodgkin's disease, sickle cell anemia, or AIDS.

Pneumonia is most often seen in infants, elderly persons, alcoholics, or persons with chronic disease. *S. pneumoniae* is by far the most common single cause of bacterial pneumonia, with as many as

500,000 cases a year occurring in the United States. In addition, the pneumococcus is responsible for up to 7 million cases of middle-ear infection, about 50,000 cases of bacteremia, and 3000 cases of meningitis, and contributes to an estimated 40,000 deaths annually. The seasonal variations and increased number of cases of all types of pneumonia occurring during influenza epidemics are shown in Figure 15-7. About 25% of the pneumonia in adults is caused by pneumococci and 75% of these cases are due to only 9 of the 84 different pneumococcal serotypes. More pneumonia is seen in the winter and generally parallels the incidence of viral respiratory infections.

✳ Diagnosis

Typical cases of pneumonia are usually diagnosed on the basis of physical examination and lung x-rays. Laboratory diagnosis entails demonstrating the pneumococci in the sputum, blood, or **cerebral spinal fluid** (CSF), by direct microscopic examination and by culturing on artificial media. Differentiation of the pneumococci from viridans streptococci is often necessary because they both produce the same type of reaction on blood agar plates. The pneumococci are distinguished from the

Predisposing factors   Factors making one susceptible to an infection or disease.
Cerebral spinal fluid   Fluid filling the ventricles and cavities of the brain and spinal column.

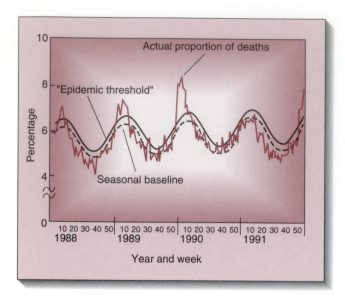

FIGURE 15-7 Percentage of all deaths attributable to pneumonia and influenza in 121 cities in the United States, January 1988 to January 1992. The sharp increases resulted from persons being predisposed to pneumonia as a result of influenzal infection. (Courtesy of Centers for Disease Control, Atlanta)

viridans streptococci by their *bile* solubility. Pneumococci contain autolytic enzymes that are activated in the presence of bile. For this test, when pneumococci are added to a solution containing bile salts, the cells dissolve, whereas in a similar test the viridans streptococcal cells do not dissolve because they lack the necessary autolysins. Alternatively, the *optochin disk test* may be used. For this test, a paper disk containing a chemical called *optochin* is placed on an agar plate that has been seeded with the test bacterium. Optochin inhibits the growth of pneumococci but not of viridans streptococci (see Plate 8). Various serologic tests are available to detect the presence of pneumococcal antibodies or antigens in the serum of the patient.

✴ Treatment

Until recently, most pneumococci have been susceptible to penicillin; and only occasionally was a resistant strain isolated. However, multiple antibiotic-resistant *S. pneumoniae* were isolated in South Africa in 1977. These strains have spread throughout the western world such that many of *S. pneumoniae* isolated in Europe and America are now resistant to penicillin. Treatment with penicillin or other appropriate antibiotics should be started as soon as pneumococcal pneumonia is suspected. The longer

treatment is delayed, the more difficult it is to cure the disease. The response to therapy, when initiated early, is often dramatic and bacteria are cleared from the system in a few hours. The effects of both sulfa drugs and penicillin on pneumonia helped justify the title of "wonder drugs" that was applied to these agents shortly after their discovery ∞ (Chapter 9, p. 127). For the first time in history, the number one killer of humans could often be cured with relative ease and rapidity. When penicillin cannot be used because of allergic reactions or resistant strains, treatments with erythromycin, chloramphenicol, or vancomycin are used.

✴ Prevention and Control

Good nursing and medical management practices are effective in minimizing or preventing the development of pneumonia in compromised hospital patients. **Prophylactic treatment** with penicillin or other antibiotics might be warranted in high-risk patients with viral respiratory disease or other conditions that might predispose the tissues to pneumococcal infections. Even though pneumonia is often caused by endogenous bacteria, exogenous transmission may also be important in a hospital or nursing home environment, and compromised patients should be isolated from pneumonia patients. In extreme cases, such as highly immunosuppressed organ transplant patients, protective isolation from all persons is necessary. Elective surgery should not be performed on a person with a respiratory infection. Hospitalized or confined patients should be required, if possible, to sit up and get out of bed periodically to prevent the pooling of fluids in the lungs. Maintaining good circulation and healthy "dry" lungs is the best preventive measure against nosocomial (hospital-acquired) pneumococcal pneumonia.

Before the development of effective antibiotics, vaccines against pneumonia were used with some positive effect. Because of the ongoing increased death rates from pneumonia in both elderly and compromised patients, plus the difficulty of treating pneumonia in many of these persons, renewed interest in antipneumococcal vaccines was stimulated. A pneumococcal vaccine (Pneumovax) has been available in the United States since 1978. The present vaccine contains the capsular antigens from 23 of the commonly encountered serotypes that are responsible for about 90% of serious pneumococcal disease. Although data regarding the usefulness of this vaccine are varied, Pneumovax is still recom-

---

Bile A fluid secreted by the liver to aid the emulsification of fats.
Prophylactic treatment A preventive treatment against infection and disease.

TABLE 15-2			
Purified Capsular Polysaccharide Antigens (Types) Contained in *Streptococcus pneumoniae* Vaccine			
1	7F	12F	19F
2	8	14	20
3	9N	15B	22F
4	9N	17F	23F
5	10A	18C	33F
6B	11A	19A	

mended for young children, elderly persons, and others with health conditions that would predispose them to serious pneumococcal disease. Table 15-2 lists the pneumococcal antigens presently included in the vaccine.

OTHER DISEASE-CAUSING STREPTOCOCCI

✷ Group B

Lancefield's group B streptococci (*S. agalactiae*) are commonly found in the genital and intestinal tracts of healthy persons. They are also widely (2–40%) found in the vagina of pregnant women. There are six capsular serotypes of *S. agalactiae* (Ia, Ib/c, Ia/c / II, III, and IV). Although asymptomatic colonization is about the same for types I, II, and III, type III is most frequently responsible for human disease. Group B streptococci may cause urinary tract, ear, and wound infections in adults, but they are most important as a cause of neonatal (newborn) infection. Newborn infants may become colonized during passage through an infected birth canal and (1.3 to 3/1000 live births) may develop serious **septicemia**, meningitis, and pneumonia, with a mortality rate of about 55%. This "early onset" form of the disease is often associated with complicated or premature deliveries. Another form of the disease, "late onset," affects infants more than 7 days of age (average 24 days). These infants are infected after birth and most commonly have meningitis and a lower (23%) mortality rate. Type III *S. agalactiae* is a leading cause of meningitis in infants under 4 months of age in the United States. Prompt treatment with penicillin is effective in reducing the number of fatal infections.

✷ Group C

Occasionally species of group C streptococci have been implicated as the cause of such infections as impetigo, abscesses, pneumonia, and pharyngitis.

CLINICAL NOTE

Group B Streptococcal Disease in the United States, 1990: Report From a Multistate Active Surveillance System

Group B streptococcal (GBS) disease is the most common cause of neonatal sepsis and meningitis in the United States. It is also an important cause of morbidity among pregnant women and adults with underlying medical conditions. Because most states have not designated GBS disease as a reportable condition, previous estimates of the incidence of GBS disease were based on studies from single hospitals or small geographic areas. This report summarizes the results of population-based active surveillance for invasive GBS disease in counties within four states that had an aggregate population of 10.1 million persons in 1990. A case of GBS disease was defined as isolation of group B streptococcus from a normally sterile anatomic site in a resident of one of the surveillance areas.

Age- and race-adjusted projections to the U.S. population suggest that >15,000 cases and >1300 deaths due to GBS disease occur each year. The projected age- and race-adjusted national incidence is 1.8/1000 live births for neonatal GBS disease and 4.0/100,000 population per year for adult GBS disease. Intrapartum chemoprophylaxis for pregnant women at risk for delivering infants with GBS disease is the most effective strategy available for prevention of neonatal disease. Development of effective GBS vaccines may prevent GBS disease in both infants and adults. Ongoing surveillance for GBS disease is important for targeting preventive measures and determining their effectiveness (*MMWR* 42:103, 1993).

Septicemia   The presence of bacteria in the blood.

Glomerulonephritis, but not rheumatic fever, has resulted from previous group C infections.

＊ Group D

Group D streptococci are significant human pathogens, causing endocarditis, sepsis, and urinary tract infection. Three species (*S. durans, S. faecalis, S. faecium*) of this group are designated as *enterococci*, and are separated from other group D organisms because of their ability to grow in media containing 6.5% NaCl. These organisms are normal inhabitants of the human gut and are resistant to a broad range of antibiotics. This group of organisms has recently been placed in their own genus, known as *Enterococcus*. Their resistance to antimicrobial therapy along with the serious nature of endocarditis due to these organisms (mortality rates range from 30 to 60%) make them of special interest in the medical community.

＊ Viridans Group

At least 19 species of non-beta-hemolytic streptococci are designated as viridans streptococci. They are the primary cause of up to 70% of all cases of bacterial endocarditis and are sometimes isolated from patients with deep wound infections, abdominal abscesses, and septicemia. These organisms are present in the normal human mouth and pharynx, and they are thought to be a contributing cause of tooth decay. Some species are very difficult to culture and will grow only where culture medium is supplemented with vitamin B_6. Most strains are susceptible to penicillin, and serious infections are often treated with penicillin in combination with an aminoglycoside. Persons with rheumatic fever damage to their heart valves should be given prophylactic antibiotics before oral or genital surgery to reduce the risk of subsequent endocarditis from these organisms.

MATERIAL FOR REVIEW

CONCEPT SUMMARY

1. The genus *Streptococcus* includes a large number of primarily commensalistic bacteria. These organisms are often found as normal flora in both animals and humans. They are gram-positive, and diseases produced by these organisms are suppurative.

2. The pathogenic streptococci are classified on the basis of a polysaccharide capsular antigen. The group A organisms are of particular concern to humans, although human infections are also caused by members of the other groups, notably groups B and D.

3. The relatively benign disease, streptococcal sore throat, is of importance primarily because of the delayed sequela associated with such infections. Most streptococcal diseases are readily treated with penicillin.

CLINICAL SUMMARY TABLE

Microorganism	Virulence Mechanisms	Diseases	Transmission	Treatment	Prevention
Streptococcus pyogenes	Hemolysins Fibrinolysin Hyaluronidase M-protein Protein F Erythrogenic toxin Pyogenic toxin Immune complexes	Sore throat Scarlet fever Skin and wound infections Toxic shock-like syndrome Puerperal fever Glomerulonephritis Rheumatic fever	Person to person Airborne contact	Penicillin Erythromycin Cephalosporin	No vaccines Isolation Good aseptic procedures Prophylactic antibiotics
Streptococcus pneumoniae	Capsule Pneumolysin	Pneumonia Pleurisy Otitis media Meningitis	Person to person Opportunists	Penicillin or Other antibiotics	Vaccine Good nursing practices Prophylactic antibiotics

STUDY QUESTIONS

1. What component of the streptococcal cell is used as an antigen in determining serologic classification?
2. Streptococcal pharyngitis is a relatively benign disease without mortality. What feature of this disease makes it important for patients to receive antibiotic therapy?
3. What is the usual source of group A streptococci that is transmitted from person to person?
4. Name the primary clinical concerns (diseases) associated with each of the following streptococci: (a) group A, (b) group B, (c) group D, and (d) viridans.
5. What is the present theory as to the basis of rheumatic fever?

CHALLENGE QUESTIONS

1. If *Streptococcus pneumoniae* is isolated from a patient's throat, does this indicate the patient has pneumonia? Explain.
2. Would the cell walls from pneumococcus make a good vaccine? Explain.
3. Why would physicians and clinicians find a broad-spectrum antibiotic to be popular for treating pneumonia?

CHAPTER

16 *Neisseria*

OUTLINE

Bacteria of the genus *Neisseria* are the fourth of the major pyogenic cocci to be presented. These organisms differ from the other pyogenic cocci both in their fundamental properties (they are gram-negative) and in the specific kinds of diseases that they produce. There are several species of *Neisseria*, and two of them have taken their species names from the disease they produce: *Neisseria meningitidis* is the leading cause of young adult meningitis, and *Neisseria gonorrhoeae* is the cause of the most frequently reported sexually transmitted disease, gonorrhea. Although these two species are similar in structural and physiological makeup, the diseases they cause are clinically quite different and are discussed separately.

BACTERIA

The *Neisseria* are gram-negative diplococci (paired cocci, Figure 16-1) which usually have a common flattened or concave side. They are a little less than 1 μm in diameter and are nonmotile. The pathogenic strains **ferment** sugars (Table 16-1).

These organisms are environmentally fragile and are readily killed by exposure to drying, chilling, sunlight, acids, or alkalies. They can be cultivated on laboratory media if special care is taken to prevent inactivation. Growth occurs best on blood or chocolate agar (a medium enriched with blood and heated, which turns the blood a chocolate color) in

| **Ferment** | Microbial production of acid from carbohydrates. |

HUMAN BODY SYSTEMS AFFECTED

Neisseria

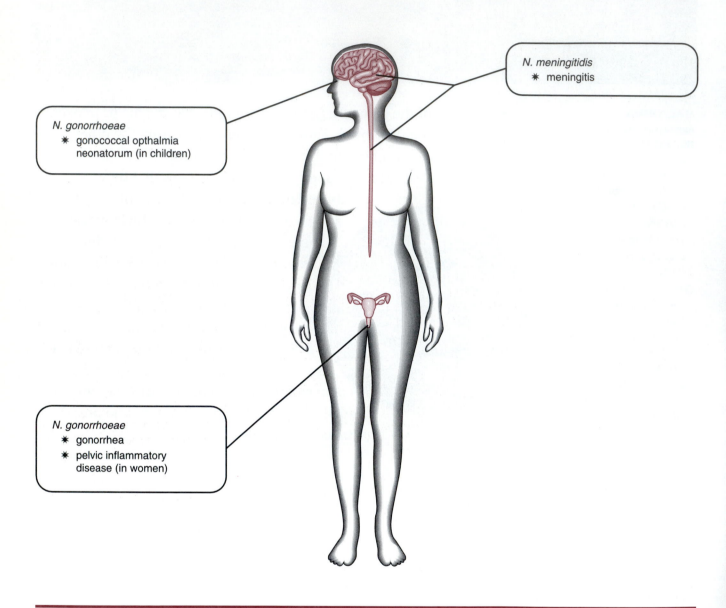

N. gonorrhoeae
* gonococcal opthalmia neonatorum (in children)

N. meningitidis
* meningitis

N. gonorrhoeae
* gonorrhea
* pelvic inflammatory disease (in women)

a CO_2-enriched environment (see Plate 12). The two species can be differentiated on the basis of sugar fermentation tests. Most *N. meningitidis* strains produce apparent capsules, whereas *N. gonorrhoeae* strains produce very slight capsules. Pili are present on *N. gonorrhoeae*. Even though the *Neisseria* possess typical gram-negative-type cell walls, they are susceptible to penicillin. *N. meningitidis* is commonly referred to as the *meningococcus* and the *N. gonorrhoeae* as the *gonococcus*. Nonpathogenic species of *Neisseria* are found on the mucosal surfaces of the human alimentary and genitourinary tracts.

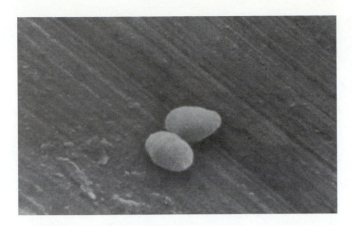

FIGURE 16-1 Scanning electron micrograph of *Neisseria gonorrhoeae* showing the typical diplococcal arrangement of the "bean-shaped" cocci (magnification 10,000×). (A. S. Klainer and I. Gies, *Agents of Bacterial Disease*. Harper & Row: Hagerstown, MD, 1973. Figure 7-1, p. 83)

MENINGOCOCCAL INFECTIONS

＊ Pathogenesis and Clinical Diseases

Based on a polysaccharide capsule, *N. meningitidis* can be divided into 13 different serogroups. Group A (sometimes called the *epidemic strain*) is most often the cause of meningococcal disease throughout the world, but in the United States groups B, C, Y, and W-135 are most often involved (see Clinical Note). The main mechanisms of pathogenicity and virulence appear to be the capsule that helps retard phagocytosis and cell wall endotoxin that is thought to be responsible for most of the toxic effects of the disease. The meningococci are widespread in the general population and are readily transmitted via the airborne route ∞ (Chapter 13, p. 192). They multiply in the nasopharyngeal area without causing any disease; an infected individual may carry these organisms for many months. Such a **carrier** state may induce an immune response in the person and cause resistance to the given serogroup with which that person is colonized. This carrier, however, serves as a reservoir from which the bacteria may be transmitted to those who live in close contact with the person. The carrier rate varies in different populations. The carrier rate is 3 to 30% in a general population in which no clinical disease is present, compared to a rate of 15 to 50% in persons who live or work around patients with clinical meningococcal diseases, and may reach 80% during an epidemic. The carrier rate in military personnel is generally higher than in other populations.

Once meningococci are colonized in the respiratory membranes, the person may experience mild sore throat and fever. Meningococci may be carried into the lymphatic system and then into the blood. If the host is unable to contain the infection at this point—and no information is available as to how many infected persons progress to this point without symptoms developing—the bacteria become deposited in various tissues, such as skin, **meninges**, joints, and lungs. In a few days lesions may develop in these tissues with the manifestation of signs and symptoms of the disease. In the dis-

TABLE 16-1

Biochemical Identification of *Neisseria* Species

Species	Acid From			
	Glucose	**Maltose**	**Sucrose**	**Lactose**
N. meningitidis	+	+	−	−
N. gonorrhoeae	+	−	−	−
N. lactamica	+	+	−	+
N. sicca	+	+	+	−
N. cinerea	−	−	−	−

Carrier A living host that is infected by an organism but does not have clinical symptoms of disease. Carriers may have asymptomatic infections or may be individuals who have recovered from the disease but continue to shed the causative agent into the environment.

Meninges Membrane coverings of the brain and spinal cord.

seminated infections, **hemorrhagic** lesions may occur in the involved tissue, along with high fever and prostration. Subcutaneous hemorrhagic lesions sometimes occur, giving the person a spotted appearance.

The disease *spinal meningitis*, more specifically called *meningococcal meningitis*, is a relatively rare outcome of the much more common meningococcal infections. Meningitis is accompanied by various **neurologic** symptoms, including headache, vomiting, and stiffness in the neck. The death rate in untreated cases may be as high as 85%. With treatment, however, the overall death rate is 8 to 10%. *Disseminated meningococcal disease* may also occur without meningitis and has a high death rate as well.

The presence of specific antibodies in the serum offers protection against the disseminated disease but apparently does not cure the carrier stage in the nasopharyngeal tissues. Most adults have acquired antibody immunity against meningococci due to previous subclinical infections, and most infants have passive immunity during the first months of life. The pathogenesis of this disease is outlined in Figure 16-2.

✳ Transmission and Epidemiology

Transmission of meningococci is usually via the airborne route. Because of the large number of carriers in the general population, transmission may occur at any time; however, transmission is more efficient from a person with a clinical infection, and a significantly higher number of cases occur in a patient's household contacts than in the general population. Children are frequently exposed early in life and a higher incidence of disseminated disease is seen in the 6- to 24-month age range. This represents the period between loss of passive immunity and acquisition of active immunity ∞ (Chapter 12, p. 174). For less obvious reasons, the next age groups that are most susceptible are the 10- to 20-year-olds. More than two-thirds of all cases occur in persons under 20 years of age. Many meningococcal disease outbreaks have been associated with military recruits. The higher number of cases in military recruits probably results from close personal contact in barracks, and the increased exposure due to the high carrier rate in military personnel.

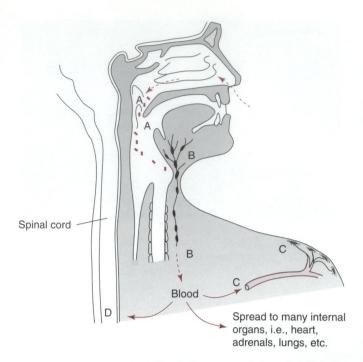

FIGURE 16-2   The pathogenesis of meningococcal infections. Stages of infection: A, Bacteria may be implanted in the nasopharynx. This phase of the infection is often asymptomatic, but in some cases may cause a sore throat. B, Bacteria may pass into the bloodstream via the lymphatic system. C, Septicemia with inflammation of blood vessels and areas of hemorrhaging in the skin. Much of the damage is caused by endotoxins. The disease may range from mild to rapidly fatal. D, Meningitis, with headache, fever, and signs of meningeal irritation (rash may or may not be present).

Sporadic cases are seen in most areas; however, epidemic meningitis does occur periodically in areas throughout the world. The sporadic cases run about 3 per 100,000 in the general population per year, whereas the rates during an epidemic may increase to several hundred per 100,000 per year. The annual case rate in the United States is presented in Figure 16-3 and shows between 2000 and 3000 cases per year.

✳ Diagnosis

Direct microscopic examination of infected tissues or fluids may reveal the presence of the meningococci and provide a rapid diagnosis. Inoculation

Hemorrhagic   Associated with bleeding. Hemorrhagic lesions of the skin often appear as red spots or blemishes. These lesions frequently become dark blue or black in color and may involve large areas of the skin.
Neurologic   Pertaining to the nervous system, either central (brain and spinal cord; CNS) or peripheral (nerves extending from the CNS to other body tissues).

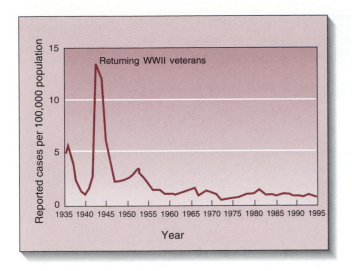

FIGURE 16-3 Reported cases of meningococcal infection per year in the United States from 1935 and 1994. (from CDC annual summaries)

onto appropriate nutrient agar is also used to demonstrate the presence of meningococci. Cerebral spinal fluid and other body fluids can be rapidly tested for the presence of *N. meningitidis* antigens. This procedure may help to provide a guide to specific therapy against the disease.

✴ Treatment

Intravenous administration of large doses of penicillin is the most effective treatment. Broad-spectrum antibiotics such as chloramphenicol and cephalosporins may be used in penicillin-allergic patients. The sulfonamides are effective if resistant strains are not causing the disease, although currently about one-half the isolates are resistant to the sulfonamides. Treatment is much more effective if started early in the infection. If meningitis is suspected in a young child, immediate treatment is essential, for the disease may progress rapidly and lead to death or permanent neurologic damage within less than 24 hours after the onset of severe symptoms.

✴ Prevention and Control

Quadrivalent vaccines have been developed against serogroups A, C, Y, and W-135. These vaccines use purified capsular antigens and a single dose produces a good antibody response. These vaccines are used in selected populations at special risk. Since 1971, military recruits have been vaccinated with serogroup C vaccine and school children in coun-

tries where epidemics occur have been vaccinated with serogroup A vaccine. No serious adverse reactions have occurred in those vaccinated, and serogroup C meningococcal meningitis has been virtually eliminated in military recruit populations. Before 1971, serogroup C was frequently the cause of meningitis in military recruits. These vaccines do not seem to be effective in children under 2 years of age. The vaccine might also be used along with antibiotics in the prophylactic treatment of household contacts of meningococcal disease.

Household contacts and hospital personnel who are exposed to clinical cases should be given prophylactic doses of the antibiotic rifampin to reduce the chance of secondary cases occurring. Meningococcal patients in the hospital should be kept in isolation.

GONORRHEA

The term *gonorrhea* was introduced by the ancient physician Galen; it means the "flow of seed" and refers to the flow of pus from the urethra of infected persons. Diseases that may have been gonorrhea were described in ancient medical writings, and in the thirteenth century, gonorrhea was recognized as a sexually transmitted disease. Not until many years later, however, was a clear distinction made between gonorrhea and syphilis, because the two diseases often occurred simultaneously in the same person. Today, gonorrhea is among the most prevalent of the sexually transmitted diseases and is the most frequent reportable disease in the United States.

Gonorrhea usually responds well to penicillin therapy; in fact, the availability of this drug in the 1940s made people believe that gonorrhea could be controlled or eliminated. Nevertheless, the liberalized sexual attitude since the 1960s, together with the development and use of oral contraceptives, caused a marked upsurge in the number of cases of gonorrhea. The epidemic of gonorrhea peaked at more than one million reported cases per year for a number of years. However, during the past several years there has been a general decline in the reported cases of this disease. The development of penicillin resistance among some strains may make future control of this disease more difficult.

✴ Pathogenesis and Clinical Diseases

Virulence factors associated with *N. gonorrhoeae* include pili that are involved with attachment to the

Serogroup B Meningococcal Disease: Oregon 1994

In Oregon, the incidence of meningococcal disease has increased substantially, more than doubling from 2.2 cases per 100,000 persons in 1992 to 4.6 per 100,000 in 1994—the highest incidence in Oregon since 1943. This incidence was almost fivefold higher than recent estimates for the United States during 1989–1991 (approximately one case per 100,000 persons annually). This report describes meningococcal disease surveillance data from 1994 and summarizes epidemiologic and laboratory data on serogroup B meningococcal disease in Oregon during 1987–1994.

During 1994, a total of 143 cases of meningococcal disease was reported to the State Health Division. In 124 cases, *Neisseria meningitidis* was isolated from a normally sterile site (confirmed cases); in four cases, gram-negative diplococci were detected in specimens obtained from a normally sterile site or in persons who had classic symptoms after contact with a confirmed case (presumed cases). Characteristic symptoms (including petechial rash and hypotension) occurred in 14 cases; however, these cases were not culture confirmed (suspected cases). Of 115 isolates for which serogroup was known, 70 (61%) were serogroup B, 40 (35%) were serogroup C, four were serogroup Y, and one was serogroup W-135. When compared with 1992 and 1993, the serogroup-specific incidence in 1994 was higher for both serogroups B and C.

Of the 70 culture-confirmed cases of serogroup B infection, 34 (49%) occurred in females. Seven (10%) cases were fatal; of these, one occurred in a child aged 2 years, and four deaths occurred in persons aged 55–88 years.

During 1987–1992, 63% (84 of 133) of cases of serogroup B occurred in children aged <5 years; in comparison, in 1994 27% (19 of 70) occurred in this age group. When compared with 1987–1992, the incidence of reported serogroup B disease in 1994 increased modestly among those aged <5 years (from 6.9 to 8.4), approximately 14-fold among those aged 15–19 years (from 0.4 to 5.4), and approximately fourfold among those aged ≥60 years (from 0.3 to 1.1).

In 1994, serogroup B cases occurred in 10 of the 36 counties in Oregon; these counties account for 83% of the total population of Oregon. The risk for disease was highest in counties in the Willamette Valley in the northwestern part of the state. Based on investigation of serogroup B cases, six (9%) were linked to other cases. Two coprimary cases (disease in a close contact within 24 hours of disease onset in a primary case) were linked to a single primary case. Four secondary cases (disease in a close contact >24 hours after disease onset in a primary case) were identified; at least two occurred in patients for whom appropriate chemoprophylaxis had been prescribed but who were noncompliant with therapy.

Of the 114 *N. meningitidis* serogroup B strains isolated in Oregon during 1993–1994, a total of 64 (56%) have been characterized at CDC by multilocus enzyme electrophoresis. Of these, 55 (86%) belong to the enzyme type-5 (ET-5) complex, a group of genetically related serogroup B meningococcal strains associated with epidemic meningococcal disease in other countries.

Editorial Note: The recent increased occurrence of serogroup B meningococcal disease in Oregon has been associated with a group of closely related strains belonging to the ET-5 clonal complex. These strains were first identified as the cause of a serogroup B meningococcal epidemic in Norway that began in 1974 and persisted through 1991. After its identification in 1974, serogroup B meningococci belonging to the ET-5 complex subsequently caused epidemics in Europe, Cuba, and South America. Endemic meningococcal disease typically is caused by a heterogeneous mix of strains. In comparison, the predominance of closely related strains, or clones, in Oregon is characteristic of epidemic disease, as is the disproportionate increase in age-specific incidence among young adults. The latter pattern has been suggested as a reliable predictor of the transition from endemic to epidemic meningococcal disease (*MMWR* 44:121, 1995).

mucosal tissues of humans. A complex cell wall structure of the gonococci appears to possess mechanisms that allow these bacteria to persist in cells (see Plate 20), induce inflammation, and avoid the immune responses of the host.

The patterns of gonorrhea vary somewhat among the male, female, and the newborn. The clinical diseases are discussed separately for these three groups.

Gonorrhea in the Male. The gonococci are usually deposited in the lower (anterior) portion of the urethra during sexual intercourse, and the bacteria attach to the surface cells of the urethra. The growing gonococci induce an acute inflammatory response in 2 to 8 days that may be accompanied by fever, a purulent discharge, and pain during urination. In most cases, the male experiences definite symptoms when infected. The infection may spread by direct extension along the ducts of the genitourinary tract with accompanying inflammation. In about 1% of the cases, the inflammation may close the urethra and prevent urination. The infection may also spread to the *vas deferens* (sperm ducts) and cause scarring and closure of this duct, which then results in infertility of the patient. In some cases, the bacteria spread into the blood and are carried to various internal organs or bone joints where inflammation (*septic arthritis*) may develop. *Pharyngitis* and rectal inflammation (*prochitis*) may occur, particularly among bisexuals and homosexuals.

Gonorrhea in the Female. Infection usually results from gonococci transmitted during sexual intercourse. The ensuing infection is of the cervical–vaginal junction and often remains at a low level; in up to 80% of the infected females it is asymptomatic. Those patients with signs of the disease experience vaginal discharge, fever, a burning sensation, and abdominal pain. The duration of the asymptomatic infections is not known, but some evidence indicates that the infection may persist for months. Women using oral contraceptives are thought to be more susceptible to gonorrhea due to changes in the pH of the vaginal mucosa induced by birth control pills.

As with males, the gonorrhea may spread into the blood and be carried to other organs in a small percentage of infected females. In women of child-bearing age, *N. gonorrhoeae* is the leading cause of septic arthritis. Rectal gonorrhea is seen in about half the infected females. Pharyngitis and urethritis may also be seen. The abdominal pains are normally associated with spread of the infection to the fallopian tubes, a serious condition known as *salpingitis*. In 20% of these women, this form of gonorrhea may induce scarring with closure of the fallopian tubes and resultant loss of fertility. The fallopian tubes are also predisposed to secondary infections by other bacteria.

Further extension of the infection into the lower

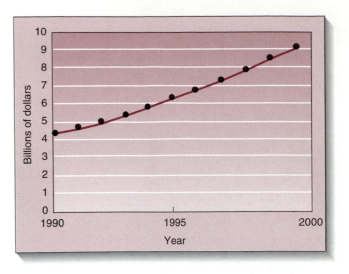

FIGURE 16-4 Total projected cost (direct and indirect) associated with pelvic inflammatory disease in the United States, 1990–2000. (Courtesy of Centers for Disease Control, Atlanta)

abdomen occurs in 10% of the women with gonorrhea and is called **pelvic inflammatory disease (PID)**. PID is an acute or chronic infection in women. Multiple etiologic agents can cause the disease, including streptococci, anaerobic bacteria, *Chlamydia* and, most important, *N. gonorrhoeae*. It is estimated that more than 275,000 women are hospitalized for PID each year in the United States. The projected medical cost associated with PID in the United States is shown in Figure 16-4. These women are at an increased risk of chronic pelvic pain, *ectopic pregnancy*, and tubal infertility. They also have a risk of disseminated infection, which can produce fatal disease. There is no single risk factor that leads to PID. However, individuals who maintain a healthy sexual behavior (limited number of sex partners, avoiding sex with high-risk partners, following procedures that reduce the likelihood of developing sexually transmitted diseases, and postponing initial sexual intercourse until older) are significantly less likely to develop PID.

Gonorrhea in Children. Persons of all ages are susceptible to gonorrhea. A common form of childhood gonorrhea is seen as an eye infection (*gonococcal ophthalmia neonatorum*) of the newborn.

Pelvic inflammatory disease (PID) A serious intra-abdominal infection in females. PID usually follows vaginal or uterine infection, and is most often caused by one or more of the sexually transmitted disease agents. Reproductive sterility is a common outcome, but in some instances death may occur from PID.
Ectopic pregnancy A pregnancy occurring "out of place"—with embryo implantation occurring outside the uterus.

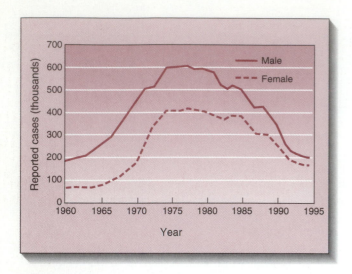

FIGURE 16-5 The number of cases of gonorrhea in males and females in the United States, 1960 to 1994. (From CDC annual summaries)

Eye infection takes place during passage through the birth canal of a mother who has gonorrhea. If not treated, severe inflammation of the eyes and even blindness may result. The serious nature of gonococcal conjunctivitis in the newborn has led to the universally accepted practice of placing a few drops of 1% silver nitrate, 0.5% erythromycin, or 1% tetracycline directly onto the conjunctiva of the infant immediately following birth. Before routine disinfection of the eyes of newborns was required, gonococcal ophthalmia was the major cause of blindness in children. Any gonococcal infection of other mucosal tissues of children is a strong indication of sexual abuse.

✳ Transmission and Epidemiology

Gonorrhea in adults is almost always transmitted by sexual contact. Increases and decreases in disease rates are directly related to major social changes such as those associated with war or with liberalized attitudes toward sex. The reported cases of gonorrhea are shown in Figure 16-5. The rate of gonorrhea increased steadily through the 1960s and into the 1970s in most parts of the world. The number of cases leveled off in the mid–1970s and decreasing numbers of cases have been noted through the 1980s. These decreases are thought to be associated, in part, with control programs directed against the more serious sexually transmitted diseases, such as AIDS. Most cases of gonorrhea are seen in the ages and social groups that are most sexually active. The highest rate of disease is seen in the 15- to 30-year-age groups, with a peak rate in the 20-

to 24-year-age groups. High rates are also seen in subpopulations of individuals who have promiscuous sexual contacts. Though nearly half a million cases of gonorrhea are officially reported in the United States each year, the actual number is thought to be three or four times greater. The number of cases of gonorrhea varies among the states and is shown in Figure 16-6.

Immunity to gonorrhea infection does not develop and a person may be infected over and over again. The probability of a male becoming infected after sexual intercourse with an infected female is about 20%. The probability of the female contracting the disease from an infected male is thought to be greater; however, it is not as well determined because many females contract asymptomatic infections that go undetected.

✳ Diagnosis

Diagnosis is much easier in males than in females. The purulent discharge is examined with the microscope for the presence of intracellular gram-negative diplococci; if positive, the diagnosis is confirmed. If the microscopic test is negative, an attempt should be made to isolate the gonococcus on a culture medium to provide a positive diagnosis. It is much more difficult to observe gonococci microscopically in smears of exudate from infected females, and this procedure is of limited diagnostic value in women. Cultures should be obtained from the cervix and anal canal and, if possible, inoculated directly onto a gonococcal selective agar medium, such as modified Thayer–Martin medium. If the specimen must be transported to the laboratory, special handling methods are required to prevent the inactivation of these delicate microorganisms. Because the diagnosis of asymptomatic infections in females is often difficult, there is currently a great need for a simple reliable test to detect these infections. Serologic tests are not well developed and are of limited value at present.

✳ Treatment

Until recently, penicillin was still the drug of choice in treating gonorrhea. Over the years, however, gonococci have developed increased resistance to penicillin, and in 1976 a strain of *N. gonorrhoeae* that was completely resistant to penicillin was discovered. Penicillin-resistant strains carry a plasmid that directs the formation of the enzyme β-lactamase (penicillinase) that is able to inactivate penicillin G, ampicillin, and cephalosporins.

These penicillin-resistant strains and other resistant gonococci now make up a sizable portion (up

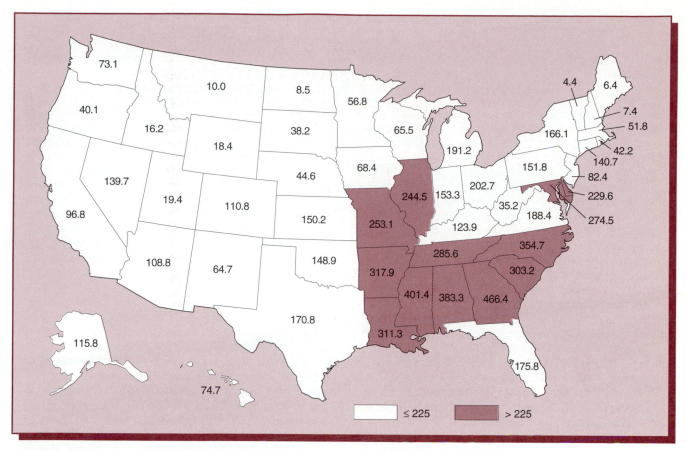

FIGURE 16-6 Reported cases of gonorrhea, per 100,000 population, within states of the United States in 1993. The year 2000 objective is ≤225 per 100,000. (From CDC annual summaries)

to 45%) of the strains causing disease. In addition, many persons who are infected with gonococci are coinfected with chlamydiae (Chapter 27). Chlamydial infections are difficult to diagnose and do not respond to penicillin treatment. Because of the above conditions, the current recommended treatments for gonorrhea are directed against both the resistant gonococci and chlamydiae. The recommended treatment for uncomplicated gonorrhea is a single intramuscular injection of ceftriaxone plus doxycycline given orally for 7 days to cover any undiagnosed chlamydial infection. Alternative treatments may use spectinomycin combined with doxycycline or a single oral dose of ciprofloxacin, cefixime, or ofloxacin plus the doxycycline regimen.

✳ Prevention and Control

No vaccines are currently available. Gonorrhea in the newborn is prevented by irrigating the eyes with a 1% solution of silver nitrate, 0.5% erythromycin, or 1% tetracycline. This procedure is required by law and has almost eliminated gonococcal ophthalmia in the newborn. If a pregnant female has gonorrhea, she should be treated before delivery to further reduce the hazard to the newborn infant.

Prevention and control of gonorrhea are complicated by the **occult** nature of the infection in the female. Extensive studies involving more than 1.5 million women visiting a wide variety of medical services indicated that the rate of infection varied from 19.7% of those visiting a sexually transmitted disease clinic to 1.4% of women who were military dependents. Overall, the rate of infection was 5.7% of all females tested. Gonorrhea in adults is preventable by avoiding exposure to the disease. Still, history has clearly demonstrated that prevention will not be accomplished by this means. The risk of exposure can be reduced by using such devices as

Occult   Hidden or unknown.

CLINICAL NOTE

Increasing Incidence of Gonorrhea: Minnesota 1994

In the United States, gonorrhea is an important cause of urethritis in men and cervicitis in women; reproductive complications include infertility and ectopic pregnancy. During 1981–1993, the annual incidence rate of gonorrhea in Minnesota declined; the average annual change in the rate of infection was −8.5%. However, in 1994, the incidence rate increased 32% (from 56 cases per 100,000 persons in 1993 to 74 cases per 100,000 in 1994). No corresponding increases occurred in rates of other reportable sexually transmitted diseases (STDs), including chlamydial infection and early syphilis. To elucidate possible explanations for the increased rate of gonorrhea in Minnesota in 1994, the Minnesota Department of Health (MDH) analyzed surveillance data for 1994 and compared it with data for 1993. This report presents the findings of the analysis.

In 1994, a total of 3346 gonorrhea cases were reported to MDH, compared with 2543 cases in 1993. From 1993 to 1994, the incidence rate of gonorrhea increased at least 30% in Minneapolis and St. Paul and in the remainder of the seven-county Minneapolis–St. Paul metropolitan area; in rural areas of the state, the rate increased 17% but remained low (i.e., <10 cases per 100,000). Six urban zip code areas accounted for 49% of all gonorrhea cases but represented only 5% of the state's population.

From 1993 to 1994, the rate of gonorrhea in Minnesota increased 14–44% for all racial/ethnic groups; the rate was highest for non-Hispanic blacks. Sex-specific rates increased approximately 30% and were similar for men and women. Age-specific rates increased 20–86% for all age groups except 10- to 14-year-olds; rates were highest among adolescents (i.e., 15- to 19-year-olds).

From 1993 to 1994, the increase in reported cases varied by reporting source. During this period, the number of gonorrhea cases reported by STD clinics increased 28% (from 1120 to 1430, respectively) and by all other sources increased 34% (from 1423 to 1916, respectively). In addition, the related increase in positive cultures for gonorrhea varied by laboratory testing source. At the two STD clinics in Minneapolis and St. Paul that accounted for most (43%) cases during 1993 and 1994, all clients were tested for gonorrhea. These clinics submitted 18,032 culture specimens to the Minnesota Public Health Laboratory (MPHL) in 1994. Although specimen collection, handling procedures, and volume of tests were unchanged at the two clinics, the percentage of cultures in 1994 that were positive for *Neisseria gonorrhoeae* increased 24% (from 6.7 to 8.3%) and 28% (from 6.0 to 7.7%). Of the five clinics that each submitted ≥1500 gonorrhea cultures to the MPHL in 1994, the proportion of positive cultures increased substantially for only one clinic. For 16 private and hospital-based laboratories, the proportion of all tests (i.e., culture and nonculture) that were positive increased from 1.7% (409 of 24,531) during the fourth quarter of 1993 to 1.9% (491 of 26,231) during the fourth quarter of 1994. From 1993 to 1994, the proportion of gonorrhea patients who were interviewed by health department staff (30%) to identify and treat sex partners remained constant.

Testing for antimicrobial resistance was performed on every fourth *N. gonorrhoeae* isolate identified at the MPHL; in 1994, a total of 443 isolates were tested. All were susceptible to ceftriaxone and ciprofloxacin, two of the recommended therapies for gonorrhea.

Editorial Note: Gonorrhea is a major cause of pelvic inflammatory disease and may play a role as a cofactor in human immunodeficiency virus transmission. During 1975–1993, the rate of reported gonorrhea decreased 65% in the United States, from a peak of 467.7 cases per 100,000 persons to 165.8 per 100,000. Despite the decline, gonorrhea rates in the United States remain the highest among developed countries.

The surveillance findings in Minnesota probably reflect a real increase in the incidence of gonorrhea because reported cases increased in all age and race groups without apparent change in program activities, reporting practices, or laboratory procedures. In addition, the proportion of positive cultures increased at the MPHL. Rates remained highest for adolescents, non-Hispanic blacks, and residents of urban areas. National surveillance data also indicate high incidence in these groups. Adolescents and young adults are at increased risk for gonorrhea because they are more likely to have multiple sex partners, to have unprotected sex, and to select partners at increased risk. In 1993, 81% of the total reported cases of gonorrhea in the United States occurred among blacks; although explanations for the high rates among blacks are undetermined, race may be a marker strongly associated with risk factors for STDs, such as low socioeconomic status, limited access to health care, poor health-care seeking behavior, illicit drug use, and residence in communities with high prevalences of STDs. In Minnesota, the concentration of gonorrhea cases in some zip code areas suggests that the disease is highly focal, and intervention should be targeted geographically (*MMWR* 44:282, 1995).

condoms and reporting cases of gonorrhea to public health officials to aid in finding and treating persons who might be serving as sources of infection. Educational programs have been tried with various age groups in attempts to reduce the incidence of gonorrhea; These programs have had varying degrees of success.

MATERIAL FOR REVIEW

CONCEPT SUMMARY

1. The gram-negative diplococci (*Neisseria*) include two species of major clinical significance: *N. meningitidis*, the etiologic agent of epidemic meningitis, and *N. gonorrhoeae*, the causative agent of gonorrhea.
2. *N. meningitidis* is often found as normal body flora and yet is responsible for a fulminant infection that causes rapid mortality in its victims. The balance between a normal flora and disease due to this organism is not well understood.
3. Gonorrhea has become such a common disease that it is literally a household word. The epidemic state of this disease and the ease with which it is generally treated belie its potential for producing life-threatening infections.

CLINICAL SUMMARY TABLE

Microorganism	Virulence Mechanisms	Diseases	Transmission	Treatment	Prevention
Neisseria meningitidis	Capsule Endotoxin	Spinal meningitis Disseminated infection	Airborne Person to person	Penicillin Broad-spectrum antibiotics Sulfonamides	Vaccines Prophylactic antibiotics
Neisseria gonorrhoeae	Attachment pili Induces inflammation Endotoxins	Gonorrhea	Sexually Mother to newborn	Penicillin Ceftriaxone & doxycycline	Avoid exposure Treat newborns

STUDY QUESTIONS

1. The *Neisseria* are pyogenic cocci, which means that they are _____ .
2. What is the usual mode of transmission and the reservoir for (a) *N. meningitidis* and (b) *N. gonorrheae*?
3. What is the basis for protection against infections due to (a) *N. meningitidis* and (b) *N. gonorrheae*?
4. List two reasons why a microorganism of relatively low virulence such as *N. gonorrheae* is of such great medical concern.
5. Why is gonorrhea more easily diagnosed in males than in females?
6. How can you account for a female infection rate due to *N. gonorrheae* of more than 5%, but a disease rate less than 0.1%?

CHALLENGE QUESTIONS

1. Today, few newborns are infected with *N. gonorrheae*, yet hospitals usually require antimicrobial compounds be added to the eyes of all newborns. Why don't hospitals simply treat only those babies whose mothers were identified as having gonorrhea?

2. Based on your current knowledge, propose a vaccine formulation for gonorrhea and identify what components would be in the vaccine.

17 Spirochetes

The spirochetes are members of a family called *Spirochaetaceae*, which includes long, slender, coiled, motile microorganisms (Figure 17-1). Three genera of spirochetes contain species that are able to cause diseases in humans: *Treponema*, *Borrelia*, and *Leptospira*. The most important of these genera in human disease is *Treponema*, which includes the etiologic agent of syphilis (*Treponema pallidum*). Other syphilis-like diseases, such as yaws, pinta, and bejel, are caused by bacteria either closely related to or identical with the syphilis spirochete. Lyme disease, which has gained prominence in recent years, is caused by the spirochete *Borrelia burgdorferi*. Two less-common diseases, relapsing fever and leptospirosis, are produced by spirochetes of the genera *Borrelia* and *Leptospira*, respectively.

TREPONEMA PALLIDUM: SYPHILIS

T. pallidum is a slender, tightly spiraled organism with a length of from 5 to 20 μm (Figure 17-1; see Plate 24). The width of these organisms is only about 0.2 μm, which makes observation with a standard bright-field microscope difficult. The treponemas are highly motile and move through fluid in a twisting motion. When a fresh, wet-mounted specimen is viewed with a dark-field microscope ∞ (Chapter 3, p. 28), the motile spirochetes are easily seen. *T. palladium* has not been grown on any artificial culture media and humans are the only natural host. Monkeys and rabbits, however, can be experimentally infected. The testes of rabbits support the growth of some syphilis spirochetes and are used to grow these bacteria for experimental and diagnostic

HUMAN BODY SYSTEMS AFFECTED

Treponema, Borrelia, and Leptospira

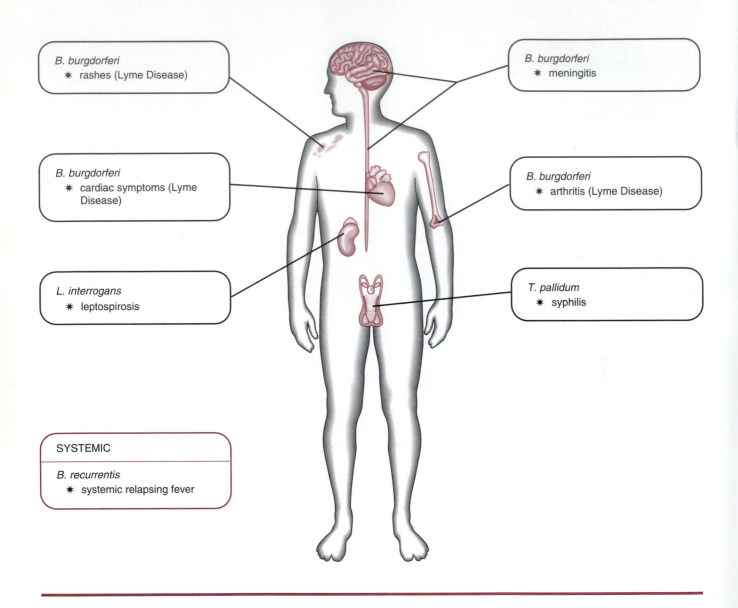

B. burgdorferi
* ✳ rashes (Lyme Disease)

B. burgdorferi
* ✳ cardiac symptoms (Lyme Disease)

L. interrogans
* ✳ leptospirosis

SYSTEMIC

B. recurrentis
* ✳ systemic relapsing fever

B. burgdorferi
* ✳ meningitis

B. burgdorferi
* ✳ arthritis (Lyme Disease)

T. pallidum
* ✳ syphilis

tests. This bacterium can be kept alive and motile for up to 2 weeks if stored at 25°C under anaerobic conditions in a special medium. *T. pallidum* is quite fragile and lives for only a short period when shed from the body. It is readily killed by disinfectants and is highly susceptible to penicillin and other antibiotics.

✳ Pathogenesis and Clinical Disease

Among sexually transmitted diseases (Table 17-1), syphilis follows only gonorrhea in frequency of reporting to the U.S. Public Health Service. However, syphilis is potentially far more serious than

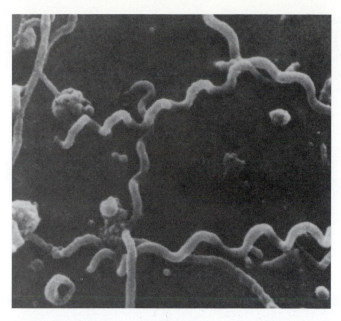

FIGURE 17-1   Scanning electron micrograph of the syphilis spirochete *T. pallidum* growing in rabbit testicular cells. (N. S. Hayes, K. E. Muse, A. M. Collier, and J. B. Baseman, *Infection and Immunity* 17:174–186, Figure 3c, with permission from ASM)

gonorrhea and is a disease that, when untreated, may go through several stages over an extended period of time with varying clinical manifestations. The stages of syphilis in adults are primary, secondary, latent, and tertiary (Figure 17-2). **Congenital** syphilis also occurs when the disease is transferred in utero from an infected woman to her fetus. Syphilis is somewhat unusual in that both specific and nonspecific antibodies are formed against the infecting organism, but neither are protective against reinfection. In fact, it appears that many of the clinical symptoms produced in syphilitic patients are the result of the host immune response to the organism.

Primary Syphilis.   Sexual contact is almost always the mode of transmission of *T. pallidum.* Treponema seem able to penetrate the mucosal tissues or gain entry through small lesions in the skin. The bacteria begin to multiply at the site of entry and are soon carried to the adjacent lymph nodes and eventually to the blood, by which they are spread throughout the body. As a rule, after 10 to 30 days, but in some cases up to 70 days, a primary lesion appears at the site of infection. This lesion, which is usually on the genital tissues, is called a

TABLE 17-1

Microbial Pathogens That Are Transmitted by Sexual Contact

Bacteria	Viruses	Fungi and Protozoa
Primarily Transmitted by Sexual Contact		
Neisseria gonorrhoeae	Human immunodeficiency virus	*Trichomonas vaginalis*
Ureaplasma urealyticum	Herpes simplex II	
Gardnerella vaginalis	Papillomavirus	
Haemophilus ducreyi	Cytomegalovirus	
Calymmatobacterium		
granulomatis		
Chlamydia trachomatis		
Treponema pallidum		
Occasionally Transmitted by Sexual Contact		
Group B streptococci	Hepatitis B virus	*Entamoeba histolytica*
Campylobacter jejuni	Hepatitis A virus	
Shigella species	Molluscum contagiosum	*Giardia lamblia*
		Cryptosporidium
		Candida albicans

Congenital   Associated with the fetal state. A congenital disease is one that is acquired by the fetus, usually a result of a preceding maternal infection such that the infant is infected at birth.

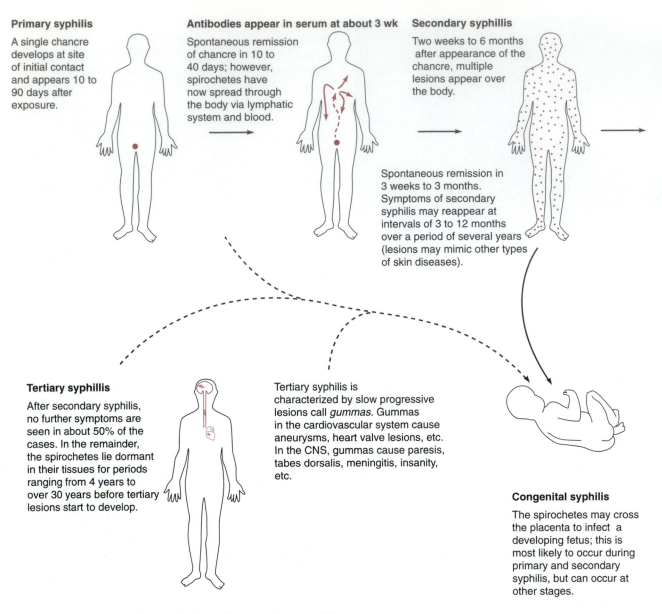

Primary syphilis

A single chancre develops at site of initial contact and appears 10 to 90 days after exposure.

Antibodies appear in serum at about 3 wk

Spontaneous remission of chancre in 10 to 40 days; however, spirochetes have now spread through the body via lymphatic system and blood.

Secondary syphillis

Two weeks to 6 months after appearance of the chancre, multiple lesions appear over the body.

Spontaneous remission in 3 weeks to 3 months. Symptoms of secondary syphilis may reappear at intervals of 3 to 12 months over a period of several years (lesions may mimic other types of skin diseases).

Tertiary syphillis

After secondary syphilis, no further symptoms are seen in about 50% of the cases. In the remainder, the spirochetes lie dormant in their tissues for periods ranging from 4 years to over 30 years before tertiary lesions start to develop.

Tertiary syphilis is characterized by slow progressive lesions call *gummas*. Gummas in the cardiovascular system cause aneurysms, heart valve lesions, etc. In the CNS, gummas cause paresis, tabes dorsalis, meningitis, insanity, etc.

Congenital syphilis

The spirochetes may cross the placenta to infect a developing fetus; this is most likely to occur during primary and secondary syphilis, but can occur at other stages.

FIGURE 17-2   Stages in the pathogenesis of syphilis.

chancre (see Plate 13). The chancre is shallow, ulcerative, has a firm base, and is relatively painless. A chancre may reach more than 1 cm in diameter and is teeming with spirochetes (Figure 17-3). In some patients, often women, the chancre goes unnoticed due to its location in the deeper passages of the genital tract. With the appearance of the chancre, the disease is in the primary state, which lasts until the chancre goes away. The spontaneous disappearance of the chancre (usually within 3 to 6 weeks) may convince a naive patient to mistakenly believe that the disease is cured. The patient may then be without symptoms for a period ranging from 2 weeks

to 6 months before the symptoms of secondary syphilis become evident.

Secondary Syphilis. The spirochetes that spread through the body from the primary site of entry become deposited in many tissues, where they slowly multiply. Secondary syphilis is usually characterized by the appearance of a generalized rash and lesions anywhere on the body (Figure 17-4). The lesions on the skin and mucous membranes may contain large numbers of spirochetes and are highly infectious. The patient may also experience such symptoms as headaches, fever, and

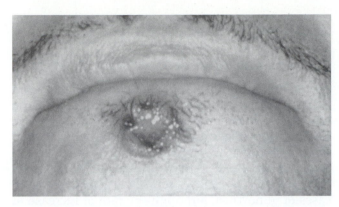

FIGURE 17-3 Primary syphilis with a chancre on the lower lip. (Courtesy of Centers for Disease Control, Atlanta)

sore throat. Lesions may also be present in bones, the central nervous system, or other organs. The skin lesions may be quite prominent; in earlier days this stage of the disease was called "great pox" to distinguish it from another common disease with smaller lesions, that is, smallpox. The secondary stage gradually subsides. Complete recovery may be slow, with some signs remaining for several years. Any one of the following conditions may occur following this phase of the disease:

1. Complete remission with no further manifestations of the disease. This occurs in about 30 to 40% of untreated patients.
2. Recurrence of secondary lesions, often in modified forms. This may occur at 3- to 12-month intervals over the next several years. These recurring lesions may mimic other types of skin diseases.
3. Progression first into latent and then into tertiary syphilis.

Latent Syphilis. This is the period following secondary syphilis. No symptoms of the disease are present during the latent period, but high levels of antibodies are in the serum. About half the patients who progress to the late latent phase have no further symptoms of syphilis. The others progress to the tertiary stage of the disease.

Tertiary Syphilis. Signs of tertiary syphilis may appear any time from 3 to over 30 years following the secondary stage. However, in persons infected with *human immunodeficiency virus* (HIV, the etiologic agent of AIDS), symptoms of tertiary syphilis may develop rapidly and appear within only a few months following primary syphilis.

Lesions, called *gummas* (Figure 17-5), develop in various tissues of the body and may be due to a hypersensitivity reaction to the small number of spirochetes that have persisted in the body. Lesions may develop in the central nervous system, resulting in *neurosyphilis*, which may cause mental changes or resemble other neurologic diseases. Neurosyphilis is a major cause of insanity. The cardiovascular system is often involved, a common

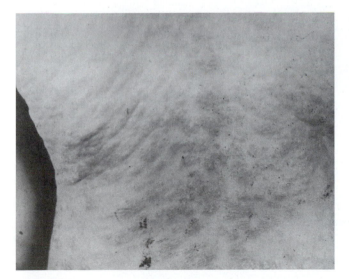

FIGURE 17-4 Secondary syphilis with rash-type lesions on the back. (Courtesy of Centers for Disease Control, Atlanta)

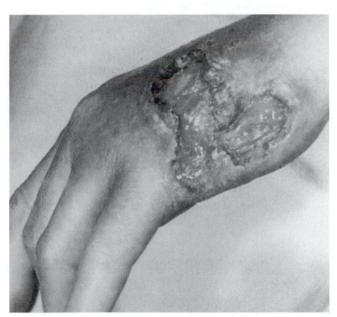

FIGURE 17-5 Tertiary syphilis with an ulcerative gumma on the hand. (Courtesy of Centers for Disease Control, Atlanta)

manifestation being the development of aneurysms in the aorta. Although in the years just prior to 1986 the number of tertiary cases of syphilis had declined, this trend was reversed in 1986, and the number of tertiary cases nearly doubled in 1986 and 1987. The reason for this increase is not known.

Congenital Syphilis.   *T. pallidum* readily passes the placental barrier and a pregnant syphilitic woman may transmit the infection to the developing fetus. Between 1950 and 1980 there was a general decline in the incidence of congenital syphilis. However, the mid-1980s saw a remarkable and continuing increase in in-utero transmission of *T. pallidum* (Figure 17-6). Congenital infection almost always occurs when the expectant mother develops primary syphilis, occurs about 90% of the time if she has secondary syphilis, and occurs around 30% of the time if the disease is latent. The fetus does not develop signs of congenital syphilis until the second trimester. The spirochetes become disseminated throughout the fetus and 30 to 40% die in utero, resulting in a spontaneous abortion or stillbirth. Over half the syphilitic infants who are alive at birth show marked congenital defects, while some may appear normal at birth but develop symptoms at later periods. The manifestations of congenital syphilis are variable but include such signs as skin lesions, enlarged spleens, anemia, pneumonia, eye damage, neurologic lesions, mental retardation, and deformed bones, teeth, and cartilaginous tissues.

✳ Transmission and Epidemiology

Generally, transmission is by sexual contact, and persons most likely to contract this disease are those who have multiple sex partners. The epidemiologic patterns of primary and secondary syphilis are similar to those seen with gonorrhea; that is, most cases occur in sexually active young adults. Overall, about 1 case of syphilis is reported to about every 10 cases of gonorrhea, although the rate of increase in the number of cases has been greater in recent years than for gonorrhea. Beginning in 1985 there was a continuing increase in the number of primary and secondary cases of syphilis reported to the Centers for Disease Control (Figure 17-7). The number of new cases reached epidemic levels in 1989, with some American cities reporting a 293% increase in the number of cases between 1989 and 1990. The increased number of cases did not occur uniformly throughout the population, and three factors have apparently been responsible for much of the increase. First, syphilis transmission increased among medically hard-to-reach groups, such as crack cocaine users and other drug users. Second, persons at increased risk for syphilis may not have considered health care a high priority. Third, declines in

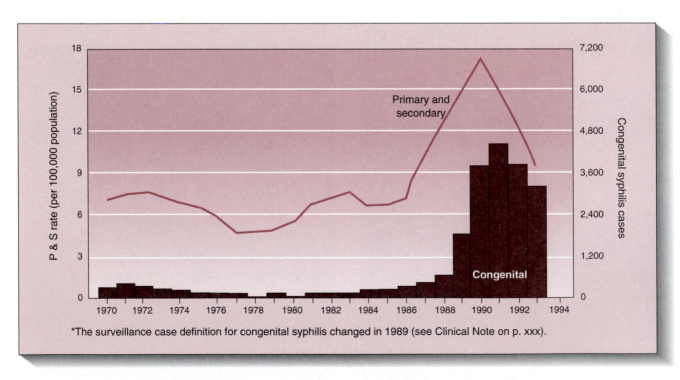

*The surveillance case definition for congenital syphilis changed in 1989 (see Clinical Note on p. xxx).

FIGURE 17-6   Congenital (under 1 year) and primary and secondary syphilis in women in the United States from 1970 to 1993. (from CDC annual summaries)

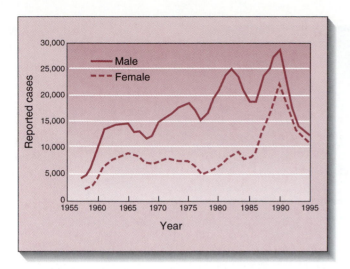

FIGURE 17-7 Reported cases of primary and secondary syphilis in males and females in the United States from 1955 to 1994. (from CDC annual summaries)

socioeconomic and education levels in certain populations have been associated with increased unemployment, drug use, prostitution, and family disruption, conditions conducive to the spread of syphilis. Since 1990 the number of cases have been decreasing.

✳ Diagnosis

The appearance of a primary chancre or secondary lesions is suggestive of syphilis; these signs, however, may be confused with lesions or skin rashes caused by allergies or other infections. Diagnosis is first made by demonstrating the presence of the spirochetes in exudate from the lesion and then by serological tests. About 75% of cases can be diagnosed by direct dark-field microscopic examination of exudate from primary or secondary lesions. The exudate can also be fixed to a slide and stained with a specific fluorescent-labeled antibody for examination with a fluorescence microscope.

Various serological tests have been used over the past 80 years to detect syphilis antibodies. Antibodies appear several weeks after the primary lesion develops and remain in those persons who continue to harbor the spirochetes through the latent and tertiary stages. The serological tests for syphilis, often referred to as **STS**, are perhaps the most widely used of all serologic tests. They are sometimes required when applying for marriage licenses, for some types of employment, entrance into a military service, and as part of prenatal examinations. About 40 million such tests are run each year in the United States.

Two general types of serologic tests are used (Table 17-2). The first detects a nonspecific antibody, called *reagin*, that reacts with antigens from several sources and is produced by some disease other than syphilis. Among others, reagin reacts with a *cardiolipin* antigen that is extracted from beef hearts. Because the cardiolipin antigen is inexpensive, it is used in routine screening tests for syphilis. Flocculation tests, the chief tests used today, consist of adding the cardiolipin antigen to the test serum. Of these, the most frequently used test is called the *VDRL* (*Venereal Disease Research Laboratories*) test and is positive when the cardiolipin forms into fine aggregates or floccules. Persons with a number of other diseases, such as hepatitis, diabetes, and malaria, or other chronic diseases, may give a positive reaction to these screening tests for syphilis; such reactions are called *biological false positives*. All sera that are positive in the screening tests must be retested to confirm the results with a more specific serologic test that detects antibodies specific for *T. pallidum*. One such specific test is called the *fluorescent treponemal antibody* (FTA) test. The FTA test uses freeze-dried *T. pallidum* that are fixed to a glass slide. The test serum is added; if specific antibodies are present, they react with the *T. pallidum*. To determine if this reaction has taken place, the slide is rinsed to remove any unattached antibodies and then covered with fluorescent-labeled antibodies against human gamma globulin. If the test is positive, the spirochete-fluorescent antibody complex fluoresces when examined under a fluorescence microscope ∞ (Chapter 3, p. 14). Nonspecific antibodies must be removed from the patient's sera before testing. This has given the name *absorbed* to the test, known as FTA-ABS.

Not all of the available tests have equal **specificity** and **sensitivity** (Table 17-2). Careful consideration must be made regarding the circumstances of

STS Serologic tests for syphilis.
Cardiolipin A phospholipid found in beef heart tissue.
Specificity The capacity of a diagnostic test to detect only truly positive specimens. A highly specific test would not have any false positive reactions.
Sensitivity The capacity of a diagnostic test to detect all positive specimens. A highly sensitive test would not have any false negative reactions.

TABLE 17-2

Serologic Tests for Syphilis (STS)

Test	Use
Nontreponemal	
Flocculation tests	
VDRL (Venereal Disease Research Laboratory)	Screen and diagnosis
Reagin tests	
PCT (plasmacrit)	Screen
USR (untreated serum reagin)	Screen
RPR (rapid plasma reagin)	Screen and diagnosis
Complement fixation tests	Rarely available
Kolmer or Wasserman	
Treponemal	
Immobilization	
TPI (*Treponema pallidum* immobilization)	Diagnosis and research
Immunofluorescence	
FTA-ABS (fluorescent treponemal antibody absorption)	Diagnosis and screen
Hemmagglutination	
TPHA (*Treponema pallidum* hemagglutination assay)	Screen and diagnosis

each patient in order to select the proper laboratory test for diagnosis.

✳ Treatment

Penicillin is highly effective in treating syphilis and there is no evidence that resistant strains of *T. pallidum* are developing. Persons allergic to penicillin can be treated with erythromycin or tetracyclines. To treat primary or secondary syphilis successfully, penicillin must be maintained continuously in the system for 7 to 10 days. In the later stages of the disease, penicillin therapy should continue over at least 21 days. Treatment of syphilis at any stage before the onset of tertiary disease will block further progression of the disease. Treatment of pregnant syphilitic women early in pregnancy usually prevents congenital disease. Treatment later in pregnancy clears the infection in the fetus and prevents the occurrence of further tissue damage. Infants suspected of having congenital syphilis should receive treatment as soon as possible.

✳ Prevention and Control

No vaccine is available. The use of condoms reduces the chance of transmission. One widely used procedure to limit the spread of syphilis is to follow up and treat all persons who have had sexual contact with known syphilitic patients. This procedure, however, is often limited by a lack of public health personnel and operating funds. In addition, educational programs try to help people understand the mode of transmission, recognize symptoms, and tell how to receive treatment. Restricting one's sexual contact to a marriage partner is the most effective means of preventing syphilis and other sexually transmitted diseases.

OTHER TREPONEMAL INFECTIONS

Three similar clinical diseases called *yaws*, *pinta*, and *bejel* result from treponemal spirochetes that are generally indistinguishable from *T. pallidum*. These diseases occur in people living under conditions of poor hygiene in the tropics or the Mideast. Transmission occurs by direct contact, not necessarily sexual, and primary infection usually happens in young children. Primary, secondary, and tertiary symptoms are seen. The exact relationship of these diseases to syphilis is not clear. It has been theorized that syphilis may have evolved from these diseases or vice versa. Yaws, pinta, and bejel are readily cured by penicillin. Infection rates decrease significantly when improvements are made in personal hygiene and living conditions.

CLINICAL NOTE

Evaluation of Congenital Syphilis Surveillance System: New Jersey 1993

To monitor disease burden and trends associated with congenital syphilis (CS), effective prevention programs require a surveillance system that identifies CS cases in an accurate and timely manner. Before 1988, comprehensive CS surveillance was difficult for health departments to conduct because documentation of infection in infants required complex and costly long-term follow-up for up to 1 year after delivery; follow-up often was incomplete, and many infected infants were not identified. To estimate the public health burden of CS more accurately and eliminate long-term follow-up of infants by health department personnel, in 1988 CDC implemented a new CS case definition. Rather than relying on documentation of infection in the infant, the new case definition presumes that an infant is infected if it cannot be proven that an infected mother was adequately treated for syphilis before or during pregnancy. During 1993–1994, the Sexually Transmitted Disease Prevention and Control Program of the New Jersey Department of Health (NJDOH) evaluated its CS surveillance system to assess the accuracy and completeness of reporting using the new case definition and to determine the personnel costs associated with identifying and classifying CS cases. This report summarizes the results of the evaluation.

New Jersey statutes mandate that all pregnant women receive a serologic test for syphilis (STS) during pregnancy or at delivery if no test was done during pregnancy. Newborns also routinely receive a STS at birth if born to a mother with a reactive STS. Laboratories are required to report all reactive STSs (including maternal, delivery, and newborn) to the NJDOH, and all such reports are investigated by NJDOH. Investigation activities include reviewing infant and maternal medical records to determine whether syphilis was previously diagnosed, reviewing laboratory results and

health department records to determine the mother's treatment status, and verifying missing information by contacting the patient and/or provider by telephone or field visit.

For this analysis, reports of all reactive STSs for newborns received by NJDOH during 1 January–31 December 1993, were reviewed manually to assess the completeness and accuracy of case classification and reporting. Infants with reactive STSs had been classified using the four categories recommended by CDC: (1) not infected, (2) syphilitic stillbirth, (3) confirmed case of CS, and (4) presumptive case of CS. Costs associated with investigation and follow-up of reactive STSs for newborns were estimated by multiplying the average time spent at each task by the hourly wage (excluding benefits) of the person performing the task. Time spent on an investigation was determined by interviewing the persons who performed the tasks.

During 1993, a total of 497 reactive STSs for newborns were reported to NJDOH. Of these reports, 266 (53%) had been classified as not infected, but reactive secondary to passive transfer of maternal syphilis antibodies from a mother adequately treated for syphilis before or during pregnancy, and 143 (29%) had been classified as presumptive cases. In addition, a total of 10 (2%) reports initially classified as not infected were reclassified as presumptive cases, and 78 (16%) reports were still under investigation.

For 1993, the estimated average cost of investigating one reactive STS for a newborn using routine surveillance methods was $183. Based on an average of 41 reactive STSs for newborns reported to NJDOH each month in 1993, the estimated costs for investigation and follow-up were $7500 per month or $90,000 per year (*MMWR* 44:225, 1995).

BORRELIA: RELAPSING FEVER AND LYME DISEASE

Borrelia organisms are responsible for some instances of relapsing fever in humans and lyme disease. In contrast with *Treponema*, the members of this genus stain with a variety of aniline dyes, and

have been cultured in vitro. A common species, *Borrelia recurrentis*, has an irregular, open spiral with several micrometers between turns. These organisms are frequently carried by wild rodents and their ticks or lice and infection is transmitted to humans when they are bitten by the insect vector. Tick-borne relapsing fever (called *endemic* relapsing fever) is seen occasionally in campers, for instance,

Endemic Referring to a disease that is always present in a population.

who may spend time in tick-infested areas. The louse-borne, or *epidemic* relapsing fever, occurs most often in persons living in poverty or under crowded impoverished conditions associated with wars or other calamities. (See Plate 14.)

✳ Relapsing Fever

Sporadic cases of relapsing fever in humans are caused by *B. recurrentis*. The clinical disease is characterized by a sudden onset with accompanying fever, headache, and muscle pain. A rash may be seen in some patients. The infection may involve many tissues of the body. The first episode usually lasts 3 to 7 days, and climaxes with severe illness involving nausea, vomiting, and prostration. The patient quickly feels better and begins to gain strength. This afebrile period usually lasts up to 10 days, after which the patient relapses with symptoms similar to those of the initial period of illness.

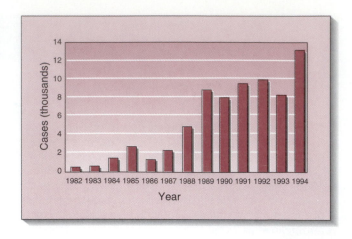

FIGURE 17-8 Reported cases of Lyme disease in the United States from 1982 through 1994. (From CDC weekly report)

CLINICAL NOTE

Common Source Outbreak of Relapsing Fever: California 1990

In August and September 1989, six persons who had each at different times spent the night in the same cabin at Big Bear Lake, San Bernardino County, California, developed sudden onset of high fever with severe headache, prostration, nausea, and vomiting. Four patients were initially considered to have viral gastroenteritis and received no antibiotic therapy. When fevers recurred, relapsing fever was suspected and confirmed serologically in two of these patients. Two of the four patients were treated with tetracycline and recovered. Two recovered without treatment after three relapses.

The fifth patient, an elderly woman, was admitted to intensive care with septic shock. Laboratory findings included spirochetes visible in a Giemsa-stained smear of peripheral blood, neutrophilic pleocytosis of the cerebrospinal fluid, peripheral leukocytosis, thrombocytopenia, and hypophosphatemia. She was treated with intravenous doxycycline and developed a probable Jarisch–Herxheimer reaction before recovering fully.

For the sixth patient, the suspected diagnosis of relapsing fever was confirmed by the identification of spirochetes in a thick blood smear. This patient was treated promptly with tetracycline and recovered rapidly.

The Santa Barbara and San Bernardino county health departments were advised of the outbreak and conducted an epidemiologic and environmental investigation of the Big Bear Lake area and the cabin used by each of the six patients. The investigators noted a large population of California ground squirrels and chipmunks and the widespread practice by visitors of feeding these animals. Inhabited ground-squirrel burrows were found under the cabin. The cabin was fumigated to kill vectors (ticks). No other cabins were known to be associated with illness. In the spring of 1990, educational literature about relapsing fever and prevention of the disease was distributed to cabin owners in the Big Bear Lake area (*MMWR* 39:579, 1990).

Epidemic Referring to a disease that occurs at greater than normal levels in a population.

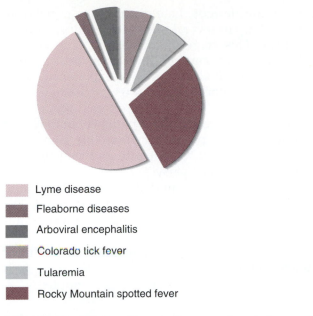

Lyme disease

Fleaborne diseases

Arboviral encephalitis

Colorado tick fever

Tularemia

Rocky Mountain spotted fever

FIGURE 17-9 Reported cases of vectorborne diseases in the United States, 1989. (Courtesy of Centers for Disease Control, Atlanta)

Patients often suffer 4–10 such relapses, usually with decreasing severity. Such relapses are the result of the selection of antigenic variants of the organism, which then multiply and repeat the symptomatic process. Treatment is with broad-spectrum antibiotics such as tetracycline and the disease is prevented by avoiding exposure to ticks or lice.

✳ Lyme Disease

This disease was first reported in Europe in the early twentieth century. The disease is named for the first reported American cases that occurred in 1975 in Lyme, Connecticut. During the 1980s and into the 1990s, increasing numbers of cases were reported throughout the United States (Figure 17-8). Lyme disease, which is tick-borne, has been reported from most states, including Hawaii, and has become the most common diagnosed tick-borne illness in the United States (Figures 17-9 and 17-10). Typical of vector-borne diseases, the occurrence of Lyme disease is seasonal and parallels increased

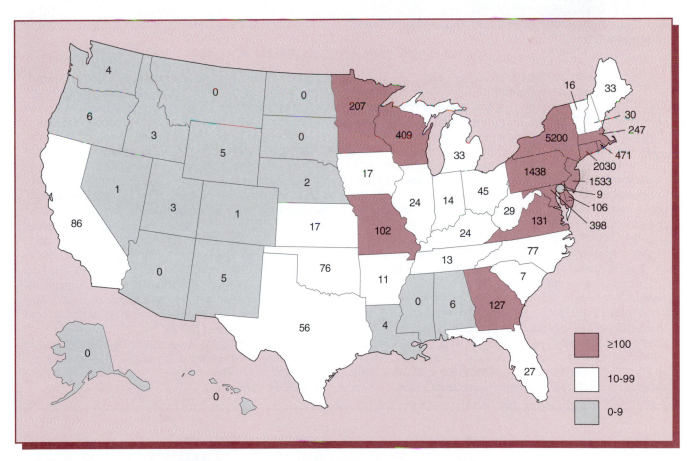

FIGURE 17-10 Reported number of Lyme disease cases by state in the United States in 1994. (from CDC weekly report)

human outdoor activity during the summer months. The causative agent of Lyme disease was first identified by Dr. W. Burgdorfer in 1982 in the United States and is called *Borrelia burgdorferi*. It is a long, narrow, motile spirochete that can be grown in complex enriched medium and is able to infect a variety of different animal hosts.

The disease is characterized by two relatively distinct clinical stages. Early disease is associated with flu-like symptoms and one or more circular rashes known as *erythema migrans* that occur at the site of the tick bite. The appearance of the rash is very helpful in making a diagnosis of Lyme disease. The rash lasts from two to four weeks and often expands in a circular fashion out from the site of infection, giving the appearance of a target with a bull's-eye. Early disease may also be associated with swollen lymph nodes, fever, headache, stiff neck, *arthralgias*,

myalgias, and fatigue. Late disease may occur from one month up to several years after early symptoms disappear. This stage of the disease is associated with neurologic, cardiac, and/or arthritic symptoms. About 60% of patients experience some arthritic discomfort and about 6% of these develop chronic arthritis. Cardiac symptoms occur in a smaller number (about 8%) of patients but may lead to serious health problems. Neurolgic signs may occur in about 15% of patients and may include meningitis and facial palsy.

Lyme disease is transmitted by ticks belonging to the genus *Ixodes* (commonly called deer ticks). The tick becomes infected by feeding on infected animals.

Therapy for Lyme disease includes penicillin or tetracycline, which, if given early in the disease, is effective in preventing late stage symptoms. A

CLINICAL NOTE

Lyme Disease: United States 1994

For surveillance purposes, Lyme disease (LD) is defined as the presence of an erythema migrans rash 5 cm in diameter or laboratory confirmation of infection with *Borrelia burgdorferi* and at least one objective sign of musculoskeletal, neurologic, or cardiovascular disease. In 1982, CDC initiated surveillance for LD, and in 1990, the Council of State and Territorial Epidemiologists adopted a resolution that designated LD a nationally notifiable disease. This report summarizes surveillance data for LD in the United States during 1994.

In 1994, 13,083 cases of LD were reported to CDC by 44 state health departments, 4826 (58%) more than the 8257 cases reported in 1993 (Figure 17-8). As in previous years, most cases were reported from the northeastern and north-central regions (Figure 17-10). The overall incidence of reported LD was 5.2 per 100,000 population. Eight states reported incidences of more than 5.2 per 100,000 (Connecticut, 62.2; Rhode Island, 47.2; New York, 29.2; New Jersey, 19.6; Delaware, 15.5; Pennsylvania, 11.9; Wisconsin, 8.4; and Maryland, 8.3); these states accounted for 11,476

(88%) of nationally reported cases. Six states (Alaska, Arizona, Hawaii, Mississippi, Montana, and North Dakota) reported no cases. Reported incidences were 100 per 100,000 in 15 counties in Connecticut, Maryland, Massachusetts, New Jersey, New York, Pennsylvania, and Wisconsin; the incidence was highest in Nantucket County, Massachusetts (1197.6).

Six northeastern states accounted for 95% of the increase in reported cases for 1994: Maryland, New Jersey, New York, Rhode Island, Connecticut, and Pennsylvania. Reported cases increased by 218 cases (121%) in Maryland, 747 cases (95%) in New Jersey, 2382 cases (85%) in New York, 199 cases (73%) in Rhode Island, 680 cases (50%) in Connecticut, and 353 cases (33%) in Pennsylvania. Reported cases remained stable in the states with endemic disease in the north-central region (Minnesota and Wisconsin) and decreased in California (36%).

Males and females were nearly equally affected in all age groups except those aged 10–19 years (males: 55%) and those aged 30–39 years (females 56%) (*MMWR* 44:459, 1995).

Arthralgia   Pain in the joints.
Myalgia   Pain in the muscles.

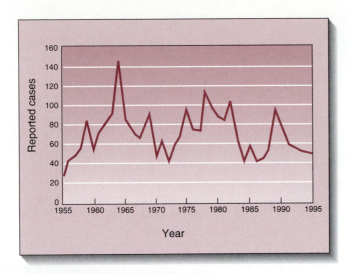

FIGURE 17-11 Reported cases of *leptospirosis* in the United States from 1955 to 1994. (Modified from CDC annual summaries)

genetic engineered vaccine is in the development stages.

LEPTOSPIRA: LEPTOSPIROSIS

The *Leptospira* are characterized by a very tight coil. One end of the organism is characteristically bent into a hook shape. Only one species, *L. interrogans*, is a pathogen, and it is found in a wide variety of wild and domestic animals. Among other tissues, the kidneys of the animals become infected and the leptospira are shed in the urine.

Humans become infected by direct or indirect contact with the urine from infected animals or with animal tissues. *Leptospirosis* is most often seen in persons who are frequently exposed to animals, such as veterinarians, farmers, abattoir workers, persons living in rodent-infested housing, and dog owners. Infections have also occurred from contact with contaminated water. In the United States, leptospirosis occurs during all seasons, but most of the cases occur during the summer months, a time of significantly increased outdoor activity.

The portal entry to the leptospira is probably through the mucosa or breaks in the skin. The organisms spread through the blood and usually the kidneys become infected; other tissues are infected as well. Although a wide variety of symptoms may be seen, most cases begin with influenza-like symptoms, including fever, chills, back pain, headache, and prostration. Less than 100 cases of human leptospirosis are confirmed by laboratory diagnosis each year in the United States (Figure 17-11), although it is thought that many more cases do occur but are not specifically diagnosed. It is possible both to isolate this spirochete on artificial media and to measure specific antibodies by various serologic tests. Isolation, however, is difficult and is infrequently accomplished.

Vaccines are used in veterinary medicine and are available to persons in high-risk occupations. Penicillin, streptomycin, and tetracyclines are effective if used early in the disease, but are generally not effective if given after about 4 days of illness.

MATERIAL FOR REVIEW

CONCEPT SUMMARY

1. The spirochete of greatest medical concern is *Treponema pallidum*, the causative agent of syphilis. This disease occurs in several distinct stages with the last stage responsible for the most serious pathology. It is a classic example of a sexually transmitted disease, although other means of transmission occur. No vaccine is available for control, but the disease responds well to antibiotic therapy.

2. Other spirochetial diseases, can be severe and include leptospirosis, yaws, pinta, relapsing fever, and Lyme disease.

CLINICAL SUMMARY TABLE

Microorganism	Virulence Mechanisms	Diseases	Transmission	Treatment	Prevention
Treponema pallidum	Avoids immune responses	Syphilis	Sexually Congenital	Penicillin Erythromycin	Avoid exposure Treat pregnant females
Borrelia recurrentis	Antigenic variants	Relapsing fever	Tickborne Animals to humans	Broad-spectrum antibiotics	Avoid ticks
Borrelia burgdorferi		Lyme disease	Tickborne Animals to humans	Penicillin Tetracyclines	Avoid vectors
Leptospira interrogans		Leptospirosis	Animals to humans	Penicillin Streptomycin Tetracyclines	Vaccine

STUDY QUESTIONS

1. Describe the features of the lesion associated with primary syphilis.
2. What structural (anatomical) features of the spirochetes are different from those common to a gram-negative bacillus?
3. During which phase of syphilis is a patient most infectious for others?
4. How does tertiary syphilis differ from latent syphilis?
5. What are the advantages and disadvantages of the nontreponemal STS?
6. Which of the spirochetes can be cultured in vitro? Which can be visualized by staining with analine dyes?
7. What are the major sources of human infection by *Leptospira*?

CHALLENGE QUESTIONS

1. The newsprint media and television tend to focus more on Lyme disease than on syphilis. Explain why this is so. Is it correct to do so?
2. Explain why the nonspecific VDRL test for symphilis is still used even though it is not completely specific and despite the availability of more specific tests like FTA.

CHAPTER

18 Anaerobic Bacteria

OUTLINE

The anaerobic bacteria of medical interest are easily separated into two groups: those with and those without spores. The spore-forming anaerobes are all members of the genus *Clostridium*; they are gram-positive bacilli, produce powerful *exotoxins*, and occur across a wide variety of habitats. The non-spore-forming anaerobes are a group of diverse and incompletely characterized microorganisms. They form a significant proportion of the normal flora of the human body. Bacteria from this category include from 80 to 90% of the bacteria present on the skin, in the mouth, or in the upper respiratory tract, and about 99.9% of bacteria present in the intestinal tract. These anaerobic bacteria include both gram-negative and gram-positive cocci and bacilli as well as spirochetes. Because these bacteria are more difficult to isolate and grow in pure culture than the nonanaerobes, they have often been ignored or missed during routine bacteriological examinations of clinical specimens. These bacteria, however, are recognized as important opportunistic pathogens that produce a significant number of infections. The two groups of bacteria are presented separately in this chapter.

CLOSTRIDIA: GENERAL CHARACTERISTICS

The genus *Clostridium* contains a large number of gram-positive, spore-forming species, several of which are able to produce disease in humans. These anaerobic bacteria range in habitat from soil to the human mucosa. They are obligate anaerobes, many of which are killed by exposure to oxygen. They have a very active metabolism, ferment a variety of sugars, and have a very short generation time. Most clostridia produce one or a variety of very poisonous protein toxins. Representative species and the diseases produced by the clostridia are presented.

HUMAN BODY SYSTEMS AFFECTED

Clostridia and Other Anaerobic Bacteria

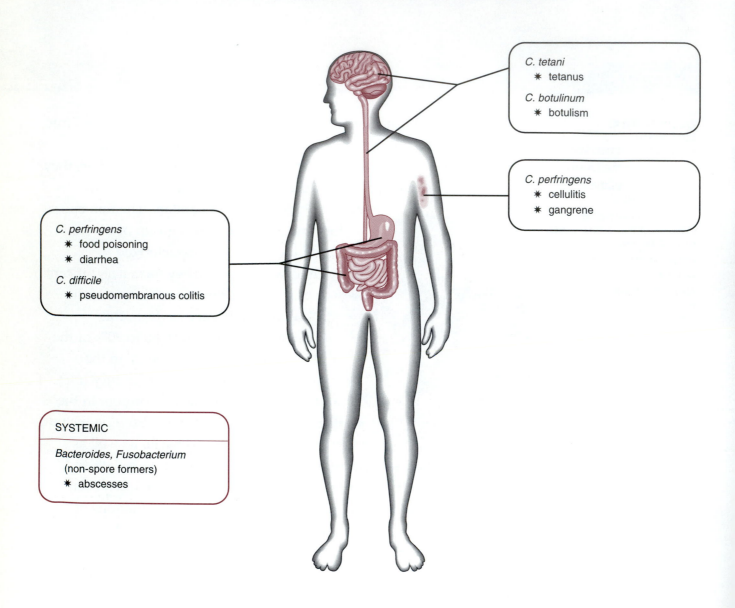

C. tetani
  * tetanus

C. botulinum
  * botulism

C. perfringens
  * cellulitis
  * gangrene

C. perfringens
  * food poisoning
  * diarrhea

C. difficile
  * pseudomembranous colitis

SYSTEMIC

Bacteroides, Fusobacterium
(non-spore formers)
  * abscesses

C. TETANI: TETANUS

Clostridium tetani, the causative agent of tetanus, is a large, gram-positive, spore-forming, motile, obligate anaerobic bacillus ∞ (Chapter 7, p. 9) (Figure 18-1; see Plate 15). It can be grown on blood agar or media containing meat. Various strains exist, but all produce the same exotoxin (tetanospasmin).

* Pathogenesis and Clinical Diseases

Because of the wide distribution of *C. tetani* (commonly found in soils and manures), wounds are

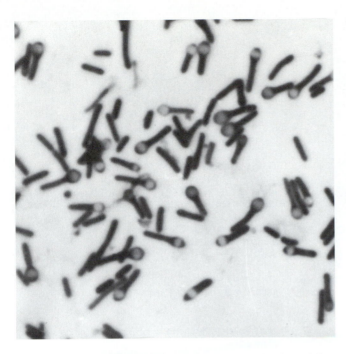

FIGURE 18-1 Photomicrograph of *Clostridium tetani.* Many have endospores at the terminal end of the bacterial cell. (Courtesy of Visuals Unlimited)

often contaminated with these spores. However, the disease of tetanus does not develop in a great majority of cases. The condition of the wound must be such that an anaerobic environment exists with some dead tissue present. These conditions allow spores to germinate, bacteria to proliferate, and toxin to be produced. Such conditions are often seen in puncture wounds produced, for instance, by nails or splinters. Yet other types of wounds or conditions resulting in tissue damage may also offer a suitable environment for the growth of *C. tetani.* One frequently encountered form of tetanus occurring in underdeveloped countries is tetanus of the umbilicus of infants born at home and treated with unsterile instruments.

The tetanus toxin is extremely potent. It is among the most toxic substances known and a small amount is able to cause the disease. The growth of *C. tetani* per se causes no tissue damage. Usually signs and symptoms of the disease begin to occur 4 to 10 days after injury but may be delayed up to several months. Symptoms are entirely the result of the toxin spreading to the spinal cord. The toxin specifically affects the **synaptic** junction of the nerves by preventing the inhibition or erasing of nerve impulses once they have crossed the synaptic junction (Figure 18-2). The nerve continues to send impulses, a condition that results in spasmodic contractions (*tetany*) of the involved muscles. Early symptoms are muscle stiffness with the muscles of the jaw often developing spasms first. This condition gives the disease its common name of "lockjaw." As the disease progresses, spasms develop in other muscles. The spasms may be brief, but they can occur frequently and cause great pain and exhaustion. In some cases, the spasms may be powerful enough to cause bones to break. Respiratory complications are common and death rates high, especially in young children and elderly persons. In nonfatal cases, recovery takes several weeks but is usually complete.

✳ Transmission and Epidemiology

Transmission and epidemiology of tetanus do not follow the pattern seen in many infectious diseases where the microbes are passed from host to host. Some situations are conducive to the development of tetanus, however. Soils or materials in contact with animal wastes are usually heavily contaminated with *C. tetani* and offer excellent sources of infection. Soil-contaminated wounds are the most frequent source of infection. Before the development of an effective vaccine, tetanus often resulted from wounds received in wars. Hundreds of thousands of cases of tetanus occurred during the Civil War, but only 12 cases were reported among U.S. troops during World War II. Less than 100 cases per year are now reported in the United States, 40% of which result in death. The numbers of reported cases in the United States are shown in Figure 18-3. Most of the cases (>70%) that occur in the United States are in adults 50 years of age or older (Figure 18-4). Older individuals are also more likely (52%) to die from tetanus than are younger patients (13%). Large numbers of cases still occur in some developing tropical countries.

✳ Diagnosis

Diagnosis is made on the basis of the clinical disease. *C. tetani* is a common contaminant of wounds and may be found in patients who do not develop

---

Synaptic   Referring to the region where the nerve impulse of a neuron is transmitted to the responsive tissue (either another neuron or a receptor cell). These impulses are transmitted in only one direction—neuron to tissue.

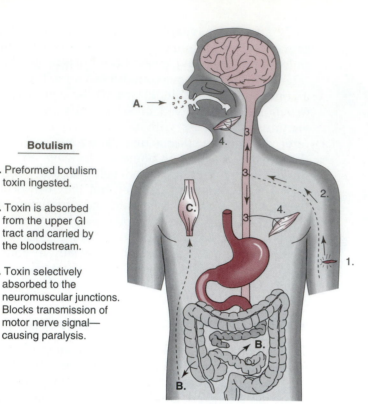

Botulism

A. Preformed botulism toxin ingested.

B. Toxin is absorbed from the upper GI tract and carried by the bloodstream.

C. Toxin selectively absorbed to the neuromuscular junctions. Blocks transmission of motor nerve signal—causing paralysis.

Tetanus

1. *C. tetani* spores introduced into an anaerobic niche of damaged tissue.

2. Tetanus toxin travels along motor neuron to spinal cord.

3. Toxin blocks the inhibition of spinal reflex.

4. Spasmatic contractions of muscles.

FIGURE 18-2 Mechanisms of action of tetanus and botulism toxins on muscle functions.

tetanus. Therefore, the isolation of the bacterium from a patient may not be diagnostic.

✳ Treatment

As soon as clinical tetanus is suspected, steps should be taken to neutralize the existing toxin and

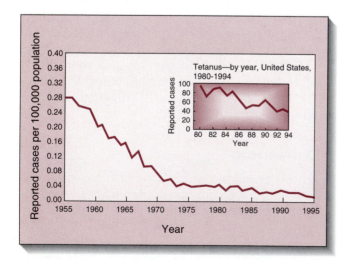

FIGURE 18-3 Reported cases of tetanus by year in the United States, 1955 to 1994. (modified from CDC annual summaries)

prevent the formation of new toxin. *Antitoxin ∞* (Chapter 12, p. 176), produced in humans, called *tetanus immune globulin*, should be administered immediately. While antitoxin cannot reverse the toxic activity of the toxin bound to nerve tissue, it will inactivate toxins in the blood. Wounds should be debrided to remove dead tissues or foreign bodies, antibiotics should be given to inhibit growth of *C. tetani*, and a tetanus toxoid booster immunization should be given to persons who have not received tetanus toxoid within the preceding 5 years. If muscle spasms occur, antispasmodic drugs should be used and respiration should be maintained by a positive-pressure breathing apparatus if necessary.

✳ Prevention and Control

At present, the major involvement with *C. tetani* by most medical personnel in developed countries concerns the prevention of this disease. Prevention through immunization has been extremely effective. In fact, approximately 95% of persons who develop tetanus have no record of immunization against the disease.

Immunization with tetanus toxoid should begin with infants 6 weeks to 3 months old. The toxoid is usually given in combination with diphtheria tox-

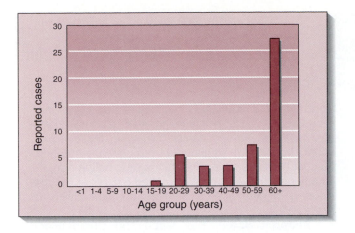

FIGURE 18-4   Occurrence of tetanus as a function of patient age in 1993. (from CDC annual summaries)

oid and pertussis vaccine as the DPT vaccine ∞ (Chapter 12, p. 173). Three doses of DPT should be given several weeks apart. A booster dose should then be given 1 and 4 years later. After the age of 15, a tetanus booster should be given every 10 years. The vaccine given to adults is called *Td* (tetanus and diphtheria toxoid) and does not contain the pertussis antigen.

When a person suffers an injury that is likely to result in tetanus and that person has no history of immunization, *passive immunization* with human antitoxin should be given as temporary protection. It should be followed by *active immunization* with the toxoid.

Neonatal umbilical cord tetanus can be prevented by actively immunizing pregnant females who have no history of previous immunizations. The newborn infant will then have natural passive immunity at the time of birth, which should reduce the chance of developing umbilical cord tetanus.

C. BOTULINUM: BOTULISM

C. botulinum is a large, gram-positive, anaerobic bacillus that produces an exotoxin that causes food poisoning. Its spores are among the most heat resistant and are able to withstand temperatures of 100°C for several hours. These spores will survive in heat-processed foods if the temperature does not reach the required level. Growth can occur in a wide variety of culture media as well as in many types of food. Eight serotypes, based on the antigenic characteristic of the exotoxin, have been identified and are designated A through H. Cases due to type G have not been reported in humans. (See Plate 18)

✳ Pathogenesis and Clinical Diseases

In most cases, botulism results from the ingestion of preformed toxin produced during the growth of *C. botulinum* in foods. Botulinum toxin is one of the most powerful known toxins, and extremely small amounts (1–2 µg) are able to cause illness or death in humans. These toxins are destroyed by heating to 100°C for a brief period.

The toxin, generally type A, B, or E, is absorbed primarily from the small intestine, passes into the blood, and is carried to the peripheral nerves where it specifically reacts at the muscle–nerve junction (Figure 18-2). The toxin produces complete paralysis of the nerve impulse by preventing the release of **acetylcholine**. Death results from the paralysis of respiratory functions. Symptoms may appear as soon as 12 to 36 hours after ingesting contaminated food or may take as long as 8 days to appear. The first symptoms are often weakness and dizziness. Double vision (*diplopia*), difficulty in speaking (*dysphonia*) and swallowing (*dysphagia*), and dilated pupils usually occur. Some abdominal distress may be experienced. Fever is rare. Muscle weakness develops, leading to paralysis as the disease progresses. When paralysis of respiratory muscles occurs, death results. The mortality rate varies between 20 and 70% and is influenced by the amount and serotype of toxin consumed, as well as the time between ingestion and the initiation of antitoxin therapy.

In 1976 it was discovered that *C. botulinum* could grow in the intestines of infants and produce enough toxin to cause serious illness—infant botulism. The spore may be in various infant foods, but honey has been implicated in several cases. Not all infants appear to be susceptible and at present it is not known just which conditions allow the intestinal tract of some infants to support the growth of *C. botulinum*. Signs of the disease in these patients may start with constipation, followed by weakness and then paralysis of the muscles of the head and neck.

Acetylcholine   A chemical messenger that functions to make a nerve–muscle junction. When this messenger is inhibited, the brain cannot signal the muscle to contract. This lack of muscle response is known as *flaccid paralysis*.

Tetanus Fatality: Ohio 1991

In August 1991, the Ohio Department of Health received a report of a fatal case of tetanus. This report summarizes the investigation of this case.

On 21 July, 1991 an 80-year-old woman sought treatment in the emergency department of a hospital in central Ohio because of a stiff jaw and difficulty in swallowing (dysphagia). On examination, she had slightly slurred speech and difficulty opening her mouth but no difficulty breathing. A wood splinter from a forsythia bush had been lodged in her left shin approximately 1 week; the wound site was erythematous and draining purulent material. The emergency room physician diagnosed tetanus and admitted the woman to the hospital. Treatment included tetanus immune globulin (3000 units), tetanus toxoid (0.5 cc) and intravenous clindamycin because of a reported history of penicillin allergy.

The patient had no history of any previous tetanus vaccinations. She had been treated at an undetermined time in the 1960s for an infected wound associated with a fractured ankle. In addition, she had sought medical care periodically for treatment of hypertension and other medical problems.

The patient's clinical status gradually deteriorated, and mechanical ventilation was required because of increasing generalized rigidity. During the ensuing 2-week period, she was treated for tremors, muscle spasms, abdominal rigidity, apnea, pneumonia, and local infection from her leg wound. Despite aggressive treatment, the patient died on 5 August.

As a result of this case, a public health nurse, serving as part of the Occupational Health Nurses in Agricultural Communities project, instituted community-wide educational activities to increase tetanus vaccination coverage among adults. Following these educational efforts, from August 1991 through July 1992, the number of adults receiving tetanus vaccination from the county health department increased 51% over the previous 12 months (79 vaccinations compared with 52, respectively).

Editorial Note: This report and others underscore the consequences of missed opportunities for vaccination. Although the patient in this report had numerous prior contacts with the health-care system, she had no history of vaccinations against tetanus. Of the 57 persons with tetanus in 1989 and 1990 for whom vaccination status was known, 45 (79%) reported never having received more than 2 doses of tetanus toxoid. In addition, of the 12 who had sought medical care for their injuries and for whom tetanus toxoid was indicated, 11 were not vaccinated.

Wounds such as that described in the patient in this report are common in persons with tetanus and may not be considered sufficiently severe by the person to warrant a visit to a health-care provider. In 1989 and 1990, only 27 (31%) of 86 persons with tetanus and a clear antecedent acute injury sought medical treatment for their wounds. Therefore, internists, family practitioners, occupational physicians, and other primary health-care providers who treat adults should use every opportunity to review the vaccination status of their patients and administer Td and other indicated vaccines as appropriate (*MMWR* 42:148, 1993).

Paralysis may proceed to the arms and legs, with death resulting from paralysis of the respiratory muscles. In cases that progress slowly, supportive medical treatment can be applied and most infants survive. In rapidly developing cases, death may result before significant signs are noted. Such death may be included under the sudden infant death syndrome (sudden crib death), a problem that has long baffled medical investigators. It is now suspected that at least a small percentage of sudden infant deaths is a result of botulism.

Occasionally *C. botulinum* is able to grow in various environmental niches, such as animal feed, carcasses of dead animals or invertebrates, and sediments in lakes or ponds. Outbreaks of botulism poisoning occur when these contaminated materials are eaten by domestic or wild animals or birds. Under certain conditions outbreaks of botulism have occurred in waterfowl, resulting in the death of hundreds of thousands of the birds. Such wildlife epidemics are commonly due to botulinum toxin types C and D.

✳ Transmission and Epidemiology

C. botulinum spores are often present on food. Production of botulinum toxin results when the microorganism grows in an anaerobic environment in foods stored at room temperature. Alkaline foods favor the growth of *C. botulinum* and the development of the exotoxin, while acidic foods, those with a pH of 4.6 or less, inhibit their growth.

Because botulism is a poisoning rather than an infectious disease, transmission from person to person does not occur. Circumscribed outbreaks, however, may result when a toxin-containing food is eaten by a number of people. Transmission of botulism generally involves home-canned foods; then the disease usually occurs only among family members. Outbreaks from improperly sterilized commercially canned food are quite rare. With commercially canned food, only four deaths from botulism poisoning have resulted in the United States in the past 45 years, whereas during this time more than 775 billion cans of food have been eaten. Yet the threat of botulism is always present, and continued monitoring of food-processing procedures and foods is needed to maintain this remarkable safety record. Usually, from 20 to 30 cases a year of food-borne botulism are reported in the United States. Nevertheless, in 1977 several significant outbreaks of botulism food poisoning resulted when restaurants illegally used home-canned sauces and relishes; the total number of cases for that year increased to over 80 (Figure 18-5). Usually, from 60 to 80 cases of infant botulism are reported each year in the United States (Figure 18-6).

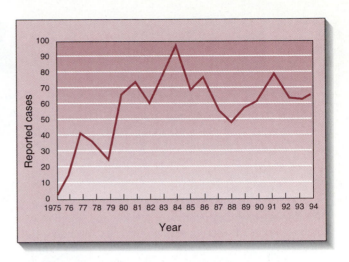

FIGURE 18-6 Reported cases of infant botulism in the United States, 1975 to 1994. (modified from CDC annual summaries)

✳ Diagnosis

Because time is important in dealing with botulism food poisoning, a preliminary diagnosis must often be made on the basis of clinical and epidemiological evidence. A diagnosis based on clinical appearance is often difficult, for early symptoms are easily confused with other diseases. Furthermore, few physicians have personally seen patients with botulism and so would not easily recognize the symptoms. When isolated cases occur, there is little reason to suspect botulism. However, when botulism is suspected, serum, stool, and **gastric washing** specimens should be collected and the suspected foods obtained. The presence of toxin in these materials can be detected by injecting extracts into mice. If the toxins are present, the mice usually die within a few days.

✳ Treatment

Once botulism is suspected, antitoxin should be given as soon as possible. Because any one of the three most common toxin serotypes may cause the disease, a *polyvalent* antitoxin containing antitoxin to types A, B, and E is used. The antitoxin will not reverse the effects of toxin already affecting the nerves but will neutralize any circulating toxin.

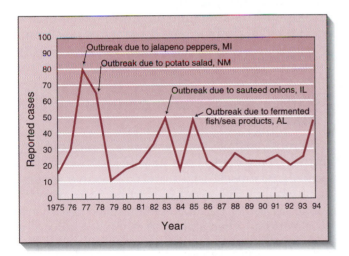

FIGURE 18-5 Reported cases of food-borne botulism in the United States, 1975 to 1994. (modified from CDC annual summaries)

Gastric washing Removal of stomach contents through a tube swallowed to allow washing of the stomach.
Polyvalent Antisera containing antibodies against more than one epitope or antigen.

CLINICAL NOTE

Food-borne Botulism: Oklahoma 1994

On 2 July, 1994, the Arkansas Department of Health and the Oklahoma State Department of Health were notified about a possible case of food-borne botulism. This report summarizes the investigation, which implicated consumption of improperly stored beef stew.

On 30 June, 1994, a 47-year-old resident of Oklahoma was admitted to an Arkansas hospital with subacute onset of progressive dizziness, blurred vision, slurred speech, difficulty swallowing, and nausea. Findings on examination included a paralytic drooping of the upper eyelid (ptosis), extraocular palsies, facial paralysis, palatal weakness, and impaired gag reflex. The patient also had partially healed superficial knee wounds incurred while laying cement. He developed respiratory compromise and required mechanical ventilation.

Differential diagnoses included wound and food-borne botulism, and botulism antitoxin was administered intravenously. Electromyography demonstrated an incremental response to rapid repetitive stimulation consistent with botulism. Anaerobic culture of the wounds were negative for *Clostridium*. However, analysis of a stool sample obtained on 5 July detected type A toxin, and culture of stool yielded *C. botulinum*. The patient was hospitalized for 49 days, including 42 days on mechanical ventilation, before being discharged.

The patient had reported that during the 24 hours before onset of symptoms he had eaten home-canned green beans and a stew containing roast beef and potatoes. Although analysis of the leftover green beans was negative for botulism toxin, type A toxin was detected in the stew. The stew had been cooked, covered with a heavy lid, and left on the stove for 3 days before being eaten without reheating. No other persons had eaten the stew (*MMWR* 44:200, 1995).

Supportive care, particularly in maintaining respiratory functions, is very important. Infants require only supportive treatments.

✳ Prevention and Control

Proper procedures in preparing home-canned foods—that is, using pressure cookers to ensure destruction of spores—are the most important preventive measures. Before being served, home-canned foods should be boiled for several minutes to destroy any toxin that might be present.

C. PERFRINGENS: CELLULITIS AND GAS GANGRENE

Different species of the genus *Clostridium* are able to grow in the anaerobic environment of damaged body tissue. These bacteria may release toxins and enzymes that produce further tissue damage to the surrounding healthy tissue, a condition known as anaerobic **cellulitis**, **myonecrosis**, or gas gangrene. *C. perfringens* is the most frequently involved *Clostridium* species producing these conditions. These clostridia are common inhabitants of the intestinal tracts of humans and animals and are found both on clothing and in the soil. *C. novyi* and *C. septicum*, or other clostridia are occasionally involved (Figure 18-7).

✳ Pathogenesis and Clinical Diseases

Numerous toxins and enzymes produced by clostridia (Table 18-1) are able to destroy tissues, particularly muscle fibers and connective tissues. Many wounds that occur in wartime or from automobile accidents become contaminated with various clostridia. When a niche of devitalized (dead or dying) tissue or foreign debris exists in a wound, it may create an anaerobic area in which the clostridia can proliferate. As the toxins and enzymes of the growing clostridia are released, the tissues adjacent to the wound are devitalized and the supply of oxy-

---

Cellulitis   Inflammation of cellular or connective tissues.
Myonecrosis   Death and destruction of muscle cells. Myonecrosis is a serious clinical condition characteristic of gas gangrene and is fatal without proper rapid treatment.

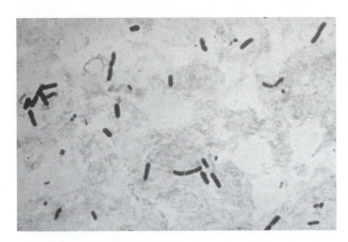

FIGURE 18-7   Specimen from a patient with gas gangrene showing large gram-positive bacilli (*Clostridium*) in a background of cellular debris. (Courtesy of Centers for Disease Control, Atlanta)

genated blood is stopped. This situation creates an expanded anaerobic area into which the clostridia can grow; the result is a further production of toxins, which again extends the area of tissue damage. Phagocytic cells of the host are essentially helpless against this infection, for the bacteria are sequestered in dead tissue out of reach of the defense mechanisms of the host.

Several forms of clostridial wound infections occur. A condition called *anaerobic cellulitis* is less severe than gas gangrene (*myonecrosis*) and does not involve the muscles. The infection spreads through subcutaneous tissues and between muscles. Gas is produced by the bacteria, causing distention of the tissues. Myonecrosis is a more severe form of the infection, with toxic destruction of adjacent muscle tissues and an ever-widening expansion of the lesion. The swollen tissues have a dark yellowish discoloration and produce a foul-smelling, dark fluid exudate. Gas is formed by the bacteria, causing some distention of the subcutaneous tissues and considerable pain. Usually, symptoms of gas gangrene begin to appear 12 to 72 hours following injury. Along with the local tissue involvement, bacteria may spread via the blood to other organs, and a generalized toxic reaction, including fever, *toxemia*, and *shock*, may be seen in the patient. Without proper treatment, death results. Gas gangrene may develop in the uterus after mechanically induced abortion, and is seen more frequently following illegal abortions induced by nonmedical practitioners. Rapidly developing gas gangrene may occur in any section of the bowel that is deprived of a normal blood supply.

✳ Transmission and Epidemiology

As with the other clostridial diseases, person-to-person transmission is not a factor in the epidemiology. Disease results when host tissues are altered so that they allow the growth of the clostridia. Gas gangrene is a serious threat to persons with traumatic injuries, bowel obstructions, bowel surgery, reduced blood supply to given tissues, and so on. Elderly persons with poor circulation are prime candidates for this type of infection. Gas gangrene is a major problem in battlefield wounds when treatment is delayed.

✳ Diagnosis

Laboratory diagnosis of clostridial wound infections is difficult and early diagnosis is made on clinical grounds.

✳ Treatment

Removal of dead tissue and debris from the wound is the first step in treatment. Penicillin or other antibiotics should be given and are helpful if all dead tissue is removed. In many cases of gas gangrene, the spread of the infection cannot be stopped unless all infected tissue is removed; this often requires amputation of the involved limb or extensive tissue removal. Hyperbaric oxygen (chambers of oxygen gas under 3 atmospheres of pressure to increase the amount of O_2 in the tissues) has been

TABLE 18-1	
Enzymes from Clostridia Associated with Pathogenesis	
Enzyme	**Function**
Lecithinase (alpha toxin)	Digests lecithin
Collagenase	Destroys connecting tissue
DNase	Disrupts DNA
Hemolysin (theta toxin)	Cardiac toxin, hemolysis

Toxemia   The presence of toxins in the blood.
Shock   A condition in which there is a sharp decrease in blood pressure accompanied by depression of a number of physiologic parameters.

used with some limited beneficial effects in treating these anaerobic infections.

✳ Prevention and Control

Prompt cleaning and surgical debridement of wounds constitute the most important preventive measure. The rapid evacuation to field hospitals by helicopter of military personnel wounded during combat has greatly reduced the incidence of gas gangrene associated with battlefield wounds. Antibiotic treatment may help prevent the development of clostridial infection in wounds.

OTHER CLOSTRIDIAL DISEASES

✳ Food Poisoning

In addition to producing gangrene, *C. perfringens* is a common cause of food poisoning. Infection is most often associated with consumption of contaminated meat dishes. The spores may survive the normal cooking process, then germinate as the meat cools, and within a few hours at warm temperatures massive numbers of bacteria develop. The ingested bacteria grow in the intestines and release toxins. These toxins cause diarrhea, cramps, and abdominal pain. Onset is 8 to 20 hours after ingestion of contaminated meat, and symptoms last about one day. Deaths do not result. In New Guinea, *C. perfringens* (type C) is responsible for a unique type of gastroenteritis referred to as *pigbel*. This condition follows the consumption of large quantities of pork and is often fatal, particularly in children.

✳ Pseudomembranous Colitis

A potentially serious form of diarrhea following some antibiotic treatments is caused by *C. difficile*. This bacterium is widespread in the hospital environment and is part of the normal flora of the intestinal tract of 20% of hospital patients, but is not able to compete with the normal bacterial flora. When antibiotic treatments reduce the normal bacterial flora, *C. difficile* is able to proliferate rapidly and secrete toxins. This may result in diarrhea or the more serious condition of pseudomembranous colitis. *C. difficile* cause about 25% of antibiotics-associated diarrheas and most cases of pseudomem-

branous colitis. Pseudomembranous colitis results from toxins that cause fluids to collect and damage the cells of the bowel. The first symptoms are abdominal pain with watery diarrhea. Mixtures of fibrin, mucus, and white blood cells accumulate in patches on the mucosa of the colon. These patches are called *pseudomembranes* (false membranes) and are the basis of the name *pseudomembranous colitis*. About one-third of the patients with this disease die, possibly from combined effects of both the primary disease for which they were being treated and from the antibiotic-induced colitis. No single class of antibiotics is responsible for this condition, and care must be used in administration of all antibacterials. Treatment includes discontinuing the antibiotic that started the problem.

✳ Bacteremia

Several species of *Clostridium* are able to produce *bacteremia* (bacteria in the bloodstream). This situation is serious not only because of the presence of the bacteria in the blood, but because clostridial bacteremia is one of the conditions that frequently accompanies certain types of cancer. When these organisms are found in the bloodstream, the physician usually begins a careful examination for the presence of a tumor.

NON-SPORE-FORMING ANAEROBES

In contrast to the clostridia, the non-spore-forming anaerobes are a rather heterogenous group of bacteria comprising a large number of species from several genera. Currently their classification is not complete. But because of the increased interest and research activity involving these microbes, names and classification schemes will undoubtedly continue to undergo frequent revisions. This group of bacteria are not characterized by distinct exotoxins or virulence properties and they are present in as many as 40% of properly collected clinical specimens. They are the most common causes of brain, lung, and abdominal abscesses (Table 18-2).

Because the non-spore-forming anaerobes are part of the normal flora of many body tissues, their incrimination as the cause of a given infection can be made only when they are isolated from a tissue that is normally free of these organisms. The source

Colitis   An inflammation of the colon.

Clostridium perfringens Gastroenteritis Associated with Corned Beef Served at St. Patrick's Day Meals: Ohio and Virginia, 1993

Clostridium perfringens is a common infectious cause of outbreaks of food-borne illness in the United States, especially outbreaks in which cooked beef is the implicated source. This report describes two outbreaks of *C. perfringens* gastroenteritis following St. Patrick's Day meals in Ohio and Virginia during 1993.

Ohio On 18 March, 1993, the Cleveland City Health Department (CCHD) received telephone calls from 15 persons who became ill after eating corned beef purchased from one delicatessen. After a local newspaper article publicized this problem, 156 persons contacted CCHD to report onset of diarrheal illness within 48 hours of eating food from the delicatessen on 16 March or 17 March. Symptoms included abdominal cramps (88%) and vomiting (13%); no one was hospitalized. The median incubation period was 12 hours (range: 2–48 hours). Of the 156 persons reporting illness, 144 (92%) reported having eaten corned beef; 20 (13%), pickles; 12 (8%), potato salad; and 11 (7%), roast beef.

In anticipation of a large demand for corned beef on St. Patrick's Day (17 March), the delicatessen had purchased 1400 pounds of raw, salt-cured product. Beginning 12 March, portions of the corned beef were boiled for 3 hours at the delicatessen, allowed to cool at room temperature, and refrigerated. On 16 and 17 March, the portions were removed from the refrigerator, held in a warmer at 120°F (48.8°C), and sliced and served. Corned beef sandwiches also were made for catering to several groups on 17 March; these sandwiches were held at room temperature from 11 a.m. until they were eaten throughout the afternoon.

Cultures of two of three samples of leftover corned beef obtained from the delicatessen yielded 10^5 colonies of *C. perfringens* per gram.

Following the outbreak, CCHD recommended to the delicatessen that meat not served immediately after cooking be divided into small pieces, placed in shallow pans, and chilled rapidly on ice before refrigerating and that cooked meat be reheated immediately before serving to an internal temperature of 165°F (74°C).

Virginia On 28 March, 1993, 115 persons attended a traditional St. Patrick's Day dinner of corned beef and cabbage, potatoes, vegetables, and ice cream. Following the dinner, 86 (76%) of 113 persons interviewed reported onset of illness characterized by diarrhea (98%), abdominal cramps (71%), and vomiting (5%). The median incubation period was 9.5 hours (range: 2–18.5 hours). Duration of illness ranged from 1 hour to 4.5 days; one person was hospitalized.

Corned beef was the only food item associated with illness; cases occurred in 85 (78%) of 109 persons who ate corned beef compared with one of four who did not. Cultures of stool specimens from eight symptomatic persons all yielded 10^6 colonies of *C. perfringens* per gram. A refrigerated sample of leftover corned beef yielded 10^5 colonies of *C. perfringens* per gram.

The corned beef was a frozen, commercially prepared, brined product. Thirteen pieces, weighing approximately 10 pounds each, had been cooked in an oven in four batches during 27–28 March. Cooked meat from the first three batches was stored in a home refrigerator; the last batch was taken directly to the event. Approximately 90 minutes before serving began, the meat was sliced and placed under heat lamps.

Following the outbreak, Virginia health officials issued a general recommendation that meat not served immediately after cooking be divided into small quantities and rapidly chilled to 40°F (4.4°C), and that precooked foods be reheated immediately before serving to an internal temperature of 165°F (74°C).

Follow-up Investigation The results of the epidemiologic and laboratory investigations suggest that the two outbreaks in this report were not related. Traceback of the corned beef in both of these outbreaks indicated that the meat had been produced by different companies and sold through different distributors. Serotyping was performed on *C. perfringens* isolates recovered from the stool samples in Virginia and on an isolate from a food sample obtained in Ohio. Six of the seven Virginia stool isolates were serotype PS86; however, the food isolate from Ohio could not be serotyped using available antisera (*MMWR* 43:137, 1994).

of these infections is usually **endogenous**, and disease results when the integrity of the tissue is altered to allow passage of these resident bacteria from their normal site of growth to other tissues. Often an anaerobic environment can be created by facultative anaerobes that use up the free oxygen

---

Endogenous From inside; endogenous infections are caused by the normal bacterial flora of the body.

TABLE 18-2

Commonly Encountered Anaerobes

| Organism | Percentage of Organisms Associated with | |
	Pulmonary Infections	Abdominal Sepsis
B. fragilis	17	91
B. melaninogenicus	34	19
Fusobacterium	38	25
Clostridium	10	66
Peptostreptococcus	49	28
Peptococcus	21	12

and thus act in a synergistic manner to allow the growth of the anaerobe. A listing of the most common pathogens, their morphology, and the infections they most often cause is given in Table 18-3.

✳ Clinical Diseases

Some general clinical problems attributed to non-spore-forming anaerobes are briefly discussed in the following paragraphs. It should be noted that these infections are often **polymicrobic**—that is, they are caused by more than one species acting simultaneously.

Infections of the Abdominal Cavity. Infections may result from passage of fecal material into the abdominal cavity as a result of condi-

tions like injury, surgery, appendicitis, or cancer. Over 90% of abdominal abscesses and infections are caused by anaerobic bacteria. Members of the genus *Bacteroides* (*B. fragilis* in particular) are most frequently involved in these infections (see Plate 17).

Infections of the Female Reproductive Tract. About 75% of the infections and abscesses of female reproductive organs are caused by anaerobes. Such conditions as abortion, surgery, cancer, extensive medical manipulations, prolonged labor, intrauterine contraceptive devices, and gonorrhea may predispose the tissue to these anaerobic infections. A serious condition known as *pelvic inflammatory disease* ∞ (Chapter 16, p. 231) is frequently associated with the presence of these bacteria.

TABLE 18-3

Major Non-Spore-Forming Bacteria Commonly Associated with Infections

Bacteria	Morphology	Associated Infections
Bacteroides	Gram-negative bacilli	Peritonitis, liver abscesses, gynecologic, pulmonary, upper respiratory, wounds, bacteremia
Fusobacterium	Gram-negative bacilli	Liver abscesses, gynecologic, pulmonary, upper respiratory, wounds, bacteremia
Propionibacterium	Gram-positive bacilli	Upper respiratory (pathogenicity uncertain) Acne
Eubacterium	Gram-positive bacilli	May be isolated from infected tissues, but pathogenic role, if any, is not known
Veillonella	Gram-negative cocci	May be isolated from infected tissues, but pathogenic role, if any, is not known

Polymicrobic Produced by more than one species of bacteria at the same time. Anaerobic infections are commonly polymicrobic.

Liver Abscesses. Non-spore-forming anaerobic bacteria may reach the liver by direct spread from adjacent tissues or from the blood. One-half or more of the liver abscesses are now known to be caused by anaerobic bacteria.

Respiratory Tract Infections. A variety of infections of the respiratory tract are caused by anaerobes. Between 75 and 95% of such diseases as pneumonitis and lung abscesses are induced by these microbes. More than 90% of patients with aspiration pneumonia (caused by breathing oral secretions down into the lungs) have cultures that are positive for non-spore-forming anaerobes.

Infections of the Skin and Muscles. Anaerobic infections may develop in skin, muscle, and connective tissue following injury, surgery, or lack of blood supply (ischemia). Human bites are a common reason for such infections.

Septicemia. Invasion of the blood system may result from any of the preceding anaerobic infections. About 10% of all blood infections detected in hospital patients are caused by non-spore-forming anaerobes.

Brain Abscesses. Bacteroides and anaerobic cocci are the most frequent etiologic agents of brain abscesses. Without proper surgical and antimicrobial intervention, these abscesses give the patient an extremely poor prognosis.

✳ Diagnosis

There are a number of clinical clues (e.g., site of infection and foul odor) to infection with non-spore-

forming anaerobes, but species identification of the isolated organism can be a difficult task. Recent efforts to provide rapid organism identification have resulted in the development of a variety of colorless enzyme substrates which become brightly colored when they have been changed by their specific enzyme. Use of these *chromogenic substrates* has facilitated rapid (within less than 24 hours) identification of the most common of the anaerobic bacteria. Other, more traditional approaches, such as fermentation reactions and chromatographic analysis, may be necessary for the more unusual anaerobic bacteria.

✳ Prevention and Treatment

Because most non-spore-forming anaerobic infections are derived from endogenous (normally present) bacteria, prevention of infection is difficult and largely depends on good medical procedures and nursing care. There are no vaccines, and it is not possible to eliminate these endogenous bacteria from the normal microbial flora of humans.

Treatment of these anaerobic infections is often difficult. Drainage and surgical removal of dead tissue, when possible, are most helpful, and antibiotic therapy is essential. Unfortunately, susceptibility to many of the available antibiotics seems to be declining among these organisms. The bacteroides are largely resistant to penicillins, and susceptibility to the cephalosporins is not uniform. Chloramphenicol, metronidazole, and clindamycin are generally effective across a wide range of these bacteria.

MATERIAL FOR REVIEW

CONCEPT SUMMARY

1. Human clostridial infections are serious, life-threatening diseases resulting from powerful exotoxins produced by these organisms.
2. Ability to limit disease due to clostridia depends on adequate use of available vaccines, proper food processing and preparation, and effective medical intervention. Such practices

have greatly reduced the frequency of these diseases in many countries of the world.

3. Human infections due to anaerobic bacteria are common. Most such infections have an endogenous source and are due to non-spore-forming anaerobes. These infections tend to be polymicrobic and serious. Proper medical help, including

surgical intervention, is frequently necessary to cure these infections.

4. Anaerobic infections have distinctive characteristics, which facilitate diagnosis. Most anaerobic infections are produced by a relatively small number of different bacterial species.

CLINICAL SUMMARY TABLE

Microorganism	Virulence Mechanisms	Diseases	Transmission	Treatment	Prevention
Clostridium tetani	Tetanus exotoxin	Tetanus	Environment to humans	Antitoxin	Vaccine
Clostridium botulinum	Botulinum exotoxin	Botulisum food poisoning	Foodborne	Antitoxins	Proper food handling
Clostridium Spp.	Exotoxins Enzymes	Cellulitis Gangrene	Soil to wound	Removal of infected tissue Penicillin	Rapid treatment of wounds
Clostridium difficile	Exotoxins	Pseudo-membranous colitis	Endogeneous	Stop antibiotic treatment	Limit use of antibiotics

STUDY QUESTIONS

1. Pathogenic anaerobic bacteria can be classified as to their Gram reaction, ability to form spores, and natural habitat. Construct a table that shows the above features for the following anaerobic bacteria: *C. tetani, C. septicum, B. fragilis,* and *Fusobacterium nucleatum.*

2. What feature of the non-spore-forming anaerobes makes them such a common cause of human infection?

3. What virulence mechanism possessed by most of the clostridia gives them the capacity to produce serious human diseases?

4. DPT vaccine has been used to reduce the incidence of tetanus. At what ages and how often should this vaccine be given?

5. List the features of infant botulism that are different from those associated with botulism in adults.

6. List the sources of *C. perfringens* that could lead to infection of a wound and possible gas gangrene.

CHALLENGE QUESTIONS

1. Some of the genes for the botulinum toxins are carried on plasmids. Should medical and research professionals be concerned of the danger that these gene might appear in *Escherichia coli* or gram-negative anaerobes commonly found in the human intestinal microflora? Explain.

2. Although there is a vaccine for botulism (rarely used), why is there no vaccine for food-borne diseases caused by *Clostridium perfringens*?

Gram-Positive Bacilli

Species from four genera of gram-positive, facultatively anaerobic bacilli—*Corynebacterium*, *Listeria*, *Erysipelothrix*, and *Bacillus*—are of continuing concern as *etiologic agents* of human disease. Although members of each genus are gram-positive, this is almost the only characteristic they share in common. *Bacillus* is a large genus with many endospore-producing, **ubiquitous** species, while only a single, nutritionally fastidious species is found in the genus *Erysipelothrix*. The diseases produced by these microorganisms range from classic diphtheria and anthrax to benign food poisoning due to *B. cereus*. Such variation in pathogenesis is reflective of the physiologic variations that occur among these four genera. Each genus is discussed in separate sections of this chapter.

CORYNEBACTERIA: GENERAL CHARACTERISTICS

Several species of corynebacteria are found in humans and animals. Diphtheria is the major disease of humans caused by corynebacteria.

Corynebacteria are narrow (0.5 to 1 µm), gram-positive bacilli that may range in length up to 5 µm (Figure 19-1). When prepared on slides for staining, the cells, which often remain attached on one side during division, are frequently oriented in *palisades* or in V- and L-shaped arrangements. When mixed together, such cellular arrangements are said to give

Etiologic agent The causative agent; the microorganism responsible for a specific disease.
Ubiquitous Present everywhere.

HUMAN BODY SYSTEMS AFFECTED

Corynebacteria, Listeria, and Bacillus

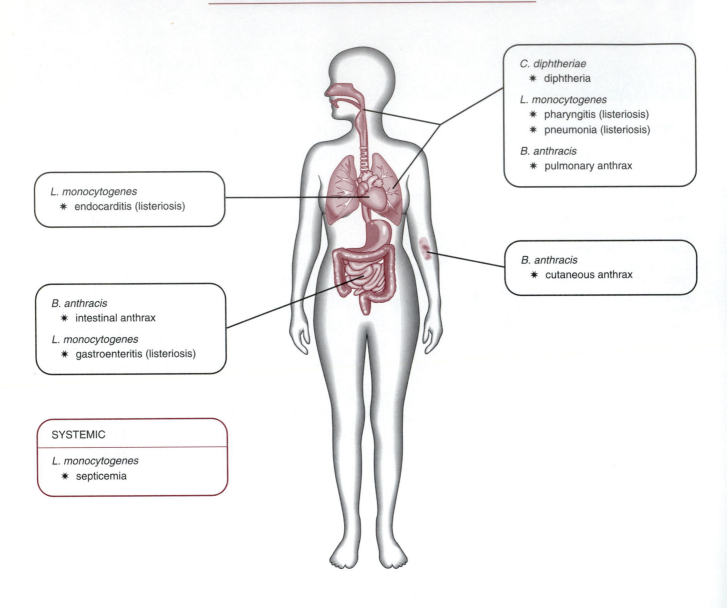

C. diphtheriae
* diphtheria

L. monocytogenes
* pharyngitis (listeriosis)
* pneumonia (listeriosis)

B. anthracis
* pulmonary anthrax

L. monocytogenes
* endocarditis (listeriosis)

B. anthracis
* cutaneous anthrax

B. anthracis
* intestinal anthrax

L. monocytogenes
* gastroenteritis (listeriosis)

SYSTEMIC

L. monocytogenes
* septicemia

the appearance of Chinese letters. The cells are *pleomorphic* (multiple shaped), often with bulging at one end that gives a club appearance (Greek *coryne* = club). They may also contain accumulations of phosphates (*metachromatic granules*) that stain differently from the other cell materials and give the stained cells a beaded appearance. Although these bacteria can be cultivated on a variety of media, in diagnostic laboratories they are grown on a selective medium ∞ (Chapter 7, p. 90) containing tellurite salts. Reduction of the tellurite causes these bacteria to produce gray or black colonies. With this procedure, three different types of the causative agent of diphtheria, *C. diphtheriae*—gravis, mitis, and intermedius—have been determined. All three types, however, are capable of producing the same

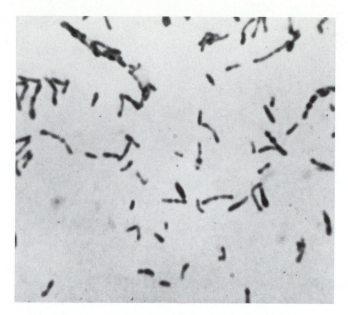

FIGURE 19-1 Micrograph of *Corynebacterium diphtheriae*. (Courtesy of Biological Photo Services)

toxin and clinical disease. *C. diphtheriae* is more resistant to drying than many vegetative bacteria and may remain viable for as long as 3 to 4 months in dried respiratory exudates.

C. DIPHTHERIAE: DIPHTHERIA

Diphtheria is caused by some strains of the bacterium *Corynebacterium diphtheriae*. This disease results when the organism produces a powerful *exotoxin* that is absorbed by various tissues within the body.

The ability of a given strain of *C. diphtheriae* to produce the exotoxin is determined by the presence of a lysogenic bacteriophage ∞ (Chapter 6, p. 77). Some bacteriophages have the ability to insert their DNA into the DNA molecule of the host bacterium, a process called *lysogeny*. The gene that directs the production of the diphtheria toxin is carried by the bacteriophage (known as *beta-corynephage*), and the toxin is therefore produced only by those corynebacteria that contain the bacteriophage genes. Species of this genus other than *C. diphtheriae* may occasionally be infected with beta-corynephage which enables them to produce diphtheria toxin, but this event seems to be quite rare. Research on diphtheria during the latter two decades of the 1800s led to the discovery of the first bacterial exotoxins, and demonstrated that such toxins are the cause of some bacterial disease. This research also led to the development of methods for treating toxic diseases with antitoxins.

✳ Pathogenesis and Clinical Diseases

Both toxin-producing and non-toxin-producing strains of *C. diphtheriae* can adhere to and colonize the mucosal tissue of the upper respiratory tract. Humans are the only natural host, and transmission is primarily from person to person by airborne droplets. The disease results entirely from the effects of the exotoxin.

Diphtheria toxin is one of the most frequently studied and best known of the bacterial exotoxins. This toxin prevents protein synthesis ∞ (Chapter 6, p. 72). The toxin specifically adheres to and is initially absorbed into cells around the growing *C. diphtheriae*. No structural damage is seen until the lack of newly synthesized proteins causes the death of the host cells. As the dead host cells accumulate, the bacteria continue to produce toxin, which causes the lesion to expand. An incubation period of several days to one week is required before the patient begins to experience clinical symptoms from diphtherial lesions. These lesions usually appear first in the tonsillar–pharyngeal area as patches of a thick fibrinous exudate containing many entrapped host cells and bacteria. This layer of exudate, called a *pseudomembrane*, adheres firmly to the epithelial surfaces (Figure 19-2). As the disease progresses, the pseudomembrane may spread upward into nasopharyngeal tissues and/or downward into the larynx and trachea. In severe cases, obstruction of the airway may occur, resulting in suffocation of the patient. The lesions remain superficial and rarely do the bacteria invade deeper tissues, although the

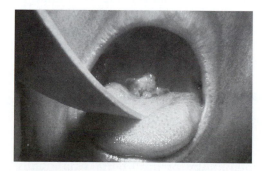

FIGURE 19-2 Pharyngeal pseudomembrane produced by infection with *Corynebacterium diphtheriae*. (Courtesy of Visuals Unlimited)

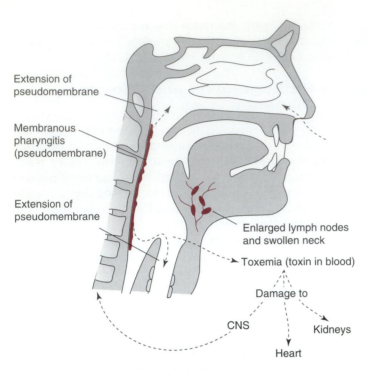

Extension of pseudomembrane

Membranous pharyngitis (pseudomembrane)

Extension of pseudomembrane

Enlarged lymph nodes and swollen neck

Toxemia (toxin in blood)

Damage to

CNS

Kidneys

Heart

FIGURE 19-3 Pathogenesis of diphtheria.

toxin may produce *necrosis* of such internal organs as liver, kidney, and adrenal glands.

Besides formation of the pseudomembrane and a sore throat, the patient experiences fever, malaise, and enlarged regional lymph nodes, resulting in the swelling of the neck. The term "bull-neck" is sometimes used to refer to this condition. The most frequent and serious damage due to toxin occurs to the heart and central nervous system (CNS), with death often resulting from damage to the heart. The disease may linger for many weeks, and due to damage to the CNS, varying levels of paralysis may occur. Aspects of the pathogenesis of diphtheria are shown in Figure 19-3. Before methods of treating or preventing this disease were available, diphtheria was a major killer of children. When poor sanitation exists, primary or secondary diphtherial lesions may occur on the skin (Figure 19-4). The toxin is also produced in these lesions and patients experience an illness similar to pharyngeal diphtheria.

✳ Transmission and Epidemiology

Recovery from diphtheria does not necessarily eliminate *C. diphtheriae* from the throat; many patients remain asymptomatic carriers for prolonged periods. The epidemiology of diphtheria was greatly

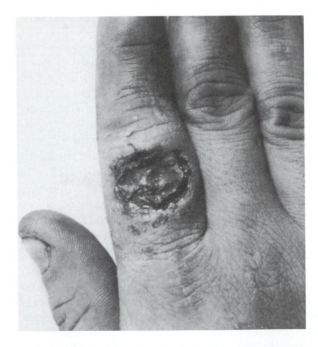

FIGURE 19-4 Necrotic cutaneous diphtherial lesion about 15 days after onset. (Armed Forces Institute of Pathology, AFIP MIS #55 997-1)

Necrosis The death and subsequent destruction of cells; usually appearing as darkened or ulcerated tissue.

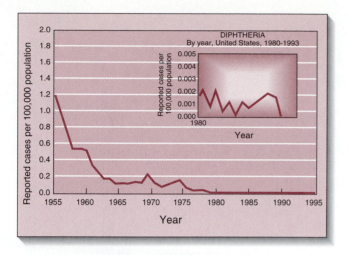

FIGURE 19-5 Reported cases of diphtheria in the United States between 1955 and 1995. (From CDC annual summaries)

altered during the past several generations in countries where widespread immunization has been practiced. This has also greatly reduced the number of healthy carriers. At present, only a few cases of diphtheria occur each year in the United States, and since 1988 all cases have been imported (Figure 19-5). Those who develop diphtheria are most often either individuals who have never been immunized or elderly persons who have lost their immune status.

✳ Diagnosis

The diagnosis of diphtheria is most frequently made on the basis of clinical or serological evidence. Nevertheless, the isolation of toxin-producing *C. diphtheriae* from the throat or lesion provides a positive diagnosis.

✳ Treatment

When diphtheria is suspected, the patient should receive passive immunization with antitoxin. Early antitoxin treatment is necessary in order to be effective because, once the toxin has bound to cell receptors, the reaction cannot be reversed. Penicillin or erythromycin should also be given to stop further growth of *C. diphtheriae* in the throat, and to prevent the patient from developing into a carrier. Persons

who are asymptomatic carriers can be freed of *C. diphtheriae* by antibiotic treatment.

✳ Prevention and Control

Diphtheria can be prevented and controlled to a large extent by active immunization with a toxoid vaccine. The vaccine is administered during the second or third month of life in combination with vaccines for tetanus and whooping cough (DPT vaccine). A booster is given a year later and again when the child enters school. Revaccination every 10 years thereafter is also recommended. Even though diphtheria has been essentially eliminated in the United States, it is important to maintain an up-to-date vaccination program. The recent widespread outbreaks of diphtheria in the Commonwealth Independent States of the former Soviet Union (see Clinical Note and Figure 19-6) underscores the ability of this disease to reoccur when vaccination programs are not properly maintained.

Unimmunized persons who may have been exposed to diphtheria can be passively immunized with antitoxin. A procedure called the *Schick test* can be used to determine if a person is susceptible to diphtheria. Here a small amount of diphtheria toxin is injected subcutaneously; if the person has no protective antibodies, an inflammatory reaction occurs at the site of toxin injection. Such a person should then be actively or passively immunized, depending on circumstances.

OTHER CORYNEBACTERIA

As already noted, other members of the genus *Corynebacterium* (*C. ulcerans, C. pseudotuberculosis*) can produce diphtheria toxin when they are infected with the beta-corynephage. Other members of this genus, which are often referred to as **diphtheroids**, are occasionally the cause of endocarditis and sepsis in immunocompromised patients (Table 19-1). In particular, one of these, known as *JK*, produces serious life-threatening disease with mortality rates as high as 75%. This organism is very resistant to most antibiotics and is found primarily in patients with artificial heart valves or intravenous catheters. Two species, *C. haemolyticum* and *C. ulcerans*, can cause pharyngitis, with symptoms similar to those produced by group A streptococci.

Diphtheroids   Gram-positive bacilli of the genera *Corynebacterium*, **Propionibacterium**, which are commonly found on body surfaces and are frequently found as contaminating bacteria in clinical specimens.

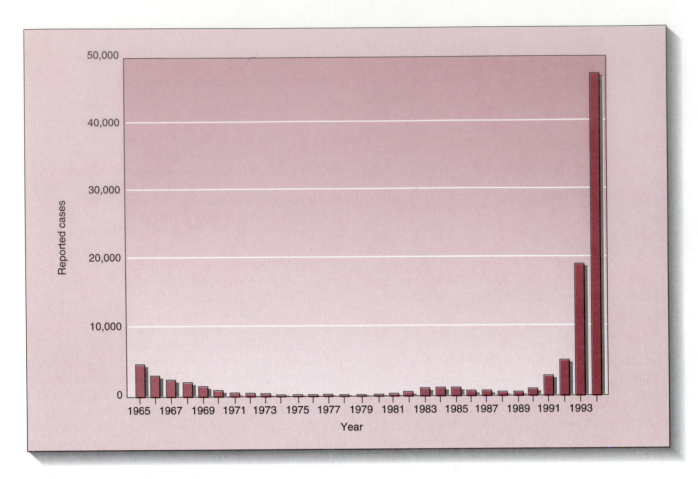

FIGURE 19-6 Reported number of diphtheria cases in the Commonwealth of Independent States of the former Soviet Union, 1965–1995. (World Health Organization)

LISTERIA: GENERAL CHARACTERISTICS

Listeria are gram-positive short bacilli or coccobacilli with a tendency to occur in short chains and produce a zone of hemolysis when grown on blood agar. Because of these traits, *Listeria* can easily be mistaken for a group A streptococcus. They survive in the environment for years, but are readily killed by antiseptics and the heat of pasteurization. *L. monocytogenes*, the only one of the four species in the genus *Listeria* to be pathogenic for humans, is found in the genital and intestinal tracts and throat of many humans and animals. It is also found in soil, water, sewage, and animal secretions, especially milk.

LISTERIA MONOCYTOGENES

The disease caused by *L. monocytogenes* is called *listeriosis* and is primarily an infection in animals, but can also occur in human fetuses, newborns, pregnant females, and immunosuppressed patients.

✳ Pathogenesis and Clinical Diseases

Once in the tissues of the host, *L. monocytogenes*, is phagocytized by monocytes and macrophages

TABLE 19-1	
Diseases Caused by *Corynebacterium* Species	
Corynebacterium	**Disease Association**
C. diphtheriae	Diphtheria
C. haemolyticum	Pharyngitis
C. xerosis	Endocarditis
C. pseudotuberculosis	Tuberculosis-like illness
C. ulcerans	Pharyngitis
JK	Endocarditis, bacteremia

Diphtheria Acquired by U.S. Citizens in the Russian Federation and Ukraine, 1994

Epidemic diphtheria has reemerged in 14 of the 15 Commonwealth of Independent States (CIS) of the former Soviet Union; during 1994, a provisional total of 47,802 cases and 1746 deaths from diphtheria were reported throughout the CIS. This report escribes one confirmed and one probable case of diphtheria acquired in countries where the disease is epidemic (Russian Federation and Ukraine) by U.S. citizens during November and December 1994.

Patient 1 A 42-year-old woman, born in Russia but living in the United States for several years, arrived in Moscow on 22 November. She had onset of fever and sore throat on 6 December and was hospitalized on 7 December with a provisional diagnosis of diphtheria. Her vaccination history was unknown, and she was not aware of contact with diphtheria patients or carriers. She was treated with 9000 international units (IU) of equine diphtheria antitoxin, antibiotics, and prednisone. On 8 December, she was transferred to a referral hospital in Helsinki; findings on examination included a pharyngeal membrane. Treatment included administration of 40,000 IU of antitoxin, penicillin G (for 6 days), and several days of roxithromycin. Toxigenic *C. diphtheriae*, biotype *gravis*, was isolated from a pharyngeal culture obtained 9 December. Follow-up cultures on 12 and 15 December were negative. Her antitoxin level was measured in Helsinki by Vero cell neutralization assay and

was >5 IU/ml; however, the level measured by an enzyme immunoassay that is specific for human antibodies was <0.03 IU/ml, indicating that the Vero cell assay was detecting recently administered equine antitoxin. She recovered fully without complications.

Patient 2 On 28 November, a 22-year-old woman from New Jersey working in Kherson in southern Ukraine since June 1994 had onset of a sore throat; she was hospitalized on 29 November with a provisional diagnosis of diphtheria. She had received five doses of diphtheria and tetanus toxoids and pertussis vaccine (DTP) during childhood and an adult formulation tetanus and diphtheria toxoids (Td) booster in August 1991. She had no recognized contact with a known diphtheria patient or carrier. Findings on examination included a tonsillar and posterior pharyngeal membrane. The patient had treated herself with ciprofloxacin for 1 day and had had at least one dose of oral penicillin before a throat culture was obtained (the culture was negative). Treatment comprised 80,000 IU of diphtheria antitoxin and a course of parenteral penicillin. A diphtheria antitoxin level of 0.2 IU/ml by Vero cell neutralization assay was detected in both a blood specimen obtained at the time of her arrival in Ukraine in June 1994 and a convalescent sample. She recovered fully without complications (*MMWR* 44:237, 1995).

∞ (Chapter 10, p. 146). However, rather than being destroyed by the phagocytic cells, this bacterium is able to multiply and then escape from the phagocytic cell and spread to adjacent cells. Thus, *L. monocytogenes* is able to spread through the tissue and remain inside the host cells where it is protected from circulating antibodies. Only when *cell-mediated-immunity* ∞ (Chapter 11, p. 164) develops can this infection be controlled.

Diseases of Fetus and Newborn. *L. monocytogenes* gains access to the blood of the fetus or the newborn from the infected mother. This may result in spontaneous abortion or stillbirth of the fetus. In the newborn, septicemia results, with lesions on the legs and trunk, and the bacterium is carried to other organs where lesions occur; the most common manifestation is meningitis.

Diseases of Adults. Pregnant females and immunocompromised persons are the most suscep-

tible; disease is rarely seen in otherwise healthy individuals. Infection is from contaminated milk products, water, domestic animals, or human carriers. The organisms colonize the throat and intestinal tract where they are phagocytized and then spread into the lymph nodes and the bloodstream while they continue to survive and multiply inside the phagocytic cells. Clinical manifestations that are seen include pharyngitis, gastroenteritis, septicemia, meningitis, endocarditis, pneumonia, and abscesses in various organs.

✳ Transmission and Epidemiology

About 2000 cases are diagnosed each year in the United States, with 400 to 500 deaths. AIDS patients have a rate of infection with *L. monocytogenes* that is 300 times greater than the rate in the general population.

＊ Diagnosis

Isolation of *L. monocytogenes* from the infected tissues is the first phase of diagnosis. Next it must be differentiated from group A streptococci. *L. monocytogenes* has a characteristic "tumbling" motility at 22 to 25°C that helps to identify it.

＊ Treatment

Early treatment with ampicillin, erythromycin, or a combination of penicillin and aminoglycosides offers the best chance for success. Treatment is often not successful.

＊ Prevention and Control

There are no vaccines. Prevention is best achieved by using only pasteurized milk or milk products and keeping pregnant females and immunocompromised patients away from known human cases and animals.

ERYSIPELOTHRIX

The single species *Erysipelothrix rhusiopathiae* may cause local cutaneous lesions (*erysipeloid*) or more serious systemic illness. Most infections are associated with animal contact and the disease most frequently occurs in those who work with animals.

BACILLUS ANTHRACIS: ANTHRAX

Anthrax is primarily a disease of animals. During the developmental years of medical microbiology

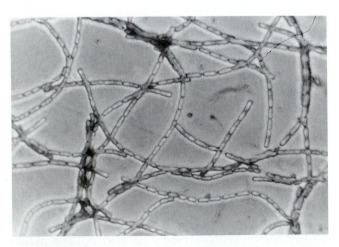

FIGURE 19-7 Photomicrograph of *Bacillus anthracis* showing both endospores and vegetative cells. (Courtesy of Biological Photo Service).

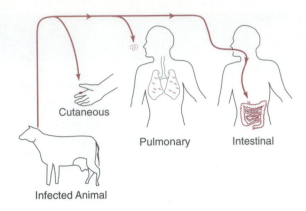

FIGURE 19-8   Three forms of anthrax that might be contracted by exposure to infected animal products.

anthrax was a frequently encountered disease in sheep and cattle. Because of the relatively large size and definite shape of the bacillus, plus the availability of experimental animals, anthrax served as an excellent model for the study of infectious diseases. Many important discoveries regarding the germ theory of disease and immunity resulted from investigations of anthrax ∞ (Chapter 1, p. 7).

Bacillus anthracis is a large, aerobic, gram-positive, spore-forming bacillus 5 to 10 μm long and 1 to 3 μm wide (Figure 19-7). The spores are highly resistant to drying and may remain viable on animal products or in the soil for years. This bacterium grows readily on laboratory media, where it may be easily confused with other members of the genus *Bacillus*.

＊ Pathogenesis and Clinical Diseases

The spores of *B. anthracis* enter the body either through abrasions of the skin, by inhalation, or by ingestion. The virulent strains possess an unusual capsule composed of D-glutamic acid that retards phagocytosis by host cells. A complex toxin causes the signs and symptoms of the disease. The three main clinical forms of anthrax are cutaneous, pulmonary, and intestinal (Figure 19-8).

Cutaneous Anthrax.   The cutaneous form of anthrax is the most common and results when spores enter the tissues through abrasions or lesions. Infection usually occurs on the exposed skin surface. A local lesion develops that rarely contains pus, but it is swollen, hemorrhagic, and forms a black scab called an *eschar* (Figure 19-9). If the infection remains localized, death rates are low. The infection spreads to the blood in about 5% of the

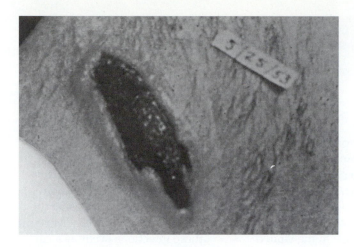

FIGURE 19-9 Cutaneous anthrax lesion on the neck, fifteenth day of disease. The patient had worked with air-dried goat skins from Africa. (Armed Forces Institute of Pathology, AFIP MIS #75-4203-9)

cases and this generalized infection is often fatal.

Pulmonary Anthrax. The pulmonary form of anthrax, which results from inhalation of spores, is seen in persons who handle contaminated animal products and is sometimes referred to as "wool-sorters' disease." The onset is sudden, with high fever and respiratory distress. Pneumonia is often followed by sepsis and in untreated cases, death usually results.

Intestinal Anthrax. The intestinal form of anthrax results from eating contaminated meat. Severe enteritis results and mortality rates are high. This form is rare and has not been reported as occurring in the United States.

✳ Transmission and Epidemiology

Transmission of anthrax among animals usually occurs from the ingestion of spore-contaminated feed. Anthrax spores may remain viable for 20 to 30 years or longer in a pasture contaminated by the remains of animals dying of anthrax. At present, relatively few cases of anthrax occur in animals in the United States. Anthrax is still found in animals in certain countries, however, and this presents a hazard to persons living in those countries who come in

contact with contaminated animal products. Since the 1950s, outbreaks in the United States have resulted from contact with contaminated hides covering bongo drums, infected pig bristles used in shaving brushes, goats' hair used in weaving, and similar items, all imported from foreign countries. Overall, the risk of contracting anthrax in the United States is slight; an average of only two cases per year have been reported in the past 10 years. Public health personnel should be aware of the threat of anthrax and be able to recognize the possible sources of infection.

✳ Treatment and Control

Anthrax can be successfully treated with penicillin or broad-spectrum antibiotics. Vaccines are available to control the disease in animals and humans of high risk. Carcasses of diseased animals should be buried deep in the soil (below the earthworm line as spores can be spread by earthworms) or burned to prevent the spread of spores. Gas sterilization or radiation may be used to decontaminate hides, wool, and related animal products.

One of the ominous aspects of anthrax is the ease with which it can be used as a deadly agent in germ-warfare. During the Gulf War there was the threat that anthrax spores might be used against the United Nations military forces. A great deal of effort was expended by the UN forces to both vaccinate and to provide barrier protection against such biological agents.

OTHER *BACILLUS* SPECIES

As noted earlier, other members of the genus *Bacillus* are able to produce human disease. Although these diseases are not as severe as those caused by *B. anthracis,* they are more common in the United States and can be serious. In otherwise compromised patients, *B. subtilis* may produce serious **endophthalmitis,** right-sided endocarditis, and even meningitis. Intravenous drug abuse is frequently an antecedent to such infections. *B. cereus* is a common cause of gastroenteritis. Eating cooked, improperly stored rice is often the cause of this illness.

Endophthalmitis Inflammation or infection of the inside of the eyeball. This is a serious condition, often requiring removal of the eyeball.

MATERIAL FOR REVIEW

CONCEPT SUMMARY

1. The most common serious infection caused by members of the genus *Corynebacterium* is diphtheria. Diphtheria is caused by lysogenic strains of the bacterium that produce a powerful exotoxin. The disease is generally held under control by appropriate immunization. The organism is frequently found in the human respiratory tract, but superficial wound infections are also known to result in the disease diphtheria.

2. *Bacillus anthracis* is responsible for the disease anthrax. The disease, though rare in the United States, is a serious, often life-threatening one acquired by associating with infected animals or contaminated animal by-products.

3. Diseases due to gram-positive bacilli other than diphtheria and anthrax are increasing in frequency and concern. Most of these infections occur as a consequence of some form of host compromise.

CLINICAL SUMMARY TABLE

Microorganism	Virulence Mechanisms	Diseases	Transmission	Treatment	Prevention
Corynebacterium diphtheriae	Exotoxin	Diphtheria	Airborne Person to person	Antitoxin	Vaccine
Listeria monocytogenes	Resists killing by phagocytes	Widespread infections	Water & milkborne Animals to humans Carriers	Various antibiotics	Pasteurization of milk
Bacillus anthracis	Capsule toxin	Anthrax	Animals to humans	Broad spectrum antibiotics	Vaccines Destroy infected animals

STUDY QUESTIONS

1. Describe the characteristics of the corynebacteria that are responsible for each of the following observations: (a) toxin production, (b) irregular staining, (c) black colonies on tellurite agar, (d) palisade formation, and (e) pseudomembrane formation.

2. Draw a diagram that shows the attachment, entry, and inhibitory action of diphtheria toxin.

3. What is the most common disease presentation in individuals infected with *Listeria*?

4. Which of the three clinical forms of anthrax is most unusual in the United States?

5. What activity commonly precedes serious infection due to *B. subtilis*?

6. What characteristic of diphtheria toxin production makes it possible for organisms other than *C. diphtheriae* to produce the disease?

CHALLENGE QUESTIONS

1. What mutations in eukaryotic cells could provide resistance to the diphtheria toxin?

2. When a person is infected with *L. monocytogenes*, antibodies to the bacteria are produced. Explain why these antibodies are not protective.

Mycobacteria and Related Microorganisms

Tuberculosis and leprosy are the major diseases of humans caused by bacteria of the genus *Mycobacterium*. Their impact is difficult to determine, but historically they would certainly rank among the most devastating of all human diseases. It has been estimated that tuberculosis was the single greatest cause of human disease and death during the past 200 years. During the 1800s as many as 30% of all deaths in the eastern United States were due to this disease. And even today there are about 2000 deaths per year due to this disease in the United States.

Although modern medical practices and improved living standards have greatly reduced the prevalence in developed nations, in some areas of the world, up to 40% of the people are infected with *Mycobacterium tuberculosis*. In the United States, 4.1% of New York City school children are infected with *M. tuberculosis* and 1% of patients admitted to general practice hospitals carry the bacillus.

It is estimated that about 2.5 million of the world's population suffer from leprosy due to *Mycobacterium leprae*. The mycobacteria are distinguished by their acid-fast staining property, which results in part from the high content of lipids in their cell walls.

M. TUBERCULOSIS: TUBERCULOSIS

The species *M. tuberculosis*, commonly referred to as the *tubercle bacillus*, is the major cause of human tuberculosis. This bacterium is a non-spore-forming bacillus measuring about 0.5 × 3 μm. It can be grown on simple culture media; in routine laboratory isolation procedures, however, best results are obtained with a medium containing egg yolk and starch, such as the Lowenstein–Jensen medium. *M. tuberculosis* is an obligate aerobe ∞ (Chapter 7, p. 99) and grows slowly. Doubling time is from 12 to 20 hours and several weeks may be required for visible colonies to develop (Figure 20-1). The organism Gram-stains poorly and is usually stained by a special procedure known as the acid-fast stain (see Plate 20). Mycobacteria are often referred to simply

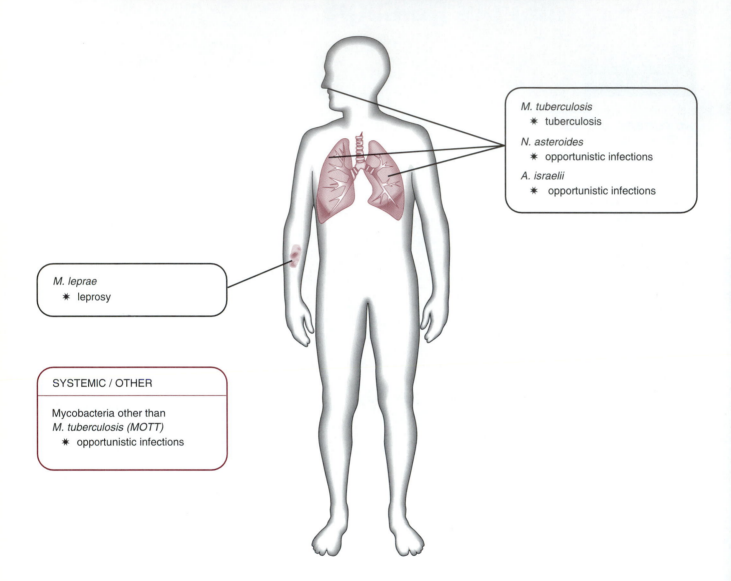

HUMAN BODY SYSTEMS AFFECTED

Mycobacteria, Nocardia, Actinomyces, and Streptomyces

M. tuberculosis
* tuberculosis

N. asteroides
* opportunistic infections

A. israelii
* opportunistic infections

M. leprae
* leprosy

SYSTEMIC / OTHER

Mycobacteria other than
M. tuberculosis (MOTT)
* opportunistic infections

as acid-fast bacteria. Experimental infections can be produced in a variety of laboratory animals.

* Pathogenesis and Clinical Diseases

Tuberculosis is a complex disease that may go unrecognized for many years in the human host. Characteristically, it may be seen in different stages.

Humans are readily infected with the tubercle bacilli, but progression of the disease depends on many subtle host and environmental factors. In some cases the bacteria are disposed of with no manifestation of the disease. However, in most cases the primary infection develops in the lungs, then becomes dormant and remains in this state for the remainder of the person's life. To provide a better understanding of the complex nature of tuberculosis, the fol-

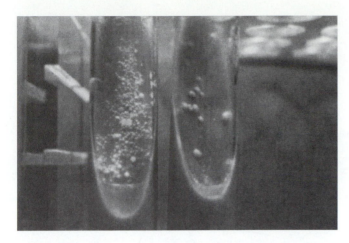

FIGURE 20-1 Colonies of *Mycobacterium tuberculosis* after several weeks growth on an agar surface. (S. S. Schneierson, *Atlas of Diagnostic Microbiology*, p. 39. Courtesy Abbott Laboratories, Abbott Park, IL)

lowing discussion divides the disease into various stages. These stages are shown diagrammatically in Figure 20-2.

Primary Tuberculosis. The primary stage of the disease results when a person becomes infected for the first time. **Aerosolized** bacteria from a person with active tuberculosis are inhaled and the bacilli that reach the **alveoli** are able to cause infection. These bacilli begin to multiply slowly and many are phagocytized by the alveolar macrophages, but these phagocytic cells are generally unable to destroy the bacilli. Consequently, many bacilli are carried by macrophages to regional lymph nodes. Phagocytosis of the bacteria, however, stimulates the development of *cell-mediated immunity* ∞ (Chapter 11, p. 164) against the mycobacterial antigens. A brisk reaction then occurs with activated macrophages concentrating around the focus of bacterial growth. The growth of the bacilli inside the macrophages is slowed and a scar tissue barrier forms around the bacilli. The resulting nodule of scar tissue and cells is called a *tubercle*. At this phase of the disease the patient develops a positive skin reaction, which is a manifestation of cell-mediated immunity against tuberculosis antigens. This process is commonly called "converting" to skin

positive. The tubercle usually prevents any further spread of the bacilli and the center of the tubercle provides a niche of dead tissue where the bacilli are protected from the host's defense mechanisms. The primary stage of the disease is usually without definite clinical symptoms. From this point, the disease may enter any one of the following stages.

Healing. The infection may be completely contained and the bacteria destroyed by the primary response. The patient would have experienced no symptoms but would have a positive reaction to the skin test. In persons not receiving adequate chemotherapy it is difficult to determine if this healing does indeed occur.

Disseminated Tuberculosis. In a small percentage of persons, mostly young children or immunologically impaired individuals, the infection spreads from the primary site of multiplication into the blood and the bacteria may be seeded throughout the body. Cellular immunity does not readily develop in some individuals and the bacilli grow in many body tissues causing a fatal infection. If dissemination occurs and hypersensitivity develops, numerous small tubercles are formed, a condition called **miliary tuberculosis**. The death rate is high from this form of the disease.

Latent–Dormant Tuberculosis. The latent–dormant stage is a continuation of the primary infection and is the usual outcome of pulmonary tuberculosis. During the primary stage, secondary foci usually develop around the initial focus and in adjacent lymph nodes. Tubercles are formed around these foci. The bacilli may live inside these tubercles for many years or for the lifetime of the infected person. Before chemotherapy was available, a general saying about tuberculosis was "Once infected, always infected."

During this stage the activated host defense mechanisms prevent the bacilli from spreading to other parts of the body, and the environment inside the tubercle protects some of the bacilli from these same defenses. This "truce" may last for the lifetime of the person or may be "broken" by the influence of various, often not well understood, changes in the physiology of the host. In past years, a vast

---

Aerosolized Suspended particles in the atmosphere.
Alveoli Small open spaces in the lung where oxygen and carbon dioxide exchange takes place.
Miliary tuberculosis A condition in which the bacilli are widely disseminated throughout the lung. Each of these bacilli serve as the focus for the development of a tubercle.

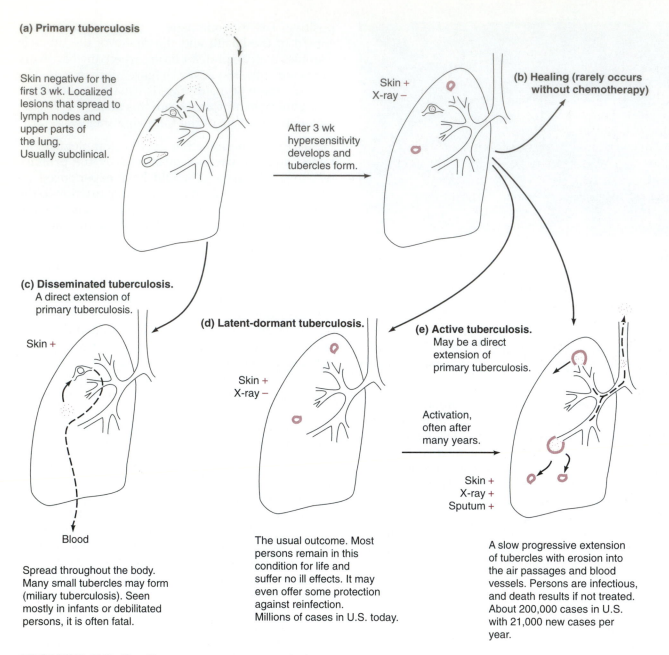

(a) Primary tuberculosis

Skin negative for the first 3 wk. Localized lesions that spread to lymph nodes and upper parts of the lung. Usually subclinical.

After 3 wk hypersensitivity develops and tubercles form.

Skin +
X-ray −

(b) Healing (rarely occurs without chemotherapy)

(c) Disseminated tuberculosis. A direct extension of primary tuberculosis.

Skin +

(d) Latent-dormant tuberculosis.

Skin +
X-ray −

(e) Active tuberculosis. May be a direct extension of primary tuberculosis.

Activation, often after many years.

Skin +
X-ray +
Sputum +

Blood

Spread throughout the body. Many small tubercles may form (miliary tuberculosis). Seen mostly in infants or debilitated persons, it is often fatal.

The usual outcome. Most persons remain in this condition for life and suffer no ill effects. It may even offer some protection against reinfection. Millions of cases in U.S. today.

A slow progressive extension of tubercles with erosion into the air passages and blood vessels. Persons are infectious, and death results if not treated. About 200,000 cases in U.S. with 21,000 new cases per year.

FIGURE 20-2 The most common stages in the pathogenesis of tuberculosis.

majority of the world's population had latent–dormant tuberculosis and some estimates place the current number of persons in the United States with latent–dormant tuberculosis at over 20 million. Although these persons have an increased resistance to reinfection due to their activated defenses against the tubercle bacilli, most active cases develop as an extension of this latent infection. So it is not generally considered advantageous to have the latent form of the disease. Persons in the latent–dormant stages remain skin test-positive, but the tuber-

cles may not be large enough to be seen by x-ray.

Secondary or Active Adult-type Tuberculosis. The secondary stage of tuberculosis may develop as a direct extension of the primary stage but most cases result from a "reawakening" of the dormant lesion. It has been estimated that greater than 90% of persons with active tuberculosis harbored the tubercle bacilli in the latent–dormant stage for a year or more. The exact mechanisms associated with **reactivation** after long periods of dormancy are not understood. It occurs most often in persons who have had their defense mechanisms compromised—for instance, young adults who become run-down, overworked, malnourished, or stressed; elderly persons with general declining health; alcoholics; persons with such diseases as AIDS, diabetes, or silicosis; and those on immunosuppressive therapy.

The previously dormant tubercle begins to expand in size and causes an enlarged central area of dead tissue and debris to form. This material is referred to as *caseous necrosis* (cheesy, dead tissue). Eventually the expanding tubercle erodes into a bronchial tube and the inner contents are expelled into the airways. At this time, the patient begins to expel large numbers of the bacilli from the respiratory tract. The fibrous and calcified walls of the tubercle then form an air-filled cavity where the tubercle bacilli may continue to grow. At this phase of the disease, healing or treatment is difficult, for the cavity wall forms a barrier not only against the body's defense mechanisms but also against chemotherapeutic agents. Surgical removal of such cavities is sometime necessary before chemotherapy can be effective. Without treatment, the tubercular lesion may continue to expand and consume the normal tissue until death results. Earlier tuberculosis was called "consumption" because of this progressive destruction or consumption of the tissues. In some cases, the adverse factors that stimulated the "reawakening" of the disease may be removed and the disease may again become stabilized in a dormant stage.

✳ Transmission and Epidemiology

Prior to the studies of tuberculosis by Robert Koch, it was generally agreed that this disease was of genetic origin. This idea probably resulted from the fact that the disease was often observed to occur in families, and among close relatives. However, it wasn't the infection per se that had genetic origins, but rather the response to the infection—that is, the manifestation of disease. Indeed, with the development of skin testing procedures it was possible to demonstrate that a majority of the population was infected with *M. tuberculosis*, although only a portion of the infected population developed disease symptoms. It has subsequently been shown that

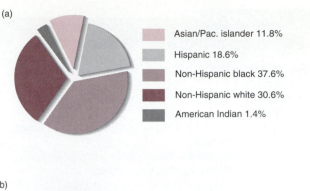

(a)

- Asian/Pac. islander 11.8%
- Hispanic 18.6%
- Non-Hispanic black 37.6%
- Non-Hispanic white 30.6%
- American Indian 1.4%

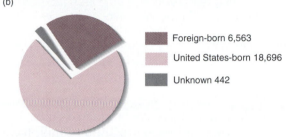

(b)

- Foreign-born 6,563
- United States-born 18,696
- Unknown 442

FIGURE 20-3 (a) Occurrence of tuberculosis by race and ethnic background in the United States in 1990. (b) Occurrence of tuberculosis among American- and foreign-born patients in the United States. (Courtesy Centers for Disease Control, Atlanta)

host factors (some of which may well be inherited) play a greater role in the outcome of infection by this organism than possibly in any other infectious process (Figure 20-3). Application of Koch's postulates ∞ (Chapter 13, p. 197) was of considerable significance in demonstrating the role of *M. tuberculosis* in this disease. In fact, because of his monumental work with *M. tuberculosis*, Koch was awarded the first Nobel Prize for medicine in 1905.

M. tuberculosis is usually found only in humans, and large numbers of bacilli may be disseminated by the airborne route from persons with active tuberculosis. A person may unknowingly transmit the bacilli for some time before being aware of having the disease. It is generally felt that prolonged close contact with an infectious person is necessary for successful transmission of this disease. Casual contact with an infectious person, such as passing on a street, would not normally result in transmission. Several groups of individuals have been identified as having a high risk for contracting tuberculosis (Table 20-1). Bovine tuberculosis, caused by *M. bovis*,

Reactivation Development of an active, spreading infection in an individual in whom a disease had been latent after an earlier infection.

TABLE 20-1

Populations at High Risk for Tuberculosis

Persons infected with human immunodeficiency virus (HIV)

Household and other close contacts with tuberculosis patients

Foreign-born persons from countries with high tuberculosis prevalence

Medically underserved, low-income populations

Alcoholics and intravenous drug users

Residents of long-term residential and care facilities

Some ethnic groups (black, Hispanic, and Native American)

Homeless persons

---

was a problem at one time because transmission was from cows to humans via contaminated milk. But inspection of cows and pasteurization of milk have almost eliminated this source of infection in the United States and many other countries.

The occurrence of mycobacterial disease due to **mycobacteria other than tuberculosis (MOTT)** is causing increasing concern. *M. kansasii, M. avium, M. avium intracellulare, M. chelonei,* and *M. fortuitum* are presently responsible for an increasing number of cases of tuberculosis (Table 20-2). Of considerable epidemiologic interest is the fact that these agents are transmitted to humans from the environment,

and human-to-human transmission rarely if ever occurs. Relatively large numbers of persons are infected with these agents but remain healthy unless there is a significant change in their resistance to disease. This fact has been dramatically demonstrated by the frequent occurrence of such mycobacterial disease in AIDS patients.

✳ Diagnosis

Tuberculosis has been the motivating factor in the development of a number of useful clinical diagnostic procedures and tools. The discovery of the stethoscope by Laennec in the early 1800s was prompted by the clinical necessity of monitoring the progress of tuberculous patients. And the first medical use of the x-ray by Roentgen was in connection with the diagnosis of tuberculosis (Figure 20-4). Present-day diagnosis of tuberculosis occurs in different stages. The first stage is *skin testing*, which provides a rapid, inexpensive procedure for screening large numbers of persons. The second stage involves chest x-rays, and the last stage is the isolation of *M. tuberculosis* from the infected individual.

Skin Testing.   The highly specific delayed-type hypersensitivity ∞ (Chapter 11, p. 164) that develops against the tubercle bacilli can be demonstrated by injecting an antigen from the bacillus into the skin (intradermal injection). This antigen is called *tuberculin* and was originally supplied as *old tuber-*

TABLE 20-2

Common Mycobacterial Agents of Disease

Agent	Communicable	Disease
M. tuberculosis	Yes	Human tuberculosis
M. bovis	Yes	Human tuberculosis
M. ulcerans	No	Cutaneous ulceration
M. kansasii	No	Pulmonary tuberculosis
M. avium-intracellulare	No	Pulmonary tuberculosis
M. marinum	No	Cutaneous lesions
M. scrofulaceum	No	Lymph node infection
M. africanum	Yes	Pulmonary tuberculosis
M. fortuitum-chelonei	No	Wound infection
M. szulgai	No	Human tuberculosis
M. leprae	Yes	Human leprosy

Skin testing   Application of an antigen directly to the skin or intradermally to determine the host immune status in relation to the antigen. A positive reaction depends on cell-mediated immunity and is a delayed hypersensitivity response.

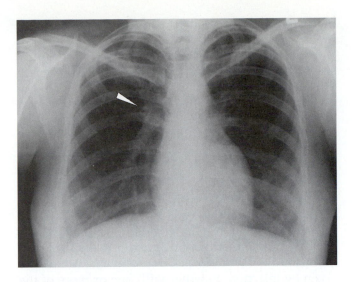

FIGURE 20-4 Chest x-ray of patient with tuberculosis (tubercle lesions indicated by arrows). (Courtesy of Photo Researchers, Inc.)

culin (OT), a crude extract from a broth culture of bacilli. Today a more refined extract called *purified protein derivative* (PPD) is used. The reference skin test—the *Mantoux test*—uses the intradermal injection of standard amounts of tuberculin. More convenient but slightly less reliable tests have been developed for routine screening programs such as the "Tine" test, which uses a disposable unit with metal tines that are covered with dried tuberculin and pressed into the skin.

Skin reactions are read 48 hours after testing; an area of redness with swelling 10 mm in diameter is considered a strong positive (Figure 20-5; see Plate 21). The skin test cannot distinguish clearly between the different stages of tuberculosis, but if someone converts to skin test-positive it shows that the person has been exposed to the disease and has progressed at least through the primary stage. Further diagnostic tests are then indicated.

Chest X-ray. Routine chest x-rays are discouraged. Such examinations are most appropriately used as a follow-up procedure on those who have converted to skin test-positive or to establish the extent of tissue damage in previously diagnosed cases of active or dormant tuberculosis.

Bacteriologic Tests. A definite diagnosis of active tuberculosis requires the isolation of the mycobacterium from the patient. The most effective method for obtaining bacterial specimens is the collection of induced sputum samples. The patient inhales a fine aerosol mist that induces deep coughing. The cough carries sputum and bacilli from the lungs to the mouth. Care should be taken to prevent health care personnel from being exposed to aerosolized bacilli during the collection of induced sputums. Special safety hoods or cubicles are recommended for this procedure (Figure 20-6).

The collected sputum is treated with sodium hydroxide, which kills most microorganisms other than mycobacteria, and, *N*-acetylcysteine, an amino acid derivative, which digests the mucus. The bacteria in the sputum are then concentrated by centrifugation, and cultures, as well as slides for microscopic examination, are prepared from the sediments. The standard acid-fast stain (Ziehl–Neelsen) provides a specific and reasonably sensitive diagnostic tool (see Plate 20). Newer stains, based on the same acid-fast principle, use fluorescent dyes such as auramine-O to enhance the microscopic examination of these specimens. The presence of acid-fast bacilli gives a rapid provisional diagnosis. The growth of *M. tuberculosis* on the inoculated culture media confirms the diagnosis.

✳ Treatment

With recognition that tuberculosis is an infectious disease, treatment became a real possibility. Over the years, isolation of patients in **sanitariums**, diet

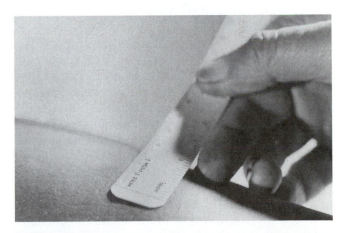

FIGURE 20-5 A positive Mantoux skin test showing an area of redness and swelling (about 15 mm in diameter) after 48 hours. (Courtesy of Centers for Disease Control, Atlanta)

Sanitarium A facility designed to care for persons and help them regain their health. In tuberculosis sanitariums patients were given healthful meals and were kept as free from stress as possible. Surgical and chemotherapeutic treatments were also available at some sanitariums.

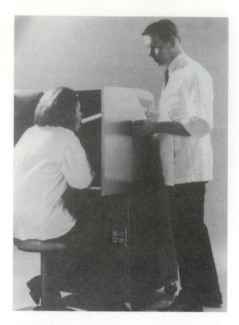

FIGURE 20-6 A safety hood used during the collection of sputum samples from tuberculosis patients. During the collection process, deep coughing is induced in the patient (seated). All airborne tubercle bacilli expelled by the patient are sucked into this safety hood where they are trapped by filters. The technician (standing) assisting the patient is protected from these airborne bacteria. (Jensen Research Laboratories)

modification, and surgery have all had a significant impact on tuberculosis therapy. However, the ability to treat tuberculosis successfully with chemotherapeutic agents (Table 20-3) has been the most notable advance in controlling this disease and has led to significant changes regarding management of tuberculosis patients. No longer is it practical to maintain sanitariums specifically for such patients. The majority of patients can be successfully treated in a relatively short time either in a general hospital that has specific facilities to handle tuberculosis, or at home. Moreover, most patients can be rendered noninfectious and then released from the hospital after some weeks. Generally, however, long-term follow-up therapy must be carried out on an outpatient basis. An important phase of the hospitalization program includes the proper motivation and education of patients so that self-medication continues and return visits to the clinic are made. The personnel staffing these clinics must be trained to understand the **chronic** nature of tuberculosis and

the need for the patients to return for regular follow-up appointments.

The antibiotic streptomycin was the first highly effective antituberculosis agent to be developed. This agent had such a profound effect on the outcome of disease that its discoverer, Selman Waksman, was awarded a Nobel Prize for its discovery. Several years later, in the early 1950s, the synthetic compounds isoniazid (INH) and p-aminosalicylate (PAS), were developed. Later the compound ethambutol (EMB) and the antibiotic rifampin were effectively used to treat tuberculosis. Additional antituberculosis agents are available for selected applications (Table 20-3).

These chemotherapeutic agents are always given in various combinations. Streptomycin, if used, is given by injection, together with one or more of the less toxic compounds INH, PAS, or EMB, which are given orally. Rifampin, combined with INH, is often administered orally throughout the entire treatment period. The time of treatment may vary, depending on the severity of the infection or the patient's reliability in taking the medication properly. The standard regimen has been daily medication for periods of 6 to 24 months. But it has now been determined that under controlled conditions short-course therapy for about 9 months can be effective. If it is difficult to control daily medication, an intermittently supervised high-dose treatment given twice weekly for the conventional time is effective. Well-regulated programs of administering the antituberculosis drugs are important. Failure to follow prescribed regimens results in the development of resistant strains of the tubercle bacillus. These strains have led to infections that cannot be readily treated.

TABLE 20-3

Agents Used to Treat Tuberculosis

Major or Primary Compounds	Minor or Secondary Compounds
Ethambutol (EMB)	Capreomycin
Isoniazid (INH)	Cycloserine
p-Aminosalicylic acid (PAS)	Ethionamide
Rifampin	Kanamycin
Streptomycin	Pyrazinamide
	Thiacetazone
	Viomycin

Chronic Lasting a long time and tending to progress slowly.

* Prevention and Control

Prompt diagnosis of active cases is the most important phase in the control of tuberculosis. Once infected individuals have been identified, they can be rendered noninfectious by proper chemotherapy. The next major problem is determining who might have become infected from the index cases. All possible contacts should be investigated, starting with those who shared common environmental air at home or work. The investigation should also extend to those who may have had less extensive contact with the patient. Appropriate diagnostic procedures should be carried out on the contacts, beginning with a history of any prior tuberculin skin tests, vaccinations, or infections. Skin tests should then be done on all contacts and follow-up x-rays and sputum specimens used if indicated. Contacts showing conversion to positive skin reactions or having other signs of the disease should be placed on prophylactic chemotherapy. It is sometimes advisable to place the more susceptible close contacts, such as young children, on prophylactic chemotherapy even if they show no signs of the disease.

This program of detection and follow-up has worked quite well in countries with relatively few active cases and with the necessary medical personnel and public health facilities. A vaccine may be advisable in countries where tuberculosis is more prevalent and medical facilities are limited. A live vaccine, containing an *attenuated M. bovis* mutant known as *bacillus Calmette-Guerin,* or BCG, has been available since 1923 and induces an increased resistance to tuberculosis, but not complete immunity. BCG vaccination has been used in some countries for many years and appears to offer some protection. But this vaccine induces hypersensitivity against the tubercle bacillus and thus renders the tuberculin skin test useless as a diagnostic aid. It is rarely used in countries where the incidence of tuberculosis is low, because it is considered more valuable to have the diagnostic usefulness of the skin test than the moderate protection provided by the vaccine.

* Present-Day Concerns

The study of tuberculosis facilitates the learning of numerous principles and concepts that have application to a broad perspective of infectious disease. However, in spite of remarkable progress in reducing the incidence of this disease from 200 cases per 100,000 population in 1900 to fewer than 9.5 per 100,000 in 1989, there are significant concerns for the future. Since the latter part of the 1980s there has been a small but persistent increase in the number of cases reported in the United States (Figure 20-7). Perhaps of more significance than the number of new cases has been the number of these cases where the organism is antibiotic-resistant. This primary resistance is not a new phenomenon, but occurrence in the United States is relatively recent.

As medical practice and patterns of disease change, new avenues of infection become available to microorganisms, including *M. tuberculosis* and various MOTT bacteria. *M. chelonei* and *M. fortuitum* are somewhat opportunistic in their occurrence, and produce wound infections most often associated with prosthetic surgery. *M. avium-intracellulare* have taken advantage of the immunologically disabled AIDS patients to produce serious, disseminated mycobacterial disease. *M. avium-intracellulare* infections are producing an ever-increasing percentage of the total mycobacterial disease in the United States. Another significant factor in increasing mycobacterial disease is the number of homeless individuals. These persons often share crowded, hygienically poor conditions, have inadequate diets, and suffer from drug or alcohol abuse; these are all circumstances that increase the inci-

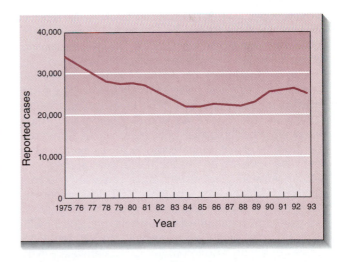

FIGURE 20-7 Reported cases of tuberculosis in the United States from 1975 to 1993. (from CDC annual summaries)

Attenuated The condition of an organism in which virulence factors have been reduced or removed.

Exposure of Passengers and Flight Crew to *Mycobacterium tuberculosis* on Commercial Aircraft, 1992–1995

From January 1993 through February 1995, CDC and state health departments completed investigations of six instances in which passengers or flight crew traveled on commercial aircraft while infectious with tuberculosis (TB). All six of these investigations involved symptomatic TB patients with acid-fast bacillus (AFB), smear-positive, cavitary pulmonary TB, who were highly infectious at the time of the flight(s). In two instances, *Mycobacterium tuberculosis* isolated from the index patients was resistant to both isoniazid and rifampin; organisms isolated from other cases were susceptible to all antituberculous medications. In addition, in two instances, the index patients were aware of their TB at the time of travel and were in transit to the United States to obtain medical care. However, in none of six instances were the airlines aware of the TB in these passengers. This report summarizes four of the investigations by CDC and state health departments and provides guidance about notification to passengers and flight crew if an exposure to TB occurs during travel on commercial aircraft.

Investigation 1 During 1993, the Minnesota Department of Health conducted an investigation of a foreign-born (i.e., born outside the United States or Canada) passenger with pulmonary TB who traveled in the first-class section of an aircraft during a 9-hour flight from London to Minneapolis in December 1992. Of the 343 crew and passengers on the aircraft, TST results were obtained for 59 (61%) of 97 U.S. citizens and 20 (8%) of 246 non-U.S. citizens. Tuberculin skin tests (TSTs) were positive for eight (10%) persons, all of whom had received bacillus Calmette–Guérin (BCG) vaccine or had a history of past exposure to *M. tuberculosis*. The investigation indicated no evidence of transmission of TB during the flight.

Investigation 2 In March 1993, a foreign-born passenger with pulmonary TB traveled on a flight from Mexico to San Francisco. This investigation included efforts by the San Francisco Department of Public Health to obtain information by mail from all 92 passengers on the flight; 17 persons could not be contacted because of invalid addresses. TSTs were positive in 10 (45%) of the 22 persons who were contacted and completed TST screening: 9 of these TST-positive persons were born outside the United States. The other was a 75-year-old passenger who may have become infected with *M. tuberculosis* while residing outside the United States or during a period when TB was prevalent in the United States. The San Francisco Department of Public Health found no conclusive evidence of transmission during this flight.

Investigation 3 In March 1994, a U.S. citizen with pulmonary TB and an underlying immune disorder who had resided long term in Asia traveled on flights from Taiwan to Tokyo (3 hours), to Seattle (9 hours), to Minneapolis (3 hours) and to Wisconsin (1/2 hour). Of 661 passengers on these four flights, 345 (52%) were U.S. residents. The Wisconsin Division of Health contacted the 345 U.S. residents and received reports about TST results from 87 (25%) persons; of these, 14 (17%) had a positive TST. All 14 persons had been seated more than five rows away from the index patient; nine of these persons had been born in Asia (including two with a known prior positive TST). Of the five who were TST-positive and U.S.-born, one was known to have had a positive TST previously, two had resided in a country with increased endemic risk for TB, and two were aged 75 years. The investigation indicated that, although transmission of TB during flights could not be excluded, the positive TSTs may have resulted from prior *M. tuberculosis* infection.

Investigation 4 In April 1994, a foreign-born passenger with pulmonary TB traveled on flights from Honolulu to Chicago (7 hours, 50 minutes) and to Baltimore (2 hours), where she lived with friends for 1 month. During that month, her symptoms intensified; she returned to Hawaii by the same route. Investigation in Baltimore determined that TST conversion had occurred in the 22-month-old child of her friends. The four flights included a total of 925 passengers and crew who were U.S. residents, of whom 755 (82%) completed TST screening; of these, 713 (94%) were U.S.-born. The investigation by CDC indicated no evidence of transmission on the flight from Honolulu to Chicago or the flight from Chicago to Baltimore. Of the 113 persons who had traveled on the flight from Baltimore to Chicago, TSTs were positive in 3 (3%), including 2 who were foreign-born. However, of the 257 persons who traveled from Chicago to Honolulu (8 hours, 38 minutes), TSTs were positive in 15 (6%), including 6 who had converted; 2 of these 6 persons apparently had a boosted immune response, while the other 4 had been seated in the same section of the plane as the index patient. Because of TST conversions among U.S.-born passengers, the investigation indicated that passenger-to-passenger transmission of *M. tuberculosis* probably had occurred.

Editorial Note: The investigations described in this report were undertaken to determine whether exposure to persons with infectious pulmonary TB was associated with transmission of *M. tuberculosis* to others traveling on the same aircraft. Two of these investigations indicated that transmission occurred. In investigation 4, transmission occurred on the return to Hawaii, the longest flight, when the index passenger was most symptomatic. All persons with TST conversions were seated in the same section of the aircraft as the index passenger, suggesting that transmission was associated with seating proximity (*MMWR* 44:137, 1995).

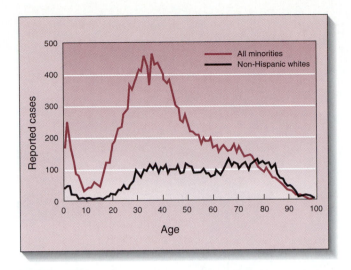

FIGURE 20-8 The frequency distribution of cases of tuberculosis in the United States by age, race, and ethnicity during 1993. (From CDC annual summaries)

M. LEPRAE: LEPROSY

Mycobacterium leprae is the causative agent of leprosy. This bacterium shares many common characteristics with *M. tuberculosis.* It is an acid-fast bacillus containing large amounts of lipid, it induces hypersensitivity, and it multiplies slowly. *M. leprae* is found in enormous numbers in certain lesions of infected persons. This feature allowed Gerhard Hansen in 1874 to make the first reliable causal association between a bacterial agent and a human disease. Leprosy is often referred to as *Hansen's disease*, partly to honor Hansen's discovery and partly to avoid the use of the unpleasant name of leprosy.

Even though this bacterium is found in greater numbers in infected tissues than any other bacterium, it has not been possible to cultivate it on artificial media. Some growth occurs when it is inoculated into the foot pad of a mouse. Today it is known that armadillos are susceptible to this bacillus. Some evidence even suggests that armadillos may be a natural nonhuman reservoir for leprosy. Thus, both mice and armadillos are being used in some limited experimental laboratory studies of leprosy.

✳ Pathogenesis and Clinical Diseases

Leprosy is probably transmitted from person to person under conditions of poor sanitation and may gain entry via the respiratory tract or skin lesions. The incubation period averages several years but may extend to 20 years or more. The major growth of the bacilli occurs in the low-temperature body tissues—that is, nose, ears, and the skin of extremities. The leprosy bacilli are easily phagocytized but not destroyed, and large numbers are found growing inside macrophages. Nerves are uniquely susceptible to infection; early symptoms of leprosy are often associated with anesthesia (lack of feeling) over an area of the body. The exact mechanism of tissue destruction is not understood but probably results from a combination of neurological damage, massive accumulation of bacilli, immunologic reactions, and *secondary infections.*

Two forms of leprosy are seen, the *lepromatous* and the *tuberculoid* (Figures 20-9 and 20-10). The lepromatous form is the most severe and is character-

dence of mycobacterial disease. Present studies show a tuberculosis prevalence as high as 6.8% among these persons.

In areas where there is a high prevalence of HIV infection, up to 30% of tuberculosis patients are also HIV positive. When HIV-seropositive persons are skin tested for reactivity to tuberculin, 15% of these people are positive. These rates of infection are so much higher than that found in the population without HIV that an associative relationship is established. It is not only clear that HIV infection can predispose persons to mycobacterial infection, it is also apparent that persons with tuberculosis have an increased likelihood of having HIV infection. This latter relationship emphasizes the importance of lifestyle in the transmission and occurrence of both of these diseases.

Lastly, there has been a continued drift of the disease into older patient populations (Figure 20-8). In 1960 tuberculosis was considered as an illness of the middle-aged and young adult patient. However, by 1985, a greater proportion of cases were seen in patients 65 years of age and older. The incidence of tuberculosis in this group has reached 34.9 cases per 100,000 and the incidence remains less than 14 per 100,000 for persons between 25 and 44 years of age.

Secondary infections Infections that occur as a consequence of other disease processes in the host.

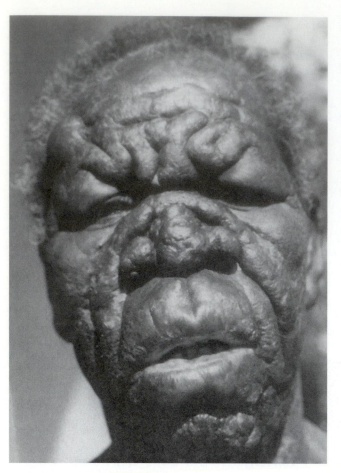

FIGURE 20-9 A patient with lepromatous leprosy. Note the loss of eyebrows and the deep furrowing that exaggerates the normal folds of the face. (Courtesy American Leprosy Missions, Bloomfield, NJ 07003)

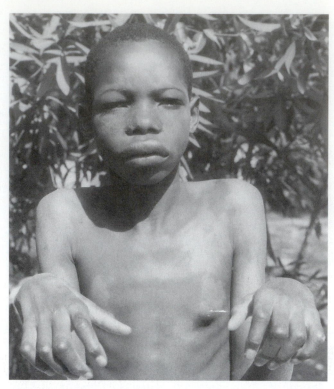

FIGURE 20-10 A patient with tuberculoid leprosy. Note the flat discolored plaques on the shoulder, chest, and hands. Swelling and contraction of the fingers results from inflammation of the nerve fibers. (Courtesy American Leprosy Missions, Bloomfield, NJ 07003)

ized by large nodular lesions. In lepromatous leprosy the immune response is impaired, limiting the formation of granulation (scar) tissue. The tuberculoid form is less severe and is associated with a normal immune response that causes granulation-type lesions; bacilli in the lesions are sparse, tissue damage is less, and response to therapy is better. Forms intermediate between lepromatous and tuberculoid are also seen. Overall, leprosy is a slowly progressing disease that often disfigures and cripples. Death usually results after a number of years and is commonly associated with secondary infections.

✳ Transmission and Epidemiology

Leprosy is generally found in tropical and subtropical areas of developing nations. Most cases seen in developed countries were contracted—sometimes years earlier—while the person resided in a tropical or subtropical area. Only about 100 cases occur each year in the United States (Figure 20-11). But the disease is still a major problem worldwide. Estimates, as of 1993, indicated that about 2.5 million people worldwide have leprosy. This is down from the 10 million who had leprosy in the 1970s.

From a historical perspective the epidemiology of leprosy offers some unexplainable paradoxes. Early reports suggested that the disease was highly contagious. During the eleventh and fifteenth centuries, for example, leprosy was widespread in Europe. A sharp decline in incidence followed in the sixteenth century. The reasons for this decline and for the comparative decrease in virulence and communicability of this disease today are not known.

✳ Diagnosis

Bacterial diagnosis is made by direct microscopic demonstration of the presence of acid-fast bacilli in scrapings of fluids from the lesions (Figure 20-12). A skin test using an antigen called *lepromin*, obtained from the heat-inactivated extracts of infected tis-

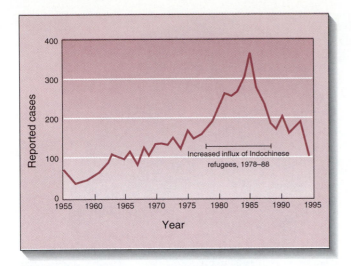

FIGURE 20-11 Reported cases of leprosy in the United States: 1955 to 1995. (Modified from CDC annual summaries)

sues, is of some value. Clinical findings are often characteristic enough for a tentative diagnosis.

✳ Treatment

Antimicrobial agents called *sulfones,* which are related to the sulfonamides ∞ (Chapter 9, p. 127), are fairly effective in arresting progression of the lesions and in allowing them to heal (Figure 20-13). These agents render patients noninfectious and allow them to return to normal daily activities as outpatients. Treatment is prolonged, for months to years, and it is not certain when or if a complete cure is ever obtained. Unfortunately, there have been reports of increasing bacterial resistance to dapsone, the leading antileprosy sulfone. The antibiotic rifampin has been shown to render patients noninfectious in just a few weeks, and when combined with sulfones a complete cure may be attained in some patients. The other antituberculosis drugs are not effective against leprosy.

✳ Prevention and Control

Only persons who have prolonged contact with leprosy patients under poor sanitary conditions seem to stand an increased risk of being infected. Using good sanitary procedures when dealing with patients is thus recommended. Young children are more susceptible than adults. It may be advisable, in some situations, to remove young children from infectious parents and place them on a course of

preventive chemotherapy. Tattooing parlors in endemic areas should be avoided, for contaminated tattooing needles have been shown to transmit leprosy. New cases of leprosy were prevented on a Pacific island by subjecting the entire population of 1500 people to a course of sulfone treatment. Such prophylactic chemotherapy might be used in other endemic areas to block the spread of this disease.

Leprosy vaccine trials have been conducted in South America. This vaccine is unusual in that it is given to patients who already have leprosy. The vaccine (BCG mixed with killed cells of *M. leprae*) is apparently able to increase the patients immune response to the active disease. Because of the slow progressive nature of leprosy, it is possible to give this vaccine after onset of the disease and still have time for the immune system to be stimulated. With the proper commitment and using current technology it might be possible to eliminate leprosy (see Clinical Note).

ACTINOMYCETES AND RELATED MICROBES

Actinomycetes and related microbes are gram-positive bacteria; some species are acid-fast and related to the mycobacteria. These organisms grow in long filaments (Figure 20-14) with extensive branching

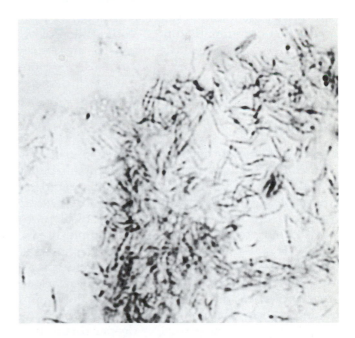

FIGURE 20-12 Micrograph of *Mycobacterium leprae* taken from a lesion and stained by the acid-fast method. (Courtesy American Leprosy Missions, Bloomfield, NJ 07003)

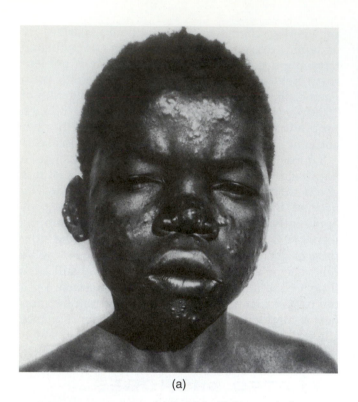
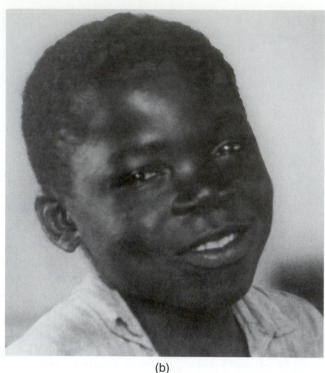

(a)   (b)

FIGURE 20-13   (a) A child with leprosy before treatment. (b) The same child after treatment with a sulfone. (Courtesy American Leprosy Missions, Bloomfield, NJ 07003)

and thus resemble the morphology of fungi (Chapter 39). They are prokaryotes, however, and possess bacterial-type cellular structure. Widespread in nature, they are found in soil and are noted for their production of antibiotics and decomposition of organic matter. Some of these organisms are associated with disease in humans and animals. The more prominent human pathogens are briefly discussed here.

✳ *Nocardia*

The most frequently encountered *Nocardia* species is *N. asteroides*. It is acid-fast and a common inhabitant of soil. Generally, it is an opportunistic pathogen and causes disease in patients with already lowered resistance due to other concurrent medical problems. Lung infection is the most common disease caused by *N. asteroides* and may be misdiagnosed as tuberculosis (see Plate 19). The infection may spread from the lungs to the blood and involve various other parts of the body, especially the brain. *Nocardia* species may also cause penetrating lesions of the dermal, subcutaneous, or deeper tissues that are localized and have connecting passages (*sinuses*)

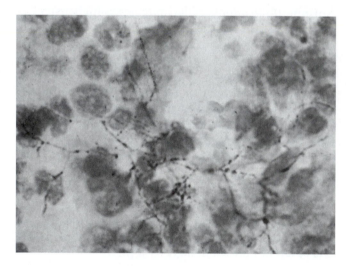

FIGURE 20-14   A pulmonary specimen containing an actinomycete among many PMNs. Note branching and beading configuration of bacteria. (Courtesy Center for Disease Control, Atlanta)

to the surface through which the infection drains. Characteristic clumps (granules) made of compact colonies of nocardia are present in the exudate. Infections are usually best treated with sulfa drugs.

Sinus   An opening or space in a tissue. Infectious processes often produce a nonhealing sinus that may connect between tissues or even to the outside.

* Actinomyces

Actinomyces is an anaerobic gram-positive, non-acid-fast, filamentous organism with or without branching. The species *Actinomyces israelii* is the major human pathogen. *A. bovis* is a common pathogen of cattle only and causes a disease called "lumpy jaw." *A. israelii* is normally found in the mouth of humans and usually acts as an opportunist by causing infections in damaged tissues. The following types of infection are produced:

1. Head and neck infections following injury to the mouth or jaw, such as tooth extractions.
2. Pulmonary infections resulting from aspiration of infectious material from the mouth.
3. Abdominal infections, probably resulting from swallowing organisms after abdominal surgery or injury.
4. Infections resulting from human bites that directly introduce the organisms into the tissues or any injury that breaks the skin; foot infections are common in some areas (Figure 20-15).

Infections are often characterized by draining abscesses with the actinomyces filaments embedded in yellowish granules in the exudate. Diagnosis is based on clinical appearance and the presence of typical organisms in the granules. Actinomyces are susceptible to various antibiotics, including penicillin, the antibiotic of choice. Surgical removal of the abscess is often necessary before successful treatment is possible.

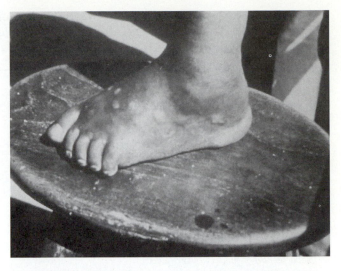

FIGURE 20-15 *Actinomyces* infection of the foot. Note nodular lesions (Courtesy Centers for Disease Control, Atlanta)

* Streptomyces

Streptomyces form long, branching filaments that segment into bead-like structures called *conidia*. Each conidium can develop into a new colony. Most streptomyces are nonpathogens and are found in the soil. Many of the commonly used antibiotics are produced from these organisms. The species *Streptomyces somaliensis*, and perhaps a few other species, causes localized swollen lesions that are distinguishable clinically from lesions caused by *Nocardia* species.

MATERIAL FOR REVIEW

CONCEPT SUMMARY

1. Members of the genus *Mycobacterium* are causative agents of two chronic diseases of great historical interest and significance: tuberculosis and leprosy. The mycobacteria are unique in that their cellular composition includes a high concentration of lipid. Because of their staining characteristics, they are known as acid-fast bacteria. They are strict aerobes.

2. Tuberculosis is also caused by a number of mycobacteria other than *M. tuberculosis* (MOTT). MOTT are not transferred from person to person but produce a disease that is similar to that caused by *M. tuberculosis*.

3. Tuberculosis therapy has been difficult and requires long periods of antimicrobial use, but has been quite effective. A number of compounds are commonly used in therapy. A vaccine called BCG is available and has produced limited results in areas where tuberculosis is a major health concern.

4. Leprosy is an age-old human disease. There are millions of cases of this disease throughout the world, but it is infrequently found in the United States. New approaches to therapy have

enabled many leprosy patients to have normal lives.

5. The actinomycetes are very similar in structure and composition to the *Mycobacteria*. They are much less often involved in human infections, but the infections are serious and often life-threatening.

CLINICAL SUMMARY TABLE

Microorganism	Virulence Mechanisms	Diseases	Transmission	Treatment	Prevention
Mycobacterium tuberculosis	Induced hypersensitivity Resists destruction by phagocytes	Tuberculosis	Airborne Person to person	Various anti- microbial agents	Prompt diagnosis & isolation of active infectious patients BCG vaccine
Mycobacterium leprae	Induced hypersensitivity	Leprosy	Person to person	Sulfones Rifampin	Good sanitation

STUDY QUESTIONS

1. What features of mycobacteria differ from those of other bacteria and may be responsible for the unusual pathogenesis of tuberculosis?
2. What changes in the organism and the host appear to be responsible for adult-type tuberculosis?
3. Describe the use of the skin tests as a means of diagnosing tuberculosis. What are the limitations to the use of this procedure?
4. Explain the advantages and disadvantages of the use of BCG vaccine.
5. Infections due to MOTT are not contagious. What characteristics of lifestyle correlate with tuberculosis due to these organisms?
6. List the two clinical forms of leprosy and the characteristics of this disease that are associated with each type.
7. Discuss the epidemiologic aspects of the transmission of leprosy.
8. What is the major physiologic distinction between the *Nocardia* and the *Actinomyces*?

CHALLENGE QUESTIONS

1. The drawback of using the BCG vaccine is that it induces hypersensitivity against the tubercle bacillus, rendering the tuberculin skin test ineffective as a diagnostic tool for tuberculosis. Is it possible to develop a vaccine that would not interfere with the tuberculin skin test? Explain.

2. Many physicians and clinicians do not believe tuberculosis is a highly contagious disease. However, they are very concerned about the increased spread of tuberculosis. How can you resolve these two observations?

CHAPTER

21 Enterobacteriaceae

OUTLINE

*E*nterobacteriaceae is a family of commonly isolated gram-negative bacilli. It is a large family containing more than 100 species of bacteria. These organisms are closely related genetically, physically, and biochemically. Over the years, various classification arrangements have been used and the names of some species were periodically changed. For example, as of 1972 there were only 12 genera and 26 species in this family, but presently there are more than 25 genera, 17 of which contain organisms of clinical significance, and the number of species exceeds 100, with additions being made regularly. The following discussion uses mainly the genus or species names. In the clinical setting these microorganisms are frequently referred to as the gram-negative or the enteric bacilli.

ENTEROBACTERIACEAE: GENERAL CHARACTERISTICS

The members of the family *Enterobacteriaceae* are relatively small (0.5 × 2 µm), non-spore-forming bacilli (Figure 21-1). Some are motile (see Plate 23), others are not. Some have capsules, but others do not. They ferment a variety of different carbohydrates and the patterns of carbohydrate fermentation are used to help differentiate and classify these bacteria. The bacterial colonies appear similar on nondifferential media. Various differential and selective media, however, are used to help in the preliminary classification of the *Enterobacteriaceae* (see Plate 25). Once these microorganisms have been classified to the genus or species level by biochemical tests, further differentiation is made by serologic tests.

Serologic identification of these bacteria is based

HUMAN BODY SYSTEMS AFFECTED

Enterobacteriaceae

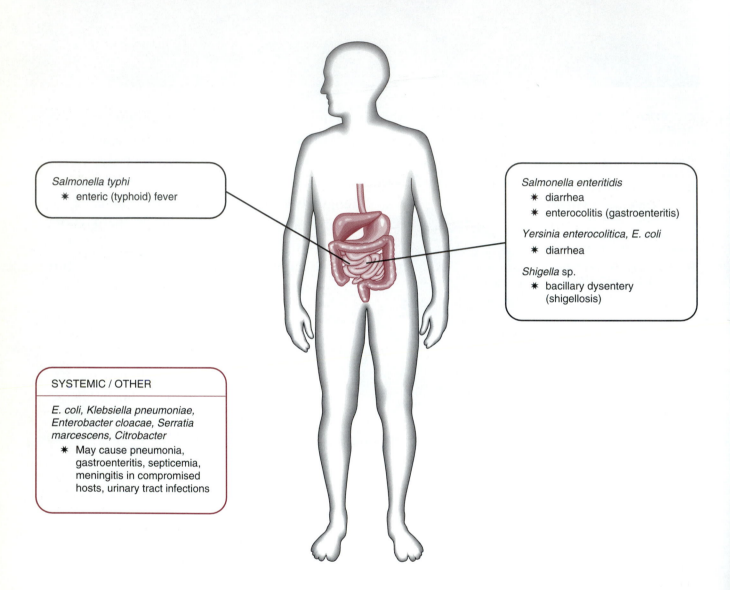

Salmonella typhi
* enteric (typhoid) fever

Salmonella enteritidis
* diarrhea
* enterocolitis (gastroenteritis)

Yersinia enterocolitica, E. coli
* diarrhea

Shigella sp.
* bacillary dysentery (shigellosis)

SYSTEMIC / OTHER

E. coli, Klebsiella pneumoniae, Enterobacter cloacae, Serratia marcescens, Citrobacter
* May cause pneumonia, gastroenteritis, septicemia, meningitis in compromised hosts, urinary tract infections

on the presence or absence of several antigenic structures (Figure 21-2). Common to all of these bacteria is a heat-stable lipopolysaccharide cell wall antigen called the *O antigen*. More than 150 known variations of this antigen are distributed among the *Enterobacteriaceae*. A second antigen, known as K, is a heat-labile capsular polysaccharide antigen that occurs in about 100 different antigenic configurations. The third common antigen is located on the flagella of motile species of bacteria. This antigen, designated H, may be present in either or both of two forms known as phase 1 or phase 2. Some 50 H antigens are found among these bacteria.

Use of these antigens provides a means for identification of individual strains of bacteria; a kind of "fingerprint" that separates serotypes within a species. Such serologic separation has considerable epidemiologic significance in helping to identify organisms associated with an epidemic outbreak or

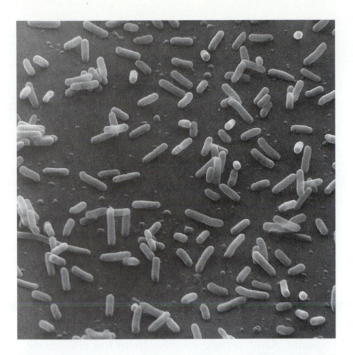

FIGURE 21-1 Photomicrograph of *Salmonella* (all *Enterobacteriaceae* appear the same when viewed with a microscope). (Courtesy of Visuals Unlimited)

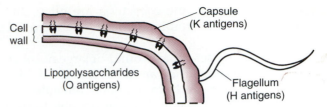

FIGURE 21-2 A section of a cell envelope of an *Enterobacteriaceae* showing the major antigens used in serotyping.

characterized by significant virulence. For example, the *Escherichia coli* K-1 capsular antigen is often found on those organisms that cause neonatal meningitis, and the *E. coli* responsible for bloody diarrhea is designated O157:H7.

People in a clinical bacteriology laboratory spend much time identifying and differentiating the species of these gram-negative bacilli. Being widespread normal inhabitants of the body, they are often found in clinical specimens and it is a rather laborious task to determine if pathogenic strains are present among the nonpathogens. When dealing with an infected patient, particularly a compromised host, it is also necessary to determine if the disease is caused by a low-virulence strain. Trying to find the involvement, if any, of these bacteria in a given clinical disease is a challenge to physicians and laboratory workers.

✱ General Pathogenesis

The *Enterobacteriaceae* inhabit primarily the large intestine of humans and animals. Some are also found in soil, water, and decaying matter, and some of these bacteria are of moderate to high virulence and are able to cause disease when they infect susceptible hosts. Members of the most virulent group are referred to as primary pathogens and include species of the genera *Salmonella*, *Shigella*, and

Yersinia—the causative agents of typhoid fever, dysentery, and bubonic plague, respectively. Most other enteric bacilli are of lower virulence and function as opportunistic pathogens; these organisms are regular inhabitants of the intestinal tract of humans or are routinely found in the general environment. However, these opportunistic species cause disease when they gain access to normally sterile body compartments or when the host defenses become compromised. The enteric genera of clinical significance are listed in Table 21-1.

Endotoxins contained in the cell walls of these enteric bacteria often play an important role in the pathogenesis of the diseases they cause. A number of species also produce exotoxins called *enterotoxins* that specifically affect the intestinal tract, causing diarrhea and fluid loss from the body. Various species of the *Enterobacteriaceae* are able to cause pneumonia and are also the most common cause of urinary tract infections. These microorganisms are now recognized as a major cause of wound infections and other *nosocomial* infections acquired by hospital patients. They may also cause severe *systemic* infections such as *bacteremia* and occasionally *meningitis*. It has been estimated that infections by these enteric bacilli may be contributing factors in about 100,000 deaths per year in the United States, and they account for about 50% of all the clinically significant bacteria isolated by hospital laboratories.

TABLE 21-1

Genera of Clinically Significant *Enterobacteriaceae*

Cedecia	Hafnia	Salmonella
Citrobacter	Klebsiella	Serratia
Edwardsiella	Kluyvera	Shigella
Enterobacter	Morganella	Tatumella
Escherichia	Proteus	Yersinia
Ewingella	Providencia	

Infections caused by these bacteria are often difficult to treat with routinely used antimicrobial agents.

Because these bacteria are found in large number in the intestinal tract, they are often transmitted by the fecal–oral route and are frequent contaminants of food and water. The enteric bacilli are able to survive for extended periods. They can be transferred in water supplies over long distances and may be found on foods or in other environmental niches where moisture is present. Freezing does not destroy these bacteria and frozen foods or ice can remain contaminated for extended periods. They are responsive to relatively low concentrations of common disinfectants and are effectively reduced in water supplies treated with small amounts of chlorine. However, their antibiotic susceptibility is unpredictable, and they are frequently resistant to common antimicrobials.

SALMONELLA: GENERAL CHARACTERISTICS

Members of the genus *Salmonella* are so closely related that they could easily be considered a single species. However, for epidemiologic and clinical purposes it is useful to characterize the following species: *S. typhi, S. choleraesuis, S. typhimurium, S. enteritidis,* and *S. paratyphi.* These species are divided among more than 1500 serotypes. Serotypes are based on the presence of specific O (cell wall) and H (flagellar) antigens. The diseases caused by *Salmonella* species are enteric fever (*typhoid fever*), bacteremia, and *enterocolitis* (*gastroenteritis*).

SALMONELLA TYPHI: ENTERIC FEVER

✳ Pathogenesis and Clinical Diseases

Enteric fever is caused by only a few of the salmonellae. Typhoid fever is the most important of the enteric fevers. Typhoid fever is caused by *S. typhi* (see Plate 24), which attaches to and penetrates the epithelial lining of the small intestine. Following penetration, the bacteria are phagocytized by macrophages. Unfortunately, they are not destroyed but are actually able to multiply within the macrophages. The macrophages carry the *S. typhi* throughout the mononuclear phagocyte system ∞ (Chapter 10, p. 146). These events occur during the first week of infection and may be accompanied by fever, malaise, lethargy, and aches and pains. During the second week, extended bacteremia is present. The foci of infection may occur in various tissues and often the gallbladder becomes infected. Bacteria may be shed from the gallbladder back into the intestinal lumen. During this time ulcerative lesions of *Peyer's patches* may develop and the patient is often severely ill with a constant fever as high as 40°C (104°F), abdominal tenderness, diarrhea or constipation, and vomiting. By the third week, in uncomplicated cases, the patient is exhausted, may still be **febrile**, but shows improvement. Death may result in up to 10% of untreated patients. After people recover from the clinical disease, *S. typhi* may continue to multiply in the gallbladder of about 3% of the patients. These persons may become chronic carriers and serve as a source of future outbreaks. The pathogenesis of typhoid fever is shown in Figure 21-3.

✳ Transmission and Epidemiology

The primary mode of transmission of the typhoid bacillus is the fecal–oral route through contaminated food or water. In developed countries, typhoid fever cases have declined significantly because of adequate water supply and sewage treatment systems. Most outbreaks of typhoid fever in the United States today are associated either with persons living in undeveloped areas or with **point-source** outbreaks due to contamination of food or beverage by a carrier. The threat of typhoid is always present when normal water and sewage systems are disrupted by disasters such as floods and earthquakes. Normally, between 400 and 600 cases of typhoid fever occur each year in the United States. The typhoid fever rate in the United States is seen in Figure 21-4.

Gastroenteritis   An infection or intoxication of the intestinal tract. Gastroenteritis is commonly characterized by nausea, diarrhea, and malaise.
Enteric fever (typhoid fever)   Systematic disease due to *S. typhi* infection.
Peyer's patches   Lymphoid tissue lining the small intestine.
Febrile   Referring to fever.
Point source   A single source from which dissemination of an infectious agent occurs.

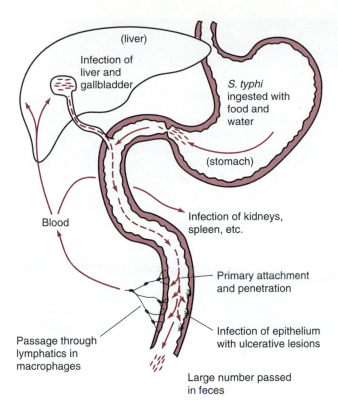

FIGURE 21-3 The pathogenesis of typhoid fever.

✳ Diagnosis

Typhoid fever, particularly in early stages, is easily confused with other diseases. A positive diagnosis depends on the isolation of *S. typhi* from the blood, feces, or other parts of the body.

✳ Treatment

Typhoid fever may be treated with either trimethoprim sulfamethoxazole, ampicillin, third-generation cephalosporins, or chloramphenicol. Chloramphenicol has traditionally been the treatment of choice, but resistant strains are now developing against this agent. Treatment must be continued for several weeks to ensure killing bacteria that became sequestered in the phagocytic cells. Carriers are best cured by daily treatment for three months. If this treatment fails, surgical removal of the gallbladder may be necessary.

✳ Prevention and Control

Proper water treatment and sewage disposal are the most important factors in controlling typhoid fever.

Pasteurization of milk and exclusion of chronic carriers as food handlers are also helpful. Killed vaccines have been used for many years but are of limited value. A recently developed oral (living attenuated) vaccine has been used with considerable success in highly endemic countries.

SALMONELLA: GASTROENTERITIS

✳ Pathogenesis and Clinical Diseases

Many serotypes of *Salmonella* are found as normal flora in the intestinal tract of animals and birds. However, when ingested by humans, salmonellae proliferate in the intestines and symptoms of gastroenteritis (enterocolitis) may begin within 18 to 36 hours. Such infections constitute a form of infectious food poisoning. Symptoms of enterocolitis include fever, nausea, abdominal pain, and diarrhea. This condition is usually self-limiting and complete recovery occurs within several days. In serious or prolonged cases, extensive dehydration may occur; this is particularly true in very young or in elderly persons. Fluid imbalance resulting from dehydration may be life threatening. Human infection by these organisms is usually limited to the **lumen** of the intestine, but on occasion may progress to the bloodstream, producing enteric fever.

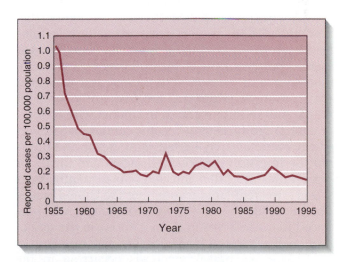

FIGURE 21-4 Reported cases of typhoid fever in the United States, 1955–1995. (Courtesy Centers for Disease Control, Atlanta)

Lumen Opening, often used to refer to the hollow space in a tube-like structure.

CLINICAL NOTE

Typhoid Fever: Skagit County, Washington, 1990

In June and July 1990, an outbreak of typhoid fever occurred in Skagit County, Washington, following a family gathering of 293 people from five states. This report provides a preliminary summary of the investigation of this outbreak by the Skagit County and Washington State departments of health.

Based on interviews of 257 attendees, 17 (6.6%) of these persons developed an illness that met the case definition for probable or confirmed typhoid fever. Blood cultures were obtained from seven case-patients and from three other symptomatic persons; four of these yielded *Salmonella typhi*. Stool specimens from nine case-patients and six asymptomatic persons yielded *S. typhi*. The 17 case-patients ranged in age from 1 to 50 years; eight were male. Fourteen were from Washington, and three were from California. The mean incubation period was 16.1 days (range: 7–27 days); mean duration of illness was 19.7 days (range: 7–35 days). Two case patients were hospitalized and treated with systemic antibiotics.

The investigation indicated that consumption of three food items served during the gathering was associated with risk for illness. A food handler who prepared one of the implicated food items had an *S. typhi*-positive stool culture and an elevated antibody titer (1:80) to the Vi antigen, suggesting chronic carriage of *S. typhi*. No other suspected carriers were identified.

To prevent secondary transmission of *S. typhi* associated with this outbreak, the county and state health departments implemented several measures from 30 July to 17 August, including (1) widely disseminating information about typhoid fever and its prevention; (2) recruiting local family members to assist with case findings and disease-control efforts by asking them to contact family members and friends who had attended the gathering; (3) culturing stool samples from household contacts of infected persons, food handlers who had worked at the gathering, and other attendees who had jobs as food handlers; (4) excluding selected persons (food handlers who worked at the gathering, attended the gathering, or cultured positive for *S. typhi*) from food handling until three consecutive negative stool cultures were obtained; and (5) instructing all other infected persons in proper hand washing and advising these persons to refrain from handling food until three consecutive negative stool cultures were obtained. No new cases of typhoid fever related to this outbreak have occurred since 4 August (*MMWR* 39:749, 1990).

✳ Transmission and Epidemiology

Salmonellosis is one of the more common infectious diseases in the United States. It is estimated that more than 2 million cases occur annually. Infections result from ingesting salmonella-contaminated foods. Poultry products, including eggs, are a common source of *Salmonella* infections, but meat and meat products, in general, are frequently contaminated. Any food that comes in contact with rodents or animal products may become contaminated. Pets or other animals may harbor salmonellae and transmit the infection directly to humans. Pet turtles or pet reptiles are often infected; consequently, their sale is often regulated throughout the United States. The increased use of widely distributed mass-produced foods may result in an increase in the rate of salmonellosis in the United States. The numbers of reported cases of salmonellosis occurring in the United States are shown in Figure 21-5.

✳ Diagnosis

Isolation of *Salmonella* species from the intestinal tract is required for a positive diagnosis.

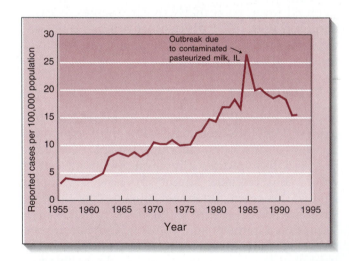

FIGURE 21-5 Reported cases of salmonellosis in the United States, 1955–1994. (Courtesy Centers for Disease Control, Atlanta)

✳ Treatment

Supportive therapy is the recommended treatment. Antibiotics are not recommended except in extreme cases of disseminated disease or cases involving

Outbreak of *Salmonella enteritidis* Associated With Homemade Ice Cream: Florida, 1993

On 7 September, 1993, the Epidemiology Program of the Duval County (Florida) Public Health Unit was notified about an outbreak of acute febrile gastroenteritis among persons who attended a cookout at a psychiatric treatment hospital in Jacksonville, Florida. This report summarizes the outbreak investigation.

On 6 September, seven children (age range: 7–9 years) and seven adults (age range: 29–51 years) attended the cookout at the hospital. A case of gastroenteritis was defined as onset of diarrhea, nausea or vomiting, abdominal pain, or fever within 72 hours of attending the cookout. Among the 14 attendees, 12 cases (in five of the children and all seven adults) were identified. The median incubation period was 14 hours (range: 7–21 hours); the mean duration of illness was 18 hours (range: 8–40 hours). Predominant symptoms were diarrhea (93%), nausea or vomiting (86%), abdominal pain (86%), and fever (86%). All ill persons were examined by a physician. *Salmonella enteritidis* (SE) (phage type 13a) was isolated from stool of three of the seven patients from whom specimens were obtained.

Eleven of the 12 ill persons had eaten homemade ice cream served at the cookout. No other food item was associated with illness. Testing of a sample of ice cream revealed contamination with SE (phage type 13a).

The ice cream was prepared at the hospital on 6 September using a recipe that included six grade A raw eggs. An electric ice cream churn was used to make the ice cream approximately 3 hours before the noon meal. The ice cream had been properly cooled, and no food-handling errors were identified. The person who prepared the ice cream was not ill before preparation; however, she became ill 13 hours after eating the ice cream. Her stool specimen was one of the three stools positive for SE (phage type 13a).

The U.S. Department of Agriculture's (USDA) Animal and Plant Health Inspection Service attempted to trace the implicated eggs back to the farm of origin. The hospital purchased eggs from a distributor in Florida. The traceback was terminated because the implicated eggs from the distributor had been purchased from two suppliers, one of whom bought and mixed eggs from many different sources. Current USDA *Salmonella* regulations limit testing of flocks to one clearly implicated flock (*MMWR* 43:669, 1994).

---

infants, elderly persons, or other immunocompromised patients.

✳ Prevention and Control

Proper cooking eliminates salmonella, and refrigeration of foods prevents their growth. Sanitary procedures in slaughterhouses help reduce the level of contamination. Some foods are routinely monitored for the presence of salmonellae.

SHIGELLA: GENERAL CHARACTERISTICS

The shigellae are primarily pathogens of humans and are not naturally found in other environments. The four species of the genus *Shigella*—*S. dysenteriae*, *S. flexneri*, *S. boydii*, and *S. sonnei*—can all cause

TABLE 21-2

Distribution of *Shigella* Species Recovered from Dysentery in the United States

Species	Group	Percentage of Total
S. dysenteriae	A	<1
S. flexneri	B	28
S. boydii	C	<1
S. sonnei	D	70

dysentery in humans. *S. sonnei* is by far the most commonly involved species, and *S. dysenteriae* causes the most severe type (Table 21-2).

> **Dysentery**   Serious diarrhea, accompanied by mucus and blood in the stool along with severe abdominal cramping.

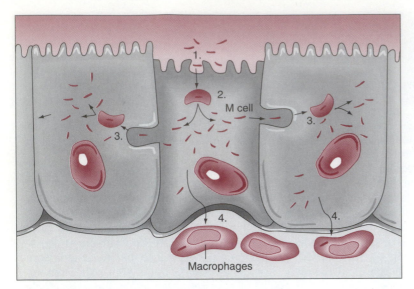

FIGURE 21-6 Invasion of intestinal epithelial cells by *Shigella*. 1. *Shigella* attach to M cells and induce their own phagocytosis. 2. *Shigella* escape the phagocytic vacuole and multiply. 3. Penetration of adjacent epithelial cells with multiplication and cell destruction. 4. *Shigella* released from infected cells are phagocytized by macrophages, thus preventing the spread to deeper tissues.

S. DYSENTERIAE (AND OTHERS): BACILLARY DYSENTERY (SHIGELLOSIS)

✷ Pathogenesis and Clinical Diseases

Following ingestion, the shigellae usually penetrate the large intestine by stimulating the endothelial cells, called M cells, that line the intestine to phagocytize them. However, these "nonprofessional" phagocytes are unable to kill the ingested bacteria and the bacteria then multiply and invade neighboring cells (Figure 21-6). Generally, penetration is not deeper than the submucosal cells. Inflammation, together with sloughing of the epithelial cells, results in ulcerative lesions. After 1 to 3 days of incubation the patient experiences a sudden onset of symptoms—abdominal cramps, fever, and diarrhea. The diarrheal stool frequently contains mucus and blood. Significant loss of water and salts may occur and in young and/or debilitated patients this dehydration and electrolyte imbalance may cause death. In otherwise healthy persons the disease is usually self-limiting and recovery occurs in 3 to 7 days. The death rate from dysentery in young children is significant in countries with poor sanitation and nutrition.

Infections due to *S. dysenteriae* are always potentially more serious than those due to other species. This organism produces a very powerful exotoxin (*shiga toxin*) that greatly increases its virulence. During a recent epidemic in Central and South America the mortality rate among those infected with this organism was between 8 and 10%. Although not endemic in the United States, this species has recently been introduced by tourists returning from Central America and Mexico. Most residents in areas where dysentery is endemic develop some immunity to the disease either through clinical or subclinical cases. Many such persons, however, remain carriers of the organism and serve as a source of infection for new susceptibles, such as visitors or newborns entering the population.

✷ Transmission and Epidemiology

Transmission is from human to human via the fecal–oral route by "fingers, food, feces, fomites, or flies." Infection can occur with as few as 10 or 10^2 bacteria (this is in contrast to 10^5–10^7 bacteria necessary to cause salmonellosis). Transmission by

---

Shiga toxin A powerful toxin produced by *S. dysenteriae* that acts on tissues of the central nervous system.

CLINICAL NOTE

Outbreak of *Shigella flexneri* 2a Infections on a Cruise Ship

During 29 August–1 September 1994, an outbreak of gastrointestinal illness occurred on the cruise ship *Viking Serenade* (Royal Caribbean Cruises, Ltd.) during its roundtrip voyage from San Pedro, California, to Ensenada, Mexico. A total of 586 (37%) of 1589 passengers and 24 (4%) of 594 crew who completed a survey questionnaire reported having diarrhea or vomiting during the cruise. One death occurred in a 78-year-old man who was hospitalized in Mexico with diarrhea.

Shigella flexneri 2a has been isolated from fecal specimens from at least 12 ill passengers. Antimicrobial susceptibility testing of representative isolates indicated resistance to tetracycline and susceptibility to ampicillin and trimethoprim sulfamethoxazole. The subsequent two cruises of the ship were canceled. Investigation of the mode of transmission is under way (*MMWR* 43:657, 1994).

Shigella species is commonly associated with poor or crowded living conditions. Most of the approximately 20,000 cases occurring annually in the United States (Figure 21-7) are associated with institutionalized individuals, where hygienic conditions may be difficult to maintain because of crowding and lack of individual capabilities. This relationship between an ability to maintain personal hygiene and the frequency of shigella infection is reflected in the age distribution of the disease in the United States, where it is seen most frequently in the pediatric population.

Shigellosis is endemic in underdeveloped countries. Historically, dysentery has been a problem in military populations and entire armies have become temporarily disabled when living under unsanitary conditions that commonly exist during wartime. People traveling from countries like the United States often contract bacillary dysentery within a short period after entering a country where dysentery is endemic.

✳ Diagnosis

Diagnosis is made by isolating shigellae from the feces or intestinal tract.

✳ Treatment

In contrast with *Salmonella* gastroenteritis, most cases of shigellosis are improved by chemotherapy. The recent development of *multiresistant* strains of *S. sonnei* (resistant to ampicillin, tetracycline, and trimethoprimsulfamethoxazole) has complicated the approach to therapy, but several available antibiotics remain effective. Oral rehydration and maintaining proper electrolyte balance is an essential component of treatment.

✳ Prevention and Control

Prevention of person-to-person transmission by following good sanitary practices is the most effective means of avoiding shigellosis. Patients with the disease should be isolated.

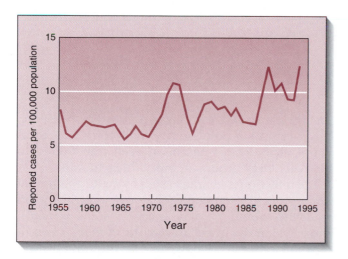

FIGURE 21-7 Reported cases of shigellosis in the United States, 1955–1994. (Courtesy Centers for Disease Control, Atlanta)

Multiresistant Bacteria that are resistant to a variety of antibiotics with different mechanisms of antimicrobial action.

Update—Multistate Outbreak of *Escherichia coli* O157:H7 Infections from Hamburgers: Western United States, 1992–1993

From 15 November 1992 through 28 February 1993, more than 500 laboratory-confirmed infections with *E. coli* O157:H7 and four associated deaths occurred in four states—Washington, Idaho, California, and Nevada. This report summarizes the findings from an ongoing investigation that identified a multistate outbreak resulting from consumption of hamburgers from one restaurant chain.

On 13 January 1993, a physician reported to the Washington Department of Health a cluster of children with hemolytic uremic syndrome (HUS) and an increase in emergency room visits for bloody diarrhea. During 16–17 January, a case-control study comparing 16 of the first cases of bloody diarrhea or postdiarrheal HUS identified with age- and neighborhood-matched controls implicated eating at chain A restaurants during the week before symptom onset. On 18 January, a multistate recall of unused hamburger patties from chain A restaurants was initiated.

As a result of publicity and case-finding efforts, during January–February 1993, 602 patients with bloody diarrhea or HUS were reported to the state health department. A total of 477 persons had illnesses meeting the case definition of culture-confirmed *E. coli* O157:H7 infection or postdiarrheal HUS. Of the 477 persons, 52 (11%) had close contact with a person with confirmed *E. coli* O157:H7 infection during the week preceding onset of symptoms. Of the remaining 425

persons, 372 (88%) reported eating in a chain A restaurant during the 9 days preceding onset of symptoms. Of the 338 patients who recalled what they ate in a chain A restaurant, 312 (92%) reported eating a regular-sized hamburger patty. Onsets of illness peaked from 17 January through 20 January. Of the 477 case patients, 144 (30%) were hospitalized; 30 developed HUS, and three died. The median age of patients was 7.5 years (range: 0–74 years).

During the outbreak, chain A restaurants in Washington linked with cases primarily were serving regular-sized hamburger patties produced on 19 November, 1992; some of the same meat was used in "jumbo" patties produced on 20 November, 1992. The outbreak strain of *E. coli* O157:H7 was isolated from 11 lots of patties produced on those two dates; these lots had been distributed to restaurants in all states where illness occurred. Approximately 272,672 (20%) of the implicated patties were recovered by the recall.

A meat traceback by a CDC team identified five slaughter plants in the United States and one in Canada as the likely sources of carcasses used in the contaminated lots of meat and identified potential control points for reducing the likelihood of contamination. The animals slaughtered in domestic slaughter plants were traced to farms and auctions in six western states. No one slaughter plant or farm was identified as the source (*MMWR* 42:258, 1993).

YERSINIA

Yersinia pestis is the most widely known species of the genus *Yersinia* and is the cause of the plague. Plague is primarily a disease of rats and other rodents, not humans. While plague has been an important disease in the past, it is currently a rare disease in developed countries, and its pathogenesis and epidemiology are quite different from the enteric pathogens discussed in this chapter. For these reasons, plague is discussed in Chapter 25.

By far the most common disease in humans caused by *Yersinia* is that caused by *Y. enterocolitica*. This bacterium is primarily associated with domestic animals, but may be transmitted to humans through contaminated foodstuffs or directly from such animals as dogs or swine. The organism is fre-

quently associated with large outbreaks of disease. This is typical of point-source food-borne infections and *Y. enterocolitica* has often been transmitted to relatively large groups of individuals in milk.

The infection is characterized by acute abdominal pain, profuse (sometimes bloody) diarrhea, and headache. Vomiting may occur in some patients. These symptoms are characteristic of many cases of appendicitis and so have led to the unnecessary removal of normal appendixes in some patients with this infection. Patients may also have an inflammation of the abdominal lymphatic structures, a condition referred to as *mesenteric adenitis*. Although symptoms may persist for several weeks, recovery is usually uneventful and complete.

Y. enterocolitica is most often isolated by plating a stool specimen on a selective medium. However, when present in low numbers, *Y. enterocolitica* can be cultured by a cold-enrichment procedure in which the culture specimen is held in saline at 4°C for up to 3 weeks. Periodic cultures are made from the cold enrichment to regular media and incubated at 35°C to recover the organism. There are more than 50 serotypes of *Y. enterocolitica*, and the distribution of these serotypes is strongly geographic. In the United States, most disease is due to serotype 03, 08, or 09.

OPPORTUNISTIC ENTERIC BACILLI

The bacteria discussed in this section are generally of low virulence and are often present as normal or transient inhabitants of the intestinal tract of humans and animals. Some are also found in water, sewage, and soil. Members of the genera *Escherichia*, *Klebsiella*, *Enterobacter*, *Serratia*, and *Citrobacter* usually ferment the sugar lactose and share a number of other properties; they are sometimes called **coliforms**.

All these opportunistic bacilli are capable of producing similar infections. When they gain access to such tissues as the urethra, bladder, lungs, wounds, or internal organs, they are capable of causing disease, particularly in a compromised host. The most frequent problems are urinary tract and wound infections. Pneumonia, gastroenteritis, septicemia, and meningitis may also result from these microbes. The more commonly encountered genera or species are discussed briefly in this section, with emphasis on some unique characteristics of each. The short, gram-negative morphology and physiologic characteristics generally associated with the enteric bacilli are typical for this group of bacteria.

✳ *Klebsiella*

The major *Klebsiella* species is *K. pneumoniae*. This nonmotile organism is characterized by a large, mucoid polysaccharide capsule. It is a cause of *primary pneumonia* (about 3% of all bacterial pneumonias) in older persons with such predisposing medical problems as chronic bronchitis, diabetes, or alcoholism. It is also a fairly common cause of septicemia as well as urinary tract and wound infec-

tions. Many antibiotic-resistant strains are found in hospitals; treatment is often difficult. *K. oxytoca* is commonly associated with nosocomial infections.

✳ *Enterobacter*

The *Enterobacter* are closely related to the *Klebsiella*, but are a more common cause of urinary tract infections. The most commonly encountered species is *E. cloacae*. These bacteria are frequently resistant to antimicrobial therapy and therefore may cause serious infections.

✳ *Serratia*

The most frequently encountered *Serratia* is *S. marcescens*. This bacterium sometimes produces red-pigmented colonies when grown at room temperatures. For many years it was considered a nonpathogen and, because of the pigmented colonies, it was used in various experiments to follow the movement of airborne particles in hospitals and other environments. It is now recognized that *S. marcescens* may cause pneumonia, cystitis (inflammation of the urinary bladder), and other serious infections in compromised hosts. Treatment is difficult because of its resistance to many commonly used antimicrobials.

✳ *Proteus, Providencia, and Morganella*

Proteus, *Providencia*, and *Morganella* are closely related members of normal bacterial flora of the bowel. Bacteria of these genera are found in water, soil, sewage, and the intestinal tracts of humans and animals. *Proteus* is characterized by exceptional motility that enables the cells to swarm across an entire petri dish within a few hours (Figure 21-8). Most clinical infections are of the urinary tract or burns. These bacteria are responsible for 10% of nosocomial infections. Treatment with antibiotics is often difficult.

✳ *Citrobacter*

The genus *Citrobacter* is closely related to *Salmonella*. *Citrobacter* bacilli have occasionally been incriminated as the cause of infections in humans, particularly urinary tract infections.

Coliform Enterobacteriaceae that are commonly normal flora of the intestinal tract.
Primary pneumonia Pneumonia that occurs as a result of infection without previous compromise to host defenses.

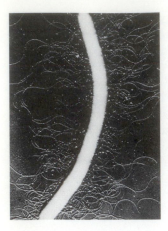 

FIGURE 21-8   (a) A peritrichous *Proteus vulgaris* cell (Courtesy of Biological Photo Service.) and (b) swarming *Proteus* on an agar plate. (Courtesy of Visuals Unlimited.)

✳ *Escherichia*

The species involved is *Escherichia coli*. Like many other enteric bacilli, *E. coli* was considered a non-pathogen for many years. It is one of the predominant facultative anaerobic bacteria of the intestinal tract and so is used as an indicator organism in determining the amount of fecal contamination in water and food. Many strains are used in experimental work in cell research. Much of the current work in molecular biology and recombinant DNA uses *E. coli*, and more is probably known about this organism than any other bacterium.

It became increasingly apparent over the past several decades that *E. coli* of varying degrees of pathogenicity exist and that these bacteria are responsible for numerous human infections. Diarrhea-causing *E. coli* are now placed into different categories based on their virulence properties. The major categories are *enterotoxigenic* (ETEC), *enteropathogenic* (EPEC), *enterohemorrhagic* (EHEC) *E. coli* and *enteroinvasive* (EIEC). ETEC strains produce enterotoxins that function like those produced by the cholera bacillus, *V. cholerae* (Chapter 23).

ETEC Strains.   ETEC carry one or more plasmids that are responsible for the production of a heat-labile toxin or a heat-stable toxin. These toxins are responsible for producing the symptoms of disease. These strains are the most frequent cause of infectious diarrhea throughout the world. Often referred to as "traveler's diarrhea," it is considered by many to be one of the nuisances of international travel. In adults it is usually self-limiting after a few days of nausea and profuse watery diarrhea. In global terms it is not so benign, and is a leading cause of child mortality throughout the world. The extent of such intestinal diseases is not known because most cases are self-limiting and a specific laboratory diagnosis is difficult. Epidemics in hospital nurseries due to ETEC have been reported and infections are probably widespread in infants living under impoverished conditions. Where infants are malnourished and supportive therapy is not given, a significant number of deaths result from dehydration and electrolyte imbalance due to *E. coli* enteritis. The use of oral rehydration therapy has produced remarkable results in reducing mortality due to *E. coli* diarrhea.

EPEC Strains.   EPEC strains cause a significant amount of the diarrhea in infants under 1 year of age in developing areas of the world, particularly in bottle-fed infants. These organisms adhere tightly to the mucosal cells of the small bowel resulting in a watery diarrhea that can become chronic.

EHEC Strains.   EHEC produce a powerful toxin that is similar to the shiga toxin produced by *Shigella dysenteriae* and is called Shiga-like toxin. The prominent serotype, designated *E. coli* O157:H7, has been associated with increasing outbreaks in the United States over the past 15 years (see Clinical Note). This microbe is usually transmitted to humans from animals via contaminated raw milk or undercooked meat. The clinical disease starts as watery diarrhea with abdominal pain about 4 days after exposure and develops into bloody diarrhea (hemorrhagic colitis) a day or two later. Recovery is usually complete within 10 days. However, in about 10% of children under 5 years, this disease develops into a condition called hemolytic–uremic syndrome. In this condition, the toxin causes direct kidney damage and is life-threatening. In recent years, outbreaks caused by EHEC have occurred in fast-food restaurants in the United States from undercooked ground beef products.

EIEC Strains.   The EIEC are very much like *Shigella*. They are able to invade the intestinal mucosa, causing dysentery-like illness. Most EIEC infections occur in children in developing countries or in travelers to these countries.

Other infections that are commonly due to *E. coli* include urinary tract infections. Strains of *E. coli* with specific attachment pili for epithelial cells of the urinary tract cause over 90% of urinary tract infections. *E. coli* may also cause bacteremia and neonatal meningitis. *E. coli* are a common cause of nosocomial infection including bacteremia, urinary tract, and wound infections.

MATERIAL FOR REVIEW

CONCEPT SUMMARY

1. The enteric bacilli consist of a large group of gram-negative bacilli that are normally found in the intestinal tract of humans and animals. They are transmitted by the fecal–oral route and usually are commensals with only a few primary pathogens, such as *Salmonella* and *Shigella*.
2. Typhoid fever, a disease caused by *Salmonella typhi*, is of a considerable significance worldwide. Most frequently transmitted in contaminated water, this organism causes a serious life-threatening illness. Proper sewage and water treatment, as well as adequate food-handling laws, are necessary to prevent the spread of this organism. Other *Salmonella* species cause disease in humans but these infections are nearly always transmitted through contamination of our environment by feces.
3. Shigellosis or bacillary dysentery is a human disease and is transmitted through human fecal contamination of food, water, or inanimate objects. The disease is severe but usually self-limiting.
4. The opportunistic enteric bacilli are common environmental inhabitants that are responsible for a high percentage of hospital-associated infections, and they are the leading cause of urinary tract infection.

CLINICAL SUMMARY TABLE

Microorganism	Virulence Mechanisms	Diseases	Transmission	Treatment	Prevention
Salmonella typhi	Not killed by macrophages Endotoxins	Typhoid fever	Fecal-oral routes Carriers	Broad spectrum antibiotics	Water and sewage treatment Vaccines Control carriers
Salmonella species	Endotoxins	Gastroenteritis	Food and waterborne	Supportive therapy	Proper cooking of foods Sanitary procedures
Shigella species	Shiga toxin Endotoxins	Dysentery	Fecal-oral routes	Oral rehydration Antibiotics	Good hygiene
Escherichia coli	Enterotoxins Attachment pili	Travelers diarrhea Hemorrhagic colitis Urinary tract infections	Fecal-oral routes Opportunists	Oral rehydration	Proper food handling and good sanitation

STUDY QUESTIONS

1. How would you respond to a statement that the *Enterobacteriaceae* are low-level pathogens of minimal medical significance?
2. What feature of the *Enterobacteriaceae* makes antibiotic therapy of their infections difficult?
3. Contrast the following terms: (a) diarrhea and dysentery; (b) food poisoning and gastroenteritis.
4. Describe the normal route of infection for each of the following organisms: (a) *Salmonella*, (b) *Shigella*, (c) *Y. enterocolitica*, and (d) *E. coli*.
5. List the cultural, morphologic, and virulence features that are common to the genera of the opportunistic enteric bacilli.

CHALLENGE QUESTIONS

1. Explain why the incubation period for typhoid fever would be longer than that for enterocolitis.
2. The EHEC strains, such as *E. coli* O157:H7 that have been responsible for outbreaks of hemorrhagic colitis (see Clinical Note), have brought food safety to the minds of the general public. One group suggests that the meat processors are responsible for providing safe meat to eat. The other group believes the food preparer is responsible for properly cooking the meat. Present arguments that support and refute the beliefs of these two groups.

CHAPTER

22 Nonfermentative Gram-Negative Bacilli

OUTLINE

A significant number of genera of gram-negative bacteria are of relatively low virulence and produce disease primarily in immune compromised hosts. The genera covered in this chapter include *Pseudomonas*, *Acinetobacter*, *Alcaligenes*, *Moraxella*, and *Eikenella*. Although these organisms produce primary infection in such hosts, their natural habitat is almost always the environment.

GENERAL CHARACTERISTICS

The organisms discussed in this chapter are all non-spore-forming, gram-negative bacilli, motile or non-motile. The most characteristically distinguishing feature of the group is that they are not fermentative. And while they utilize carbohydrates, they do so oxidatively without producing the acid or alcohol end products of fermentation.

A common feature of this group of genera is that they produce similar infections, and they are not known for a particular type of disease. It is also characteristic of these bacteria that they are commonly responsible for *nosocomial* (hospital-acquired) rather than community-acquired infec-

HUMAN BODY SYSTEMS AFFECTED

Pseudomonas, Acinetobacter, and Eikenelle

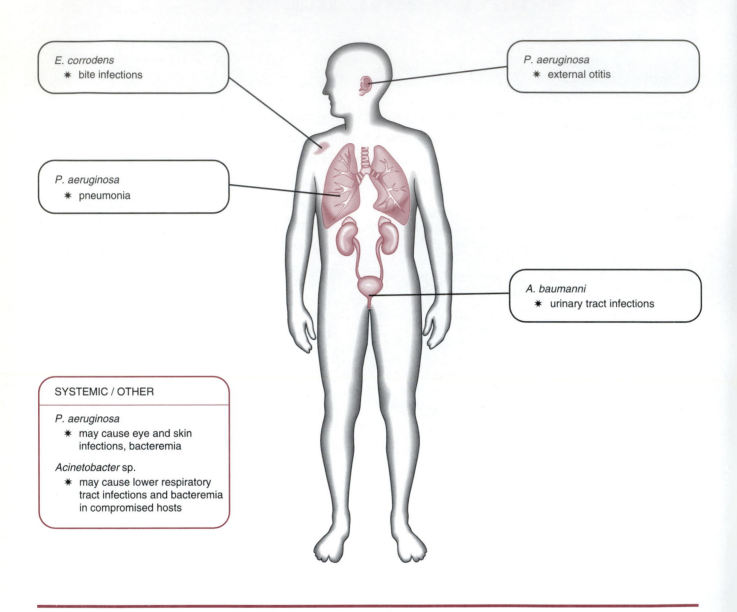

E. corrodens
  ✳ bite infections

P. aeruginosa
  ✳ external otitis

P. aeruginosa
  ✳ pneumonia

A. baumanni
  ✳ urinary tract infections

SYSTEMIC / OTHER

P. aeruginosa
  ✳ may cause eye and skin infections, bacteremia

Acinetobacter sp.
  ✳ may cause lower respiratory tract infections and bacteremia in compromised hosts

tions. About 15% of all clinical bacterial isolates are from these genera. Therefore, they are most likely to be found in urinary tract infections (as a consequence of catheterization) or burn infections (as a consequence of a reduction in the immune status of the patient), and bacteremia or pneumonia (in patients with reduced phagocyte function). Microbi-ologists frequently refer to these bacteria as "water bugs" because they can often be found in any amount of standing moisture, and they pose a particular patient care problem due to this association. Therapy is often difficult because of high levels of antibiotic resistance. Of the various genera (Table 22-1) of bacteria that fit the above description, only

TABLE 22-1

Nonfermentative Gram-Negative Bacilli of Clinical Significance

Achromobacter	Alcaligenes	Moraxella
Acinetobacter	Eikinella	Pseudomonas
Agrobacterium	Flavobacterium	

those most frequently involved in clinical disease are presented in this chapter.

PSEUDOMONAS

There are more than 300 species within the genus *Pseudomonas*, but of these only a few are commonly associated with human disease (Table 22-2). *P. pseudomallei* and *P. mallei* differ from others of this genus in that they are primary pathogens and are extremely virulent. *P. pseudomallei* is endemic in Southeast Asia, but infections outside of that area are uncommon. The three pseudomonads most frequently seen in human disease are *P. cepacia*, *P. maltophelia*, and *P. aeruginosa*. About two-thirds of all the clinically significant isolates of nonfermentive gram-negative bacilli are *P. aeruginosa*.

P. aeruginosa is a motile bacillus that is usually about 2 μm long. It is an obligate aerobe and grows well on most culture media. More than 90% of *P. aeruginosa* isolates produce a blue-green, water-soluble pigment (*pyocyanin*), which diffuses into the culture medium (see Plate 31). This pigment was first observed on bandages covering infected wounds. The growing colonies give off a sweet odor variously described as grape-like or corn tortilla-like. *P. aeruginosa* grows well at 42°C and is oxidase positive.

At least one *exotoxin* produced by *P. aeruginosa* has been identified as an important virulence factor. This toxin, known as *exotoxin A*, has the same mechanism of action as *diphtheria toxin* and is able to prevent protein synthesis by infected host cells. A variety of other bacterial products, such as elastase, proteases, endotoxic cell wall, and occasionally a mucoid capsule, probably account for the virulence associated with this organism. The organism is commonly found on plants, in areas where there is any collected water, and occasionally as transient flora in the human intestine. This organism is so ubiquitous in the environment that no open wound, burn, or immunocompromised patient is free from exposure.

As is characteristic for the gram-negative nonfermenters, *P. aeruginosa* infection most often occurs in a hospital setting. Several clinical conditions are highly correlated with *P. aeruginosa* infection, including cystic fibrosis, burns, urinary catheters, cancer chemotherapy, or any other condition that lowers the patient's immune responses. From any such infection, the patient may develop *P. aeruginosa* pneumonia or bacteremia. These conditions are serious and have mortality rates on the order of 60 to 70%. It is estimated that more than 90% of deaths in cystic fibrosis patients are due to *P. aeruginosa*. *P. aeruginosa* is a regular cause of external *otitis* (outer ear infection), sometimes referred to as "swimmers ear"; *P. aeruginosa* infection of the eye is a serious condition that may lead to perforation of the cornea with subsequent loss of the eye. Pseudomonas infection of the skin can occur in persons who bathe in contaminated hot tubs.

P. aeruginosa infections are very resistant to therapy. The remarkable metabolic capabilities and diverse plasmids associated with these organisms have resulted in the development of a high level of resistance to a broad range of antimicrobials. A number of new antibiotics have been specifically designed for treatment of pseudomonal infections, but as yet the ideal antipseudomonal compound has not been discovered.

Control and prevention of pseudomonal infections rely on proper aseptic techniques when dealing with burns and open wounds. *Pseudomonas* species are able to grow in water with minimal nutrients and are found in such places as water baths used for heating baby bottles, hot tubs, vases for fresh-cut flowers, and water sumps in humidi-

TABLE 22-2

Pseudomonas Species Associated with Human Disease

Species	Disease
P. aeruginosa	Opportunistic
P. cepacia	Opportunistic
P. mallei	Glanders
P. pseudomallei	Melioidosis
P. maltophilia	Opportunistic
P. fluorescens	Opportunistic
P. stutzeri	Opportunistic
P. putrifaciens	Opportunistic
P. acidovorans	Opportunistic
P. paucimobilis	Opportunistic
P. diminuta	Opportunistic

fiers. Certain precautions, such as not allowing fresh-cut flowers in critical areas of a hospital and routine monitoring of water held in baths and air systems, help reduce the hazard of these infections. *Pseudomonas* species are among the most resistant vegetative bacterial cells to chemical disinfectants.

ACINETOBACTER

Organisms from this genus have undergone repeated taxonomic and nomenclatural changes. Recent DNA homology studies have hopefully brought some order to this group and it may be possible to rely on the present taxonomy into the foreseeable future. There are presently six species in the genus, two of which are regularly responsible for human disease—*Acinetobacter baumannii* and *A. johnsonii*. These organisms, while fitting the description for gram-negative, nonfermenting bacilli, are not closely related to the pseudomonads, but are part of the family Neisseriaceae, which also includes the *Neisseria* species ∞ (Chapter 16, p. 225). Acinetobacter are *oxidase* negative, nonmotile, non-spore-forming coccobacilli (Figure 22-1) about 1.0×0.7 μm. They are normally found in soil and water and on moist skin areas (axilla, groin, etc.) of humans. Although they are capable of producing primary infections, most of the infections are seen in immune-compromised hospital patients. These bacteria cause about 1% of nosocomial infections.

Infections by *A. baumannii* are far more common than those due to *A. johnsonii* and most infections

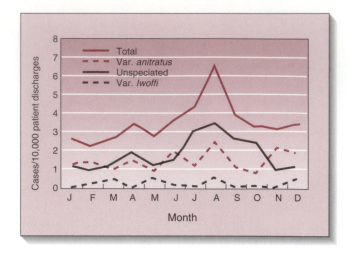

FIGURE 22-2 Seasonal variation in the occurrence of *Acinetobacter* infections. (Courtesy Centers for Disease Control, Atlanta)

are of the lower respiratory tract. Urinary tract infections and bacteremia following intravenous catheterization also occur. Most unusually, there is a marked seasonal occurrence associated with infections due to these organisms (Figure 22-2). Control of these infections depends largely on good patient care procedures.

ALCALIGENES

The genus *Alcaligenes* is similar in morphology and biochemical characteristics to bacteria of the other genera listed in Table 22-1. They are, however, unable to break down any of the sugars commonly used for bacterial identification. They are obligately aerobic and motile. The natural habitat of the most commonly isolated species, *A. faecalis*, is similar to that of pseudomonads. *A. faecalis* is occasionally found in wounds, ear infection, blood, or urine of infected patients. Patients suffering from these infections are somewhat more easily treated than those infected with the organisms previously discussed in this chapter.

MORAXELLA

Moraxella species are gram-negative coccobacilli that are nonmotile and oxidase-positive. As with the other genera in this chapter, they are unable to fer-

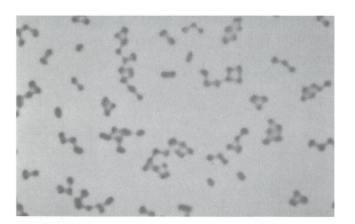

FIGURE 22-1 Photomicrograph of *Acinetobacter*. Note the paired coccobacillary structure of the cells. (Courtesy J. M. Matsen, University of Utah)

Oxidase An enzyme used as one identification test for bacteria; negative means the bacterium does not produce a respiratory enzyme, positive means it does.

ment glucose. These organisms grow well on supplemented laboratory medium.

There has been, and continues to be, some disagreement regarding the proper classification of these species. The clinical status of this group is also under question, although these bacteria have been isolated from a variety of clinical conditions such as conjunctivitis, septic arthritis, bacteremia, and urethritis. Reports of such isolations have not always correlated with patient status, and there are many who feel that these organisms are commensals or, at best, low-grade opportunists. They complicate the work of the diagnostic laboratory because they are difficult to classify. They are somewhat unusual among this group of nonfermenters because they are susceptible to penicillin.

EIKENELLA

The genus *Eikenella* has only a single species, *E. corrodens*. This organism is of interest here because it fits the description of the organisms under discussion and because of the rather unusual infections with which it is associated. *E. corrodens* is somewhat difficult to culture in the laboratory, and grows best on blood supplemented medium in a 10% CO_2 atmosphere. It is part of the normal flora of the mucous membranes of the human respiratory tract.

C L I N I C A L   N O T E

Pseudomonas cepacia at Summer Camps for Persons with Cystic Fibrosis

Pseudomonas cepacia (PC) is a multidrug-resistant, gram-negative bacillus that causes chronic colonization and infection of the respiratory tract of persons with cystic fibrosis (CF). PC colonization is usually difficult to eradicate with antimicrobial therapy and, in some patients, infection is associated with rapid decline in pulmonary function, increased hospitalization, and earlier death. Previous studies have suggested person-to-person transmission of PC both within and outside of hospitals. However, possible transmission of PC at CF summer camps—sites for physical and psychosocial therapy for many patients—has not been well characterized. To assess the risk for PC transmission in this setting, in 1987 and 1990, the CF Foundation and CDC conducted epidemiologic investigations in four CF summer camps in Michigan, Ohio, Utah, and Ontario, Canada. This report summarizes the results from Michigan.

In June 1987, 55 previously known PC-negative patients who were PC-culture negative immediately before attending camp attended a week-long CF camp with 36 other campers known to be colonized or infected with PC. The camp was staffed by 79 volunteer medical, paramedical, and laypersons who served as counselors and administered respiratory therapy and chest physiotherapy to campers. To determine the incidence of sputum conversion from PC-negative to PC-positive among campers, sputum or throat cultures were performed on all participants on arrival at, daily during, and within 3 months after camp.

To determine exposures of PC-negative campers to PC-positive patients or to particular camp staff and potential environmental sources of PC at camp, two investigators visually monitored campers' activities and administered a daily written questionnaire to each camper and/or the camper's counselor. None of the 55 CF campers with initially PC-negative sputum had PC-positive sputum cultures on departure from camp. However, five (9%) were PC-positive on their first follow-up culture within 2–13 weeks after camp. None were exposed to PC outside of the camp setting during this period. All five had reported close contact with PC-positive patients at camp, including participating in the same activities together for most of the day (four patients), hugging (three), lip-to-cheek kissing (one), and sharing toothpaste or finger food with (two) a PC-positive camper.

PC isolates from all five converters had the same ribotype (i.e., the restriction fragment-length polymorphism banding patterns were identical or had one- or two-band differences) as isolates from one or more PC-colonized campers and different from those of control isolates from other CF campers from other summer camps or CF centers. Of the five converters, three had PC with the same ribotype as that of isolates from PC-colonized campers with whom they had reported close contact.

Of 22 environmental cultures, three lake water samples grew PC. All three had an identical ribotype distinct from any of the PC isolates from campers (modified from *MMWR* 42:456, 1993).

Most infections are in wounds that have become contaminated with human saliva. This bacterium is the most common cause of infections resulting from human bites or in wounds resulting from striking teeth, appropriately referred to as "clenched fist injuries." Such infections may be serious, with extensive cellulitis, osteomyelitis, or even bacteremia. The organism is very susceptible to penicillin, which is used as the treatment of choice.

MATERIAL FOR REVIEW

CONCEPT SUMMARY

1. The nonfermentative gram-negative bacilli are environmental organisms of low virulence that are not uncommonly found in agents of human disease in compromised hosts. Antibiotic therapy cannot be empirically applied to disease caused by this group of bacteria and is often difficult because of resistance among these bacteria.

2. These organisms are frequently nonreactive with a variety of sugars and other metabolic substrates. This nonreactivity often makes identification and phenotypic classification of these bacteria very difficult.

CLINICAL SUMMARY TABLE

Microorganism	Virulence Mechanisms	Diseases	Transmission	Treatment	Prevention
Pseudomonas aeruginosa	Exotoxin A Endotoxins Proteases	Wound infection Pneumonia Outer ear infection Skin infection	Environment to humans Opportunists	Resistant to most antibiotics New antibiotics	Proper aseptic techniques

STUDY QUESTIONS

1. Describe the most common types of infections due to the nonfermentative bacteria.
2. What host–parasite relationship is demonstrated by infection and pathogenesis by the nonfermentative bacteria?
3. What feature of *Acinetobacter* species may lead to their being confused with *Neisseria* species?
4. Which of the genera presented in this chapter are aerobic?

CHALLENGE QUESTIONS

1. *Pseudomonas aeruginosa* is usually thought of as an opportunistic bacterium. Does this mean the bacterium is a lesser threat to cause disease than bacteria described in previous chapters? Explain.

CHAPTER

23

Vibrio, Campylobacter, and Helicobacter

OUTLINE

T he genera *Vibrio, Helicobacter,* and *Campylobacter,* at one time considered to be but one genus, are closely related in morphology and biochemistry. These genera are composed of slightly curved gram-negative bacilli, and there are many similarities in the diseases they produce. *Vibrio* species are free-living and marine water is their most common natural habitat, while the *Campylobacter* and *Helicobacter* species are found only in animal hosts.

VIBRIO: GENERAL CHARACTERISTICS

The genus *Vibrio* are members of the family *Vibrionaceae,* which includes gram-negative, motile, oxidase positive, facultative anaerobes that utilize glucose as a source of energy. These organisms are widely distributed in surface waters throughout the world. Of the more than 20 species of *Vibrio,* 11 have been responsible for human infections. These infections have generally produced one of two types of disease: localized infection, such as the gastroenteritis of classical cholera, or a severe disseminated bacteremic disease only recently recognized.

HUMAN BODY SYSTEMS AFFECTED

Vibrio, Campylobacter, and Helicobacter

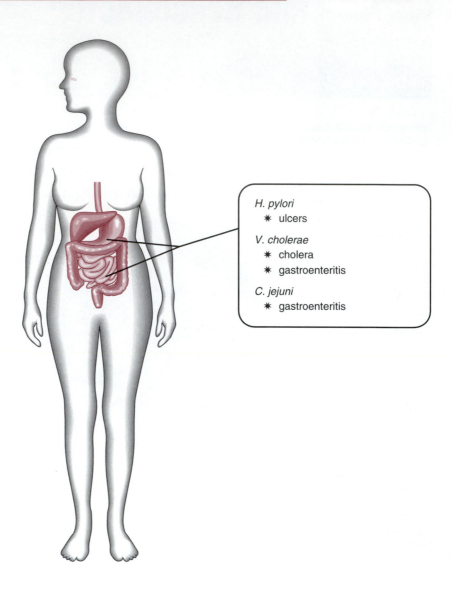

H. pylori
  ✳ ulcers

V. cholerae
  ✳ cholera
  ✳ gastroenteritis

C. jejuni
  ✳ gastroenteritis

V. CHOLERAE: CHOLERA

Vibrio cholerae, the causative agent of human **cholera**, is similar to the enteric bacilli in many respects. This bacterium can be grown on simple media and can be differentiated from the enteric bacilli by using selective and differential media (see Plate 28). The bacilli are slightly curved on initial isolation, giving comma-shaped cells (Figure 23-1; the first name given to this bacterium was *Kommabacillus*). Based on the outer lipopolysaccharides, *V. cholerae* can be divided into more than 100

Cholera   An acute enterotoxin-mediated infection caused by *Vibrio cholerae*.

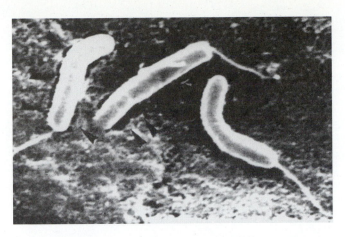

FIGURE 23-1 Electron micrograph of *Vibrio cholerae* on intestinal epithelial cells. Note the polar flagellum and the curved bacilli. (Courtesy J.S. Teppema et al., *Infection and Immunity* 55:2093. Reprinted with permission from ASM.)

serotypes. Of these, serogroup O1 is responsible for epidemic cholera, although some organisms from the non-O1 groups are also capable of producing a cholera-like illness. *V. cholerae* has also been subclassified into various biotypes. The biotypes *classical* and *El Tor* (hemolytic) are responsible for the more serious epidemics of cholera, although in recent years a different serogroup (O139) has been associated with epidemic cholera.

✳ Pathogenesis and Clinical Disease

The disease of human cholera, often referred to as *Asiatic cholera*, can be devastating, especially in crowded populations with poor sanitary and medical facilities. The term *cholera* itself means any conditions characterized by violent diarrhea. The bacterium is transmitted via contaminated food or water. The organisms attach to and proliferate in the small intestines but do not invade the tissues (see Plate 29). *V. cholerae* produces a powerful heat-labile *enterotoxin*. This toxin, composed of two proteins (A and B), is bound to intestinal cell receptors by part B. Part A then enters the cell and activates an enzyme known as *adenyl cyclase*. This enzyme acts as an internal messenger within the cell and the cell responds by secreting large volumes of water and dissolved salts (**electrolytes**) into the lumen of the intestine. Many liters per day of fluid are hypersecreted from the body with resultant acidosis,

dehydration, shock, and often death. Virulent strains also produce two minor enterotoxins called *zot* and *ace.*

The incubation time averages 1 to 2 days and the onset is abrupt. The initial signs are vomiting and diarrhea. The solids in the intestinal tract are purged early in the disease and the subsequently voided fluid is watery, without odor, and contains such electrolytes as sodium chloride, potassium, and bicarbonate. The patient develops sunken eyes and cheeks and the skin becomes wrinkled. Death results from rapid dehydration and resulting electrolyte imbalance.

No direct tissue damage occurs. *V. cholerae* may remain in the intestinal tract for several weeks after recovery in some untreated survivors. The pathogenesis of cholera is outlined in Figure 23-2. Without treatment, death rates are known to exceed 60%. With treatment, death rates may be reduced to less than 1%.

✳ Transmission and Epidemiology

In epidemic cholera, transmission follows the typical fecal–oral routes, with carriers and clinical cases serving as sources of infection. In normal persons the infectious dose is high (10^8 to 10^{10} organisms), but in persons with reduced stomach acidity, or if the organism is ingested with foods that reduce the stomach pH, the infectious dose is considerably lower and as few as 10^3 organisms can result in disease.

Cholera is endemic in regions of India and Bangladesh, and periodic epidemics erupt throughout Asia and Africa. Occasional outbreaks occur in other parts of the world. Because of today's rapid international travel, cholera could break out in any part of the world. The current cholera epidemic began in Asia in 1961 and reached Peru in early 1991. This epidemic spread to other South and Central American countries (Figure 23-3), and as of late 1994, over 1,000,000 cases and over 9000 deaths had been associated with this epidemic (Table 23-1). Cases of cholera have been reported in North Americans following travel to these countries. In late 1992, an epidemic of cholera caused by strain O139 started in southern Asia and continues to spread (Figure 23-4). It is estimated that as of late 1994, over 100,000 cases of cholera have occurred in this epidemic (see Clinical Note).

During the mid-1800s major outbreaks of cholera

Electrolytes   Sodium, potassium, chloride, and bicarbonate ions present in the body fluids.

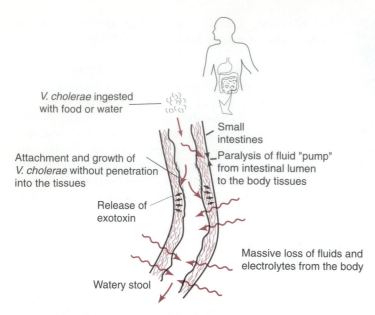

FIGURE 23-2   The pathogenesis of cholera showing the toxin-mediated effects of *V. cholerae* on the intestinal mucosa.

occurred along the Mississippi and Missouri river valleys of the United States. However, no cases of cholera were reported in the United States between 1911 and 1973. In 1973 a single case occurred in Texas. Then in 1977 a case developed in Alabama and in 1978 eleven cases were reported in Louisiana. Subsequently, cases have continued to occur along the Gulf Coast. Their sources were shown to be improperly cooked shellfish. The source of infection of these shellfish is not known, but evidence now indicates that *V. cholerae* is widespread along the Gulf Coast and provides the

FIGURE 23-3   The spread of *Vibrio cholerae* strain O1 in Central and South America from January 1991 to November 1994. (*MMWR* 44:215, 1995)

TABLE 23-1

Cholera Cases and Deaths Reported to the Pan American Health Organization: Western Hemisphere, January 1991 to September 1994

Country	Number of Cases	Number of Deaths
Peru	625,259	4,396
Ecuador	86,429	992
Colombia	27,481	370
Mexico	23,947	351
Guatemala	52,990	655
El Salvador	26,817	125
Panama	3,636	82
Brazil	134,425	1,359
Bolivia	35,203	695
Chile	146	3
United States	158	1
Honduras	3,537	112
Nicaragua	13,627	338
Total	**1,033,655**	**9,479**

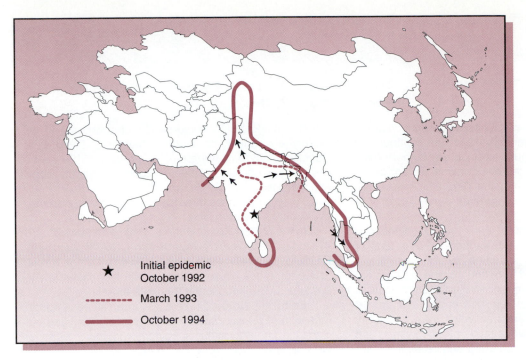

FIGURE 23-4 The spread of *Vibrio cholerae* strain O139 in southern Asia from October 1992 to October 1994. (*MMWR* 44:215, 1995)

source of a number of sporadic cases of cholera each year in the United States. During 1993 and 1994, 22 and 47 cases of cholera were reported in the United States, respectively. Of these, 65 (95%) were associated with foreign travel.

✳ Diagnosis

Cholera is generally diagnosed by clinical observation, although isolation of *V. cholerae* from the patient is needed to confirm diagnosis. Direct fluorescent antibody tests on stools from patients have been successfully used in making a rapid diagnosis.

✳ Treatment

Effective treatment of cholera, if initiated soon enough, is amazingly simple. Inasmuch as death results from loss of fluids and not tissue damage, the patient can be maintained by replacing the lost fluids and electrolytes. For severe cases, intravenous **isotonic electrolyte** solutions may be given until the patient recovers, often within hours. However, fluid balance can usually be maintained by oral administration of electrolyte solutions containing glucose. Fluid therapy is continued until antibodies are produced to neutralize the enterotoxin. Oral therapy alone can be highly effective, especially if the initial fluid loss is not excessive. Such a procedure is appropriate for clinics in rural areas that are staffed with paramedical personnel. Under proper treatment death rates may be reduced to less than 1%. Such **oral rehydration therapy (ORT)** should be used in treatment of all enterotoxic diarrheas. The results of its use have been remarkable, and literally tens of thousands of lives are saved each year by its use. Some medical observers have suggested that use of simple ORT solutions may represent the greatest single medical advance in history. Tetracyclines are effective and help eliminate the organism from the intestinal tract, but are not used routinely in endemic areas.

✳ Prevention and Control

Proper sewage treatment and water purification systems are the most important preventative measures. Countries with adequate systems have few, if

---

Isotonic electrolytes Common metabolic ions (Na$^+$, K$^+$, Cl$^-$) in the same concentration as found inside the cell.
Oral rehydration therapy (ORT) The supplying of water, glucose and electrolytes to compensate for their loss due to cholera diarrhea.

Update: *Vibrio cholerae* O1: Western Hemisphere, 1991–1994, and *V. cholerae* O139: Asia, 1994

The cholera epidemic caused by *Vibrio cholerae* O1 that began in January 1991 has continued to spread in Central and South America. In southern Asia, the epidemic caused by the newly recognized strain *V. cholerae* O139 that began in late 1992 also has continued to spread. This report updates surveillance findings for both epidemics.

From the onset of the *V. cholerae* O1 epidemic in January 1991 through 1 September 1994, a total of 1,041,422 cases and 9642 deaths (overall case-fatality rate: 0.9%) were reported from countries in the Western Hemisphere to the Pan American Health Organization. In 1993, the numbers of reported cases and deaths were 204,543 and 2362, respectively. From 1 January through 1 September 1994, a total of 92,845 cases and 882 deaths were reported. In 1993 and 1994, the number of reported cases decreased in some countries but continued to increase in several areas of Central America, Brazil, and Argentina.

The epidemic of cholera caused by *V. cholerae* O139 has affected at least 11 countries in southern Asia. *V. cholerae* O139 produces severe watery diarrhea and dehydration that is indistinguishable from the illness caused by *V. cholerae* O1 and appears to be closely related to *V. cholerae* O1 biotype El Tor strains. Specific totals for numbers of *V. cholerae* O139 cases are unknown because affected countries do not report infections caused by O1 and O139 separately; however, >100,000 cases of cholera caused by *V. cholerae* O139 may have occurred.

In the United States during 1993 and 1994, 22 and 47 cholera cases were reported to CDC, respectively. Of these, 65 (94%) were associated with foreign travel. Three of these were culture-confirmed cases of *V. cholerae* O139 infection in travelers to Asia.

Editorial Note: Cholera is transmitted through ingestion of fecally contaminated food and beverages. Because cholera remains epidemic in many parts of Central and South America, Asia, and Africa, healthcare providers should be aware of the risk for cholera in persons traveling in cholera-affected countries—particularly those persons who are visiting relatives or departing from the usual tourist routes, because they may be more likely to consume unsafe foods and beverages (*MMWR* 44:215, 1995).

---

any, outbreaks of cholera. Rapid detection, isolation, and treatment of patients are also important.

Persons traveling in countries where cholera exists should avoid consuming uncooked fruits or vegetables, raw seafood, and nonsterilized beverages. Considering the outbreaks along the Gulf Coast in the United States, care should be taken to properly cook shellfish taken from these waters.

A killed vaccine has been used for many years but is not highly effective. Promising field tests are under way with a live, genetically engineered vaccine using a strain of *V. cholerae* from which the genes for part A of the cholera toxin, and the *zot* and *ace* toxins have been deleted.

OTHER DISEASE-CAUSING VIBRIOS

Recent recognition that the non-O1 vibrios are responsible for a variety of clinical illness has prompted considerable interest in this group of bacteria. These organisms are all from marine habitats and some are true *halophiles*, requiring increased sodium chloride concentrations for growth ∞ (Chapter 7, p. 100).

The most common illness produced by the non-O1 serotypes of *V. cholerae* is gastroenteritis. This illness can be mild to severe, and is nearly always associated with the ingestion of seafood. A recent study showed that more than 80% of such victims had recently consumed oysters or crabs and that only 10% of these patients could not associate their illness with seafood. Most of the patients (96%) had diarrhea, 82% had cramping, and 70% suffered headaches; more than 50% were hospitalized. The illness appeared to result from the production of an enterotoxin similar to cholera toxin.

Vibrio species other than *V. cholerae* are also implicated in human disease (Table 23-2). *V. parahaemolyticus* infection is acquired by eating improperly cooked or stored seafood. The illness is accompanied by a watery diarrhea which lasts for a number of days. It is now recognized that about half the cases of diarrhea during the summer in Japan are caused by *V. parahaemolyticus*.

TABLE 23-2

Vibrio Species Pathogenic for Humans

Species	Growth in NaCl			Fermentation of		Human Disease
	0%	3%	8%	Sucrose	Salacin	
V. cholerae	+	+	−	+	−	Cholera
V. parahaemolyticus	−	+	+	−	−	Gastroenteritis
V. vulnificus	−	+	−	−	+	Bacteremia, cellulitis
V. fluvialis	−	+	+/−	+	+	Gastroenteritis
V. mimicus	+	+	−	+	−	Gastroenteritis
V. alginolyticus	−	+	+	+	−	Cellulitis
V. damsela	−	+	−	−	−	Cellulitis

Vibrio vulnificus is a halophilic vibrio found in warm seawaters. It has long been known that persons with liver disease occasionally become infected with this organism. It is now recognized that persons who have open wounds or who injure themselves in a marine setting run the risk of developing severe *V. vulnificus* cellulitis, which may lead to septicemia. The usual source of bacteremia due to this organism is from ingestion of contaminated seafood, such as raw oysters (see Clinical Note). In approximately 50% of patients with bacteremia, the disease is fatal.

Other species of *Vibrio* listed in Table 23-1 are occasionally involved in human diseases. As more and more attention is given to this group of bacteria, it becomes apparent that they are increasingly involved in human infections.

CAMPYLOBACTER

Bacterial species now belonging to the genus *Campylobacter* were classified as vibrios for many years. The term campylobacter means "curved rod" in Greek and describes the shape of these bacteria. They are gram-negative, motile, microaerophilic microorganisms and are best isolated on a selective agar medium in an atmosphere of reduced oxygen concentration (5 to 10%) and 10% CO_2. Microscopically, they are often seen in pairs or short chains, giving them a typical "gull wing" appearance (Figure 23-5). Although they grow well on most laboratory media, they grow slowly, and are easily missed if other bacteria are present. Laboratories often use highly selective media incubated at 42°C in order to isolate campylobacter from clinical material.

✳ Pathogenesis and Clinical Diseases

Long recognized as part of the intestinal microbial flora of many mammals and birds, these bacteria have been incriminated as a cause of abortions in domestic animals. Since the 1970s, it has been recognized that campylobacter cause enteritis and are frequently disseminated in humans. The most common human infection due to campylobacter is acute gastroenteritis. Of the species shown in Table 23-3, *C. jejuni* is most often the etiologic agent involved. This organism is usually acquired from food, milk, or contact with infected animals. The disease is accompanied by fever, bloody diarrhea, headache,

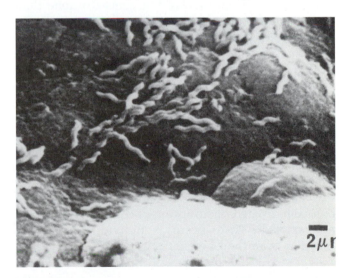

FIGURE 23-5 Scanning electron micrograph of *Campylobacter jejuni* on the intestinal epithelium of a colonized moose. Bar = 2 μm. (Courtesy A. Lee et al., *Infection and Immunity* 51:536. Reprinted with permission of ASM.)

TABLE 23-3		
Campylobacter Species and Associated Human Diseases		
Species	Growth at 42°C	Clinical Significance in Humans
C. fetus	–	Systemic infection/bacteremia
C. jejuni	+	Acute gastroenteritis
C. coli	+	Proctitis/gastroenteritis
C. hyointestinalis	–	None
C. laridis	+	Rare in humans
C. sputorum	–	None

and abdominal pain. It is usually self-limited, lasting 6–10 days, although chronic cases are known. Systemic infection, usually due to *C. fetus* subspecies *fetus* is known to occur, particularly among immunocompromised persons.

A recent observation has extended our appreciation of the pathogenesis of this group of bacteria. Evidence is increasing that *C. fetus* subsp. *fetus* infection may induce spontaneous abortion in humans. This organism has long been known to cause fetal loss in lower animals, but has recently been associated with similar pathology in pregnant humans.

✳ Transmission and Epidemiology

Although the *Campylobacter* species of bacteria are less commonly known by the general public than are *Salmonella* and *Shigella* species, they appear to be responsible for as many cases of infectious enteritis as the other two combined. Intestinal infection occurs in patients of all ages; it is common in young children and may account for up to 30% of all acute diarrheas in children under 8 months of age living in underdeveloped countries. Under conditions of poor sanitation, transmission is by the fecal–oral route from person to person. The presence of these organisms in the normal gut flora of domestic animals and poultry also widely disseminates them in our environment. Most reported cases (60%) are associated with consumption of contaminated milk, and sporadic cases occur in individuals who are infected from their house pets.

✳ Diagnosis

Diagnosis is made by observing campylobacter among blood cells in fecal specimens, or by culture of the organism. Systemic cases are nearly always diagnosed by culture.

✳ Treatment and Control

Even though most cases of campylobacter infection are self-limiting, therapy is usually recommended to reduce the length and severity of the disease. A number of antibiotics, including erythromycin or the quinolones, are effective agents when given properly. Control is best achieved by good hygiene, with special emphasis on avoiding transmission of microorganisms from animals to food and water consumed by humans.

HELICOBACTER: GENERAL CHARACTERISTICS

A comma-shaped bacterium first isolated from a stomach biopsy in 1982 was initially called *Campylobacter pylori*. However, it was found to differ from other *Campylobacter* species and has been reclassified to a new genus, *Helicobacter*, and given the species name *H. pylori*. While comma-shaped bacteria had long been observed on the stomach lining, before the 1980s they had not been associated with any disease condition or studied in detail. Evidence accumulated during the 1980s to indicate that *H. pylori* was the cause of gastritis (inflammation of the stomach lining), as well as stomach and duodenal ulcers. And current evidence is accumulating to indicate that stomach cancer may be associated with chronic *H. pylori* infection. This knowledge has drastically changed the methods of treating gastric (stomach) and peptic (duodenal) ulcers; they are now treated as infectious diseases rather than as physiological disorders.

Vibrio vulnificus Infections Associated with Raw Oyster Consumption: Florida, 1981–1992

Vibrio vulnificus is a gram-negative bacterium that can cause serious illness and death in persons with preexisting liver disease or compromised immune systems. From 1981 through 1992, 125 persons with *V. vulnificus* infections, of whom 44 (35%) died, were reported to the Florida Department of Health and Rehabilitative Services (HRS). This report summarizes data on these cases and presents estimates of the at-risk population in Florida.

The infections generally occurred each year from March through December and peaked from May through October. Seventy-two persons (58%) had primary septicemia, 35 (28%) had wound infections, and 18 (14%) had gastroenteritis. In patients with primary septicemia, 58 infections (81%) occurred among persons with a history of raw oyster consumption during the week before onset of illness. The mean age of these persons was 60 years; 51 (88%) were male. Fourteen (78%) of the patients with gastroenteritis also had raw oyster-associated illness. Their mean age was 49 years; seven (50%) were male.

Of the 40 deaths caused by septicemia, 35 (88%) were associated with raw oyster consumption. Nine of these deaths occurred in 1992. The case-fatality rate from raw oyster-associated *V. vulnificus* septicemia among patients with preexisting liver disease was 67% (30 of 45), compared with 38% (5 of 13) among those who were not known to have liver disease.

Results of the 1988 Florida Behavioral Risk Factor Survey (BRFS) were used to estimate the proportions of the Florida population who ate raw oysters, and the proportion of the population who ate raw oysters and who believed they had liver disease (e.g., cirrhosis). These estimates were used in conjunction with case reports and population data from the Florida Office of Vital Statistics to estimate the risk for illness and death associated with *V. vulnificus*.

BRFS and state population data indicate that approximately 3 million persons in Florida eat raw oysters; of these, 71,000 persons believe they have liver disease. Based on the number of cases reported to the Florida HRS during 1981–1992, the annual rate of illness from *V. vulnificus* infection for adults with liver disease who ate raw oysters was 72 per 1 million adults—80 times the rate for adults without known liver disease who ate raw oysters (0.9 per 1 million). The annual rate of death from *V. vulnificus* for adults with liver disease who ate raw oysters was 45 per 1 million—more than 200 times greater than the rate for persons without known liver disease who ate raw oysters (0.2 per 1 million) (*MMWR* 42:405, 1993).

H. PYLORI: GASTROINTESTINAL INFECTIONS

H. pylori is a microaerophilic, curved, gram-negative bacillus. It has polar flagella and displays rapid, darting motility. *H. pylori* can be cultured on standard media, such as chocolate agar, when incubated under microaerophilic conditions as used for *Campylobacter*; it may take 4 to 7 days for colonies to develop. An important characteristic is its ability to produce a urease enzyme that rapidly breaks down urea.

✳ Pathogenesis and Clinical Diseases

H. pylori is one of the few microbes adapted to live in the stomach of humans. This ability may be due in part to the production of the enzyme urease, which splits ammonia from urea and creates an alkaline microenvironment around the microbe. *H. pylori* appears to specifically adhere to the stomach mucosa, and the ammonia and possible toxins it produces may cause injury to the adjacent gastric epithelium. This leads to inflammation and is associated with such clinical manifestations as pain, gas, painful digestion, foul breath, vomiting, and nausea. Further tissue damage leads to ulcers in the stomach and duodenum. On the other hand, many persons harbor this microbe in their stomachs and remain asymptomatic.

✳ Transmission and Epidemiology

H. pylori is found primarily in humans. The mode of transmission has not been demonstrated. It is difficult to find these microbes in feces, but they have been isolated from saliva. Many people appear to be infected with *H. pylori* and have no clinical disease.

However, current evidence indicates that gastric colonization with *H. pylori* is prerequisite to the development of most stomach and duodenal ulcers and also greatly predisposes a person to stomach cancer.

✳ Diagnosis

Direct diagnosis is done by endoscopic examination, during which a biopsy of the stomach mucosa is obtained. The biopsy can be examined microscopically for curved gram-negative bacilli, cultured for *H. pylori*, and also tested directly for the urease enzyme. A "breath test" gives a strong indication of infection; this is done by having the patient ingest radiolabeled urea and then measuring the amount of labeled carbon dioxide (CO_2) exhaled. The labeled CO_2 results from the breakdown of urea

by the urease enzyme produced by *H. pylori*. Several serologic tests for *H. pylori* are currently being evaluated.

✳ Treatment

Combinations of drugs have been effective in treating helicobacter. Current treatments usually include bismuth salts plus metronidazole in combination with either amoxicillin or tetracycline. Cure rates are over 90% and relapse rates are low.

✳ Prevention and Control

No vaccine is available and until the mode of spread is understood no control measures are known.

MATERIAL FOR REVIEW

CONCEPT SUMMARY

1. The genus *Vibrio* is represented by several highly virulent species that cause a variety of human diseases. Most of the organisms in the genus are halophilic, curved, gram-negative, motile bacilli.

2. Asiatic cholera is a disease known for its great human plagues. This fecally transmitted disease can cause death in its victims within hours of infections. Infections have resulted from eating improperly cooked shellfish or drinking contaminated water.

3. *Campylobacter* produces a painful, sometimes serious gastrointestinal disease. Usually transmitted from domestic animals, the disease is generally self-limiting.

CLINICAL SUMMARY TABLE

Microorganism	Virulence Mechanisms	Diseases	Transmission	Treatment	Prevention
Vibrio cholerae	Enterotoxin	Cholera	Foodborne Waterborne	Oral rehydration	Vaccine Water & sewage treatment
Campylobacter jeiuni	Uncertain	Gastroenteritis	Animals to humans Fecal-oral route	Oral rehydration Antibiotics	Good hygiene
Helicobacter pylori	Urease enzyme Possible toxins	Stomach and duodenal ulcers	Not determined	Bismuth salts plus metronidazole plus amoxicillin	No known procedures

STUDY QUESTIONS

1. Describe the mechanism of action of cholera toxin.
2. Why is cultural diagnosis of cholera a questionable approach in highly endemic areas?
3. List four reasons why oral rehydration therapy is such a valuable therapeutic procedure.
4. What is the difference between O1 and non-O1 vibrios?
5. Compared to the more classic *Shigella* and *Salmonella* gastroenteritis, how common is *Campylobacter* gastroenteritis?

CHALLENGE QUESTIONS

1. *Vibrio cholerae* does not survive in conditions similar to those of the human stomach. How then can it infect the small intestines?
2. Many people who never develop ulcers still harbor the bacterium *Helicobacter pylori*. What characteristics might explain this observation?
3. Often combinations of antibiotics are used to eliminate *H. pylori*. Why is such chemotherapy used?

CHAPTER

24 Haemophilus and Bordetella

OUTLINE

The two genera, *Haemophilus* and *Bordetella*, discussed in this chapter are composed of small, fastidious, gram-negative bacilli. Representatives from both genera are capable of producing human respiratory disease. These diseases primarily afflict children and can be serious, even life-threatening. Based on current classification methods, these two genera are not closely related, and so each is discussed separately in this chapter.

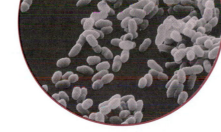

HAEMOPHILUS: GENERAL CHARACTERISTICS

Bacterial species of the genus *Haemophilus* (sometimes spelled *Hemophilus*) require special growth factors that are found only in blood and other body fluids. The name *Haemophilus* means "blood-loving" (Greek *haemo* = blood; *philus* = loving). *Haemophilus influenzae* is the major disease-producing species of this genus. Several other species, such as *H. ducreyi* and *H. aphrophilus*, are virulent but are less common agents of serious human infection.

H. INFLUENZAE INFECTIONS

H. influenzae is a small (1 × 0.3 μm), gram-negative coccobacillus (Figure 24-1). It requires blood or blood products—specifically hemin, called *X factor*,

HUMAN BODY SYSTEMS AFFECTED

Haemophilus and Bordetella

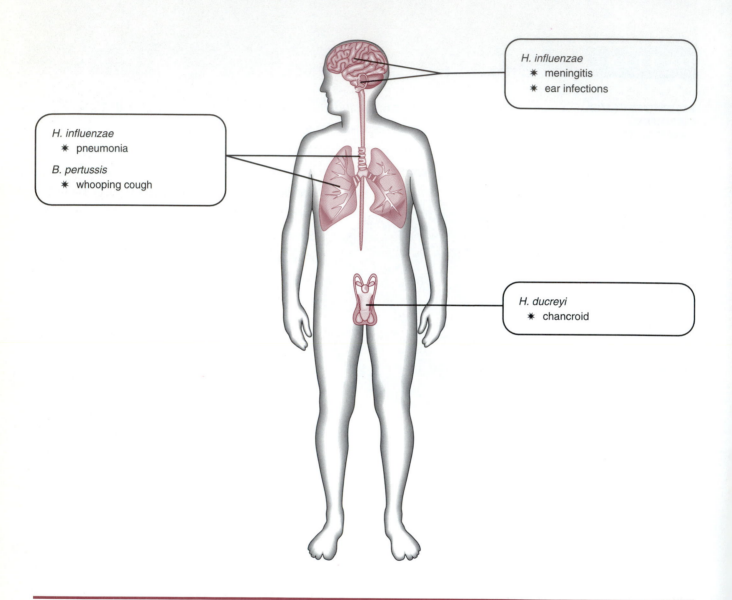

H. influenzae
  * meningitis
  * ear infections

H. influenzae
  * pneumonia

B. pertussis
  * whooping cough

H. ducreyi
  * chancroid

and NAD, a coenzyme called *V factor*—for growth in artificial media (Table 24-1).

H. influenzae can be divided into a variety of subtypes based on metabolic reactions and antigenic capsular polysaccharides. There are six capsular types (a–f); essentially all serious systemic disease is due to type b. The type b capsule is a polyribose-ribitol phosphate and is the primary component of the *H. influenzae* type b (Hib) vaccine. The capsule helps retard phagocytosis ∞ (Chapter 4, p. 13). Other virulence factors are less well defined, although this bacterium can produce an enzyme that specifically degrades protective IgA molecules that may be produced by the host. These traits may contribute significantly to the virulence of this bacterium. This bacterium does not survive well outside the body and is found only in humans.

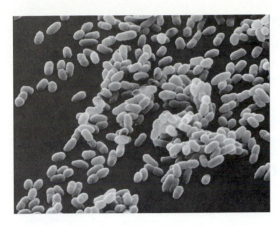

FIGURE 24-1 A scanning electron micrograph of *Haemophilus influenzae* (15,000×). (Courtesy of Visuals Unlimited.)

✳ Pathogenesis and Clinical Diseases

H. influenzae colonizes in the respiratory tract and as many as 50% of young children may be carriers. Only a small number of persons who carry this bacterium develop clinical disease, and disease is most common in young children. Thus, *H. influenzae* generally functions as an opportunist. Clinical disease often occurs as a nose and throat infection (*nasopharyngitis*). In some persons, the infection may spread to the sinus and middle ear or develop into pneumonia. In a small percentage of cases the bacteria spread to the meninges (see Plate 33).

In the 1980s, *H. influenzae* was the leading cause of invasive bacterial disease among children in the United States and was also the most common cause of bacterial meningitis, and an estimated 8000 to 10,000 cases of *Haemophilus* meningitis occurred annually, primarily among children under 5 years of age. Even with treatment, the mortality rate from this infection is about 5%, and a significant number (25 to 35%) who recover have permanent, residual damage to the central nervous system. In addition to bacterial meningitis, *H. influenzae* is responsible for other invasive diseases, including *epiglottitis*, sepsis, *cellulitis*, pneumonia, and septic arthritis. Patients with altered host defenses, such as those patients without a spleen or who have sickle cell disease or Hodgkin's disease, have a greatly increased risk of serious outcome from Hib infection. Patients with Hib epiglottitis pose a true medical emergency because they are in danger of suffocation due to obstruction of the trachea. Other, less serious infections due to *H. influenzae* include ear infection (*otitis media*), infectious conjunctivitis (pink eye), and nasopharyngitis. Most of these latter infections are caused by organisms that do not have a capsule and seemingly pose little threat to life.

✳ Transmission and Epidemiology

The *Haemophilus* bacteria are transmitted by respiratory droplets ∞ (Chapter 13, p. 192) and are widespread in humans; most persons develop active antibody immunity against them before reaching adulthood. Newborns receive passive immunity from their mothers and hence valuable protection during the early months of life ∞ (Chapter 12, p. 176). Because antibody to the capsular polysaccharide is protective, as long as infants have this transplacental antibody from their mothers they are at little risk of serious *H. influenzae b* infections.

TABLE 24-1

Identification of *Haemophilus* Species

Species	Requirement for		Production of		
	V factor	**X factor**	**Urease**	**Indole**	**Beta-hemolysis**
H. influenzae	+	+	+/−	+/−	−
H. parainfluenzae	+	−	+/−	−	−
H. hemolyticus	+	+	+	+/−	+
H. aphrophilus	−	−	−	−	−
H. ducreyi	−	+	−	−	−

Epiglottitis Infection of the tissue that normally covers the trachea during swallowing.
Cellulitis Inflammation of connective tissue.

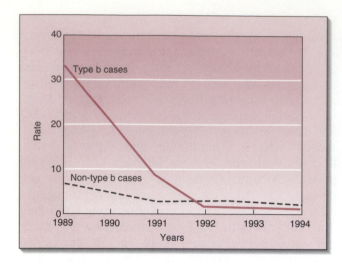

FIGURE 24-2   Rate of *Haemophilus influenzae* type b invasive disease and non-type b cases per 100,000 children under 5 years of age living in four regions of the United States.

However, as this antibody begins to be lost, there is a definite and increasing risk of meningitis until the child develops its own antibody. The epidemiology of invasive Hib infection is currently (1995) changing rapidly in countries where vaccination programs are carried out. Since the introduction of the Hib conjugated vaccines in 1990, the cases of invasive Hib infection in the United States have decreased over 95% among children under 5 years of age (Figure 24-2). *Haemophilus* infection also occurs in adults and in the era before the use of Hib vaccines represented only a small portion of the total cases. However, with the marked decrease in Hib infections in young children due to vaccination, more cases are now reported in older persons than in those under 5 years of age. In 1994, about 1100 cases of invasive Hib infection were reported in the United States, and only 29% of these were in children less than 5 years of age. Most haemophilus infections occur on a **sporadic** basis, except for conjunctivitis, which is highly contagious, and may occur as an epidemic. *H. influenzae* pneumonia may occur in conjunction with an epidemic viral respiratory disease, such as influenza. The name *H. influenzae* was first applied to this bacterium because early studies mistakenly thought it was the primary cause of influenza.

✳ Diagnosis

Diagnosis of *H. influenzae* meningitis can often be made by direct microscopic observation of the bacteria in the spinal fluid. Bacteria can be cultured on chocolate agar and then specifically identified. Some cases are diagnosed through serological procedures that can detect free bacterial capsular antigens in body fluids like urine or spinal fluid.

✳ Treatment

Prompt, proper therapy is essential in life-threatening circumstances (e.g., meningitis, sepsis, epiglottis), and not only reduces mortality but also limits serious sequelae. Delay in treatment greatly reduces the chance for therapeutic success. Ampicillin is considered the drug of choice if the organism is susceptible. However, recent reports indicate that as many as 35% of *H. influenzae b* are now resistant to this antibiotic. A recent effort to overcome the developing antibiotic resistance among *H. influenzae* has produced an unusual combination antibiotic called *augmentin*. Augmentin is a combination of ampicillin and a beta-lactamase inhibitor known as clavulanic acid. This combination extends the qualities of ampicillin by using the clavulanic acid to block ampicillin-inactivating enzymes before they can bind to the active ampicillin molecule. Other antibiotics, such as chloramphenicol and a cephalosporin called *cefotaxime*, may be necessary in cases where ampicillin or augmentin are not effective. Sensitivity tests to determine the most effective chemotherapeutic agent are generally completed on all strains isolated from serious disease conditions.

✳ Prevention and Control

Currently, the prevention and control of Hib infections are based on the use of successful *conjugated vaccines* ∞ (Chapter 12, p. 174). These vaccines use the capsular polysaccharide from type b *H. influenzae* strains. However, this polysaccharide antigen alone does not induce a good immune response in children under 18 months of age. The original Hib vaccine contained only purified capsular polysaccharide, and while it induced good immunity in children over 2 years of age, it was ineffective in children under 18 months of age. Continued efforts to improve the Hib vaccine have resulted in several

Sporadic   Occurring at irregular intervals. Outbreaks or occurrences of sporadic infections are very difficult to predict.

Progress Toward Elimination of *Haemophilus influenzae* Type B Disease among Infants and Children: United States, 1987–1993

Editorial Note: This report documents the continued decline in the incidence of all *Haemophilus influenzae* (Hi) and *Haemophilus influenzae* type b (Hib) disease in children aged <5 years in the United States. National surveillance monitors the occurrence of Hi disease, which in the past was primarily caused by Hib organisms; the decline in incidence monitored by national surveillance most likely reflects a decline in Hib disease associated with use of Hib conjugate vaccines. In addition, the laboratory-based surveillance system provided direct evidence of a decline in Hib disease, which coincided with introduction and use of Hib conjugate vaccines for children aged 18 months in 1988 and infants aged 2 months in 1990.

Based on findings from the National Health Interview Survey, in 1992, 67% of children aged 12–23 months had received at least one dose of Hib vaccine, and 36% had received three or more doses. Despite this incomplete level of vaccination coverage, surveillance indicates a decline of more than 90% in disease incidence, probably reflecting an unexpected additional benefit of conjugate vaccine use—elimination of carriage—resulting in reduced exposure to the pathogen and decrease in disease incidence even among unvaccinated persons. The decrease in incidence of Hib disease among persons aged 5 years in laboratory-based surveillance sites also is most likely a result of decreased carriage and transmission of the organism by infants and children. The availability of Hib conjugate vaccines, which are efficacious in children and reduce carriage, make feasible the goal of elimination of Hib disease among children aged <5 years by 1996.

Achievement of the 1996 goal to eliminate Hib disease requires participation by all levels of the health-care provider system in collection of surveillance data (i.e., rapid identification, assessment, and prompt reporting of all cases) and optimal use of this information to prevent increased disease incidence among poorly vaccinated populations. To optimize surveillance, case reports should ideally satisfy four criteria. First, because Hib vaccines protect against serotype b organisms only, serotype should be determined and reported for all invasive Hi isolates. Second, to identify persons and groups at risk for Hib disease, vaccination status of all children with invasive Hib disease should be assessed. Third, to evaluate the possible role of incomplete or ineffective vaccination in persons with Hib disease, the date, vaccine manufacturer, and lot number for each Hib vaccination should be determined. Fourth, important measures of morbidity and mortality associated with Hi infections should be reported and should include information on the type of clinical syndrome, specimen source (e.g., cerebrospinal fluid, blood, or joint fluid), and the outcome from disease. CDC is working with state health departments to optimize collection, compilation, and analysis of Hi surveillance data (*MMWR* 43:147, 1994).

new versions that became available in the early 1990s. These new vaccines, called conjugate vaccines, overcame the limited infant immune response to polysaccharide antigens by linking a bacterial capsular antigen from type b *H. influenzae* to a carrier protein. These Hib conjugated vaccines can be effectively given to infants as young as 2 months of age, thus overcoming the major concern associated with the epidemiology of *H. influenzae* disease, that is, the immunity of infants between the ages of 3 and 15 months who are no longer under the protection of maternal passive immunity.

Currently, four different Hib conjugated vaccines are licensed in the United States (Table 24-2). It is generally recommended to give the first vaccination at 2 months of age, with two additional vaccinations given 2 months apart and a booster vaccination at 12–18 months of age. In 1993 a vaccine was licensed that combines a Hib conjugate vaccine with the DTP vaccine. Because of the high success rate with these

TABLE 24-2

Licensed *Haemophilus influenzae* Type b Conjugate Vaccines for Use Among Children

Trade Name	Protein Carrier
ProHIBiT	Diphtheria toxoid
HibTITER	Mutant *Corynebacterium diphtheriae* toxin protein
PedvaxHIB	*Neisseria meningitidis* outer membrane protein complex
ActHIB	Tetanus toxoid

Hib vaccines, Hib disease in children under 5 years of age is now listed as a vaccine-preventable disease. A goal had been set to eliminate Hib infections in young children in the United States by 1996 (see Clinical Note).

✳ Other Disease-Causing *Haemophilus* Species

H. ducreyi is commonly isolated from patients suffering from a sexually transmitted disease known as *chancroid*. This infection is characterized by the development of a small papule or pustule at the point of infection, usually the genitalia. The pustule ruptures and forms an ulcer similar to the chancre observed with syphilis. The ulcer (commonly there are multiple ulcers) is painful and often accompanied by tender, swollen, and suppurative regional lymph nodes. Genital ulcers are most frequently seen in male patients. There is often a high correlation between this disease and contact with female prostitutes. Specific diagnosis is made by isolating *H. ducreyi* from the lesion or lymph nodes. Treatment with sulfonamide or tetracycline is generally effective.

BORDETELLA PERTUSSIS: WHOOPING COUGH

Bordetella pertussis is the causative agent of **pertussis**, which is commonly called *whooping cough*. Historically, pertussis has been one of the prominent childhood diseases and before the advent of an effective vaccine was a frequent cause of death in young children. Vaccination has greatly reduced the number of cases of pertussis in developed countries.

Morphologically, *B. pertussis* is similar to *H. influenzae*. It differs from *Haemophilus*, however, in that it is a strict aerobe, does not require specific blood components for growth, and will grow on various types of culture media. Agar containing blood (see Plate 32), potato starch, charcoal, and the antibiotic cefalexin is the culture medium of choice. This bacterium survives for only a short time when expelled from the body in respiratory secretions.

✳ Pathogenesis and Clinical Disease

B. pertussis is aerosolized from the throat of a person with whooping cough and is transmitted to others by the airborne route. This bacterium selectively attaches to the ciliated epithelial cells of the respiratory tract and growth is limited to the superficial tissues. After an incubation period of 10 days, generalized signs of an upper respiratory infection occur, such as sneezing, runny nose, and coughing (the term *catarrhal* refers to such signs). This first stage, or **prodrome**, lasts a week or two. The second stage progresses into episodes (*paroxysms*) of uncontrollable coughs. Each paroxysm may consist of 5 to 20 rapid coughs, with the patient unable to breathe between coughs. At the end of the paroxysm a forced inspiratory breath causes the "whooping" sound. This coughing and whooping form the basis for the common name for this disease. Such prolonged coughing may lead to anoxia (decreased oxygen in the blood), expelling of mucus, and vomiting. The second stage may continue for 1 to 6 weeks. The third stage may include some coughing during convalescence and may last for several more weeks. Various toxins produced by *B. pertussis* are thought to induce the accumulation of mucoid materials, destruction of tracheal cilia, and extensive coughing. Respiratory distress and secondary bacterial pneumonia also contribute to the seriousness of many cases of whooping cough, particularly in young children.

✳ Transmission and Epidemiology

B. pertussis is found only in humans and is transmitted, in most cases, only by persons with an active infection. Forty-five percent of cases occur in children less than 1 year of age and greater than 80% occur in children under 10 years (Figure 24-3). Up to 90% of the unimmunized household contacts of a clinical case may develop whooping cough. Pertussis is found worldwide and has no seasonal distribution. Widespread immunization in the United States has caused a steady decline in the number of cases. For example, in 1950 there were 120,000 cases and 1100 deaths due to pertussis com-

---

Posttussive   After coughing.
Prodrome   An early phase of infection leading to disease. The prodromal phase of disease is usually not associated with specific diagnostic symptoms, but may include a variety of nonspecific host changes.

CLINICAL NOTE

Transmission of Pertussis from Adult to Infant: Michigan, 1993

During 1993, a total of 6586 pertussis cases was reported in the United States, including 675 (10%) cases among persons aged >19 years. However, the total number of cases probably was substantially higher because only an estimated 10% of all pertussis cases are reported; underreporting is greater among adults, who often have only a mild cough. This report summarizes the investigation of two cases of pertussis in which transmission occurred from an adult resident of Massachusetts who was visiting the residence of an infant in Michigan.

Patient 1. On 11 October 1993, a 4-month-old boy in Ann Arbor, Michigan, developed a mild cough. His symptoms gradually worsened during the following 2 weeks with paroxysms of cough associated with whooping, vomiting, and apnea. A culture of a nasopharyngeal specimen obtained on 25 October yielded *Bordetella pertussis*. On 8 September, the infant had received one dose of diphtheria and tetanus toxoids and pertussis vaccine (DTP). Treatment for the infant and chemoprophylaxis of his household contacts with erythromycin were initiated on 27 October and continued for 2 weeks. Although hospitalization was not required, the infant had a persistent cough for 3 months. The infant had not attended group day care.

Patient 2. During 17–20 September 1993, the infant's 47-year-old aunt from Cambridge, Massachusetts, had visited his home. During the visit, she developed a mild cough. On return to Massachusetts, her cough worsened, and she developed paroxysms of cough with inspiratory whoop and **posttussive** apnea. On 23 September, bronchitis was diagnosed, and she was treated with codeine. On 29 September, therapy with a 7-day course of erythromycin and a steroid inhalant was initiated. Because of the history of close contact with her nephew—in whom a culture-confirmed case of pertussis had been diagnosed—on 17 November, a serum sample was obtained for immunoglobulin G (IgG) antibody to pertussis toxin; the serum IgG concentration was >30 g/mL. Inhalant therapy was discontinued, and a 14-day course of erythromycin was prescribed for the patient and three household contacts.

Editorial Note: The investigation of the two cases described in this report indicates the continuing occurrence of pertussis in adults and that adults can be a source of pertussis infection for susceptible infants. Health-care providers should consider pertussis in the differential diagnoses for acute cough of 7 days' duration in adults, particularly if the cough is paroxysmal and associated with posttussive vomiting and/or whooping. Although the cough associated with pertussis in adults generally is mild, the case in this report indicates that classic symptoms of pertussis (e.g., paroxysms of cough, posttussive whoop, and apnea) can occur in adults (*MMWR* 44:74, 1995).

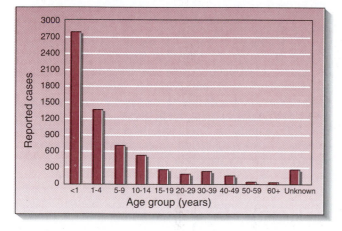

FIGURE 24-3 Reported cases of pertussis in the United States by age group during 1993. (Courtesy Centers for Disease Control, Atlanta)

pared to about 3600 cases and only a few deaths in 1994. Similar decreases are seen in other countries where wide-scale immunization is used. Reported cases in the United States are shown in Figure 24-4 and have been increasing slightly in recent years. However, many cases are atypical, particularly in partially immunized persons, and go unreported; it is estimated that only about 10% of pertussis cases are reported.

✳ Diagnosis

Much of the diagnosis depends on physician observation of the patient. However, fluorescent-tagged antibodies can be used for a rapid and specific identification of *B. pertussis* obtained directly from nasopharyngeal swabs (swabs passed through the

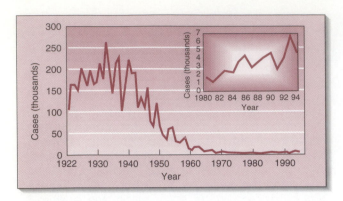

FIGURE 24-4 Reported cases of pertussis (whooping cough) in the United States, 1965–1994. (CDC *Annual Summary*)

nostril into the nasopharynx). Diagnosis should also include attempts to culture the bacterium on agar plates inoculated directly with the nasopharyngeal swab.

✳ Treatment

Although no antibiotic is completely successful against *B. pertussis*, the antimicrobials erythromycin or trimethoprim sulfamethoxazole are the most effective chemotherapeutic agents. The earlier antibiotic therapy is initiated the more effective is the treatment. Even if administered late, however, use of antibiotics helps to reduce the spread of the organism to others who may come in contact with the patient. Removal of respiratory secretions, oxygen therapy to aid breathing, and general supportive measures are of value in treating patients with severe symptoms.

✳ Prevention and Control

A killed whole-cell pertussis vaccine has been widely used for many years and is associated with a decline in the number of cases of whooping cough. The vaccine is usually given in combination with tetanus and diphtheria toxoids as the DPT vaccine. Little passive immunity to this disease is transferred to newborn infants; therefore, vaccination should be initiated as soon as possible. The first immunization of a series of three should be given at about 6 weeks of age and the other two at bi-monthly intervals. Booster immunizations should be given at about 15 to 18 months of age and again just before starting school. When a person under 4 years of age who has

---

CLINICAL NOTE

Health Officials Tout Whooping Cough Vaccines: Bethesda, Maryland, 1995

Three new whooping cough vaccines have been shown to be more effective and to have fewer side effects than current vaccines in large clinical trials conducted in Italy and Sweden. Officials at the National Institutes of Health announced the findings Thursday and told reporters that they will attempt to accelerate approval for use of the new vaccine in the United States.

Whooping cough is a very contagious respiratory disease that infects more than 50 million people worldwide and causes about 350,000 deaths annually. A vaccine against the disease, also called pertussis, is one of the routine childhood inoculations now used in the United States.

In the drug trials in Italy and Sweden, the three new vaccines were found to protect more than 84% of vaccinated children and caused fewer side effects than the vaccine now used in the United States.

The new vaccine uses only parts of the pertussis bacteria to prompt the body to become immune. The vaccine now in use in the United States is a whole-cell vaccine that uses the entire killed pertussis organism.

Dr. Anthony S. Fauci, the director of the National Institute of Allergy and Infectious Diseases at NIH, said his agency is working closely with the Food and Drug Administration and with vaccine manufacturers to ensure that the new pertussis vaccines are available in the United States as soon as possible.

In the meantime, Fauci recommended that parents continue to inoculate children against whooping cough using the currently available vaccine. "While whole-cell pertussis vaccines have saved tens of thousands of lives and have long been the cornerstone of childhood immunization programs in the United States," said Fauci, the new vaccines "promise to become the new gold standard in pertussis immunization."

The clinical trials of the new vaccines began in 1992 and involved more than 25,000 European children (*Deseret News*, 13 July 1995).

been immunized is exposed to someone with whooping cough, a booster injection should be given. When exposed, unimmunized children and all household contacts should be given prophylactic treatments with erythromycin or trimethoprim sulfamethoxazole for a period of at least 14 days.

There have been numerous anecdotal accounts of mild to severe reactions in children who have received pertussis vaccine. However, several recent carefully controlled studies have clearly demonstrated the safety of the DPT vaccine. These studies led the National Immunization Practices Advisory Committee to issue a strong statement in 1991 favoring the use of this vaccine and to continue the recommendation that the first DPT immunizing dose should be given to children between 6 and 8 weeks of age.

Research efforts to provide an alternative to the whole-cell vaccine have led to the development of an acellular vaccine. Clinical trials have shown this acellular vaccine to have fewer side effects and to be more effective than the current vaccine. Efforts are under way to have this new vaccine licensed in the United States (see Clinical Note).

MATERIAL FOR REVIEW

CONCEPT SUMMARY

1. *Haemophilus* is a genus of fastidious gram-negative bacilli that are responsible for a wide variety of clinical diseases. These organisms are particularly serious pathogens of children.
2. Because of some problems associated with use of pertussis vaccine, there has been a worldwide decrease in the use of DPT. Reduction in immunization use has been accompanied by a corresponding increase in the reported number of cases of whooping cough.

CLINICAL SUMMARY TABLE

Microorganism	Virulence Mechanisms	Diseases	Transmission	Treatment	Prevention
Haemophilus influenzae	Capsule Anti IgA enzyme	Meningitis Ear infections Epiglottitis Pneumonia	Airborne Person to person	Ampicillin	Vaccine
Bordetella pertussis	Attachment to cilia Toxins	Whooping cough	Airborne Person to person	Respiratory support Antimicrobials	Vaccine

STUDY QUESTIONS

1. What virulence feature is almost always associated with serious *Haemophilus* infection?
2. Discuss the observation that serious *Haemophilus* infection usually occurs between the ages of 6 months and 6 years.
3. What determines the age at which the Hib vaccine should be administered?
4. What explanation can you provide to account for the observation that antibiotic therapy has only limited effect on the course of pertussis?
5. How do you justify the use of DPT in view of the toxic response that occurs in some persons?

CHALLENGE QUESTIONS

1. Why would poor nutrition, alcoholism, diabetes, or cancer be predisposing factors to *Haemophilus influenzae* pneumonia?

Brucella, Francisella, Pasteurella, and Yersinia

Organisms presented in this chapter are a mixture of families, genera, and species that may be related only in that they produce human disease. The sporadic occurrence of disease due to these organisms should not suggest that they are not serious. Many of the infections discussed have a serious morbidity and significant mortality.

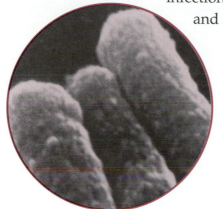

BRUCELLA: BRUCELLOSIS

Four of the six species of *Brucella* (Table 25-1) produce disease in humans. These organisms have their normal habitat in lower animals and human infections are **zoonotic**. Humans become infected by contact with contaminated animal products or ingestion of contaminated unpasteurized milk or milk products. The resulting pathological condition is termed *brucellosis*. Of the three most common *Brucella* species, *B. abortus* normally causes infections of cattle, *B. melitensis* is usually found in sheep or goats, and *B. suis* commonly infects swine. Humans are susceptible to all three species.

Zoonotic	Referring to a disease of animals that is transmitted to humans.

HUMAN BODY SYSTEMS AFFECTED

Brucella, Francisella, Pasteurella, and Yersinia

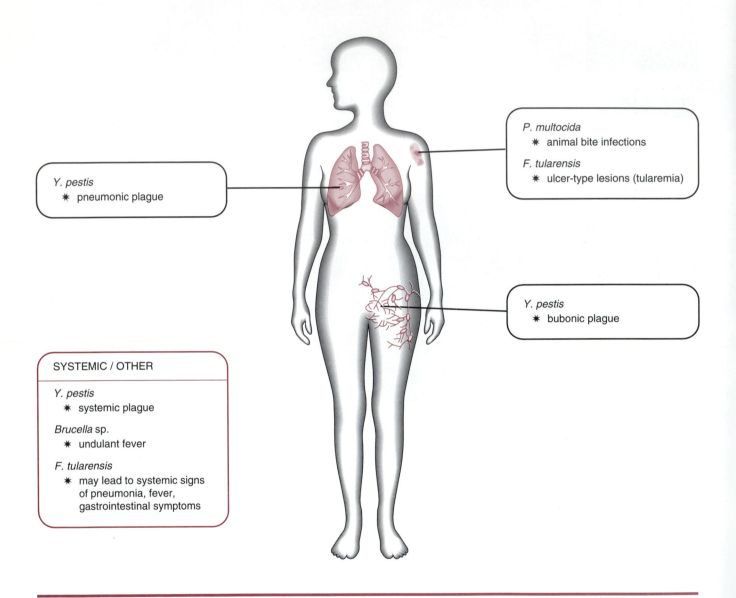

P. multocida
  ✳ animal bite infections

F. tularensis
  ✳ ulcer-type lesions (tularemia)

Y. pestis
  ✳ pneumonic plague

Y. pestis
  ✳ bubonic plague

SYSTEMIC / OTHER

Y. pestis
  ✳ systemic plague

Brucella sp.
  ✳ undulant fever

F. tularensis
  ✳ may lead to systemic signs of pneumonia, fever, gastrointestinal symptoms

Brucellae are small, gram-negative, pleomorphic, coccobacillary-shaped bacteria (Figure 25-1). Because they grow slowly on artificial culture media, incubation for several days to several weeks is sometimes necessary before colonies develop. These microorganisms may survive from several days to several weeks when shed in the body fluids of an animal.

✳ Pathogenesis and Clinical Disease

The brucellae organisms enter the body via lesions or cuts, ingestion, or inhalation. The bacteria are readily phagocytized by white blood cells; however, they are able to survive inside both polymorphonuclear cells (PMNs) and macrophages and much of

TABLE 25-1

Common Reservoir Host and Human Infections of *Brucella* Species

Species	Reservoir Host	Human Disease
B. abortus	Cattle	Brucellosis (undulant fever)
B. canis	Dogs	Brucellosis
B. melitensis	Goats	Brucellosis (Malta fever)
B. neotomae	Rodents	None
B. ovis	Sheep	None
B. suis	Swine	Brucellosis (Bang's disease)

the pathogenesis of brucellosis is associated with this intracellular survival.

Bacteria are carried with the phagocytic cells through the lymphatic system to the blood and into such organs of the mononuclear phagocyte system ∞ (Chapter 10, p. 146) as the liver and spleen, which may, in turn, become enlarged. Circulating antibodies are produced but are unable to neutralize the bacteria sequestered inside the white blood cells.

The onset of clinical symptoms of brucellosis is usually gradual and often occurs weeks or months after exposure. Clinical symptoms are quite generalized and include fever, weakness, malaise, body ache, headache, and sweating. The fever may occur in cycles, with febrile periods alternating with afebrile periods. The fever pattern has prompted the use of the name **undulant fever** for this disease. When cell-mediated immunity develops, the body is better able to contain or eliminate these infections. Recovery from the disease is gradual and partial disability is often prolonged. Some infections are not apparent or go unrecognized. Brucellosis induces abortions in animals, an important factor in the economics of the cattle and swine industries. This is not, however, an important feature of the human disease.

✳ Transmission and Epidemiology

Large numbers of brucellae organisms are shed in urine, placental fluids, milk, and other secretions of infected animals. Transmission among animals occurs by direct contact with contaminated materials. Similarly, transmission to humans is by contact with contaminated animal products. Thus, infections are most often seen in persons who work with

animals or in meat-processing plants. Between 1980 and 1987, 1187 cases of brucellosis were reported in the United States. More than half were in persons working in meat-processing jobs. But the number of cases of brucellosis has decreased significantly in the past three decades (Figure 25-2) due to intensive control programs with domestic animals. Currently, many states are certified as being free of brucellosis. Still, a slight increase in the number of cases has occurred in the past few years among persons working with cattle. Drinking of unpasteurized milk is a major transmission route of brucellosis in some parts of the world.

✳ Diagnosis

Various clinical findings in persons most likely to be exposed are suggestive of the disease in humans.

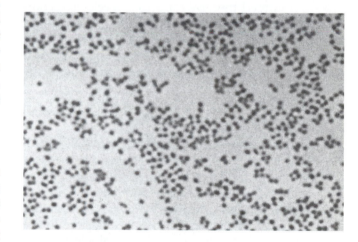

FIGURE 25-1   Photomicrograph of *Brucella suis*.

Undulant fever   One of several common names applied to the disease *brucellosis*. *Bang's disease* is also commonly used to refer to infection by the *Brucella*.

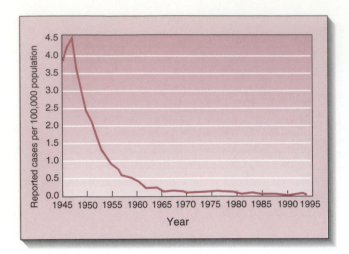

FIGURE 25-2 The reported cases of brucellosis in humans in the United States, 1945–1994. (Modified from CDC annual summaries)

Diagnosis is confirmed by cultivation of *Brucella* species from the blood or tissues of the patient. However, culture is difficult and is positive in only about 20% of cases. Most clinicians rely on the development of humoral antibody to confirm their diagnosis. The serologic test is often difficult to interpret because many veterinarians and meat industry workers may have high antibody titers because of repeated exposure to the organisms.

✳ Treatment

Tetracyclines are the most effective antibiotics and generally cure the clinical disease within a few days. Prolonged therapy for 3 to 4 weeks is recommended, however, to ensure killing of the bacteria sequestered in the white blood cells and to prevent relapses. Streptomycin is sometimes used in conjunction with tetracyclines to help prevent recurrences of this disease.

✳ Prevention and Control

Pasteurization of milk ∞ (Chapter 8, p. 112) is effective in preventing transmission of brucellosis. Most control efforts are directed against the animal reservoir and extensive serologic and skin testing of domestic animals is required by law in most states. Infected animals are destroyed. An effective living vaccine is available for animals. Serologic tests or

certification of vaccination is required to transport cattle into brucellosis-free areas or across state lines. These measures have resulted in a general decline in the number of brucellosis infections in both humans and animals.

FRANCISELLA TULARENSIS: TULAREMIA

Francisella tularensis is the cause of the disease **tularemia**. Tularemia, like brucellosis, is a zoonotic disease. However, in this instance, the animal reservoir is wild animals, and transmission is extended beyond direct contact and ingestion of contaminated foods to also include vector transmission. *Francisella tularensis* is a small (0.5 × 0.3 μm), gram-negative coccobacillus. Cysteine, one of the 20 amino acids ∞ (Chapter 5, p. 58), and blood or serum must be present in culture media before this bacterium will grow and even then growth is slow.

✳ Pathogenesis and Clinical Disease

Tularemia is a widespread disease in many species of animal and insects. Humans become infected

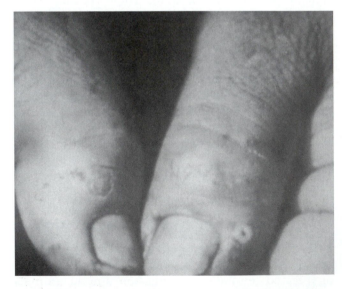

FIGURE 25-3 Lesions at the site of exposure to *Francisella tularensis*. (Centers for Disease Control, Atlanta)

Tularemia A zoonotic disease of rabbits caused by *Francisella tularensis*.

from bites by an infected insect, by handling blood or tissues from infected animals, or by drinking contaminated water. An **ulcer**-type lesion usually develops at the site of inoculation or bite (Figure 25-3) and regional lymph nodes become swollen and painful. Systemic signs, such as dizziness, headache, chills and fever, sweating, and prostration, develop. Gastrointestinal symptoms may result from ingesting contaminated water or meats. *Pneumonia* may also be part of the clinical picture, either as a result of airborne exposure or as an extension of the systemic disease. The lesion heals in 4 to 7 weeks without treatment and complete recovery may require 3 to 6 months. Some relapses may occur, for *F. tularensis* is able to remain sequestered inside certain host cells and not be removed by normal host defense mechanisms. Unless treated, a death rate of 10% is possible, but death rates are less than 1% with antibiotic therapy.

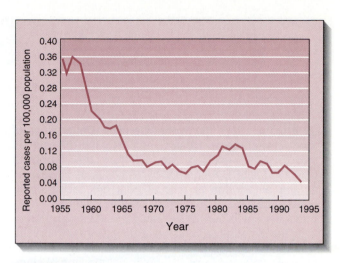

FIGURE 25-4 The reported cases of tularemia in the United States, 1955–1994. (Courtesy Centers for Disease Control, Atlanta)

✳ Transmission and Epidemiology

Most cases occur in rural areas and among persons who come in contact with wild animals or who are routinely exposed to ticks, deerflies, or other biting arthropods. Rabbits and rodents are the most common animal sources of human infection and transmission occurs during handling or dressing of an infected animal. Tularemia is sometimes referred to as "rabbit fever" or "deerfly fever." Open streams of water may become contaminated from infected beavers or muskrats. Infections can be maintained in ticks by the passage of *F. tularensis* from the female to her offspring (transovarian passage). No human-to-human transmission happens under natural conditions and the occurrence of diseases in humans is sporadic. From 100 to 200 cases are reported each year in the United States (Figure 25-4).

✳ Diagnosis

Growth of *F. tularensis* occurs slowly, even on specialized culture media, and isolation of the microbe from clinical specimens is difficult. Diagnosis is usually made on the basis of clinical symptoms and a history of possible exposure. Serologic agglutination tests are available; a sharp rise in antibodies against *F. tularensis* is needed to confirm the diagnosis.

✳ Treatment

Streptomycin, the treatment of choice, is highly effective against *F. tularensis*. It is more effective than some other broad-spectrum antibiotics in preventing recurrent diseases.

✳ Prevention and Control

The best methods of preventing tularemia are to avoid possible infected animals, especially sick rabbits, and to take precautions to reduce the chance of being bitten by arthropods. When working with *F. tularensis*, care should be taken to prevent laboratory-acquired infections. Because of the widespread nature of this disease in wild animals and arthropods, it cannot be controlled by the eradication of these natural hosts. A live attenuated vaccine is available to persons who have a high risk of exposure, including laboratory personnel working with this microorganism, sheepherders, and those who work with wild animals or process hides or other products from these animals.

PASTEURELLA

There are presently four species in the genus *Pasteurella* and, as with brucellosis and tularemia, their related diseases are zoonotic. The bacteria are

---

Ulcer An area of tissue erosion. Skin ulcers are typical of a variety of diseases such as syphilis and tularemia.

CLINICAL NOTE

Brucellosis Outbreak at a Pork Processing Plant: North Carolina, 1992

During 1992, the North Carolina Department of Environment, Health, and Natural Resources (NCDEHNR) received reports from the Sampson County Health Department of 18 cases of brucellosis among employees at a local pork processing plant; onsets of illness occurred from November 1991 through September 1992. Clinical features and serologic testing of all patients were consistent with brucellosis, and *Brucella suis* was isolated from blood samples obtained from 11 persons at the time of acute illness. Two patients were hospitalized. All of the affected employees had documented exposure to the kill floor of the plant. In March 1993, plant employees requested that CDC's National Institute for Occupational Safety and Health (NIOSH) evaluate occupational transmission of brucellosis at the facility.

The NIOSH investigation was conducted during May–June 1993 and included a questionnaire survey, serologic testing, and an industrial hygiene survey. Serologic status was determined using the standard tube agglutination (STA) test. The 2-mercaptoethanol (2-ME) test was also used to assist in differentiating recent or persistent infection from past infection with low-titered antibody. A case of brucellosis was defined as an STA titer 160:1 and either (1) two or more symptoms (fever, chills, headache, myalgia/arthralgia, fatigue, anorexia, sweats, weight loss, and weakness) during the preceding 12 months or (2) a positive 2-ME test (2-ME titer 20:1).

Of the 156 workers in the kill division, 154 (99%) participated in the survey; of these, 30 (19%) met the case definition for brucellosis, including 16 (53%) with previously unrecognized cases. Twelve of these 16 had been symptomatic. Within the kill division, risk for brucellosis was highest among workers in the head (33%) and red offal (25%) departments. Twenty-nine of the 30 employees with cases reported a history of having been cut or scratched at work, compared with 102 of 124 employees without cases.

NIOSH investigators distributed educational material concerning swine brucellosis to all kill floor employees, notified participants of their individual results by mail, and met with individual employees to supplement the mail notifications. Information about swine brucellosis was provided to local physicians. NIOSH staff recommended that the plant process only brucellosis-free swine. In addition, NIOSH staff provided recommendations to management and employees concerning personal protective equipment usage (i.e., rubber gloves and face shields), the need to maintain the kill floor at negative pressure with respect to the contiguous building, and the importance of ongoing education.

The plant processes approximately 8000 swine per day, and the animals originate in at least 10 states. NIOSH and NCDEHNR are working with the U.S. Department of Agriculture (USDA) to determine the possible source of infected swine processed at the plant (*MMWR* 43:113, 1994).

small, gram-negative, coccobacillary forms that have bipolar staining. They all grow on normal laboratory media at 37°C.

Pasteurella multocida is the most frequently recovered species of the genus, and gets the name multocida as a consequence of both its broad host range (many animals can be infected) and the serious consequence of infection. In poultry *P. multocida* causes a serious disease known as *fowl cholera*, and in other domestic animals it is responsible for a frequently fatal hemorrhagic septicemia. The organism can be found in the upper respiratory tract of normal dogs and cats. Human infection often follows a bite or scratch by one of these animals. The onset of disease is rapid with marked swelling and pain at the site of the injury. Fever usually occurs and is accompanied by swollen regional lymph nodes. If treated, the infection can be fairly easily terminated, or it may go on to produce septicemia with systemic and life-threatening consequences. Infections are treated with penicillin, tetracyclines, or a cephalosporin.

YERSINIA PESTIS: PLAGUE

The bacterial species, *Yersinia pestis*, is the cause of *epidemic plague*, which was also known as the "black death" during the Middle Ages. Due to improved

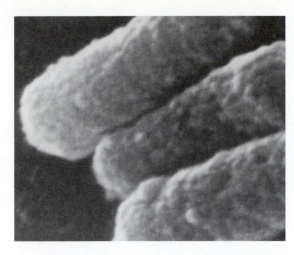

FIGURE 25-5 Scanning electron micrograph of *Y. pestis* magnified 40,000×. (T.H. Chen and S.S. Elberg, *Infection and Immunity* 15:972-977, Figure 4, with permission from ASM)

living conditions, and perhaps changes in other factors influencing host–parasite relationships, plague is no longer the devastating disease it once was. Sporadic cases still occur in the western United States and it remains endemic in Asia.

Y. pestis (Figure 25-5), a gram-negative coccobacillus, shows *bipolar* staining that produces a "safety pin" appearance when viewed with an optical microscope. It can easily be grown on common laboratory media, and grows best at 30°C. It is nonmotile.

✳ Pathogenesis and Clinical Disease

Y. pestis is able to multiply in a variety of different mammalian and insect hosts and occurs in three transmission cycles. The first, or natural cycle, is found in a wild rodent reservoir and is called *sylvatic plague*. Sylvatic plague is transmitted among rodents by fleas. The most frequent carriers of plague are squirrels, mice, prairie dogs, and chipmunks. Some animal species die of the disease, whereas others experience subclinical infections.

The second cycle occurs when the plague is spread to urban rodents, mainly rats, that live in close proximity to humans. This cycle is called *urban* or *domestic plague*. Urban rats frequently die of plague, which causes their fleas to seek new hosts

and results in an increased chance of spread to other animals as well as humans.

The third cycle, called *human plague*, starts when humans are bitten by an infected flea from either the sylvatic or urban cycles. Generally, the flea bite occurs on the legs and the bacteria spread to regional lymph nodes in the groin, where extensive multiplication and swelling occur. The swollen lymph nodes are called *buboes*, especially in older medical writings, and this form of the disease is called *bubonic plague* (Figure 25-6). The buboes usually appear less than a week after the flea bite. Fever, chills, nausea, malaise, and pain may precede and accompany the buboes. The spread of the bacteria is not stopped by the lymph nodes and so bacteremia results. The presence of the *Y. pestis* in the blood is called *septicemic plague*. Massive involvement of blood vessels occurs, resulting in purpuric (purple) lesions in the skin. This manifestation was responsible for the "black death" title earlier applied to the disease. Bacterial **emboli** may become trapped in the lungs (see Plate 27), where the lesions erode into

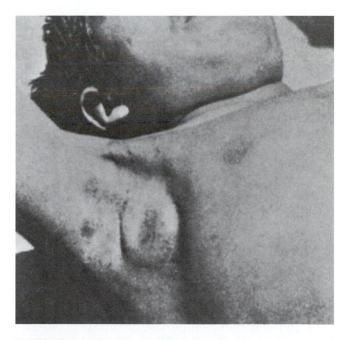

FIGURE 25-6 Swollen axillary lymph nodes (bubo) in a patient with plague. (Armed Forces Institute of Pathology, AFIP MIS #219900(7-B))

Bipolar Referring to a dye that stains the ends (poles) of cells more than the center, giving a "safety pin" appearance.
Embolus Bacteria or other foreign bodies that block a vessel.

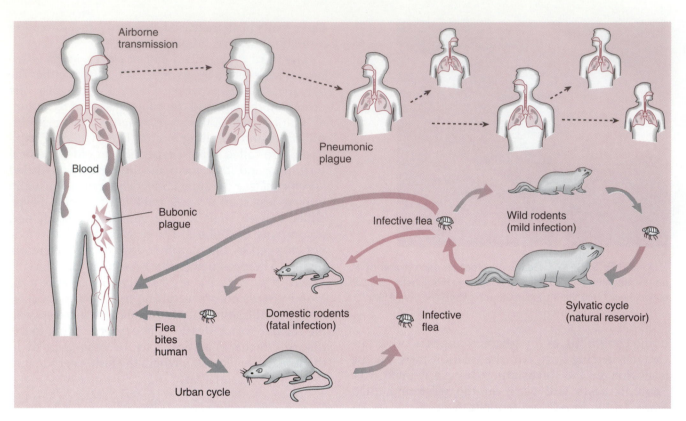

FIGURE 25-7   The epidemiology and pathogenesis of plague.

the air sacs and cause *pneumonic plague*. Pneumonic plague gives an added dimension to this disease, for the bacteria can be readily transmitted from the patient to other persons by the airborne route. Those contracting pneumonic plague in this way rapidly develop severe signs of the disease and die within 2 or 3 days. The epidemiology and pathogenesis of plague are shown in Figure 25-7. The death rate from untreated bubonic and septicemic plague is 50 to 75%, whereas that from pneumonic plague is close to 100%. Some persons do develop mild nonfatal cases of plague.

✳ Transmission and Epidemiology

Plague has occurred in pandemics throughout history and was one of the most devastating diseases of mankind. The first well-documented pandemic occurred in A.D. 550 and resulted in an estimated 100 million deaths over a 60-year period. The next major pandemic took place in the fourteenth century when it is estimated that 25% of the population of Europe died. Smaller epidemics continued until about 1800; then a general decline set in. The last major epidemic happened in China at the end of the 1800s. During the nineteenth century, plague was carried to most parts of the world, including the West Coast of the United States, by rat-infested ships. Today, most cases of plague are reported in southeast Asia. Virtually all plague seen today results from flea bites and not from airborne transmission from person to person. In the fall of 1994 an outbreak of plague in western India included both bubonic and pneumonic cases. More than 5000 persons were affected in the outbreak that was sustained by pneumonic transmission and spread by refugees fleeing from the epidemic center.

Sylvatic plague still exists in the western United States in over 50 species of rodents and their fleas. Most cases seen in the United States occur among persons who live in rural areas or who camp in the West. Although the number of human cases is small, the area of sylvatic plague has been expanding and now includes all of the western United States (Figure 25-8). The numbers of cases of bubonic plague diagnosed each year in humans in the United States are shown in Figure 25-9. Urban plague from domestic rats in seaport cities remains a possible threat, but no such outbreaks have occurred for many years.

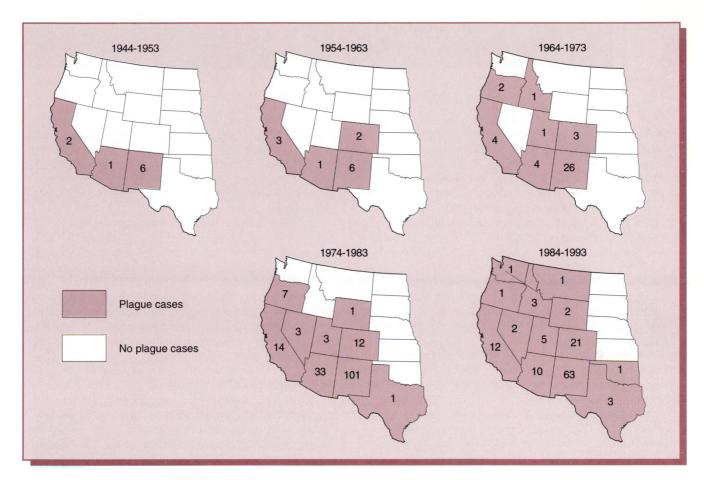

FIGURE 25-8 Number of human plague cases reported, by state and decade, in the United States, 1944–1993. (*MMWR* 43:243, 1994)

✳ Diagnosis

Preliminary laboratory diagnosis of plague may be made by direct microscopic examination (see Plate 26) of smears of fluids from lymph nodes or lesions. The appearance of gram-negative, bipolar-staining coccobacilli is suggestive of *Y. pestis*. A confirmed diagnosis can be made by culturing the bacteria and by serologic tests. Extreme care is needed to prevent laboratory-acquired infections when working with *Y. pestis*.

✳ Treatment

Y. pestis is highly susceptible to streptomycin, tetracyclines, and chloramphenicol. Early treatment is extremely important, especially for pneumonic plague victims because treatment after the first day may not be successful. Proper early treatment of bubonic plague reduces the mortality to less than 10%.

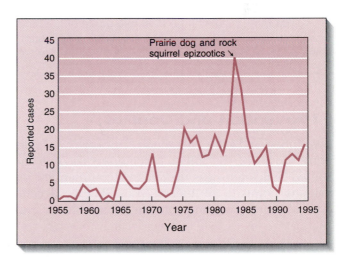

FIGURE 25-9 Reported cases of human plague in the United States, 1955–1994.

Human Plague: United States, 1993–1994

From 1944 through 1993, 362 cases of human plague were reported in the United States; approximately 90% of these occurred in four western states with endemic disease (Arizona, California, Colorado, and New Mexico). During each successive decade of this period, the number of states reporting cases increased from three during 1944–1953 to 13 during 1984–1993, indicating the spread of human plague infection eastward to areas where cases previously had not been reported. In 1993, health departments in four states reported 10 confirmed cases of human plague to CDC; one case has been confirmed during 1994. This report summarizes information about the 11 cases of human plague reported during 1993–1994 and describes epidemiologic and epizootic trends of plague in the United States.

In 1993, the 10 confirmed cases of human plague were reported from New Mexico (six cases), Colorado (two), Texas (one), and Utah (one). Persons with plague infection were aged 22–96 years (median: 55.5 years); five were aged 67 years. Six cases occurred among men. Five cases occurred during June–August, three during March–May, and two during September–November. Seven persons were exposed at their homesites, and one (a veterinarian) was exposed at work; exposure sites could not be determined for two cases. Seven cases were bubonic plague; two, primary septicemic; and one, primary pneumonic. Nine of the 10 patients recovered with antibiotic therapy; one patient died.

For three patients, the probable mode of transmission was flea bite (based on the presence of an inguinal bubo or a recollection of flea bites). Two patients (including the veterinarian) were infected by domestic cats with visible signs of plague infection (i.e., oral lesions and a swollen tongue). For five cases (including the fatal case), the probable mode of transmission could not be determined; however, evidence of plague infection in local animal populations was detected in association with three of these cases.

In 1994, plague infection has been confirmed in a 56-year-old resident of Inyo County, California, who had onset of illness on 1 January (the first report in California of a human plague case during winter since 1928). The patient lived in a county where plague was known to be endemic. In addition, he had recently worked in a subterranean gold mine and slept in a cabin at the minesite; signs of rodent activity were found in the mine shaft and the cabin outbuildings.

Editorial Note: The findings in this report emphasize the increasing importance of two related trends in the epidemiology of human plague in the United States: (1) increased peridomestic transmission and (2) the role of domestic cats as sources of human infection. Peridomestic transmission is especially important in the most highly plague-endemic states of Arizona, Colorado, and New Mexico, where rapid suburbanization has resulted in increasing numbers of persons living in or near active plague foci. Domestic cats that are permitted to roam freely in areas where plague occurs in rodents are at increased risk for infection and, therefore, increase the risk for peridomestic transmission to humans. Before 1977, domestic cats were not reported as sources of human plague infection; however, since 1977, cats have been identified as the source of infection for 15 human plague cases (*MMWR* 43:242, 1994).

✳ Prevention and Control

Both living and killed plague vaccines are available. Three doses of the killed vaccine are used in the United States and this dosage is recommended for persons going to southeast Asia. Plague vaccinations were effective in protecting U.S. military personnel during the Vietnam War. In general, plague can be controlled by improved living conditions that reduce close contact between humans and rodents. Efforts to prevent the importation of rats by ship or airplane have been quite effective and are a necessary phase of plague control.

MATERIAL FOR REVIEW

CONCEPT SUMMARY

1. Zoonotic disease agents are able to produce infection in both animals and humans. They are characterized by a wide range of bacteria of which *Brucella*, *Francisella*, and *Pasteurella* are typical.
2. A classic zoonotic disease, brucellosis is usually transmitted to humans through contact with contaminated animals or animal products. Individuals in the animal industry, particularly males, have an increased risk of getting the disease. Meat, milk, and meat products have been implicated in the spread of infection. Several species exist. In the United States the most common cause of human disease is due to *Brucella abortus*.

3. Tularemia, resulting from infection by *F. tularensis*, is a not uncommon infection of humans, particularly in some areas of the world. Because of its normal habitat—rodents—it is most often a disease found or contracted in a rural setting. Antibiotics have been successful in treating the infection.
4. The plague bacillus, *Yersinia pestis*, is transmitted by the bite of infected fleas. Since *Y. pestis* is able to multiply in a variety of different mammalian and insect hosts, plague exists in three cycles: sylvatic (wild rodent → flea → wild rodent); urban (urban rodent → flea → urban rodent); and human (flea → human → human).

5. An extremely dangerous form of the plague (pneumonic plague), which is spread by the airborne route, has been very rare in recent times. The plague is endemic in the United States. Sylvatic cases that occur each year are usually associated with outdoor activities in endemic (western U.S.) areas. Antibiotic therapy is effective in treatment and preventing transmission of the disease among human hosts.

CLINICAL SUMMARY TABLE

Microorganism	Virulence Mechanisms	Diseases	Transmission	Treatment	Prevention
Brucella species	Survive in phagocytes	Brucellosis (undulant fever)	Animal to human	Tetracyclines	Animal vaccination Pasturization of milk Destroy infected animals
Francisella tularensis	Not known	Tularemia	Animal to human Insectborne	Streptomycin	Vaccine Avoid infected animals
Yersinia pestis	Rapid growth Systemic toxicity	Plague	Rodent to human Fleaborne Human to human	Various antibiotics	Vaccine Avoid wild rodents

STUDY QUESTIONS

1. What is the method by which *Brucella* infects humans?
2. What aspect of *Brucella* infection promotes chronic disease?
3. What features characterize the epidemiology of zoonotic diseases?

4. What is the source of most human infections due to *P. multocida*?
5. Describe the normal route of infection for *Yersinia pestis*.
6. Compare and contrast the following cycles of "the plague":

(a) sylvatic plague, (b) domestic (urban) plague, and (c) human plague.

CHALLENGE QUESTIONS

1. Explain why patients with pneumonic plague should be handled with the utmost care. What steps would you recommend be taken, and why do you recommend them?

2. Explain why plague acquired by inhaling airborne droplets from an infected person is more deadly than plague acquired from the bite of a flea.

CHAPTER

26 Mycoplasma and Legionella

OUTLINE

There are few similarities between the genera *Mycoplasma* and *Legionella*. They both contain species that cause human pneumonia and are commonly treated with the same antibiotic, but they are vastly different in their epidemiologic, morphologic, and clinical properties. Mycoplasma pneumonia is common in young adults. This primary pneumonia, while serious, is rarely life-threatening, and the organism must be transmitted from person to person. *Legionella* is a much more rare cause of pneumonia that occurs most often in compromised elderly patients, is frequently life-threatening, and is not communicated among patients.

Scientists seeking to resolve the role of these two organisms in causing diseases have been puzzled by both. For many years mycoplasmas were thought to be viruses, while the etiologic role of *Legionella* defied characterization because these organisms could not be grown on the usual bacteriologic media or stained with the usual aniline dyes. In some ways, both groups of organisms were an enigma to researchers seeking to determine their roles in human health. They have been included in a common chapter primarily for convenience.

MYCOPLASMAS

The mycoplasmas are a group of both free-living and parasitic microorganisms that are unique in that they possess no cell wall. They are widespread and are found as natural flora in the mouth, throat, and genitourinary tract of mammals and birds. There are three genera of mycoplasmas: *Acholeplasma*, which includes the free-living forms, *Urea-plasma* with but two species, and *Mycoplasma* with 69 species. The *Ureaplasma* and *Mycoplasma* contain only parasitic agents, but of these, *M. pneumoniae*, *M. hominis*, and *U. urealyticum* are the only recognized human pathogens.

Mycoplasma cells, except for the lack of a cell wall, have the same intracellular components and metabolic activities as other bacteria. The cell membrane differs from that of other bacteria and the cell

HUMAN BODY SYSTEMS AFFECTED

Mycoplasma and Legionella

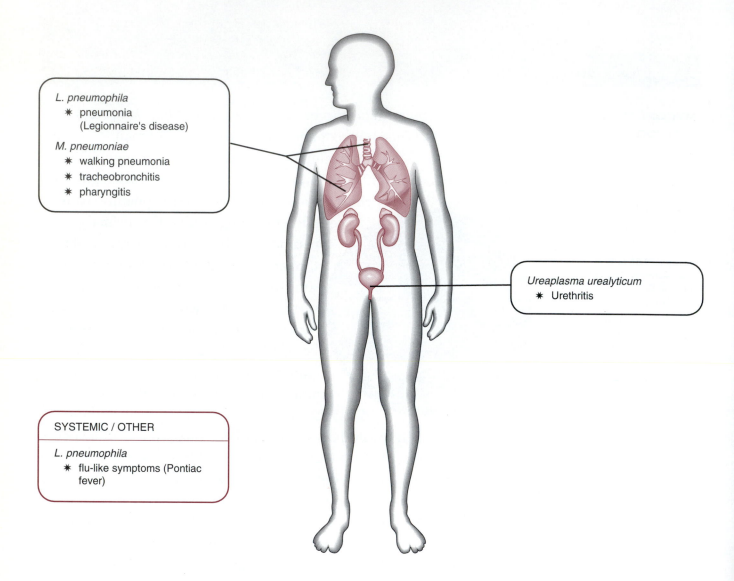

L. pneumophila
* pneumonia
  (Legionnaire's disease)

M. pneumoniae
* walking pneumonia
* tracheobronchitis
* pharyngitis

Ureaplasma urealyticum
* Urethritis

SYSTEMIC / OTHER

L. pneumophila
* flu-like symptoms (Pontiac fever)

shapes are variable (*pleomorphic*) due to the flexibility of this membrane (Figure 26-1). The mycoplasmas and ureaplasmas are the smallest free-living bacteria, with the spherical-shaped cells being about 300 nm in diameter. These pleomorphic cells may form filament- or branch-shaped cells as well. Multiplication is by binary fission, which sometimes takes the form of fragmentation into groups of *daughter cells*. Mycoplasmas can be grown in enriched agar media and produce small colonies

Pleomorphic   Having many morphological shapes. Pleomorphic bacilli often appear as long, drawn-out cells or as short coccobacilli. These figures are often found in the same microscopic field.
Daughter cell   A cell that is produced by binary fission or budding directly from another cell.

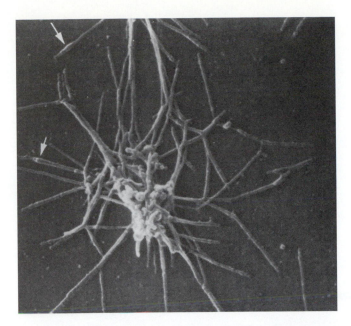

FIGURE 26-1 Scanning electron micrograph of a small colony of *Mycoplasma pneumoniae* showing a filamentous shape around the periphery and pleomorphic shapes in the center. Bulbous swellings (arrows) are seen in some individual cells. (K.E. Muse, D.A. Powell, and A.M. Collier, *Infection and Immunity* 13:229–237, Figure 1b with permission from ASM)

about 0.5 mm in diameter that have the appearance of a fried egg (Figure 26-2). These bacteria are facultative anaerobes or aerobes.

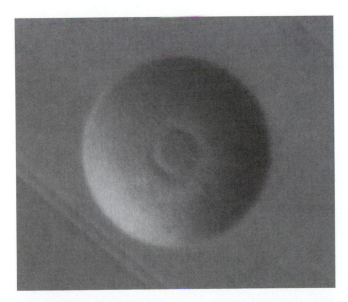

FIGURE 26-2 Figure of *Mycoplasma hominis* colony showing the "fried egg" colony appearance. This unusual colony form is produced by the colony growing down into the medium as well as across the surface of the agar.

✳ Pathogenesis and Clinical Diseases

Various diseases caused by the mycoplasmas have been diagnosed in mammals and birds. The first mycoplasma to be discovered was isolated from cattle suffering from a disease called *pleuropneumonia*; consequently, other mycoplasmas have often been referred to as pleuropneumonia-like organisms or PPLOs. The best-known disease of humans caused by mycoplasmas is called *primary atypical pneumonia*; the causative microbe is *Mycoplasma pneumoniae*. This form of pneumonia, which is characterized by cough, chest pain, fever, pharyngitis, and bronchitis, is called "atypical" because it differs from the "typical" pneumococcal-type bacterial pneumonia. It is usually less severe than typical bacterial pneumonia and is also called *walking pneumonia*, because the patient rarely requires hospitalization. *M. pneumoniae* is also responsible for tracheobronchitis and pharyngitis. The symptoms of pharyngitis are similar to those produced by group A streptococci or by respiratory viral infections. Respiratory infections due to *Mycoplasma* are often protracted and associated with long-term illness. Resistance to commonly used (beta-lactam) antibiotics complicates therapy.

Two other mycoplasmas, *M. hominis* and *Ureaplasma urealyticum*, are commonly found as part of the microbial flora of the genitourinary tract of humans. *M. hominis* is a common cause of postpartum septicemia in women. *U. urealyticum* is found more often in sexually active persons and is transmitted primarily by sexual contact. It is a common cause of *urethritis* and in women may cause postpartum fever, spontaneous abortions, or low birth weights in infants. Table 26-1 lists the agents and diseases due to these organisms.

✳ Transmission and Epidemiology

Primary atypical pneumonia is widespread and may account for 20% of pneumonias in urban pop-

TABLE 26-1	
Mycoplasma and _Ureaplasma_ Causing Diseases in Humans	
Species	**Disease**
Mycoplasma pneumoniae	Primary atypical pneumonia
Mycoplasma hominis	Pelvic inflammatory disease: postpartum fever
Ureaplasma urealyticum	Nongonococcal urethritis

ulations. It is chiefly responsible for pneumonia in persons between 5 and 30 years of age. Outbreaks are often associated with schools, families, summer camps, or military barracks and are common during the summer and fall months. The organisms seem to be transmitted efficiently by aerosol from person to person during the acute stages of the disease, which lasts for 1 to 2 weeks.

Women may be asymptomatic carriers of *U. urealyticum* and transmission is primarily by sexual contact. The incidence of infection due to *M. hominis and U. urealyticum* corresponds to sexual activity. As many as 50% of sexually active individuals are colonized by these organisms.

✳ Diagnosis

Pneumonia caused by mycoplasmas can be differentiated from viral or rickettsial pneumonias by clinical and laboratory tests. DNA probes are now available for direct detection of *M. pneumoniae* in clinical specimens. Various serologic tests are available to detect specific antibodies. Mycoplasmas can be cultured by using enriched media.

✳ Treatment

Antibiotics of the tetracycline and erythromycin groups are effective in reducing the severity and duration of mycoplasma infections. Antibiotics such as penicillin and the cephalosporins that affect only

bacterial cell walls are obviously of no value against microoganisms that lack this structure.

✳ Prevention and Control

The best prevention is to avoid contact with persons suspected of having the disease. However, this is difficult because of the prolonged period of infectivity of persons either in early or recovery phase of the disease. Antibiotics can be taken prophylactically to prevent the spread of pneumonia to members of a group or family where a known index case has been diagnosed. To prevent urethritis caused by *U. urealyticum*, sexual partners should be treated simultaneously.

LEGIONELLA: LEGIONNAIRES' DISEASE

The disease now called the *Legionnaires' disease* or legionellosis is a pneumonia-like illness that went unrecognized as a specific disease until the late 1970s. Attention was dramatically focused on it in the summer of 1976 when some 5000 members of the American Legion attended a convention in Philadelphia. Within 2 weeks of the convention's close, many of these Legionnaires complained of chills, fever, and muscle aches. Over the next week, 12 of the Legionnaires died of a pneumonia-like illness. During the ensuing weeks, 170 of those

---

CLINICAL NOTE

Outbreaks of *Mycoplasma pneumoniae* Respiratory Infection: Ohio, 1993

From 15 June through 5 September, acute respiratory illness (ARI) characterized by acute onset of cough and fever occurred among 47 (12%) of 403 staff members and clients of a sheltered workshop for developmentally disabled adults in Ohio. The median age of patients was 35 years (range: 20–60 years); seven (15%) required hospitalization, and 31 (66%) had radiographic evidence of pneumonia.

Thirty-eight persons had laboratory evidence of *Mycoplasma* infection: All had convalescent-phase serum antibody titers for *Mycoplasma* 32 by complement fixation (CF), 22 (58%) had CF titers of 128, and four (11%) had a fourfold rise in CF titers. *M. pneumo-*

niae was isolated from nasopharyngeal secretions of two of eight patients with available specimens. Serologic and microbiologic studies were negative for acute viral and non-*Mycoplasma* bacterial infections.

Although no deaths occurred among persons with laboratory-confirmed cases, one workshop participant who had not been evaluated for *Mycoplasma* infection died on 30 June from complications of pneumonia.

Beginning 6 August, persons with ARI were excluded from work until completion of at least 3 days of antimicrobial therapy. No cases of *M. pneumoniae* have been identified since 5 September (*MMWR* 42:931, 1993).

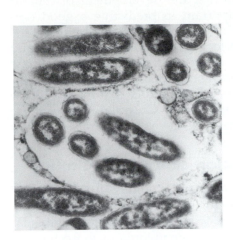
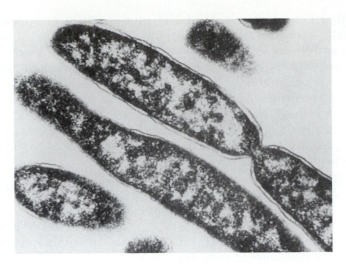

FIGURE 26-3 Electron micrographs of *Legionella pneumophila.* (Courtesy Centers for Disease Control, Atlanta).

attending the convention were hospitalized with pneumonia and 29 deaths occurred.

An extensive investigation was immediately launched to find the cause of this newly recognized disease. Tests were run with all known pathogenic microorganisms or toxins that might be responsible and all tests were negative. After several months of continued research a previously unidentified bacterium (*Legionella pneumophila*) was found to be the culprit.

The causative bacteria are gram-negative bacilli and grow in guinea pigs, in the yolk sac of embryonated eggs, and on special nutrient media containing activated charcoal and extra amounts of iron and the amino acid cysteine. They do not stain well with the Gram stain; therefore special stains, such as the Gimenez silver-impregnation stain, are used to demonstrate the presence of these cells (see Plate 34). When grown in yolk sacs, the cells have both coccoid and bacillary forms. The shape is pleomorphic (Figure 26-3) when grown on culture media but is predominantly bacillary. These bacteria are quite stable and may remain viable for up to one year in water. Evidence now indicates that these bacteria survive in water and soil as intracellular parasites of protozoa.

No close relationship has been found between these bacteria and any of the previously characterized bacteria. Thus a new family, *Legionellaceae,* has been established, which contains but one genus *Legionella* and at present over 30 species (Table 26-2). There are a variety of serogroups associated with some of the species and only about half of the known species have been associated with human

disease. The most frequent cause of legionellosis is the species *Legionella pneumophila.*

✳ Pathogenesis and Clinical Disease

Two forms of disease due to the legionellae occur. A rather mild, common, flu-like disease that occurs throughout the year in epidemic outbreaks is known as *Pontiac fever.* This disease has a short incubation period of one to two days and is self-limiting. The mortality rate for patients with Pontiac fever is less than 1% and the disease occurs in otherwise healthy individuals. Legionnaires' disease is much more severe than Pontiac fever and occurs in sporadic cases and most often in individuals who are compromised by age, respiratory illness, immunosuppression, or other conditions.

The clinical presentation of severe Legionnaires' disease most closely resembles lobar pneumonia and begins 2 to 10 days after exposure. Some early

TABLE 26-2	
***Legionella* Species Associated with Human Disease**	
L. bozemanii	L. maceachernii
L. feeleii	L. micdadei
L. dumoffii	L. pneumophila
L. hackeliae	L. wadsworthii
L. longbeachae	L. jordanis
L. birminghamensis	L. tuconensis
L. oakridgensis	L. cincinnatiensis

symptoms are diarrhea, weakness, headache, muscle aches, malaise, anorexia, and a dry cough. An increasing temperature to 40°C may be seen after several days, along with prostration and pulmonary consolidation. Some patients experience stupor and show signs of involvement of the kidneys and liver. The primary involvement, however, appears to be in the lungs. The duration of the disease is from 5 to 16 days. Most cases are seen in older or immunosuppressed persons, although all ages may be afflicted. Risk factors include alcoholism, advanced age, smoking, and immunosuppressive therapy. Most sporadic cases occur in renal transplant or cancer patients.

Humoral immunity ∞ (Chapter 11, p.157) plays only a limited role in resistance to Legionnaire's disease. Some patients who have expired from this infection have been found to have high antibody titers. It appears that the development of cellular immunity is important in predicting a good outcome following infection. This is consistent with the observation that legionellae are able to survive and multiply inside of phagocytic monocytes. Unless treated, 20% of those patients requiring hospitalization die. Evidence now indicates that many mild or *subclinical* cases may occur. Only about 5% of those exposed to the microbe develop disease.

✳ Transmission and Epidemiology

Person-to-person transmission does not seem to be involved. When clusters of cases occur, it appears that exposure was from a common environmental source. The organisms are ubiquitous in fresh water and are readily isolated from streams, water found in air-conditioning systems, or in domestic water

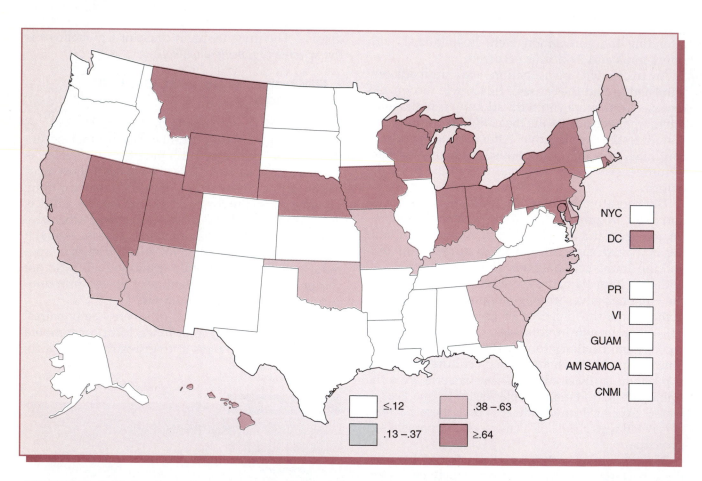

FIGURE 26-4 Reported cases of legionellosis, per 100,000 population, in the United States and territories in 1993. (Courtesy Centers for Disease Control, Atlanta)

Subclinical   Occurrence of infection with such mild symptoms that no diagnostic signs are evident. Antibody responses are usually as strong as with clinical infections.

supplies. Airborne dissemination via dust or water sprayed in air-conditioning systems or from showers has been suggested.

Once the causative bacterium was isolated, thus providing a known antigen, it was possible to study serum samples that had been collected from patients who had suffered from unknown types of pneumonia before 1976. Many of these samples contained specific antibodies against *L. pneumophila*, indicating that this disease has been around for many years. Since the initial recognition of this disease in 1976, numerous outbreaks have been identified in all parts of the United States and in many foreign countries. Many isolated cases occur as well as some clusters of infection. Several hundred severe cases are diagnosed each year in the United States (Figure 26-4). In addition, an estimated 30,000 to 50,000 additional mild or subclinical cases are thought to also occur yearly. It is presently determined that between 3 and 6% of all hospital-associated pneumonias are due to *L. pneumophila*.

✳ Diagnosis

A tentative diagnosis is based on symptoms when no other causative agent can be demonstrated. Cultivation of *L. pneumophila* from the lung tissues or blood confirms the diagnosis. Specific laboratory tests also rely on serologic procedures that show specific increases in, or high levels of, antibodies against the *L. pneumophila*. The presence of *L. pneumophila* in sputum or lung tissues can sometimes be demonstrated by direct microscopic examination using special stains, by specific fluorescent-labeled antibodies, or by the use of DNA probes. Detection of legionella antigen in the urine is also a useful diagnostic test.

✳ Treatment

Erythromycin has been the most effective antibiotic in treating this disease. Often recovery is prompt and dramatic following the use of this agent. Therapy usually lasts for about 2 weeks. Prompt treatment greatly decreases the death rate. Rifampin may also be effective.

✳ Prevention and Control

No vaccines are avialable. Elimination of possible contamination of water used in some air-conditioning systems may reduce the number of cases. Avoiding exposure to high concentrations of dust from such activities as soil excavation could be helpful. Protecting compromised patients from excessive dust or aerosolized water is recommended. As more is learned about the epidemiology of this disease, additional preventative measures can be developed.

C L I N I C A L   N O T E

Update—Outbreak of Legionnaires' Disease Associated with a Cruise Ship, 1994

On 15 July 1994, CDC was notified by the New Jersey State Department of Health of six persons with pneumonia who had recently traveled to Bermuda on the cruise ship *Horizon*. In conjunction with local and state health departments, an investigation was initiated; as of 10 August, a total of 14 passengers had Legionnaires' disease (LD) confirmed by either sputum culture (one patient), detection of antigens of *Legionella pneumophila* serogroup 1 (Lp1) in urine by radioimmunoassay (seven patients), or fourfold rise in titer of antibodies to Lp1 between acute- and convalescent-phase serum specimens (six patients). Possible cases in 28 other passengers with pneumonia that occurred within 2 weeks after sailing aboard the *Horizon* are under investigation. Cases have occurred from nine separate week-long cruises during 30 April–9 July 1994.

To identify the source of *Legionella* sp., a case-control study was conducted, and environmental sampling of the ship's water system was performed. Exposure to the whirlpool baths was strongly associated with illness (odds ratio = 16.4%; 95% confidence interval = 3.7–72.3). Cultures taken from a sand filter used for recirculation of whirlpool water yielded an isolate of Lp1; this isolate and the clinical isolate had matching monoclonal antibody subtyping patterns.

A variety of interventions were completed, including hyperchlorination of the ship's potable water supply, removal of the whirlpool filters, and discontinuation of the whirlpool baths. Following completion of these interventions, on 30 July the *Horizon* resumed its weekly sailing schedule from New York City to Bermuda (*MMWR* 43:574, 1994).

MATERIAL FOR REVIEW

CONCEPT SUMMARY

1. Mycoplasma are unique bacteria without cell walls. They produce a number of diseases in animals and humans. Primary pneumonia is common among children and young adults. Older individuals seem somewhat refractory, probably due to antibodies from earlier infection.

2. Legionnaires' disease is due to a bacillus normally present in water environments. Individuals with pulmonary compromise are usually the victims of this organism. Frequently nosocomial, the infection represents upward of 5 to 6% of all pneumonias in hospitalized individuals. Diagnosis is difficult because the organism is characteristically refractile to the aniline stains used in early diagnosis.

CLINICAL SUMMARY TABLE

Microorganism	Virulence Mechanisms	Diseases	Transmission	Treatment	Prevention
Mycoplasma pneumoniae	Attachment to cilia	Primary atypical pneumonia	Airborne Person to person	Tetracyclines Erythromycin	Avoid infected persons
Legionella pneumophila	Survive in phagocytes	Legionnaires disease	Waterborne	Erythromycin	Protect compromised patients

STUDY QUESTIONS

1. How important is *M. pneumoniae* as a cause of human pneumonia?

2. What feature of *Legionella* would make it difficult to prevent exposure to these bacteria?

3. Why did Legionnaires' disease "suddenly" appear as a cause of human disease in the 1970s?

4. What is the primary clinical symptom associated with Legionnaires' disease?

5. What risk is there to family members and friends of a patient with Legionnaires's disease?

CHALLENGE QUESTIONS

1. *Legionella pneumophila* normally survives in water. However, it can be cultivated only on special nutrient media. Explain these observations.

CHAPTER

OUTLINE

27 Chlamydiae and Rickettsiae

GRAM –

20μm

PROKARYOTE

The two groups of bacteria presented in this chapter are not closely related, but they share a number of characteristics. Both contain zoonotic species, with animals both harboring and transmitting the disease in some instances. Both of these groups of bacteria must live in a host cell, since they lack the metabolic ability to grow outside of host cells. Both are very small and have gram-negative-type cell walls. Diseases due to rickettsiae are generally much more severe than those due to chlamydiae, although the latter are by far the most common and contribute extensively to human misery and suffering.

CHLAMYDIAE: GENERAL CHARACTERISTICS

Chlamydiae are **obligate intracellular parasites** that depend completely on the host cell for energy, because they lack the ability to produce their own ATP molecules ∞ (Chapter 5, p. 53). They go through a unique developmental cycle inside cytoplasmic vacuoles of the host cell (Figure 27-1). The basic structure, called an *elementary body*, is 0.2 to 0.4 μm in diameter. It is infectious, metabolically inert, and enters the host cell by phagocytosis.

Obligate intracellular parasite   Organism that depend on a living cell for their nutrition, growth, and development.

HUMAN BODY SYSTEMS AFFECTED

Chlamydiae and Rickettsiae

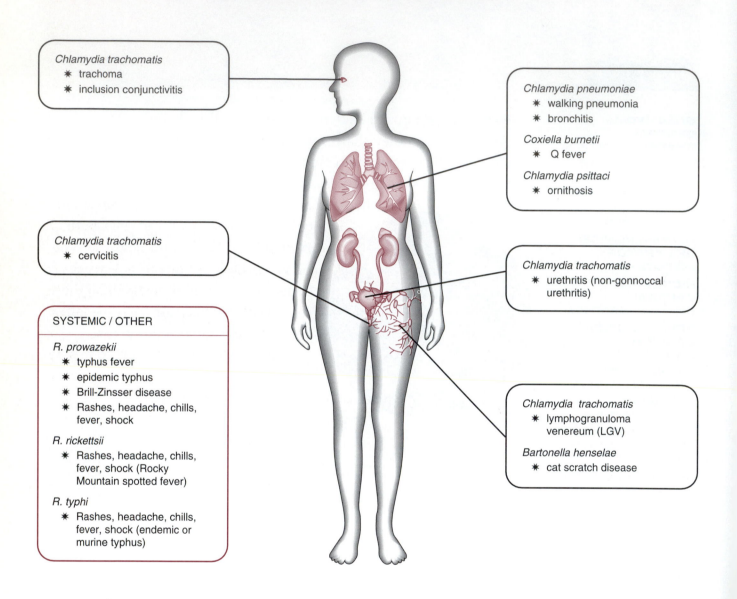

Chlamydia trachomatis
* trachoma
* inclusion conjunctivitis

Chlamydia pneumoniae
* walking pneumonia
* bronchitis

Coxiella burnetii
* Q fever

Chlamydia psittaci
* ornithosis

Chlamydia trachomatis
* cervicitis

Chlamydia trachomatis
* urethritis (non-gonnoccal urethritis)

SYSTEMIC / OTHER

R. prowazekii
* typhus fever
* epidemic typhus
* Brill-Zinsser disease
* Rashes, headache, chills, fever, shock

R. rickettsii
* Rashes, headache, chills, fever, shock (Rocky Mountain spotted fever)

R. typhi
* Rashes, headache, chills, fever, shock (endemic or murine typhus)

Chlamydia trachomatis
* lymphogranuloma venereum (LGV)

Bartonella henselae
* cat scratch disease

When present in phagosomes, chlamydiae prevent the formation of phagolysosomes ∞ (Chapter 10, p. 146). Before chlamydial replication can occur, the elementary body changes into a larger structure, 0.7 to 1.0 μm in diameter, that is called a *reticulate body*. The reticulate body is not infectious, is metabolically active, survives only intracellularly, and is the replicating form of chlamydiae. Reticulate bodies divide by binary fission to fill and enlarge the vacuole. The reticulate bodies next change into the smaller, infectious elementary bodies that are not able to multiply. The growth of chlamydiae may cause the death and the breakup of the host cell, which results in the release of the elementary bodies. The released elementary bodies enter new host cells and the cycle is repeated.

Three species of the genus *Chlamydia*—*C. trachomatis*, *C. psittaci*, and *C. pneumoniae*—cause dis-

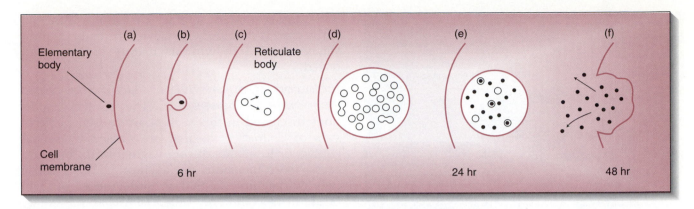

FIGURE 27-1 Stages in the replication of chlamydiae: (a) attachment of elementary body (EB) to cell membrane; (b) phagocytosis of EB; (c) EB changes into a reticulate body (RB) that divides by binary fission; (d) continued replication of RBs inside of vacuole; (e) RBs change into EBs; and (f) lysis of cell membrane and vacuole with release of EBs.

eases in humans. *C. trachomatis* is further divided into 15 serovars designated alphabetically A through L.

C. TRACHOMATIS: TRACHOMA AND OTHER DISEASES

For a single species, *C. trachomatis* causes a remarkably wide variety of clinical diseases. This fact probably stems from slightly different variations in virulence or host tissue affinity for the many different serovars of this microbe. This diversity is also a function of the different routes of transmission and

the degree of natural resistance of the persons being infected. In many cases, the chlamydiae are apparently able to persist in the cells of the host for prolonged periods without causing disease; then, at a later date, clinical disease may occur. Because chlamydiae inhabit the host cells and are therefore not readily exposed to the body's immune system, the immune response is weak and a person can be infected repeatedly.

✳ Disease Categories

Three categories of diseases of humans caused by *C. trachomatis* are listed in Table 27-1: (1) trachoma (caused by serovars A, B, Ba, and C); (2) a complex of diseases transmitted by direct (primarily sexual)

TABLE 27-1

Diseases for Which Chlamydiae Have Been Implicated as Etiologic Agents

Chlamydia	Normal Host	Disease in Humans
C. psittaci	Birds	Psittacosis, ornithosis
C. trachomatis serovars A, B, Ba, C	Humans	Trachoma
Serovars B, D–K	Humans	Nongonococcal urethritis (NGU), inclusion conjunctivitis, infant pneumonia
Serovars L_1–L_3	Humans	Lymphogranuloma venereum
C. pneumoniae	Humans	Pneumonia

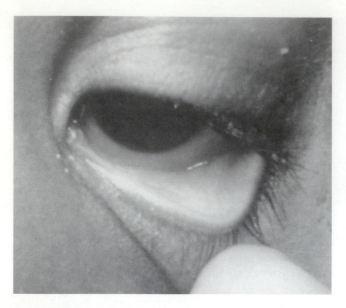

FIGURE 27-2   Trachoma with infection of the eyelid (conjunctivitis). (Centers for Disease Control, Atlanta)

and indirect routes (caused by serovars B and D through K); and (3) lymphogranuloma venereum (caused by serovars L_1, L_2, and L_3).

Trachoma.   *Trachoma* (Figure 27-2) is an infection of the eyelid, more serious than typical *conjunctivitis*. Scarring of the eyelid often results in inturned eyelids. The inturned eyelashes cause abrasion and ultimately ulceration of the cornea. Cases of varying degrees of severity appear and range from asymptomatic infections to those showing extensive scarring of the cornea with resultant blinding. This disease is widespread in areas where poverty, overcrowding, and unsanitary conditions exist. Humans are the only hosts; infection is spread by direct and indirect contact. Trachoma is most prevalent in Asia and Africa, but occurs throughout the world. An estimated 400 million people worldwide suffer from this disease and of them, up to 20 million may become blind following repeated infections over 15 to 20 years. The number of persons with severely impaired vision or blindness in countries where trachoma is prevalent definitely impairs socioeconomic progress. Trachoma ranks among the world's major infectious disease problems. Most trachoma seen in the United States appears in Appalachia or on Indian reservations in the Southwest. This disease is rarely seen in persons living under modern sanitary conditions. If treated early, cure is obtained. However, in many circumstances treated individuals readily become reinfected.

Inclusion Conjunctivitis, Urethritis, and Associated Diseases.   The clinical conditions discussed in this section are caused by serovars of *C. trachomatis* that are passed by sexual contact or to infants during vaginal birth. In many cases, these infections are asymptomatic or cause only mild symptoms that go undiagnosed. A form of urethritis commonly called *nongonococcal* (NGU) or *nonspecific urethritis* (NSU) is now recognized as a common sexually transmitted disease. *C. trachomatis* can be isolated from approximately 60% of males suffering from NGU. This infection seems to be widespread but is often without symptoms. As many as 10% of cervical specimens taken from asymptomatic females who undergo routine medical examinations contain chlamydiae.

The organism is found most often in sexually active female adolescents and can be found in as many as 27% of pregnant adolescents. It is one of the most common sexually transmitted pathogens, and although as many as 50% of infected females are asymptomatic, others commonly develop **cervicitis**, pelvic inflammatory disease, urethritis, or inflammation of the fallopian tubes (salpingitis). Salpingitis is often followed by infertility or a marked increase in tubal (ectopic) pregnancies. Ten percent of women with a single occurrence of salpingitis become infertile and after a third infection, 75% may be infertile. There is some evidence of ascending infection in pregnant women. Studies have shown that chlamydia-infected women deliver a stillborn child or a child that suffers a neonatal death ten times as often as uninfected women.

It is now recognized that *Chlamydia trachomatis* infection of neonates occurs at birth during vaginal delivery through an infected cervix. Infants who are exposed to *C. trachomatis* at birth may manifest their infections as conjunctivitis or pneumonitis. Infants born to untreated, culture-positive mothers have an overall chlamydia infection rate varying from 23 to 70%, including conjunctivitis in 17 to 50% and pneumonia in 11 to 20%. Other infants may exhibit isolated nasopharyngeal carriage or positive serology as their only evidence of infection. Multiple sites have yielded positive culture, including conjunctiva, nasopharynx, rectum, and vagina.

Cervicitis   Inflammation of the cervix.

CLINICAL NOTE

Chlamydia Prevalence and Screening Practices: San Diego County, California, 1993

Chlamydia trachomatis is the most common bacterial sexually transmitted disease in the United States and causes an estimated 4 million infections annually. Approximately 70% of infected women have few or no symptoms, and asymptomatic infection in women can persist for up to 15 months. Infection can progress to involve the upper reproductive tract and may result in serious complications. To identify women who may have chlamydial infections, CDC has recommended routine testing based on age, risk behavior, and clinical findings—especially in clinics and group practices that provide reproductive health care to adolescent and young women. This report describes the prevalence of chlamydial infections among patients visiting the family-planning clinic service of the San Diego County Department of Health Services from July 1989 through June 1993 and summarizes the findings of a survey in May 1993 that assessed chlamydia screening, reporting, and treatment practices for women who attended primary-care community-based clinics and group practices in San Diego County.

Prevalence of Chlamydial Infections   The San Diego County Department of Health Services provides family-planning services in San Diego County in six public health centers. Each clinic follows a written protocol that requires screening of all clients during their initial visit and recommends screening for clients during annual visits—particularly for those whose sexual behavior increases their risk for infection. From March 1989 through February 1991, endocervical specimens were tested at the San Diego County Public Health Laboratory using the Chlamydia Antigen ELISA (Ortho Diagnostic Systems, Inc., Raritan, New Jersey); beginning March 1991, specimens were tested using the MicroTrak EIA (Syva, San Jose, California). The proportion of women screened was determined using data from annual family-planning clinic-service utilization

reports. Test results and demographic and limited clinical information were obtained from the laboratory's chlamydia-test database.

During July 1989–June 1993, approximately 95% of family-planning clients were tested for *Chlamydia* during their initial visit, and 70% were tested during their annual visit. Of 11,044 specimens tested, 91% were obtained during routine testing of clients without symptoms. The prevalence of chlamydial infections decreased from 10.0% during July–December 1989 to 1.9% during January–June 1993, a decline of 81.0%.

During July 1989–June 1993, the prevalence of chlamydia among black women was 8.5%, more than 1.5 times that among Hispanic (5.3%) and white (4.5%) women. During the 4-year period, the prevalence declined minimally among black women and steadily among white and Hispanic women. Prevalence was inversely related to age, with the highest prevalence among women aged <20 years (8.4%); among women aged <20 years, the prevalence decreased from 9.9% during July 1989–December 1990 to 4.8% during January 1992–June 1993, a 51.5% decline.

Editorial Note: The decline in the prevalence of chlamydial infections among women receiving family-planning clinic services in San Diego County during 1989–1993 was consistent with patterns in other areas. For example, findings from a screening demonstration project in Public Health Service Region X (Alaska, Idaho, Oregon, and Washington) indicated that, among the approximately 70,000 women screened annually in public and private family-planning clinics, the prevalence declined from 9.3% in 1988 to 4.2% in 1993. The prevalence also decreased among women attending family-planning clinics in Wisconsin, where a statewide selective screening program has been operated since 1986 (*MMWR* 43:366, 1994).

Sporadic cases may occur in adults whose eyes become infected through contact with contaminated towels, fingers, and similar items. Transmission can also occur in unchlorinated swimming pools. This form of conjunctivitis, known as *inclusion conjunctivitis*, is much less severe than trachoma and is readily treated with erythromycin or tetracycline.

Lymphogranuloma Venereum (LGV).   Lymphogranuloma venereum is transmitted by sexual

contact and is caused by distinct chlamydial serovars (Table 27-1). A variety of nonspecific symptoms may be experienced in the early stages of illness, with lesions on the skin and mucous tissues of the genital organs. This condition is followed by the characteristic signs of enlarged and painful lymph nodes in the inguinal area (buboes). The enlarged lymph nodes may break (suppurate) and drain. Patients often have fever, nausea, headache, and conjunctivitis or skin rash. If untreated, the

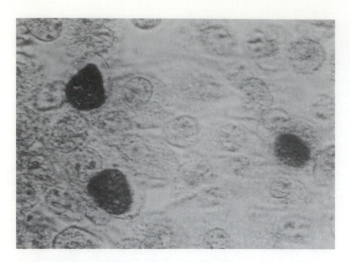

FIGURE 27-3   Chlamydia infected cells in cell culture. Dark inclusions are aggregates of chlamydiae. (Courtesy Centers for Disease Control, Atlanta)

disease can lead to permanent lymphatic or rectal blockage.

✳ Diagnosis

C. trachomatis infections are diagnosed by observing *inclusion bodies* (the vacuoles filled with the chlamydiae) (Figure 27-3) in cells scraped from infected tissues, by culturing the chlamydiae in cell cultures or embryonated eggs, and by showing a rise in specific antibodies. Specific fluorescent antibodies and DNA probes are also available to help detect the presence of chlamydiae in clinical specimens. Recently, highly sensitive polymerase chain reaction amplification probe tests have been developed.

✳ Treatment

Tetracyclines are effective agents in treating these chlamydial infections. Erythromycin is used for pregnant females to avoid teeth staining of the fetus. Ofloxacin is also effective for adults. For sexually transmitted infections, the sex partner should also be treated. Treatment is often extended to include possible coexisting gonococcal infection ∞ (Chapter 16, p. 229).

✳ Prevention and Control

Vaccines have been tried but generally do not induce a high enough level of protection to warrant their use. Proper hygiene and sanitation are the most effective means to preventing eye infections. Some persons may remain chronic carriers and serve as a reservoir of infections, a factor that makes control difficult. The chances of sexually transmitted infections are reduced using the same precautions used for other STDs ∞ (Chapter 16, p. 233; Chapter 17, p. 244).

C. PSITTACI: *PSITTACOSIS*

Psittacosis is a zoonotic disease, and *C. psittaci* is able to infect a wide range of birds and animals. This microbe was first recognized as an infection of psittacine birds, such as parrots and parakeets; hence the name *psittacosis*. Today the disease is also known as **ornithosis** because it infects many species of birds besides those of the *Psittacine* family. Recent studies have also demonstrated the organism in a variety of domestic animals from which human infection is possible.

Many tissues of the bird are infected and a wide variety of clinical signs may be seen. Diarrhea is a common finding in infected birds and chlamydiae are shed in the feces. The organism remains viable in dried avian feces for several months. Many birds have latent infections that may develop into acute diseases when stresses such as crowding or shipping occur. Ornithosis can be a major problem in the shipping and holding of pet birds and in the poultry industry.

Persons who work or live closely with birds stand the greatest risk of being infected. Exposure is usually by the airborne route via contaminated dust. Symptoms in humans are varied and may be subclinical or mild and simply passed off as a common minor respiratory disease. From 100 to 200 severe respiratory infections due to *C. psittaci* are reported each year in the United States (Figure 27-4). Some deaths result, but early treatment with antibiotics usually reduces the mortality rate.

Inclusion bodies   Intracellular inclusion often found in the cell as a consequence of infection. These inclusions are usually masses of developing microorganisms being formed within the parasitized cell.
Ornithosis   A disease associated with birds from the *Ornithine* family.

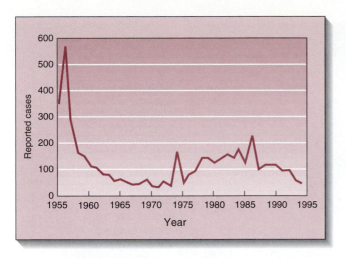

FIGURE 27-4   Reported cases of human psittacosis in the United States, 1955–1994. (Courtesy Centers for Disease Control, Atlanta)

C. PNEUMONIAE: RESPIRATORY INFECTIONS

In recent years, an antigenically distinct chlamydia, called *C. pneumoniae*, has been recognized. *C. pneumoniae* infections appear to be fairly prevalent in many parts of the world. The organism spreads from person to person and causes infections of the respiratory tract. Peak transmission occurs in young adults and usually results in bronchitis, sinusitis, or mild pneumonia. About 10% of *walking pneumonia* and 5% of bronchitis in young adults are caused by this agent. It is estimated that there are as many as 300,000 cases of pneumonia each year in the United States due to this organism. Fatal infections have been seen in infants and elderly where other underlying conditions exist.

RICKETTSIA: GENERAL CHARACTERISTICS

Rickettsiae are a group of small, gram-negative bacteria that traditionally have been considered separately from the typical bacteria. In fact, the rickettsiae have many characteristics, such as methods of laboratory cultivation and modes of transmission, that suggest a close relationship to the viruses. For many years the rickettsiae were considered a separate group of microorganisms positioned between the bacteria and viruses and the study of rickettsial diseases was usually included in the general subject area of virology. It is now well established, however, that rickettsiae are small obligate intracellular parasitic bacteria.

Rickettsiae are about 0.3 μm in diameter and up to 1.0 μm in length (Figure 27-5). Their shape ranges from *pleomorphic* to coccobacillary to bacillary. Their metabolic activities are highly dependent on energy received from living host cells and all but one (*Rochalimea quintana*) are able to multiply only inside the living host cells and are thus obligate intracellular parasites. They are grown in the laboratory in cell cultures, embryonated eggs, or in animals, much like viruses ∞ (Chapter 28, p. 370). Most rickettsiae are readily inactivated once they leave the host. Therefore, their transmission depends on direct contact or vector transmission and all rickettsial diseases except Q-fever require an arthropod vector for successful transmission between hosts.

Most rickettsiae and rickettsial diseases share some common features. These features are described first, followed by a discussion of several of the more important rickettsial diseases of humans (Table 27-2).

✳ Pathogenesis

Except for the agents of trench fever and epidemic typhus, rickettsiae are zoonotic disease agents for

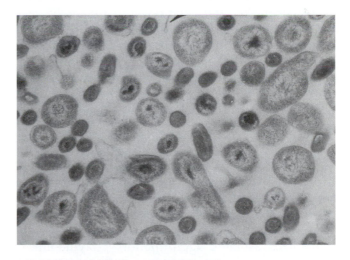

FIGURE 27-5   Transmission electron micrograph of the rickettsia *Coxiella burnettii* showing the pleomorphic nature of this microorganism. (T.F. McCaul and J.C. Williams, *Journal of Bacteriology* 147:1063–1076, figure 1b, with permission of ASM)

TABLE 27-2

Diseases Caused by Rickettsiae

Disease	Reservoir	Vector	Rickettsial Species
Trench fever	Human	Human louse	*Rochalimaea quintana*
Q fever	Lower animals	Tick	*Coxiella burnetii*
Rocky Mountain spotted fever	Tick	Tick	*Rickettsia rickettsii*
Epidemic typhus	Human	Human louse	*Rickettsia prowazekii*
Endemic typhus	Rodent	Flea	*Rickettsia typhi*
North Asia tick typhus	Rodent	Tick	*Rickettsia sibirica*
Boutoneuse fever	Rodent	Tick	*Rickettsia conorii*
Rickettsial pox	Mouse	Mite	*Rickettsia akari*
Scrub typhus	Rodent	Mite	*Rickettsia tsutsugamushi*
Ehrlichiosis	Lower animals	Tick	*Ehrlichia chaffeensis*
Cat scratch disease	Domestic cat	none	*Bartonella henselae*

which humans are only *accidental hosts*. The rickettsiae are usually introduced into the tissues by the bite of an arthropod and have a predilection for the cells that line the small blood vessels (endothelial cells). These bacteria multiply and spread along the blood vessels of the body. The signs and symptoms result from inflammation and swelling of the small blood vessels. This condition may cause blockage or a reduced blood flow to some tissues and some leakage of blood into the surrounding tissues. The leakage of blood in the skin produces the spots and rashes seen with most rickettsial diseases (Figure 27-6). Headache, chills, and fever are due to the generalized inflammation; and stupor, delirium, and shock may occur due to alterations in blood flow to the brain and other vital organs. Symptoms may last for several weeks. Death rates may be as high as 70% with epidemic typhus or less than 1% with rickettsial pox. Lifelong immunity usually results after recovery.

* Diagnosis

Most rickettsial diseases are diagnosed by patient history and clinical signs or in the laboratory by measuring an increase in specific antibodies between the acute and convalescent sera. Such serologic tests as latex agglutination, enzyme-linked immunosorbent assay, and immunofluorescence are routinely used. An older, less specific test, called the Weil–Felix test, is seldom used today. Laboratory isolation of rickettsiae is difficult and is rarely done.

* Treatment

Rickettsial diseases respond well to treatment with tetracyclines and chloramphenicol. Death rates are drastically reduced when chemotherapy is applied.

* Prevention and Control

Two general methods are used to prevent rickettsial diseases. First, vaccines developed to fight some rickettsial diseases may be used on persons who have a high risk of being exposed to a specific rickettsial disease. The second method is to eliminate or

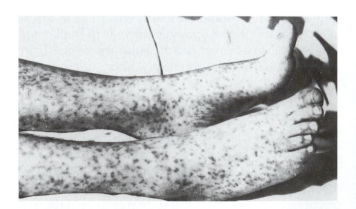

FIGURE 27-6 Rash of Rocky Mountain spotted fever consists of generally distributed, sharply defined purpuric macules involving the palms and soles. The rash may be gangrenous in regions such as the scrotum. (Armed Forces Institute of Pathology, AFIP 67987-3)

Accidental host An organism accidentally invaded by a pathogan.

reduce the animal reservoir and avoid or reduce contact with the arthropod vector of a given rickettsia.

SPECIFIC DISEASE-CAUSING RICKETTSIA

✳ *R. prowazekii*: Epidemic Typhus

Epidemic typhus is caused by the species *Rickettsia prowazekii*, named after Howard Ricketts and S. von Prowazek, two early investigators of this disease. Both scientists died following accidentally acquired laboratory infections of typhus. Without treatment, death rates from epidemic typhus may be as high as 70%.

The typhus rickettsia infects humans and the body louse of humans. The louse becomes infected when it feeds on infected humans and it leaves when the human body temperature significantly increases or decreases from normal. The louse can move only a short distance when searching for a new host. The louse will die of typhus in 1 to 3 weeks, but during this period the rickettsiae proliferate in its digestive tract and are excreted in the feces. When the infected louse infests and bites a new human host, louse feces are deposited on the skin. The louse bite causes itching, which is scratched, and the scratching forces the contaminated feces into the bite wound, thus initiating a new infection.

Because of the short distance the louse can travel from human to human, epidemic typhus is associated with conditions of poor hygiene where humans are crowded together. These conditions are found in times of war, flooding, or other major disruptions of normal human activities. Such conditions have existed often enough over the years to have allowed epidemic typhus to be a major killer of humankind. Armies of the past were frequently stricken with typhus fever and the outcome of many military campaigns was determined not so much by the strategies of the generals as by the epidemics of typhus fever. In 1489 during the Spanish siege of Granada, for example, 3000 Moorish troops died in battle and 17,000 died of typhus; in 1528, as the French army was on the verge of victory at Naples, typhus struck down 30,000 French troops and the tide of battle changed, resulting in a French defeat. During his campaign to conquer Moscow, Napoleon's army of 500,000 troops was reduced to fewer than 200,000 by disease. About 180,000 died of typhus. The last major epidemic typhus outbreaks occurred in southeastern Europe during World War I and in Russia just afterward. In the Russian epidemic an estimated 30 million people contracted the disease and over 3 million deaths resulted. The mode of transmission of epidemic typhus is shown in Figure 27-7.

Epidemic typhus has greatly decreased over the past 60 years. A few limited outbreaks occurred during World War II, but effective use of DDT as a delousing agent and vaccination of military personnel have generally controlled this disease.

Some humans who have recovered from typhus can apparently carry the rickettsiae in their tissues for the remainder of their life. Some of these carriers may experience a mild case of clinical typhus, called *Brill–Zinsser disease*. These persons could serve as a focus of a new epidemic if they were part of a

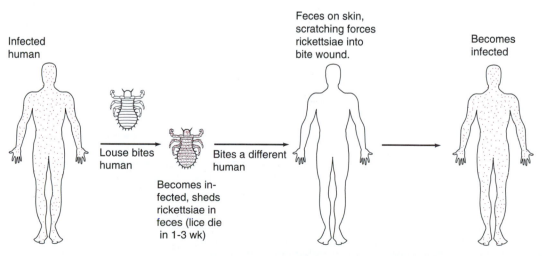

FIGURE 27-7 The transmission of epidemic typhus from human to human by the body louse.

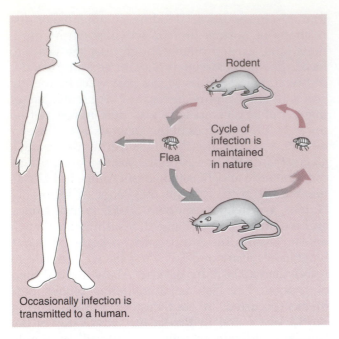

FIGURE 27-8 The transmission cycle of endemic typhus involves rodents and fleas.

crowded, deprived environment where body lice are present.

* *R. typhi*: Endemic or Murine Typhus

Endemic or *murine typhus*, which is caused by *Rickettsia typhi*, is milder than epidemic typhus. Only sporadic cases are seen in humans; the death rate is less than 5% among untreated cases. The natural infection is found in rats, mice, and other rodents and is sporadically transmitted to humans by fleas. The disease in rodents and fleas is mild or subclinical and *R. typhi* may be carried in these hosts as a latent infection. Infection may also be passed from female fleas to offspring through the eggs; this is called *transovarian passage*. The rickettsiae are shed in the feces of the flea and are transmitted to humans through the skin, respiratory tract, or conjunctiva. Persons who live, work, or play around rodent-infested areas stand the greatest risk of infection. No human-to-human transfer occurs. Endemic typhus is found worldwide. In the United States it appears mostly in the southeastern states; generally, fewer than 75 cases per year are reported

in this country. Deaths are rare when chemotherapy is applied. The transmission of endemic typhus is seen in Figure 27-8, and Figure 27-9 shows cases reported in the United States.

* *R. rickettsii*: Rocky Mountain Spotted Fever

This disease was first recognized around the year 1900 in the Rocky Mountains—hence the name. Yet it is found throughout North and South America and in Russia. Currently, the greatest number of cases in the United States occur in the southeastern regions (Figure 27-10). The causative agent, called *Rickettsia rickettsii*, is primarily a parasite of ticks. This rickettsia infects many tissues of the tick. The eggs of the female tick may be infected and so the infection is passed through the ovary directly to her progeny. The rickettsiae are in the saliva of the tick and are transmitted to humans or animals by a tick bite. Ticks may become infected by feeding on infected animals or by transovarian passage from tick to tick without the involvement of an animal reservoir. No human-to-human transfer occurs. The transmission of Rocky Mountain spotted fever is shown in Figure 27-11. Death rates vary in different areas, suggesting that strains of varying degrees of virulence exist. Without treatment, death rates may be as high as 25%; with treatment the death rate is about 4%. Early treatment is important in reducing

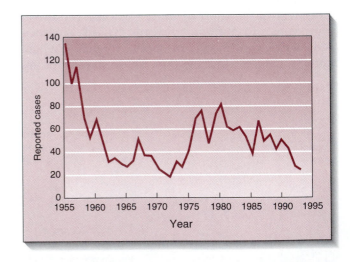

FIGURE 27-9 Reported cases of endemic typhus in the United States, 1955–1994. (Courtesy Centers for Disease Control, Atlanta)

Transovarian passage A process whereby eggs are infected by an etiologic agent present in the female. When the eggs hatch, the insect is infected and capable of transmitting the agent to a new host or, if female, to her own eggs.

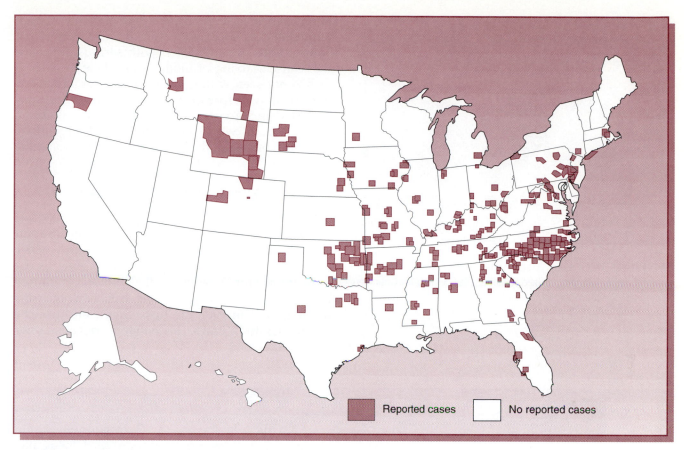

FIGURE 27-10   Distribution of tick-borne Rocky Mountain spotted fever in the United States, by county, in 1993. (Courtesy Centers for Disease Control, Atlanta)

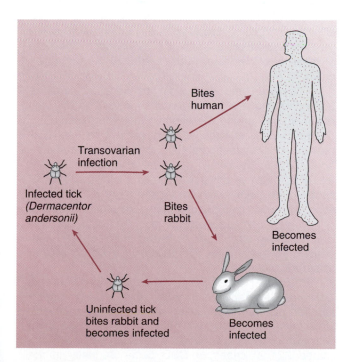

FIGURE 27-11   The transmission of Rocky Mountain spotted fever.

the death rate. There are presently 400 to 500 reported cases of Rocky Mountain spotted fever in the United States each year (Figure 27-12).

✳ *Coxiella burnetii*: Q Fever

A rickettsia called *Coxiella burnetii* is the cause of Q-fever. The name *Q-fever* comes from "Query" fever, for the cause of this disease was not known for some time. *C. burnetii* is more stable than other rickettsiae and is able to survive for long periods outside the host, allowing transmission by indirect means. Natural infections are found in many species of ticks and apparently can be spread by tick bites to birds and mammals in which asymptomatic infections develop. Rickettsiae are shed in saliva, urine, and milk. High concentrations are present in the placenta and amniotic fluids of infected animals. These rickettsiae remain viable on drying and may be carried on dust particles by the airborne route. Humans may be infected by the bite of ticks, by ingesting contaminated milk, by direct contact with infected tissues, or by the airborne route. The airborne infection results in pneumonitis, whereas

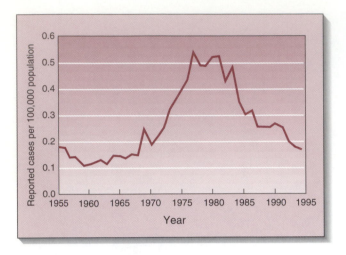

FIGURE 27-12 Reported cases of Rocky Mountain spotted fever in the United States, 1955–1994. (Courtesy Centers for Disease Control, Atlanta)

other routes may produce a nondescript disease with chills, malaise, and fever similar to influenza. Some strains are associated with serious infections that cause scarring of the liver and heart valves. A rash is not seen with Q-fever. Asymptomatic infections seem widespread in cattle, and rickettsiae may be found in unpasteurized milk. Q-fever is found throughout the world and occurs most frequently in persons working in the livestock or meat-producing industries.

✳ Other Rickettsial Diseases

In addition to the preceding varieties, other rickettsial diseases are found in different geographic areas throughout the world. Scrub typhus, which occurs over a wide area of southeast Asia, is the most important. A disease called *ehrlichiosis*, caused by species of the genus *Ehrlichia*, has been occurring in Japan and in 1987 the first cases were detected in the United States. Recently, a rickettsial-like

CLINICAL NOTE

Encephalitis Associated with Cat Scratch Disease: Broward and Palm Beach Counties, Florida, 1994

On 14 August 1994, the Broward County Public Health Unit of the Florida Department of Health and Rehabilitative Services was notified of three children from Pompano Beach who were hospitalized with encephalitis attributed to cat scratch disease (CSD). All three children (aged 5, 6, and 11 years) were previously healthy and had no histories of seizure disorders or diagnoses of CSD. This report summarizes the investigation of these cases.

On 12 and 13 August, during a 26-hour period, each child entered the emergency department of the same hospital with sudden onset of generalized seizures, coma, and respiratory depression requiring intubation and assisted ventilation. Two of the children had focal lymphadenopathy (axillary and epitrochlear) on admission; cervical lymphadenopathy developed in the third child during hospitalization. Clinical examinations and laboratory tests ruled out some causes of encephalopathy, including viral infections (e.g., herpes simplex and arboviruses), metabolic disorders, and toxic ingestions.

On 5 and 27 September, additional cases of CSD encephalitis were confirmed in a 9-year-old boy and a 3-year-old girl from the same area (Fort Lauderdale and Boynton Beach). CSD lymphadenopathy had been diagnosed in both children approximately 3 weeks before the onsets of seizure and coma. Although the girl had been treated with successive courses of amoxicillin/clavulanate potassium and trimethoprim sulfamethoxazole before the onset of CSD encephalitis, both of these cases were clinically similar to the first three cases.

Case and contact investigations identified exposure (e.g., handling and petting) to stray kittens as the only common link among the affected children; histories of overt scratches or bites were not elicited. Indirect fluorescent-antibody testing at CDC detected elevated antibody titers to *Bartonella henselae*, the etiologic agent for CSD, in all five patients. Microscopic examination of lymph node biopsies was consistent with CSD for the two children with lymphadenopathy on presentation.

During hospitalization (range: 11–17 days), all children received supportive care and antibiotic and anticonvulsant therapy. All five children recovered without apparent sequelae.

Editorial Note: CSD is caused by infection with *Bartonella* (formerly *Rochalimaea*) *henselae*, an organism that has been associated with bacillary angiomatosis in immunocompromised persons. CSD is associated with exposure to cats infected with *B. henselae*. An estimated 22,000 cases of CSD occur annually in the United States (*MMWR* 43:909, 1994).

microbe called *Bartonella henselae*, which can be grown in artificial media, has been associated with *cat scratch disease* (CSD). CSD has been recognized for many years as a systemic, febrile illness, with swollen lymph nodes that occurs following cat scratches. About 22,000 cases occur each year in the United States. *B. henselae* can be isolated from healthy cats, and patients with CSD develop specific antibodies against this microbe (see Clinical Note). CSD is best treated with rifampin. The most common rickettsial diseases of humans are listed in Table 27-2.

MATERIAL FOR REVIEW

CONCEPT SUMMARY

1. The chlamydiae are obligate intracellular parasites and are parasites of humans or animals. Numerous diseases occur from these organisms, including trachoma, LGV (a sexually transmitted disease), and a serious but usually not fatal form of pneumonia. These diseases respond well to therapy.

2. The rickettsiae are obligate energy parasites of animal cells and are accidental parasites of humans. They are transmitted to humans by vectors and produce serious generalized infections characterized by high fever and a rash. Members of the genus *Coxiella* may be found in domestic animals and can be transmitted to humans by the airborne route. These infections respond well to antibiotic therapy and vaccines are available for some diseases.

CLINICAL SUMMARY TABLE

Microorganism	Virulence Mechanisms	Diseases	Transmission	Treatment	Prevention
Chlamydia trachomatis	Block formation of phagolysosomes Intracellular growth	Trachoma Inclusion conjunctivitis Urethritis Lymphogranuloma venereum	Human to human Direct contact Sexually	Tetracyclines	Proper hygiene Barrier protection
Rickettsia prowazekii	Growth in blood vessels	Epidemic typhus	Louseborne Human to human	Tetracyclines	Control body lice
Rickettsia typhi	Same as above	Endemic typhus	Fleaborne Rodent to human	Tetracyclines	Avoid wild rodents
Rickettsia rickettsii	Same as above	Rocky mountain spotted fever	Tickborne Animal to human	Tetracyclines	Avoid ticks
Coxiella burnetii	Unknown	Pneumonitis Generalized illness	Animal to human	Tetracyclines	Pasteurization of milk

STUDY QUESTIONS

1. How does trachoma differ clinically from inclusion conjunctivitis?
2. What is the most common chlamydia disease in the United States?
3. Define the nature of the risk to a newborn infant who is born to a woman infected with *C. trachomatis*.
4. What is the host range for *C. psittaci*?
5. Construct a table listing the rickettsial agents of infection that shows the common features of clinical presentation, transmission, and culture of these organisms.
6. Briefly outline the effect epidemic typhus has had on military operations.
7. Describe the relationship between Brill–Zinsser disease and epidemic typhus.
8. List the ways that members of the genus *Coxiella* differ from the organisms of the genus *Rickettsia*.

CHALLENGE QUESTIONS

1. Identify what types of characteristics established the rickettsiae as bacteria rather than viruses.

28 Viruses

Before the germ theory of disease was established, people believed that many diseases were caused by poisons. The Latin term for poison is *virus*. Because they couldn't be propagated on artificial culture media or observed with standard optical microscopes, virus particles went undiscovered during the so-called Golden Age of Microbiology in the late 1800s. Yet during this time it was recognized that many diseases were caused by agents that had not been identified and such unidentified agents were still referred to as "virus." Eventually, it was shown that the causative agents of some diseases could pass through filters that would hold back bacterial-sized cells, and by the 1930s it was possible to crystallize these agents. This latter procedure showed that they were particulate agents and not chemical poisons. By this time, however, the term virus had become permanently associated with these agents and the original meaning was, to a large extent, lost. A great deal of information regarding the structure and functions of viruses has been obtained in the past several decades. This chapter presents a brief summarization of the major characteristics and functions of viruses that infect animals.

INTRODUCTION

Viruses are distinctly different from cells. They possess no independent metabolic capabilities and are thus totally dependent on living host cells (*obligate intracellular parasites*) to supply all the needed energy and building blocks for replication. Upon entering a cell, a virus essentially "pirates" the cellular machinery of its host and directs the cell to make new viruses. The newly replicated virus particles are then released to infect other cells, a process that usually damages or destroys the host cell.

All forms of life seem to have specific viruses that parasitize their cells. There are viruses of animals, plants, insects, bacteria, algae, fungi, and so on. The viruses that attack bacteria are called *bacteriophages* or just *phages* and are much easier to work with in laboratory experiments than animal or plant viruses ∞ (Chapter 6, p. 77). Thus, many significant discoveries on the nature of viruses were made using phage. Fortunately, most viruses, regardless of the types of hosts they attack, function by many similar mechanisms and the information obtained through studies with phage has aided in studies of animal viruses.

STRUCTURE OF VIRUSES

Viruses range in size from about 20 to 300 nm in diameter and generally consist of a small amount of genetic material (the viral **genome**) and a protein coat (the **capsid**), which contains the genome. A given virus contains only one molecule of one type of nucleic acid, which can be either DNA or RNA. Some simpler, so-called *naked*, viruses consist only of the single molecule of nucleic acid and a protein coat. Other viruses may possess an *envelope* over the protein coat. In addition, some may have internal proteins and/or small projections that are called peplomers or spikes (Figure 28-1). Viruses are often very specific in the kinds of cells they can infect and the external proteins functions to "recognize" appropriate host-cell receptors. The complete infectious virus particle, regardless of its structure, is called a *virion*. Five basic morphological shapes—spherical, cylindrical, brick, bullet, and tailed—are observed and most of these shapes (except tailed-shaped) are seen among the viruses infecting animals (Figure 28-2).

THE VIRAL GENOME

Unlike cells, in which the genetic material is always double-stranded DNA, viruses have a genome that can be either DNA or RNA. Furthermore, the nucleic acid molecule may be either *single-stranded* (ss) or *double-stranded* (ds). The genome structure, along with the method of its replication in the host cell, is used to classify the viruses into six groups, as discussed later in the chapter.

Most animal viruses contain fairly small amounts of genetic information, which means that they are limited in the number of different types of protein molecules that can be synthesized under their direction. For example, many viruses that cause disease in humans contain from 5 to 15 genes in their nucleic acid. These genes code for the bare minimum the virus needs to be replicated. This may be very little, as in the case of some DNA viruses in which the

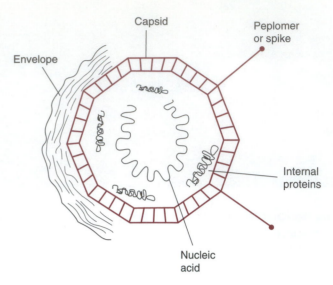

FIGURE 28-1 A schematic outline of a virus. The protein coat is called the capsid. An envelope is present on some viruses and not on others. Peplomers (spikes) are present on some viruses. One molecule of nucleic acid is contained inside the capsid. Some viruses may have additional internal proteins. The entire virus is called a virion.

host cell machinery is already capable of replicating and transcribing the viral genome. Essentially, only the proteins needed to construct the viral capsid—plus whatever other internal proteins or external structures, such as spikes, that are part of the virion—need to be encoded. In the case of RNA viruses, however, the viral genome would also need to code for enzymes that enable viral RNA to be replicated.

THE CAPSID (PROTEIN COAT)

Most animal viruses contain fairly small amounts of genetic information, which means that they are limited in the number of different types of protein molecules that can be synthesized under their direction. Many viruses must construct their protein coat, the capsid, out of just one or a few different types of polypeptides. The individual protein molecules that make up the building blocks of the capsid are called *structure subunits*. Viral capsids are symmetrical. Many capsids are constructed as regular icosahedrons—that is, an enclosed shell of 20 triangu-

---

Genome   The genetic information (DNA or RNA) in a virus or cell.
Capsid   The external protein structure of a virus and is composed of capsomers.

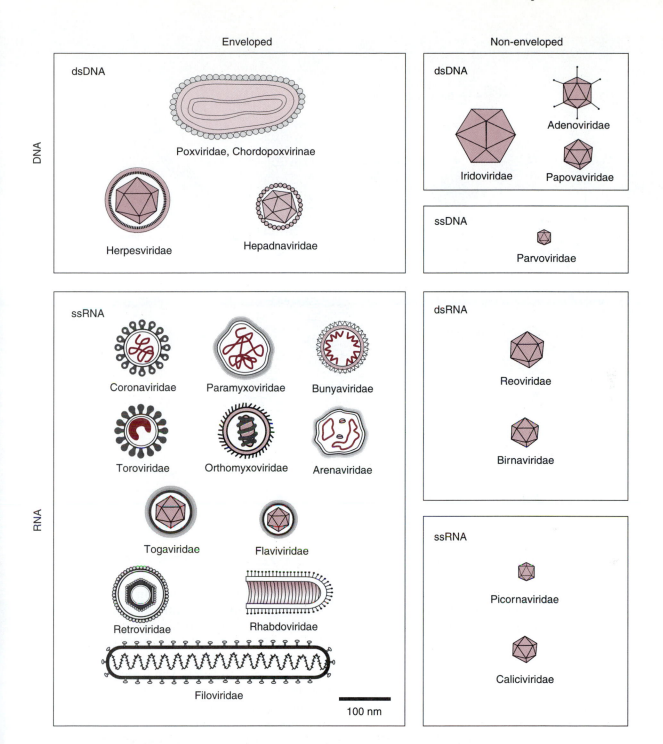

FIGURE 28-2   The basic shapes and relative sizes of some representative families of viruses. (Courtesy C. M. Fauquet and Springer-Verlag, Vienna)

lar faces, 12 vertices, and 30 edges (Figures 28-3 and 28-4). To form the icosahedral capsid, the structure units must first form into clusters of five, called *pentamers*, and clusters of six, called *hexamers*. These clusters are called *capsomers*. The number and arrangement of the capsomers in an icosahedron are restricted geometrically. The 12 vertices must be pentamers and one of the smallest possible icosahedrons is composed of only the 12 pentamers. As the icosahedron increases in size, hexamers are added between the pentamers in regular increments. The number of pentamers remains constant at 12, how-

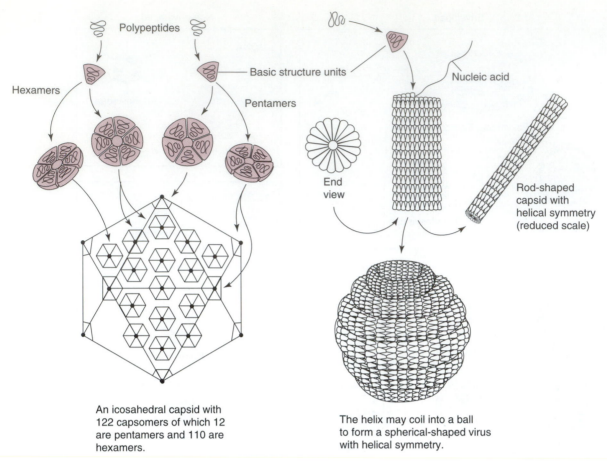

An icosahedral capsid with 122 capsomers of which 12 are pentamers and 110 are hexamers.

The helix may coil into a ball to form a spherical-shaped virus with helical symmetry.

FIGURE 28-3 The formation of viral capsids with cubical and helical symmetry from polypeptides that form the basic structure subunits.

ever. The smallest number of hexamers that will fit is 20—that is, one for each triangular face of the icosahedron. The result is a virus with a total of 32 capsomers (12 pentamers and 20 hexamers). Other increments of capsomers that can form into icosahedrons are 42, 72, 92, 122, 162, 252, and so on. The number of capsomers of a given virus is constant and serves as a useful characteristic in classifying and identifying viruses. Icosahedral viruses are said to have *cubical symmetry*. The nucleic acid is packaged inside the icosahedral capsid.

Other viruses have *helical symmetry*, which refers to an arrangement of structure units connected side by side in a continuous ribbon that spirals into a tubular helix. The nucleic acid is connected to each structure unit much like a string that is connected along a row of beads. The tubular helix may remain extended to form a cylindrical-shaped virus or it may coil into a ball to form a spherical-shaped virus (Figures 28-3 and 28-4). A few larger viruses have more complex structures.

CULTIVATION OF VIRUSES

Viruses can be propagated only in living host cells and a major activity in a viral laboratory is to provide a suitable supply of such cells. The main sources of cells for the propagation of animal viruses are intact animals, embryonated eggs, and organ, tissue, or cell cultures. The injection of viruses into susceptible living animals gives useful information on the pathogenesis of viral diseases but is usually not a useful method of producing large amounts of viruses for vaccines or experimental studies. When dealing with human viruses, an attempt is made to find an experimental animal that will support the multiplication of the virus and produce a disease similar to that seen in humans. Such attempts are not always successful, however, for many viruses are host specific; that is, human viruses will multiply only in human cells, cat viruses in cat cells, and so forth. Although it is possible to get some specific human viruses to multiply in such

FIGURE 28-4   Models of viruses made from repeated copies of identical interconnecting structure subunits (plastic triangles) to form capsids with either cubical or helical symmetry.

primates as monkeys or chimpanzees, the experimental procedure is expensive and cumbersome. If there is no chance that the virus will cause death, permanent damage, or undue discomfort, human volunteers may be used in virus research. An example is research on the common cold; here human volunteers have been used. Even though research using intact hosts has many limitations, it is often the only means available to study some diseases.

Embryonated eggs provide an inexpensive, easy-to-handle, sterile container full of a variety of living cells (Figure 28-5). Some animal viruses will multiply in embryonated eggs; others will not. Often, embryonated eggs are used to produce large amounts of viruses for vaccines. A small hole can be drilled through the eggshell and virus can be injected into the appropriate embryonic cavity or membrane. Generally, embryos between 7 to 10 days of age are used.

The development of methods to grow living animal cells routinely in test tubes (in vitro) greatly accelerated research work with animal viruses by making it possible to discover many new viruses, grow large amounts of viruses, and develop sensitive quantitative assays for many viruses. The impact of cell-culturing procedures is dramatized by the fact that in the late 1940s, when these procedures first came into wide use, only about 35 viruses associated with human diseases had been discovered. During the following 15 to 20 years, 500 additional viruses associated with humans were identified and tremendous advances were made in our understanding of the nature of viruses. *Organ culture* is a procedure in which a section of an organ is taken from an intact host and the cells of that organ are kept alive by immersing them in a nutrient fluid in a test tube. This procedure is difficult to carry out and has limited use. The terms *tissue culture* and *cell culture* are often used interchangeably. In the technical sense, however, tissue culture means implanting tissue fragments into test tubes and allowing cells to grow out from these fragments. Cell culture is the most widely used method of growing cells in vitro. A wide variety of tissues can be used, but tissues from embryos generally work best. To produce cell cultures, the tissue is cut into small pieces and treated with a *proteolytic* (protein-splitting) enzyme; this process causes the cells to separate into a suspension of single cells or small clumps of several cells. These cells are then washed, suspended in a nutrient medium, and placed in specially cleaned glass or plastic containers in an incubator. The cells settle onto and adhere to the surface and begin to divide. Usually after several days a continuous layer, one cell deep (*monolayer*), will form over the surface. These cells can again

Embryonated eggs   Fertilized eggs in which the embryo has been permitted to grow to a recognizable stage of development.

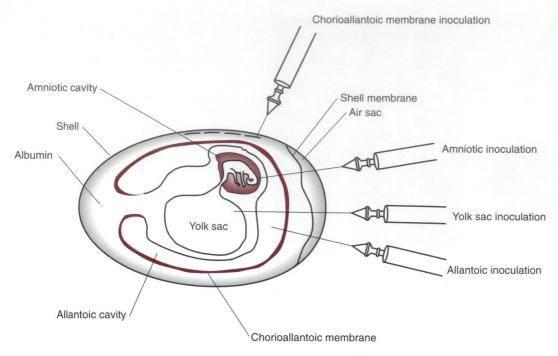

FIGURE 28-5 An embryonated egg showing the routes of inoculation.

be treated with a proteolytic enzyme and the cells of the monolayer will disassociate into a suspension of single cells. The cells can then be passed to new containers where they will continue to multiply. Enough cells can be obtained from one container to seed two or three new containers. A procedure for setting up a cell culture is shown in Figure 28-6.

Some cells can be used to seed new cultures only five or six times, others up to a hundred, and yet other cells adapt so well to cell cultures that they can be passed indefinitely. One particular cell culture line, called *Hela cells*, was started from human cancer tissue in 1952 and is still being passed. It is now possible to grow some cell cultures in fluid suspensions without forming a monolayer and to grow monolayers of cells on large numbers of small beads in a container. Both of these procedures greatly increase the density of cells that can be grown in a given container.

Much current work in virology uses cell cultures for the detection, propagation, and measurement of viruses. When a suspension of viruses is placed on a monolayer, the cells become infected and are usually destroyed. If the virus suspension is sufficiently dilute, a small number of viruses will be widely spread out across the surface, resulting in isolated areas of cell destruction in the otherwise continuous monolayer. Each area of destroyed cells is called a **plaque** (Figure 28-7) and represents the effects of a single virus. Thus, plaques can be useful for counting the number of viruses in a given sample.

MULTIPLICATION OF VIRUSES

When studying the subject of viral multiplication, the virus should be viewed as a segment of genetic information that becomes inserted into the "genetic pool" of the host cell. The only contribution of the virus is a single molecule of nucleic acid—that is, the viral genome—and, in some cases, an enzyme for the transcription of this nucleic acid. All other components involved in the multiplication of the virus are supplied by the host cell. Thus, while cells grow and divide, viruses are said to *replicate*. Viral replication can be likened to a factory process in which individual viral parts are made separately and later assembled into the virions, which are released from the cell.

Plaque An area of cell destruction caused by the propagation of a single virus in a monolayer of cells.

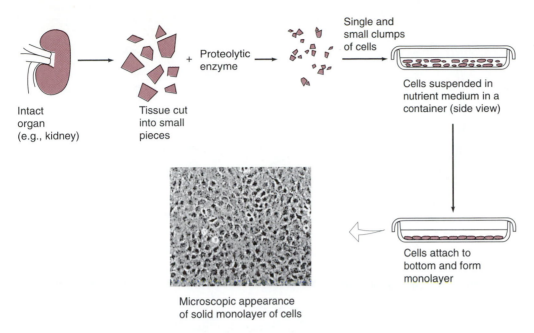

Intact organ (e.g., kidney)

Tissue cut into small pieces

+ Proteolytic enzyme

Single and small clumps of cells

Cells suspended in nutrient medium in a container (side view)

Cells attach to bottom and form monolayer

Microscopic appearance of solid monolayer of cells

FIGURE 28-6   A procedure for setting up a cell culture and creating a cell monolayer.

* Steps in Viral Replication

The replication of animal viruses can be divided into the following general steps (Figure 28-8):

1. *Attachment* to the cell surface. This is a specific reaction and only cells that have the correct receptor site can be infected by a specific virus. This phenomenon accounts for the ability of a virus to infect only a certain animal species or only a given tissue within the infected animal.

2. *Penetration* into the cytoplasm. After attachment has occurred, the plasma membrane

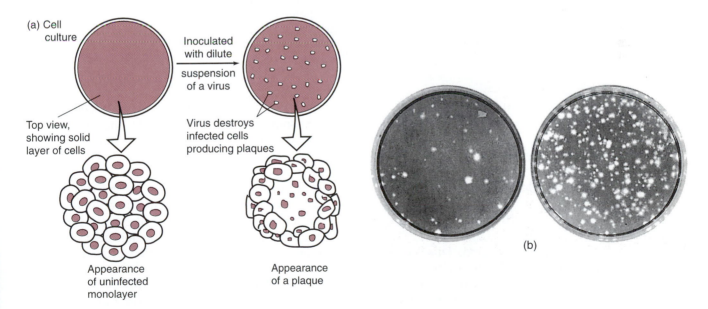

(a) Cell culture

Inoculated with dilute suspension of a virus

Top view, showing solid layer of cells

Virus destroys infected cells producing plaques

Appearance of uninfected monolayer

Appearance of a plaque

(b)

FIGURE 28-7   **(a)** A procedure for inoculating a cell culture (in a petri dish) with viruses. Each plaque represents a single virus' destruction of the monolayer cells. Thus, the number of plaques can be used to estimate the number of viruses in the original sample solution. **(b)** Photos of plaques on monolayer cultures.

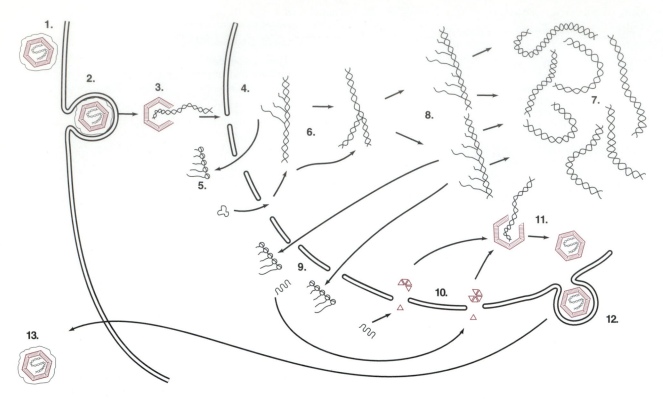

FIGURE 28-8   Some of the steps in the replication of a ds-DNA virus (e.g., a herpesvirus). (1) Specific attachment. (2) Penetration. (3) Release of viral nucleic acid and its migration to the cell nucleus. (4) Early transcription of viral DNA, directed by cellular enzymes. (5) Translation of viral *m*RNA and synthesis of an enzyme needed to replicate viral DNA. (6) Replication of viral DNA. (7) Synthesis of many viral DNA molecules. (8) Late transcription of viral DNA. (9) Translation of late viral *m*RNA with the synthesis of basic structure units. (10) Formation of capsomers. (11) Assembly of capsid around viral DNA. (12) Budding of virus from the nuclear membrane and picking up of an envelope. (13) Release of virus from the cell. The herpesviruses are able to produce about 500 new viruses in 10 hours from each infected cell.

invaginates into the cytoplasm and forms a vacuole around the virus. This process is called **pinocytosis**.

3.   *Uncoating* (release of viral nucleic acid). The viral capsid is broken down by cellular enzymes and the nucleic acid is released. This procedure may occur in either the nucleus or the cytoplasm, depending on the type of virus.

4.   *Transcription* of viral nucleic acid. This process occurs in the cytoplasm for some viruses and in the nucleus for others. The various modes of transcription are discussed later in this chapter.

5.   *Translation* of viral-directed mRNA. Cellular ribosomes, tRNA, amino acids, energy, and so on, are used in this step to bring about the synthesis of the proteins needed for the synthesis of new viral particles.

6.   *Replication* of viral nucleic acid. One or more proteins produced under the direction of the viral genome function as enzymes for directing the synthesis of new viral nucleic acid molecules. The original viral nucleic acid molecule must serve as a template.

7.   *Assembly* of virus particles. Viruses do not replicate by dividing as cells do, but are assem-

Pinocytosis   A process analogous to phagocytosis where the ingested material is a liquid or very small particle.

bled from pools of the viral nucleic acid and protein structural subunits. Many copies of a virus may be forming simultaneously within a single cell. Assembly may occur in the nucleus or cytoplasm, or partly in one and partly in the other area, depending on the type of virus. Assembly of the protein structure subunits into capsomers and then into capsids was described earlier. As the capsid forms, it encloses the viral nucleic acid. This process is not efficient, however, and often the capsid forms without enclosing the nucleic acid; this is called an empty virus particle. Often enough viral building blocks are produced to make 10,000 to 20,000 new virus particles per infected cell. Because of the inefficient assembly, however, only 200 to 300 particles will be properly assembled. Yet the entire process may take only an hour or two and several hundred "offspring" in this time provide an effective means of reproduction.

8. *Release* of viruses. Some viruses are released when the cell disintegrates as a result of the damage produced by the replication process. These viruses would have no envelopes. Other viruses migrate to a cell membrane and bud out through the membrane. The membrane pinches off and remains attached to the virus, thus forming the envelope. With some viruses, viral-directed proteins are formed and become embedded in the cell membrane and thus incorporated into the envelope.

Throughout all other biological systems in nature protein synthesis is directed by the same reliable mechanism of information contained in double-stranded (ds) DNA molecules being transcribed into single-stranded (ss) mRNA molecules, which, in turn, direct the synthesis of polypeptides through the process of translation ∞ (Chapter 6, p. 69). These reactions are summarized as ds-DNA → mRNA → protein.

When dealing with viruses, however, not only is the conventional ds-DNA → mRNA → protein pathway used but so are a variety of other modified pathways. Most viruses of animals can be divided into at least the following six general classes, based on the type of nucleic acid they contain and on the pathways used to express their genetic information. This broad-based classifying convention is outlined below:

Class 1: Viruses with ds-DNA The flow of information is ds-DNA → mRNA → protein. This is the classical pathway seen in all high-

er forms of life. An example of this class of viral multiplication is outlined in Figure 28-8.

Class 2: Viruses with ss-DNA The flow of information is ss-DNA → ds-DNA → mRNA → protein. The ss-DNA must first be changed into ds-DNA, which is done by cellular enzymes after the viral nucleic acid enters the cell. Afterward the flow of information is the same as in class 1.

Class 3: Viruses with ds-RNA The flow of information is ds-RNA → mRNA → protein. This class of viruses presents a unique problem because ds-RNA is not found in normal cells. Thus, no enzyme is present in a cell to direct the transcription of ds-RNA molecules. To solve this problem, these viruses direct the synthesis of a special enzyme that will transcribe mRNA from the ds-RNA molecule. This enzyme is packaged inside the viral capsid along with the ds-RNA.

Class 4: Viruses with ss-RNA of the same polarity as mRNA (called ss-RNA⁺) The information flow is simply mRNA → protein. The viral nucleic acid acts directly as mRNA once inside the cell. Additional steps (not covered here) are needed to replicate and amplify the viral genome.

Class 5: Viruses with ss-RNA of the opposite polarity from mRNA (called ss-RNA⁻) The information flow is ss-RNA⁻ → mRNA → protein. As in class 3, this is a unique situation in which a function not found in a normal cell must be carried out—that is, the transcription of mRNA from a ss-RNA molecule. It is therefore necessary to provide a specific enzyme to accomplish this task. These viruses carry this special enzyme and also direct its formation as part of their replication process. An example of this class is outlined in Figure 28-9.

Class 6: Viruses with ss-RNA⁺ and a special enzyme called reverse transcriptase or RNA-dependent-DNA polymerase. Some viruses of this class are associated with the induction of cancer in animals and others cause acquired immunodeficiency syndrome (AIDS). These viruses change the genetic information on their ss-RNA molecule into a ds-DNA molecule or, in other words, move backward from the normal flow of genetic information. The information flow is ss-RNA → ss-DNA → ds-DNA → mRNA → protein. The enzyme reverse transcriptase

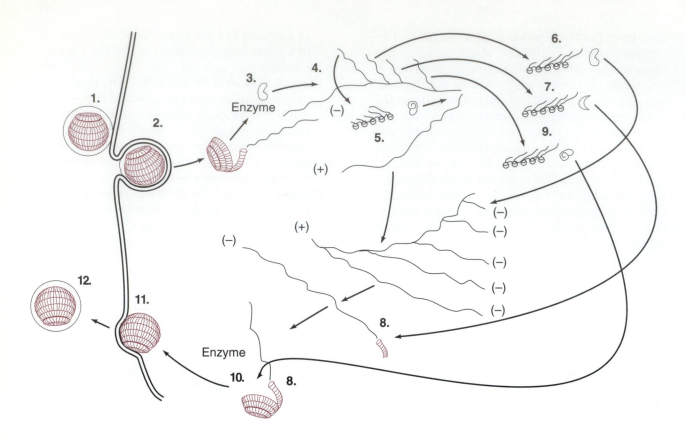

FIGURE 28-9 An interpretation of some of the major steps in the replication of a ss-RNA⁻ virus (i.e., a virus with a ss-RNA molecule of opposite polarity from the *m*RNA and thus it must carry its own enzyme for transcribing *m*RNA). All steps occur in the cytoplasm. (1) Attachment. (2) Penetration. (3) Uncoating with release of viral RNA and enzyme. (4) Transcription of viral RNA to *m*RNAs by the enzyme carried by the virus. (5) Translation of a viral *m*RNA with the formation of an enzyme that will direct the synthesis of a complementary copy of the complete viral RNA. (6) Synthesis of an enzyme that directs the synthesis of negative ss-RNA copies off the positive copies synthesized in step 5; these negative copies are identical to the original viral RNA. (7) Synthesis of structure units. (8) Formation of helical capsid around the viral RNA. (9) Synthesis of the enzyme to be packaged in the virus. (10) Packaging of enzyme in the virus. (11) Budding capsid and picking up of an envelope. (12) Release of complete virion. Hundreds of viruses are made in each infected cell.

is needed for the step from ss-RNA to ss-DNA. Normal cellular enzymes are able to direct the other functions. When the ds-DNA is formed, it may insert into the DNA of the host cell. Under certain conditions that are not well understood some of the information in this inserted DNA may be transcribed and direct changes in the cell that may cause cancer. Under other conditions the inserted DNA may be transcribed and direct the formation of new virus particles.

CLASSIFICATION OF VIRUSES

The classification of viruses into related taxonomic groups has been more difficult than with the higher forms of life. However, in the past 35 years significant progress has been made in our understanding of the nature of viruses. This has made it possible to classify viruses into families of related members according to chemical and morphologic characteristics, as well as on the broader criterion of mode of replication.

✳ Some Rules of Nomenclature

Virus families are designated with the suffix *-viridae* and in formal taxonomic usage the first letter of the name is capitalized and the name is underlined or printed in italics; for example, *Poxviridae* and *Herpesviridae*. Some families are subdivided into subfamilies, but the most used subdivision is the genus. Viral genera are designated with the suffix *-virus*, with the first letter capitalized and the name underlined or italicized; for example, *Enterovirus*. However, at the species level it has not been practical to use the typical binomial (double) Latin names ∞ (Chapter 2, p. 16). Viral species are most often referred to in the vernacular form and are not capitalized, underlined, or italicized; for example, polio virus and herpes simplex virus. Subspecies are designated by a letter or number after the genus level name; for example, influenza A virus or enterovirus 71.

Table 28-1 lists the major families of animal viruses based on the following characteristics; type of nucleic acid, symmetry of capsid, presence or absence of an envelope, size of virion, and number of capsomers, if the symmetry is cubical.

TABLE 28-1

The Classification into Families of Viruses That Infect Animals

Family	Type of Nucleic Acid	Symmetry of Capsid	Envelope	Size(nm)	Number of Capsomers
Picornaviridae	RNA	Cubical	−	24–30	32
Caliciviridae		"	−	35–39	32
Reoviridae		"	−	60–80	32
Togaviridae		"	+	60–70	32
Flaviviridae		"	+	40–50	32
Orthomyxoviridae		Helical	+	80–120	−
Paramyxoviridae		"	+	150–300	−
Rhabdoviridae		"	+	60 × 180[a]	−
Filoviridae		"	+	80 × 800[b]	−
Coronaviridae		"	+	60 − 220	−
Bunyaviridae		"	+	80 − 110	−
Retroviridae		Uncertain	+	80–100	−
Arenaviridae		"	+	50–300	−
Parvoviridae	DNA	Cubical	−	18–26	32
Papovaviridae		"	−	45–55	72
Adenoviridae		"	−	70–90	252
Hepadnaviridae		"	+	40–50	?
Herpesviridae		"	+	120–200	162
Iridoviridae		"	+	130–300	1500
Poxviridae		Complex	−	230–300	−

[a] Bullet-shaped.
[b] Filamentous.

MATERIAL FOR REVIEW

CONCEPT SUMMARY

1. Viruses are submicroscopic structures containing both nucleic acid and protein. These structures are not capable of self-replication but require living host cells for synthesis and development. Hundreds of viruses have been identified. They are known to be responsible for many human, plant, and animal diseases and yet in some cases appear to exist within the host cell without harming it.

2. Viruses infect specific host cells. This specificity is determined by viral protein surface structures and corresponding host cell receptors. For convenience in studying viruses, they are frequently cultivated in vitro in cell culture systems.

3. A variety of viral replication methods have evolved. The exact nature of these replicative processes depends on the type of nucleic acid in the virus genome. There are at least six variations of virus replication, although the procedures involved in attachment, penetration, release of nucleic acid, and so on, are similar among all viruses.

STUDY QUESTIONS

1. Write a short paragraph that describes the general properties of viruses.

2. What are the two most common forms of symmetry of viral capsids?

3. Define each of the following terms: (a) capsid, (b) organ culture, (c) cell culture, (d) tissue culture, and (e) plaque.

4. Diagram the process of nucleic acid replication for each of the six classes of viruses illustrated in the chapter.

5. What feature of possible host cells and virus accounts for host specificity?

6. Describe the process of viral maturation (virus assembly).

CHALLENGE QUESTIONS

1. Explain each word in the term *obligate intracellular parasite* when talking about viruses. Does this term differ when talking about the chlamydiae and rickettsiae?

2. Based on the description of viruses in this chapter, provide arguments for and against the statement, "Viruses are living organisms."

The outcome of the multiplication of viruses in an animal host may vary from rapidly progressing destruction of tissues with resulting disease and death to a completely inconsequential relationship in which the viruses multiply at a low level or are dormant and cause no damage or disease. Some general effects of viruses on individual cells and the general characteristics of viral infections are discussed in this chapter.

EFFECTS OF VIRUSES ON CELLS

Some viruses have a profound effect on the host cell, for they redirect most of the cellular metabolic processes to the production of new viral components. Such a redirection usually results in the destruction of the infected cell. Other viruses may set up a less dramatic relationship in which they do not seriously interfere with the cell's metabolic processes and redirect only a small percentage of the cellular components into the production of new virus particles. Some viruses are able to insert their nucleic acid into the DNA of the host cell. In some such cases, no influence is manifested and the viral genome simply replicates and "rides along" with the DNA of the host cell. In other cases, the viral genome that becomes inserted into the host cell DNA is partially transcribed and imparts specific traits to the host cell. These three general types of

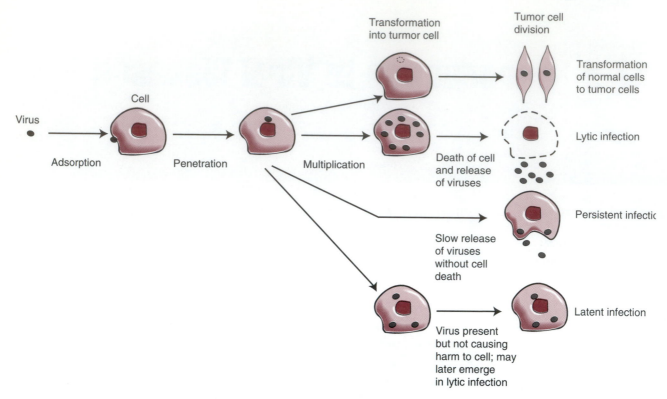

FIGURE 29-1 Some possible effects of viruses on animal host cells. (Biology of Microorganisms, 7/E by Brock, Madigan/Martinko/Parker, © 1994. Adapted by permission of Prentice-Hall, Inc., Upper Saddle River, NJ.)

effects—cell destruction, cell alteration, and cell transformation—are described below and summarized in Figure 29-1.

✳ Cell Destruction

Many viruses enter the host cell and promptly rearrange the cellular components into those needed for the assembly of viruses. These viruses are often able to turn off the metabolic pathways of the cell. The result is a rapid degradation of the cellular components, with the resultant disintegration of the cell; this is called a *lytic* infection. This process may require 15 minutes for some viruses to several hours for many others. During this process, changes may be observed in the infected cells. Aggregates of viral materials or cell debris may collect in areas of the cell and are called *inclusion bodies*. Some viruses cause the cytoplasmic membranes of adjacent infected cells to fuse together to form multinuclear giant cells called *syncytia*. Characteristic syncytia or inclusion bodies occur with some viral infections

and serve as useful diagnostic aids (Figure 29-2). Usually within a few days these *cytocidal* (cell-destroying) viruses have caused enough damage that symptoms of the clinical disease begin to appear in the host. Common acute viral diseases such as rabies, influenza, and measles are caused by these cytocidal viruses.

✳ Cell Alterations

Some viruses enter the host cells and replicate without significantly interfering with normal cellular functions. Only about 1% of the cellular components are "pirated" for the production of new viruses. These viruses are gradually released from the cell by *budding* through the cell membrane without causing apparent damage. With such infections a balance may be maintained for years between the host cell and the infecting virus, a situation known as a *persistent* or *chronic* infection. Because of the lack of observable changes, this type of virus–cell relationship has been difficult to study. There is evi-

Budding The act of release for some viruses. The replicated virus becomes enveloped by host membrane as it leaves an organelle or the cell.

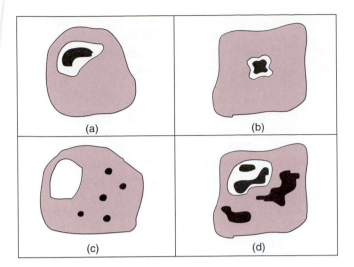

FIGURE 29-2 Some examples of the types of inclusion bodies formed when different viruses infect cells. (a) Intranuclear inclusions from a herpesvirus infection. (b) A rosette-type intranuclear inclusion from an adenovirus infection. (c) Intracytoplasmic inclusions (Negri bodies) from rabies virus infection. (d) Both intranuclear and intracytoplasmic inclusions in the same cell from a measles virus infection.

dence, however, that after a time the host's immune mechanisms may react against these persistently infected cells. The immune reaction or other effects of the virus may lead to the degeneration of the infected tissue.

Certain DNA viruses are able to enter a cell and integrate their DNA into the DNA of the host cell. This condition may allow the viral genome to be carried by the host cell without being expressed for long periods of time and even be passed to progeny host cells; this condition is called *viral latency*, also known as a *latent* infection. At intermittent intervals this latently carried viral genome may direct the formation of new infectious virus particles, resulting in the recurrence of clinical manifestations long after the original infection. This phenomenon of viral latency is characteristic of the herpesviruses (Chapter 31).

✳ Cellular Transformation

A special type of alteration, called **transformation**, may occur following the insertion of viral nucleic acid into host cell DNA. Transformed cells are cancer cells that multiply rapidly and do not respond to the normal mechanisms that control cellular proliferation. The causes of many types of cancer are not known, but animal experiments have shown that certain types are caused by viruses. Well-documented studies have demonstrated that a number of ds-DNA viruses are able to produce tumors in animals. The *papovaviruses* are primarily tumor-producing viruses and certain strains are able to produce tumors in mice, rabbits, and other experimental animals. These tumors are most effectively produced when the papovaviruses are injected into newborn animals. The only tumors known to be induced in humans by a papovavirus are common warts. Some human adenoviruses (Chapter 30) have been shown to produce tumors when injected into baby rodents, but there is no evidence of adenovirus-induced cancer in humans. Herpesviruses (Chapter 31) are able to produce cancer in their specific animal hosts. Some herpesviruses of humans are suspected of causing cancer, and hepatitis B virus has a strong association with human liver cancer.

At present, there is much interest in the cancer-producing capabilities of the RNA-containing *retroviruses*. Retroviruses, which have ss-RNA genomes, are the members of the class 6 viruses ∞ (Chapter 28, p. 375). As early as 1910 it was shown that a retrovirus called *Rous sarcoma* virus could produce cancer in chickens. At first the induction of cancer by these ss-RNA viruses presented a dilemma, for it is impossible for the ss-RNA molecule to become inserted into the ds-DNA molecules of the host cells. However, this dilemma was resolved when the phenomenon of *reverse transcription* was discovered in the 1960s. (The ability to carry out reverse or *retro* transcription from RNA to DNA is the basis for the current name of this group of viruses.) Retroviruses contain an enzyme called *reverse transcriptase*, which enables them to change the genetic information in a ss-RNA molecule into genetic information in a ds-DNA molecule. The ds-DNA molecule is able to insert into the host cell DNA to induce the cellular transformation. In the past few decades retroviruses that cause cancer in humans, mice, cats, baboons, and other animals have been found.

Not all retroviral infections result in cell transformation. The **human immunodeficiency virus (HIV)**, which causes AIDS, is a retrovirus but is able to produce *lytic* (destructive) infections in some cells. Other cells infected with HIV may per-

Transformation The morphological changes that occur to some virus infected cells that may make them cancerous.

mit long-term, chronic infection with a rather gradual release of the virus. Integration of the viral DNA into host DNA is possible with this virus, but direct tumor formation is not a usual consequence of these infections.

CELL DEFENSE: INTERFERON PRODUCTION

Considering the rapid multiplication of cell-destroying viruses, it would seem that massive numbers of cells of the infected host could be destroyed in a relatively short time. However, along with host defense mechanisms, such as phagocytosis and antibody formation, cells have a special built-in defense mechanism that produces a group of substances called **interferons** that limit the proliferation of viruses. Interferons are a group of proteins produced by cells of vertebrates following infections by viruses or following stimulation by some antigens. Three distinct types of interferons have been identified and are called *alpha*, *beta*, and *gamma*. About 13 different varieties of alpha and one or two varieties of beta interferon have been found. Alpha and beta interferons are induced by any virus in most infected cells. Only one variety of gamma interferon has been identified and it is produced only by lymphocytes following antigenic stimulation. Gamma interferon seems to function much like some of the lymphokines ∞ (Chapter 11, p. 157). In fact, as our knowledge of interferons increases, it appears that interferons and lymphokines represent a large group of immunomodulating proteins and the full actions and interactions of these proteins are as yet not fully understood.

Interferons induce protection against most viruses, not only the type of virus that stimulated their production, and are thus nonspecific in their antiviral action. It has been hoped that interferons might prove to be effective antiviral chemotherapeutic agents; in fact, much effort has been made to use interferons in treating viral infections. A major limitation has been the species specificity of interferons; that is, interferons produced by human cells will induce protection only in other human cells. Therefore, it was not possible to produce

interferon in such hosts as experimental animals or embryonated eggs for use in treating humans. In the 1970s, methods were developed for producing some interferon in human white blood cells and fibroblasts grown in cell culture systems. This interferon was effective in treating some human viral infections, but only small amounts could be produced, which made treatments very expensive. Most applications were limited to research trials. Another limitation is the short duration of protection offered by interferon, which is about 2 weeks.

In the 1980s, the genes for human interferons were cloned into bacteria and yeast cells using genetic engineering technology ∞ (Chapter 6, p. 81). This made large amounts of human interferons available for research and clinical trials. There has been much optimism that interferons would function as the long sought after effective antiviral therapeutic agents. While these genetically engineered interferons have shown some effects in treating certain viral infections, the hoped for dramatic effect on treating many viral diseases has as yet not been realized. Much research work is still going on with various forms and combinations of interferons, which may yet yield effective way of using these proteins for antiviral treatments. Some interferons have also shown promise in treating certain forms of cancer.

Most evidence indicates that interferons function in intact animal hosts as important rapidly mobilized natural defense mechanisms against viral infections. Interferons probably play a vital role in slowing down the destructive effects of rapidly multiplying viruses during the early stages of the infection, which allows the host to survive long enough for the humoral and cell-mediated immune responses to be mobilized to specifically react against the invading virus.

The mode of interferon production and action is outlined in Figure 29-3. Interferons are proteins that are stable and readily diffuse out of the cells in which they are produced. When they reach adjacent cells, they attach to specific receptors on the cell membrane and induce a series of *cascading* reactions that eventually result in blocking the replication of viruses in these cells. In these reactions, at least three new enzymes are induced, which catalyze the

---

Interferon   A chemical substance produced by cells in response to an infection by most viruses.
Cascading   Referring to a series of reactions that follow one another.

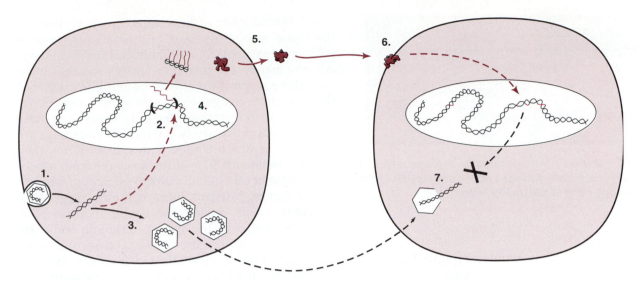

FIGURE 29-3   A simplified outline of the production and action of interferon. (1) Viral or chemical inducer of interferon enters the cell. (2) The virus induces derepression of the interferon-producing gene. (3) New viruses produced and released from cell. (4) mRNA for interferon synthesized. (5) Interferon produced and secreted from cell. (6) Interferon binds to membrane of adjacent cells. (7) The binding triggers secondary reactions that induce mechanisms that interfere with the replication of any viruses that might enter this cell.

synthesis of short segments of nucleotides. These in turn activate endonuclease enzymes that then destroy viral mRNAs, which prevents the synthesis of proteins that are needed to produce the infecting virus. Some of these intermediate proteins have now been produced through genetic engineering and may also prove to be effective agents for treating viral infections.

SIGNIFICANCE OF VIRAL DISEASES

Many of the infectious diseases of humans are caused by viruses. Fortunately, today, most are not serious; that is, they are not life-threatening and they do not require hospitalization. Still, they cause discomfort and incapacitation and are disruptive overall to normal human activities. Many serious problems still result from viral infections and as patterns of human behavior change, a new environment may be created that allows a new virus to enter the human population and be transmitted to large numbers of persons with devastating results; such an example is HIV, which causes AIDS.

On the other hand, several serious viral diseases have been either eradicated or greatly reduced through active immunization programs. *Smallpox*, once the greatest killer of all viral diseases, has been completely eradicated. Such diseases as poliomyelitis, measles, and rubella, which up until the past few decades caused many deaths and left thousands crippled or impaired, have been greatly reduced in number.

Viruses are not cells ∞ (Chapter 28, p. 367) Therefore, antibiotics and other "wonder drugs," which generally operate by disabling cellular machinery and processes, are ineffective in treating viral infections. Yet some antiviral compounds are now available to treat some viral infections ∞ (Table 9-3, p. 136). However, most treatments against viruses are symptomatic, using procedures that reduce the discomfort and maintain the proper physiological functions of the body—that is, they treat the symptoms, not the cause. Recovery results primarily from the patient's own natural defense mechanisms. Procedures or activities that would weaken the host, such as malnourishment, lack of sleep, and immunosuppressive drugs, should be avoided. Vaccines are very effective in preventing some viral infections and a new generation of vaccines against additional viral disease is being developed using genetic engineering technology. Viral infections may cause sufficient tissue damage to predispose a patient to secondary bacterial infections. In such cases, chemotherapy to combat the bacterial complications would be appropriate.

The impact of viral diseases on humans is most

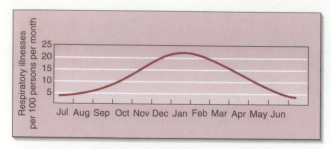

FIGURE 29-4 An approximate representation of the number of respiratory illnesses during each month of the year in persons living in the United States.

apparent with infections of the respiratory tract. It is estimated that about 75% of all infectious diseases are of the respiratory tract and about 80% of these respiratory diseases are caused by viruses. A great majority of these viral infections are of the upper respiratory tract and cause such clinical diseases as common colds (*rhinitis*) and sore throats (*pharyngi-*

tis). A fair number of infections, however, involve the lower respiratory tract and may cause serious problems, particularly in infants and small children. Approximately 150 different viruses are associated with these respiratory infections and it is estimated that, on the average, each person will experience about six such infections per year. Because of these infections, approximately 100 million work days are lost per year in the United States. In addition, they are responsible for 50% of the absences from school and a third of all patient visits to general practitioners or clinics. Most of these common respiratory infections occur during the cooler months, with peak incidences in the winter, although some are seen year-round (Figure 29-4). The second most prevalent type of viral disease involves the gastrointestinal tract. Over 50% of the cases of gastroenteritis in humans are caused by viruses; and viral gastroenteritis is a major cause of death of undernourished infants.

The viruses causing diseases of humans will be discussed in the following chapters.

MATERIAL FOR REVIEW

CONCEPT SUMMARY

1. Intracellular multiplication of viruses may result in one or more of several effects on the host cell. These responses vary from rapid and complete destruction of the host cell to long-term coexistence within the cell with no apparent adverse effect.

2. Cell transformation is a unique response to viral infection in which the host cell is not injured but is changed from its normal function. Viral-induced tumors are the result of this type of host–parasite interaction.

3. Response of the host cell to viral infection includes the production of interferons, which increase the protection of other cells to viral replication.

STUDY QUESTIONS

1. List the possible outcomes of viral infection of a host cell.
2. What aspect of host cell–virus interaction is essential if cellular

transformation is to be the result of virus infection?
3. Why haven't scientists been able to produce interferon in

experimental animals for use in treatment of human disease?
4. Write a paragraph on your own experience with viral infections.

CHALLENGE QUESTIONS

1. Koch's postulates ∞ (see Chapter 13, p. 197) have been used to determine the agent responsible for a disease. Would the application of his postulates be

useful in determining viral disease ideology? Explain.

500nm

VIRUS

In terms of clinical significance the adenoviruses and poxviruses provide an interesting contrast. The major human disease caused by a poxvirus is smallpox, a disease of great historical significance that has been responsible for untold human misery and suffering. Among the first known viral diseases, smallpox was the first for which immunization was available. It has now been completely eradicated as a result of extensive worldwide disease preventive efforts. On the other hand, adenoviruses are a large group of viruses whose clinical significance is still being described, and for which only limited immunization is possible. These viruses cause some of the most common respiratory diseases of humans, but have also recently been implicated as a source of human disease ranging from conjunctivitis to gastroenteritis. The properties and clinical role of these two groups of DNA viruses are presented in this chapter.

ADENOVIRUSES

Adenoviruses are widespread in nature and currently about 47 distinct serotypes have been isolated from humans. They are designated types 1, 2, and so on, to 47. Additional adenoviruses are also found in various avian and nonhuman mammalian species. The adenoviruses were so named because they were first isolated from *tonsils* and *adenoids*. Many adenoviruses have been shown as the cause of some common upper respiratory infections of humans and some are associated with diarrheal diseases in infants. Less frequently, they may cause infections in other tissues. The adenoviruses have

Adenoids   Nasopharyngeal lymphoid tissue.

Human Body Systems Affected

Adenoviruses and Poxviruses

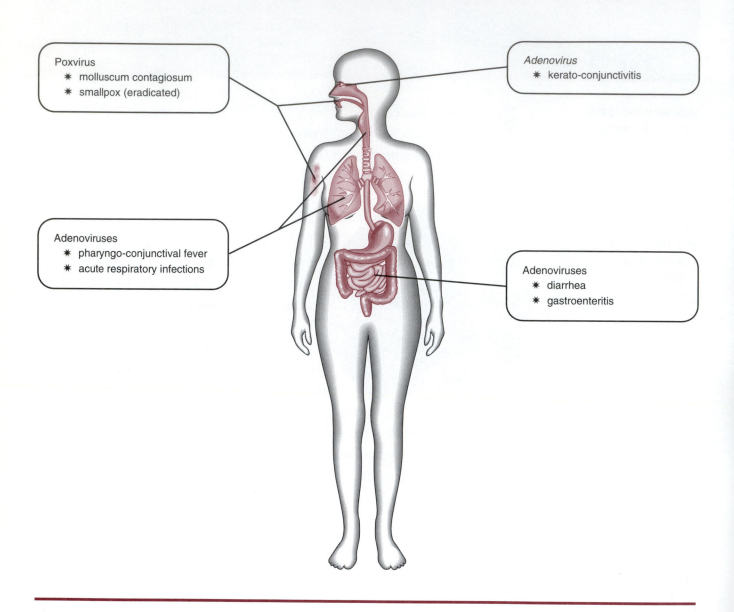

Poxvirus
* molluscum contagiosum
* smallpox (eradicated)

Adenovirus
* kerato-conjunctivitis

Adenoviruses
* pharyngo-conjunctival fever
* acute respiratory infections

Adenoviruses
* diarrhea
* gastroenteritis

the ability to remain sequestered in lymphatic tissues and tissues of the intestinal tract long after the primary exposure. About 50% of surgically removed tonsils and adenoids are found to contain some of these latent adenoviruses.

Adenoviruses have a distinct appearance and can often be identified by direct observation with an electron microscope (Figure 30-1). The capsid is 60 to 90 nm in diameter and has cubical symmetry with 252 capsomers, characteristic spikes, and no envelope. These viruses contain ds-DNA and can be cultivated in various types of cell cultures. Adenoviruses are quite stable and are able to survive for long periods once expelled from a host.

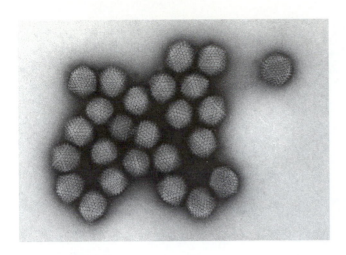

FIGURE 30-1 Transmission electron micrograph of adenoviruses magnified 150,000×. (Courtesy Visuals Unlimited)

✳ Pathogenesis and Clinical Diseases

Acute Respiratory Diseases.
Close to 100% of the children living under crowded conditions become infected with adenoviruses early in life. The infectivity rate is less in children living in less crowded conditions, but is still 50% or more. About half the infected children develop definite clinical symptoms. The most common clinical disease is an *acute respiratory infection* with such common symptoms as sore throat, cough, runny nose (*coryza*), and headache. Tonsillitis and pneumonia may be seen occasionally. Studies have shown that about 5% of acute respiratory illness in children is due to adenovirus. The most common serotypes in children are types 1 through 7. Adenovirus infections are also seen in adults but generally are less frequent and less severe.

A unique form of acute respiratory disease caused by adenoviruses has been seen in military recruit training centers. Typically, outbreaks among the recruits are caused by type 4, 7, and 21 adenoviruses. The reason for this unique epidemiologic pattern among military recruits is not known, but may be associated with assembling large numbers of young persons from diverse backgrounds and subjecting them to the crowding and "stresses" of military life.

Pharyngoconjunctival Fever and Keratoconjunctivitis.
Adenovirus-induced *pharyngoconjunctival fever* is a disease most often caused by types 3, 7, and 14 and characterized by fever, *conjunctivitis* (inflammation of the eyelids), and sore throat. This disease occurs in sporadic outbreaks associated with swimming pools, summer camps, and small lakes. It usually involves children and young adults. All evidence indicates that transmission is by direct contact with contaminated water. Several outbreaks have occurred in swimming pools that were not properly chlorinated. The incubation period is 6 to 9 days.

Keratoconjunctivitis is an infection transmitted by adenovirus-contaminated instruments or medications in ophthalmological (eye) clinics. The adenoviruses are quite stable and often survive disinfection procedures that use alcohol and many other common chemical disinfectants. This syndrome starts with inflammation of the eyelid and spreads to the cornea. Damage to the cornea may lead to impaired vision. The incubation period is 8 to 10 days.

Gastroenteritis.
Besides their extensive multiplication in the respiratory tract, adenoviruses are also able to grow in and are shed from the intestinal tract. It is becoming increasingly apparent that some of the adenoviruses—for example, adenovirus 40 and adenovirus 41—are major causes of diarrhea in children. Although the full extent of these diseases is not known, in the United States it appears that about 15% of the cases of viral gastroenteritis of children that require hospitalization are caused by adenoviruses. Some scientists have suggested that adenoviral gastroenteritis is a common cause of mortality among children in developing countries.

✳ Transmission

Spread is from person to person by direct and indirect contact. The fecal–oral route is probably a major means of spread in children and the airborne route affects all age groups. Contact with contaminated materials is the major route of transmission for eye infections. Immunocompromised persons such as organ transplant recipients or AIDS patients are at increased risk of infection due to adenoviruses. Recovery from infection provides a solid, long-term immunity against reinfection by the same type. Maternal antibody provides short-term protection

---

Keratoconjunctivitis An infection of the eye surfaces that leads to inflammation of the conjunctiva as well as the cornea of the eye.

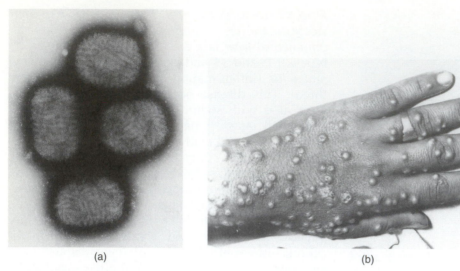

(a)   (b)

FIGURE 30-2   (a) An electron micrograph of poxviruses. (b) Lesions (pox) on the hand of a patient with smallpox. (a. Courtesy of Visuals Unlimited. b. Armed Forces Institute of Pathology, AFIP MIS #CA 44271-3)

for many newborn infants. However, antibody protection is type specific.

✳ Diagnosis

Adenoviruses can be isolated on a variety of cell cultures. Isolation should be correlated with a specific rise in antibody titer before an adenovirus can be considered the cause of a specific disease, for adenoviruses are frequently isolated from healthy persons. Adenovirus respiratory infections in the general population cannot be distinguished on clinical grounds from the many respiratory infections caused by a variety of other infectious agents. The specific laboratory tests required to identify an adenovirus infection are generally not done.

✳ Treatment

No treatment is available other than measures taken to relieve symptoms.

✳ Prevention and Control

Only in training centers for military recruits are measures taken to prevent adenovirus infections. A live *attenuated vaccine* against types 4, and 7 when given by the oral route to military recruits, has suc-

cessfully controlled the epidemics of acute respiratory disease syndrome caused by these viruses. Eye clinics should use proper methods of sterilization and asepsis to prevent the spread of adenoviruses.

POXVIRUSES

Poxviruses are large, 250 × 300-nm, brick-shaped, DNA-containing viruses. These are the most complex virus structures. Many animal species have their own specific poxvirus infections, usually in the form of lesions (pocks) on the skin. The major poxvirus of humans is the *variola virus*, which causes the disease of smallpox (Figure 30-2).

✳ Smallpox

Smallpox has been one of the most devastating diseases of humans. History is replete with reports of epidemics in which more than 50% of the inhabitants of a city or country developed smallpox, with as many as half the infected dying of the disease.

A new milestone in medical science was achieved with the eradication of smallpox from the world. It marks the first time that medical science has been able to eradicate a major infectious disease. Because future workers in medicine will not encounter

---

Attenuated vaccine   A vaccine made from living microorganisms that have been selected for their lack of virulence.

Outbreak of Pharyngoconjunctival Fever at a Summer Camp: North Carolina, 1991

On 19 July 1991, the Communicable Disease Section of the North Carolina Department of Environment, Health, and Natural Resources (DEHNR) was notified that an outbreak of acute upper respiratory illness had occurred in campers and counselors at a 4-week summer camp. Manifestations of the illness included pharyngitis, cough, fever to 104°F (40°C), headache, myalgia, malaise, and conjunctivitis. On 2 August, the DEHNR was notified of a similar outbreak during a second 4-week session at the camp. The epidemiologic investigation, initiated by the DEHNR on 7 August, identified the cause as pharyngoconjunctival fever (PCF) associated with infection with adenovirus type 3. This report summarizes findings from the investigation.

Approximately 700 persons swam each day in a 1-acre, man-made pond that had a maximum depth of 10 feet. Well water was continuously pumped into the pond at multiple sites through pipes located 1 foot below the surface of the water; the water overflowed, through a spillway, into an adjacent river. An automatic chlorination system treated the water before it entered the pond. The pond water was turbid, and plants grew in the bottom of the pond.

Every camper swam at least once during the 4 weeks; 158 (90%) of 175 swam one or more times per day. The attack rate for campers who swam daily (74 [48%] of 154) did not differ significantly from that for campers who swam less than once per week (11 [65%] of 17). The attack rate for staff who swam was higher than that for staff who did not swim (10 [77%] of 13 versus 13 [54%] of 24) and increased with increased frequency of swimming. The attack rate for nonswimmers was 54% (13 of 24); for infrequent swimmers (i.e., those who swam once per week or less), was 75% (six of eight); and for frequent swimmers (i.e., those who swam three or more times per week), was 80% (four of five). Of the 221 campers and staff members interviewed, 75 reported whether they had shared a towel with another person. Towel sharing increased the risk for illness.

Of viral cultures (nasopharyngeal and throat swabs) obtained from 25 ill persons, 19 grew adenovirus serotype 3. Convalescent geometric mean titers (GMT) to adenovirus for persons with cases during sessions one and two (GMT 14 and GMT 28, respectively) were each significantly higher than the GMT of persons not meeting the case definition (GMT 6). Bacterial analysis of grab samples of water obtained from the pond yielded 80 colonies per 100 cc of fecal coliforms, 200 colonies per 100 cc of enterococcus, and 9000 colonies per 100 cc of staphylococcus. A concentrated sample of pond water drawn approximately 6 feet below the surface yielded adenovirus serotype 3. Residual chlorine was not detectable.

One week after the end of the second session the pond was drained, and most counselors left. No further outbreaks were reported following the second session; however, all subsequent sessions during the summer and fall were of maximum 1-week duration.

Editorial Note: The illness described in this outbreak is consistent with PCF, a syndrome caused by adenovirus (especially serotypes 3 and 7). As in previous reports, three routes (person to person, fomites, and water contact) probably transmitted the virus in this outbreak. Because of the turbidity of water in soil-bottom reservoirs, chlorination is ineffective (*MMWR* 41:343, 1992).

---

smallpox, the pathogenesis, transmission, diagnosis, and control of this disease are not discussed. Instead a historical synopsis of the control of smallpox is presented, with emphasis on those concepts that may be helpful in understanding the control of infectious disease in general.

✳ Control of Smallpox

Before the nineteenth century, smallpox was controlled in a limited way by purposely inoculating persons who had not yet contracted it with a mild form of the disease. The severe form of smallpox was called *variola major* and the mild form *variola minor* or *alastrim*. Recovery from either form conferred immunity against both forms. Inoculation with variola minor usually caused mild symptoms with a fatality rate of "only" 1 to 2%. Many considered it an acceptable risk compared to that of contracting and dying of the major form of smallpox. This practice of purposely inoculating persons with variola minor, called *variolation*, was pursued in Africa and Asia from ancient times and was used to a limited extent in some areas of North America during the eighteenth century.

The most significant contribution toward the

eventual control of smallpox was the development of the vaccination procedure by Edward Jenner in 1796 ∞ (Chapter 1, p. 4), using the cowpox virus. Afterward, the practice of vaccination against smallpox became widespread in most developed countries and a steady decline in the total number of cases occurred over the ensuing years. The last cases of smallpox seen in the United States were in the lower Rio Grande Valley in 1949. The final struggle with smallpox began in 1966 when the World Health Organization embarked on a program of complete eradication of this disease. Much of the program's success was made possible by the development of a **freeze-dried** form of vaccine that was stable without refrigeration and was very effective, and the methods of administration were easy-to-learn and inexpensive. These developments made it possible to carry out effective vaccination programs in remote areas of the world. As a result of this concentrated eradication program, by 1972 all countries in North and South America were free of smallpox. By 1975 only Bangladesh, Ethiopia, and surrounding areas reported cases. After a slight reversal in early 1977, this eradication program moved to its successful conclusion. The last person to have a naturally acquired case of smallpox was a hospital cook in Somalia who came down with the disease in October 1977. After a 2-year waiting period in which no new cases occurred, the world was officially declared free of smallpox in October 1979 (see Clinical Note). Actually, the last known cases of smallpox occurred from accidental infections in a research laboratory in England in 1978. By 1992, only two reference laboratories, one in the United States and one in Russia, possessed cultures of the smallpox virus.

Although smallpox was a severe and highly contagious disease, it possessed features that permitted its successful eradication. First, the disease was easily recognized and few if any subclinical cases occurred. This factor was a help in detection and diagnosis. Second, the virus did not remain in the body as a latent infection after the patient recovered from the clinical disease. Third, no regular nonhuman host could act as a *reservoir* of the virus. Fourth, an effective, easily administered vaccine was available, and, fifth, the virus did not readily mutate.

The most notable result of this eradication program in countries where smallpox was not normally found was the discontinuation of routine smallpox vaccination. The smallpox vaccination used a living virus, called the *vaccinia virus*, and this virus caused a mild infection in the form of a single pock at the site of inoculation. However, in some persons, mostly young children with compromised host defense mechanisms, the vaccinia virus might spread beyond the single pock to cause moderate-to-severe infections. Before smallpox vaccination was discontinued in the United States in 1971, about 500 children per year developed a serious vaccine-associated disease and about a dozen deaths would occur. Thus, for many years in the United States and many other countries, many more children suffered or died from the ill effects of the vaccine than from the disease itself.

While the vaccinia virus is no longer used as a vaccine against smallpox, it is finding a possible new use in current recombinant DNA technology. Because of the nature of the poxvirus DNA, it is possible using genetic engineering methods to add multiple foreign genes to this DNA. Thus, it is possible to engineer a vaccinia virus that is both able to carry antigens from a variety of different microorganisms and is also attenuated so it cannot cause any direct illness other than a mild single pock. Such an engineered vaccinia virus can be made stable by freeze-drying, is easy to administer, is relatively inexpensive to produce, and it could be used as a vaccine against a variety of diseases that are still prevalent in children of developing countries.

✳ Other Poxvirus Infections

A variety of poxvirus diseases common among rodents, domestic animals, and some primates occasionally cause disease in humans. These diseases often have interesting names such as monkeypox, ORF, tanapox or cowpox. They have not been well studied and reservoirs and sources of transmission to humans have not been totally documented. One poxvirus disease, *molluscum contagiosum*, is transmitted from person to person. The lesions are small, pink, wart-like tumors that present most commonly on the face and back. This disease can be sexually transmitted and there has been an increase in the disease among sexually active young adults. The lesions of molluscum contagiosum will spontaneously regress in two or three years.

Freeze-dried   Referring to those materials frozen, after which the water is removed without melting.
Reservoir   A site where microbes persist and retain their ability for infection.

MATERIAL FOR REVIEW

CONCEPT SUMMARY

1. Over 40 types of adenoviruses have been discovered. These DNA viruses are responsible for a variety of infections in humans that may be severe but are usually self-limiting. Most common adenovirus infections occur in the respiratory tract, but conjunctivitis and intestinal infections are not infrequent.

2. Smallpox, a life-threatening infection by the DNA pox virus, has been eradicated from the ranks of human illness. Careful, thorough, and prolonged immunization efforts have apparently rid the world of this dreaded disease.

CLINICAL SUMMARY TABLE

Microorganism	Virulence Mechanisms	Diseases	Transmission	Treatment	Prevention
Adenoviruses	47 serotypes Latency	Upper respiratory Diarrhea Conjunctivitis	Airborne Contact	Symptomatic	Aseptic procedures Vaccine (in military)
Poxviruses		Smallpox	Contact		Vaccination

STUDY QUESTIONS

1. What property of the adenoviruses make them good candidates to produce latent viral infections?

2. The fact that available adenovirus vaccine is limited to types 4, 7, and 21 suggests a number of epidemiologic considerations. List two factors that relate to this situation.

3. What features of adenovirus and adenovirus infection reduces the likelihood of the development of a general vaccine for these agents?

4. What features of the pathogenesis of smallpox made immunization such an effective means of control?

5. Not everyone in the United States has been immunized against smallpox. Why were immunizations suspended before everyone had an opportunity to be immunized?

CHALLENGE QUESTIONS

1. By the end of 1996, all remaining stocks of the smallpox virus are to be destroyed. However, two groups have argued this point. One group agrees that remaining stocks of the virus should be destroyed, eliminating one of the worst viral diseases on earth. The other group suggests that it should not be destroyed because we may learn something about viral transmission and replication by continuing to study the virus. What do you think?

31 Herpesviruses

500nm

VIRUS

Members of the family *Herpesviridae* are widespread in nature and infect many animal species. These viruses are usually host specific; that is, the human herpesviruses cause natural infections only in humans and not in other animals and vice versa. On rare occasions, monkey herpesviruses have caused fatal infections in humans.

The common diseases of humans caused by herpesviruses are cold sores or fever blisters, genital herpes, chickenpox, infectious mononucleosis, and cytomegalovirus infection. Exposure to these viruses generally occurs in the first decade of life in those living in conditions of poor hygiene and the resulting infection may be apparent or subclinical. An important characteristic of these viruses is their ability to remain sequestered inside certain cells long after the primary infection is resolved. These latent infections may persist for the life of the person with no further clinical symptoms or they may cause sporadic recurrences of clinical symptoms. The human herpes viruses—herpes simplex, varicella–zoster, cytomegalo, Epstein–Barr, and other recently discovered human herpesvirus—are discussed in this chapter.

HERPESVIRUSES: GENERAL CHARACTERISTICS

The capsid of the herpesviruses is about 100 nm in diameter, has cubical symmetry with 162 capsomers, and is surrounded by an envelope (Figure 31-1). There is a space between the envelope and capsid that contains a number of viral enzymes and other proteins that assist in the replication process. The herpesviruses are fairly unstable. After being shed from the body, they become inactivated within hours at room temperature and are readily destroyed by drying, soaps, and disinfectants. The nucleic acid is double-stranded (ds) DNA and the viruses multiply in the nucleus of the host cell (Figure 28-8). Replication of the viral genome is under the control of enzymes encoded on the viral

HUMAN BODY SYSTEMS AFFECTED

Herpesviruses

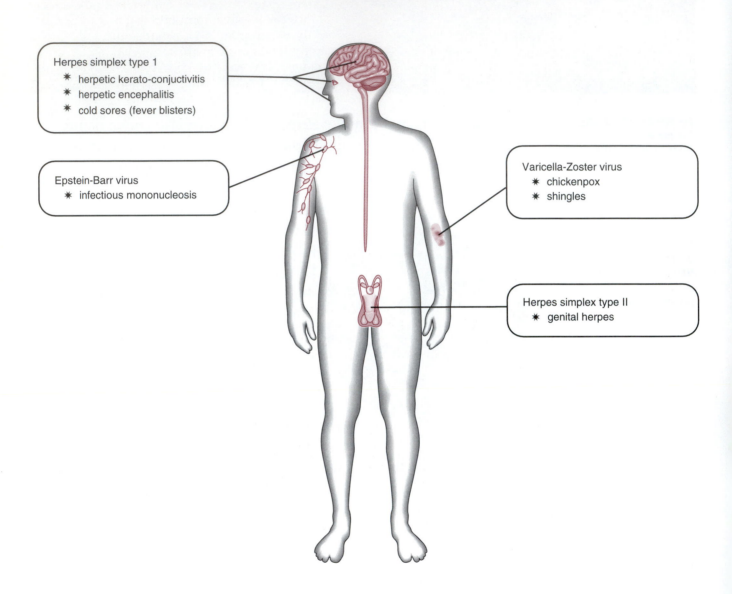

Herpes simplex type 1
* ✳ herpetic kerato-conjuctivitis
* ✳ herpetic encephalitis
* ✳ cold sores (fever blisters)

Epstein-Barr virus
* ✳ infectious mononucleosis

Varicella-Zoster virus
* ✳ chickenpox
* ✳ shingles

Herpes simplex type II
* ✳ genital herpes

DNA. These enzymes, such as *DNA polymerase* ∞ (Chapter 6, p. 68), *thymidine kinase,* and *ribonucleotide reductase,* are different from those produced by human cells and are the targets for antiviral compounds used in treatment of herpesvirus infections.

Under certain conditions the viral DNA is able to integrate into the host cell DNA to establish a *latent infection.* This association of foreign DNA with the DNA of host cells has provided theoretical evidence that herpesviruses may induce cancer. Some forms

Thymidine kinase   An enzyme needed to produce the DNA nucleotide thymidine phosphate.
Ribonucleotide reductase   An enzyme needed to synthesize deoxyribonucleotides for DNA snythesis.

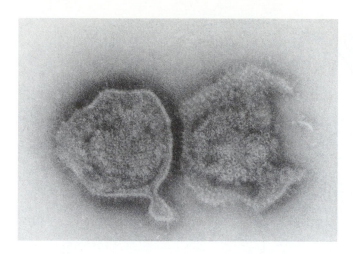

FIGURE 31-1 Transmission electron micrograph of herpesviruses showing icosahedral capsids (partially disrupted) surrounded by envelopes (magnified 160,000×). (Courtesy Robley C. Williams, University of California, Berkeley)

of cancer in lower animals are associated with these viruses and several lines of evidence associate herpesviruses, particularly the Epstein–Barr virus, with certain forms of cancer in humans. Women who have cervical herpes infection have a significantly greater incidence of cervical cancer than uninfected women.

HERPES SIMPLEX VIRUSES

Two serotypes of herpes simplex viruses (HSV) have been identified. Type 1 is generally associated with infections of the upper half of the body and type 2 with infections of the genitourinary tract and surrounding tissues. Primary infection usually occurs on the mucosal–epithelial surfaces of the body. Following initial infection, the neurons that innervate the area become infected. The primary site of infection is characterized by a lesion, while infection of the neurons leads to latent infection. Both types may cause disseminated infections in infants and compromised patients.

✳ Pathogenesis and Clinical Diseases

Cold Sores or Fever Blisters. Among the most common of all human infections are *cold sores* or *fever blisters* (herpes labialis), which are usually

caused by the type 1 HSV (Figure 31-2). The recurring lesions on the lips are the clinical manifestation of a complex chronic interaction between the virus and the host. Most newborn infants are not readily infected, possibly as a result of passive immunity that offers some protection against primary infection. Once the passive immunity is gone, the infant is highly susceptible to primary infection. Susceptibility tends to decrease somewhat as the child gets older. However, in conditions of poor sanitation as many as 90% of the population has been infected before adulthood. Persons living under conditions of improved sanitation experience about a 50% infectivity rate. The primary infection is often asymptomatic or is not diagnosed as herpes. Symptoms are seen in 10 to 15% of the cases from 2 to 12 days after being exposed to the virus. The primary lesions may appear as small **vesicles** in the throat, mouth, or nose and go relatively unnoticed. The most noticeable form of primary infection involves the lips, mouth, and gums (*gingivostomatitis*), in which the vesicles rupture and develop into ulcerative lesions. Fever, pain, and irritability usually persist for about 1 week, followed by gradual healing during the second week.

Recovery is associated with a rise in antibodies against the virus. During the primary infection, however, the virus passes along nerve fibers to regional **ganglia**. In the case of gingivostomatitis,

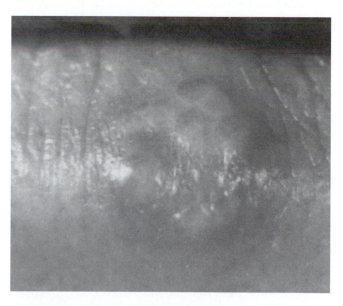

FIGURE 31-2 Herpes simplex fever blister on lower lip 2 days after onset. (Centers for Disease Control, Atlanta)

Vesicle A blister-like structure that contains a clear serous fluid.
Ganglion Major nerve trunks connecting the peripheral nerves to the CNS.

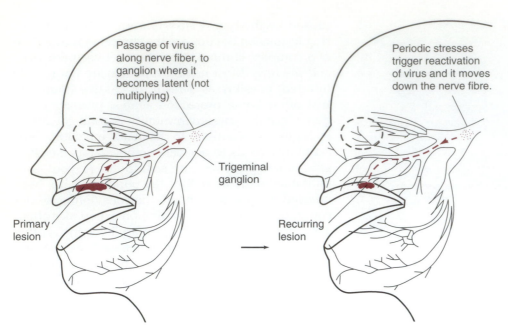

FIGURE 31-3 Aspects of the pathogenesis of primary and recurring herpes infections of the lips.

the *trigeminal ganglion* is commonly involved and the virus becomes sequestered in a latent form in this tissue. While in this latent form, the virus cannot be detected by ordinary means. It causes no symptoms and is not affected by antibodies. Periodically, in from 20 to 30% of the general population, these latent viruses become activated and move down the nerve fiber to cause recurring skin lesions at the site of the original infection (Figure 31-3). The frequency of these recurring lesions varies from person to person, ranging from once every few years to about once a month. Various stressful stimuli, such as excessive sunlight, fever, cold winds, emotional stress, and hormonal changes, apparently trigger the reactivation of the virus. As the virus moves down the nerve fibers, it passes directly into the skin cells without becoming exposed to the host's antibodies. The antibodies usually prevent the virus from spreading systemically to other tissues of the body but are unable to prevent the recurring lesions. Recurrent infections are generally less severe than primary infection. Primary and recurring infections may also occur in the eyes, causing a disease known as *herpetic keratoconjunctivitis*. Lesions on the cornea are most serious, for the accumulating scar tissue may lead to vision impairment.

Occasionally, almost any tissue of the body may become infected with these viruses. Primary and recurring infections may occur on any cutaneous area of the body. Traumatic injury may provide a portal of entry for primary infections that may develop in both children and adults. Such infections have been seen in wrestlers (herpes gladiatorum) due to skin abrasions, or in persons following burns, or on the thumb of a thumb-sucking child or the finger of a dentist (herpetic whitlow). Children with *eczema* may acquire a serious herpes infection over large areas of the body (eczema herpeticum). Herpes simplex viruses may also infect the central nervous system, causing a severe, and often fatal, infection (herpetic encephalitis).

Genital Herpes. A very common sexually transmitted disease is *genital herpes* (see Plate 38). Over 80% of these infections are caused by HSV type 2. In females the vesicles usually occur in the mucosal tissue of the vulva, vagina, or cervix, but any of the genital or surrounding tissues may be involved. These vesicles ulcerate, producing shallow lesions. The symptoms may include malaise, urinary retention, local pain, fever, vaginal discharge, and tender, swollen inguinal lymph nodes. In males the vesicles and subsequent ulcerations

Eczema Inflammation of the skin, often associated with scaling, papules, crusting, and serous discharge.

develop on the penis and surrounding tissues. These infections heal spontaneously in 2 to 4 weeks. Recurrent infections are common but are generally milder than the primary infection. All ages seem susceptible to genital herpes, but most infections are seen in young adults; it is estimated that about 500,000 cases of genital herpes occurs per year in the United States, and up to 60% of persons in some socioeconomic groups are infected.

A disseminated form of herpes occasionally occurs in newborns, particularly premature infants, and here the virus spreads throughout the body. These infections are usually fatal. Infections are acquired most often during birth as the infant passes through an infected birth canal. Disseminated herpes may also occur in patients of any age when they become immunosuppressed. HSV type 1 can cause genital herpes, and HSV type 2 may cause cold sores. HSV type 2 is also associated epidemiologically with an increased prevalence of cervical cancer.

✳ Transmission

HSV type 1 is transmitted in saliva by direct contact or by indirect contact with such objects as eating utensils, toothbrushes, and drinking glasses. Kissing an infected person leads to easy transmission. A typical mode of transmission is for an adult with an active recurring lesion to kiss or fondle a young child. Large numbers of viruses are present in the vesicular fluid and exudate from the ulcerative lesions. Transmission within family units is difficult to prevent. Some persons may be intermittent subclinical shedders of HSV, which further complicates attempts to prevent transmission of this agent. In general, transmission is much more efficient among persons living in crowded, unsanitary conditions.

HSV type 2 is spread primarily by sexual contact. HSV type 2 can be isolated from 10 to 15% of the patients who visit clinics for sexually transmitted diseases. Infected females may transmit this virus to infants during delivery. Such neonatally acquired HSV type 2 infections are generally serious and may result in death of the infant.

✳ Diagnosis

Most diagnoses are made on the clinical appearance. Disseminated infections and atypical compli-cated cases often require a laboratory diagnosis. This diagnosis involves isolating HSV from the lesion early in the illness. HSV multiplies well in some commonly used cell cultures. HSV can be detected in cell cultures after about 24 hours by the use of specific *monoclonal antibodies* ∞ (Chapter 11, p. 161) tagged with fluorescent dyes or enzymes that can differentiate HSV-1 from HSV-2. A correlation of *antibody titers* (a significant rise) is often needed to confirm a diagnosis, for these viruses may be found in normal-appearing tissues. HSV produces a characteristic *inclusion body* in the nucleus of infected cells; such intranuclear inclusions may be seen on microscopic examination of stained cells that are scraped from a suspected lesion. Direct fluorescent antibody tests or the use of specific DNA probes on biopsied tissues may also give a rapid diagnosis of a HSV infection (see Plate 36).

✳ Treatment

The herpetic infection of the cornea was the first viral infection to be routinely treated by chemotherapy. The chemical *iododeoxyuridine*, used on corneal infections, interferes with the replication of herpesviruses by inhibiting DNA synthesis.

In the 1970s, a compound called *trifluridine* became available to treat HSV keratitis. A third antiviral compound, called *adenine arabinoside*, has also been licensed for use in the United States. A fourth compound, called *acyclovir*, is currently available and has advantages over the earlier developed agents in that acyclovir is less toxic to humans than the other agents. It is currently the drug of choice in treating herpesvirus infections. An ointment of acyclovir effectively treats HSV keratitis. When given intravenously it helps treat serious disseminated herpes infections. In cream form it is able to shorten the duration of recurrent fever blisters and has some inhibitory effects on genital herpes when given topically, orally, or intravenously. Acyclovir is not effective in curing the latent state of HSV.

✳ Prevention and Control

Whenever possible, persons with active herpetic lesions should avoid contact with young children. Children with primary lesions should be isolated from other children until the lesions have healed. The virus can be shed from crusted lesions as well

---

Antibody titer   The concentration of antibody present in the blood serum.

Vaccine for Genital Herpes Shows Promise in NIAID Study

The National Institute of Allergy and Infectious Diseases (NIAID) reports that an experimental vaccine has safely reduced by a third the number of genital herpes outbreaks in people with previous recurrences of the disease. According to NIH officials, the research provides the first clinical trial evidence that a vaccine used as therapy can modify the course of a chronic viral infection.

The recombinant vaccine, which was supplied by the Biocine Company (Emeryville, CA), contains the glycoprotein D of the herpes simplex virus type 2 (HSV-2) and alum. More than 25 million Americans are infected with the disease.

During the year of the study, vaccine recipients reported 36% fewer confirmed herpes outbreaks per month and close to a 33% lower average number of recurrences than did placebo recipients.

"The study is a step forward in our search for better treatments for people with genital herpes," says Anthony S. Fauci, M.D., director of NIAID. "The research shows that an immunotherapeutic vaccine,

one designed to enhance the body's immune response to HSV-2, may modify the course of this disease."

In the study, 98 patients with documented genital herpes infections that recurred 4–14 times per year enrolled at one of two sites—NIH (Bethesda, MD) or the University of Washington in Seattle. Comparable numbers of men and women enrolled at each site.

In the study, the vaccine significantly reduced the monthly rate and total number of outbreaks, and it prolonged the time to the first recurrence. An NIH spokesperson says that these findings suggest that the vaccine altered the immune response to HSV-2.

"While the vaccine was less effective than we had hoped, it proves that the concept of a therapeutic vaccine is possible and encourages us to continue to pursue potentially more effective formulations of a vaccine," says Stephen E. Straus, M.D., chief of the laboratory of clinical investigation for NIAID and lead researcher on the study. "Such studies are now underway in our NIAID clinics." (*Genetic Engineering News*, 15 June 1994)

---

as vesicles and persons should be considered infectious until the normal epithelial tissues reappear. Persons with HSV type 2 genital herpes should refrain from sexual contact when active lesions are present. Condoms may not be fully protective against transmission. Pregnant females with genital herpetic infections may need to have a Caesarean delivery to help avoid infecting the newborn. Dentists and other medical personnel who come in contact with oral secretions of patients should wear gloves.

Evidence suggests that antibodies are able to prevent the primary infection but not the recurring lesions. This finding has stimulated research on a possible vaccine that could be given to young children before they lose their passive immunity. If the primary infection were prevented in children, then it would follow that the recurring lesions later in life would also be prevented. Using advances in genetic engineering, several approaches are being used to develop vaccines against herpes simplex

virus infections (see Clinical Note). However, no vaccines are currently available for general use.

VARICELLA–ZOSTER VIRUS

The varicella–zoster virus (VZV) is the causative agent of both **chickenpox** (varicella) and **shingles** (zoster), hence the double name. For many years it was thought that these two diseases were unrelated. Now it is known that chickenpox is the acute primary form of the disease complex, whereas shingles is a delayed recurrent form of the same infection (Figure 31-4). Only a single serotype of the VZV exists.

✳ Pathogenesis and Clinical Diseases

Chickenpox.   The VZV virus enters the respiratory tract by the airborne route. The virus then

---

Chickenpox   A highly contagious childhood disease involving skin lesions caused by the varicella–zoster virus.
Shingles   A severe vesicular eruption resulting from latent viral infection of the dorsal root ganglia by the varicella–zoster virus.

apparently passes through the lymphatic system and is disseminated throughout the body via the bloodstream, and a latent infection develops in the neuroganglia. After an incubation period of 10 to 20 days, the virus is found to be multiplying in numerous foci of the deep skin and in the throat. General discomfort, together with a sore throat, may occur a day or so before the appearance of a macular (flat) rash on the skin. Within 24 hours the rash develops into vesicles. The vesicles form into *pustules* within the next 24 hours. The lesions next form scabs that fall off after a few days. Successive crops of vesicles develop and all stages of the lesions may be seen simultaneously (Figure 31-5). After about 7 days the lesions begin to disappear. Scarring usually does not result without secondary bacterial infection. The intense itching often causes scratching that, in turn, may lead to secondary bacterial infection with accompanying inflammation and scarring.

Chickenpox rarely has serious complications in otherwise healthy children. Occasionally a serious condition called *Reye's syndrome* develops during the recovery period and consists of diffuse metabolic and neurologic malfunctions. Chickenpox, however, can be severe or fatal in newborn infants, children with leukemia, and other immunosuppressed patients. Infections in adults are more severe than in children, with varicella pneumonia being a much more frequent complication. Re-

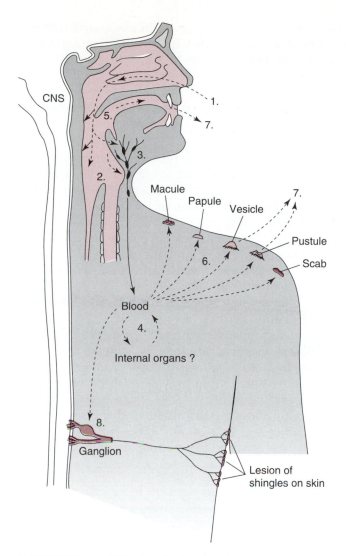

FIGURE 31-4 Pathogenesis of varicella–zoster: (1) Virus enters by airborne route. (2) Early viral multiplication in upper respiratory tract; no symptoms and patient not infectious. (3) Passage of virus into regional lymph nodes. (4) Passage of virus to blood and spread to various internal organs. (5) Several days before pock formation virus may be found in respiratory tract; patient may be infectious. (6) Virus deposited in the epithelial tissues with resultant pock formation; this occurs 14 to 21 days after exposure. (7) Virus shed from the respiratory tract and from the pocks. (8) Virus becomes sequestered in neural ganglia and remains latent. At a later time the virus may become active and moves along the sensory nerves to cause the localized lesions of shingles.

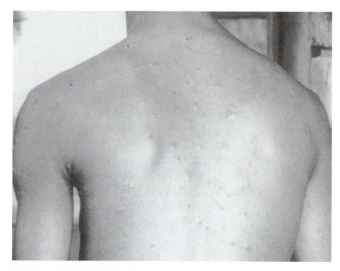

FIGURE 31-5 Chickenpox three days after onset of rash. Lesion in different stages of development can be seen. (Centers for Disease Control, Atlanta)

Pustule A small blister that is filled with pus.
Reye's syndrome A syndrome of organ changes that may occur following a viral infection. Most noticeable changes occur in mental status, and liver enlargement.

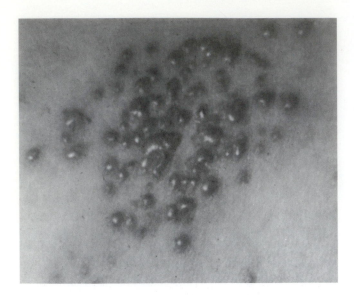

FIGURE 31-6 Vesicular lesions of shingles. (Centers for Disease Control, Atlanta)

covery confers lifelong immunity to reinfection. Nevertheless, the immune mechanisms do not remove the viruses that become sequestered in the ganglia.

Shingles. Shingles is a recurrent clinical manifestation of a latent–dormant infection with the VZV. Analogous to the latent infections with the herpes simplex virus, the VZV is able to become sequestered in neuroganglia following a typical acute chickenpox infection. About 10 to 20% of persons who are infected with VZV develop shingles at some time in their lives. The VZV may lie dormant for many years before it becomes reactivated. More than 65% of the cases occur in persons over 45 years of age; however, reactivation may occur at any age. Usually a person has only a single attack of shingles, but second episodes occasionally develop. The mechanisms of reactivation are not understood. One theory is that it may be associated with a waning of the acquired immunity. Shingles is much more common in persons who are immunosuppressed or who have irradiation or injuries to the spine. After the virus becomes activated, it moves down the sensory nerve leading from the infected ganglion. This step is accompanied by an abnormal sensation and/or pain over the area served by the involved nerve (dermatome). Several days later the eruption occurs (Figure 31-6) and is confined to the involved dermatome. The distribution is often in a band (zoster = girdle) on one side of the body with an abrupt margin at midline (Figure 31-7). The skin involvement progresses through the rash, vesicular, pustular, and crusting stages, often fusing

together into large lesions. Pain associated with the neurologic involvement may be severe for 1 to 4 weeks. Recovery occurs in 2 to 5 weeks with pain persisting (postzoster neuralgia) in some patients for an additional period of time. The skin of the trunk and face are the areas most often affected.

✳ Transmission

Chickenpox is one of the most communicable of all childhood diseases and over 90% of children become infected before the age of 10. The airborne route is the major means of transmission. Viruses are present in respiratory secretions, possibly a day or more before the appearance of a rash. Children are generally isolated after the rash appears; however, the VZV may have already been spread by respiratory droplets ∞ (Chapter 13, p. 192) to schoolmates and friends during the *prodromal* (onset) period. Viruses are also found in the vesicular fluid and pustular exudate during the first week of illness. Viruses are readily spread from this source by both direct and indirect contact and outbreaks are much more prevalent in the winter and spring (Figure 31-8).

Shingles is not transmitted from person to person; however, viruses are shed from the zoster lesions and susceptible children are able to contract chickenpox from this source. Thus, the reactivation

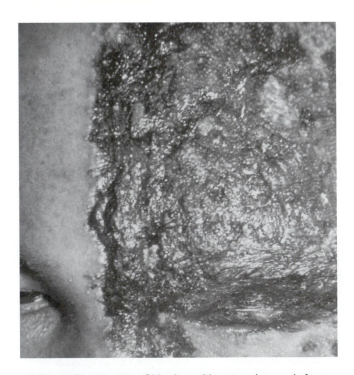

FIGURE 31-7 Shingles with extensive scab formation in a college-age student about 12 days after onset. Classical midline distribution is seen.

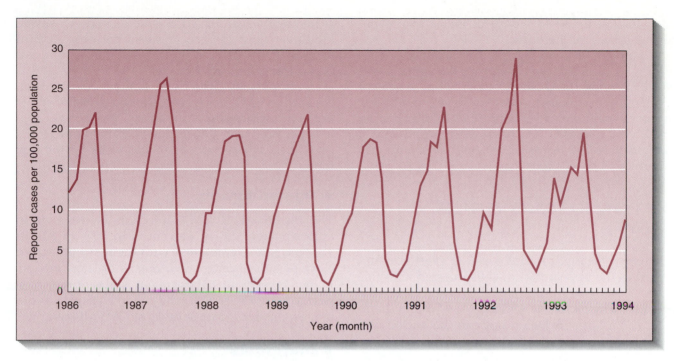

FIGURE 31-8 Reported cases of chickenpox per month in the United States, 1986–1993. Seasonal variations are consistently seen with peak incidence rates in March, April, and May. A total of 134,722 cases were reported in 1993.

of the VZV in zoster patients serves as a source that reintroduces the virus into a susceptible population year after year.

✳ Diagnosis

Chickenpox is usually diagnosed on clinical appearance. Occasionally, an atypical case may require laboratory diagnosis. Clinically, chickenpox resembles smallpox; when both diseases were present, it was important to make a rapid differential diagnosis. Since the eradication of smallpox, it is no longer important. Serologic tests are sometimes done on immunocompromised patients and on pregnant females near term who have no history of prior infection but have recently been exposed to chickenpox. This helps determine if they should receive passive immunization with zoster immune globulin (ZIG).

Diagnosis of zoster may be difficult in the early stages. Even in the later stages it may occasionally be confused with other skin infections. In such cases, it may be necessary to isolate and identify the virus by laboratory procedures.

✳ Treatment

Acyclovir is effective in stopping the progression of the disease when treatment is started just after the rash first appears. This treatment is especially recommended for immunosuppressed patients. Treatment with acyclovir will not prevent viral latency and treated individuals may subsequently develop shingles. Symptomatic relief to relieve itching may be obtained with ointments. Secondary bacterial infections should be treated with antibiotic ointments.

✳ Prevention and Control

A living attenuated vaccine was licensed for use in the United States in 1995, after having been used effectively for a number of years in Japan and Europe. A single injection of the vaccine should be given to children 1 to 12 years of age who have not had chickenpox. For persons over 12 years of age, two injections 4 to 8 weeks apart are recommended. The vaccine is not recommended for children under 1 year of age.

Passive immunization with ZIG, obtained from persons who have recently recovered from shingles, is sometimes used to treat near-term pregnant women who are seronegative to VZV antibodies but have been exposed to chickenpox. ZIG is also administered to newborn infants of mothers who contract chickenpox just before or just after delivery and to immunosuppressed patients who become exposed, but have no prior immunity to the VZV. Persons with chickenpox or zoster should be isolated for the first 5 to 7 days of illness to prevent further spread of the virus.

A number of years ago it was noted that children who received aspirin during the acute stages of chickenpox (and some other viral diseases) appeared more likely to develop Reye's syndrome. In recent years the use of aspirin in children during acute viral diseases has been discouraged and a marked decrease in cases of Reye's syndrome has occurred.

CYTOMEGALOVIRUS (CMV)

The term *cytomegalo* means a cell of great size and describes the appearance of cells infected with this herpes virus. CMV multiplies slowly and causes the host cell to swell in size. Cells up to 40 μm in diameter with a large inclusion body in the nucleus are produced.

✳ Pathogenesis and Clinical Diseases

Most CMV infections are inapparent and the virus is able to persist in the body for long periods. In much of the world, the primary infection occurs early in life and remains subclinical in the infected person for the remainder of his or her life. The virus is shed periodically in such body fluids as saliva, urine, breast milk, cervical secretions, and semen from persons with these *inapparent infections*.

Several clinical diseases are caused by the cytomegalovirus. Congenital infections are acquired when the developing fetus becomes infected by virus crossing the placental tissues from a mother who acquired a primary CMV infection during the pregnancy. Between 0.5 and 2.5% of all newborn infants are infected. Some of these congenitally infected infants are born with severe deformities and die shortly after birth. About 1% of neonatal deaths are due to this virus. Some infants survive with serious physical and mental defects. Other congenitally infected infants survive without any obvious tissue damage. CMV is currently the leading cause of viral congenital defects. The second clinical form of this disease is an activation of an inapparent infection. It may be seen in both children and adults and is associated with prolonged immunosuppressive therapy for such procedures as organ transplants or cancer. This form of infection is frequently seen in AIDS patients and is a severe life-threatening disease in these persons. A third form of CMV disease is post-transfusion or post-organ transplant mononucleosis-like illness and is seen in persons who had not previously been exposed to CMV but received blood transfusions or organ transplants that contained CMV. Normal adolescents may also develop a mild mononucleosis-like illness when exposed to this virus for the first time.

✳ Transmission

The likelihood of CMV infection increases from approximately 2% of newborns to 50% of adults. This increase is consistent with the requirement for close personal contact as a means of transmission. In the United States, three major routes of transmission (congenital, sexual, and organ transplantation) are responsible for most cases of CMV disease.

✳ Diagnosis

CMV may be cultivated from the infected tissues or secretions in cell cultures. Cells taken from the site of infection may show the characteristic cytomegalic cells and the presence of CMV antigens can be detected using immunofluorescent procedures with monoclonal antibodies. Rapid diagnostic procedures that employ DNA probes are also available.

✳ Treatment

Chemotherapeutic agents that show some effects on other herpesvirus infections have shown little if any effect on CMV infections. Interferon treatments have possibly helped reduce the incidence and severity of reactivation of CMV infections in recipients of kidney transplants. A compound called ganciclovir (closely related to acyclovir) is used to treat infections in immunocompromised persons.

✳ Prevention and Control

Because of the large number of inapparent cases in the general population, no specific control measures are available. A live attenuated vaccine is available and genetically engineered component vaccines are being developed. Immunization of sero-negative women as well as children could greatly reduce the incidence of CMV infection. Condom use and appropriate sexual abstinence would greatly reduce the incidence of this disease.

Inapparent infection A subclinical infection. The immune response is triggered in most inapparent infections but disease is not evident.

EPSTEIN–BARR VIRUS

Epstein–Barr virus (EBV) is a widespread herpesvirus that has been incriminated as the possible cause of several forms of cancer, and is also the cause of *infectious mononucleosis.*

✳ Pathogenesis and Clinical Disease

Infection with EBV is acquired by the oral route. Viral multiplication possibly occurs in the throat and **parotid glands**. The virus also apparently passes through the lymphatic system and into the blood where the principal target is the B cells. Most of the interactions between the viruses and B cells do not destroy the B cells but *transform* them. These transformed B cells contain many copies of the EBV DNA, proliferate rapidly, and have EBV antigens on their cell membrane. These transformed cells also secrete a variety of immunoglobulins unrelated to the EBV. Some of these immunoglobulins are the **heterophile antibodies** that are useful in the diagnosis of infectious mononucleosis. T cells in large numbers are mobilized against the B cells that have EBV antigens on their membranes. The transformed B cells and T cells account for the increased number of *atypical lymphocytes* (see Plate 37) that are seen in infectious mononucleosis patients. The accumulation of these cells, along with the induced inflammation caused by the T cells destroying the infected B cells, account for the tonsillitis, swollen spleen, and enlarged lymph nodes associated with infectious mononucleosis. Specific antibodies against the EBV also develop and immunity against reinfection occurs, yet multiple copies of the EBV DNA persist in a few B cells. The immune system is not able to rid the body of these infected cells and a few of these cells shed virus. Some evidence suggests that virus may also continue to multiply intermittently in the parotid glands and be shed in the saliva, which is a principal means of spread.

Infectious mononucleosis is very mild or even subclinical in young children. From adolescence on, clinical disease occurs in about 50% of those who are infected for the first time with the EBV. Most typical clinical diseases are seen in persons between 15 and 35 years of age. The incubation period may range from 3 to 7 weeks. The onset is gradual, lasting up to 1 week, and may consist of such general symptoms as malaise, headache, fatigue, and low-grade fever. The acute phase may last 1 to 3 weeks, but it is longer in some and is characterized by intermittent high fever, generalized weakness, severe sore throat, malaise, swollen lymph nodes, enlarged spleen, occasionally a rash, and a greatly increased number of abnormally appearing lymphocytes (monocular cells) in the blood. This last characteristic is the basis for the name infectious mononucleosis. The convalescent phase, with accompanying malaise and weakness, may be prolonged. Death or serious complications are rare in otherwise normal patients. Complications range from splenic rupture and respiratory obstruction to neurologic disease such as *Guillain–Barré* syndrome. However, in persons on immunosuppressive treatment for organ transplants, on anticancer chemotherapy, or with some genetic defects, infectious mononucleosis may progress unchecked by the immune system, and death may result. A form of cancer known as Burkitt's lymphoma is associated with EBV infection in areas of the world with endemic malaria.

A condition sometimes referred to as *chronic EBV disease* has been described. Patients with chronic EBV disease have persistent or relapsing unexplained fatigue lasting for months. Sometimes this is associated with slight fever, sore throat, muscle and bone pains, and headaches. Some evidence suggests that this condition is a result of a chronic EBV infection.

✳ Transmission and Epidemiology

It now appears that most people, particularly in areas where crowding and poor sanitation exist, contract this disease early in life and experience subclinical infections. The typical clinical disease usually occurs in adolescents and young adults who escaped infection earlier in life. About 70% of adults have been infected and the virus is intermittently shed in the oral secretions of an estimated 90% of the infected population.

Parotid gland   Salivary gland located at the back of the jaw.
Heterophile antibody   An antibody that reacts with antigens other than those responsible for induction. Human heterophile antibodies will hemagglutinate sheep red blood cells.
Atypical lymphocytes   Lymphocytes that are abnormal, usually enlarged with a foamy nucleus.
Guillain–Barré syndrome   A progressive loss of myelin from nerve fibers caused by the immune system's response to a viral infection.

Herpes Gladiatorum at a High School Wrestling Camp: Minnesota, 1989

In July 1989, the Minnesota Department of Health (MDH) investigated an outbreak of herpes simplex virus type 1 (HSV-1) dermatitis (herpes gladiatorum) in participants at a Minnesota wrestling camp. The camp was held 2–28 July and attended by 175 male high school wrestlers from throughout the United States. The participants were divided into three wrestling groups according to weight (group 1, lightest; group 3, heaviest). During most practice sessions, wrestlers had contact only with others in the same group. The outbreak was detected during the final week of camp, and wrestling contact was subsequently discontinued for the final 2 days.

A case was defined as isolation of HSV-1 from involved skin or eye or the presence of cutaneous vesicles. To identify cases, a clinic was held at the camp to obtain viral cultures and examine skin lesions. Additional clinical data were obtained from review of emergency department records at the facility where all affected wrestlers were referred for medical care. A questionnaire was administered to wrestlers by telephone following the conclusion of camp.

Clinical and questionnaire data were available for 171 (98%) persons. The mean age of these participants was 16 years (range: 14–18 years); 153 (89%) were white; 137 (80%) were high school juniors or seniors. The median length of time in competitive wrestling was 4 years.

Sixty (35%) persons met the case definition, including 21 (12%) who had HSV-1 isolated from the skin or eye. All affected wrestlers had onset during the camp session or within 1 week after leaving camp. Two wrestlers had a probable recurrence of HSV, one oral and one cutaneous, during the first week of camp.

Lesions were located on the head or neck in 44 (73%) persons, the extremities in 25 (42%), and the trunk in 17 (28%). Herpetic conjunctivitis occurred in five persons; none developed keratitis. Associated signs and symptoms included lymphadenopathy (60%), fever and/or chills (25%), sore throat (40%), and headache (22%). Forty-four (73%) persons were treated with acyclovir.

Thirty-eight (22%) wrestlers interviewed reported a past history of oral HSV-1 infection. The attack rate was 24% for wrestlers who had reported a past history or oral herpes and 38% for wrestlers without a history of oral herpes. Twenty-three percent of affected wrestlers continued to wrestle for at least 2 days after rash onset. Athletes who reported wrestling with a participant with a rash were more likely to have confirmed or probable HSV-1 infection.

Editorial Note: Herpes gladiatorum (cutaneous infection with HSV in wrestlers and rugby players) was first described in the mid-1960s. In 1988, an outbreak of herpes gladiatorum was reported among three Wisconsin high school wrestling teams. In a national survey of 1477 trainers of athletes, approximately 3% of high school wrestlers were reported to have developed HSV skin infections during the 1984–85 season. Lesions occur most often on the head and neck. Primary infection may cause constitutional symptoms with fever, malaise, weight loss, and regional lymphadenopathy. Ocular involvement includes keratitis, conjunctivitis, and blepharitis. Transmission occurs primarily through skin-to-skin contact. Autoinoculation may lead to involvement of multiple sites. Previous infection with HSV-1 may reduce the risk of acquiring herpes gladiatorum (*MMWR* 39:69, 1990).

The infection is widespread in young persons in tropical and subtropical areas as well as lower socioeconomic groups in temperate climates. People in such groups are frequently infected before school age and develop a subclinical or generalized infection that is not recognized as infectious mononucleosis. These persons may become lifelong carriers of the virus. The typical clinical disease is seen most often in adolescents and young adults; the highest incidence occurs in young persons who have had a more sheltered early life. When these individuals experience greater exposure during high school or college, they are prime candidates for the clinically recognizable form of infectious mononucleosis. Because the exchange of saliva by intimate oral contact is a most efficient means of transmitting this disease, the synonym "kissing disease" is appropriate; however, transmission may occur by other less direct routes such as sharing drinking glasses, drinking straws, and beverage bottles. It is estimated that about 100,000 college undergraduates contract infectious mononucleosis each year in the United States.

✳ Diagnosis

The enlarged lymph nodes and increased number of mononuclear cells, along with the compatible generalized symptoms in persons of the appropriate age, strongly suggest infectious mononucleosis. The diagnosis is confirmed by showing a rise in antibodies. Tests are now available to measure antibodies specifically against the EB virus. A nonspecific antibody test, called the *heterophile antibody test*, has been helpful in diagnosing this disease for many years. This test uses sheep or horse red blood cells as the antigen and by the third week of illness 80% of the infectious mononucleosis patients develop heterophile antibodies that will agglutinate these red blood cells.

✳ Treatment

There is no treatment other than supportive therapy. Bed rest is recommended during the acute phase and is usually spontaneous as a result of the generalized weakness of the patient. Contact-type physical activities should be avoided, for the swollen spleen could rupture if bumped.

✳ Prevention and Control

No vaccine against EBV infections is available, but some are being developed. Isolation of patients is of

little value. Persons who have recently recovered from mononucleosis should not donate blood. Due to the widespread distribution of this virus in asymptomatic carriers, only a recluse could purposely avoid being infected—a price most young people are not willing to pay to prevent this generally nonfatal disease.

OTHER HERPESVIRUSES

In recent years a herpesvirus now called *human herpesvirus 6* (HHV-6) has been incriminated as the cause of the common childhood disease of *roseola* (exanthem subitum). This disease is characterized by fever and a flat rash of 3 to 5 days duration. The HHV-6 appears to be widespread in most human populations. Almost all children show evidence of having been infected between the time they lose their passive immunity and by 5 years of age.

In 1990, the seventh human herpesvirus was isolated from CD4 T cells from healthy individuals and is called *human herpesvirus 7*. Serologic studies have shown that most children have been infected with this virus by 2 years of age. This widespread virus has as yet not been associated with any specific clinical disease.

In late 1994, yet another human herpes virus was found, which is associated with a type of cancer called *Kaposi's sarcoma*. Kaposi's sarcoma is found in about 20% of AIDS patients. Some evidence suggests that this herpesvirus, called *Kaposi's sarcoma-associated herpesvirus*, may be the cause of Kaposi's sarcoma in AIDS patients.

MATERIAL FOR REVIEW

CONCEPT SUMMARY

1. Infection due to the DNA herpesvirus group is of considerable attention and concern today. These viruses are responsible for a wide variety of disease conditions in both humans and animals. They cause benign, latent infections, such as cold sores, and extensive life-threatening infections, such as generalized herpes.

2. Herpesviruses are widely known because of current interest in their role as agents of a sexually transmitted disease caused by herpes simplex virus type 2 and because of infectious mononucleosis due to the Epstein–Barr virus.

3. Herpes viruses are among the few viruses for which a specific antiviral therapy has been developed.

4. Although not generally well known, cytomegalovirus infection is extremely common. This agent is responsible for serious, often fatal disease in compromised patients.

CLINICAL SUMMARY TABLE

Microorganism	Virulence Mechanisms	Diseases	Transmission	Treatment	Prevention
Herpes simplex	Latency	Cold sores Conjunctivitis Genital herpes Encephalitis	Direct and Sexual contact	Acyclovir	Avoid contact
Varicella-zoster	Latency	Chickenpox Shingles	Airborne	Acyclovir	Vaccine
Epstein-Barr	Latency	Infectious mononucleosis	Oral contact	Symptomatic	None

STUDY QUESTIONS

1. Briefly describe the host–parasite relationship commonly associated with herpesvirus infections.

2. Why isn't antibody to herpesvirus type I and type II protective against recurrence of the disease?

3. What is the most common site of infection for type II herpes?

4. Why would a viral-component vaccine be particularly useful against herpesviruses?

5. What is the likely source of an outbreak of chickenpox in a community that is apparently free of the virus?

6. What is the biggest risk factor associated with cytomegalovirus infection?

CHALLENGE QUESTIONS

1. Why would there be possible opposition to approving a living vaccine against herpes simplex virus infections?

2. Chickenpox in children is usually a relatively harmless disease. Why do health officials say that chickenpox in adults can be very serious?

500nm

VIRUS

Although the clinical disease of hepatitis was recognized in ancient times, it was not until the early 1940s that sufficient evidence existed to distinguish at least two distinct types and to suspect that the etiologic agents were viral. One form of hepatitis was thought to be passed from person to person by usual means, particularly the fecal–oral route. The other form was thought to be transmitted only by the injection of body fluids, such as blood, serum, and plasma, or the use of contaminated needles or syringes. Up until the late 1960s little progress was made in isolating and characterizing the viruses causing hepatitis by standard methods; since then, however, new methods have yielded significant information on the nature of hepatitis. This new information is markedly changing our understanding of these diseases; at present, our knowledge is expanding rapidly but is still incomplete.

INTRODUCTION

The disease passed from person to person was previously referred to as *infectious hepatitis* but is now referred to as **hepatitis A**. The form transmitted by injection was called *serum hepatitis* but it is now referred to as **hepatitis B**. When it became possible to specifically diagnose both hepatitis A and B by serologic tests, it was discovered that some cases of hepatitis were not caused by either hepatitis A or

> **Hepatitis A**   Inflammation of the liver. This condition is due to the hepatitis A virus; also called *infectious hepatitis*.
> **Hepatitis B**   Inflammation of the liver. This condition is due to the hepatitis B virus; also called *serum hepatitis*.

HUMAN BODY SYSTEMS AFFECTED

Hepatitis Viruses

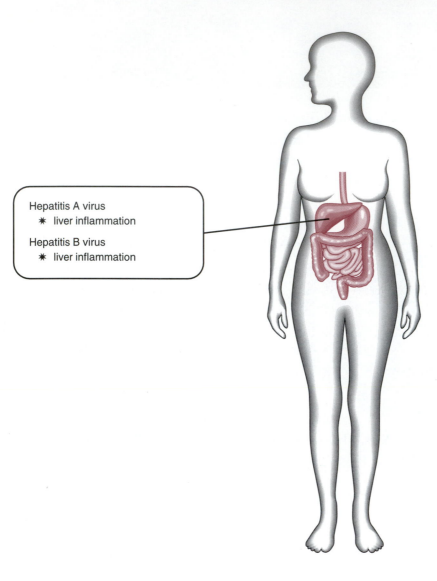

Hepatitis A virus
* ✳ liver inflammation

Hepatitis B virus
* ✳ liver inflammation

hepatitis B viruses but were apparently caused by other, as yet unidentified viruses. These cases were referred to as non-A non-B (NANB) hepatitis. As new hepatitis viruses are discovered they are removed from the NANB category and given a letter designation—hepatitis C, D, and so on. The hepatitis viruses are related by pathology and not by virus structure or composition. They constitute a rather broad mixture of both RNA and DNA, spherical and icosahedral, enveloped and nonenveloped viruses. Table 32-1 lists the general characteristics of the hepatitis viruses.

HEPATITIS A VIRUS

For many years all attempts to grow the hepatitis A virus (HAV) in cell cultures failed, but such propa-

TABLE 32-1

Characteristics of the Different Hepatitis Viruses

Virus	Size (nm)	Type of Nucleic Acid	% of Hepatitis Caused	Vaccine Available	Mean Incubation Time (days)	Main Mode of Transmission
HAV	40	RNA	50	+	25	Fecal–oral
HBV	42	DNA	41	+	75	Sexual, Transfusion
HCV	30	RNA	5	−	50	Transfusion
HDV	40	RNA	<1	−	35	Transfusion
HEV	30	RNA	<1	−	?	Fecal–oral

gation is now possible. Marmosets, monkeys, and chimpanzees can be experimentally infected, but, in general, early experimental work involved human volunteers. In 1973, virus-sized particles were observed via the electron microscope in the feces of patients with hepatitis A. This virus is a picornavirus (Chapter 35) that has been classified as enterovirus type 72. The HAV has single-strand (ss)RNA, is very stable, and only one serotype has been found.

✳ Pathogenesis and Clinical Disease

The hepatitis A virus enters the body orally. Most evidence indicates that the virus first multiplies along the intestinal epithelium and may result in anorexia (loss of appetite), malaise, intermittent fever, followed by nausea, vomiting, and diarrhea. In an undetermined percentage of infected persons the virus passes to the blood and spreads to the liver. Inflammation of the liver produces the classical signs of hepatitis, such as **jaundice** (yellowness of skin due to presence of bile pigments), dark urine, and pale, offensive smelling feces. The signs of hepatitis appear about 1–2 weeks after the initial intestinal symptoms, recovery takes 4 to 6 weeks or longer, and the general weakness continues even longer. The disease is acute and does not develop a chronic phase as is seen in cases of hepatitis due to hepatitis B virus. Death rates are less than 0.1%. The disease is mild in children and usually inapparent.

It is now suspected that the HAV might be widespread in tropical and underdeveloped areas, where most persons become infected early in life and develop only the mild or subclinical infection. In such environments, clinical infections are most often seen in adult "outsiders" who enter the area. The extent of virus spread in countries with improved sanitation is difficult to determine at present.

✳ Transmission and Epidemiology

The fecal–oral route appears to be the major means of transmission. Outbreaks have been traced to contaminated water supplies and to food vendors who are carriers. Outbreaks are common in mental institutions where sanitary practices are difficult to maintain, and infection rates are high in homosexual males. About 25,000 cases of hepatitis A are reported each year in the United States (Figure 32-1). An inexpensive ELISA serological test ∞ (Chapter 12, p. 181) is now available to measure antibodies against the HAV. Using this test, it is possible to study and gain a greater insight into the epidemiology of this disease.

✳ Diagnosis

Routine serological tests are available to detect the virus and measure antibodies. Diagnosis is based on both serology and clinical signs associated with inflammation of the liver. Liver enzymes called

---

Jaundice   A symptom due to an increase in bilirubin in the blood. Jaundiced patients often develop yellow-brown skin color.
Transaminase   A normal tissue enzyme that can transfer an amino group from one amino acid to another. Increased levels of serum transaminase indicate tissue injury.

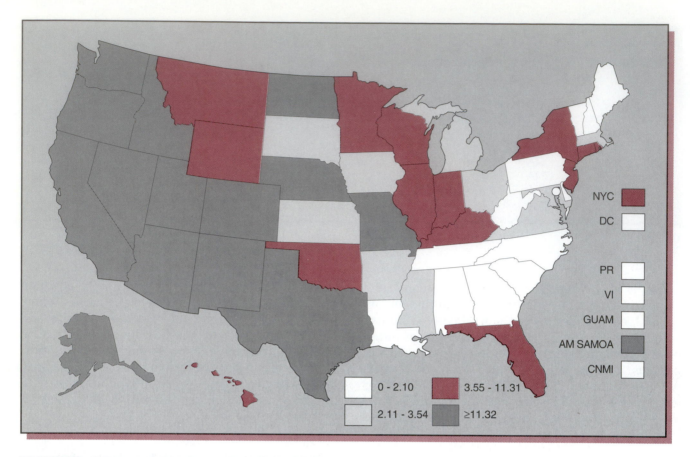

FIGURE 32-1 National distribution and occurrence of hepatitis A. Reported cases per 100,000 population, 1993. (Courtesy Centers for Disease Control, Atlanta)

transaminases, which are released into the blood during liver inflammation, can be measured as an indirect sign of viral hepatitis.

✳ Treatment

Treatment of hepatitis is symptomatic and often prolonged rest is required before complete recovery occurs.

✳ Prevention and Control

In 1995, an inactivated hepatitis A vaccine (called Havrix) was licensed for use in the United States (see Clinical Note). This vaccine has been used previously in other countries and induces nearly complete immunity in adults after one dose. A booster 6 to 12 months following the first injection is recommended to prolong protection. This vaccine is recommended for persons at risk who live in the United States and for those traveling to countries where hepatitis A is endemic. Passive immunization ∞ (Chapter 12, p. 173) with pooled human gamma globulin has been the standard procedure for offering short-term protection to persons at risk. It

would now be expected that active immunization with the inactivated vaccine would greatly reduce the use of the less effective gamma globulin treatments. Gamma globulin should still be used to protect unimmunized persons who have had known recent exposure to HAV.

HEPATITIS B VIRUS

The hepatitis B virus (HBV) differs from other known viruses and has been classified into a separate family called *Hepadnaviridae*. It cannot be grown in cell cultures, and higher primates are the only susceptible experimental hosts. In spite of the inability to readily grow this virus in the laboratory, a significant amount of information has been obtained by studying concentrations of viruses obtained from blood of infected humans. The complete virion is 42 nm in diameter and is sometimes called a Dane particle (named after D. S. Dane, who first characterized it). The virion has two concentric protein coats. The inner capsid or core surrounding the nucleic acid is 27 nm in diameter and is made of a different type of protein than the outer coat or envelope. The major protein comprising the outer coat is called the *hepatitis B surface antigen* (HBsAg) and the core protein is called the *hepatitis B core antigen* (HBcAg). The HBV contains a unique DNA molecule that is part double-stranded and part single-stranded. The DNA and HBcAg are produced and assembled in the nucleus of infected cells and then passed into the cytoplasm. The HBsAg is produced in the cytoplasm and then added over the core protein. In chronic infections, great excesses of HBsAg are produced and form into 22-nm-diameter spherical and filamentous particles. These particles contain mostly protein and are released in large numbers into the blood. The presence of this HBsAg in the blood is a valuable diagnostic aid in the detection of persons who may be carriers of HBV. The structure of HBV is shown in Figure 32-2.

✳ Pathogenesis and Clinical Disease

The clinical picture of hepatitis B is somewhat similar to that of hepatitis A except the incubation period is longer—6 weeks to 6 months—and the disease is more severe, with an overall death rate of about 1%. When adults become infected, most recover completely after several months. Some 5 to 10% develop chronic forms of hepatitis that in a few cases progress to either *cirrhosis* or cancer of the liver. Some patients become asymptomatic carriers of the infection and contain large amounts of HBsAg in their serum; this condition may last from

6 months to a lifetime. In developing countries where HBV is widespread, most adults have been infected. These infections usually occur in early childhood and many (10 to 50%) become carriers of the virus. About one-third of these will eventually develop chronic forms of hepatitis, and have a significantly increased risk of liver cancer.

✳ Transmission and Epidemiology

In the past, the most readily detected outbreaks of hepatitis B were usually transmitted by contaminated blood or blood products that were used for the treatment or immunization of patients. These cases were generally quite severe. The death rates may be from 10 to 20% in persons acquiring hepatitis B through blood transfusions, partly because of the large doses of virus received and partly because of the weakened or compromised conditions of the persons needing the transfusion. Contaminated needles, syringes, and similar items may readily transmit this virus, and infectivity rates, as measured by the presence of HBsAg in the serum, may exceed 50% among drug addicts who use and share unsterilized needles and syringes. The infection is also transmitted by close contact with blood, mucus, and semen (similar to the AIDS virus, Chapter 37). In the United States about half of the cases are transmitted by sexual contact and is a particular problem for homosexual males. Infants readily become infected if the mother is infected. About 15,000 cases of hepatitis B are reported in the United States each year, but it is estimated that 200,000 to

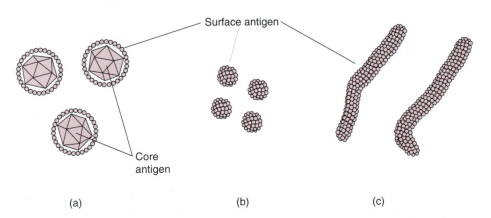

FIGURE 32-2   The various particles associated with HBV: (a) 42-nm complete virion (Dane particle); (b) 22-nm-diameter spheres composed only of surface antigen; (c) 22-nm-diameter filaments composed only of surface antigen.

Cirrhosis   A progressive disease of the liver characterized by scar tissue development and ultimate liver failure.

300,000 new infections actually occur annually and an estimated 1.25 million persons have chronic HBV infection and are potentially infectious to others. Of these chronically infected persons approximately 4500 die from chronic liver disease each year.

HBsAg was originally discovered during a study of blood proteins of the Australian aborigines when it was thought that this protein was a common but unique component of their blood; so at first it was called the Australia antigen. With the development of serologic methods to detect the presence of HBsAg in blood in the late 1960s and early 1970s, a better study of the epidemiology of hepatitis B was possible. It is now known that this infection is widespread in certain cultures, with most cases being subclinical. The infection is most widespread among populations in Asia, in primitive areas, in mental institutions, and in areas where sanitation practices are inadequate. The infectivity rates run from close to 100% in some groups, less (20 to 30%) in others with improving conditions, compared to about 0.1% in the United States.

✳ Diagnosis

Diagnosis of hepatitis, in general, is made as described for hepatitis A. When the disease occurs in a patient who has received blood transfusions or other blood products, hepatitis B is suspected. Various serologic tests to detect the presence of hepatitis B antigens and antibodies in the serum are now available and are being used as specific tools in the diagnosis of both the acute carrier and chronic stages of this disease.

✳ Treatment

Hyperimmune gamma globulin may offer some passive protection to individuals who have been exposed to HBV-contaminated materials. Interferon has shown promise in treating some severe cases, but generally treatment is symptomatic.

✳ Prevention and Control

The major efforts to prevent hepatitis B in developed countries involve the avoidance of contaminated instruments when cutting tissues or injecting medication and transfusing blood that may contain HBV. Personnel working with hepatitis patients or blood products in the laboratory should be extra careful so as to reduce the chances of contracting hepatitis B. Sensitive serologic tests such as radioimmunoassays or ELISA tests ∞ (Chapter 12, p. 181) have now made it possible to detect and eliminate most HBV-contaminated blood products.

Federal regulations now require that all blood samples used for transfusion, as well as blood products given by injection, be tested for HBV. Such serologic tests help to reduce the cases of posttransfusion hepatitis but are irrelevant to the control of HBV infections that are widespread in areas with low standards of hygiene or where transmission is primarily by sexual contact or parenteral drug abuse.

In 1987, HBsAg produced through genetic engineering became available as a vaccine. This has made a hepatitis B vaccine available to population groups at risk. High-risk groups for whom vaccination is recommended include the following:

1. Persons with occupational risks, such as health-care workers and public-safety workers
2. Clients and staff of institutions for the developmentally disabled
3. Hemodialysis patients
4. Recipients of certain blood products
5. Household contacts and sex partners of HBV carriers
6. Adoptees from countries where HBV infection is endemic
7. International travelers
8. Injecting drug users
9. Sexually active homosexual and bisexual men
10. Sexually promiscuous heterosexual men and women

Because of the short- and long-range major health consequences of HBV infection and the inability to reach many of the high-risk groups with vaccination programs, it has recently been recommended that HBV genetically engineered vaccine become a part of routine vaccination schedules for all infants. Under the program, infants born to mothers who test positive for HBsAg would receive an intramuscular dose of HBV vaccine and an injection of hepatitis B immune globulin within 12 hours of birth followed by a second dose of vaccine at 1 to 2 months and a third dose between 6 and 18 months of age. Infants born to mothers who have not been tested for HBsAg, but come from a high risk group, would be vaccinated with HBV vaccine within 12 hours of birth with two follow-up doses as above. All other infants should receive the first dose of vaccine within 2 months (preferably within the first few days) of birth, with two follow-up doses as above. Work is also progressing on cloning the gene for HBsAg into vaccinia viruses for the production of a live vaccine that could be economically feasible for use in developing countries where hepatitis B is so widespread; an estimated 300 million people are infected in those countries.

CLINICAL NOTE

Food-Borne Hepatitis A:
Alaska, Florida

From 1983 through 1989, the incidence of hepatitis A in the United States increased 58% (from 9.2 to 14.5 cases per 100,000 population). Based on analysis of hepatitis A cases reported to CDC's national Viral Hepatitis Surveillance Program in 1988, 7.3% of hepatitis A cases were associated with food-borne or water-borne outbreaks. This report summarizes recent food-borne-related outbreaks of hepatitis A in Alaska and Florida.

Alaska   Between 18 June and 20 July 1988, 32 serologically confirmed hepatitis A cases among persons who resided in or had visited Peters Creek, Alaska (population 4000), were reported to the Alaska Department of Health and Social Services. Patients ranged in age from 1 to 54 years (median: 13 years). Between 8 July and 14 August, 23 additional (secondary) cases occurred among household contacts of the original patients.

To examine potential sources of infection, the Alaska Department of Health and Social Services conducted a case-control study of the first 14 reported patients and 22 asymptomatic household members. All 14 patients and 7 (32%) household members had consumed an ice-slush beverage purchased from a local convenience market between 23 May and 10 June. No other food-consumption or exposure category (including social events, restaurants, grocery stores, or international travel) was statistically associated with illness. The 18 other patients had also consumed the ice-slush beverage.

The ice-slush beverage mixture was prepared daily with tap water from a bathroom sink using utensils stored beside a toilet. All five employees of the market denied having hepatitis symptoms; four of these were tested and were negative for IgM antibody to hepatitis A virus (IgM anti-HAV). The fifth employee, who was one of the two persons who prepared the ice-slush beverage, refused to be tested. However, a household contact of this employee had had serologically confirmed hepatitis A in early June and reported that the employee had been jaundiced concurrently with her illness.

Florida   In August 1988, the Alabama Department of Public Health noted an increase in cases of serologically confirmed hepatitis A in persons living in several areas of the state. Within 6 weeks before onset of illness, most affected persons had eaten raw oysters harvested from coastal waters of Bay County, Florida. The Florida Department of Health and Rehabilitative Services (FDHRS) contacted state health departments in neighboring and other states about hepatitis A cases in July or August 1988 in persons who had attended events serving seafood within 10–50 days of becoming ill. The 61 persons who were identified resided in five states: Alabama (23 persons), Florida (18), Georgia (18), Hawaii (one), and Tennessee (one). Patients ranged in age from 8 to 60 years (median: 31 years); all were white, and 49 (80%) were male. Fifty-nine (97%) had eaten raw oysters; one, raw scallops; and one, baked oysters. All the oysters and scallops were traced to the same growing area of Bay County coastal waters. The median incubation period between consumption of raw oysters and onset of illness was 29 days (range: 16–48 days).

To further study oyster consumption as a potential risk factor for hepatitis A, the FDHRS conducted a case-control study using uninfected eating companions of patients as controls. Fifty-three patients who had serologically confirmed hepatitis A and 64 controls were interviewed by telephone; 51 (96%) of the patients and 33 (52%) of the controls had eaten raw oysters. Consumption of other seafoods (i.e., clams, mussels, and shrimp) was not statistically associated with illness.

The implicated oysters apparently had been illegally harvested from outside approved coastal waters of Bay County. Sources of human fecal contamination were identified near oyster beds unapproved for harvesting and included boats with inappropriate sewage disposal systems and a local sewage treatment plant with discharges containing high levels of fecal coliforms.

Editorial Note: The outbreaks reported here illustrate two principal modes of transmission associated with food-borne hepatitis A outbreaks: (1) contamination of food during preparation by a foodhandler infected with hepatitis A virus and (2) contamination of food, such as shellfish, before it reaches the food service establishment (*MMWR* 39:228, 1990).

OTHER HEPATITIS-PRODUCING VIRUSES

One of the distressing discoveries in hepatitis research was the failure to eliminate post-transfusion hepatitis by eliminating blood that contains HBsAg. It became apparent that other hepatitis viruses were present. Such hepatitis has been referred to as *non-A non-B hepatitis* (NANB). Because of the HBV serologic screening tests used on all blood donated for transfusion, most of the post-transfusion hepatitis that does occur in the United

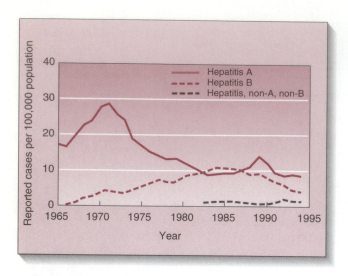

FIGURE 32-3 Reported cases of hepatitis in the United States, 1965–1994. (Courtesy Centers for Disease Control, Atlanta)

States now is NANB (Figure 32-3). An RNA-containing, 30-nm enveloped virus that appears to be associated with a large proportion of NANB hepatitis cases has been identified and named *hepatitis C virus* (HCV). More than one serotype of HCV seems

to exist. Efforts are now under way to develop reliable tests to detect HCV in blood products. Of the diagnosed hepatitis C cases in one report, 6% were transmitted by blood transfusions, 50% were found in parental drug users, and 40 to 50% had no identifiable source of transmission.

Another form of viral hepatitis that has been characterized is called *hepatitis D*. The causative agent is a 36-nm enveloped, spherical virus with a single-stranded RNA genome and has a protein coat composed of HBsAg. It is called the *delta virus*. The delta virus is a defective virus, which means it is able to replicate only in association with another virus that acts as a helper virus and supplies essential components needed by the defective virus. The HBV acts as a helper virus by supplying its surface antigen (HBsAg) as a protein coat for the delta virus. This means that only persons who have infections with HBV can be superinfected with the delta virus. The person may be a chronic carrier of HBV with no clinical signs of hepatitis before the infection with the delta virus, but after the superinfection with the delta virus the hepatitis becomes clinically active and even *fulminating*. The epidemiology of hepatitis D is not clearly understood; however, preliminary evidence suggests it could be a major problem in intravenous drug users and in some primitive cultures. When these two viruses

CLINICAL NOTE

Hepatitis A Vaccine Licensed

The new hepatitis A vaccine will be marketed under the name Havrix by Smith Kline Beecham Pharmaceuticals of Philadelphia, Pennsylvania. The product is formulated as a suspension containing inactivated particles of hepatitis A virus (HAV) that are adsorbed onto the surface of aluminum particles. The newly licensed product, which has been used in more than 40 countries, contains no blood- or plasma-derived components, unlike immune globulin, which is a blood-derived product used for passive, short-term prevention of hepatitis for individuals who may have been exposed to the virus.

When the hepatitis A vaccine was studied in Thailand in 40,000 individuals, it produced nearly complete immunity in adults after a single dose, according

to the manufacturer. Thus, 5 days after inoculation, up to 98% of recipients seroconverted and had produced specific antibodies against HAV. One month after vaccination, the vaccine elicited antibodies in 96% of subjects studied. A booster is recommended 6 to 12 months after the initial injection to prolong protection.

"This vaccine is a major advance in the prevention of hepatitis A, which is a worldwide problem," says FDA Commissioner David Kessler. In the United States, about 23,000 cases of the disease are reported annually, although an estimated 100,000 to 150,000 Americans are believed to become infected each year. Worldwide about 10 million individuals become infected with the virus each year (*ASM News* 61:324, 1995).

Fulminating Running a rapid course with continued worsening of patient condition.

Hepatitis E among U.S. Travelers, 1989–1992

Outbreaks of hepatitis E (i.e., enterically transmitted non-A, non-B hepatitis) have occurred in some parts of the world and have generally been related to contaminated water supplies. Until recently, when research-based serologic tests were developed to test for antibody to hepatitis E virus (anti-HEV), no serologic test was available to identify HEV infection, and diagnosis depended on a history of exposure in an appropriate epidemiologic setting and the exclusion of other causes of viral hepatitis. During 1989–1992, acute HEV infection was documented among six persons in the United States who had returned from international travel. This report summarizes CDC's serologic documentation of acute HEV infection—presumed to have been acquired during international travel—in two of these persons.

Patient 1 On 23 February 1991, a woman from Denver traveled to Rosarito Beach, Mexico, for 1 day. On 17 March, she developed headache and nausea, and on 23 March, became jaundiced. A serum specimen obtained on 23 March demonstrated a serum aspartate aminotransferase (AST) level of 2100 U/L (normal: 0–35 U/L), an alkaline phosphatase level of 516 U/L (normal: 110–295 U/L), and a total bilirubin level of 7.5 mg/dL (normal: 0–1 mg/dL). Physical examination was normal except for jaundice. Tests for serologic markers for hepatitis A, B, and C were negative, and an ultrasonogram of the liver was normal. Serum samples obtained on 18 April and 31 May were positive for anti-HEV by fluorescent antibody (FA) blocking assay (titers of 1:512 and 1:128, respectively) and by a Western blot assay.

The patient had no underlying medical problems and denied excessive alcohol consumption, injecting-drug use (IDU), blood transfusions, or contact with anyone known to have hepatitis during the 6 months before onset of her illness. Although the source of infection for this patient was not clearly established, she reported drinking margaritas with crushed ice at two restaurants and eating salsa and chips while in Mexico; she denied drinking water or eating other uncooked food. The patient recovered fully.

Although her three traveling companions also consumed margaritas with ice, they did not become ill, and serum samples from all three were negative for anti-HEV.

Patient 2 From mid-June through the end of July 1989, a male college student traveled in Pakistan, Nepal, and India. Before his trip, he received prophylactic immune globulin. After his return to the United States, he developed nausea, fever, epigastric discomfort, and marked fatigue. Physical examination revealed scleral icterus and a mildly tender and enlarged liver. Serum samples included an AST of 2256 U/L (normal: 9–53 U/L), total bilirubin of 6.4 mg/dL (normal: 0.2–1.4 mg/dL), and an alkaline phosphatase of 258 U/L (normal: 30–125 U/L). Although tests for serologic markers for hepatitis A, B, and C were negative, anti-HEV was detected by FA blocking assay at a titer of 1:1024.

The patient denied a history of alcohol abuse, IDU, blood transfusions, or known contact with anyone diagnosed with hepatitis. The patient reported that during his trip abroad he did not boil his drinking water (he treated the water with iodine), and he swam in the Ganges River. The patient recovered fully (*MMWR* 42:1, 1993).

(HBV and delta) occur together, the disease tends to be particularly severe with a significant increase in mortality.

More recently, a fifth hepatitis virus, designated *hepatitis E virus* (HEV), has been identified. HEV is a 27- to 35-nm, icosahedral, RNA virus. It appears to be spread by the typical fecal–oral route. Hepatitis E is usually mild and self-limiting except in women in the third trimester of pregnancy, where the death rate is about 10%. Most cases are seen in developing countries in Asia, Africa, and Central and South America. Only a few imported cases have been reported in the United States. Prevention is by good sanitation.

MATERIAL FOR REVIEW

CONCEPT SUMMARY

1. Hepatitis has long been one of the most common human diseases. Today the more widely known hepatitis A form has given way in interest to hepatitis B, which is transmitted through body excretions and is associated with increased health risks to hospitalized patients and personnel. Other viruses, called hepatitis C, D, and E also cause hepatitis.

2. Effective vaccines are now available for the prevention of hepatitis A and B.

3. Diagnosis of both hepatitis A and B, as well as the other types of hepatitis, is effectively accomplished by recently developed serologic procedures. Use of these tests has greatly reduced the risk of transfusion-associated hepatitis B.

CLINICAL SUMMARY TABLE

Microorganism	Virulence Mechanisms	Diseases	Transmission	Treatment	Prevention
Hepatitis A virus		Hepatitis A	Fecal-oral route	Symptomatic	Vaccine
Hepatitis B virus	Latency	Hepatitis B	Blood transfusion	Symptomatic	Vaccine
			Sexual	Interferon	Test blood

STUDY QUESTIONS

1. What characteristic of the hepatitis viruses significantly delayed the development of diagnostic serological tests?

2. Describe the essential differences between transmission of hepatitis A and hepatitis B viruses.

3. What reasons can you suggest for the extremely high rate of HBV infections in Asian populations?

4. List the primary risk factors associated with HBV transmission, and suggest a mechanism that may be useful to interrupt each.

CHALLENGE QUESTIONS

1. A research student forgets to label two electron microscope photographs. One is of the hepatitis A virus (HAV) and the other photo is of the hepatitis B virus (HBV). How could you determine which virus is which?

2. Which hepatitis vaccines (inactivated A or genetically engineered B) should be given to each of the following individuals? Explain why that vaccine should be used. (a) Emergency medical technician (EMT), (b) traveler to a developing nation, and (c) nurse.

33 Orthomyxoviruses

500nm

VIRUS

The family *Orthomyxoviridae* contains the influenza viruses. Various animal species become infected with their own specific strains of influenza and, in some cases, the animal strains may produce mild infections in humans. The human strains of influenza viruses cause some of the more explosive and severe viral respiratory infections. The term "flu" is used in daily speech to refer to a wide variety of infections, ranging from mild respiratory infections, such as the common cold, to various forms of enteritis ("intestinal flu"). Yet when used correctly, the terms flu or influenza should describe the very characteristic disease discussed in this chapter.

INFLUENZA VIRUSES

Three general types of influenza viruses have been identified based on differences in their internal protein and are designated types A, B, and C. Type A causes the major influenza epidemics of humans, type B causes moderate outbreaks, and type C is relatively insignificant. The following discussion is limited primarily to type A influenza.

Several features of the influenza virus need to be understood when studying the unique epidemiologic patterns of influenza. Features of the influenza virus are shown in Figure 33-1. The inner core of the virion contains a helical capsid that surrounds a ss-RNA molecule. The RNA molecule is somewhat unique in that it is made up of eight loosely connected segments and each segment contains the genetic information for the formation of a specific viral component. Of importance to the following discussion are the two protein components, known

417

HUMAN BODY SYSTEMS AFFECTED

Orthomyxoviruses

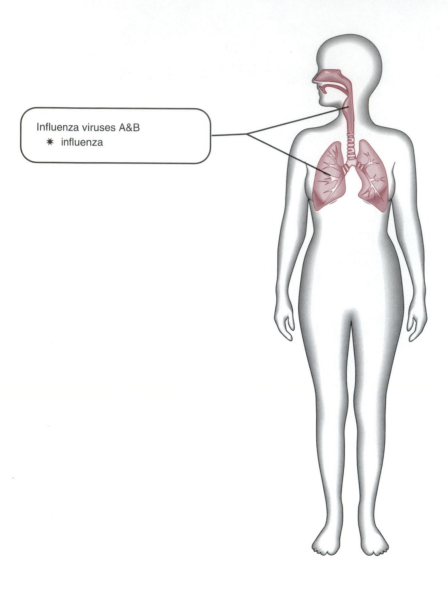

Influenza viruses A&B
  ✳ influenza

as *peplomers* or spikes, that are embedded in the lipid envelope and project out from the virus. These protein spikes are effective antigens, and the antibodies that form against them provide immunity to the host. One spike serves as the attachment site between the virus and the host cells and also causes **hemagglutination** (clumping or agglutination of RBCs); thus it is called the *hemagglutinin* or simply the *H antigen*. The other spike is an enzyme that dissolves a component of mucus called *neuraminic acid*.

Peplomer   A protrusion from the surface of a virus.
Hemagglutination   A process by which an antibody or virus particle forms bridges between red blood cells such that they become stuck together.

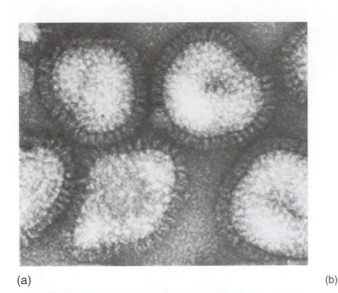

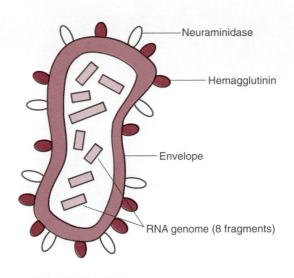

(a)

(b)

FIGURE 33-1   (a) Electron of influenza virions. (b) Schematic outline of an influenza virion. The H and N antigens are spikes that protrude from the envelope. The RNA is in eight segments and is located inside a helical capsid. (Biology of Microorganisms, 7/E by Brock, Madigan/Martinko/Parker, © 1994. Adapted by permission of Prentice-Hall, Inc., Upper Saddle River, NJ.)

This protein is called *neuraminidase* or the *N antigen*.

Among the influenza viruses that infect humans, three distinct H antigenic subtypes exist and are designated H1, H2, and H3. Fourteen distinct H antigens are found in influenza viruses that infect animals and include H1, H2, H3, and on to H14. A total of nine N antigen subtypes (N1–N9) have been found, but only N1 and N2 are routinely found in viral isolates from humans. Antibodies formed against the H and the N antigens are important protective antibodies.

Minor mutations occasionally occur in the RNA segments that control the configurations of the antigenic determinants of the H and N antigens. When minor genomic changes like this occur, the result is a phenomenon called *antigenic drift*. When antigenic drifts occur, the viral antigens produced by the mutated genome are altered just enough to cause some recognition difficulty for the antibodies in the general population, and limited outbreaks of influenza may develop.

On the other hand, when a major mutation occurs such that a *new* subtype of the H or the N antigen appears in a strain of influenza virus, the phenomenon is called *antigenic shift*. The origin of the antigenic shifts has been theoretically attributed to reassortment of RNA segments between two different influenza subtypes that are simultaneously infecting the same host. When an antigenic shift occurs, the alterations in the viral antigens are so great that antibodies already in the population are unable to recognize and defend against it. Thus, the virus is able to spread without restrictions, and a **pandemic** of influenza may result. Subtypes of type A influenza viruses are given designations that tell where and when they were first isolated, and the antigenic composition. For example, A/Hong Kong/68(H3N2) would be the influenza A subtype isolated in Hong Kong in 1968 with the H3 and N2 antigens.

✳ **Pathogenesis and Clinical Disease**

Humans are infected primarily by the airborne route. The viruses specifically attach to the ciliated epithelial cells of the respiratory tract for the initiation of the infection. As viruses are released from the few cells that are initially infected, many adja-

Antigenic drift   Small changes in the antigenic structure of a virus.
Antigenic shift   Major changes in the antigenic structure of viruses with segmented genomes (e.g., influenza virus.)
Pandemic   A worldwide epidemic.

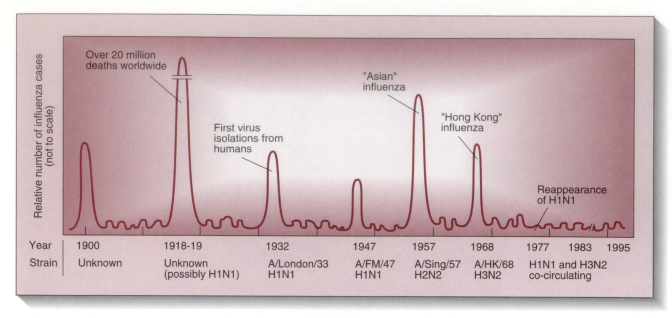

FIGURE 33-2 Time line showing when some of the major outbreaks and antigenic shifts associated with type A influenza occurred during the 1900s. Minor outbreaks occurred between the major outbreaks.

cent cells become infected and the infection spreads along the epithelium of the respiratory tract. Widespread inflammation results. Usually the infection is limited to the upper respiratory tract, but, in severe cases, the lower respiratory tract may also be involved. The influenza virus rarely spreads to the blood or deeper body tissues; however, toxic products from the virus infection are absorbed into the blood and are responsible for the generalized symptoms of influenza. Influenza has a distinct clinical picture with an onset of 1 to 3 days after exposure. A sore throat, cough, possible hoarseness, and nasal discharge are the localized signs. The systemic manifestations caused by the toxic materials include headache, fever, chills, generalized muscular aches, and, in severe cases, *prostration*. Gastrointestinal symptoms may also be present. Recovery is usually spontaneous after 4 to 7 days. Deaths are rare in otherwise healthy individuals, but increased death rates are noted among the very young, elderly, and chronically ill when an epidemic of influenza is in an area. Influenza weakens the natural defense mechanisms of the respiratory tract and so secondary bacterial pneumonia is a common complication during influenza outbreaks and is responsible for many of the resulting deaths. Reye's syndrome ∞ (Chapter 31, p. 399) may also develop in some children recovering from influenza.

✳ Epidemiology

There are few if any parallels today to the epidemiology of influenza. When major antigenic shifts occur in virulent strains, a worldwide pandemic may occur and may involve hundreds of millions of persons. Such pandemics are accompanied by significant increases in the overall death rate. Often such outbreaks force the closing of schools and factories and generally disrupt the normal flow of human activities. Fortunately, such major outbreaks do not occur often; only six have occurred in the past 100 years and the last one was in 1968. Less extensive outbreaks occur every 2 to 4 years as a result of antigenic drift.

Figure 33-2 shows the general epidemiological patterns of influenza during the past century. The most devastating pandemic occurred in 1918–1919. This strain of influenza virus was highly virulent, particularly in young adults for some yet unexplained reason, and approximately 500,000 deaths occurred in the United States and about 20 million deaths occurred throughout the world. As the world population developed antibody immunity to this virus, the rate of disease decreased. The technology of 1918 was not able to isolate the virus responsible for this great pandemic but indirect evidence suggests it may have had H1N1 antigenic

Prostration Exhaustion and loss of strength.

composition. In 1931 the first influenza virus was isolated from swine. It was noted, however, that serum taken from persons who had recovered from the 1918 influenza pandemic did contain antibodies that specifically reacted, to a limited extent, with the swine influenza virus. Whether any further relationship existed between these viruses could not be determined.

In 1932–33, epidemics occurred around the world and at that time the first influenza virus from humans was isolated. The isolates from this period had the antigenic makeup of H1N1 and over the next 20 years antigenic drifting occurred in these viruses with limited epidemics resulting every 3 to 4 years; rather extensive epidemics occurred in 1947. Originally, the 1932 and 1947 isolates were considered to be different antigenic subtypes, but more detailed studies have now shown them to be similar enough to be both classed as H1N1 subtypes. The H1N1 subtype generally disappeared from the world population in the early 1950s. Major antigenic shifts occurred in 1957 in both the H and N antigens and resulted in a new subtype with H2N2 antigens. A prototype of this subtype is A/Japan/57 (H2N2) and is commonly known as the "Asian" influenza virus. Because no antibody immunity was present in the world population against either the H2 or N2 antigen, this subtype of influenza spread very rapidly and within a year was causing major pandemics in most parts of the world. Epidemics of H2N2 subtype occurred every three years or so for the next ten years. In 1968, a major shift occurred in the H antigen, creating subtype A/Hong Kong/68 (H3N2), also known as the "Hong Kong" influenza virus. Like its predecessor, the Asian flu, the Hong Kong flu spread rapidly from Asia to most parts of the world causing major pandemics. No new major antigenic shifts have occurred from 1968 to 1995. The subtype H1N1, however, made a reappearance in 1977 and has caused some epidemics through the 1980s in persons born since 1950 and who were thus not previously exposed to this subtype. The H3N2 and H1N1 subtypes have continued to co-circulate in human populations during the 1980s and into the 1990s.

A subtype designated A/New Jersey/76 (H1N1) that was isolated in 1976 deserves special mention. This subtype was apparently the same as or closely related to the swine influenza virus. This virus was isolated from recruits at a military camp in New Jersey in January 1976. Because it was first detected in young adults and was related to "swine flu," there was a great deal of concern that it might be a return of the same strain that caused the great pandemic of 1918. To prevent a possible recurrence of such a devastating pandemic, with the possibility of millions of deaths, a massive immunization program was undertaken. By late summer and early fall of 1976 large segments of the population of the United States were being vaccinated. Fortunately, the A/New Jersey/76 subtype was not the deadly strain of 1918; it was probably a random transmission of the flu from infected swine to humans.

Limited outbreaks of type B influenza occur in most populations every 4 to 5 years.

✳ Diagnosis

Preliminary diagnosis is made on the basis of clinical and epidemiological observations. The diagnosis is confirmed by demonstrating the presence of the virus in throat washings. The virus can be grown in cell cultures or in embryonated eggs. Changes in antibody levels can be readily demonstrated by hemagglutination inhibition tests ∞ (Chapter 12, p. 181).

✳ Treatment

Unlike most viral diseases, early therapy of type A influenza can be accomplished with amantadine hydrochloride or rimantadine. These compounds prevent the uncoating of the virus in host cells and so restrict its replication. Rimantadine is less toxic to the patient than amantadine. These compounds may also be used prophylactically in high-risk persons who are in danger of being exposed to influenza. Unfortunately, these compounds are not effective against either type B or C viruses. In all cases of influenza, supportive treatment of the symptoms may be helpful in comforting the patient and antibacterial chemotherapy may be needed to treat secondary bacterial infections.

✳ Prevention and Control

Because of the increased mobility of today's world population, the influenza virus is apparently able to spread most effectively from country to country.

Quarantine Separation of humans or another animal from the general population when these individuals have or have been exposed to a communicable disease.

Influenza Activity: New York, 1994–95 Season

Influenza activity in the United States during the 1994–95 influenza season began in the Northeast, and during late January, spread to other regions of the country. This report describes influenza outbreaks in nursing homes in New York.

The first influenza outbreak reported to CDC during the 1994–95 season occurred in a 300-bed skilled-nursing facility in Long Island, New York. On 30 November 1994, eight residents on one 20-bed corridor developed influenza-like illness (ILI) (i.e., fever ≥100°F [≥38°C] and cough). On 1 December, nasopharyngeal swab specimens from these eight residents were submitted for rapid antigen testing; within 5 hours after transport to the laboratory, influenza type A was detected by enzyme immunoassay in six specimens. On the evening of 1 December, 293 of the 299 residents in the facility each received 100 mg of amantadine hydrochloride as treatment for the eight ill residents and as prophylaxis against influenza A infection for the other 285 residents. Most (285 [95%]) residents had received influenza vaccine before the outbreak. On 2 December, as part of the nursing home's contingency plan for influenza outbreaks, amantadine dosages were modified for individual residents based on estimated creatinine clearance, and prophylaxis was continued for 14 days. Other outbreak-control measures included confining ill residents to their rooms for at least 72 hours after the initiation of amantadine treatment and prophylaxis, confining all residents to their individual units, suspending group activities, and minimizing the assignment of nursing staff to multiple units. The amantadine dosage subsequently was discontinued for five residents and reduced for 13 residents because of side effects (pri-marily confusion and agitation); for most patients, side effects resolved within 48 hours of dosage adjustment.

During the first 48 hours of amantadine prophylaxis and treatment, six additional residents developed ILI. Of the 14 residents who developed outbreak-associated ILI, five subsequently developed clinical pneumonia. During the 2-week period of amantadine prophylaxis, sporadic cases of febrile respiratory illness occurred in other units of the facility; however, there was no clustering of cases.

Tissue culture of all eight nasopharyngeal specimens yielded influenza type A(H3N2). These isolates were further characterized at CDC; all were antigenically similar to the A/Shangdong/09/93 strain included in the 1994–95 influenza vaccine.

Influenza surveillance in New York State indicated increasing activity beginning in late November 1994. From 1 December 1994 through 11 February 1995, outbreaks associated with influenza type A(H3N2) in 46 other nursing homes were reported to the New York State Department of Health (NYSDOH); of these, 16 were reported from nursing homes in Long Island. For all 16 facilities, influenza type A infection was documented by rapid antigen detection; in 13 facilities, amantadine was administered as an outbreak-control measure. Outbreaks in five other nursing homes were caused by influenza type B, and in two nursing homes by influenza types A and B. Based on findings of virologic surveillance in New York, influenza occurred in persons in all age groups during the 1994–95 season. Of the 385 influenza virus isolates reported by laboratories in New York this season, 332 (86%) have been type A (*MMWR* 44:132, 1995).

Quarantine measures seem of little value. The major means of controlling outbreaks is through a killed vaccine. This type of vaccine is moderately effective in that it confers protection to about 80% of those who receive it; this protection lasts from 6 months to a year. The vaccine contains the toxic products of the virus and often induces mild systemic symptoms of influenza, such as headache, fever, and muscle aches. Currently, purified vaccines having only the H and N antigens are being used to a limited extent and do not contain the toxic by-products. The most important aspect of influenza immunization is to use the proper subtype of virus for the production of the vaccine. Each year in the United States, the composition of the vaccine is adjusted so it contains antigens against the predominant types of viruses circulating in the world. When a major antigenic shift occurs, the vaccines then in stock may be of no value.

Continual worldwide surveillance programs monitor for the appearance of new subtypes. When a new subtype is found and is determined to be of significant virulence, a race begins between humans and virus, with the goal being to produce large amounts of vaccine and vaccinate a significant proportion of the population before the viral pandemic arrives. The first time this race was run was with the advent of the Hong Kong flu of 1968. Unfortunately, delays occurred in the production of the vaccine and the pandemic had circled the earth before the vaccine had become available. The next attempt was during the "swine flu" episode of

CLINICAL NOTE

Update—Influenza Activity: United States and Worldwide, 1994–95 Season, and Composition of the 1995—96 Influenza Vaccine

In collaboration with the World Health Organization (WHO) and the international network of collaborating laboratories and with state and local health departments in the United States, CDC conducts surveillance to monitor influenza activity and to detect antigenic changes in the circulating strains of influenza viruses. This report summarizes surveillance for influenza in the United States and worldwide during the 1994–95 season and describes the composition of the 1995–96 influenza vaccine.

United States Influenza activity began in the Northeast in late November 1994 and from late January to early February spread to other regions of the country. Activity peaked during March and continues to decline.

From 27 November 1994 through 14 January 1995, regional or widespread influenza activity was reported only from northeastern states. Regional activity was first reported outside this area for the week ending 21 January, and by 11 February regional or widespread activity had been reported from every region in the country. Based on reports from state and territorial epidemiologists, peak activity occurred the week ending 11 March 1995, when 26 states reported either regional or widespread activity. The number of states reporting regional or widespread activity has declined every week since 12 March. For the week ending 8 April, four states reported regional activity, and none reported widespread activity.

Of total deaths reported through CDC's 121-city mortality surveillance system, the proportion attributed to pneumonia and influenza exceeded the epidemic threshold for 11 of the 27 weeks from 2 October 1994 through 8 April 1995. Pneumonia and influenza deaths exceeded the epidemic threshold for 2 consecutive weeks twice during this interval.

Of the 3423 influenza virus isolates reported to CDC from WHO collaborating laboratories in the United States through 8 April, a total of 2654 (78%) were type A and 769 (22%) were type B. Of the 1337 type A viruses that have been subtyped, 1318 (99%) were type A(H3N2) and 19 (1%) were type A(H1N1).

Worldwide Influenza activity has occurred at low to moderate levels in most parts of the world. Although a few countries reported epidemic activity, sporadic activity or localized outbreaks were reported more frequently. Influenza activity was usually associated with cocirculation of influenza A(H3N2) and influenza B viruses. Influenza A(H1N1) activity was reported only in association with sporadic cases. Influenza A(H3N2) viruses were first detected during October in Europe and North America. Outbreaks associated with influenza A(H3N2) were subsequently reported in the People's Republic of China, Finland, Hungary, Italy, Spain, the United Kingdom, and the United States. Although influenza A and influenza B cocirculated, influenza A(H3N2) viruses predominated in Canada, Finland, France, Italy, Spain, and the United States.

Influenza A(H1N1) viruses have been reported in association with sporadic activity from Canada, China, Hong Kong, the Netherlands, Norway, Poland, Singapore, Switzerland, Thailand, the United Kingdom, and the United States during the 1994–95 season.

Composition of the 1995–96 Vaccine The Food and Drug Administration Vaccines and Related Biologicals Advisory Committee (VRBAC) has recommended that the 1995–96 trivalent influenza vaccine for the United States contain A/Johannesburg/33/94-like (H3N2), A/Texas/36/91-like (H1N1) and B/Beijing/184/93-like viruses. This recommendation was based on the antigenic analysis of recently isolated influenza viruses and the antibody responses of persons vaccinated with the 1994–95 vaccine (*MMWR* 44:292, 1995).

1976. In this case, the vaccine was prepared in time, but this subtype of virus was of low virulence and no significant outbreak occurred. Whether this stratagem will work in aborting a major pandemic in the future due to an antigenic shift of a virulent subtype remains to be determined. Various projects are now underway, using genetic engineering technology, to develop an influenza vaccine that contains combinations of the commonly encountered H and N antigens. It is important that individuals with increased risk of serious outcome from influenza infection receive adequate immunization.

This would include elderly persons and anyone who has heart or respiratory health problems.

In the United States, information is collected each week from about 120 selected cities on the deaths due to pneumonia and influenza (Figure 15-4). When deaths exceed the expected range in a given area, an investigation is undertaken to determine if an influenza epidemic is occurring and what subtype might be present. Frequent updates are reported in the *Morbidity and Mortality Weekly Reports* (see the Clinical Notes).

MATERIAL FOR REVIEW

CONCEPT SUMMARY

1. Influenza is the most common disease due to orthomyxoviruses. While common in occurrence, this disease may have serious consequences.
2. An unusual genetic arrangement in the genome of this virus facilitates the repeated development of viruses with altered antigenic structures. These "new" viruses are the basis of periodic pandemics of influenza.
3. Vaccines are available against the influenza virus and consist of the most recent and common antigenic variations of the virus.

CLINICAL SUMMARY TABLE

Microorganism	Virulence Mechanisms	Diseases	Transmission	Treatment	Prevention
Influenza A	Multi-serotypes Neuraminidase Hemagglutinins Toxins	Type A influenza	Airborne	Amantadine Rimantadine	Vaccine

STUDY QUESTIONS

1. What is the chemical composition and the physical structure of the influenza H and N antigens?
2. Discuss the epidemiologic significance of antigenic drift and antigenic shift in the influenza virus.
3. If type A influenza is treatable with amantadine hydrochloride, why isn't this compound commonly used in the early stages of an epidemic?

CHALLENGE QUESTIONS

1. Why is it so hard to develop a vaccine for influenza?
2. If a patient had recovered from influenza caused by an H2N3 virus, to which of the following viruses would he/she be most susceptible: H2N1, H1N3, or H1N1?

34 Paramyxoviruses

The family *Paramyxoviridae* consists of related viruses that cause some common diseases of humans. These viruses are structured somewhat like the orthomyxoviruses, but, compared to the influenza virus, they are genetically stable so that new mutant serotypes do not periodically develop. Many common viral respiratory diseases, as well as such distinct clinical diseases as mumps and measles, are caused by paramyxoviruses. The parainfluenza, respiratory syncytial, mumps, and measles paramyxoviruses are discussed in this chapter. Rubella, which is caused by a togavirus, is also described here because its clinical manifestations are much like measles.

PARAINFLUENZA VIRUSES

The parainfluenza viruses, like all paramyxoviruses, contain ss-RNA and have helical symmetry. The helical capsid is wound in a loose sphere that is somewhat irregular in shape and ranges from 100 to 300 nm in diameter. Parainfluenza viruses possess an envelope with *hemagglutination* and *neuraminidase* activities. These viruses are unstable and do not survive long outside the host.

Four serotypes (designated type 1, 2, 3, and 4) of parainfluenza viruses infect humans. The infections are seen primarily in infants and young children. Illnesses range from subclinical to mild upper respiratory infections to croup or pneumonia. The viruses rarely spread to the blood. Most cases are seen during the "respiratory disease season" from late fall to early spring. Many children are infected during the first years of life so that by age 10 most (over 80%) children have antibodies against the four

HUMAN BODY SYSTEMS AFFECTED

Paramyxoviruses

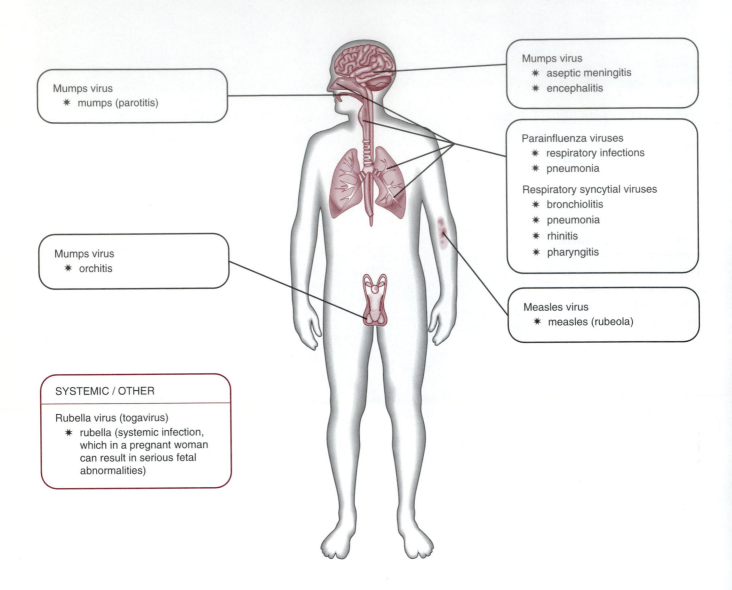

Mumps virus
* mumps (parotitis)

Mumps virus
* orchitis

SYSTEMIC / OTHER

Rubella virus (togavirus)
* rubella (systemic infection, which in a pregnant woman can result in serious fetal abnormalities)

Mumps virus
* aseptic meningitis
* encephalitis

Parainfluenza viruses
* respiratory infections
* pneumonia

Respiratory syncytial viruses
* bronchiolitis
* pneumonia
* rhinitis
* pharyngitis

Measles virus
* measles (rubeola)

---

types of parainfluenza viruses. Reinfections may occur at all ages, but they are much milder than the primary infection and present symptoms of the "common cold." These viruses cause about 15 to 30% of the acute respiratory diseases seen in children under 10 years of age. Treatment is symptomatic; no vaccines are available.

RESPIRATORY SYNCYTIAL VIRUSES (RSV)

The RSV virions vary in size from 100 to 300 nm and the envelope contains neither hemagglutinin nor neuraminidase activity. Two types exist, designated A and B. The RSV are the major cause of respiratory illness in infants during the first few months of life and the major cause of *bronchiolitis* and *pneumonia* in infants less than 1 year of age. Outbreaks occur annually from fall to early spring. Infection readily spreads within family units and among infants in common environments. Prevention of nosocomial infections in pediatric units requires the vigilance of all medical personnel. Infections can be severe and death may result in 1 to 4% of infants requiring hospitalization. In compromised infants mortality can rise to 15%. The infections are mild in older children and are limited to the upper respiratory tract (*rhinitis* and *pharyngitis*). Close to 100,000 infants with RSV infections are hospitalized per year in the United States and several thousand deaths occur. Because of the seriousness of this disease in infants, a rapid diagnosis is sometimes important; this can be done by detecting RSV antigen in respiratory aspirate using specific, labeled monoclonal antibodies. Aerosol treatment with the antiviral agent ribavirin may be effective in reducing the severity of this disease. Supportive treatment to maintain adequate respiratory functions and to prevent secondary bacterial infections is important. No vaccine is available.

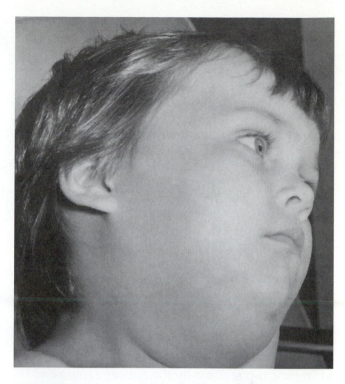

FIGURE 34-1   Child with diffuse swelling as a result of mumps infection. (Centers for Disease Control, Atlanta)

MUMPS VIRUS

Mumps has been one of the common communicable diseases of children and young adults. It is not as contagious as measles or chickenpox and outbreaks are generally limited to fewer cases. Along with the well-recognized swelling of the salivary glands, mumps may also involve other glands or the central nervous system (CNS). In countries where vaccinations are used, the number of cases have decreased markedly over the past several years.

The mumps virus is a typical paramyxovirus and is inactivated quite readily (within hours) once expelled from the body. Only one serotype is known and humans are the only natural host. This virus can be grown in a variety of different cell cultures and in embryonated eggs. *Primates* can be experimentally infected.

✷ Pathogenesis and Clinical Disease

Humans are infected by inhaling infected *droplet nuclei* that contain the virus. This disease has an incubation period of 12 to 35 days and during this time the virus passes from the site of entry in the respiratory tract into the lymphatic system and then into the blood. The virus is deposited from the blood into the meninges and various organs, such as the salivary glands, ovaries, testes, mammary glands, and pancreas. Several days to a week before the onset of symptoms, and for about an equal length of time after the onset of symptoms, the virus may be found in the saliva, urine, stool, and blood. In about a third of these patients the infection remains subclinical or at least the infected person experiences none of the classical symptoms of mumps. In the other two-thirds the most readily recognizable sign is the swollen salivary glands (*parotitis*) (Figure 34-1). Swelling occurs only on one

---

Bronchiolitis   Inflammation of the bronchioles.
Primate   An order of upright, bipedal mammals of which monkeys are an example.
Droplet nuclei   The small dehydrated particles of mucus and microbes that remain suspended in the air following coughing, sneezing, or speaking.

side (unilateral) in 25% of the cases. Several days to a week or more following the onset of parotitis, involvement of other glands may become apparent. Inflammation of the testes (*orchitis*) is rare in prepubertal males (under 11 to 13 years of age), whereas 20 to 30% of males over this age may experience orchitis. Usually orchitis is unilateral and all current evidence indicates that infertility is rarely caused by this infection. Inflammation of the ovaries and mammary glands is seen in a smaller percentage of infected females and permanent damage rarely, if ever, occurs. The most serious complication from mumps results from infection of the central nervous system. Mumps was one of the most common causes of *aseptic* (nonbacterial) *meningitis* and *encephalitis*. Most patients with CNS involvement recover without permanent complications. Deafness may result in a small percentage of cases and death occasionally occurs.

✳ Transmission and Epidemiology

Mumps is transmitted by the airborne route and by direct contact with saliva. No documentation of transmission by virus-infected urine or mothers' milk has been made. Those with subclinical cases, however, are able to transmit the disease. Transmission is not as effective as with diseases like measles, chickenpox, or influenza. Most cases occur in the winter or spring, but tend to be sporadic and may occur throughout the year. Since the introduction of a living attenuated vaccine in the late 1960s, the overall incidence of mumps has decreased (Figure 34-2). In 1994, an all-time low of about 1370 cases of mumps was reported in the United States. Recovery from mumps generally confers lifelong immunity.

✳ Diagnosis

Typical cases of mumps with swelling of the salivary glands can be readily diagnosed on the basis of clinical appearance. Atypical cases require laboratory tests to confirm that they are mumps. The virus can be isolated from the throat, saliva, urine, or spinal fluid through cell cultures or embryonated eggs. Various serologic tests are available to measure specific mumps antibodies.

CLINICAL NOTE

Update—Respiratory Syncytial Virus Activity: United States, 1994–95 Season

Respiratory syncytial virus (RSV), a common cause of winter outbreaks of acute respiratory disease, causes an estimated 90,000 hospitalizations and 4500 deaths each year from lower respiratory tract disease in both infants and young children in the United States. Outbreaks occur annually throughout the United States, and community activity usually peaks within 1 month of the national peak. RSV activity in the United States is monitored by the National Respiratory and Enteric Virus Surveillance System (NREVSS), a voluntary, laboratory-based system. This report presents provisional surveillance results from the NREVSS for RSV during 2 July–9 December 1994, and summarizes trends in RSV from 1 July 1990 through 1 July 1994.

Since 1 July 1990, a total of 105 hospital-based and public health laboratories in 47 states have participated in the NREVSS and have reported weekly to CDC the number of specimens tested for RSV by the antigen detection and virus isolation methods and the number of positive results. Widespread RSV activity is defined by the NREVSS as the first of 2 consecutive weeks when at least half of participating laboratories report any RSV detections. This definition generally indicates a mean percentage of specimens positive by antigen detection in excess of 10%. During the previous four seasons, from 1 July 1990 through 1 July 1994, onset of widespread RSV activity began in November and continued an average of 24 weeks until April or mid-May. The peak in activity occurred each year from mid-January through mid-February. For the current reporting period (2 July–9 December 1994), 85 laboratories in 43 states reported results of testing for RSV. Since 12 November, more than half of the participating laboratories reported detections of RSV on a weekly basis, indicating the onset of RSV activity for the 1994–95 season (*MMWR* 43:920, 1994).

Encephalitis   Inflammation of the brain. Encephalitis is most often due to viral infection.

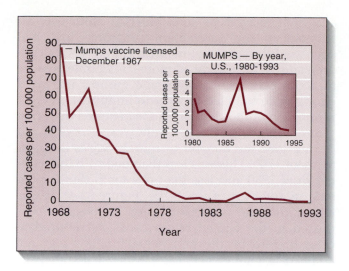

FIGURE 34-2 Reported cases of mumps by year in the United States, 1968–1993.

* Prevention and Control

Children with mumps should be kept from associating with other children for about 2 weeks. A living attenuated vaccine has been available since 1968 in the United States. This vaccine is quite effective and induces immunity in over 95% of the recipients. The vaccine has few side effects. The vaccine should be given as part of the routine childhood vaccination series at about 15 months of age. It is usually given as part of a triple vaccine known as MMR, which stands for measles, mumps, and rubella. Mumps vaccines have been used effectively in military recruit training centers to control troublesome outbreaks in that environment. Significant decreases in mumps have been observed in areas where the vaccine has been widely used.

MEASLES (RUBEOLA) VIRUS

Measles (also called *rubeola*) is one of the most contagious diseases of humans and before the development of an effective vaccine in the mid 1960s, 98% of the population of the United States would have contracted it by the age of 18. Even though measles was considered a disease that everyone had to get and "get it over with," it is a serious disease and a lead-ing cause of death in many undernourished children in developing countries where vaccination is not carried out. Cases of measles have declined significantly in the past 25 years in countries where vaccination programs are applied.

Only one serotype of measles virus exists. The viral morphology is much like that of the other paramyxoviruses. This virus can be adapted to grow in cell cultures and embryonated eggs. It has a short survival time outside the body, but does remain viable for a long enough time in droplet nuclei to be spread effectively by the aerosol route.

* Pathogenesis and Clinical Disease

The measles virus enters by the airborne route and is taken up by the draining lymph ducts of the respiratory tract. The virus passes into the blood and is deposited throughout the body. The incubation period is 9 to 12 days before the onset of prodromal signs of fever, cough, runny nose (*coryza*), and conjunctivitis. At this stage of the disease red *macules* or ulcer-type lesions, called *Koplik spots*, appear on the inside of the cheek. After about 3 additional days, the rash appears on the skin and continues to spread and intensify over the next 1 or 2 days (Figure 34-3). General symptoms may be severe and such complications as secondary bacterial infections may occur. The CNS is infected and in about 1 out of every 1000 cases of measles the encephalitis is severe enough to cause noticeable signs. About 15% of those with severe signs of encephalitis die; others may have permanent effects, such as epilepsy, hearing loss, and personality changes.

A rare degenerative neurological disease called *subacute sclerosing panencephalitis* develops in a few children or adolescents several years after a measles infection. All current evidence indicates that this disease is caused by measles viruses that have remained latent in the CNS.

In the mid-1970s a new dimension in the clinical picture of measles appeared. The immunity of some children who had been vaccinated waned or their vaccinations were of insufficient potency to produce full immunity; consequently, a significant number of cases of atypical measles occurred. These atypical cases resembled various other diseases and it was necessary to use serologic tests to confirm that they were measles.

Macules Reddish-pink skin spots.

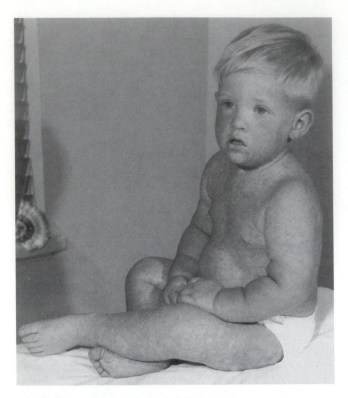

FIGURE 34-3   Measles rash on a child three days after onset. (Centers for Disease Control, Atlanta)

✳ Transmission and Epidemiology

Transmission is by droplet nuclei expelled from the respiratory tract for a few days before to a few days after the onset of the rash. The disease is highly infectious. When measles is introduced into an isolated population that has not had this disease, and no acquired immunity is present, almost 100% of the inhabitants contract it. In such cases, the mortality rate is high and may reach 10% if local health care is not available and the general nutritional state of the population is poor. Such an epidemic occurred on the Hawaiian Islands during early colonization by Europeans and Americans.

Before the introduction of vaccination, measles epidemics would occur every 2 to 5 years in a given area, with most cases happening during the winter and early spring. This epidemiologic pattern of measles has changed significantly since the introduction of vaccines in the mid-1960s. Within a few years after initiation of vaccination programs the total number of cases in the United States decreased 90%. Through the 1970s, however, the number of cases remained at around 20,000 to 30,000 cases per year. During this time the age distribution of cases shifted, with more outbreaks occurring in teenagers

and college-aged persons. Such a shift was attributed to the vaccination programs that had decreased the spread of measles among young children and, in turn, allowed more susceptible persons to accumulate in the older age groups. Increased efforts were initiated in 1979 to immunize a larger proportion of the children as they entered their first year of schooling. This program was successful and during 1982 the number of measles cases dropped to about 1700. This was less than one-fifth the number from 1980 and less than one-tenth the number seen in any year of the 1970s (Figure 34-4). It was hoped that the number of cases of measles contracted in the United States could be reduced to close to zero by the mid-1980s, but this was not attained. A total of 1495 cases was reached in 1983, then the numbers increased somewhat up to the year 1990, when a sharp increase to 27,786 cases occurred. Renewed efforts to increase vaccination coverage have reduced measles cases to all-time lows; 312 cases were reported in 1993. Measles is a life-threatening disease in malnourished children. Therefore, in many of the developing countries of the world this disease remains a serious health-care consideration. Many of the cases seen in the United States are contracted outside of the country.

✳ Diagnosis

Because of its characteristic appearance and epidemiology, measles is usually diagnosed clinically without laboratory tests. The appearance of Koplik

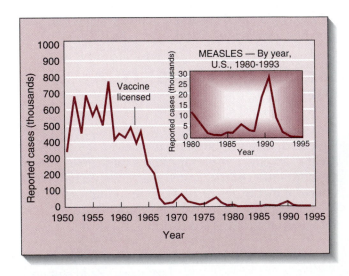

FIGURE 34-4   Reported cases of measles by year in the United States, 1950–1994.

CLINICAL NOTE

Interstate Measles Transmission from a Ski Resort: Colorado, 1994

During 1 April–25 May 1994, a chain of measles transmission began in Breckenridge, Colorado, and extended into nine additional states; a total of 247 measles cases were reported, representing 36% of all U.S. measles cases reported to the National Notifiable Diseases Surveillance System (excluding those reported from U.S. territories) through 2 July (week 26). The source of exposure was unknown but is believed to have been an out-of-state tourist who probably visited Breckenridge during March, because (1) no measles cases had previously been reported in Colorado during 1994, and (2) the only common exposure appeared to have been at a ski resort visited by many out-of-state travelers. Persons associated with spread of measles from Breckenridge were predominately school- and college-aged. This report summarizes the investigation of this chain of interstate measles transmission.

A total of 15 measles cases with rash onset during 4–21 April occurred in Breckenridge. Persons with measles ranged in age from 16 years to 46 years (median: 27.6 years). All cases met the CDC measles clinical case definition; 12 were serologically confirmed. All 15 ill persons either lived in Summit County (Breckenridge) or three neighboring counties (Arapahoe, Chaffee, and Park) or worked in tourism-related services in or near Breckenridge. Twelve of the 15 ill persons are believed to have been exposed to the unidentified source, and three cases resulted from secondary transmission. Two cases occurred among high school students; no further transmission in schools was reported. Interstate transmission of measles occurred through four out-of-state travelers and a Silver Thorn, Colorado, resident—all of whom had visited Breckenridge during 18–25 March. All five visitors are believed to have been exposed to the unidentified source. Two persons (a 46-year-old Texas resident [rash onset: 16 April] and a 29-year-old Missouri resident [rash onset: 4 April]) developed measles on return home but have not been linked to additional cases. The other three persons—an Illinois resident, a Maryland resident, and the Silver Thorn resident—became sources for further transmission.

Illinois   A 14-year-old, unvaccinated, female high school student returned home to Jersey County, Illinois; she developed a rash on 4 April. The student was identified as the source of an outbreak involving 51 unvaccinated persons (age range: 1–24 years; median: 18 years; last rash onset: 3 June) in her community—which was associated with a Christian Science college in the county. She also was identified as the source of an outbreak involving 156 persons (age range: 4–25 years; median: 15 years; rash onsets: 17 April–15 May) at the Christian Science boarding high school she attended in St. Louis County, Missouri. After several unvaccinated persons from other states visited the school during the outbreak, six additional cases occurred. Five persons developed measles on return home (two persons to Maine and one each to California, New York, and Washington); the California patient was the source of exposure for a sibling. No further transmission associated with these six cases is known.

Maryland   A 25-year-old woman returned home to Baltimore County, Maryland; she developed a rash on 4 April. The woman was the source of exposure for her 56-year-old father, who had rash onset on 21 April.

Michigan   A 25-year-old Silver Thorn man visited his family in Wayne County, Michigan; he developed rash on 17 April. The man was identified as the source of an outbreak involving 12 persons (age range: 9 months–37 years; median: 24 years; rash onsets: 17 April–18 May) who were exposed at a wedding and a restaurant. One additional case (rash onset: 16 April) was reported in a 12-year-old Chicago resident who had visited Wayne County. No further transmission associated with the Michigan or Chicago cases is known (*MMWR* 43:627, 1994).

---

spots is useful in diagnosing measles during early stages of the disease. But laboratory tests are needed to make a specific diagnosis in the cases of atypical measles in partially immunized children. Virus isolation can be made during the acute phase of the disease; however, most laboratory diagnoses are made by showing a significant increase in antibodies to measles virus between the acute or convalescent sera.

✻ Treatment

No specific antiviral treatment is available. Antibiotics may be given to treat secondary bacterial infection.

✻ Prevention and Control

The control of measles at present relies almost entirely on the use of an attenuated vaccine. This

CLINICAL NOTE

Outbreaks of Rubella among the Amish: United States, 1991

From 1 January through 19 April 1991, at least nine outbreaks of rubella, involving more than 400 cases, have been reported in Amish communities in the United States. These outbreaks have been reported from Mecosta and Montcalm counties, Michigan; Allegany, Cattaraugus, Chautauqua, and St. Lawrence counties, New York; Geauga, Knox, and Trumbull counties, Ohio; and Lawrence County, Tennessee. In addition, serologically confirmed cases of rubella have been reported from Amish communities in six Pennsylvania counties, suggesting widespread rubella activity among the Amish in Pennsylvania. In general, cases have occurred among unvaccinated children and young adults.

In 1990, three linked outbreaks causing an estimated 171 cases occurred in Amish communities in Minnesota, New York, and Ohio. No cases of congenital rubella syndrome (CRS) associated with these outbreaks have been reported. However, during 1990, rubella outbreaks not involving Amish communities occurred among unvaccinated adolescents and adults in the western United States; as a result, for 1990, at least 16 confirmed or compatible indigenous CRS cases and six additional provisional cases occurred and have been reported to the National Congenital Rubella Syndrome Registry.

Editorial Note: Because interstate and intrastate travel to other Amish communities is common among the Amish population, state and local health departments and clinicians should be alerted to the risk for local outbreaks of rubella among Amish communities. Many rubella infections cause only mild illness; therefore, outbreaks may remain unreported unless active surveillance for cases is conducted. In addition, active surveillance should be conducted for cases of CRS that may result from large outbreaks of rubella. Amish communities should be alerted to the risk for rubella outbreaks, the consequences of rubella infection during the first trimester of pregnancy, and the importance of increasing vaccination levels in their communities, especially among women of childbearing age and children.

During the past 5 years, outbreaks of other vaccine-preventable diseases, such as measles and pertussis, have been reported from Amish communities. Although vaccination coverage among the Amish is low, some health departments report that, with vigorous effort, many Amish will accept vaccination (*MMWR* 40:264, 1991).

---

vaccine is given subcutaneously to young children after they have lost all passive immunity from their mothers, usually between 12 and 15 months of age ∞ (Chapter 12, p. 175). The measles vaccination is normally administered as part of the MMR triple vaccine. One vaccination, when given to children over one year of age, offers lifelong protection. For children vaccinated under one year of age, or for those who received inadequate protection (as determined by antibody tests) from the first immunization, it is recommended that a second immunization be given before the child enters school at 5 or 6 years of age. All current evidence indicates that effective vaccination programs directed at preschool-age children can control measles. It is important that persons traveling outside of the United States be properly immunized.

RUBELLA VIRUS

Rubella is a mild disease that is of little direct concern in children or adults, but it may have disastrous effects on the developing fetus should the mother become infected early in pregnancy. The terms *German measles* or *three-day measles* are also used for this disease; however, these terms often lead to confusion with regular measles and it would be better to use the name rubella consistently when referring to this disease.

The virus that causes rubella defied isolation by standard procedures through the 1950s but was finally isolated in the early 1960s. The feature that made isolation difficult was the inability of this virus to cause observable changes in cell cultures (*cytopathic effect*) or death in embryonated eggs even though the virus will proliferate in these systems. Only a single serotype of rubella has been found. The rubella virus is a member of the family *Togaviridae* (Chapter 38). These are 40- to 80-nm, spherical viruses with single-stranded RNA and an envelope.

✳ Pathogenesis and Clinical Disease

This disease has the same pathogenesis as many other diseases; that is, the virus is inhaled into the respiratory tract, followed by passage through the lymphatic system into the blood. After an incubation period of around 14 days, the signs and symptoms produced are usually trivial and include swollen lymph nodes, mild fever, and often a slight rash lasting for 2 to 3 days.

When a pregnant female is infected, the rubella virus may infect the developing fetus and cause a disease known as *congenital rubella syndrome*. This syndrome may include any of the following effects: cataracts with partial or complete blindness, loss of hearing, heart defects, mental retardation, or generalized tissue damage. The chance that the developing baby will develop serious damage due to congenital rubella is much greater if infection occurs during the first trimester of pregnancy. Severe damage results in about 50% of the fetuses infected during the first month; serious effects on the fetus are rare if the mother contracts rubella after the fourth month of pregnancy.

It is theorized that the rubella virus has such damaging effects on the fetus because of its mild nature. More virulent viruses, if they infected the fetus, would cause severe damage that would result in fetal death and lead to spontaneous abortion. On the other hand, the rubella virus, because of its mild effect, simply slows cell growth or damages or kills a limited number or cells, which, in turn, leads to malformation of the tissues as they differentiate during early fetal development.

✳ Transmission and Epidemiology

Rubella is primarily transmitted by the respiratory route, with the patient being infectious several days before to about a week after the onset of the rash. Before the widespread use of the vaccine, moderate epidemics of rubella would occur every 6 to 9 years and major epidemics at greater intervals of up to 30 years. The last major epidemic in the United States

was in 1964 when about 500,000 cases occurred. As a result of this epidemic, thousands of children were born with congenital rubella syndrome. Rubella babies present a special problem in transmission, for they continue to shed virus in saliva, urine, and other body secretions for up to 6 years after birth.

The reported cases of rubella per year in the United States are seen in Figure 34-5. Prior to the development of the vaccine, a majority of the cases occurred in young children. However, since the widespread use of the vaccine the ratio of cases in older persons has increased. In recent years the greatest increases have been in persons over 15 years of age; the exceptions to this were in certain religious communities where vaccinations are not used. An all-time low of 192 cases of rubella was reported in 1993. Currently, only a few cases of congenital rubella syndrome are reported per year in the United States.

✳ Diagnosis

The major problem in diagnosing rubella centers around the concern of exposure or potential exposure of females during the early stages of pregnancy. The expense and time involved in a laboratory diagnosis are not warranted in routine nonpregnant cases. However, when a susceptible female in the first trimester of pregnancy has been exposed to a possible rubella patient, it is important to know whether the exposure was indeed to rubel-

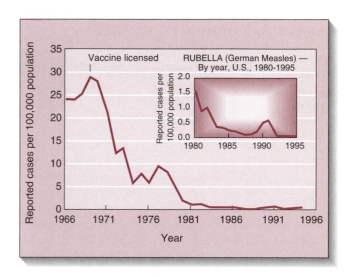

FIGURE 34-5   Reported cases of rubella by year in the United States, 1966–1994.

Cytopathic effect   The visible effect that virus infection has on cells.

la. Laboratory serologic tests and viral isolation can determine if the infection is caused by the rubella virus. Routine serologic tests are available to determine whether females of child-bearing age have antibodies against rubella.

✳ Treatment

No direct treatment is available.

✳ Prevention and Control

Once discoveries in the early 1960s showed that the rubella virus could be grown in cell cultures or embryonated eggs, rapid progress was made in developing a rubella vaccine. A living vaccine was first licensed in 1969 and appears to induce a high degree of immunity after a single dose. Mild infection sometimes results from the vaccine, but in nonpregnant persons it is of little consequence. There is no evidence, from 20 years of observation, that this living vaccine causes damaging infections in the developing fetus. As a precaution, however, it is still recommended that conception should be delayed until at least 3 months after a female has been vaccinated and pregnant females should not be vaccinated.

The rubella vaccine is generally given as part of the MMR vaccine at 15 months of age. It is also recommended to immunize girls in their early teens and older females of childbearing age who have no antibodies against rubella.

CONCEPT SUMMARY

1. The RNA-containing paramyxoviruses are responsible for several of the more commonly known childhood diseases. These infections include croup (epiglottitis), mumps, measles, and a not uncommon pneumonia in newborns caused by respiratory syncytial virus.
2. Diseases due to these viruses have been so widespread that although any given infection may be mild in its symptoms, the small percentage of infections with serious consequence were, in fact, quite common. This situation has spawned the development of excellent vaccines that should be used to limit the occurrence of rubella, measles, and mumps.

CLINICAL SUMMARY TABLE

Microorganism	Virulence Mechanisms	Diseases	Transmission	Treatment	Prevention
Parainfluenza viruses	4 serotypes	Respiratory infections	Airborne	Symptomatic	Avoid exposure
Respiratory syncytial viruses	2 serotypes	Croup and pneumonia in infants	Airborne	Symptomatic Ribavirin	Avoid exposure
Mumps virus		Mumps	Airborne Contact	Symptomatic	Vaccine
Rubeola virus		Measles	Airborne	Symptomatic	Vaccine
Rubella virus		Rubella Congenital rubella syndrome	Airborne Congenital	Symptomatic	Vaccine

STUDY QUESTIONS

1. What is the normal route of transmission of the paramyxoviruses?
2. In what age group is infection with the mumps virus likely to cause the most severe disease?
3. Why are so many individuals unsure as to whether or not they have had mumps, while almost everyone who has had measles is sure of this infection?
4. What procedure is available to prevent persons from developing mumps? measles? rubella?
5. If rubella is such a benign disease, why has there been a major effort to ensure immunization against this disease?
6. What aspect of epidemiology has resulted in shifting the most likely age for mumps and measles from childhood to older ages?

CHALLENGE QUESTIONS

1. At an international sports competition, an outbreak of the measles occurs. This competition involves over 500 athletes, managers, and coaches from 36 countries. As head of the local health agency overseeing the competition, what would you do to minimize and prevent spread of the disease?
2. In the United States, rubella control involves immunizing all children at the age of 15 months using the MMR vaccine. In England, only young girls are immunized just before they enter childbearing age. Which approach is preferable and why?

35 Picornaviruses

500nm

VIRUS

The family *Picornaviridae* includes large numbers of viruses that infect humans and animals. They are among the smallest viruses and contain ss-RNA, hence the name *pico* ("small") plus RNA (Figure 35-1). Two genera of picornaviruses are found in humans: the *rhinoviruses*, which primarily inhabit the nasal cavity, and the *enteroviruses*, which are found mainly in the alimentary tract. The rhinoviruses can only proliferate in the superficial tissues of the upper respiratory tract and are one of the frequent causes of the common cold. The enteroviruses have traditionally been subdivided into the poliovirus, coxsackievirus, and echovirus groups, but since 1970 newly discovered members have just been designated as enteroviruses, followed by a number. Generally they produce mild or asymptomatic infections of the intestinal tract but may also cause respiratory or systemic infection. Enteroviruses are also responsible for infections of the central nervous system. The picornaviruses are stable and able to survive for long periods in sewage, water, and foods.

RHINOVIRUSES

Rhinoviruses grow best at 33°C and have thus adapted to grow in the cooler superficial tissues of the nasal cavity, thereby inducing the infections referred to as **coryza** or the "common cold." These infections usually do not spread beyond the nasal cavity and tissue damage is usually slight to moderate. Much of the symptomatology of the common cold is due to the release of histamine from the patient's mast cells ∞ (Chapter 11, p. 167)—a reaction stimulated by virus-induced tissue damage.

Coryza   Acute inflammation of the nasal mucosa accompanied by profuse nasal discharge.

HUMAN BODY SYSTEMS AFFECTED

Picornaviruses

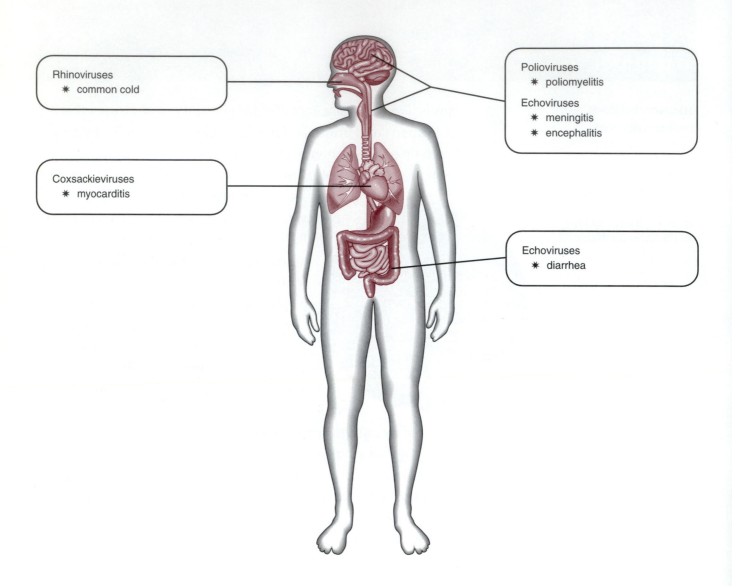

Rhinoviruses
  * common cold

Polioviruses
  * poliomyelitis

Echoviruses
  * meningitis
  * encephalitis

Coxsackieviruses
  * myocarditis

Echoviruses
  * diarrhea

The released histamine induces such symptoms as increased mucous secretions, watery eyes, and sneezing. To date, more than 100 different rhinovirus serotypes have been identified.

Common colds occur year-round but are most frequent during the cooler months and it has been postulated that such factors as crowding in buildings and low relative humidity may increase the rate of infection. Attacks of the common cold occur repeatedly, partly because of the large number of serotypes of rhinoviruses. Another factor is that serum antibodies are not protective. The relatively short duration of immunity that is conferred to the superficial mucosal tissues seems to depend on the presence of IgA antibody. Various other viruses are also able to cause common colds.

The common cold appears to be transmitted by relatively close contact. Large respiratory droplets

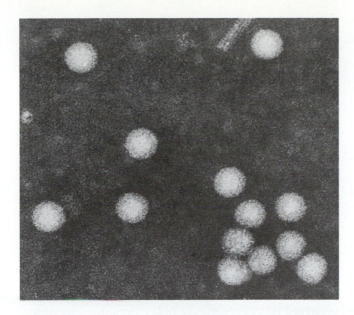

FIGURE 35-1 Transmission electron micrograph of polioviruses magnified 300,000×. (Courtesy Robley C. Williams, University of California, Berkeley)

are relatively easily transferred between individuals and onto the hands and fingers of infected persons. It appears that hand to hand or hand to inanimate object (*fomite*) transmission of the virus with subsequent self-inoculation is a common mode of transmission. The incubation period is relatively short, 2–4 days, symptoms usually last for about seven days, although cough may persist for 1–2 weeks in some patients.

No specific treatment for the common cold is available; *antihistamines* and other medications, however, may be used to relieve symptoms. Controlled studies have failed to show any beneficial effects of vitamin C in treating and preventing colds. Vaccines are not available.

POLIOVIRUSES

Three serotypes of polioviruses have been identified and at present the diseases caused by these viruses constitute minor problems in developed countries. The widespread use of effective vaccines has greatly reduced the number of cases of *poliomyelitis*, or just polio, in the United States. The conquest of polio is one of the great success stories of modern medical research. Although the disease of *paralytic polio* is no

longer a threat in the developed countries, persons involved in health fields need to understand both the principles involved in the "rise and fall" of this disease during the first 65 years of the twentieth century and how these principles may still apply in developing countries. It is also important to understand why continued emphasis on immunization is needed to prevent possible future recurrences of the disease.

✳ Pathogenesis and Clinical Diseases

The poliovirus generally enters the body via the oral route. The virus first proliferates in the throat and small intestines and then passes through the draining lymph nodes and enters the blood, thus becoming widely disseminated (Figure 35-2). On occasion the virus passes into the central nervous system (CNS) but infects only certain motor nerve cells of the spinal cord or brain. Varying degrees of paralysis may result, depending on the location of the destroyed nerve cells. Paralysis of lower limbs results from infection of the anterior horn cells of the spinal cord, whereas infection in the brain stem, called *bulbar poliomyelitis*, may cause death due to respiratory or cardiac failure. It is now known that less than 1% of those infected with the poliovirus show signs of CNS involvement. In over 99% of the cases the infection is limited to tissues other than the CNS and the infected persons may experience no or only minor nonspecific symptoms of a respiratory or intestinal tract infection. The presence of circulating antibodies is very effective in preventing the spread of the virus through the blood to the CNS.

An illness called *postpolio syndrome* has been characterized in recent years. The symptoms may include fatigue, weakness, pain, and loss of muscle mass, and usually involve the muscles that were affected when the patient previously had paralytic polio. This syndrome typically occurs 25 years or longer after the original attack. The mechanisms causing this syndrome are not known.

✳ Epidemiology

The epidemiology of paralytic polio is closely associated with the sanitary conditions of a population in a paradoxical way—that is, paralytic disease is rarely seen in populations living under conditions

---

Antihistamine Drug used to counteract allergic reaction by blocking histamine-induced reactions.
Paralytic polio Infection with poliovirus resulting in paralysis in the patient.

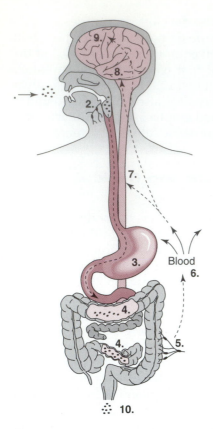

FIGURE 35-2 The pathogenesis of poliomyelitis. (1) Ingestion of virus. (2) Some multiplication may occur in tonsils and lymph nodes of upper respiratory tract with common cold-type symptoms. (3) Virus resistant to stomach acids and digestive fluids. (4) Multiplication of virus in Peyer's patches and other lymphoid cells along the intestinal tract with clinical symptoms of enteritis. (5) Viruses drain into regional lymph nodes, which stimulates antibody response. (6) Viruses may pass into blood. (7) In a small percentage of infections, the viruses may cross the blood–brain barrier into the CNS. Motor nerves (anterior horn cells) are specifically infected and result in paralysis of certain muscles. (8) Lower brain centers may be infected causing bulbar polio. (9) Motor cortex may be infected causing widespread paralysis. (10) Large amounts of virus shed in feces.

of poor sanitation. The paradox is explained in the following manner. Polioviruses are widespread in populations where poor sanitary conditions prevail and all persons possess specific polio antibodies. An infant born into such a culture possesses sufficient passive immunity ∞ (Chapter 12, p. 176) to prevent the virus from passing through the blood to the CNS. These circulating IgG-type antibodies, however, will not prevent infections of the intestinal tract; under unsanitary conditions the infants are invariably infected with the polioviruses before they lose their passive immunity. The natural

intestinal infection then stimulates lifelong natural active immunity. Only when sanitary conditions begin to improve is it possible for infants to avoid the primary infection long enough for the passive immunity to disappear. Yet these infants will generally contract the natural disease fairly early in life and about 1% will develop the paralytic disease.

The change from poor sanitary conditions to improved conditions began to occur during the latter part of the nineteenth century in the United States and in some European countries. At that time paralytic polio was first recognized as a specific disease; and because it occurred mainly in young children, it was called *infantile paralysis*. As sanitary conditions continued to improve, it became possible for more and more children to avoid primary exposure for longer and longer periods and in the first decades of the 1900s more and more paralytic polio occurred in progressively older persons. The annual polio rate in the United States reached its highest level in the early 1950s and then dropped sharply after the introduction of the killed vaccine (Figure 35-3).

✳ Prevention and Control

Polio has been controlled through the development of effective vaccines. A killed vaccine, containing all three serotypes, was first introduced in 1955 and induced adequate immunity when given as a series of three injections over a 3- to 6-month period, followed with booster injections every 2 or 3 years. By the early 1960s, a living attenuated vaccine became available and has generally replaced the use of killed vaccine in the United States and many other

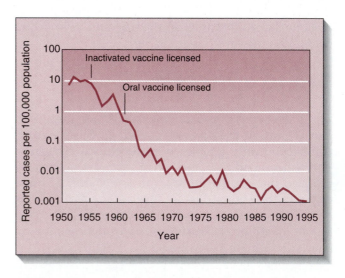

FIGURE 35-3 Reported cases of paralytic poliomyelitis by year in the United States, 1951–1994.

CLINICAL NOTE

Mass Vaccination with Oral Poliovirus Vaccine: Asia and Europe, 1995

The theme of World Health Day, 7 April 1995, is "Target 2000—A World Without Polio." In conjunction with World Health Day activities, 18 geographically contiguous countries in Europe, Central and South Asia, and the Middle East are conducting coordinated National Immunization Days (NIDs) with oral poliovirus vaccine (OPV). The World Health Organization (WHO) has designated this effort "Operation MECACAR" (MEditerranean, CAucasus, and Central Asian Republics). This report describes the efforts of this campaign and summarizes polio surveillance data for 1994.

To maximize the geographic area covered and the number of children targeted simultaneously for mass vaccination with OPV, Operation MECACAR has been committed to by adjoining countries in Europe (Armenia, Azerbaijan, Bulgaria, Georgia, and Turkey), Central Asia (Kazakhstan, Kyrgyzstan, Tajikistan, Turkmenistan, and Uzbekistan), South Asia (Pakistan), and the Middle East (Afghanistan, Iran, Iraq, Jordan, Lebanon, Syria, and one national identity [Palestine]). Approximately 56 million children aged <5 years have been targeted to receive two doses each of OPV. Efforts have been planned and will be coordinated under the direction of both the European Regional Office (EURO) and the Eastern Mediterranean Regional Office (EMRO) of WHO.

Participating countries in each region have provisionally reported a high proportion of the total polio cases in their respective regions in 1994. In EURO, participating countries reported 200 (95%) of the 211 cases reported in the region, including Uzbekistan with 117 cases; Tajikistan, 28; Turkey, 23; Azerbaijan, 17; Turkmenistan, 6; Armenia, 5; and Kazakhstan, 4. Participating countries in EMRO reported 669 (69%) of the 973 cases reported in the region, including Pakistan with 520; Iran, 80; Iraq, 63; Jordan, 4; and Lebanon, 2.

Some of the countries in these regions previously have conducted NIDs, including Azerbaijan (1993 and 1994), Lebanon (1994), Iran (1994), Syria (1993 and 1994), Pakistan (1994), and Uzbekistan (1994), while others conducted subnational immunization days. Based on the desirability of scheduling mass vaccination campaigns simultaneously and during the low polio incidence season, either the first round (EURO) or the second round (EMRO) of NIDs has been scheduled during 24 March–29 April. Countries participating in Operation MECACAR are planning to repeat NIDs in 1996 and 1997.

Editorial Note: Since 1988, when the World Health Assembly (the governing body of WHO) adopted the goal of global polio eradication by the year 2000, substantial progress has been made toward this goal. In particular, during 1994 the Western Hemisphere was certified free of wild poliovirus by an international certification commission. From 1988 through 1994, reported polio declined 82%, with particular progress in the Western Pacific Region of WHO, including China, Philippines, and Vietnam; polio-free zones are emerging in Western Europe, Southern and Northern Africa, and the Arabian peninsula.

The technical basis for achieving worldwide polio eradication already exists; persistent impediments to the eradication objective for the year 2000 include insufficient political will and inadequate resources. Operation MECACAR is supported by a coalition of organizations that includes WHO, United Nations Children's Fund (UNICEF), other bilateral and multilateral organizations, and Rotary International, which provided the funds for the OPV vaccine needed by member countries of the European Region to conduct NIDs in 1995 (*MMWR* 44:234, 1995).

countries. The living vaccine provides long-lasting immunity after a series of three administrations and it can be taken orally, thus saving time and money and avoiding the discomfort of hypodermic injections.

Even though a small risk is involved with the use of the living vaccine, in that 1 out of approximately every 3 million persons receiving it may develop paralytic disease, the overall success of the polio vaccines has been phenomenal. Paralytic cases were reduced from a high of 57,879 in 1952, to 31 in 1970 and 9 in 1980. About one-half the cases in the 1970s

in the United States resulted from infections with vaccine strains in immunosuppressed children. This prompted the remanufacture of a small supply of killed vaccine to be used for vaccinating children who may be immunosuppressed. This procedure has resulted in further reductions of paralytic polio; in 1993 only three cases were confirmed; two were vaccine associated, and one was imported. Many areas of the world are now free of polio. Major campaigns are underway under the direction of the World Health Organization to eradicate poliomyelitis from the world (see Clinical Note) and signifi-

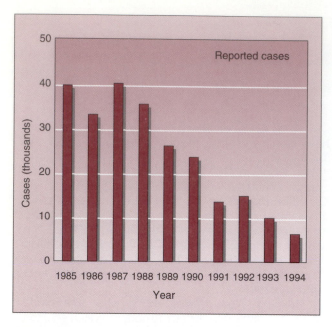

FIGURE 35-4 Worldwide reduction in the number of reported cases of poliomyelitis since the start of the global eradication campaign in 1985.

cant progress has been made towards this goal (Figure 35-4).

COXSACKIEVIRUSES AND ECHOVIRUSES

The coxsackieviruses are so named because they were first isolated in 1948 from children living in the town of Coxsackie, New York. The children were thought to have mild cases of polio. The first echoviruses were discovered several years later during field trials of the killed polio vaccine. It was customary in these trials to examine many vaccinated children to determine if they were shedding the polioviruses. Besides finding some polioviruses and coxsackieviruses, a group of previously unidentified viruses was discovered. Because these new viruses were isolated from the intestinal tract, produced a cytopathic effect in cell cultures, were isolated from humans, and were not associated with any apparent diseases, they were given the name *echo*, which is an acronym for *e*nteric *c*ytopathic

CLINICAL NOTE

Aseptic Meningitis: New York State and United States, Weeks 1–36, 1991

During April–October 1991, several state health departments noted increased reports of aseptic meningitis. This report summarizes findings from epidemiologic investigations of and surveillance efforts for aseptic meningitis in New York state and elsewhere in the United States.

New York In New York, information on cases is collected by local health units and forwarded to the New York State Department of Health (NYSDOH), using the Council of State and Territorial Epidemiologists' (CSTE) case definition for surveillance. From January through August 1991, 636 cases of aseptic meningitis were reported to the NYSDOH (excluding New York City), a 153% increase over the average number of cases reported during the same 8-month period for 1987–1990.

Preliminary data from 11 state laboratories, which perform viral isolation, showed increased isolations of coxsackieviruses, echovirus 30, and enteroviruses not yet typed. During June–July 1991, the Nassau County Medical Center detected echovirus 30 in 12 (67%) of 18 patient specimens from which nonpolio enteroviruses were isolated; during 1990, echovirus 30 was isolated from one (2%) of 57 patient specimens.

United States From the reporting period ending 24 August 1991 through the period ending 12 October 1991, reports of aseptic meningitis nationally have exceeded historical limits for each 4-week reporting period. Cases of aseptic meningitis are not reportable in five states (Connecticut, Idaho, New Jersey, Oregon, and Washington); however, among states with reporting requirements, 8415 cases were reported during the first 36 weeks of 1991, compared with an average of 2992 cases reported during weeks 1–36 of 1986–1990. The highest rates were reported from Vermont and Rhode Island (34.3 and 29.1 cases per 100,000 persons, respectively). In Vermont, reported cases increased 10-fold over baseline from April through July. States reporting elevated rates of aseptic meningitis were concentrated in the eastern United States, particularly in New England and among the mid-Atlantic states.

Outbreaks were reported in Massachusetts, Ohio, and other states. For example, in Massachusetts, echovirus 30 was isolated from specimens from seven patients involved in a communitywide outbreak. In Ohio, a middle school football coach, a student manager, and three members of the team developed aseptic meningitis during an 8-day period in September; in this outbreak, an enterovirus (not yet identified) was isolated from two patients. The local health department initiated an education campaign that promoted handwashing and discouraged the shared use of drinking vessels and open ice buckets (*MMWR* 40:773, 1991).

human orphan viruses. Since that time, however, many of the echoviruses have been shown to cause a variety of diseases.

About 30 different serotypes of coxsackieviruses and 33 serotypes of echoviruses have been identified. Differences between these two groups of viruses are slight and the current procedure is not to classify new isolates into either of these groups but simply call them *enteroviruses* and give them a numerical designation—for instance, enterovirus 70. These viruses are transmitted by both the fecal–oral and the airborne routes. Generally, the enterovirus infections are subclinical or mild with generalized symptoms. Occasionally, symptoms may be more severe. Wide varieties of disease illnesses are produced by these viruses. They range in severity from rather benign colds and diarrhea to fairly severe and even life-threatening cases of aseptic meningitis, encephalitis, and myocarditis. Fevers, with and without skin rashes, pneumonia, hepatitis, and conjunctivitis can be caused by coxsackieviruses, echoviruses, or other enteroviruses. They are very easily transmitted, with the highest rate of infection occurring in late summer. Enterovirus 72 causes hepatitis A and was discussed in Chapter 32.

MATERIAL FOR REVIEW

CONCEPT SUMMARY

1. The picornaviruses are a large group of small RNA viruses that are responsible for both mild upper respiratory infections and severe neurological or cardiovascular disease.

2. The picornaviruses are easily transmitted by the fecal–oral route, and most of these viruses multiply in the intestine prior to further dissemination within the patient.

CLINICAL SUMMARY TABLE

Microorganism	Virulence Mechanisms	Diseases	Transmission	Treatment	Prevention
Rhinoviruses	100 + serotypes	Common colds	Airborne Contact	Symptomatic	Wash hands Avoid exposure
Polioviruses	Affinity for CNS	Poliomyelitis	Fecal-oral route	Symptomatic	Vaccines
Enteroviruses (Coxsackie Echo)	Numerous serotypes	Respiratory Diarrhea Myocarditis Encephalitis	Fecal-oral route Airborne	Symptomatic	Good sanitation

STUDY QUESTIONS

1. Describe the picornaviruses in general taxonomic terms.
2. What rationale can you give for the present lack of a "common cold" virus vaccine?

3. To be immune to poliovirus it is necessary to receive all three serotypes in a vaccine. What does this tell you about the nature of the serogrouping antigen?

4. The name "enterovirus" refers to what characteristic of the echo and coxsackieviruses?

CHALLENGE QUESTIONS

1. There are two forms of the polio vaccine, the killed vaccine and the live, attenuated one. Some physicians and clinicians believe the live attenuated vaccine should be discontinued. Why do they believe this?

2. Postpolio syndrome is an illness that has been recently characterized. Although the mechanism causing the syndrome is not known, try to suggest possible causes for this illness.

36 Rhabdoviruses

This chapter discusses only one virus and one human disease caused by that virus. There are a number of viruses in the family *Rhabdoviridae* and they infect various vertebrate and invertebrate hosts; however, only one, the rabies virus, causes serious disease in humans. In almost all cases, rabies produces an invariably fatal disease of humans and is a major animal health problem throughout the world. This chapter details the epidemiology of this virus and the efforts made to prevent rabies throughout the world.

RABIES VIRUS: RABIES

An enveloped, helical ss-RNA virus, the rabies virus has a unique and distinctive shape. The helix is wound in a rod shape with one end tapered, which gives the virion the characteristic "bullet" shape of the *Rhabdoviridae* (Figure 36-1). The rabies virus appears capable of infecting and causing serious disease in most mammals. Most infected animals die, but some are able to carry and shed the virus for prolonged periods. The disease of rabies in humans was recognized and reported in ancient times as a disease transmitted by the bite of a mad (rabid) dog. Once signs of rabies begin to appear in humans, indicating infection of the central nervous system (CNS), the disease progresses in almost all cases until death results.

Rhabdoviruses

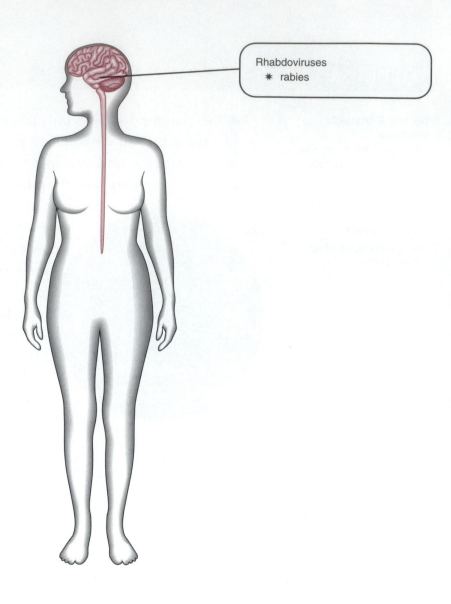

Rhabdoviruses
* rabies

---

* Pathogenesis and Clinical Disease

Generally, humans are infected by the bite of an infected animal that introduces the virus into the tissues. Infection can also occur from saliva of an infected animal that is deposited on mucous membranes or on open sores or abrasions. The virus may remain at the site of entry for weeks, where some local multiplication may occur. The virus then spreads from the local site of entry, moving along nerve fibers to the regional ganglia and the CNS. The incubation time—that is, the period between the time of the bite and the onset of signs of the disease in the CNS—varies greatly, depending on the site and the severity of the bite as well as other host factors. The incubation period may range from a week to months and even a year or more (average is 1 to 2 months). The incubation period is usually

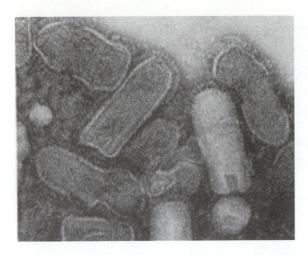

FIGURE 36-1 Electron micrograph of a rhabdovirus (vesicular stomatitis virus), showing the envelope and characteristic bullet shape of this family of viruses. (*Biology of Microorganisms, 7/e*, by Brock, Modigan/Martinko/Parker. © 1994. Reprinted with permission of Prentice Hall, Inc., Upper Saddle River, NJ.)

longer when the bite occurs on an extremity, such as an arm or leg, and shorter when the bite occurs about the neck and head.

The virus fails to progress from the site of entry to the CNS in approximately half or more of the persons bitten by an infected animal. Once the virus enters the CNS, however, it spreads rapidly throughout these tissues and then spreads down the peripheral nerves to other tissues. In particular, the rabies virus proliferates in the salivary glands and may be found in high concentrations in the saliva. When this stage is reached, the overt signs and symptoms of the disease begin with fever, headache, and loss of appetite. Then they progress to convulsions, excessive flow of tears and saliva, insomnia, anxiety, muscle spasms triggered by swallowing, and sometimes maniacal behavior. Ascending paralysis may be seen and death sometimes results after 2 to 6 days because of respiratory or cardiac failure. In other cases, patients lapse into a coma, followed by death. Except in a few known cases, clinical rabies has invariably resulted in the death of the patient.

The pathogenesis of rabies in animals has many similarities and some differences from the disease in humans. The route of infection in animals is often by the bite of another infected animal, but disease may also result from eating a diseased animal. In many animals the virus passes to the CNS during the long incubation period. In dogs, cats, and related animals the virus appears in various excretory glands, especially the salivary glands, several days before the onset of signs of CNS involvement. Once signs of the CNS infection occur, the animal usually

dies within a relatively short time. A small percentage of these animals have been found to shed rabies virus in their saliva for up to 2 years without developing the clinical disease. Bats are often infected with rabies virus and appear to tolerate the infection much better than most other animals. Some evidence indicates that bats are able to carry and shed the rabies virus for prolonged periods without showing signs of the disease.

✳ Transmission and Epidemiology

Normally the route of transmission is by the bite of an infected animal (Figure 36-2). This route works extremely well, for the irritation of the infection in the CNS drives the animal into a frenzy that results in the biting of other animals. Such biting occurs at the time when high concentrations of virus are found in the saliva. Rabies in humans and in animals like cows or horses is usually a dead end because the infection is not conveniently transmitted to other hosts by biting; however, the saliva from infected humans or nonbiting animals may be infectious. Although only a few cases of rabies are

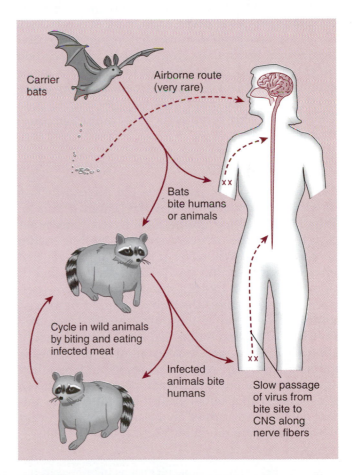

FIGURE 36-2 Modes of transmission of rabies viruses among animals and to humans.

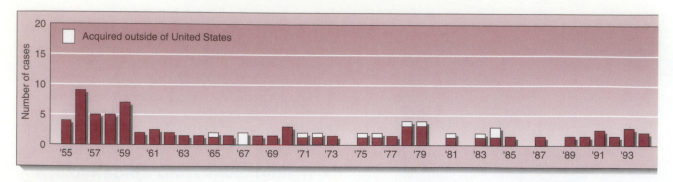

FIGURE 36-3 Reported rabies cases in humans by year in the United States, 1955–1994. (Modified from CDC annual summaries)

seen in humans each year in the United States (Figure 36-3), many wild animals are infected (Figure 36-4). Rabies in humans is much more frequent in many developing countries, and an estimated 15,000 cases occur worldwide. Some studies have suggested that from 1 to 2% of all deaths in some cultures are due to rabies.

Rabies may be transmitted orally through eating infected meat and may be an important means of transmission among wild carnivores. Airborne transmission in bat caves, while rare, has occurred. Six cases have been documented in persons receiving corneal transplants from donors who had died of undiagnosed rabies. Rabies in cattle is a major economic problem in Latin America and is transmitted by the bite of infected vampire bats. Bats, in general, probably serve as the most important reservoir of the rabies virus.

✷ Diagnosis

The most important priority is the rapid detection of rabies in an animal that has bitten a person. Whether the biting animal is infected determines directly the type of treatment that the bite victim should receive. Whenever possible, an animal that bites a person should be apprehended or killed. If domestic animals like dogs or cats do not show definite signs of rabies, the animal should be placed under observation at the public health laboratory for 8 to 10 days and monitored for the appearance of clinical signs of rabies. Wild animals or domestic animals showing signs of rabies should be killed and either the intact animal or the head taken immediately to a public health laboratory. The animal should be wrapped in a plastic bag to prevent possible contamination and kept cool if possible.

The laboratory worker examines the brain cells for the appearance of clusters of viral materials (*inclusion bodies*) in the cytoplasm called *Negri bodies* ∞ (Chapter 29, p. 381). Negri bodies are rapidly and specifically identified by using specific fluorescent antibodies. A positive diagnosis can be made on most infected animals within a few minutes via this procedure. As a backup test, brain tissue is injected into suckling mice that will develop rabies within the next 1 to 3 weeks if the virus is present. Cell cultures may also be used to isolate the virus.

A preliminary diagnosis in humans is based on clinical signs and symptoms and on a possible history of exposure to rabies. A confirmed diagnosis may be made by showing the presence of rabies virus antigens in corneal impressions or skin biopsies (taken from the nape of the neck) using fluores-

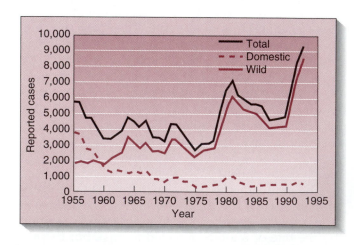

FIGURE 36-4 Reported rabies cases in wild and domestic animals by year in the United States and Puerto Rico, 1955–1994.

Negri bodies   Cytoplasmic inclusions in brain cells infected with the rabies virus.

cent antibodies. Diagnosis is confirmed after death by demonstrating the presence of rabies virus in the brain tissue by the fluorescent antibody method and by mouse or cell culture inoculations.

✱ Treatment

Treatment of clinical rabies has generally been unsuccessful. A few exceptions have occurred. The most notable was a 6-year-old boy in Ohio who was bitten by a rabid bat in 1970. He went through the Pasteur treatment described below but still developed clinical rabies. Through close medical supervision and such standard medical interventions as a tracheotomy to control respiration, drainage of cerebrospinal fluid to prevent excessive pressure on the brain, anticonvulsive drugs, and monitoring of heart rhythm, this patient was able to survive complications that killed most other known rabies patients. Apparently his partial immunity and his own defense mechanisms were able to fight off the rabies virus infections and his recovery was complete.

✱ Prevention and Control

A great deal of effort and resources go toward preventing and controlling rabies and in many countries these programs have been successful. Cases of human rabies in the United States have been reduced to a few cases per year out of approximately 10,000 persons exposed to rabies each year. In developing countries, hundreds of thousands are exposed to rabies each year.

Prevention and control of rabies are carried out in the following areas:

1. Control in domestic animals by vaccination and quarantine
2. Vaccination of persons who have a high risk of being exposed
3. Postexposure *immunoprophylaxis* of those exposed.

Before the initiation of strict control measures, most human rabies were contracted by the bite of domestic cats or dogs. Leash laws, licensing of dogs, and vaccinations have reduced the cases of rabies in these animals to an insignificant number in many countries and eliminated rabies completely in such island nations as Great Britain, Australia, and Japan. Rabies is prevented from reentering these rabies-free countries by the strict quarantine of imported animals. Still, in many parts of the world rabies is widespread in wildlife and no effective methods are available to control or eliminate this reservoir of the virus. The major animal reservoirs in the United States are bats, skunks, foxes, and raccoons; yet virtually any wild animal may be infected and so any animal exhibiting abnormal behavior should be suspect. In some countries, domestic dogs and cats are still the major source of exposure for humans; for example, in 1986 in Mexico 8483 rabid dogs were reported and this made up 96% of all reported animal rabies in Mexico that year. Currently, experiments are underway to vaccinate wild animals with bait that is mixed with a live, harmless pox virus into which the rabies virus antigen has been genetically engineered. Should this procedure work, it would offer a means of possibly reducing the incidence of rabies in wild animal populations.

Several types of vaccine are available for animals. A living attenuated vaccine is generally used to vaccinate dogs and cats. The only vaccines currently available for humans in the United States are, first, a killed purified human diploid cell vaccine (HDCV) and second, a killed-absorbed rhesus lung diploid cell vaccine (RVA). These vaccines are free of most adverse reactions that were associated with the crude rabies vaccines that were used before the 1980s. Preexposure vaccination with 3 injections of HDCV or RVA over 4 weeks is recommended for veterinarians, animal control officers, rangers, and other people who are in frequent contact with wild animals.

The prevention of rabies primarily concerns how people are handled following their exposure to rabies. Possibly because of the long incubation period, rabies is the only disease in which the initiation of active vaccination after exposure is successful. This procedure was developed by Pasteur in 1884 when he prevented the development of rabies in a boy who had been severely bitten by a rabid dog. Pasteur injected progressively more potent preparations of rabies-infected rabbit spinal cord extracts into the boy daily over a 21-day period. The treatment worked and was referred to as the Pasteur treatment. Some modifications were made in the Pasteur treatment over the years, and it continued to be used up to the introduction of the HDCV in the early 1980s.

Local treatment of the wound is helpful in preventing rabies. The wound should immediately be

---

Immunoprophylaxis   Prevention of disease through treatment with a vaccine or antiserum.

CLINICAL NOTE

Mass Treatment of Humans Exposed to Rabies: New Hampshire, 1994

On 22 October 1994, the laboratory of the New Hampshire Division of Public Health Services (NHDPHS) diagnosed rabies in a kitten that had been purchased from a pet store in Concord, New Hampshire. On 19 October the animal had developed seizures, then died of unknown causes during the night of 20–21 October. Approximately 665 persons received rabies postexposure prophylaxis because of exposure to this kitten and other cats from the same pet store. This report summarizes the epidemiologic investigation of the source of the infection and follow-up care of humans and animals potentially exposed to rabies.

Because the pet store did not keep records for kittens acquired for sale, the kitten's origin and date of arrival were unknown. However, on 26 September, a group of kittens reported to have included the rabid kitten was examined by a veterinarian and given health certificates, in accordance with state law, before being offered for sale by the pet store. The kitten was sold on 5 October and kept by its owners until its death. On 22 October, rabies was diagnosed in the kitten by fluorescent antibody testing at the NHDPHS laboratory. At CDC, genetic typing of the rabies virus isolated from the kitten indicated that it was a variant associated with raccoons. The investigation could not determine whether the kitten was infected with rabies before, during, or after its stay in the pet store; two other kittens sold by the pet store during the same period as the infected kitten died of unknown causes at their new homes but were unavailable for testing for rabies.

On 12 October, a raccoon captured in Henniker (a suburb of Concord), where the kitten was suspected to have originated, tested positive for rabies. Subsequent investigation indicated that the raccoon may have had direct contact with three feral kittens acquired by the pet store on 20 September. All three feral kittens developed signs of respiratory illness and died during approximately 4–6 October—a period overlapping that during which the rabid kitten was in the store. None of these three kittens were available for testing for rabies and all were younger than the minimum age (3 months) recommended for rabies vaccination.

From 19 September through 23 October (the last date any potentially exposed kittens were in the pet store), a minimum of 34 kittens had been offered for sale by the store. In addition to the infected kitten, 33 other kittens were included in the investigation: 27 were located and tested negative for rabies, and five died of unknown causes but were unavailable for testing (including the three feral kittens); one kitten was quarantined at the owner's request, and its status is unknown.

Because of limitations in the store's records regarding the origins and sale destinations of the kittens, local news media assisted in alerting community residents about the potential exposures to rabies at the store. The kittens had been allowed to roam freely throughout the store, which was frequented by children from child-care centers and a nearby school. As a result, NHDPHS and two major health-care facilities screened approximately 1000 persons who responded to media alerts and referred to private sector health-care providers for definitive evaluation of those persons who might need rabies postexposure treatment. NHDPHS gave medical providers an algorithm to determine the necessity for recommending rabies postexposure treatment. Rabies postexposure treatment, consisting of one dose of rabies immune globulin and five doses of rabies vaccine, was initiated for approximately 665 persons.

Editorial Note: This incident of rabies associated with a pet store resulted in the largest number of persons ever reported to have received rabies postexposure treatment as a result of potential contact with a point source in the United States. The costs associated with the public health response to exposure to the rabid kitten in New Hampshire are unprecedented in the United States. The overall estimated cost was $1.5 million, including expenditures for rabies immune globulin and vaccine ($1.1 million), laboratory testing of animals ($4200), and investigation by NHDPHS and CDC personnel ($15,000). This cost is nearly 15-fold higher than that ($105,790) associated with rabies postexposure treatment of 70 persons after a single case of rabies occurred in a domestic dog in California in 1981 (*MMWR* 44:484, 1995).

cleansed thoroughly and flushed with soap and water. The patient should then be taken to a physician to receive antiserum treatment. A **hyperimmune globulin** known as *human rabies immune globulin* is now available and should be instilled in the depths of the wound and infiltrated around the

Hyperimmune globulin   Globulin containing high levels of antibody to a specific antigen. These globulins are prepared by repeatedly immunizing animals or persons with the antigen. Blood is then collected and the globulins are obtained.

wound. Next, postexposure immunization with HDCV or RVA should be initiated with intramuscular injections on days 0, 3, 7, 14, and 28.

The treatment regimen should be correlated with the diagnostic program on the animal that inflicted the bite. If tests show that the animal is not infected, the immunization program should be stopped to avoid further discomfort to the patient. For this reason, it is important to apprehend the biting animal. If the animal is not apprehended, the victim usually has little choice but to go through the immunization program. About 10,000 postexposure immunizations are given in the United States annually.

MATERIAL FOR REVIEW

CONCEPT SUMMARY

1. A bullet-shaped, helical RNA rhabdovirus is responsible for the disease rabies. This is a common disease among lower animals and in humans in foreign countries, but its occurrence is limited in people living in the United States.

2. Rabies is unusual in its mode of transmission (usually from the bite of an infected animal) and devastating in its consequences.

3. Control of rabies is best accomplished by the immunization of domestic animals and individuals at high risk. Vaccines with risk of serious side effects have largely been replaced by a more useful diploid cell vaccine.

CLINICAL SUMMARY TABLE

Microorganism	Virulence Mechanisms	Diseases	Transmission	Treatment	Prevention
Rabies virus		Rabies	Animals to humans Biting	None	Post-exposure vaccination Vaccinate animals Animal control

STUDY QUESTIONS

1. What three preventive procedures should be taken for an individual who is bitten by a rabid animal?

2. What is the most likely source of human exposure to rabies in the United States?

3. What features of rabies give this disease characteristics of an occupational disease?

4. Describe the physical properties of the rabies virus.

CHALLENGE QUESTIONS

1. Why is the incubation period for rabies longer when the bite occurs on the leg and shorter when it occurs on the neck?

2. In some years there are no reported cases of human rabies. Does this mean no one was exposed to the virus? Explain.

Human Immunodeficiency Virus and AIDS

500nm

VIRUS

The appearance of a new and strange disease at the beginning of the 1980s puzzled physicians and researchers. Several cases of an unusual immunological deficiency were reported in the United States and other places around the world. The deficiency was quite complex, since victims often died from one of several opportunistic infections. Epidemiological studies suggested these infections were a new syndrome. Close examination of patients showed that they contained unusually low numbers of certain lymphocytes. Without these cells, individuals could not mount an effective attack on invading infectious agents. For these reasons, the immune deficiency was termed **acquired immunodeficiency syndrome** by U.S. health officials. The acronym **AIDS** soon became a household word. AIDS may ultimately prove to be among the most devastating medical experiences in history. In terms of human life lost, AIDS may not exceed the devastation of smallpox, plague, tuberculosis, or malaria, but in terms of economic demand, social change, and commitment of medical resources it may eclipse all previously known maladies. This chapter presents a review of our current understanding of AIDS.

HUMAN IMMUNODEFICIENCY VIRUS: GENERAL CHARACTERISTICS

Today we know that AIDS is caused by the *human immunodeficiency virus* (HIV). HIV is a single-stranded (ss) RNA virus which belongs to a group of viruses known as *retroviruses*. Retroviruses are commonly associated with animal tumors ∞ (Chapter 29, p. 381). Like other retroviruses, HIV depends on the presence of a reverse transcriptase enzyme to transcribe the viral RNA genome into DNA. In retroviral infections, once the viral DNA is produced, it readily integrates into the host cell

HUMAN BODY SYSTEMS AFFECTED

Opportunistic Pathogens in HIV-Infected Person

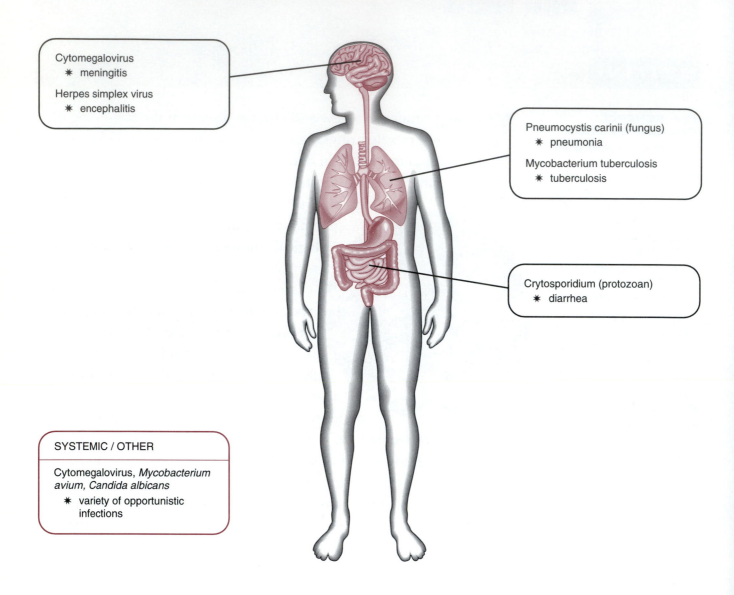

Cytomegalovirus
* meningitis

Herpes simplex virus
* encephalitis

Pneumocystis carinii (fungus)
* pneumonia

Mycobacterium tuberculosis
* tuberculosis

Crytosporidium (protozoan)
* diarrhea

SYSTEMIC / OTHER

Cytomegalovirus, *Mycobacterium avium, Candida albicans*
* variety of opportunistic infections

chromosome. In this *provirus* state, the viral DNA of most retroviruses is able to provide genetic messages that *transform* the host cell. This type of transformed cell is characteristic of tumor cells. In HIV infections, however, the DNA transcribed from viral RNA does not induce tumors, but remains in many of the infected cells, especially activated CD4 T cells ∞ (Chapter 11, p. 156). The HIV readily multiply, causing cell destruction and the release of large numbers of viruses (see Plate 75). The CD8 T cells or cytotoxic T cells destroy the virus-infected cells and new CD4 T cells are produced.

Provirus   A complete viral genome that has been integrated into the host genome.

Early in an infection, a person has HIV disease. In these individuals, the daily battles between T cells and virus result in a draw because infected cells are destroyed and replaced by more T cells. As the years pass, it becomes more difficult for the immune system to replace infected T cells and the number of such cells dwindles. Without these cells, secondary infections characteristic of AIDS develop.

The virus appears to be infective for the chimpanzee but not for other animals, and the chimpanzee does not develop symptoms of AIDS. Two different virus isolates, called HIV 1 and HIV 2, have been isolated and are antigenically different. HIV 1 is the original isolate, made in 1983, and is the predominant isolate found in most parts of the world. HIV 2 was first isolated in 1986, and is found predominantly in West Africa. HIV 2 does not seem to be as virulent as HIV 1, and the development of AIDS in infected persons is significantly delayed compared to persons infected with HIV 1.

Periodically, reports of other HIV occur. However, their relationship to the present HIV epidemic is not clear. There are two major serogroups of HIV viruses. These are designated as M and O groups. The M group contains eight subspecies named A through H. Each subspecies occurs in a specific geographic area, with B being the most common subspecies in the United States. There are 30 subtypes within group O. The rapid rate of HIV evolution makes it difficult to generate a clear taxonomy for HIV, especially since it is likely that new subspecies will continue to appear.

PATHOGENESIS AND CLINICAL CONDITIONS

AIDS was first recognized in the United States in 1981. The accumulated deaths from AIDS has rapidly increased since that time (Figure 37-1). Realization that a new disease was developing was first made indirectly. A few individuals with the same, but unusual, clinical conditions raised questions as to the etiology of their disease. Each of these patients developed a rather rare form of cancer known as *Kaposi's sarcoma* (Figure 37-2; see Plate 40), suffered from unexplained weight loss, fever, **lymphadenopathy**, and an unusual type of pneumonia caused by a fungus, *Pneumocystis carinii* ∞ (Chapter 40, p. 508). The other feature in common with each

of these patients was that all were homosexual males. The clinical characteristics of these early patients set the diagnostic pattern associated with AIDS. Most patients suffer from malaise, diarrhea, shortness of breath, lymphadenopathy, and infection by one or more low-grade pathogens. What was not realized at that time was that these patients were near the end of a long-term, chronic infection now called HIV disease, and that AIDS is the terminal clinical condition for HIV-infected persons.

IMMUNODEFICIENCY AND OPPORTUNISTIC INFECTIONS

As explained in Chapter 11, ∞ (p. 155) normal immunologic health is dependent on a variety of lymphocytes. These cells communicate with each other in order to provide optimal host resistance to infectious disease. Some of these cells, such as the CD4 lymphocytes, produce lymphokines ∞ (Chapter 11, p. 157) that are needed to activate many of the cells involved in the immune response. CD4 cells are sometimes referred to as T-helper cells. One kind of cell activated by CD4 cells is the CD8 cell. These are cytotoxic T cells, and when activated they are able to destroy human (host) cells that are infected with viruses or other intracellular parasites. Destruction of the infected host cell exposes the parasite (virus, fungus, bacteria) to effective host protective measures such as antibody or phagocytosis.

The numbers 4, 8, and so on, associated with these cells represent the kind of antigens found on the surface of the cell. HIV is able to bind to the type 4 antigen and infect cells that have that antigen on their surface. HIV is a *lymphotropic* virus that preferentially infects T-helper cells (CD4 cells). Once infected, these cells are eventually destroyed and the reduction in the number of CD4 cells decreases the amount of lymphokines that are necessary to maintain an effective immunologic response by the host. Other cells, such as *glial cells* (located in the brain), macrophages, vaginal lining cells, and retinal cells, also have type 4 antigen on their surface and can be infected by HIV. However, infection of these cells may not lead to cell death, but may serve as a continuing source of HIV release from the cells into body fluids. These viruses are produced by persons with HIV disease and provide the means

Lymphadenopathy Enlargement of the lymph nodes.
Lymphotropic Having a preference for lymphoid cells.
Glial cells Specialized brain cells surrounding the nerve cells (neurons).

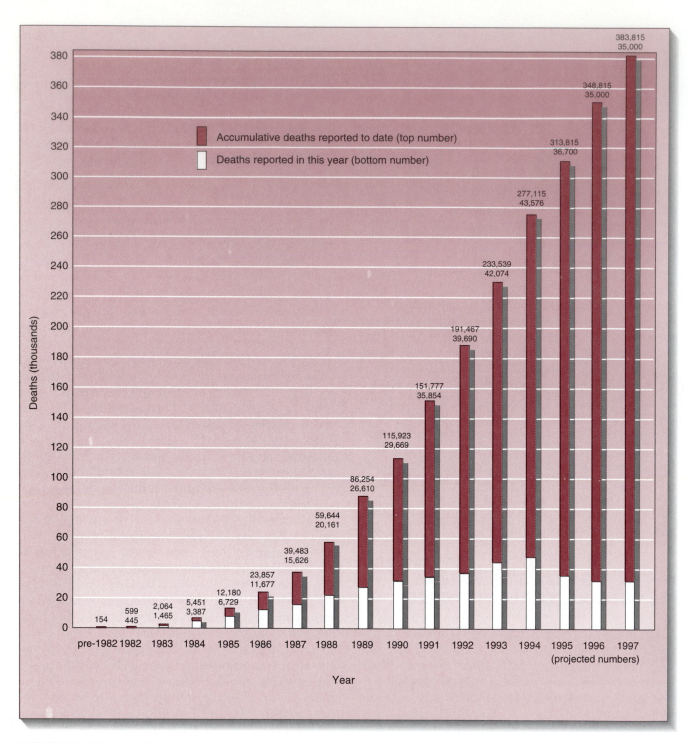

FIGURE 37-1   Cumulative deaths from AIDS in the United States between 1982 and 1994. Projected for 1995 through 1997.

whereby other individuals may become infected through contact with the body fluids or tissue of an HIV-positive person.

The normal ratio of CD4 helper T cells to CD8 T cells is just slightly greater than one. The end result of destruction of helper T cells by HIV is to lower

this ratio to values as low as 0.2. Individuals with such small numbers of helper T cells are unable to maintain a proper immunologic response and they become highly susceptible to a variety of infectious agents characteristic of individuals with immuno-suppressed T-cell functions. Humoral immune

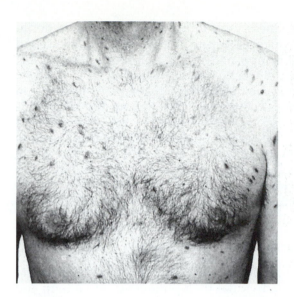

FIGURE 37-2 Patient with Kaposi's sarcoma. Note dark pigmented areas of skin, characteristic of this disease. (Courtesy Burroughs Welcome Co., Research Triangle Park, NC)

response may also be diminished in these patients. The complications associated with immunodeficiency include infection by microorganisms not commonly found in other patients. Serious life-threatening diseases due to viruses such as cytomegalovirus and herpes simplex virus, bacteria such as *Mycobacterium avium* and *Salmonella*, fungi such as *Candida albicans*, *Cryptococcus neoformans*, and *Pneumocystis carinii*, and parasites such as *Cryptosporidium* have become characteristic of AIDS patients. In spite of aggressive therapy, nearly 90% of AIDS patients die from opportunistic infections. At present, the single most important risk factor in HIV disease is the development of tuberculosis caused by *Mycobacterium tuberculosis* ∞ (Chapter 20, p. 285).

There is good evidence that the body immune surveillance system is an important part of a cancer control mechanism. An increasingly defective immune system associated with HIV disease significantly increases the risk of some types of cancer. Several cancers, such as invasive cervical cancer, squamous cell carcinoma, progressive leukoencyphalopathy, lymphoma, and Kaposi's sarcoma, occur more commonly in persons with AIDS than in individuals without this disease. Kaposi's sarcoma is manifested primarily by multiple vascular nodules in the skin and other organs (Figure 37-2; see

Plate 40). It is a disease that is rare in young adults but afflicts as many as 36% of people with AIDS. Recent findings have suggested that Kaposi's sarcoma may be associated with the activation of an unidentified herpesvirus ∞ (Chapter 31, p. 406).

In addition to predisposing individuals to opportunistic infections and some cancers, HIV also seems to potentiate some illnesses caused by other etiologic agents. For example, as noted in Chapter 17, ∞ (p. 241) tertiary syphilis normally takes years for development. In patients with AIDS, however, this phase of syphilis may be expressed after only a few months. Treatment of any of these complicating illnesses is made particularly difficult due to the lack of host response in AIDS patients. The basic immune deficiency, coupled with complicating factors of associated diseases, has resulted in a condition that has a mortality rate that is essentially 100%.

NEUROPATHIC EFFECTS

In addition to being lymphotropic, HIV is also *neurotropic*. Infection of the central nervous system by HIV seems to be facilitated by infected macrophages that cross the blood–brain barrier carrying with them the infectious virus. Neurologic abnormalities in these patients fall into two general categories. One is a progressive mental deterioration (dementia) that often begins with relatively benign symptoms such as forgetfulness, slurring of speech, and loss of memory or balance. Some of these symptoms occur in as many as 50% of HIV-infected patients. In later stages of dementia, patients express profound fatigue, bladder and bowel incontinence, headache, loss of speech, seizures, and even coma. When the brain is examined at autopsy, it appears that as many as 80% of AIDS patients have some neurologic lesions.

The second type of neurologic symptom that commonly occurs in HIV-positive patients is a change in the peripheral nervous system known as *neuropathy*. Peripheral neuropathy may result in sensations of burning, numbness, pins and needles, or frank pain in the feet, legs, arms, or hands. The extent to which peripheral neuropathy may be mediated by opportunistic infectious agents such as cytomegalovirus or by toxic drugs used to treat the HIV infection is not entirely understood. Various neurologic manifestations of HIV disease are listed in Table 37-1.

Neurotropic Having a preference for the nervous system.
Neuropathy A disorder that affects the nervous system.

TABLE **37-1**

Neurological Manifestations Associated With Direct HIV Infection of the Nervous System

AIDS dementia complex
Asymptomatic infection
Acute encephalitis
Aseptic meningitis
Vacuolar myelopathy
Inflammatory demyelinating polyneuropathy
Radiculopathy
Mononeuropathies
Distal sensory neuropathy

STAGES OF HIV DISEASE

The Centers for Disease Control (CDC) have divided individuals with HIV disease into four groups, depending on the stages of their disease (Figure 37-3). The entire course of the disease may vary, being as short as 2 to 3 years in some persons (1 to 3%) and as long as 10 to 15 years in others.

✳ Acute HIV Infection

Three to 8 weeks following the initial infection with HIV, most individuals experience a brief, acute, self-limited illness with symptoms similar to mononucleosis, such as fever, sore throat, headache, and swollen lymph nodes. The symptoms usually last less than 4 weeks. During the *acute HIV infection*, high levels of virus are produced in the patient and the virus is distributed throughout the body. It is also during this phase that the first antibodies to the HIV are released into the bloodstream. Because individuals may have large amounts of virus in their body tissue prior to the development of antibodies, there is a risk that they may unknowingly transmit HIV to another person during this time.

✳ Asymptomatic Stage

The symptoms of the acute HIV infection disappear and the individual then progresses into a long and variable period without symptoms, the *asymptomatic stage*. Some individuals (about 30%) do not develop an acute illness following HIV infection, and enter the asymptomatic phase directly. These individuals, while clinically well, may experience a decrease in helper T cells and have circulating antibodies to HIV. Asymptomatic individuals are very important epidemiologically because of their ability to transmit the virus to others.

The actual number of HIV antibody-positive persons is unknown, but is presently estimated to be

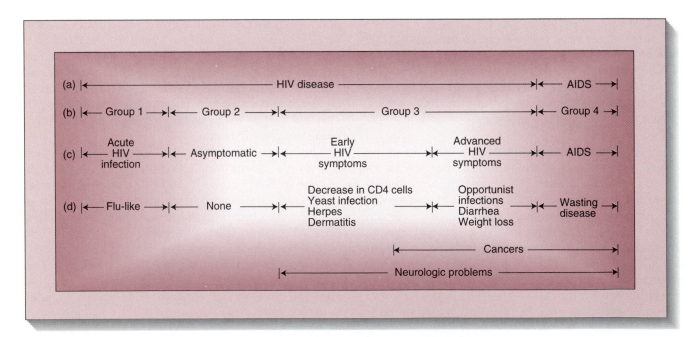

FIGURE 37-3   Time line showing sequence of events in patients infected with HIV: (a) Category of diagnosis; (b) CDC patient categories, where groups correspond to (c) patient condition, (d) symptoms of disease in HIV-infected individuals (time of events varies; line represents a 15-year time between infection and death).

more than 2 million in the United States. The proportion of these persons who will ultimately develop symptomatic disease is estimated to be in the order of 40 to 50% over a 10-year period. Whether or not all HIV-infected persons will ultimately develop AIDS is unknown, but it appears that the disease ultimately affects nearly 100% of those with HIV disease.

✳ Early and Late HIV Symptoms

Persons in the asymptomatic phase of HIV disease live a relatively normal life without HIV-associated illness. However, during this time there is a continuing decrease in the number of CD4 cells. This change in immune status eventually leads to a condition where the body is no longer able to effectively resist opportunistic infections. Early HIV symptoms are often constitutional, such as fatigue, weight loss, and so on, but ultimately HIV overwhelms the immune system, leading to repeated episodes of serious infections by viruses, bacteria, and fungi.

✳ AIDS

Patients in this stage have all of the conditions associated with the early and late symptoms plus a combination of opportunistic infections and cancers that

result from severe immunosuppression. Prior to 1993 AIDS cases were reported on the basis of the patients' opportunistic infections and cancer. This definition of the disease did not include a direct consideration of the change in the immune system and eliminated a number of persons with the disease from inclusion in the statistics. The 1993 CDC definition of the disease included not only persons with unusual opportunistic infections, but those with fewer than 200 CD4 cells per cubic millimeter of blood (normal is between 800 and 1200 per cubic millimeter). This change in the definition of persons with AIDS significantly increased the number of persons considered to have AIDS.

TRANSMISSION AND EPIDEMIOLOGY

Although AIDS was first described in the United States and has subsequently been found in most countries throughout the world (Figure 37-4), it appears that HIV infection first occurred in Africa and was later transmitted to the western world. By late 1994, over 441,000 cases of AIDS had been reported in the United States. Most of the cases in the United States have occurred in males, and AIDS is currently the leading cause of death in young

FIGURE 37-4 Estimated number of global HIV infections projected by the World Health Organization in 1995.

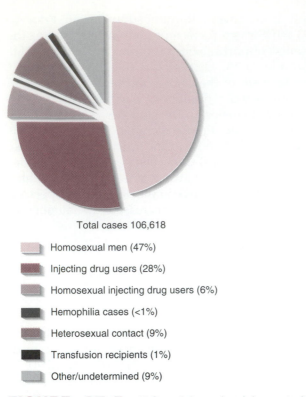

Total cases 106,618

Homosexual men (47%)

Injecting drug users (28%)

Homosexual injecting drug users (6%)

Hemophilia cases (<1%)

Heterosexual contact (9%)

Transfusion recipients (1%)

Other/undetermined (9%)

FIGURE 37-5 U.S. adult and adolescent AIDS cases by transmission category. (Courtesy Centers for Disease Control and Prevention)

adult males in the United States. Of the cases reported in men in the United States in 1993, 53% were in homosexual or bisexual males, 32% were in intravenous drug users, and 9% were in persons infected by heterosexual contact (Figure 37-5). Increased numbers of cases are now occurring in females as a result of heterosexual transmission; in 1994 about 18% of all reported cases were in women. In the United States, women generally acquire HIV from infected sexual partners or from illicit IV drug use (Figure 37-6). Children obtain the virus in utero or at the time of birth from infected mothers, from treatment for hemophilia, or through sexual abuse.

The estimated number of infected, pregnant women varies throughout the world, but in some areas of Central and East Africa as many as 2 to 15% of pregnant women show evidence of infection. Between 30 and 65% of their children will be infected either prior to or at birth. The AIDS epidemic in some parts of Africa is most alarming, with estimates of infection as high as 25% of individuals between the ages of 20 and 40 years. In Central Africa, the ratio of men to women with AIDS is about 1:1. This reflects primarily heterosexual transmission of the disease in Africa, where frequency of occurrence is related to number of sexual partners

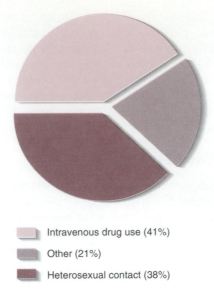

Intravenous drug use (41%)

Other (21%)

Heterosexual contact (38%)

FIGURE 37-6 Sources of HIV infection in U.S. women. (Courtesy Centers for Disease Control and Prevention)

more than to sexual preference. The percentage of HIV cases occurring in women throughout the world is shown in Figure 37-7.

Although much controversy has existed regarding possible mechanisms for the spread of this virus, it is becoming increasingly clear that even though the virus has been isolated from human milk, saliva, tears, urine, cerebrospinal fluid, vaginal fluid, semen, and blood, there is almost no risk of transmission except through contact with the latter three fluids. It is also now known that casual contact will not transmit the virus, and that it cannot be transmitted through food prepared by an HIV-positive person or among children who attend school or play together. With the exception of those whose lifestyles place them at increased risk (male homosexuals, bisexuals, female prostitutes, or in-

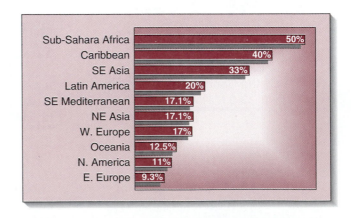

FIGURE 37-7 Percentage of total HIV cases who are women. (Courtesy Centers for Disease Control and Prevention)

TABLE 37-2

Guidelines to Prevent Transmission of AIDS

Health-Care Workers

General
- Use appropriate barrier precautions (gloves, masks, eyewear, gowns).
- Immediately wash any skin surfaces exposed to possibly contaminated body fluid.
- Use precautions to prevent personal injury from sharp objects (needles, scalpels, instruments) used in any medical procedure.
- Use procedures that minimize need for mouth-to-mouth resuscitation.
- If you have open or weeping lesions, do not participate in direct patient care.
- Implement universal blood and body fluid precautions for all patients.

Dental
- Consider blood, saliva, and gingival fluid from all patients as infective.
- Use appropriate barrier precautions (gloves, protective eyewear, etc.) for all contact with oral surfaces.
- Sterilize handpieces after use with each patient.
- Carefully clean blood and saliva from all materials that are used in the mouth. Disinfect all devices before they are placed in patient's mouth.
- Carefully wrap all equipment surfaces that are difficult to disinfect with impervious paper, plastic wrap, or foil and replace after each patient.

Laboratory
- Specimens should be placed in leak-proof containers.
- Persons processing blood or body fluid specimens should wear gloves. Masks and eye protection should be used when appropriate.
- Use safety cabinet for procedures with risk of generating droplets.
- Use mechanical pipetting devices.
- Limit use of needles and syringes. Use proper discard procedures for all sharp objects.
- Immediately decontaminate all spills. Regularly decontaminate work surfaces.
- Implement universal body fluid precautions for all patients.

Other
- Strictly follow established precautionary procedures for handling patients and patient materials in areas where there is increased risk: hemodialysis center, emergency rooms, nursing care facilities, ophthalmology and optometric centers.

Individual

If you are HIV-positive:
- Use barrier (condom) protection for sexual partner.
- Refrain from donating blood, sperm, body tissues, or organs.
- Recognize that you can transmit HIV to others even if you remain asymptomatic.
- Properly disinfect any surface accidentally contaminated by bleeding.
- Notify responsible personnel of your HIV status when seeking medical or dental assistance.

If you are HIV-negative:
- Limit sexual contacts to a monogonous relationship.
- If your sexual partner is HIV-positive, use barrier (condom) protection.
- Do not share needles, toothbrushes, razors, or any item that may be contaminated with blood.
- Avail yourself of HIV testing if there is a possibility that you have been exposed to the virus.

jecting drug abusers), the only individuals who are at significantly increased risk are medical and dental health professionals. Specific guidelines have been developed by the CDC to reduce the risk to this group of individuals. An abbreviated list of guidelines is given here as Table 37-2.

Prior to the development of diagnostic antibody tests, HIV was occasionally transmitted through contaminated blood or blood products to hospitalized recipients. However, present screening of blood donors to rule out high-risk individuals, and serologic testing procedures used in the United States make transmission by this mode extremely rare.

DIAGNOSIS

Laboratory diagnosis is used for two primary reasons—either to detect HIV contamination of blood

Human Immunodeficiency Virus Transmission in a Household Setting: United States

Transmission of human immunodeficiency virus (HIV) has been reported in homes in which health care has been provided and between children residing in the same household. CDC has received reports of HIV infection that apparently occurred following mucocutaneous exposures to blood or other body substances. This report summarizes the findings of the epidemiologic and laboratory investigations, which underscore the need to educate persons who care for or are in contact with HIV-infected persons in household settings where such exposures may occur.

A 5-year-old child whose parents were both HIV-infected tested negative for HIV antibody in 1990 and July 1993 but tested positive in December 1993. In February 1994, all other close household contacts of the child tested HIV-antibody negative.

From January through December 1993, when the child was likely to have become infected, the child's parents were the only known HIV-infected persons with whom the child had any contact. During this period, the child lived with both parents until the father's death as the result of acquired immunodeficiency syndrome (AIDS) in May 1993. The child continued to live with the mother, who had AIDS, until 8 days before the child's last negative antibody test in July 1993. The child then lived in foster care.

The child had several opportunities for contact with HIV-infected blood and exudative skin lesions. Based on the mother's medical records and history, from March through August 1993 the mother had recurrent, purulent, exudative skin lesions on her face, neck, torso, buttocks, and extremities. She frequently scratched the lesions until they bled, left the lesions uncovered, and discarded onto the furniture or the floor the gauze and tissues used to wipe the exudate. During periods when the mother's skin lesions were uncovered and draining, the child frequently hugged and slept with the mother. In addition, the child intermittently had scabs from impetigo and abrasions that the mother sometimes picked off and caused to bleed. When the mother had intermittent gingival bleeding, she periodically shared a toothbrush with the child. From January through May 1993, the child had no known contact with the father's blood or body fluids, although the child sometimes used his toothbrush.

No other situations were identified in which the child potentially may have been exposed to HIV-infected blood or had contact with an HIV-infected person. There were no known HIV-infected persons in either the foster home or the school, and the child had no known contact with blood in these settings. Based on interviews and medical record reviews, no household members at either the parents' home or foster home engaged in injecting-drug use. Based on history and physical examination, sexual abuse of the child was believed to be unlikely. During 1993, the child had no injections, blood transfusions, vaccinations, or invasive dental or medical procedures.

Proviral DNA from peripheral blood mononuclear cells obtained from the mother and the child was amplified by polymerase chain reaction. By direct sequencing, the two DNA fragments encompassing 343 nucleotides of the V3 and flanking regions of the gene encoding the HIV-1 envelope glycoprotein (gp120) were genetically similar, differing by only 2.6%. No specimen was available from the child's father.

Editorial Note: The findings of the investigations described in this report indicate the transmission of HIV as the result of contact with blood or other body secretions or excretions from an HIV-infected person in the household. Exposures occurred after the source-patient had developed AIDS; consequently, relatively high HIV titers may have been present in their blood (*MMWR* 43:347, 1994).

---

or other medical products that are to be used for transfusion; or to determine whether or not a person is infected. Three general approaches are used: (1) isolation of the virus; (2) demonstration of HIV antigens or nucleic acids; and (3) detection of specific antibodies.

✳ Isolation of the Virus

Virus isolation can be made from lymphocytes, bone marrow, blood, or other fluids containing infected cells; however, this procedure is usually carried out only in a research setting and has little direct application to patient care or disease prevention.

✳ Demonstration of HIV Antigens or Nucleic Acids

Antigen detection is possible by several methods. ELISA procedures ∞ (Chapter 12, p. 181) using high titer human antibodies against HIV can be used to capture and detect HIV antigens in blood and other body fluids. In other tests, fluorescent-labeled ∞ (Chapter 3, p. 29) monoclonal antibody ∞ (Chapter 11, p. 161) kits are available to rapidly and specifically detect the presence of HIV antigens in blood cells or other cells. HIV nucleic acids can be detected in different cells and tissues with commercially available enzyme-conjugated nucleic acid probes ∞ (Chapter 7, p. 93). Polymerase chain reaction procedures are being developed to amplify HIV nucleic acid, and provide an exquisitely sensitive method for detecting the presence of the virus.

✳ Detection of Specific Antibodies

A variety of commercial kits are available and are used to detect the presence of HIV antibodies in serum. Many of the newer kits use genetically engineered HIV antigens. ELISA serologic tests are most often used. A drawback of using antibody tests to detect infected persons or contaminated blood is the so called "detection window" that exists in recently infected persons who are carrying the virus in their blood, but have not yet developed antibodies (in some cases such a condition can exist for months). When antigen detection methods are used this "window" is not present.

TREATMENT

Increasingly effective therapy of HIV-associated opportunistic infections (Table 37-3) has nearly doubled the life expectancy of HIV-infected persons

---

TABLE 37-3

Some Common Opportunistic Diseases Associated with HIV Infection and Possible Therapy

Causative Agents	Possible Treatments
Protozoa	
Cryptosporidium muris	Investigational only
Isospora belli	Trimethoprim sulfamethoxazole (Bactrim)
Toxoplasma gondii	Pyrimethamine and leucovorin, plus sulfadiazine, or Clindamycin, Bactrim
Fungi	
Candida sp.	Nystatin, clotrimazole, ketoconazole
Coccidioides immutis	Amphotericin B, fluconazole, ketoconazole
Cryptococcus neoformans	Amphotericin B, fluconazole, itraconazole
Histoplasma capsulatum	Amphotericin B, fluconazole, itraconazole
Pneumocystis carinii	Trimethoprim sulfamethoxazole (Bactrim, Septra), Pentamidine, Dapsone
Bacteria	
Mycobacterium avium complex (MAC)	Rifampin + ethambutol + clofazimine − ciprofloxacin +/− amikacin; clarithromycin & azithromycin (both investigational)
Mycobacterium tuberculosis (TB)	Isoniazid (INH) + rifampin + ethambutol +/− pyrazinamide
Viruses	
Cytomegalovirus (CMV)	Ganciclovir, Foscarnet
Epstein–Barr	Acyclovir
Herpes simplex	Acyclovir
Papovavirus J-C	None
Varicella–zoster	Acyclovir, Foscarnet
Cancers	
Kaposi's sarcoma	Local injection, surgical excision or radiation to small, localized lesions

since the mid-1980s. However, in spite of this excellent effort, there is still no totally effective therapy for the underlying HIV infection itself. Although several strategies have been directed against the HIV infection, such as immune stimulants and interferon, little positive effect has, as yet, been produced by these methods. A variety of antiviral compounds including zidovudine (AZT), didanosine, lamivudine, atevirdine, and stavudine have been approved for treatment of AIDS patients. Although they have improved survival time, they are not able to cure patients. An intensive effort is under way to develop other antiviral agents that might be effective against HIV infections.

PREVENTION AND CONTROL

Because the modes of transmission of the AIDS virus and the segments of the population that are at high risk are well known (Table 37-4), the major tool presently available to reduce the spread of AIDS is education. Educational efforts have already caused

---

TABLE 37-4

Adult/Adolescent AIDS Cases by Sex and Exposure Categories Through December 1994, United States

	Total No.	(%)
Male Exposure Category (88.5%)		
Men who have sex with men	228,152	(61)
Injecting drug use	82,131	(22)
Men who have sex with men and inject drugs	29,891	(8)
Hemophilia/coagulation disorder	3,545	(1)
Heterosexual contact:	10,641	(3)
Sex with injecting drug user	4,719	
Sex with person with hemophilia	25	
Sex with transfusion recipient with HIV infection	239	
Sex with HIV-infected person, risk not specified	5,659	
Receipt of blood transfusion, blood components, or tissue	3,823	(1)
Other/undetermined	19,116	(5)
Total male AIDS cases	377,299	(100)
Female Exposure Category (13%)		
Injecting drug use	27,948	(48)
Hemophilia/coagulation disorder	97	(0)
Heterosexual contact:	21,021	(36)
Sex with injecting drug user	11,039	
Sex with bisexual male	1,798	
Sex with person with hemophilia	242	
Sex with transfusion recipient with HIV infection	389	
Sex with HIV-infected person, risk not specified	7,553	
Receipt of blood transfusion, blood components, or tissue	2,819	(5)
Other/undetermined	6,589	(11)
Total female AIDS cases	58,474	(100)
Total cases	435,773	

major adjustments in how health-care personnel handle patients and patient materials (see Table 37-2). The challenge now is to extend appropriate educational programs to individuals whose lifestyles and sexual conduct place them at high risk of HIV infection.

HIV is not an extremely stable virus outside of host cells. It is relatively easily destroyed by a variety of disinfectants (Table 37-5). Individuals who have need to disinfect possibly contaminated objects can do so by using undiluted household bleach.

VACCINE

Considerable effort and resources have been used to develop a useful HIV vaccine, but at present there is no acceptable vaccine available for general use. Problems associated with the development of a vaccine are numerous and include biological, economic, and ethical factors. Just finding a population of individuals willing to test potential HIV vaccine poses a serious problem.

From a biological perspective, development of an HIV vaccine is a difficult task. Considering that HIV-infected persons develop relatively high anti-HIV antibody titers and that these antibodies do not stem the progress of the disease raises a question regarding possible effectiveness of any vaccine. The rapid mutations associated with replication of all retroviruses makes the selection of appropriate antigens for a vaccine very difficult. There is good evidence that the antigens of the virus normally change within a given individual over the course of HIV disease. This raises a serious question regarding the ability of science to ever find just the right vaccine antigen. Perhaps such a universal antigen

TABLE 37-5	
Agents Effective Against Human Immunodeficiency Virus	
Agents (freshly prepared)	**Recommended Concentration**
Sodium hypochlorite (household bleach)	Full strength (no dilution)
Chloramine-T	2%
Sodium oxychlorosene	4 mg/mL
Sodium hydroxide	30 mM
Glutaraldehyde	2%
Formalin	4%
Paraformaldehyde	1%
Hydrogen peroxide	6%
Propiolactone dilution	1:400
Nonoxynol-9	1%
Ethyl alcohol	70%
Isopropyl alcohol	30–50%
Lysol	0.5–1%
NP-40 detergent	1%
Chlorhexidine gluconate/ ethanol mix	4/25%
Chlorhexidine gluconate/ isopropyl mix	0.5%/70%
Tincture of iodine/isopropyl	1/30–70%
Betadine	0.5%
Quarternary ammonium chloride	0.1–1%
Acetone/alcohol mix	1:1
pH of 1 or 13	
Heat, 56°C for 10 minutes	

does not even exist. Another biologic problem is the possibility that some antibodies may actually help HIV infect monocytes. Such antibodies could possibly increase risk of infection in HIV-exposed persons.

MATERIAL FOR REVIEW

CONCEPT SUMMARY

1. AIDS is a recently discovered, life-threatening, viral disease with worldwide distribution. The infection is readily transmitted horizontally through sexual contact and by indiscriminate use of IV drug para-

phernalia or vertically from infected pregnant females to their offspring.

2. AIDS is caused by the human immunodeficiency virus (HIV). HIV is an RNA retrovirus that is lympho- and neurotropic.

HIV reduces the CD4 to CD8 lymphocyte ratio, which seriously diminishes T cell-based immunity. These patients are highly susceptible to a broad range of pathogenic microbes.

3. AIDS can be diagnosed on the

basis of clinical symptoms, by isolation of the virus or by serological testing. Careful screening of blood donors and testing for anti-HIV antibody have essentially eliminated risk of virus transmission through blood or blood products.

4. AIDS poses serious economic concerns for the health-care system.

CLINICAL SUMMARY TABLE

Microorganism	Virulence Mechanisms	Diseases	Transmission	Treatment	Prevention
Human immunodeficiency virus	T-cell infection	AIDS	Sexual Blood transfusion Intravenous drug use	None effective	Avoid exposure Blood testing

STUDY QUESTIONS

1. Describe the process of HIV replication in a host cell.
2. What characteristic of the human immune response creates the "diagnostic window" in currently used laboratory diagnostic procedures for AIDS?
3. List the sociological factors presently associated with transmission of HIV.
4. List the medical factors presently associated with transmission of HIV.
5. Discuss the observation that AIDS patients suffer from "unusual" opportunistic infections.

CHALLENGE QUESTIONS

1. Would a live, attenuated AIDS vaccine be useful? Explain.
2. If such a vaccine were developed, how could it be tested?

38 Other Viruses

500nm

VIRUS

This chapter discusses the following families of viruses: *Togaviridae, Flaviviridae, Bunyaviridae, Reoviridae, Arenaviridae, Coronaviridae, Filoviridae, Parvoviridae,* and the so-called Norwalk group of viruses. Though unrelated morphologically or genetically, these groups of viruses cause a variety of significant diseases of humans—from the mild common cold-type infections of coronavirus to the deadly hemorrhagic fever of Ebola virus.

TOGAVIRUSES AND FLAVIVIRUSES

The *Togaviridae* and *Flaviviridae* families contain about 250 related viruses. The viruses in both families contain RNA, have cubical symmetry, and pos-

sess an envelope. These viruses constitute a major portion of a group of viruses that have traditionally been called arthropod-borne viruses and referred to as *arboviruses*. Arboviruses are able to multiply in such bloodsucking arthropods as mosquitoes, ticks, and gnats, as well as in many different vertebrate

Arboviruses   A general term describing viruses transmitted from arthropods to vertebrates.

HUMAN BODY SYSTEMS AFFECTED

Miscellaneous Viruses

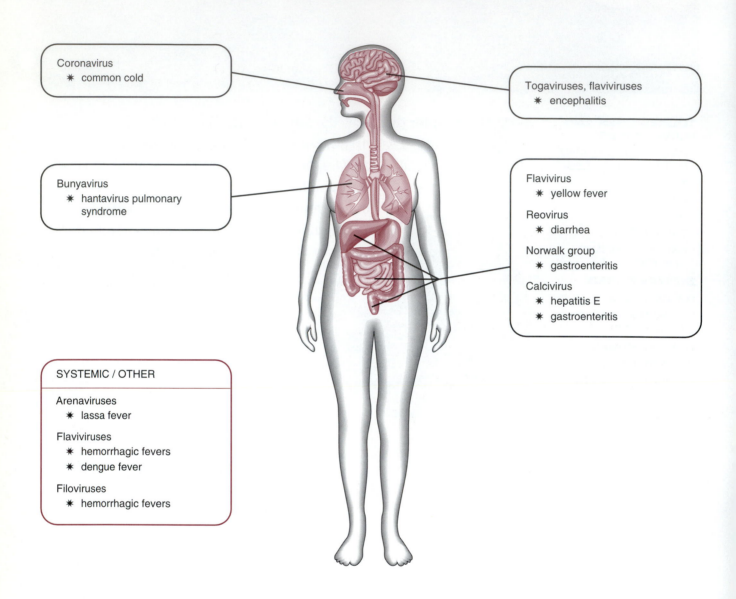

Coronavirus
 * common cold

Togaviruses, flaviviruses
 * encephalitis

Bunyavirus
 * hantavirus pulmonary syndrome

Flavivirus
 * yellow fever

Reovirus
 * diarrhea

Norwalk group
 * gastroenteritis

Calcivirus
 * hepatitis E
 * gastroenteritis

SYSTEMIC / OTHER

Arenaviruses
 * lassa fever

Flaviviruses
 * hemorrhagic fevers
 * dengue fever

Filoviruses
 * hemorrhagic fevers

---

hosts, including mammals, birds, and reptiles. These viruses are passed from vertebrate hosts to arthropods and vice versa during the taking of blood meals by the arthropods. Mosquitoes are the main vector involved in transmission. The rubella virus is morphologically similar to the togaviruses, but its pathogenesis and epidemiology are similar to that of the paramyxoviruses ∞ (Chapter 34, p. 432).

* Pathogenesis and Clinical Diseases

Many arboviruses are maintained in nature with only slight or no observable effects on their natural arthropod or vertebrate host. Humans are accidentally infected (*accidental host*) when bitten by an infected insect and are usually a dead end in the transmission chain. Most human infections are mild

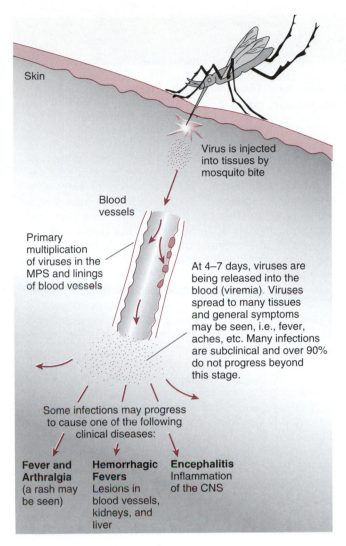

Skin

Virus is injected into tissues by mosquito bite

Blood vessels

Primary multiplication of viruses in the MPS and linings of blood vessels

At 4–7 days, viruses are being released into the blood (viremia). Viruses spread to many tissues and general symptoms may be seen, i.e., fever, aches, etc. Many infections are subclinical and over 90% do not progress beyond this stage.

Some infections may progress to cause one of the following clinical diseases:

Fever and Arthralgia
(a rash may be seen)

Hemorrhagic Fevers
Lesions in blood vessels, kidneys, and liver

Encephalitis
Inflammation of the CNS

FIGURE 38-1 The general pathogenesis of arbovirus infections.

or subclinical; some arboviruses, however, are able to cause serious disease. Figure 38-1 shows an outline of the pathogenesis and general categories of these arbovirus diseases. The viruses first multiply in the cells lining the blood vessels (endothelial cells) and the cells of the mononuclear phagocyte system (MPS) ∞ (Chapter 10, p. 146). In 4 to 7 days viruses are being released into the blood. Most human infections are subclinical or relatively mild with generalized symptoms of fever, aches, and chills. Moreover, most do not progress beyond this mild or subclinical stage. A dozen or so viruses, however, are able to cause serious diseases that progress beyond the generalized symptoms to produce the syndromes listed next.

Fever and Arthralgia. The least serious of the clinical arbovirus diseases are characterized by the sudden onset of fever, headache, swollen lymph nodes, conjunctivitis, and excruciating pains in the back, muscles, and joints. A rash may or may not occur. Diseases of this category are primarily seen in subtropical and tropical areas. Deaths are rare. *Dengue fever* (caused by a flavivirus) is the most frequently experienced disease of this type and increased numbers of cases have been seen in Mexico and the Caribbean islands in the 1970s and 1980s. Limited outbreaks that occurred in southern Texas in 1980 were the first cases reported in the United States since 1945. Since 1980, about 30 imported cases have been detected per year in various areas of the United States.

Hemorrhagic Fevers. A few flaviviruses damage the blood capillaries, resulting in subcutaneous hemorrhaging and bleeding from body openings. This type of hemorrhagic disease, which is often tick-borne, is found in many parts of the world but not in the United States. A second type of hemorrhagic fever causes lesions of the liver and kidneys: it is the serious disease of *yellow fever*. Yellow fever was once one of the major diseases of humans, with devastating outbreaks occurring in crews of sailing ships in tropical areas, among military troops, or in construction crews. Work on the Panama Canal was stopped in the late 1800s because of yellow fever. Walter Reed, a U.S. medical officer studying yellow fever in Cuba during the Spanish–American War, first demonstrated that this disease was transmitted by the *Aedes aegypti* species of mosquito. Control of the *Aedes aegypti* mosquito led to control of yellow fever, although the disease is still prevalent in some tropical areas. An effective vaccine is available and persons traveling in endemic areas should be vaccinated. Death rates from yellow fever are about 10%.

Encephalitis. Various togaviruses and flaviviruses are able to cause *encephalitis*. This disease begins with generalized symptoms and then after several days develops into encephalitis. The symptoms of encephalitis include drowsiness and stiff neck; severe cases may progress to confusion, paralysis, coma, and death. The term *sleeping sickness* is sometimes applied to these diseases but they should not be confused with the protozoal disease of African sleeping sickness. Some residual effects, such as mental retardation, deafness, and blindness,

---

Arthralgia Severe pain in joints.
Hemorrhagic Referring to bleeding or blood accumulation.

Dengue Fever among U.S. Military Personnel: Haiti, September–November, 1994

Since 19 September 1994, approximately 20,000 U.S. military personnel have been deployed to Haiti as part of Operation Uphold Democracy. To monitor the occurrence of mosquito-borne illnesses (including dengue fever [DF] and malaria) among deployed military personnel, on 19 September the U.S. Army established a surveillance system for febrile illness. Before deployment, all military personnel were instructed to take antimalarial chemoprophylaxis, either chloroquine phosphate (500 mg weekly) or doxycycline (100 mg daily). This report summarizes surveillance findings for 19 September–4 November.

U.S. military personnel who developed a febrile illness with no apparent underlying cause and reported to a military outpatient clinic were referred to the U.S. Army's 28th Combat Support Hospital in Port-au-Prince for admission and evaluation, including serial blood smears for malaria, blood specimens for virus isolation, and serologic studies. Because dengue virus is the principal flavivirus known to be endemic in Haiti, a probable case of DF was defined as detection of antiflavivirus immunoglobulin M (IgM) antibodies. A confirmed case was defined as isolation of dengue virus.

During 19 September–4 November, a total of 106 military personnel who had febrile illnesses were evaluated. Onset of illness began as early as 7 days after deployment, and the weekly number of cases peaked during the week ending 22 October. Of the 106 patients, 24 had an illness compatible with DF (i.e., fever, headache, myalgia and/or arthralgia, with or without rash). Dengue-like illnesses occurred in personnel stationed in both urban and rural areas of Haiti. One patient with probable DF had hemorrhage from a duodenal ulcer. Another had onset of fever, myalgias, and thrombocytopenia after returning to the United States.

As of 10 November, preliminary laboratory results were available for 48 febrile patients. Of these, antiflavivirus IgM was detected in 11 (23%), and dengue virus was isolated from three additional patients (dengue type 1 [one patient] and dengue type 2 [two patients]). Confirmatory testing of specimens from these patients and other febrile personnel is ongoing. Repeated malaria smears were negative for all patients.

The detection of DF cases among U.S. troops in Haiti prompted the following interventions: (1) use of personal protective measures against biting insects (e.g., insect repellant and bed nets) was reemphasized among unit commanders; (2) routine ultralow volume spraying of troop areas with insecticide (i.e., malathion) was implemented; and (3) common larval habitats of *Aedes aegypti* mosquitos (e.g., discarded automobile tires) were identified and eliminated where possible (*MMWR* 43:845, 1994).

may result. These infections occur in all parts of the world and several thousand sporadic cases are seen in the United States during the summer months. Rates are highest when large numbers of mosquitoes are present, as during extra wet summers. Mosquito-borne encephalitis is the only type generally found in the United States. Tick-borne, as well as mosquito-borne, cases appear in other countries. The major types seen in the United States are Eastern, Western, St. Louis, and Venezuelan equine encephalitis. Eastern equine encephalitis is the most severe and causes frequent deaths. The others are less severe in humans, death rates being around 1%. The term *equine* is attached to the name of these diseases because they were first recognized in horses. Horses are often infected and the disease may be quite severe with relatively high death rates. Equine encephalitis can be a major problem to the equine industry because it forces the cancellation of races, rodeos, and horse shows and results in the death of valuable animals.

✴ Transmission and Epidemiology

Togaviruses and flaviviruses, as well as other arboviruses, are transmitted in a complex ecosystem that involves the bloodsucking arthropod vectors and animal reservoirs. The viruses are usually passed in these natural hosts with few consequences. Problems result when these infections "spill over" from these natural hosts into humans and horses. This spillover occurs when environmental conditions permit a large buildup in the numbers of insect vectors and natural animal hosts and when the viruses are introduced into this environment. Such conditions frequently occur in swampy areas and in other areas that have extra wet summers.

✴ Diagnosis

The generalized nonspecific arbovirus infections usually go undiagnosed. The severe infections are tentatively diagnosed on clinical and epidemiological findings. Confirmed diagnoses are made by viral isolation and/or by showing specific increases in serum antibody levels.

✱ Treatment

No specific treatment for these virus infections is available.

✱ Prevention and Control

The major control methods are to prevent exposure to mosquitoes or ticks. Eradication or reduction in the number of mosquitoes is routinely carried out in many areas during the summer. Immunization is required for persons traveling to areas where yellow fever is present. Vaccines against other viral encephalitides have been developed but are not licensed for human use in the United States. Such vaccines, however, are widely used for the immunization of horses.

BUNYAVIRUSES

The family *Bunyaviridae* is made up of over 200 interrelated viruses, most of which are arboviruses. These are 80- to 110-nm-diameter, single-stranded RNA viruses with helical symmetry and an envelope. These viruses are often found in rodents and are transmitted by such arthropods as sandflies, ticks, and mosquitoes. Such diseases as *Korean hemorrhagic fever*, *California encephalitis* (see Figure 38-2), *sandfly fever*, and *Rift Valley fever* are caused by bunyaviruses.

The virus that causes Korean hemorrhagic fever has been of special interest in recent years. This virus infected many United Nations troops during the Korean War in the early 1950s. However, the virus was not isolated until 1978 and was given the name *hantavirus*. Hantaviruses are carried as asymptomatic infections in wild rodents, and evidence of hantavirus infections have been found in many areas of the world.

Hantavirus infection of rodents in the United States had been suspected for some time before 1993. Then in 1993 an outbreak of severe respiratory disease with high mortality occurred (14 deaths out of the first 18 cases) in the four-corners area of Arizona, Colorado, New Mexico, and Utah, and was shown to be caused by a hantavirus. An intensive investigation of this outbreak demonstrated that the virus was present primarily in deer mice and was shed in their saliva, urine, and feces. Humans are infected by coming in direct contact with infected excreta from the mice; most often by

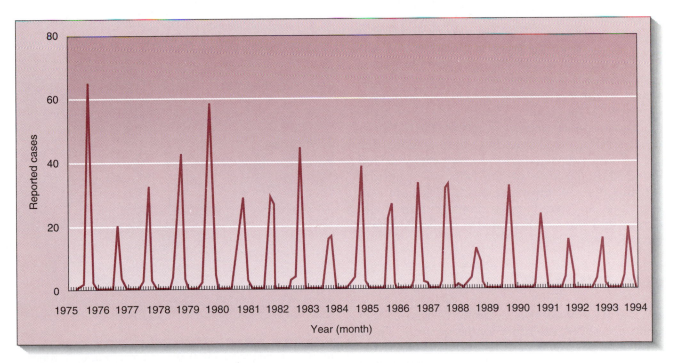

FIGURE 38-2 Reported cases of viral encephalitis due to the California serogroup arbovirus. Seasonal occurrence is distinctly demonstrated. (Courtesy Centers for Disease Control, Atlanta)

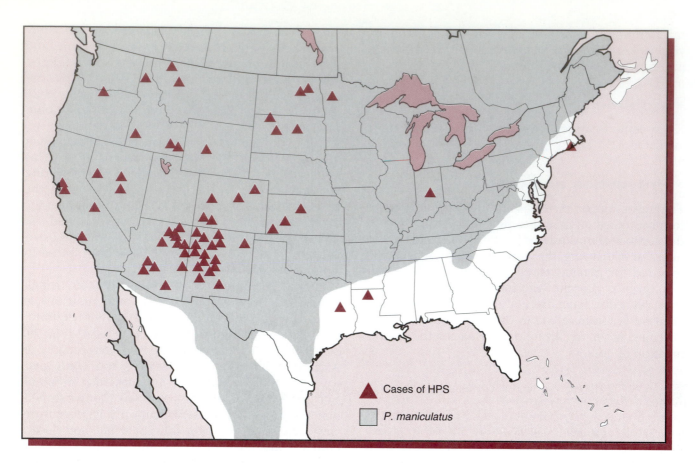

FIGURE 38-3   The distribution of deer mice (*Peromyscus maniculatus*) and the number of recognized cases of hantavirus pulmonary syndrome (HPS) in the United States as of March 1994.

inhalation of contaminated dust. No arthropod vector seems to be involved and no person-to-person transmission has been observed. Up to mid-1994, 83 cases of *hantavirus pulmonary syndrome* had been diagnosed in the United States and 45 of these patients died. Most cases have occurred in the western states, but isolated cases have been diagnosed in eastern Texas, Louisiana, Florida, and Rhode Island (Figure 38-3). Because of the extensive publicity given this disease, most persons at risk now take precautions to avoid open contact with wild rodents, and no further sharp outbreaks have occurred.

REOVIRUSES

Reoviridae (Figure 38-4) is a family of viruses that contain a ds-RNA molecule as the genome. These viruses also possess a double-layered protein coat and are about 70 nm in diameter. The first viruses of this family were isolated from humans in 1959 and were found in both the respiratory and intestinal

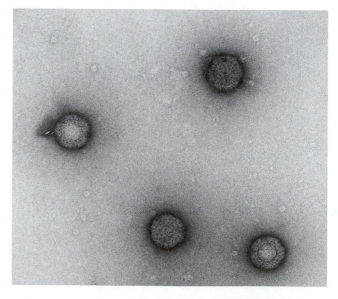

FIGURE 38-4   Transmission electron micrograph of reoviruses magnified 110,000×. (Courtesy Robley C. Williams, University of California, Berkeley)

tracts. However, they were not associated with any specific diseases and were thus referred to as orphan viruses. The acronym "reo," from *r*espiratory, *e*nteric, *o*rphan viruses, was used to name these viruses. These viruses are now classified as the genus *Reovirus*. Three serotypes have been identified. They infect humans as well as other animals, and their role in any specific human disease remains unclear.

A second type of reovirus, now called *orbivirus*, is an arbovirus that is widespread in insects and may be transmitted to humans or animals by the bite of the virus-carrying tick. Colorado tick fever virus is the only known orbivirus to cause disease in humans. The symptoms are much like those of dengue fever except that no rash is produced.

A third genus of the family *Reoviridae*, called *Rotavirus*, was discovered in 1973. The double-layered protein coat of these viruses, when viewed with an electron microscope, gives the characteristic appearance of a wheel and the term rotavirus is derived from the Latin word "rota," which means wheel. Seven serogroups have been identified. Most human infections are caused by group A rotaviruses. Other groups of rotaviruses are frequently associated with diarrheal diseases of various other animal species. Initially, rotaviruses were detected

only by electron microscope examination of tissues and feces. It is now possible to grow these viruses in cell cultures.

Rotaviruses are now recognized as the most common cause of severe viral diarrhea in infants and young children in all countries. To put the importance of rotaviruses in perspective, it is conservatively estimated that worldwide about 18 million cases of severe diarrhea in children are caused by these viruses each year and that this is a major cause of death in infants. Figures for the United States indicate that about 1 million cases occur each year, and 65,000 of these require hospitalization. Ninety percent of children have been infected by age 3. Rotavirus illness is characterized by vomiting, diarrhea, and dehydration. Treatment is directed primarily at replacing body fluids and electrolytes that are lost from vomiting and diarrhea. Often hospitalization is required. Diagnosis can be made by serologic tests that are able either to detect the presence of the rotavirus antigen in feces or to measure an increase in antibody levels between acute and convalescent sera. Diagnosis can also be made by demonstrating the presence of the characteristic wheel-shaped virus particles in fecal samples examined with an electron microscope. Vaccines are currently under development, but until they are

CLINICAL NOTE

Multistate Outbreak of Viral Gastroenteritis Associated with Consumption of Oysters: Apalachicola Bay, Florida, December 1994–January 1995

On 3 January 1995, the Florida Department of Health and Rehabilitative Services (HRS) was notified of an outbreak of acute gastroenteritis associated with eating oysters. The subsequent investigation by HRS has identified 34 separate clusters of cases, many of which were associated with oysters harvested during 29–31 December from 13 Mile Area and Cat Point in Apalachicola Bay. Oysters were shipped to other states, but additional clusters of illness associated with these oysters have been reported only in Georgia. Most of these oysters were served steamed or roasted. This report summarizes the preliminary findings of the ongoing investigation of this outbreak.

On 4 January, Apalachicola Bay was closed to harvesting even though levels of fecal coliforms in the water and in the oyster meat were within acceptable limits. The preliminary investigation identified no gross

breaches of sanitation; however, during the holiday season, the bay was used heavily by recreational boaters and commercial fishermen. Clusters of cases identified since the bay was closed prompted concern regarding the continued marketing of these oysters as unshelled and as shucked product both in Florida and other states.

Following the detection of cases associated with oysters from Apalachicola Bay, enhanced surveillance detected three additional clusters of cases in Florida and two in Texas initially linked to oysters harvested in Galveston Bay. As a result, on 13 January, Galveston Bay was closed to harvesting.

Norwalk-like viruses have been detected by electron microscopy in stool specimens from 7 of 11 persons who ate oysters from Apalachicola Bay (*MMWR* 44:37, 1995).

Hantavirus Pulmonary Syndrome: United States, 1993

In June 1993, a newly recognized hantavirus was identified as the etiologic agent of an outbreak of severe respiratory illness (hantavirus pulmonary syndrome [HPS]) in the southwestern United States. Since this problem was recognized, sporadic cases have been identified from a wide geographic area in the western United States. This report summarizes the epidemiologic characteristics of HPS cases reported to CDC from 1 May through 31 December 1993.

Through 31 December, 53 persons with illnesses meeting the surveillance case definition of HPS have been reported to CDC. Patients' ages have ranged from 12 years to 69 years (median age: 31 years), and 32 (60%) were aged 20–39 years; 30 (57%) were male. Twenty-six (49%) were American Indians; 22 (42%), non-Hispanic whites; four (8%), Hispanic; and one (2%), non-Hispanic black. Thirty-two (60%) patients died; persons with fatal cases and persons with nonfatal cases were similar in age, sex, and race.

Cases have occurred in residents of 14 states. Of the 34 (64%) persons who were residents of Arizona, Colorado, or New Mexico, illness occurred in 25 (74%) during April–July 1993 and in one before 1993. In comparison, of 19 cases reported from other states, five (26%) had onset of illness during April–July 1993, and seven (37%) had onset before 1993. All patients either lived in rural areas or had visited rural areas during the 6 weeks before onset of illness.

The etiology of HPS was initially identified by serology, polymerase chain reaction (PCR), and immunohistochemistry. Additional cloning and sequencing of virus ribonucleic acid (RNA) from human autopsy tissues indicated that all three of the RNA segments of this new virus were unlike those of any known hantavirus; the new hantavirus is most closely related to the Prospect Hill strain of hantavirus.

In November 1993, the etiologic hantavirus associated with HPS was isolated from tissues of a deer mouse (*Peromyscus maniculatus*) trapped in New Mexico in June 1993 near the residence of a person with confirmed HPS. Lung material from this animal was twice passed in uninfected laboratory deer mice and then adapted to Vero E6 cell cultures. The genetic sequence of the 139-nucleotide PCR product from the isolated virus was identical to PCR products amplified from this rodent in June 1993 and from lung tissue of the associated patient. At the same time, the U.S. Army Medical Research Institute of Infectious Diseases isolated the virus from specimens from a person in New Mexico and from a rodent in California. Muerto Canyon virus has been proposed as the name for this virus, following standard conventions for naming zoonotic viruses after a nearby geographic feature (*MMWR* 43:46, 1994).

---

available, the best control measure involves good sanitation to reduce transmission by the fecal–oral route.

NORWALK GROUP OF VIRUSES

Viruses of the Norwalk group cannot be cultivated and have not been well characterized. They are round, nonenveloped, RNA-containing, 27-nm virions, and seem to be related to the caliciviruses. Some studies have suggested that these viruses may be associated with about one-third of all outbreaks of *viral gastroenteritis* in some population groups (see Clinical Note). Infections are associated more with gastroenteritis in older persons and less with infections in infants and young children.

Transmission is by the fecal–oral route and has been associated with contaminated water and shellfish. About 50% of the population have antibodies to these viruses by the time they reach 50 years of age. Diagnosis is made by examining feces for viral particles with an electron microscope, or for viral antigens using enzyme-labeled antibodies, or by amplifying the viral RNA using reverse transcription and polymerase chain reaction. Humoral antibody responses can also be measured.

CALICIVIRUSES

Caliciviruses have been isolated from humans and other animals. They are nonenveloped, 30- to 35-nm RNA viruses that have cup-shaped depressions on their surface (calices = cups). Caliciviruses cause

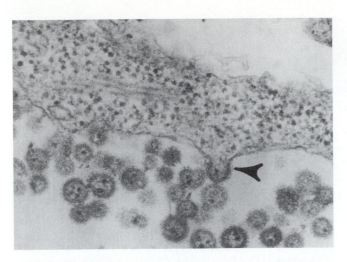

FIGURE 38-5 Transmission electron micrograph of granule-containing arenaviruses. Arrow points to a virus budding from the infected cell. (Centers for Disease Control, Atlanta)

gastroenteritis in persons of all ages and account for about 3% of diarrhea in children that requires hospitalization. The highest rates of infection occur between 3 months and 6 years and most persons have been infected by age 12. Caliciviruses may be associated with sharp outbreaks of gastroenteritis in institutional settings. Hepatitis E may be caused by a calicivirus (Chapter 32).

ARENAVIRUSES

In the late 1960s accumulated evidence demonstrated that some previously unclassified viruses shared a common morphology. These RNA viruses are pleomorphic, with diameters ranging between 100 and 130 nm. They contain dense, ribosome-sized (20-nm) particles that give the appearance of sand particles when viewed by the electron microscope (Figure 38-5). This is the basis for the name *Arenaviridae* (*arenosus* = sandy) for this family of viruses.

These viruses are normally found in various wild rodents where they cause persistent, lifelong infections, usually without acute clinical signs. The viruses are shed in the feces and urine of the rodent and contaminate food or water. Humans coming in major contact with the contaminated materials may contract an acute viral infection. Human-to-human transfer readily occurs. No vaccines are available against any of the arenavirus infections. Proper sanitation and antirodent measures are the most effective means of prevention. There are several major arenavirus infections of humans.

Lymphocytic Choriomeningitis. Lymphocytic choriomeningitis virus is widespread in domestic house mice throughout the world. In rural areas about 1% of the human inhabitants show serologic evidence of having been infected. The illness is usually nondescript or asymptomatic and rarely is a specific diagnosis made.

Argentinean and Bolivian Hemorrhagic Fevers. Argentinean and Bolivian hemorrhagic fevers occur in limited geographic areas of the respective countries and are caused by closely related arenaviruses (called the Junin and Machupo viruses). Most outbreaks occur in farmworkers during the harvesting of crops when they apparently come in contact with contaminated urine and feces of infected rodents. These diseases are quite severe in humans, causing widespread damage to the linings of blood vessels and capillaries. The result is internal hemorrhaging and death rates between 15 and 20%.

Lassa Fever. Lassa fever was first characterized in 1969 when this hemorrhagic disease was dramatically brought to the attention of the world through extensive press coverage of an outbreak among American missionaries in Nigeria, Africa. The first case was reported in a missionary nurse and the disease was transmitted to two attending nurses. The **index case** and one of the attending nurses died. Clinical specimens were flown to the United States, where two laboratory personnel working with these specimens contracted the disease. One died and the other was saved by receiving passive immunization from serum taken from the original attending nurse, who had recovered from Lassa fever. Lassa fever rapidly developed the label of a severe new West African killer virus disease. In the intervening years this disease was studied in greater detail, using stringent safety precautions, and was found to be present in a variety of wild rodents in Nigeria, Liberia, and Sierra Leone. Both mild and severe cases occur among humans. The mild cases usually go undetected, whereas those severe enough to require hospitalization have a death rate of 30 to 66%.

Index case First known case of an infectious disease.

Update: Outbreak of Ebola Viral Hemorrhagic Fever: Zaire, 1995

As of 25 June 1995, public health authorities have identified 296 persons with viral hemorrhagic fever (VHF) attributable to documented or suspected Ebola virus infection in an outbreak in the city of Kikwit and the surrounding Bandundu region of Zaire; 79% of the cases have been fatal, and 90 (32%) of 283 cases in persons for whom occupation was known occurred in health-care workers. This report summarizes characteristics of persons with VHF from an initial description of cases and preliminary findings of an assessment of risk factors for transmission.

A case was defined as confirmed or suspected VHF in a resident of Kikwit or the surrounding Bandundu region identified since 1 January. The median age of persons with VHF was 37 years (range: 1 month–71 years); 52% were female. Based on preliminary analysis of 66 cases for which data were available, the most frequent symptoms at onset were fever (94%), diarrhea (80%), and severe weakness (74%); other symptoms included dysphagia (41%) and hiccups (15%). Clinical signs of bleeding occurred in 38% of cases.

Potential risk factors for intrafamilial transmission were evaluated for secondary cases within households of 27 primary household cases identified through 10 May. A primary household case was defined as the first case of VHF in a household; household was defined as persons who shared a cooking fire at the onset of illness in the primary household case. Among 173 household members of the 27 primary household cases, there were 28 (16%) secondary case patients. The risk for developing VHF was higher for spouses of the primary household case patients than for other household members (10 [45%] of 22 compared with 18 [14%] of 151) and for adults (aged ≥18 years) than for children (24 [30%] of 81 compared with four [4%] of 92).

Needle sticks or surgical procedures during the 2 weeks before illness were reported for two of the 27 primary household case patients and none of 28 secondary case patients. Of the 28 secondary case patients, 12 had direct contact with blood, vomitus, or stool of the ill person during hospitalization (i.e., later stages of illness), and 17 simultaneously shared the same hospital bed. Of 78 household members who had no direct physical contact with the person with the primary household case patient during their clinical illness, none developed VHF.

Editorial Note: The incidence of VHF related to Ebola virus in Kikwit has diminished following the institution of interventions, including (1) training of medical and relief personnel on the proper use of protective equipment, (2) initiation of aggressive case finding; and (3) educational measures in the community (e.g., pamphlets and public announcements). However, cases continue to occur, and each case has the potential to be a source for additional infections. Therefore, ongoing measures, including continued intensive surveillance, training activities, and public education, are necessary to contain the epidemic (*MMWR* 44:468, 1995).

CORONAVIRUSES

The human coronaviruses were first discovered in the mid-1960s and were shown to be a frequent cause of common cold-type infections. The structure of these viruses resembles that of the paramyxoviruses except that the spikes are more prominent, with knob-like structures on the ends. When viewed by the electron microscope, these projections give the appearance reminiscent of the solar corona—hence the name *Coronaviridae* for this family of viruses.

Usually the human viruses cannot be isolated on regular cell cultures, but require a more difficult procedure that uses organ cultures of tracheal ciliated epithelium. Consequently, the amount of research that can be carried out on these viruses is limited. All current evidence indicates that the coronaviruses cause about 15% of the mild upper respiratory tract infections, such as the common cold. All ages are infected and immunity seems to be short-lived; a person can be reinfected with the same serotype periodically throughout life. Two different serotypes have been identified.

FILOVIRUSES

The filoviruses are filamentous (rod-shaped), enveloped, RNA viruses that are tentatively classified in a new family called *Filoviridae*. Two filoviruses have been identified. The first, called the *Marburg virus*, was isolated from persons in Marburg, Germany, who contracted an unrecognized disease while working with tissues from African monkeys. Seven of the 25 patients who contracted this disease died. The disease is characterized by fever, hemorrhages, diarrhea, vomiting, myalgia, pharyngitis, and conjunctivitis. Mortality is about 25%. Some sporadic cases have been seen in Africa. The natural reservoir for this infection has not been identified.

The second filovirus is called the *Ebola virus* and was discovered in 1976 when a hemorrhagic fever broke out in hospitals in Zaire and the Sudan. Over 500 people were infected, with death rates from 53 to 88%. Another outbreak occurred in Zaire in 1995 and of the initial 93 cases, 86 (92%) died. This infection is readily spread from person to person and is a major hazard to health-care and laboratory personnel working with Ebola hemorrhagic fever patients. The natural reservoir for this virus is not known.

PARVOVIRUSES

Parvoviruses are small (18–25 nm), nonenveloped viruses that contain a single-stranded DNA molecule. The major target cells for these viruses are the rapidly growing *hematopoietic* cells in the bone marrow that form into the blood cells. It has been known for some time that certain parvoviruses cause important diseases in dogs and cats. However, in recent years it has been discovered that *parvovirus B19* causes widespread infection in humans. In most persons these infections are asymptomatic or cause only a minor illness with fever, aches, and a slight rash. However, in persons with preexisting sickle cell disease or thalassemia (hereditary hemolytic anemia), parvovirus B19 can cause life-threatening anemia. AIDS patients and other immunosuppressed persons also develop severe anemia when infected with this virus. In addition, parvovirus infection of pregnant women can cause severe anemia, congestive heart failure, and death of the fetus.

This virus seems to be spread primarily by the respiratory route and serologic evidence indicates that up to 60% of adults have been infected. Laboratory diagnosis is difficult. Because of the widespread nature of this virus and its potential to cause serious disease in the immunosuppressed and fetuses, efforts are underway to develop a vaccine. The major antigens of parvovirus B19 have been successfully cloned by genetic engineering into systems where large amounts can be produced. Vaccines made from these antigens are now undergoing evaluation trials.

MATERIAL FOR REVIEW

CONCEPT SUMMARY

1. Togaviruses are widespread in nature and some cause important diseases of humans and animals.
2. An arenavirus causes Lassa fever, a killer disease found in Africa.
3. Coronaviruses cause about 15% of the common colds.
4. Rotaviruses are a frequent cause of severe infantile diarrhea.
5. Norwalk viruses are associated with gastroenteritis in older persons.

Hematopoietic   Referring to the origins of blood cells from stem cells.

CLINICAL SUMMARY TABLE

Microorganism	Virulence Mechanisms	Diseases	Transmission	Treatment	Prevention
Togavirus and *Flaviviruses*		Fevers Arthraligia Hemorrhagic fever Encephalitis	Animal to insect to human	Symptomatic	Insect eradication Vaccines
Bunyaviruses		Korean Hemorrhagic fever Encephalitis Hantavirus infection	Animal to insect to human Animal to human	Symptomatic	Avoid contact
Reoviruses		Colorado tick fever Diarrhea	Tickborne Fecal-oral route	Symptomatic	Tick control Good sanitation

STUDY QUESTIONS

1. Many of the viruses presented in this chapter are capable of producing serious, life-threatening illness. What feature of these viruses limits their discussion to only a few paragraphs?

2. What are the major encephalitis viruses in the United States?

3. What epidemiologic feature is shared by the togaviruses?

4. How are the coronaviruses, rotavirus, and Norwalk viruses transmitted?

5. What features are similar among the Ebola, Lassa fever, Marburg, and Bolivian hemorrhagic fever virus diseases?

CHALLENGE QUESTIONS

1. Several different viruses cause hemorrhaging, often resulting in death of the host. Since viruses cannot replicate unless they are in a host, what advantage is gained by the hemorrhagic viruses often killing their hosts?

39 Fungi

Fungi are eukaryotic organisms existing both as single cells, as in the case of yeast, and as multicellular filaments, as seen in the molds (moulds is an alternative spelling) and mushrooms. The fungi do not carry out photosynthetic activities and depend on preformed organic matter for their nourishment. Since they are heterotrophic, several species are pathogens and some of these can cause serious illnesses in humans and other animals. Fungal infections are perhaps more common in plants where they cause millions of dollars in crop losses. This chapter summarizes the major fungal (mycotic) infections of humans.

THE FUNGI

The term **mycology** refers to the study of fungi and is derived from the Greek term *mykes*, which means "mushroom." Fungi are widespread, with approximately 200,000 species having been identified. Most fungi are involved in the decomposition of organic matter and play an important role in the recycling of organic compounds in nature. Only about 150 species of fungi are associated with diseases in humans or animals. Most fungal diseases of humans are caused by only 10 to 15 different fungi.

Mycology	The study of fungi.

Fungi

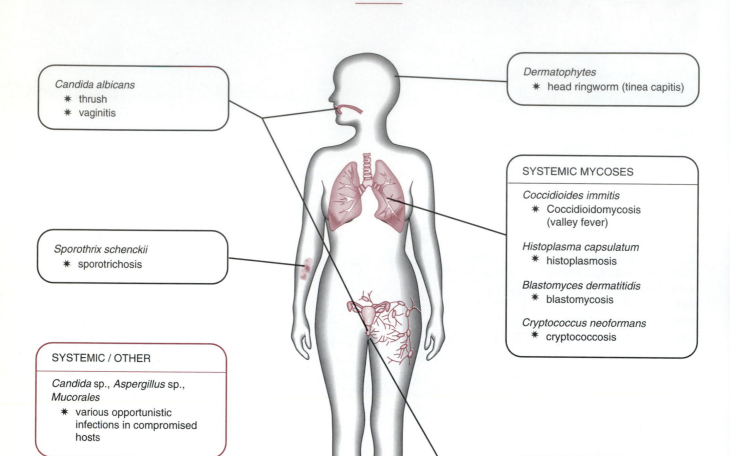

Candida albicans
* thrush
* vaginitis

Dermatophytes
* head ringworm (tinea capitis)

SYSTEMIC MYCOSES

Coccidioides immitis
* Coccidioidomycosis (valley fever)

Histoplasma capsulatum
* histoplasmosis

Blastomyces dermatitidis
* blastomycosis

Cryptococcus neoformans
* cryptococcosis

Sporothrix schenckii
* sporotrichosis

SYSTEMIC / OTHER

Candida sp., *Aspergillus* sp., *Mucorales*
* various opportunistic infections in compromised hosts

Dermatophytes
* jock itch (tinea cruris)
* athlete's foot (tinea pedis)

MORPHOLOGY AND REPRODUCTION

✳ Yeasts

Yeasts are single oval or spherical fungal cells ranging from 3 to 20 μmin diameter. As a rule, they reproduce by *budding*, which first entails dividing the nucleus, followed by passage of one nucleus to a bud that "balloons out" from the wall of the *mother cell* (Figure 39-1a). A wall forms between the bud and the mother cell. The bud separates from the mother cell and becomes a daughter cell. Many daughter cells form from a single mother cell. Initially, the daughter cell is smaller than the mother cell, but it gradually increases in size and, in turn, produces its own buds (Figure 39-1a).

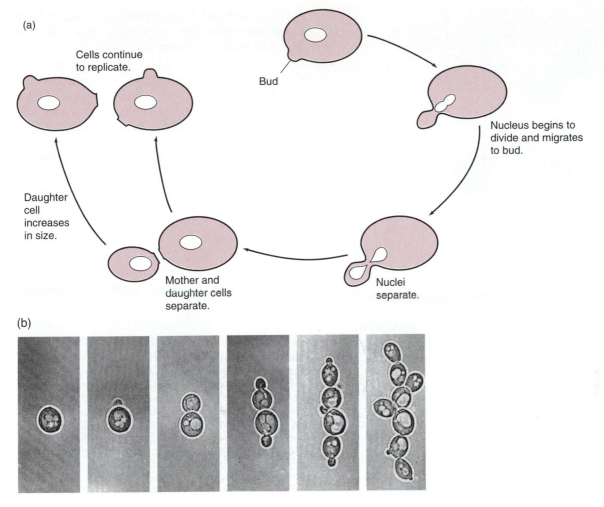

(a)

Cells continue to replicate.

Bud

Nucleus begins to divide and migrates to bud.

Daughter cell increases in size.

Mother and daughter cells separate.

Nuclei separate.

(b)

FIGURE 39-1   Replication of yeast cells by budding. (a) Schematic of process. (b) Photomicrograph of budding yeast cells.

✳ Molds

The growth of a mold (see Plate 41) usually starts with the germination of a reproductive element called a **conidium**, which sends out a filament that grows by elongation at its tip. This filament is the basic structure of growing molds and is called a *hypha*. Many branches of hyphae are formed and a mass of hyphae is called a *mycelium*. Some hyphae grow above the surface of the *substrate* and are called *aerial hyphae* or *aerial mycelia* (Figure 39-2). Other hyphae grow into the surface to absorb nutrients and are called the *vegetative hyphae*. The hyphae vary from about 2 to 10 μm in diameter, depending on the species of mold. Many nuclei are contained within the hyphae; in many species crosswalls called *septa* are located at frequent intervals along the hyphae.

Molds reproduce by developing conidia on the aerial hyphae. Conidia act as "seeds" for new colonies of molds. A reproductive cycle of one type of mold is shown in Figure 39-3. Many variations are seen in the morphology of the mycelia and conidia and the reproductive strategies of molds. These features are useful in identifying the different species. Some general structural variations of common molds are shown in Figure 39-4.

A typical mold colony is able to produce many

Conidium   An asexual reproductive unit in the fungi.
Septum   A crosswall or membrane between compartments. Molds are classified into two groups by the presence or absence of septa.

FIGURE 39-2 Photomicrograph of *Aspergillus* species showing aerial hyphae and sporangia. (Courtesy Centers for Disease Control, Atlanta)

reproductive structures, and each structure may produce many conidia. These conidia are easily disseminated through the air; therefore, mold conidia are carried to virtually every unprotected environmental habitat on the earth. This type of reproduc-

tion is called **asexual**. Some fungi carry out a form of **sexual reproduction** in which two different reproductive bodies connect and haploid cells from each body fuse to form diploid cells; the reproductive elements formed from sexual reproduction are called **spores**.

Certain species of fungi are able to grow as either the yeast or the filamentous form—a trait called *dimorphism* (Figure 39-5). Under certain growth conditions, the yeast form will develop, whereas the filamentous form is produced under other conditions. This phenomenon is important in the diagnosis of some fungal infections because certain pathogenic fungi exist in the yeast phase when growing in body tissues but change to the filamentous form when growing on an artificial laboratory medium. Most fungi can be cultivated on artificial media in the laboratory using variations of the same basic methods used to cultivate bacteria.

CLASSIFICATION OF FUNGAL DISEASES

Fungal diseases are called *mycoses*. Mycoses of humans may be divided into the following four groups, based on the level of penetration of the infection into the body tissues:

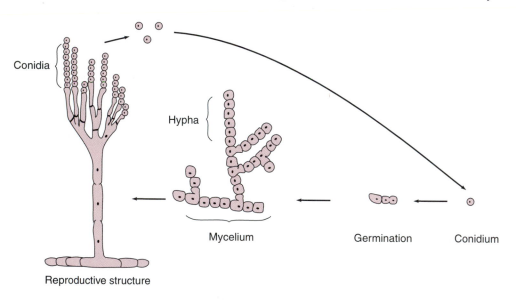

Conidia

Hypha

Mycelium

Germination

Conidium

Reproductive structure

FIGURE 39-3 An asexual reproduction cycle of a conidial-type fungus.

Sexual reproduction Reproduction through the union of sexual structures (or gametes) that produces sexual spores.
Spore A reproductive unit resulting from sexual reproduction.
Dimorphism The property of having two morphologic shapes. Some fungi are dimorphic and exist as either yeasts or molds, depending on their growth environment.

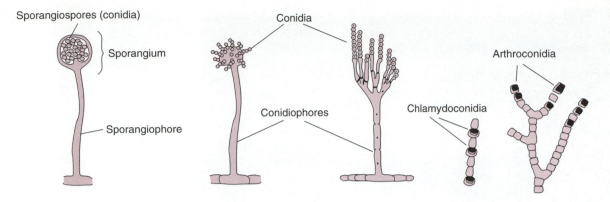

FIGURE 39-4   Some of the different types and arrangements of fungal spores.

1. *Superficial mycoses*, diseases caused by fungi that grow only on the surface of skin and hair
2. *Cutaneous mycoses* or *dermatomycoses*, including such infections as athlete's foot and ringworm, in which fungal growth occurs only in outer layers of skin, in nails, or in hair shafts
3. *Subcutaneous mycoses*, fungal infections that are able to penetrate below the skin and involve the subcutaneous, connective, and bone tissues
4. *Systemic* or *deep mycoses*, fungal infections that are able to infect internal organs and become widely disseminated throughout the body

Other fungi, such as some species of *Aspergillus* and *Candida*, may cause infections in compromised or immunosuppressed individuals. These fungi are considered to be *opportunisitic*. Table 39-1 lists the most common clinically significant fungi.

SUPERFICIAL MYCOSES AND DERMATOMYCOSES

The superficial mycoses are of minor importance because the infections are limited to hair surface or to the surface of the skin. The resulting tissue damage is minimal. These diseases are seen most often in warm climates. The lesions appear as scaly or pigmented areas on the skin or as nodules on the shafts of hair. Treatment involves removing the skin scales with a cleansing agent and removing the infected hair. Good hygiene generally prevents these infections.

The fungi that cause dermatomycoses are able to infect only the epidermis, hair, or nails (see Plate

43). About 30 different species of the genera *Epidermophyton*, *Microsporum*, and *Trichophyton*, collectively referred to as *dermatophytes*, cause these infections. Dermatomycoses are known by such lay terms as "athlete's foot," "jock itch," and "ringworm." The term **tinea**, along with the area of the body involved, is also used when referring to these infections. For example, *tinea capitis* is an infection of the scalp (Figure 39-6), *tinea corporis* is an infection on the body, *tinea cruris* (jock itch), is in the groin area, and *tinea pedis* (athlete's foot; Figure 39-7) is on the foot.

Because many dermatophytes may cause similar types of infections, specific diagnosis can be made only by laboratory tests. Although these diseases do not cause death and are rarely serious, they may produce uncomfortable symptoms and sometimes unsightly lesions. The hyphae of the dermatophytes

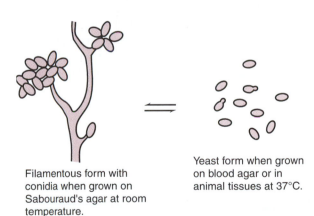

Filamentous form with conidia when grown on Sabouraud's agar at room temperature.

Yeast form when grown on blood agar or in animal tissues at 37°C.

FIGURE 39-5   Dimorphism of the fungus *Sporothrix schenckii.*

Tinea   Worm-like. Cutaneous mycotic infections are often called "ringworm." This name derives from the concentric circles formed in the skin of some infected patients.

TABLE 39-1

Most Common Clinically Significant Fungi

Classification	Body Site	Fungi
Superficial	Skin	*Malassezia furfur*
		Exophiala werneckii
Dermatophyte	Hair	*Piedraia hortae*
		Trichosporon beigelii
	Skin, hair	*Microsporum* sp.
	Skin, nails	*Epidermophyton* sp.
	Skin, hair, nails	*Trichophyton* sp.
Subcutaneous	Subcutaneous/ lymphatic	*Sporothrix schenckii*
		Cladosporium sp.
		Phialophora sp.
		Fonsecaea sp.
Systemic	Respiratory	*Histoplasma capsulatum*
		Coccidioides immitis
		Fusarium sp.
		Penicillium sp.
	Respiratory/ subcutaneous	*Blastomyces dermatitidis*
	Respiratory/CNS	*Cryptococcus neoformans*
Opportunistic	Various	*Aspergillus* sp.
		Mucor sp.
		Candida sp.
		Rhizopus sp.

grow into the keratinized tissues of the epidermis, into the hair shaft, or into fingernails or toenails. In young children, growth of these fungi in the epidermis of the scalp moves outward in concentric circles. The term "ringworm" was applied to this type of lesion years ago when it was thought that the lesion was caused by worms coiled under the skin. The lesions on the feet that characterize *tinea pedis* (athlete's foot) often start between the toes as small fluid-filled vesicles. The vesicles rupture, leaving shallow lesions that itch and may become secondarily infected with bacteria. Infections of the nails cause puffy–chalky lesions (Figure 39-8). Infections may persist for years in some persons if not treated, whereas in others the cure is spontaneous. Reinfections may occur because typical antibody-type immunity does not seem to be effective in preventing such superficial infections.

✳ Transmission and Epidemiology

Dermatophytes are usually parasites of humans and animals. Transmission is apparently from person to person by direct or indirect contact with bits of

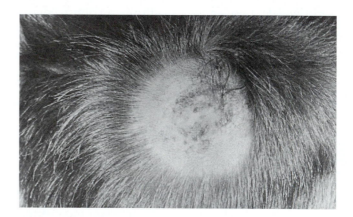

FIGURE 39-6 *Tinea capitis*, or ringworm infection, on the scalp of a child. (Courtesy Centers for Disease Control, Atlanta)

sloughed-off tissues that contain the fungus. *Tinea capitis* is seen mostly in children, *tinea pedis* in adolescents and adults. Some dermatomycoses are transmitted from pets and other domestic animals to humans. Infections are found worldwide and a significant number of people have been or are cur-

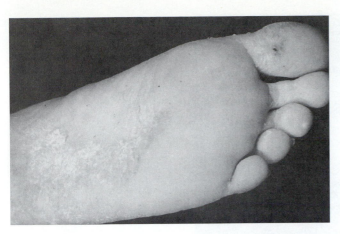

FIGURE 39-7 Dermatomycosis of the foot (athlete's foot). (Courtesy Centers for Disease Control, Atlanta)

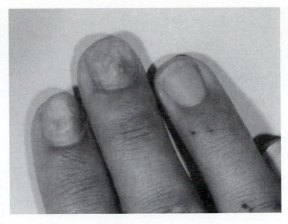

FIGURE 39-8 *Tinea unguium*, infection of the fingernails by dermatophyte.

rently infected. Clusters of ringworm infection may occur in elementary school-age children.

✳ Diagnosis

Dermatomycoses are usually diagnosed by clinical signs and symptoms. Microscopic examination of tissue scrapings shows the presence of hyphae (Figure 39-9). Tissue scrapings can be inoculated onto an agar medium and after incubation at room temperature for a week or two, fungal colonies develop (see Plate 42). Specific species can be identified by the gross appearance of the colony and by microscopic examination of the structure of the fungus. This procedure uses Sabouraud's agar medium, which contains antibiotics and a low pH to inhibit bacterial growth.

✳ Treatment

Numerous ointments, powders, and solutions are available as nonprescription treatments. Those containing tolnaflate are effective in providing symptomatic relief; when combined with good hygiene, they may help to produce a cure of skin infections due to dermatophytes. An effective systemic treatment is the antibiotic griseofulvin. When taken orally, it accumulates in the **keratin** tissues and exhibits a fungistatic effect. Treatment must be continued long enough to allow the infected tissues to be sloughed off. For infections of the scalp and skin, treatment for 2 to 3 weeks is required. Longer peri-

ods are needed for infections of nails. Infected nails may be removed surgically if prolonged treatment with griseofulvin is not practical. A newer group of antifungal agents called *azoles* (e.g., ketoconazole, miconazole, itraconazole) are now frequently used to treat dermatomycoses and other fungal infections (Figure 9-11). These compounds can be obtained without a prescription for topical application, and some are now available that can be taken orally ∞ (Chapter 9, p. 134).

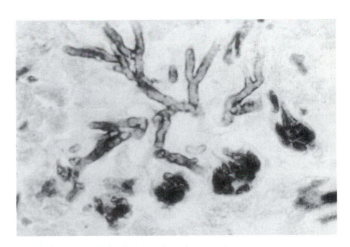

FIGURE 39-9 Microscopic appearance of fungal hyphae in infected tissue. (S.S. Schneierson, *Atlas of Diagnostic Microbiology*, p. 45. Courtesy Abbott Laboratories, Abbott Park, IL)

Keratin An insoluble protein found in skin epithelium, hair, and nails; sometimes used to refer to skin.

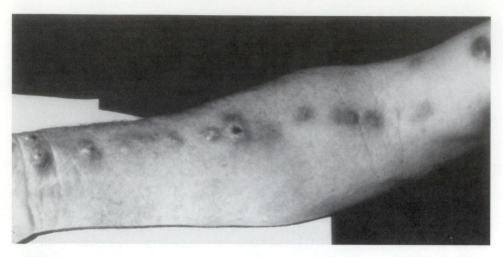

FIGURE 39-10   Sporotrichosis on the arm. (Courtesy Centers for Disease Control, Atlanta)

SUBCUTANEOUS MYCOSES

Some fungi that are normal inhabitants of the soil or organic matter are able to cause infections when introduced into the skin. These infections tend to remain in the adjacent subcutaneous tissues but on occasion may spread to deeper tissues. The most common type of subcutaneous mycosis is caused by the fungus *Sporothrix schenckii* and the disease is called *sporotrichosis* (Figure 39-10). This infection is seen in all parts of the world and occurs most often in gardeners, farmers, or other workers who come in contact with soil. An ulcerative lesion develops at the site of inoculation; then infection may spread to regional lymph nodes, where swelling occurs. Diagnosis is made by culturing the fungus from lesion exudate. This fungus is dimorphic, with a yeast phase at 36°C and a mycelial phase at room temperature (see Figure 39-5). Treatment is often difficult, but doses of potassium iodide over a 4- to 6-week period are fairly effective.

Other forms of subcutaneous mycoses occur primarily in the tropics and subtropics. The two major forms, *chromoblastomycosis* and *maduromycosis*, are caused by several different fungal species. Lesions are usually on the feet or lower extremities where the fungi have the most likely chance of entering traumatized tissues. The lesions of chromomycosis appear as ulcerative, warty, cauliflower-like growths. The lesions of maduromycosis are deeper and purulent, with openings draining to the outside. Treatment is often difficult but usually includes surgical removal of diseased tissues in the case of maduromycosis, with chemotherapy being used to control secondary bacterial infections. The lesions of maduromycosis appear similar to the subcutaneous lesions caused by *Nocardia*, *Actinomyces*, or *Streptomyces* ∞ (Chapter 20, p. 287); it is important to determine which agents are causing the disease, for it can be treated with antibiotics if bacterial agents are responsible.

SYSTEMIC MYCOSES

The systemic mycoses occur in two basic forms. The most prevalent form is a subclinical or mild respiratory infection, whereas the other is a severe disseminated infection involving many tissues. Unless treated, the disseminated form is usually fatal. The fungi that cause systemic mycoses live in the environment as saprophytes. The conidia are inhaled into the respiratory tract, where an acute, self-limiting pneumonitis may result. This initial infection is generally mild and passed off as a common bacterial or viral respiratory disease. In a great majority of cases, the infection terminates after this mild pulmonary phase. In a relatively small number of persons, often those with compromised defense mechanisms, a chronic form of the disease develops. Slowly progressing, purulent, or granulomatous pulmonary lesions develop and resemble the lesions of tuberculosis in many ways. Systemic mycoses of the lungs are often misdiagnosed as tuberculosis. These lesions may extend directly into the tissues around the lungs, or the fungi may be carried via the blood to any organ of the body, where secondary lesions will develop.

TABLE 39-2

Antifungal Agents

Compound	Used for
Amphotericin B	Severe systemic mycotic disease
Flucytosine	Systemic disease, in combination
Griseofulvin	Dermatophytes
Imidazoles (ketoconazole/miconazole)	Dermatophytes, some systemic disease
Nystatin	Topical application, all fungi
Potassium iodide	*Sporothrix schenckii*
Tolnaftate	Dermatophytes
Triazoles (fluconazole/itraconazole)	Severe systemic mycotic disease

✳ Transmission

No human-to-human transmission is known to occur. The fungi growing in soil or animal droppings produce conidia that are carried by the airborne route to humans. The major systemic mycoses are discussed below; some have worldwide distribution, whereas others are limited to specific geographic areas.

✳ Diagnosis

Most fungi that induce systemic mycoses are dimorphic and diagnosis is aided by observing the yeast form of the fungus in tissue specimens. The fungi may be cultivated on various nutrient agars, in which either the yeast or filamentous forms may be produced; serologic tests are available for some infections. Delayed hypersensitivity is induced and antigens prepared from the fungi can be used in skin tests for some ∞ (Chapter 11, p. 164). A serological test, called the *exoantigen test*, can be used to specifically identify the species of fungus causing a systemic mycoses. The fungus is grown in culture and a saline extract is made from the mycelium. This extract is then tested against a known specific antiserum in an immunodiffusion system. Also, DNA probes are now available to specifically identify these fungi.

✳ Treatment

The chemotherapeutic agent amphotericin B is the treatment of choice for most systemic mycoses ∞

(Chapter 9, p. 134). Unfortunately, use of amphotericin B is associated with serious (but reversible) side effects. Efforts to increase the availability of effective antifungal agents have resulted in several additional agents (Table 39-2). *Flucytosine* is an antimetabolite and is usually used along with other antifungal compounds. In the 1970s, the azole compounds such as miconazole and ketoconazole were introduced and have been effective against some fungal infections. Most recently a group of similar compounds known as the triazoles have been developed, of which two, *fluconazole* and *itraconazole*, have been very effective in treating some of the most difficult mycotic diseases. Surgical removal of large pulmonary lesions may be useful in some cases.

✳ Prevention and Control

No vaccines are routinely available. Avoiding areas like bird roosts and caves where conidia are most likely to be found may be wise.

SELECTED SYSTEMIC MYCOSES

✳ Coccidioidomycosis

This disease is caused by fungus *Coccidioides immitis*. The fungus is found in some desert areas and is prevalent in the southwestern United States and in some areas of Central and South America. *C. immitis* grows as a mold on the soil and produces **arthroconidia** (see Plate 46), which are carried with air

---

Arthroconidium   A fungal conidium formed as the cell walls increase in thickness. Each of the cells in the hyphae may become an arthroconidium. These conidia are very resistant to drying.

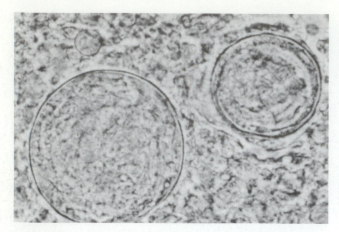

FIGURE 39-11 Spherules of *Coccidioides immitis* in sputum from patient. (Courtesy Centers for Disease Control, Atlanta)

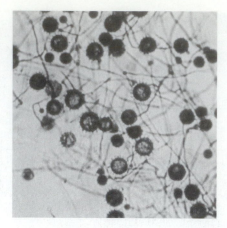

FIGURE 39-12 Photomicrograph of *Histoplasma capsulatum*, mycelial phase. (Courtesy Centers for Disease Control, Atlanta)

currents into the pulmonary spaces, where they cause infection. As many as 50 to 80% of the people living in some areas of the central valleys of California have a positive skin test to this agent, which indicates that they have had at least the mild respiratory form of this disease. Coccidioidomycosis is called "valley fever" by residents of California's San Joaquin Valley. A small percentage of those persons infected develop the disseminated disease. *C. immitis* grows in spherical forms in tissues (Figure 39-11).

✳ Histoplasmosis

The fungus *Histoplasma capsulatum* (Figure 39-12) causes the disease of *histoplasmosis*. This fungus grows well in soil enriched by bird or bat droppings and produces infectious conidia. This fungus is found worldwide, but in the United States it is most prevalent in the Ohio and Mississippi river valleys. Skin testing shows that up to 80% of the population in some areas give evidence of having been infected. Infection results from inhaling the conidia. Mild lung infections usually result and go unnoticed, but on healing they leave small, thin-walled calcified nodules that are easily mistaken for tuberculosis lesions on chest x-rays. This fungus is dimorphic and grows in the form of yeast in tissues. Disseminated infections are rare.

✳ Blastomycosis

This disease is caused by the fungus *Blastomycoses dermatitidis*. It probably grows in the soil as a mold and produces conidia that infect humans by the airborne route. This fungus is found worldwide, but

most reported diseases occur in the central river valleys of the United States. Along with common mild respiratory infection and the less-common disseminated infection, skin lesions may also be produced. *B. dermatitidis* grows as a budding yeast in human tissues (Figure 39-13).

✳ Cryptococcosis

Cryptococcosis occurs throughout the world and is caused by the yeast *Cryptococcus neoformans* (see Plate 45). This yeast has been isolated from soil and habitats of pigeons and exists only in the yeast form. Most infections are associated with mild respiratory tract involvement. In compromised persons, however, the infection may spread through the body, with a characteristic involvement of the central nervous system (Figure 39-14).

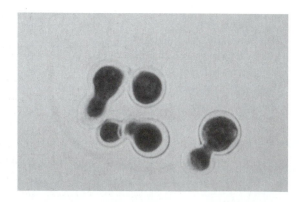

FIGURE 39-13 Budding yeast phase cells of *Blastomycoses dermatitidis*. (Courtesy Centers for Disease Control, Atlanta)

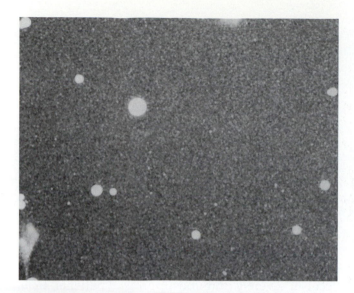

FIGURE 39-14 *Cryptococcus neoformans.* India ink preparation of spinal fluid from patient with cryptococcal meningitis.

OPPORTUNISTIC FUNGAL INFECTIONS

The number of individuals kept alive in immunocompromised conditions continues to increase. These persons are at considerably increased risk of infection, and the variety of causative agents responsible for opportunistic disease in them continues to expand. For no group of microorganisms is this more apparent than with the fungi. Many organisms, traditionally considered as saprophytic or commensalistic, manage to obtain a foothold in the promising soil of the immunocompromised host. Many of the current mycotic diseases are the result of this relationship.

The nature of the immunocompromise determines to some extent the genus of those fungal agents most likely to be responsible for infection (Table 39-3). Thus, patients with bone marrow transplants seem uniquely susceptible to *Fusarium* infection; children on long-term steroid therapy often develop infections due to *Rhodotorula*; the *Trichosporon* species often infect those patients with tumors of the blood, and patients with leukemia are frequently victims of *Mucor* or *Rhizopus*. However, the most frequent cause of mycotic infection regardless of the kind of immunocompromise is *Candida*.

✳ Candidiasis

Candida is a genus of true yeasts that are not dimorphic, although chains of yeast cells, each measuring

4–6 µm, sometimes link together, forming structures known as *pseudo* (false) *hyphae* (see Plate 44). The yeast-like fungus *C. albicans* (Figure 39-15a) is the most common of these microbes and is often part of the normal microbial flora of the skin and of the mucous membranes of the mouth, vagina, and intestinal tract. Inflammation of these epithelial surfaces may occur following various predisposing conditions. Inflammation of the mouth, called *thrush*, may occur in newborn infants, who become infected during birth, or in immunocompromised persons (Figure 39-15b). Yeast *vaginitis* is common and is frequently seen during pregnancy or in diabetics. Candidiasis of the skin may occur where the skin is damp or irritated, such as between the upper legs or under the arms. *C. albicans* is the most common cause of diaper rash. Candidiasis seems to be more prevalent in persons on broad-spectrum antibiotic therapy, because many normal indigenous bacteria are destroyed, leaving niches into which *C. albicans* can grow. Members of this genus may on occasion produce serious disease, such as endocarditis, septicemia, protracted urinary tract infection, including kidney and lung infection, esophagitis, and other soft tissue infection; even with proper treatment, which may be difficult, a

TABLE 39-3

Fungi Most Commonly Associated with Specific Immunocompromise

Immune Compromise	Opportunistic Fungi
Leucopenia (bone marrow failure)	*Candida* sp.
	Aspergillus sp.
	Phycomyces sp.
Cellular immunity (tissue transplants)	*Candida*
	Cryptococcus
	Coccidioides
	Histoplasma
Diabetes	*Zygomyces*
	Rhizopus
	Mucor
	Absidia
Steroid therapy	*Zygomyces*
Malignancy (leukemia, lymphoma, Hodgkin's disease)	*Candida*
	Cryptococcus
	Histoplasma
AIDS	*Candida*
	Cryptococcus
	Histoplasma

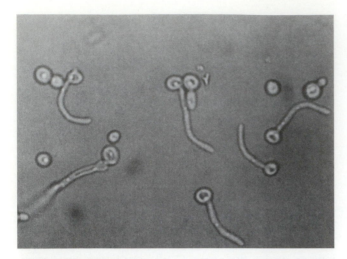

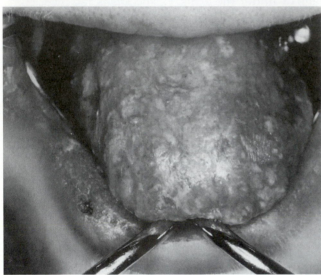

FIGURE 39-15 *Candida albicans.* (a) Germtubes by *Candida albicans.* (Courtesy W. H. Flemming III, *Journal of Clinical Microbiology* 5:236) (b) Candidiasis on the tongue and lips. (From: Council on Dental Therapeutics, American Dental Association)

high mortality rate is associated with such infections.

Nosocomial infection due to the *Candida* are increasing. In most of these patients (94%) "normal" colonization of the mucosa precedes infection and disease. *Candida* species have become among the most common causes of septicemia in hospitalized patients, and account for 10% of such diseases. These infections have a high (50%) mortality rate and usually occur as a result of some immunocompromising condition in the patient; for example, hemodialysis patients are 18 times more likely to develop systemic candidiasis than are nonhemodialysis patients. Although *C. albicans* remains the most common cause of serious candidiasis, other *Candida* species are becoming more and more com-

mon as a cause of disease (see Table 39-4). Candidiasis can be treated with imidazoles, various ointments, or with antifungal antibiotics, such as nystatin and amphotericin B.

✳ Aspergillosis

Species of the genus *Aspergillus* (Figure 39-3) are widespread in nature and several species are known to cause infections in humans. When persons with compromised immune defense mechanisms encounter large concentrations of aspergillus conidia, infections may result. Respiratory infections are the most common and lesions containing masses of mycelia may develop in the lungs or bronchi. Lesions may also develop in the ear canal, sinuses, and subcutaneous tissues. Systemic *aspergillosis* may occur in severely immunosuppressed patients, such as those with leukemia or Hodgkin's disease. Treatment is not always successful, but amphotericin B seems to be the most effective chemotherapeutic agent.

✳ Mucormycosis

Mucormycosis refers to fairly rare diseases produced by a variety of common fungi of the order *Mucorales* (Figure 39-16). These infections are seen in severely immunocompromised patients. The fungi may penetrate the respiratory or intestinal mucosa or enter through breaks in the skin. Localized lesions may develop, followed by spread to the blood and dissemination to all organs. Death often results from a combination of the predisposing illness and the fungal infection.

MYCOTIC DISEASE IN AIDS PATIENTS

Cell-mediated immunity is the major host defense system against mycotic infections. The primary immune deficit created by infection with HIV occurs in the cell-mediated system (Chapter 37). Therefore, it is not surprising that individuals with AIDS often have serious and life-threatening my-

TABLE 39-4	
Candida Species That May Cause Diseases in Humans	
Candida albicans	*Candida lusitaniae*
Candida glabrata	*Candida parapsilosis*
Candida guilliermondii	*Candida rugosa*
Candida kefyr	*Candida tropicalis*
Candida krusei	*Candida viswanathii*

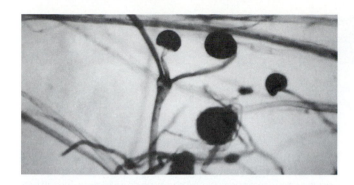

FIGURE 39-16 Photomicrograph of *Mucor species*. (Courtesy Centers for Disease Control, Atlanta)

cotic diseases. Fungal diseases that have occurred infrequently in the past are now becoming commonplace because of the large number of HIV-infected persons. This is clearly evident in those geographical areas of the United States where *Histoplasma* infection is common. In some areas, 80% of the people have serologic evidence of present or past infections and in these areas up to 20% of AIDS patients experience life-threatening disease due to *Histoplasma*.

Several mycotic agents are common among AIDS patients. These patients have repeated mucocutaneous infections such as vaginal candidiasis (76% have recurrent disease) and oropharyngeal infection with recurrence in 31% of these patients. Between 1980 and 1989 the number of cases of meningitis due to *Cryptococcus* increased by 500%. Two-thirds of this increase was due to infections in AIDS patients. These patients are extremely susceptible to this yeast, which may grow to concentrations of 10^6 organisms per ml of spinal fluid, and such infections are very difficult to treat.

Mycotic disease in AIDS patients is a reminder of the continuing change in infectious diseases occurring in a society that is able to prolong human life in the immune-compromised state.

CLINICAL NOTE

Update—Coccidioidomycosis: California, 1991–1993

Coccidioidomycosis is an infection caused by the fungus *Coccidioides immitis*, which resides in the soil in some areas of Arizona, California, Nevada, New Mexico, Texas, and Utah. Infection can occur when airborne, infective arthroconidia are inhaled. Symptomatic coccidioidomycosis, which occurs in approximately 40% of all infections, has a wide clinical spectrum, including mild influenza-like illness, severe pneumonia, and disseminated disease. Beginning in 1991, the number of cases of coccidioidomycosis reported annually to the California Department of Health Services (CDHS) increased dramatically. This report summarizes the occurrence of coccidioidomycosis in California during 1991–1993.

In 1991, 1200 cases of coccidioidomycosis were reported to CDHS, compared with an annual average of 428 reported cases during 1981–1990. The number of reported cases continued to increase during 1992 (4516 cases) but declined during 1993 (4137 cases). During 1991–1993, most (70%) cases in California were reported from Kern County in the San Joaquin Valley, where the incidence of coccidioidomycosis is high; in contrast, during 1981–1990, Kern County accounted for 52% of all cases. Coccidioidomycosis surveillance data are reported to CDHS by the counties as weekly case counts only.

Editorial Note: The public health impact of coccidioidomycosis in California during 1991–1993 was substantial. For example, based on a review of medical records in Kern County alone, coccidioidomycosis accounted for approximately $45 million in direct costs of hospitalization and outpatient care during that period. Factors potentially associated with the ongoing outbreak of coccidioidomycosis in California include weather conditions (e.g., protracted drought followed by heavy rains) conducive to the growth and spread of *C. immitis*, activities that disturb the soil and facilitate airborne spread of the organism, and a large and increasing population of susceptible persons. These factors illustrate the association between environmental and demographic factors and the emergence of some infectious diseases.

During 1991–1993 and previously, the number of coccidioidomycosis cases probably has been underreported. In Kern County, unlike other counties in California, the local health department is the diagnostic laboratory for virtually all coccidioidomycosis serologic tests from suspected cases in the county and ensures that they are reported to CDHS. Although actual rates of coccidioidomycosis are probably higher in Kern County than in other California counties, the link between the diagnostic laboratory and case reporting in the county enhances coccidioidomycosis surveillance when compared with areas that rely primarily on health-care providers to report new cases (*MMWR* 43:421, 1994).

CLINICAL NOTE

Aspergillus Endophthalmitis in Intravenous-Drug Users: Kentucky

Since May 1989, three cases of *Aspergillus* endophthalmitis (inflammation inside the eye), a potential cause of irreversible vision loss, have been reported in Louisville, Kentucky. All three patients were intravenous-drug users (IVDUs).

The patients—ages 40, 32, and 24 years—had onsets of illness in May, August, and September, respectively. Two were male. Except for IV-drug use, no risk factors for this fungal endophthalmitis were identified. All three patients were seronegative for antibody to human immunodeficiency virus. Infection was unilateral in each patient and responded to treatment with amphotericin B and flucytosine following vitrectomy. All patients had some degree of permanent vision loss. In each case, diagnosis was made by culture of specimens obtained at vitrectomy; each culture grew *Aspergillus flavus*. None of the patients had systemic or other localized signs of infection.

Two of the patients lived in the same apartment complex. The third lived approximately 1 mile away. Each denied contact with the others. Common IV-injection practices include diluting drugs with tap water and filtering this mix through cotton or cigarette filters. All patients injected cocaine and a combination of pentazocine and tripelennamine.

Editorial Note: Fungal endophthalmitis is a recognized complication of IV-drug use. Infection results from hematogenous spread after nonsterile injection. Fungal endophthalmitis can develop slowly over weeks and occurs more frequently with *Candida* than with *Aspergillus*. Other risk factors associated with *Aspergillus* endophthalmitis include the use of antibiotics, corticosteroids, and immunosuppressive therapy. Vision loss can be limited in some patients by aggressive antibiotic and surgical treatment.

Because *Aspergillus* species are ubiquitous molds and the patients reported here used common injection practices, the source of infection is difficult to determine. However, one possible explanation for the geographic and temporal clustering of these cases is a contaminated drug supply. Physicians should consider the diagnosis of fungal endophthalmitis in IVDUs with signs of intraocular infection (*MMWR* 39:48, 1990).

MATERIAL FOR REVIEW

CONCEPT SUMMARY

1. Fungi are eucaryotic, nonphotosynthetic, often multicellular organisms, some of which cause disease in humans and animals. Forms causing disease in humans include the yeasts and simple multi celled forms called molds. They do not have bacterial-type cell walls, and infections are often difficult to treat. They are ubiquitous in the environment and cause much economic loss through their destructive growth processes.

2. Mycotic infections range from the benign athlete's foot to serious systemic disease that is frequently fatal. These diseases are far more common in compromised hosts than in normal individuals.

STUDY QUESTIONS

1. List the reproductive structures found in fungi.
2. Make a table listing the systemic fungi and showing the diseases produced, cultural characteristics (dimorphism, etc.), unique morphologic features, and recommended treatments.
3. What ecological feature of *Coccidioides immitis* is responsible for its restricted geographic occurrence?
4. Characterize the role of *Candida* as etiologic agents of disease.
5. List four normally saprophytic fungi that can cause disease in humans.

CHALLENGE QUESTIONS

1. Why do the multicellular fungi such as the molds often grow as a mycelium made up of many long, thin hyphae? (*Hint*: Think about growth in the soil.)
2. Why do so many molds, including those that cause disease, produce sexual and asexual spores and conidia on aerial hyphae?

40 Protozoa

500µm

EUKARYOTE

The microorganisms presented in this chapter are animal-like in their structure and function. They are the smallest of the animal parasites and are normally included in the discipline of medical parasitology. Medical parasitology includes the study of parasites from four large phyla. These phyla are Protozoa (of the kingdom Protista), the Platyhelminthes (flatworms), the Nematodes (roundworms), and the Arthropoda. The worms and arthropods are multicellular organisms, usually macroscopic in size; in fact, some, like tapeworms, may attain lengths of several meters. It is not within the scope of this book to cover the diseases caused by these multicellular types of parasites. The protozoa are briefly reviewed in this chapter because of their small size and because their pathogenesis parallels that of the bacteria and fungi.

GENERAL CHARACTERISTICS

Protozoa are eukaryotic cells and have many of the intracellular components characteristic of higher forms of life. They range in size from 5 µm to more than 100 µm in diameter, and most protozoa have some form of active locomotion. Locomotion is an important feature used to group them into major

subdivisions. Protozoa are found in soil, in most bodies of water, and in many higher forms of life, either as commensals or parasites. Of approximately 50,000 species, relatively few are able to cause diseases in humans and most protozoa are beneficial contributors to the various biological cycles in nature. Some of the most common disease-causing prototozoa, which will be discussed below, are outlined in Table 40-1.

HUMAN BODY SYSTEMS AFFECTED

Protozoa

Naegleria
* encephalitis

Trypanosoma sp.
* trypanosomiasis (African sleeping sickness)

Leishmania sp.
* leishmaniasis

Giardia lamblia
* giardiasis

Cryptosporidium, Balantidium coli
* diarrhea

Entamoeba histolytica
* amebic dysentery

Trichomas vaginalis
* trichomoniasis (vaginitis)

SYSTEMIC / OTHER

Plasmodium sp.
* malaria

Toxoplasma gondii
* toxoplasmosis

Protozoal and other parasitic diseases are common in underdeveloped, tropical, and subtropical areas. In many such areas most of the population is infected with a variety of parasites. The overall effects of these diseases on the general well-being of the inhabitants of these areas are of major importance. In industrialized nations, particularly in temperate-climate regions, parasitic diseases are of relatively minor importance. There are, however,

several protozoal diseases of significance in the United States. These have traditionally included diarrhea due to *Giardia*, congenital disease due to *Toxoplasma*, and vaginitis due to *Trichomonas*; and in recent years marked increases in pneumonia due to *Pneumocystis* and severe diarrhea due to *Cryptosporidium* species have occurred in AIDS patients.

The diagnosis, treatment, and control of protozoal diseases differ in some ways from those used

TABLE 40-1

Some Common Disease-causing Protozoa

Type	Species	Relationship	Disease	Source
Amebas	*Entamoeba histolytica*	Intestinal pathogen	Amebic Dysentery	Human feces
	E. hartmannii	Intestinal commensal		Human feces
	E. coli	Intestinal commensal		Human feces
	E. polecki	Intestinal commensal		Human feces
	Endolimax nana	Intestinal commensal		Human feces
	Iodamoeba butschlii	Intestinal commensal		Human feces
	Naegleria fowieri	CNS pathogen	Encephalitis	Freshwater swimming
	Dientamoeba fragilis	Intestinal pathogen		Human feces
Ciliate	*Balantidium coli*	Intestinal pathogen		Pig feces
Flagellates	*Trichomonas hominls*	Tissue commensal		Human
	T. vaginalis	Tissue pathogen		Human
	Chilomastix mensnili	Intestinal commensal		Human feces
	Giardia lamblia	Intestinal pathogen	Giardiasis	Mammalian feces
	Leishmania donovani	Tissue pathogen	Leishmaniasis	Sandfly bite
	L. tropica	Tissue pathogen		Sandfly bite
	L. braziliensis	Tissue pathogen		Sandfly bite
	Trypanosoma gambiense	Tissue pathogen	Sleeping sickness	Fly bite
	T. rhodesiense	Tissue pathogen	Sleeping sickness	Fly bite
	T. cruzi	Tissue pathogen	Chagas disease	Bug bite
Sporazoa	*Plasmodium falciparum*	Tissue pathogen	Malaria	Mosquito
	Plasmodium malariae	Tissue pathogen	Malaria	Mosquito
	Plasmodium ovale	Tissue pathogen	Malaria	Mosquito
	Plasmodium vivax	Tissue pathogen	Malaria	Mosquito
	Isospora balli	Intestinal pathogen		Human feces
	Sarocystis bovicanis	Tissue pathogen		Poorly cooked meat
	Toxoplasma gondii	Tissue pathogen		Cat feces
	Cryptosporidium species	Intestinal pathogen		Animal feces

for other microbial diseases. The protozoa are not easily cultured on artificial media or inoculated into experimental animals or cell culture systems for isolation. Serologic tests are not as useful, because high levels of antibody do not readily form against many protozoa that infect the intestinal tract or other superficial tissues. Some infections, however, do stimulate an antibody response, and serologic tests are now available to aid in the diagnosis of these protozoal diseases. The diagnosis of most protozoal diseases depends mainly on demonstrating the presence of the parasite by microscopic methods. The feces are examined to determine intestinal infections and the blood and/or other tissues are examined for systemic infections. Scrapings of mucosal tissues or biopsies of infected organs may also be examined. Because of the large size and distinct shapes of the protozoa, direct microscopic identification is the routine method of diagnosis. The symptoms of many protozoal diseases are quite general and often are not used solely as a basis for a specific diagnosis.

No vaccines are available against protozoal diseases and chemotherapy is not as specific or readily available as it is against prokaryotic microbes; however, a few compounds are fairly effective against certain protozoal diseases. Toxic side effects are common with antiprotozoal drugs.

Some disease-producing protozoa go through a life cycle that involves more than one kind of animal host. In many of these situations, such as with malaria, humans are *intermediate hosts* and not *defin-*

Intermediate host   A host in which the parasite goes through asexual reproduction. The intermediate host is often an essential step in the life cycle of the parasite.

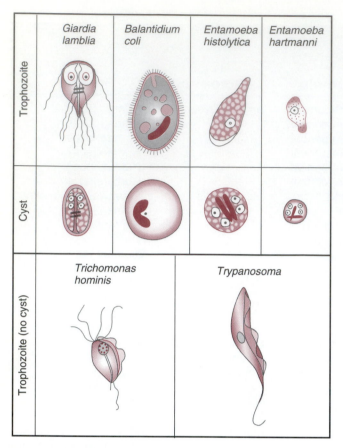

	Giardia lamblia	*Balantidium coli*	*Entamoeba histolytica*	*Entamoeba hartmanni*
Trophozoite				
Cyst				
Trophozoite (no cyst)	*Trichomonas hominis*		*Trypanosoma*	

FIGURE 40-1 Trophozoite and cyst stages of some protozoa. All except *Entamoeba hartmanni* are human pathogens.

itive hosts, which are the hosts in which the parasite goes through sexual reproduction (also referred to as *primary hosts*); and with some infections such as toxoplasmosis, humans are terminal (sometimes called *blind*) hosts because there is no way the parasite can be transmitted from them.

Many protozoa exist in two basic forms: the active, growing form, called the *trophozoite,* and the dormant, resistant form, called the **cyst** (Figure 40-1). The trophozoite form proliferates in the tissue and causes the damage that results in the clinical disease. The cyst is able to survive in the external environment and is the form of the protozoa that is usually transmitted from host to host by indirect routes. Some protozoa go through intermediate development stages in bloodsucking insects and the infected insect may serve as the vector in transmitting the disease.

In 1985, an extensive classification scheme was proposed for the protozoa and includes various phyla, subphyla, classes, and so on. However, to simplify the presentation in this chapter, the protozoa will be placed in four traditional groups based on the form of locomotion they use. Genus and species names within these groups will be used. The four groups are *sarcodina* (the amebas), *ciliophora* (the ciliates), *mastigophora* (the flagellates), and *sporozoa*. Some selected common diseases caused by protozoa are discussed below.

SARCODINA (THE AMEBAS)

The sarcodina contains protozoa that move by ameboid action—that is, by extending a section of their cytoplasm in one direction and then causing the remainder of the cytoplasm to flow into the extension. This extension is called a *pseudopodium* (false foot). The amebas are members of the sarcodina and six species from three genera are parasites of humans. Only one species, *Entamoeba histolytica*, however, causes widespread or major diseases in humans. The disease is called *amebiasis*.

✳ Amebiasis

The pathogenesis of amebiasis is outlined in Figure 40-2. Only the cyst stage of *E. histolytica* is infectious. The trophozoite stage readily dies once out of the body; if it survives long enough to be swallowed, it will be destroyed by stomach acids. Once the cyst, which contains four nuclei, passes into the small intestine, excystment occurs; that is, the cyst changes into four small trophozoite forms of the ameba. These trophozoites are carried to the colon where they continue to grow and divide by binary fission, thereby producing large numbers of offspring. These trophozoites release an enzyme that lyses tissues. This trait is the basis for the species name *histolytica* (Greek *histos* = tissue; *lysis* = dissolving). The histolytic enzyme allows the amebas to penetrate into the intestinal mucosa, where subsurface lesions develop. These lesions may coalesce into extensive ulcerative areas, thus leading to severe dysentery with stools containing bloody mucus (see Plate 46). In a small percentage of the cases, the amebas may penetrate into the mesenteric

Trophozoite An active, vegetative stage in the life cycle of a protozoan.
Cyst A dormant environmentally resistant life stage of some protozoa.

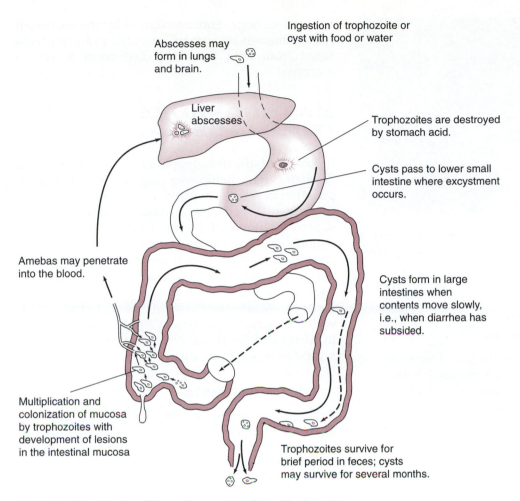

Abscesses may form in lungs and brain.

Ingestion of trophozoite or cyst with food or water

Liver abscesses

Trophozoites are destroyed by stomach acid.

Cysts pass to lower small intestine where excystment occurs.

Amebas may penetrate into the blood.

Cysts form in large intestines when contents move slowly, i.e., when diarrhea has subsided.

Multiplication and colonization of mucosa by trophozoites with development of lesions in the intestinal mucosa

Trophozoites survive for brief period in feces; cysts may survive for several months.

FIGURE 40-2   The pathogenesis of amebic dysentery.

venules and lymphatics and are disseminated to various internal organs. With disseminated amebiasis, abscesses usually develop in the liver (4 to 5% of cases) but may also develop in the brain, lungs, heart, or other tissues. Death may result from the disseminated disease. Many individuals that become infected with this ameba develop only mild diarrhea or no apparent illness.

When a person is experiencing the acute symptoms of intestinal amebiasis, called *amebic dysentery*, the contents of the intestinal tract pass rapidly through the system. Under such conditions the amebas do not have time to develop into the cyst stage and normally only the noninfectious trophozoites are released. As the body begins to establish an equilibrium with the amebas, possibly by developing neutralizing antibodies against the lytic enzymes, the lesions heal and the contents of the intestinal tract move more slowly, allowing time for the development of cysts. Thus persons showing no signs of the disease may be infectious. A large

segment of the adult population in developing countries where amebiasis is prevalent lives in semibalance with this ameba; yet they serve as a source of infection to susceptible persons. Some amebiasis is found in the United States, with most cases occurring in the South.

Specific diagnosis is made by observing the trophozoite or cyst stage of the ameba in the feces (Figure 40-3). Serologic tests are helpful in diagnosing disseminated amebiasis (see Plate 47). Treatment usually requires a combination of drugs to eliminate the amebas from both the tissues and the intestinal lumen. Metronidazole, emetine, or chloroquine are used for tissue-invasive amebas, and diloxanide or iodoquinal are used for the lumen-dwelling amebas. Tetracycline, in combination with the above drugs, is helpful to control secondary bacterial infections. Control of amebic dysentery is best achieved by following good sanitary practices, particularly in the handling and treatment of human fecal wastes.

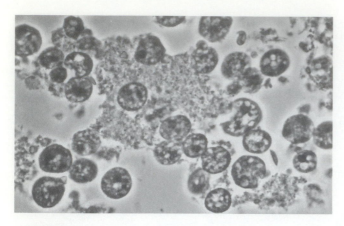

FIGURE 40-3   Microscopic appearance of cysts of *Entamoeba histolytica* in fecal material. (Courtesy Photo Researchers, Inc.)

❋ Infections Due to Free-Living Amebas

Two amebas that are normally found in water or moist organic matter have been responsible for serious human infection. Amebas of the genus *Naegleria* live in freshwater lakes or streams and can produce serious encephalitis in swimmers. Though rare, the disease is usually fatal. *Acanthamoeba* species live in soil and fresh or even brackish water. There has been a significant increase in infections due to these organisms during recent years. These infections most frequently occur in individuals who wear contact lenses, but who do not properly care for these lenses. Because *Acanthamoeba* are ubiquitous, they can easily contaminate homemade lens-cleaning solutions. The wearing of contaminated lenses may lead to infection of the cornea resulting in reduced vision or even *enucleation* of the eye. *Acanthamoeba* are also able to cause skin lesions.

CILIOPHORA (THE CILIATES)

Protozoa of the ciliophora group are surrounded by many fine cilia that beat in rhythmic patterns to propel the organism ∞ (Chapter 4, p. 44). Only one species, *Balantidium coli*, causes disease in humans. This organism is a large (50 to 100 μm in length), ovoid-shaped cell. Its normal habitat is the intesti-

nal tract of hogs. Transmission is by the fecal–oral route. Diseases in humans are rare and symptoms range from mild intestinal discomfort to severe diarrhea.

MASTIGOPHORA (THE FLAGELLATES)

Protozoa of the mastigophora group, also called *flagellates*, have whip-like flagella ∞ (Chapter 4, p. 44) that serve as their organs of locomotion (Figures 40-4, 40-5). Some mastigophora inhabit the superficial tissues of the intestinal or genital tract of mammalian hosts. Others require a bloodsucking arthropod for part of their life cycle, with the other part of their life cycle in the blood and internal tissues of the mammalian host. Two relatively mild diseases, *trichomoniasis* and *giardiasis*, found worldwide, the serious diseases of *trypanosomiasis*, found in Africa and South America, and *leishmaniasis*, are typical mastigophoral diseases.

❋ Trichomoniasis

This disease is caused by *Trichomonas vaginalis*, which is a globular-shaped cell about 15 μm in length with four anterior flagella and a short **undulating membrane** (see Figure 40-1). It exists only in

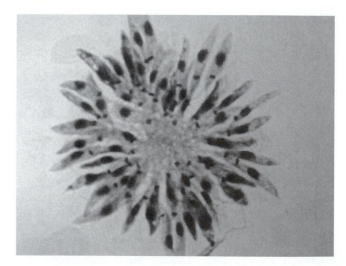

FIGURE 40-4   *Leishmania tropica* in artificial culture. Note prominent flagella. (F. Barlow and T. Minnick, *ASM News*, Vol. 51, July 1985. Reprinted with permission of ASM)

Enucleation   Removal of the entire eyeball from its socket.
Undulating membrane   An external membrane found on the surface of some protozoa. This membrane may be of various sizes and appears to be capable of wave-like movement.

CLINICAL NOTE

Flesh-Eating Parasite Afflicts 12 Million

La Nueva Estrella, Mexico—Marin Gomez pouted as the village health-care worker pointed to the yellow-and-red scabs forming over the boy's left ear, the mark of an insect-borne parasite slowly eating his flesh.

"I treated it with antibiotics, but it didn't do any good. I've never seen anything like this," said Liandro Hernandez. "If he doesn't get some medicine for this, he will lose his ear."

Gomez, 9, has been diagnosed with leishmaniasis. Though treatable, the $40 cost of the drug that combats the disease is a small fortune for the people of La Nueva Estrella.

The disease has infected dozens of the 10,000 Chiapas peasants who fled an army crackdown in February on guerrillas leading an indigenous uprising, now in its 19th month.

Most of the peasants have returned to their homes, but health workers say at least 80 contracted leishmaniasis while hiding in the jungle.

Eighteen soldiers stationed in the jungle also were treated for the disease, Chiapas State Health Department spokesman Milton Hernandez Moguel said.

"There are probably hundreds more cases," said Dr. Gabriel Garcia, who works with Health and Education Services Promoters, an independent group based 60 miles west of here in San Cristobal de las Casas.

Pronounced "leesh-muh-NIGH-uh-sis," the disease is caused by protozoa transmitted by a sandfly—a bloodsucking insect slightly larger than a pinhead that, in Chiapas, breeds in jungles.

Dozens of American troops contracted the disease while fighting the Persian Gulf War in 1991, said Dr. Richard Titus, an expert at Colorado State University in Fort Collins. Drawings in Egyptian tombs and on Maya temples suggest the disease is an ancient scourge.

The World Health Organization estimates that 12 million people have leishmaniasis, which comes in many forms. One type of the disease is highly lethal, but the form here is treatable and begins with painful stings that fester for months. The sores sometimes disappear within 6 months, after the body's defense mechanisms learn to combat the parasite's effects, although the parasite itself may remain in the body for a lifetime. In La Nueva Estrella, afflicted peasants will miss the critical corn harvest and bean-planting seasons. "Those who have the sores can't work in the fields because it bothers them to walk," said Hernandez, the village health worker. "The ulcers are disfiguring, and when they heal they leave a burn-like scar," said Dr. Richard Pearson, an expert at the University of Virginia in Charlottesville. The Health Department's Hernandez said the disease is under control. But doctors who tromp the winding muddy roads to visit the poor, isolated villages dispute that. "Government health workers hardly ever enter these areas, so they don't register the cases," said Dr. Marcos Arana of the social services umbrella organization Conpaz. "We think this is a serious problem." (AP/*Deseret News*, 24 August 1995)

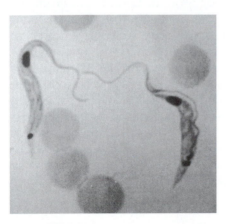

FIGURE 40-5 Photomicrograph of *Trypanosoma brucei*, the cause of sleeping sickness.

the trophozoite stage and inhabits the vagina and urethra. Transmission is usually by sexual contact. A large number of cases are asymptomatic. Males having the clinical infection may experience some irritation of the urethra, with a slight discharge and pain during urination. In the female *T. vaginalis* may cause the pH of the vagina to become slightly alkaline. This condition may be accompanied by a foul odor, a slight discharge, itching, and burning sensations. These symptoms may be caused in part by secondary bacterial infections resulting from the altered pH. This is one of the most common of the sexually transmitted diseases, with an estimated 5 to 7 million cases occurring annually in the United States.

Diagnosis is made by direct microscopic examination of smears from the vaginal or urethral discharge. Treatment of individuals and their sex partners by oral administration of the drug metronidazole effectively cures this infection.

✳ Giardiasis

This disease is caused by the flagellate *Giardia lamblia*. *Giardia* has a distinctive appearance (that resembles a human face) because of its teardrop-shaped cell, with four pairs of flagella, and two nuclei (see Figure 40-1). The cell is 9 to 16 μm in length and exists in both the trophozoite and cyst stages. The cyst is able to persist for prolonged periods outside the body and transmission is by the typical fecal–oral route. Infection is in the upper small intestines; many infections are asymptomatic or go undiagnosed and are passed off as minor intestinal disturbances. Following ingestion of the cyst, gastroenteritis and diarrhea may result, along with dark, greasy, foul-smelling feces, considerable abdominal discomfort, and flatulence. Symptoms may continue for 2 to 3 weeks with gradual improvement. The infection may become chronic in some patients, with continued intermittent bouts of diarrhea. Infection is often accompanied by anorexia, weight loss, and nausea. *Giardia* is the most common intestinal pathogenic protozoal infection of humans in the United States. Infections have occurred among campers and hikers who drank untreated water. Even water from municipal sources has been implicated as a source of infection.

Giardiasis is a recognized problem among children attending day-care centers. It is commonly associated with individuals who are deficient in immunoglobulin A and is frequently diagnosed in homosexual men. Diagnosis is made by demonstrating the parasite in feces or **duodenal** aspirates, or by the detection of giardial antigens in feces using specific enzyme-labeled monoclonal antibodies. Treatment with furazolidone, quinacrine, or metronidazole is frequently effective. This disease may not respond to treatment, however, and may be difficult to eliminate.

✳ Trypanosomiasis

Three species of the genus *Trypanosoma* cause serious diseases in humans. *T. brucei gambiense* and *T. brucei rhodesiense* cause *African sleeping sickness* and *T. cruzi* causes South American trypanosomiasis or *Chagas' disease*. The morphology of these protozoa is shown in Figure 40-5.

African sleeping sickness is found only in geographic areas of Africa where the *tsetse* (glossina) fly lives. The tsetse fly is a necessary link in the life cycle of *T. brucei gambiense* and *T. brucei rhodesiense* and functions as the vector for transmission to humans and animals. Cattle, swine, various wild animals, and humans are the major hosts and serve as a reservoir for these protozoa. When a person is bitten by an infected tsetse fly, the trypanosomes cause a local lesion at the site of the bite. The protozoa then spread and become lodged in the lymph nodes and produce a chronic infection. In some cases, the trypanosomes spread to the central nervous system (CNS) and result in the well-recognized symptoms of African sleeping sickness. During this stage, the patient becomes *somnolent* and eventually goes into a coma and dies. African sleeping sickness has a significant impact on the economy of the African continent, for large areas of otherwise productive land cannot be inhabited by humans due to the presence of infected animals and tsetse flies. An estimated 20,000 cases per year occur in Africa.

T. cruzi is found in South and Central America and is transmitted by **reduviid (triatomine) bugs**. Dogs, cats, and various wild animals serve as the reservoir of infection. The triatomine bugs bite humans at night and defecate when they feed. *T. cruzi* are in the feces, which contaminate the bite wound or other skin abrasions or are carried by fingers to the mucosa of the mouth or nose or to the conjunctiva of the eyes to cause the infection. Persons living in huts with dirt floors or walls are most likely to become infected, especially if they sleep on the floor or ground. Lesions are produced by the parasite at the site of the bite. The protozoa then spread through the body. Acute diseases occur in some persons, especially children, involving the heart and CNS, and result in a 10% death rate. Many infections are nondescript and may remain chronic for years. Chronic disease, seen mostly in adults, leads to heart diseases or gastrointestinal involvement. It is estimated that 15 to 20 million infected people live in Latin America and of these 50,000 die per year from this disease. Infections can also be transmitted congenitally and by blood transfusions.

Duodenal   Pertaining to the portion of the small intestine that attaches directly to the stomach. In adults the duodenum is about 25 cm in length.
Somnolent   Drowsy or sleepy.
Reduviid (triatomine) bug   A small biting bug.

CLINICAL NOTE

Malaria among U.S. Military Personnel Returning From Somalia, 1993

U.S. military personnel were first deployed to Somalia in late December 1992 as part of Operation Restore Hope. From the time of deployment through April 1993, malaria was diagnosed in 48 personnel who had onset of illness while in Somalia. In addition, through late June, malaria was diagnosed in 83 military personnel following their return from Somalia. This substantial number of cases has reinforced concerns regarding malaria prophylaxis, the estimated risk for infection, and the need for prompt recognition and treatment of malaria in military personnel. This report summarizes the occurrence of malaria in returning personnel and underscores for health-care providers the importance of considering malaria in the diagnostic evaluation of military personnel returning from Somalia and in other persons who have traveled to malarious areas.

Malaria infections were documented in 21 Marine and 62 Army personnel, all of whom had onset of illness after returning to the United States. Of the 62 Army personnel, 55 (89%) were stationed at Fort Drum, New York; approximately 60% of all Army troops sent to Somalia originally were stationed at Fort Drum. Detailed investigations have been completed for 32 (58%) of the Army personnel stationed at Fort Drum and all 21 Marines. Of these 53 persons, 43 (81%) had been stationed south of Mogadishu. *Plasmodium vivax* was detected in 41 (77%) of the cases, *P. falciparum* in 9 (17%), a mixed vivax and falciparum infection in 2 (4%), and *P. ovale* infection in 1.

Mefloquine was used for malaria prophylaxis by 38 persons and doxycycline by 15 persons. Because of the reportedly low frequency of vivax and ovale malaria in Somalia, terminal prophylaxis with primaquine to prevent relapses of vivax or ovale malaria following departure from Somalia had not been recommended for Army personnel. Although terminal prophylaxis had been recommended for Marine and Navy personnel, only 8 of the 15 Marines with vivax or ovale malaria had completed terminal prophylaxis. Use of prophylaxis, including terminal prophylaxis, was not supervised after arrival in the United States, and compliance was reportedly low.

Manifestations of illness included a history of fever and chills (100%), headache (97%), gastrointestinal symptoms (72%), myalgia and/or arthralgia (69%), lumbosacral pain (63%), and upper respiratory symptoms (59%). Patients with falciparum malaria had onset of symptoms an average of 34 days (range: 10–86 days) after return to the United States and 18 days (range: 0–58 days) after discontinuation of prophylaxis; patients with vivax malaria had onset at intervals of 60 days (range: 12–119 days) after return to the United States and 42 days (range: 0–102 days) after discontinuation of prophylaxis. The patients were ill an average of 4 days (range: 0–23 days) before seeking medical attention. In 13 (25%) patients, the diagnosis of malaria was delayed for 3 or more days after initial medical contact (*MMWR* 42:524, 1993).

Control is by use of insecticides to eliminate the bugs from human dwellings.

Diagnosis of trypanosomal diseases is generally made by demonstrating the presence of the protozoa in the blood, lymph node aspirates, or spinal fluid. Moderately effective serologic tests are now becoming available. African trypanosomiasis can be effectively treated during the early stages, before CNS involvement, with the drugs suramin or pentamidine. CNS infections are treated with eflornithine. There is no completely effective drug for treating South American trypanosomiasis, but a new drug called allopurinol has shown promise in treating the acute stage of this disease.

✳ Leishmaniasis

Several species of the genus *Leishmania* cause infections in tropical and subtropical regions around the world. A wide variety of mammals, including humans, serve as the natural reservoirs for the leishmanial parasite. Sandflies are the vectors of transmission. The following general types of clinical diseases are seen: (1) visceral leishmaniasis, caused by *L. donovani*, (2) cutaneous leishmaniasis, caused by several species, and (3) mucocutaneous leishmaniasis, caused by subspecies of *L. braziliensis*. Visceral leishmaniasis is severe with high mortality. The other forms of leishmaniasis are less severe and

FIGURE 40-6   Reported cases of malaria in the United States, 1930–1994. Of all the reported cases in 1980, 81% were among foreign civilians. (Modified from CDC annual summaries)

are usually limited to lesions of the skin (see Plate 96) and mucous membranes. An estimated 12–20 million people are infected worldwide and about 400,000 new infections occur each year. Diagnosis is by observing the parasite in aspirates of infected tissues; serologic and DNA probe tests are being developed.

Control measures are directed at elimination of, or avoiding exposure to, the sandfly and reducing the number of reservoir animal hosts. Treatments include drugs containing antimony as well as pentamidine or amphotericin B.

SPOROZOA INFECTIONS: MALARIA

Of all the infectious diseases of humans, malaria is probably the most important as far as total number of cases, deaths, and debilitation are concerned. Possibly hundreds of millions of people worldwide are infected with malarial parasites, resulting in more than 1 million deaths each year. The sporozoa includes four species of the genus *Plasmodium*, which cause most infections in humans: *P. vivax*, *P. malaria*, *P. falciparum*, and *P. ovale*. Most malaria is found in the subtropics or tropics. Malaria cases seen in the United States usually involve persons who were infected while visiting or residing in endemic areas outside the country. During military operations like the Vietnam War a marked increase

in malaria was seen in the United States (Figure 40-6).

✳ Pathogenesis and Clinical Disease

The severity of clinical symptoms produced varies among the different species. Generally, infections with *P. falciparum* are the most severe. The continued destruction of red blood cells, with concurrent damage to the capillaries, mononuclear phagocyte system ∞ (Chapter 10, p. 146), and various internal organs, may lead to the death of the patient; this happens most often in young children. In many adults, the immune response finally stops the replication of the parasites, no further symptoms are experienced, and these persons are immune to reinfection by the specific species involved. In some patients, the infection is somewhat suppressed and develops into a chronic disease. Occasionally, symptomatic diseases may reoccur in these chronically infected persons.

✳ Transmission and Epidemiology

The female *Anopheles* mosquito is the primary host for plasmodia; humans serve as an intermediate host. Thus, the major mode of transmission of the malarial parasite to humans is by the bite of the anopheles mosquito, although a few cases may be transmitted by other means. For example, occasionally a chronically infected female may congenitally transmit malaria to her offspring. Blood transfusions or unsterilized paraphernalia used in the injection of illicit narcotics may also transmit malaria. Persons who have had malaria or who have, within the past three years, visited areas where malaria is endemic, are requested not to give blood for transfusions.

The life cycle of the plasmodia is quite involved, with different stages of development occurring in the human and the mosquito. A simplified general outline of this life cycle is shown in Figure 40-7 and described in more detail below.

The mosquito becomes infected by taking a blood meal from an infected person. Within the mosquito, the plasmodial *gametocytes* go through a sexual reproduction cycle. This results in the accumulation of large numbers of infectious *sporozoites* in the salivary glands of the mosquito. The sporozoites are then inoculated into a susceptible host as the mosquito feeds. The sporozoites are rapidly filtered from the blood and specifically infect liver cells. The parasite matures through a series of stages in the liver cells and after a week or so parasitic forms

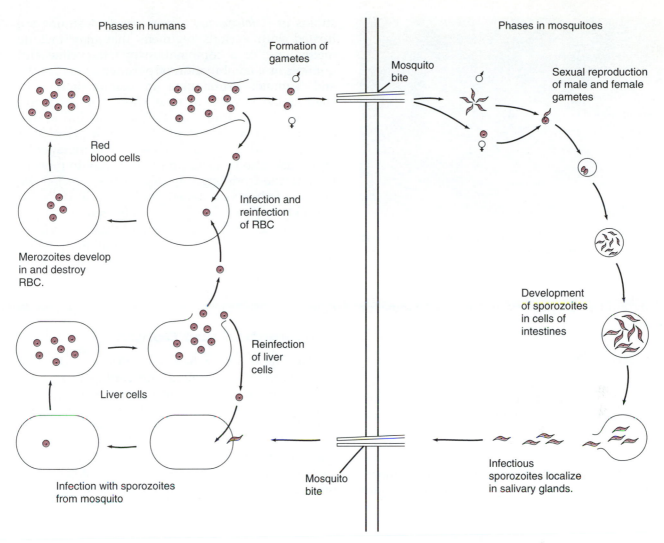

FIGURE 40-7   A simplified outline of the phases of the life cycle of a plasmodium parasite in humans and mosquitoes.

called **merozoites** are released into the blood. The merozoites infect red blood cells and go through another series of developmental stages that end with the bursting of the red blood cells, which release large numbers of new merozoites and some gametocytes. These merozoites infect more red blood cells and the cycle is repeated. The cycles of infection in the red blood cells become synchronized and with each cycle more and more cells become infected and then burst. When enough of the red blood cells burst and release cellular debris, parasites, and by-products, the onset of a malarial *paroxysm* is triggered. A paroxysm is a periodic sudden recurrence of symptoms. The paroxysms of malaria start with chills that last for 15 to 60 minutes. During this time the patient may have a headache, vomit, and generally feels nauseated. As the sensation of chilling stops, a high fever develops and lasts for several hours. It may be accompanied by a severe headache, increased nausea and vomiting, profuse sweating, and often mild delirium. The paroxysm lasts 8 to 12 hours and terminates as the exhausted patient goes to sleep. On awakening, the patient usually feels relatively well. The growth of the parasite inside the red blood cells occurs at fixed intervals and this factor determines the frequency of paroxysms. The interval is 48 hours for *P. vivax* and *P. ovale*; it varies between 32 and 48 hours for *P. fal-*

Merozoite   The asexually reproducing form of the malarial parasite. The merozoite is infective for human red blood cells.

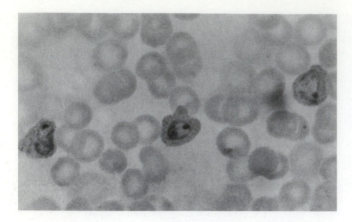

FIGURE 40-8 Malaria-infected human red blood cells. (Courtesy Abbott Laboratories, Chicago)

ciparum; and paroxysms occur every 72 hours for *P. malariae*.

✳ Diagnosis

Laboratory diagnosis is made by observing the malarial parasites in stained slides of blood examined with a microscope (Figure 40-8).

✳ Treatment

All clinical cases of malaria are treated with chloroquine, except those caused by chloroquine-resistant strains of *P. falciparum*. These resistant strains are treated with various regimens that may include melfoquine, or a combination of tetracycline and quinine, or a combination of pyrimethamine, sulfa, and quinine. Primaquine in combination with chloroquine is used to treat chronic malaria and relapses. These drugs can terminate a clinical disease, prevent recurrent attacks, and, if given long enough, result in a cure. These chemotherapeutic agents can also be used prophylactically to prevent the parasite from growing once it has infected the tissue. Medications taken orally on a daily or biweekly regimen can be quite effective when followed rigidly. This type of prophylaxis is widely used by Peace Corps volunteers or military personnel who must enter malaria-infested areas. During World War II and the Vietnam War, malaria constituted one of the major problems in military operations.

✳ Prevention and Control

Perhaps more effort has been made to develop methods for the treatment and control of malaria than for any other infectious disease. Although significant progress has been made, as yet no "magic bullet" or "wonder drug" is available as an easy cure of this disease. Control of the *Anopheles* mosquito has been one of the major methods used to control or reduce the number of cases. Malaria was markedly reduced in many countries in the 1950s

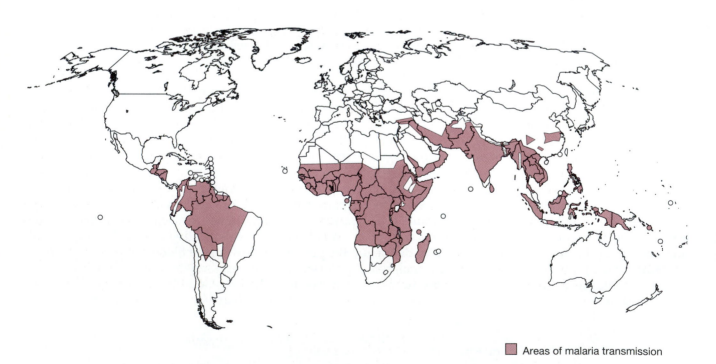

■ Areas of malaria transmission

FIGURE 40-9 World regions of endemic malaria.

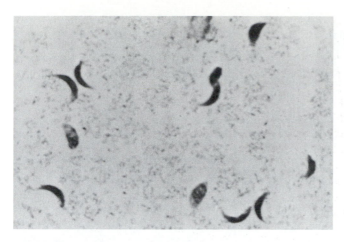

FIGURE 40-10   Micrograph of the crescent-shaped *Toxoplasma gondii*. (S.S. Schneierson, *Atlas of Diagnostic Microbiology*, p. 69. Courtesy Abbott Laboratories, Abbott Park, IL)

and 1960s when DDT was widely used to control mosquito populations. However, by the 1970s many mosquitoes were developing resistance to DDT and its use was also being reduced because of other environmental concerns. Also, some of the malarial parasites had developed resistance to the commonly used antimalarial drugs. All of these factors have led to a resurgence of malaria in many developing countries. For example, only 18 cases of malaria were reported in Sri Lanka in 1972, but by 1980 the number of cases had risen to one million.

Although research is in progress to develop a vaccine against malaria, none is currently available. Therefore, individuals traveling to malaria-infested areas (Figure 40-9) should inquire about receiving chemoprophylactic chloroquine. In addition, use of mosquito nets, insect repellents, or other methods of avoiding contact with mosquitoes are of some value in protecting persons during short visits to such areas.

SPOROZOA INFECTIONS: TOXOPLASMOSIS

In contrast to many protozoal diseases found mainly in tropical regions and under unsanitary conditions, toxoplasmosis occurs in all parts of the world and can be transmitted in cosmopolitan populations. Research has shown that from 25 to 50% of the world's population have been, or are now, infected with the toxoplasmosis parasite, *Toxoplasma gondii* (Figure 40-10). Yet this disease in humans went virtually unrecognized until the 1940s and only a handful of cases had been specifically diagnosed during the 1950s.

Only in the past several decades has the life cycle of *T. gondii* begun to be understood. The primary hosts are various felines, with domestic cats the chief transmitters to humans (Figure 40-11). Cats become infected by eating mice, birds, or raw meat; they can also be infected by the feces of other cats. The *T. gondii* goes through a sexual phase of reproduction in the intestinal tract of cats. Reproductive structures called *oocysts* are shed in the feces and in 3 to 4 days, if warm and damp, eight infectious sporozoites develop within each oocyst. This mature oocyst may remain viable for up to 1 year and infect humans and animals when ingested.

✳ Pathogenesis and Clinical Disease

Once ingested, the *T. gondii* oocyst releases its infectious sporozoites. No sexual reproduction of the parasite occurs in nonfeline animals or humans. However, the sporozoites are able to proliferate by binary fission and circulate throughout the body. The parasite penetrates into cells and continues to proliferate until the cell bursts. As this process continues, clinical disease may result with such symptoms as fever, weakness, respiratory illness, myocarditis, or an infectious mononucleosis-like illness with swollen lymph nodes. The disease is often asymptomatic or is passed off as some other disease. Only lately has a diagnosis of toxoplasmosis even been considered when dealing with such nondescript illnesses.

As the immune mechanisms of the host respond, *T. gondii* remains inside the infected cells and forms aggregates of several thousand parasitic cells that become enclosed in a fine membrane to form a cyst. The cyst evokes no further response from the host (although it can serve as a source of infection to another animal eating the infected tissue). It does not cause any tissue damage and the enclosed parasites are protected from the host's defense mechanisms. Thus a near ideal parasitic relationship is established and may persist for the lifetime of the host.

In a great majority of cases, toxoplasmosis is mild and self-limiting. However, toxoplasmosis may be a severe disease in AIDS patients or other immunosuppressed individuals and in developing fetuses.

✳ Transmission and Epidemiology

Although a *T. gondii* cyst may cause no adverse symptoms in its host, such cysts themselves can serve as a source of infection to a new host that

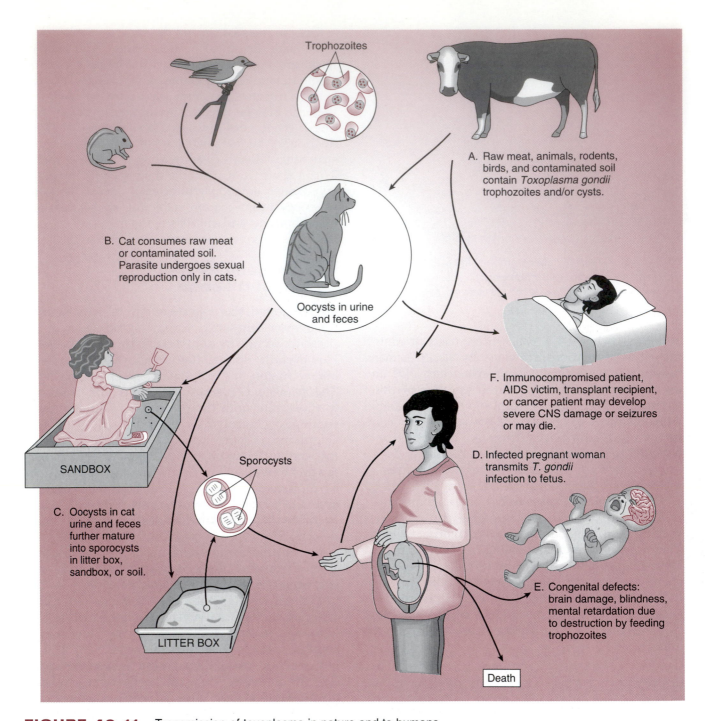

Trophozoites

A. Raw meat, animals, rodents, birds, and contaminated soil contain *Toxoplasma gondii* trophozoites and/or cysts.

B. Cat consumes raw meat or contaminated soil. Parasite undergoes sexual reproduction only in cats.

Oocysts in urine and feces

SANDBOX

Sporocysts

C. Oocysts in cat urine and feces further mature into sporocysts in litter box, sandbox, or soil.

LITTER BOX

F. Immunocompromised patient, AIDS victim, transplant recipient, or cancer patient may develop severe CNS damage or seizures or may die.

D. Infected pregnant woman transmits *T. gondii* infection to fetus.

E. Congenital defects: brain damage, blindness, mental retardation due to destruction by feeding trophozoites

Death

FIGURE 40-11 Transmission of toxoplasma in nature and to humans.

ingests the tissue. Undoubtedly this is a major means of transmission to various carnivorous birds or animals in nature. It has also been well established that humans may become infected by eating raw or rare meat. Other less common avenues of entry into humans include (1) unhygienic handling of infected cats or (2) inhalation of air or dust contaminated with cat feces.

A major concern is *congenital toxoplasmosis*. This

condition occurs when a pregnant female develops a primary case of toxoplasmosis by being in contact with cats or by eating raw or rare meat. During the systemic phase of the infection *T. gondii* is able to pass the placental barrier and infect the developing fetus. Generally, the tissue of the brain or eyes are involved. Some infants die in utero; others are born with serious CNS defects and die shortly after birth. In some cases, infected infants show no signs of dis-

Cryptosporidium Infections Associated with Swimming Pools: Dane County, Wisconsin, 1993

In March and April 1993, an outbreak of cryptosporidiosis in Milwaukee resulted in diarrheal illness in an estimated 403,000 persons. Following that outbreak, testing for *Cryptosporidium* in persons with diarrhea increased substantially in some areas of Wisconsin; by 1 August 1993, three of six clinical laboratories in Dane County were testing routinely for *Cryptosporidium* as part of ova and parasite examinations. In late August 1993, the Madison Department of Public Heath and the Dane County Public Health Division identified two clusters of persons with laboratory-confirmed *Cryptosporidium* infection in Dane County (approximately 80 miles west of Milwaukee). This report summarizes the outbreak investigations.

On 23 August, a parent reported to the Madison Department of Public Health that her daughter was ill with laboratory-confirmed *Cryptosporidium* infection and that other members of her daughter's swim team had had severe diarrhea. On 26 August, public health officials inspected the pool where the team practiced (pool A) and interviewed a convenience sample of patrons at the pool. Seventeen (55%) of 31 pool patrons interviewed reported having had watery diarrhea for 2 or more days with onset during July or August. Eight (47%) of the 17 had had watery diarrhea longer than 5 days. Four persons who reported seeking medical care had stool specimens positive for *Cryptosporidium*.

On 31 August, public health nurses at the Dane County Public Health Division identified a second cluster of nine persons with laboratory-confirmed *Cryptosporidium* infection while following up case reports voluntarily submitted by physicians. Seven of the nine ill persons reported swimming at one large outdoor pool (pool B). Because of the potential for disease transmission in multiple settings, a community-based matched case-control study was initiated on 3 September to identify risk factors for *Cryptosporidium* infection among Dane County residents.

Laboratory-based surveillance was used for case finding. A case was defined as *Cryptosporidium* infection that was laboratory-confirmed during 1 August–11 September 1993 in a Dane County resident who was also the first person in a household to have signs or symptoms (i.e., watery diarrhea of 2 or more days' duration). During the study interval, 85 Dane County residents with stool specimens positive for *Cryptosporidium* were identified. Sixty-five (77%) persons were interviewed; 36 (55%) had illnesses meeting the case definition. Systematic digit-dialing was used to select 45 controls, who were matched with 34 case patients by age group and telephone exchange. All study participants were interviewed by telephone using a standardized questionnaire to obtain information on demographics, signs and symptoms, recreational water use, child-care attendance, drinking water sources, and presence of diarrheal illness in household members.

The median age of ill persons was 4 years (range: 1–40 years). Reported signs and symptoms included watery diarrhea (94%), stomach cramps (93%), and vomiting (53%). Median duration of diarrhea was 14 days (range: 1–30 days). Swimming in a pool or lake during the 2 weeks preceding onset of illness was reported by 82% of case patients and 50% of controls. Twenty-one percent of case patients and 2% of controls reported swimming in pool A. Fifteen percent of case patients and 2% of controls reported swimming in pool B. When persons reporting pool A or B use were excluded from the analysis, the association with recreational water use was not statistically significant. Child-care attendance was reported for 74% of case patients aged <6 years and 44% of controls. Two case patients reported child-care attendance and use of pool A or pool B. No case patients reported travel to the Milwaukee area during the March–April outbreak, and no associations were found between illness and drinking water sources.

To limit transmission of *Cryptosporidium* in Dane County pools, state and local public health officials implemented the following recommendations: (1) closing the pools that were epidemiologically linked to infection and hyperchlorinating those pools to achieve a disinfection, (2) advising all area pool managers of the increased potential for waterborne transmission of *Cryptosporidium*, and (3) maintaining laboratory-based surveillance in the community to determine whether transmission was occurring at other sites (e.g., child-care centers and other pools).

On 27 August, pool A was closed and hyperchlorinated for 18 hours; on 3 September, pool B closed early for the season. Because many control measures were initiated less than 1 week before many pools closed for the season (after 5 September), their impact on transmission could not be evaluated adequately (*MMWR* 43:561, 1994).

ease but carry the toxoplasma parasite as a latent infection. Later in life the parasite may begin to proliferate, causing tissue damage. Other congenitally infected infants may remain symptomless throughout life. About one of every 1000 children born in the United States has congenital toxoplasmosis, which makes this disease one of the major congenital diseases.

✳ Diagnosis and Treatment

Diagnosis is made by examining tissues or fluids for the cysts of *T. gondii* and using serologic tests. Treatment is effective only during the systemic phase of the disease. A mixture of three different sulfa drugs in combination with the antimalarial drug pyrimethamine is used. In pregnant patients, spiramycin should be substituted for pyrimethamine.

✳ Prevention and Control

There is no vaccine against *T. gondii*, but avoiding rare or raw meats and handling domestic cats hygienically are good preventive measures. If a woman has acquired toxoplasmosis with the resultant antibody immunity before becoming pregnant, the developing fetus is protected against congenital infection. About 30% of the women of childbearing age in the United States have antibodies against this parasite. Some health officials are recommending serologic tests for toxoplasmosis during early pregnancy and suggest that those who are negative should avoid contact with cats or their feces, and also avoid rare or raw meat during their pregnancy. Such procedures should be followed by all pregnant women who have not been shown to have antibodies against toxoplasmosis.

OTHER PROTOZOAL INFECTIONS

Two ubiquitous protozoa of the genera *Cryptosporidium* and *Pneumocystis* have increasingly been recognized as causes of serious morbidity in immunocompromised patients. Cryptosporidia are clearly sporozoa, but the taxonomic position of *Pneumocystis* is less settled. It has traditionally been classified as a protozoan. However, recent studies have shown it to also have some characteristics of certain fungi, and it may represent a unique transitional position between the fungi and the protozoa.

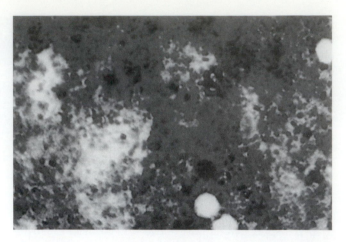

FIGURE 40-12 Photomicrograph of cryptosporidium parasites found in a human intestinal specimen.

The taxonomy of the genus *Cryptosporidium* is not well established, but the isolates from humans and cattle are given the species name *C. parvum*. Many domestic animals harbor this organism in their intestinal tract, and it appears that most humans have had one or more infections. Infectious oocysts are shed in the feces of infected animals and humans and transmission is by the typical fecal–oral routes. Cryptosporidia cause a profuse watery diarrhea that appears to be self-limited in the otherwise healthy host. However, in individuals with immune deficiencies, such as AIDS, the diarrhea is severe and often prolonged. In AIDS patients the infection may be associated with malnutrition, malabsorption, and significant weight loss and may last for many months. In developed countries from 1–4% of gastroenteritis in small children is caused by this protozoan; in developing countries the rate is 4–10%. About 15% of AIDS patients in the United States have chronic infection.

Diagnosis is by showing the oocysts in feces (Figure 40-12) using an acid-fast stain or an immunofluorescent monoclonal antibody test. Recovery is spontaneous in immunocompetent persons and treatment is generally not required, except for severe cases in children that may require rehydration therapy. AIDS patients are treated to reduce symptoms. No specific treatment is available.

Pneumocystis carinii is found worldwide and infects many different animals, and most humans show serological evidence of having been infected by the age of 4. Infections are without symptoms in immunocompetent hosts. This is a classical opportunistic microorganism and pneumonia due to

P. carinii has been recognized for many years as a complication of organ transplantation or cancer chemotherapy. However, as the epidemic of AIDS developed, this disease became the most common serious complication in these patients. Diagnosis of this infection is difficult and frequently requires removal of fluids or tissues directly from the infected lung. Trimethoprim sulfa and pentamidine are effective therapeutic agents. However, even with therapy as many as 30% of patients die, and this disease is the leading direct cause of death among AIDS patients.

MATERIAL FOR REVIEW

CONCEPT SUMMARY

1. The protozoa are a diverse group that includes both free-living forms and animals parasites. A number of protozoa require a second animal host in addition to humans in order to complete their development. Such life cycles provide a variety of alternatives for exercising control measures against these organisms.

2. Protozoal diseases range in seriousness from the relatively benign giardiasis and trichomoniasis to the very common and severe diseases of malaria and amebic dysentery. Protozoan diseases are not only among the most common worldwide but are also among the most serious. Therapy of such infections remains limited and is often ineffective.

STUDY QUESTIONS

1. List the morphologic and physiologic features of the protozoa that separate these organisms from the bacteria.
2. What features of pathogenesis associated with *E. histolytica* are responsible for the severity of disease due to this organism?
3. Make a list of protozoal diseases that occur in the United States.
4. What features of protozoa limit possible antimicrobial therapy in humans?
5. Diagram and label the life cycle of *Plasmodium*.
6. Describe the sporozoa–host relationship.

CHALLENGE QUESTIONS

1. Why are there so few cases of trypanosomiasis and leishmaniasis in the United States?
2. Many of the protozoan diseases discussed in this chapter are relatively rare in the United States. Do you believe they require less attention than other diseases discussed? Explain.

Glossary

A

abscess   A localized accumulation of pus.

accidental host   An organism accidentally invaded by a pathogan.

acetylcholine   A chemical messenger that functions to make a nerve-muscle junction. When this messenger is inhibited, the brain cannot signal the muscle to contract. This lack of muscle response is known as *flaccid paralysis*.

acquired immunity   Type of specific immunity that develops as a result of exposure to an antigen.

active immunity   A condition in which antibodies are actively produced within the host.

active site   That portion of an enzyme where the substrate binds.

acute   A current, rapid, short manifestation of disease symptoms.

acute inflammatory response   Immediate mechanisms by which host defenses wall off or destroy invading microbes and repair damaged tissue.

adenoids   Nasopharyngeal lymphoid tissue.

adenosine triphosphate (ATP)   A high-energy molecule that is the "energy currency" for all cells.

aerosolized   Dispersed particles suspended in air.

allergen   A normal substance that triggers an allergic response in a sensitized person.

alveoli   Small open spaces in the lung where oxygen and carbon dioxide exchange takes place.

amoeboid   Referring to a crawling movement by certain cells.

anionic   Having negative electrically charged groups.

antibiotic   Antimicrobial agent that is the natural product of living organisms.

antibiotic resistance   A condition in which a microbe is unaffected by the presence of a compound used for antimicrobial therapy.

antibody titer   The concentration of antibody present in the blood serum.

antigen   A chemical molecule that can stimulate antibody production in an animal.

antigen-presenting cell   A cell that presents peptide fragments associated with major histocompatibility class II proteins.

antigenic   Capable of stimulating antibody synthesis.

antigenic determinant   Those molecular groups to which an antibody reacts; also called an epitope.

antigenic drift   Small changes in the antigenic structure of a virus.

antigenic shift   Major changes in the antigenic structure of viruses with segmented genomes (e.g., influenza virus.)

antihistamine   Drug used to counteract allergic reaction by blocking histamine-induced reactions.

antimicrobial   Capable of killing or stopping the growth of a microbe.

antiseptic   An agent that usually is applied directly to living tissue to reduce the likelihood of infection or sepsis.

arboviruses   A general term describing viruses transmitted from arthropods to vertebrates.

arthralgia   Pain in the joints.

arthroconidium   A fungal spore formed as the cell walls increase in thickness. Each of the cells in the hyphae may become an arthroconidium. These conidia are very resistant to drying.

arthropods   A large group of invertebrate animals, many of which have biting or sucking mouthparts, such as the mosquito and other insects.

aseptic technique   Methods used to minimize the chances of contamination or infection by microorganisms from the environment.

attachment site   The location at which an organism attaches itself to host tissues. This is usually determined by specific receptor sites on the host cell and pathogen.

attenuated   The condition of an organism in which virulence factors have been reduced or removed.

attenuated vaccine   A vaccine made from living microorganisms that have been selected for their lack of virulence.

atypical lymphocytes   Lymphocytes that are abnormal, usually enlarged with a foamy nucleus.

autoimmunity   A hypersensitivity whereby individuals are sensitive to antigens on cells of their own body.

autotroph   An organism that obtains its carbon from CO^2 and energy from light or inorganic compounds.

axillary   Referring to the armpit.

B

bacilli   Bacteria that are elongated or rod-shaped.

bacteremia   Presence of dividing bacteria in the blood.

bacterial flora   Bacteria that normally live on or in the body but do not normally cause disease.

bacterial spore   An environmentally stable and persistent structure formed by some bacteria as protection against adverse conditions.

bacteriophage   A virus that infects bacterial cells; also called a phage.

bacteriostatic   Capable of preventing bacteria from multiplying.

bile   A fluid secreted by the liver to aid the emulsification of fats.

biopsy   The process of removing tissues from a patient for diagnostic examination.

bipolar   Referring to a dye that stains the ends (poles) of cells more than the center, giving a "safety pin" appearance.

blood agar Agar to which blood was added before the agar solidified.

booster A second or subsequent dose of antigen given to increase the level of immunity.

botulism A disease caused by toxins produced by the bacterium *Clostridium botulinum*.

broad-spectrum Effective against more than one kind of bacterium; usually suggesting antibacterial activity against both gram-positive and gram-negative bacteria.

bronchiolitis Inflammation of the bronchioles.

budding The act of release for some viruses. The replicated virus becomes enveloped by host membrane as it leaves an organelle or the cell.

C

capillary tube A glass tube with a very small opening running through the tube. The opening is usually less than 1 mm in diameter.

capsid The external protein structure of a virus and is composed of capsomers.

capsule A polysaccharide layer surrounding bacterial cells peripheral to the cell wall.

capsule An amorphous, gel-like material produced by some bacteria and collected around the outside of the cell.

cardiolipin A phospholipid found in beef heart tissue.

carrier A living host that is infected by an organism but does not have clinical symptoms of disease. Carriers may have asymptomatic infections or may be individuals who have recovered from the disease but continue to shed the causative agent into the environment.

carrier rate The number of individuals infected with a pathogen.

cascading Referring to a series of reactions that follow one another.

catheter A tube, usually of rubber or plastic, that can be placed into a body cavity (e.g., urethra) or a blood vessel to allow easy drainage of body fluids or the placement of medications into the body.

cationic Having positive electrically charged groups.

cell-mediated immune response A response in which immune cells (T-helper, T-cytotoxic) react against specific foreign antigens.

cellulitis Inflammation of cellular or connective tissues.

centrifugation A technique that spins particles or cells into a highly concentrated mass.

cerebral spinal fluid Fluid filling the ventricles and cavities of the brain and spinal column.

cervicitis Inflammation of the cervix.

chemotaxis Movement of a cell toward a chemical influence. Protective phagocytic cells are attracted by chemicals released by damaged body cells.

chemotherapeutic agent A chemical substance that is harmful to microorganisms and can be used to treat a specific disease.

chickenpox A highly contagious childhood disease involving skin lesions caused by the varicella-zoster virus.

cholera An acute enterotoxin-mediated infection caused by *Vibrio cholerae*.

chronic Lasting a long time and tending to progress slowly.

cirrhosis A progressive disease of the liver characterized by scar tissue development and ultimate liver failure.

clonal expansion The proliferation of selected lymphocytes.

clonal selection The process whereby exposure to an antigen stimulates a specific lymphocyte clone to divide and differentiate.

codon Three sequential mRNA bases that match a three-base sequence in the DNA code designating a specific amino acid in a protein.

coliform Enterobacteriaceae that are commonly normal flora of the intestinal tract.

colitis An inflammation of the colon.

combination therapy Treatment with more than one antibiotic at a time; often used when there is doubt as to the cause of the disease, or when there is likelihood that a microorganism may become resistant to one antimicrobial if used alone.

communicable Capable of being transmitted from one person to another; a synonym for *contagious*.

competitive inhibitor A compound that competes with a substrate for position at the active site on an enzyme, but that cannot be changed by the enzyme. These compounds prevent normal, essential enzyme function.

complementary pairing A structural relationship between nucleotide bases that allows adenine to bond with thymine and guanine to bond to cytosine.

compromised Reduced resistance with more susceptibility to infection.

concentration gradient The difference in the concentrations of ions or molecules on each side of a biological membrane.

congenital Associated with the fetal state. A congenital disease is one that is acquired by the fetus, usually a result of a preceding maternal infection such that the infant is infected at birth.

conidium An asexual reproductive unit in the fungi.

conjugate Attach; proteins to which a specific substance or molecule has been attached are said to be conjugated to the added molecule or substance.

contagion The passing of disease between individuals.

contrast The ability to see an object against the background.

corticosteroid A steroid derivative used to suppress some allergic reactions.

coryza Acute inflammation of the nasal mucosa accompanied by profuse nasal discharge.

cyst A dormant environmentally resistant life stage of some protozoa.

cystitis An inflammation of the bladder.

cytochrome An electron carrier in the electron transport system.

cytokine A soluble protein that regulates specific immune functions.

cytopathic effect   The visible effect that virus infection has on cells.

D

daughter cell   A cell that is produced by binary fission or budding directly from another cell.

debredment   Removal of dead tissue and foreign matter from a wound by scrapping.

denature   Altering of the tertiary or secondary structure of a protein.

desensitized   Referring to a person who is no longer sensitive to an allergen.

detergent   A compound similar to soap (not a disinfectant) that is used as a cleaning agent because of its ability to emulsify dirt.

differential staining   Using more than one dye, which enables identification bacterial components.

dimorphism   The property of having two morphologic shapes. Some fungi are dimorphic and exist as either yeasts or molds, depending on their growth environment.

diphtheroids   Gram-positive bacilli of the genera *Corynebacterium*, **Propioni bacterium**, which are commonly found on body surfaces and are frequently found as contaminating bacteria in clinical specimens.

disease   A change in the state of health of the body resulting in inability to carry out all normal functions.

disease reservoir   A natural source of disease agent. Such reservoirs may be sick patients, asymptomatic carriers, animals, recovered patients, or environmental sources.

disinfectant   An agent that usually is used to kill microorganisms on inanimate objects.

disinfection   A process used to destroy harmful microorganisms but usually not including bacterial endospores.

droplet nuclei   The small dehydrated particles of mucus and microbes that remain suspended in the air following coughing, sneezing, or speaking.

duodenal   Pertaining to the portion of the small intestine that attaches directly to the stomach. In adults the duodenum is about 25 cm in length.

dye   A colored chemical compound used to stain microorganisms provide better contrast.

dysentery   Serious diarrhea, accompanied by mucus and blood in the stool along with severe abdominal cramping.

E

ectopic pregnancy   A pregnancy occurring "out of place"-with embryo implantation occurring outside the uterus.

eczema   Inflammation of the skin, often associated with scaling, papules, crusting, and serous discharge.

edematous   Referring to an accumulation of body fluids.

electrolytes   Sodium, potassium, chloride, and bicarbonate ions present in the body fluids.

electron acceptor   An atom or molecule (such as oxygen) that is relatively easily reduced by accepting electrons and protons.

electron transport system   A series of steps that oxidize NADH and FADH^2 by transferring a hydrogen atom from them to an oxygen atom.

electron-dense   Contrast provided for electron microscopy using specific metals or salts.

electrophoresis   A technique used to separate proteins or nucleic acids using an electrical field.

embolus   Bacteria or other foreign bodies that block a vessel.

embryonated eggs   Fertilized eggs in which the embryo has been permitted to grow to a recognizable stage of development.

empirical therapy   Treatment given on the basis of experience, and not as a result of susceptibility testing.

encephalitis   Inflammation of the brain. Encephalitis is most often due to viral infection.

endemic   Referring to a disease that is always present in a population.

endocarditis   Infection of the tissues lining the inside of the heart or valves in the heart.

endogenous   From inside; endogenous infections are caused by the normal bacterial flora of the body.

endophthalmitis   Inflammation or infection of the inside of the eyeball. This is a serious condition, often requiring removal of the eyeball.

endosome   A membrane compartment in eukaryotic cells that processes proteins.

endospore   A bacterial spore associated with resistance to environmental inactivation.

endotoxin   A toxin derived from the cell wall of gram-negative bacteria.

enteric fever (typhoid fever)   Systematic disease due to *Stoyphi* infection.

enterotoxin   An exotoxin produced by a variety of bacteria that is absorbed through the intestinal mucosa. Most such toxins cause nausea, vomiting, and/or diarrhea.

enucleation   Removal of the entire eyeball from its socket.

envelope   A membrane surrounding the protein coat of some viruses.

enzyme   A catalyst that reduces the energy needed to start a chemical reaction and may increase the rate of the reaction without being used up in the reaction.

epidemic   The occurrence of a common disease such as influenza in greater than expected numbers; or the occurrence of a rare disease even in small numbers.

epidemiology   The study of mechanisms and factors involved in the spread of disease within a population.

epiglottitis   Infection of the tissue that normally covers the trachea during swallowing.

epithelium   A cellular layer covering internal and external body surfaces that lacks blood vessels.

erythema   Redness of the skin resulting from dilation of the capillaries.

etiologic agent   The causative agent; the microorganism responsible for a specific disease.

eutrophication The process by which a body of water becomes so rich in dissolved nutrients that the resultant algal growth depletes the water of oxygen supplies needed by other organisms.

exoenzyme An enzyme made in the cell and then transported outside where it is functional.

exotoxin A soluble toxin often secreted into the environment, including host tissue.

expectoration Saliva and other fluids in the mouth that are expelled by spitting.

exudate A secretion from vessels that collects in body tissues and spaces, or from the tissue that is discharged outside the body.

F

fastidious Requiring special nutrient supplementation in order to grow.

Fc region The base portion (end opposite the antigen-binding site) of the Y-shaped antibody.

febrile Referring to fever.

ferment Microbial production of acid from carbohydrates.

fermentation The conversion of pyruvic acid into such products as alcohol and organic acids.

fluorescent-tagged antibody An antibody that has a fluorescent molecule attached.

freeze-dried Referring to those materials frozen, after which the water is removed without melting.

fulminating Running a rapid course with continued worsening of patient condition.

G

gamma globulin That portion of blood serum containing antibody.

ganglion Major nerve trunks connecting the peripheral nerves to the CNS.

gangrene Tissue death due to the loss of a blood supply.

gastric washing Removal of stomach contents through a tube swallowed to allow washing of the stomach.

gastroenteritis An infection or intoxication of the intestinal tract. Gastroenteritis is commonly characterized by nausea, diarrhea, and malaise.

gene pool The total genetic information present in a population at a specific time.

generation time The length of time for a population to double in number; for bacteria, generation time is usually measured in minutes.

genetic code The three-base sequences (codons) in the mRNA that specify a specific amino acid.

genetic engineering The transfer of genetic material from one organism to another organism.

genome The genetic information (DNA or RNA) in a virus or cell.

genotype The total genetic information in the cell. Normally only selected parts of the genotype are used at any one time. Thus, cells may have much greater genetic capa-

bility than is observable at any given time.

germ theory The theory that disease is caused by microorganisms.

germination The process whereby a bacterial endospore develops into a vegetative

glial cells Specialized brain cells surrounding the nerve cells (neurons).

globulin A group of proteins found in human serum. The gamma globulins are antibody proteins and are produced by B lymphocytes.

glycolysis A series of reactions involving the catabolism of glucose, resulting in the production of pyruvic acid.

glycoprotein A protein that has simple sugars (monosaccharides) attached at various locations.

growth rate The number of generations of a species within a given length of time.

Guillain-Barré syndrome A progressive loss of myelin from nerve fibers caused by the immune system's response to a viral infection.

H

heat-sensitive Capable of being damaged or destroyed by heat.

heavy chain An antibody polypeptide of higher molecular weight.

hemagglutination A process by which an antibody or virus particle forms bridges between red blood cells such that they become to be stuck together.

hematopoietic Referring to the origins of blood cells from stem cells.

hemolysis The breaking (lysis) of blood cells. Staphylococci often cause hemolysis of the cells in blood agar.

hemolytic disease of the newborn A type II hypersensitivity where maternal anti-Rh antibodies cross the placenta and bind to Rh antigens of fetal red blood cells; also called erythroblastosis fetalis.

hemorrhagic Associated with bleeding. Hemorrhagic lesions of the skin often appear as red spots or blemishes. These lesions frequently become dark blue or black in color and may involve large areas of the skin.

hepatitis A Inflammation of the liver. This condition is due to the hepatitis A virus; also called *infectious hepatitis*.

hepatitis B Inflammation of the liver. This condition is due to the hepatitis B virus; also called *serum hepatitis*.

heterophile antibody An antibody that reacts with antigens other than those responsible for induction. Human heterophile antibodies will hemagglutinate sheep red blood cells.

heterotroph An organism that requires organic compounds as a source of carbon.

histamine An important chemical mediator of both the inflammatory response and allergies.

host An organism that supports the growth of another organism.

host range The kinds of hosts that can be infected by a specific microorganism. Host range is determined by receptors on both the host cells and the microorganism. A parasite with a broad host range can infect many kinds of cells.

humoral   Referring to body fluids, such as blood.

humoral antibody response   The production of antibodies to foreign antigens.

humoral immunity   Defenses involving antibodies that attack microbes in the body fluids.

hydrophilic   Capable of dissolving in water (water-loving).

hydrophobic   Incapable of dissolving in water (water-fearing).

hyperimmune globulin   Globulin containing high levels of antibody to a specific antigen. These globulins are prepared by repeatedly immunizing animals or persons with the antigen. Blood is then collected and the globulins are obtained.

hypersensitivity   An immune response that causes an individual to overreact to the presence of an antigen, resulting in an allergic condition.

hypersensitivity   Referring to the immune system overreacting to a specific antigen; also called *allergy*.

I

immune complex   An antibody-antigen complex that is usually eliminated from the body by phagocytic cells.

immunity   All of the mechanisms used by the body as protection against microorganisms and other foreign agents.

immunoprophylaxis   Prevention of disease through treatment with a vaccine or antiserum.

inanimate   Not capable of self-movement; usually, but not always, nonliving. Inanimate objects involved in disease transmissions are called *fomites*.

inapparent infection   A subclinical infection. The immune response is triggered in most inapparent infections but disease is not evident.

inclusion bodies   Intracellular inclusion often found in the cell as a consequence of infection. These inclusions are usually masses of developing microorganisms being formed within the parasitized cell.

index case   First known case of an infectious disease.

infection   The growth of a microorganism on or in a host.

infectious disease   A disease, such as tuberculosis or chicken pox, caused by a living organism, which in many cases can be transmitted from person to person.

inguinal   Referring to the body region near the junction of trunk and thighs.

initiation complex   An arrangement of the 30S ribosome subunit, initiator tRNA, and mRNA in a configuration such that the 50S ribosome can attach, making the synthesis of protein possible.

innate immunity   Inborn or natural immunity; mechanisms of resistance to infection that are not acquired after birth.

inoculation   A process whereby microorganisms are placed into or on culture media.

inorganic   A material lacking carbon, such as, compounds consisting of sulfur, nitrogen, or phosphorus.

interferon   A chemical substance produced by cells in response to an infection by most viruses.

intermediate host   A host in which the parasite goes through asexual reproduction. The intermediate host is often an essential step in the life cycle of the parasite.

intoxication   The ingestion of a microbial toxin that causes a disease.

ionic concentration   The concentration of ions in a solution.

isotonic electrolytes   Common metabolic ions (Na^+, K^+, Cl^-) in the same concentration as found inside the cell.

J

jaundice   A symptom due to an increase in bilirubin in the blood. Jaundiced patients often develop yellow-brown skin color.

K

keratin   An insoluble protein found in skin epithelium, hair, and nails; sometimes used to refer to skin.

keratoconjunctivitis   An infection of the eye surfaces that leads to inflammation of the conjunctiva as well as the cornea of the eye.

Krebs cycle   A series of reactions where pyruvic acid is completely oxidized and hydrogen is transferred to an appropriate carrier molecule, such as NAD.

L

lancefield system   A classification of streptococci based on the carbohydrate antigens present in the wall of the cell.

ligature   Suture; material (often silk) used to close an open wound. Commonly referred to as "stitches."

light chain   An antibody polypeptide of lower molecular weight.

lipid   A hydrophobic substance, such as a fat, phospholipid, and steroid.

lumen   Opening, often used to refer to the hollow space in a tube-like structure.

lymph   A fluid collected from tissues of the body and transported in lymphatic ducts to the venous blood.

lymphadenopathy   Enlargement of the lymph nodes.

lymphokine   A cytokine secreted by lymphocytes.

lymphotropic   Having a preference for lymphoid cells.

lysogenic   The ability of a bacteriophage to integrate its DNA into the bacterial chromosome. Under these conditions the cell does not produce new bacterial viruses, but may carry out functions under direction of the virus DNA.

lytic cycle   The series of steps by which a phage replicates and bursts (lyses) the host cell.

M

M-protein   A major virulence factor of streptococci. Located on the cell surface, this antigen can be used to separate group A streptococci into over 80 serotypes.

macrophage   A white blood cell found in tissues that phagocytizes foreign material.

macules   Reddish-pink skin spots.

major histocompatibility complex (MHC)   Cell surface proteins that are essential to self-recognition and immune responses.

megakaryocyte   Giant multinucleated cell from which platelets arise.

memory B cell   A population of long-lived B cells that act in a secondary antibody response.

meninges   Membrane coverings of the brain and spinal cord.

meningitis   An inflammation of the lining of the brain or spinal cord.

merozoite   The asexually reproducing form of the malarial parasite. The merozoite is infective for human red blood cells.

metabolic   Related to the chemical processes occurring within a cell; often associated with energy-producing and energy-using processes.

metabolism   The sum of all chemical reactions going on in a cell or microorganism.

methicillin-resistant   Resistant to methicillin, a penicillinase-resistant antibiotic. Methicillin-resistant organisms create a difficult treatment problem because of the limited number of antibiotics that can be selected from for use.

microbial contamination   The addition, usually unintended, of viable bacteria to a previously microbial-free environment, or the addition of unwanted organisms to an environment that contains known or desirable microorganisms.

microplate   A plastic plate containing up to 96 separate wells.

microtubule   A protein tubule that forms part of the structure of eukaryotic flagella and cilia.

miliary tuberculosis   A condition in which the bacilli are widely disseminated throughout the lung. Each of these bacilli serve as the focus for the development of a tubercle.

molecular weight   The sum of the atomic weights of the atoms in a molecule, or compound.

monoclonal antibody   A population of antibodies with identical variable regions.

mucus   A thick secretion produced by mucous cells that covers mucous membranes. Mucus contains a polysaccharide called *mucin* along with a variety of salts.

multiresistant   Bacteria that are resistant to a variety of antibiotics with different mechanisms of antimicrobial action.

mutagen   A physical or chemical agent that increases the rate of mutation.

mutant   A cell or organism containing a mutation.

mutation   Any change in the normal DNA base sequence.

myalgia   Pain in the muscles.

mycology   The study of fungi.

mycotic   Referring to a fungal infection or disease.

myonecrosis   Death and destruction of muscle cells. Myonecrosis is a serious clinical condition characteristic of gas gangrene and is fatal without proper rapid treatment.

N

natural killer cell   A lymphocytes that can recognize and destroy microbe-infected and cancer cells.

necrosis   The death and subsequent destruction of cells; usually appearing as darkened or ulcerated tissue.

negative staining   Using dye to stain the background but not the cells.

Negri bodies   Cytoplasmic inclusions in brain cells infected with the rabies virus.

neurologic   Pertaining to the nervous system, either central (brain and spinal cord; CNS) or peripheral (nerves extending from the CNS to other body tissues).

neuropathy   A disorder that affects the nervous system.

neurotropic   Having a preference for the nervous system.

normal microbial flora   Microorganisms normally found on or in the body that do not cause disease in the normal host.

nosocomial   Relating to a hospital.

nucleotide   A nitrogenous compound found in DNA and RNA as well as in a variety of coenzymes (NAD) and ATP.

numerical aperture   A measure of the quantity of light entering a lens.

nutrients   Ingredients used by a living organism to facilitate growth, including carbon and energy sources as well as essential vitamins or minerals.

O

obligate intracellular parasite   Organism that depend on a living host for their nutrition, growth, and development.

occult   Hidden or unknown.

oral rehydration therapy (ORT)   The supplying of water, glucose and electrolytes to compensate for their loss due to cholera diarrhea.

organelle   An anatomically distinct structure common to eukaryotic cells.

organic   Composed of carbon and hydrogen atoms.

ornithosis   A disease associated with birds from the *Ornithine* family.

osmotic pressure   Hydrostatic pressure on a biological membrane created by unqual concentrations of solute molecules on each side of the membrane.

osteomyelitis   Microbial infection of bone.

oxidase   An enzyme used as one identification test for bacteria; negative means the bacterium does not produce a respiratory enzyme, positive means it does.

oxidized   The condition of a molecule to which electrons (hydrogens) have been removed.

P

pandemic   A worldwide epidemic.

paralytic polio   Infection with poliovirus resulting in paralysis in the patient.

parasite A microorganism that derives nutrients from its host.

parenteral Given into the body by injection or through a catheter. Many medicines cannot be given orally (by mouth) because they taste bad, are destroyed by stomach acid, or are not absorbed from the intestine.

parotid gland Salivary gland located at the back of the jaw.

passive immunity A condition in which antibodies are passed from one host to another.

pasteurization The process of using mild heat to kill pathogens.

pathogen A microorganism capable of causing disease in a host.

pathogenic Referring to the ability to cause disease.

pathogenicity The ability to cause disease.

pelvic inflammatory disease (PID) A serious intra-abdominal infection in females. PID usually follows vaginal or uterine infection, and is most often caused by one or more of the sexually transmitted disease agents. Reproductive sterility is a common outcome, but in some instances death may occur from PID.

pentamer Having five sides or five subunits.

peplomer A protrusion from the surface of a virus.

peptide fragment A part of a protein.

peptidoglycan A large, complex molecule primarily made up of sugar derivatives and amino acids that forms the rigid structural portion of the bacterial cell wall.

perioperative Near the time of an operation.

periplasmic space The region between the cytoplasmic membrane and the outer membrane where many catabolic reactions occur in gram-negative bacteria.

Peyer's patches Lymphoid tissue the lining of the small intestine.

phagocytosis A process by which cells ingest particulate matter (such as microbes) from their environment. When this process is carried out by specific host cells it provides a valuable defense against infection.

pharyngitis Infection of the pharynx. The pharynx is the upper portion of the throat. This disease is usually accompanied by enlarged lymph nodes, erythema, and pain.

phenetic Based on the visible features of organisms and their apparent ability to modify their environment.

phenotype The observable characteristics of an organism.

phospholipid An important structural molecule in all biological membranes.

phosphorylation The addition of phosphorus to a molecule. This process is usually accompanied by a transfer of a relatively large amount of energy.

photosynthetic Capable of using light as a source of energy for an organism.

phylogenetic Referring to the evolutionary relatedness of a group of organisms.

pinocytosis A process analogous to phagocytosis where the ingested material is a liquid or very small particle.

plaque An area of cell destruction caused by the propagation of a single virus in a monolayer of cells.

plasma cell An antibody-secreting cell, derived from B cell, clonal expansion.

plasmid A small, circular DNA molecule found in some bacteria that is independent of the bacterial chromosome.

plasmid Circular, extrachromosomal DNA that can be transferred between cells. Plasmids replicate independently of chromosomal DNA and provide genetic information that will be expressed in addition to that of the chromosome.

pleomorphic Having many morphological shapes. Pleomorphic bacilli often appear as long, drawn-out cells or as short coccobacilli. These figures are often found in the same microscopic field.

pneumonia A condition in which a fluid exudate collects in the air spaces of the lung.

point mutation A mutation where one base has been substituted by another.

point source A single source from which dissemination of an infectious agent occurs.

polymer A relatively large molecule composed of repeating subunit molecules (monomers).

polymicrobic Produced by more than one species of bacteria at the same time. Anaerobic infections are commonly polymicrobic.

polymorphic Having many different varieties.

polysaccharide A very large sugar molecule, composed of smaller sugars units.

polyvalent Antisera containing antibodies against more than one epitope or antigen.

posttussive After coughing.

predisposing factors Factors making one susceptible to an infection or disease.

primary pneumonia Pneumonia that occurs as a result of infection without previous compromise to host defenses.

primary response The response to a first exposure of a host to an antigen.

primate An order of upright, bipedal mammals of which monkeys are an example.

prodrome An early phase of infection leading to disease. The prodromal phase of disease is usually not associated with specific diagnostic symptoms, but may include a variety of nonspecific host changes.

promoter The sequence of bases on the DNA template that signals the start point for transcription.

prophage The phage DNA integrated into the bacterial chromosome.

prophylactic treatment A preventive treatment against infection and disease.

prophylaxis Use of an antimicrobial to prevent infection from occurring.

prosthetic An artificial or manufactured substitute for a failed body structure, such as an artificial knee joint or heart valve. Other long-term artificial materials that may be implanted for a variety of reasons are also referred to as prosthetic devices.

prostration Exhaustion and loss of strength.

protein receptor Membrane-bound proteins on the external surface of cells that recognize specific antigenic determinants.

protein synthesis   The process of transcription and translation by which the information on the DNA is made into proteins.

proton motive force   The state of a membrane resulting from the transport of protons by the action of an electron transport system.

provirus   A complete viral genome that has been integrated into the host genome.

pure culture   A culture containing only a single species of microorganism.

purulent   Associated with the production of pus.

pustule   A small blister that is filled with pus.

pyoderma   Infection of the skin by a pus-producing (pyogenic) bacterium.

Q

quarantine   Separation of humans or another animal from the general population when these individuals have or have been exposed to a communicable disease.

R

reactivation   Development of an active, spreading infection in an individual in whom a disease had been latent after an earlier infection.

reduced   The condition of a molecule to which electrons (hydrogen) have been added.

reduviid (triatomine) bug   A small biting bug.

replication   The process by which a cell produces an exact replica of its chromosome.

reservoir   A site where microbes persist and retain their ability for infection.

Reye's syndrome   A syndrome of organ changes that may occur following a viral infection. Most noticeable changes occur in mental status, and liver enlargement.

ribonucleic acid (RNA)   One of the two kinds of nucleic acid.

ribonucleotide reductase   An enzyme needed to synthesize deoxyribonucleotides for DNA snythesis.

ribosomal RNA (rRNA)   The type of RNA used, along with proteins, to build the structure of ribosomes.

ribosome   A ribonucleic acid-protein complex used by cells for the synthesis of new protein.

S

sanitarium   A facility designed to care for persons and help them regain their health. In a tuberculosis sanitariums patients were given healthful meals and were kept as free from stress as possible. Surgical and chemotherapeutic treatments were also available at some sanitariums.

sanitation   Any cleaning technique that physically removes microorganisms.

sebaceous gland   A skin structure that secretes oily substances.

secondary infection   Infections that occur as a consequence of other disease processes in the host.

secondary response   An immune response to second or later exposure of a host to an antigen.

selective toxicity   The quality possessed by a substance that can damage or destroy a living organism in the presence of another organism that remains unaffected.

seminal fluid   A secretory fluid that carries sperm.

sensitivity   The capacity of a diagnostic test to detect all positive specimens. A highly sensitive test would not have any false negative reactions.

septicemia   The presence of bacteria in the blood.

septum   A crosswall or membrane between compartments. Molds are classified into two groups by the presence or absence of septa.

sequela   A condition or illness directly relating to an earlier condition but developing some time after the first illness.

serotype   A microorganism with a unique antigen such that it induces antibody specific for that organism.

sexual reproduction   Reproduction through the union of sexual structures (or gametes) that produces sexual spores.

Shiga toxin   A powerful toxin produced by *S. dysenteriae* that acts on tissues of the central nervous system.

shingles   A severe vesicular eruption resulting from latent viral infection of the dorsal root ganglia by the varicella-zoster virus.

shock   A condition that is often life-threatening in which there is a sharp decrease in blood pressure accompanied by depression of a number of physiologic parameters.

simple staining   Adding a single dye to cells, followed by rinsing off the slide to remove excess dye.

sinus   An opening or space in a tissue. Infectious processes often produce a nonhealing sinus that may connect between tissues or even to the outside.

sinusoid   A small open cavity through which body fluids travel. Sinusoids greatly increase the space available for fluid, causing the fluid to go slowly through these areas.

skin testing   Application of an antigen directly to the skin or intradermally to determine the host immune status in relation to the antigen. A positive reaction depends on cell-mediated immunity and is a delayed hypersensitivity response.

somnolent   Drowsy or sleepy.

specificity   The capacity of a diagnostic test to detect only truly positive specimens. A highly specific test would not have any false positive reactions.

sporadic   Occurring at irregular intervals. Outbreaks or occurrences of sporadic infections are very difficult to predict.

spore   A reproductive unit resulting from sexual reproduction.

sputum   Expectorated matter, such as mucus, derived from the lower respiratory tract.

staphylococcal colitis   An inflammation of the intestine (colon) caused by *Staphylococcus aureus*.

stationary phase   A period during which the number of new cells produced is equal to the number that die. This is not a static condition, and microbes continue to metabolize, multiply, and die during this phase.

sterile The state of being free of microorganisms.

sterilization A process that destroys all living organisms.

sternotomy Incision into or through the sternum.

STS Serologic tests for syphilis.

subclinical Occurrence of infection with such mild symptoms that no diagnostic signs are evident. Antibody responses are usually as strong as with clinical infections.

sulfhydryl Referring to a sulfur bonded to a hydrogen (-SH) in a molecule.

superantigen An antigen that stimulates an extreme immune response.

superinfection A second infection that develops in addition to a previous infection; an infection that occurs as a result of antibiotic treatment.

suppuration An accumulation of white blood cells resulting in the formation of pus.

symbiosis the co-habitation of two different types of organisms.

synaptic Referring to the region where the nerve impulse in a neuron is transmitted to the responsive tissue (either another neuron or a receptor cell). These impulses are transmitted in only one direction-neuron to tissue.

syndrome A number of symptoms occurring together that characterize a specific disease.

synergistically Capable of working together. Two organisms are synergistic when they are able to produce a host response greater than the sum of the effects they produce when acting alone.

systemic Referring to a bodywide infection.

T

T cell receptor A protein receptor on T cells that is specific for a foreign peptide fragment/MHC complex.

T lymphocytes A type of white blood cell involved in an acquired immune response.

T-cytotoxic cell A T-lymphocyte that can destroy infected or abnormal host cells.

T-helper cell A T-lymphocyte that interacts with macrophages and B cells and to initiate an immune response.

temperate Referring to a bacteriophage that does not cause immediate lysis of the infected bacterial cell.

template A DNA molecule or sequence used as the pattern for making a new nucleotide polymer during replication or transcription.

terminator A sequence of bases on the DNA template that signals the end of transcription.

thymidine kinase An enzyme needed to produce the DNA nucleotide thymidine phosphate.

thymus An endocrine gland located behind the breastbone near the throat. It is large in childhood and decreases in size in adults.

tincture An alcoholic solution; a tincture of iodine is a solution of alcohol that contains iodine.

tinea Worm-like. Cutaneous mycotic infections are often called "ringworm." This name derives from the concentric circles formed in the skin of some infected patients.

titer The concentration of specific antibody present in the serum of an individual.

topical Applied locally, such as to the surface of the skin.

total magnification The magnification of the objective lens multiplied by the magnification of the ocular (eyepiece).

toxemia The presence of toxins in the blood.

toxin A poisonous substance produced by some microorganisms.

toxoid A modified toxin that has lost its toxic properties but retains its antigenic properties.

transaminase A normal tissue enzyme that can transfer an amino group from one amino acid to another. Increased levels of serum transaminase indicate tissue injury.

transcription RNA synthesis; the process by which information encoded on the DNA is copied onto an RNA molecule.

transformation The morphological changes that occur to some virus infected cells that may make them cancerous.

transovarian passage A process whereby eggs are infected by an etiologic agent present in the female. When the eggs hatch, the insect is infected and capable of transmitting the agent to a new host or, if female, to her own eggs.

trophozoite An active, vegative stage in the life cycle of a protozan.

tularemia A zoonotic disease of rabbits caused by *Francisella tularensis*.

U

ubiquitous Present everywhere.

ulcer An area of tissue erosion. Skin ulcers are typical of a variety of diseases such as syphillis and tularemia.

ultracentrifugation A technique to separate particles or cell organelles using very high-speed centrifugation.

undulant fever One of several common names applied to the disease *brucellosis*. *Bang's disease* is also commonly used to refer to infection by the *Brucella*.

undulating membrane An external membrane found on the surface of some protozoa. This membrane may be of various sizes and appears to be capable of wave-like movement.

urethra The tube through which urine passes from the bladder to the outside of the body during urination.

V

vaccination A process by which small amounts of infective material, or material similar to that which is infective, is introduced into individuals to increase their resistance to disease; sometimes used in the general sense to refer to the process of immunization.

vaccine A preparation of killed or weakened microorganisms (or their products) that can be used to immunize against disease.

vegetative Referring to the nonspore stage of a bacterium.

vehicle Any object or substance that can carry microorganisms from one host to another.

vesicle A blister-like structure that contains a clear serous fluid.

viable Living or capable of reproducing.

virulence factor A structural or physiological character that enables a microbe to cause infection and disease.

W

wavelength The distance between crests of a light wave.

Z

zoonotic Referring to a disease of animals that is transmitted to humans.

Color Plate Reference

1. Algal Bloom

2. Antibiotic disc inhibition of growth on agar plates

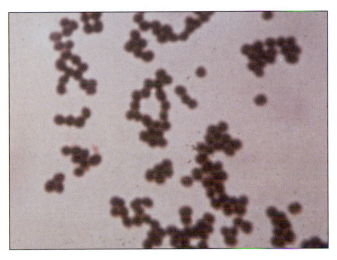

3. Gram stain of staphylococcus from culture/gram positive

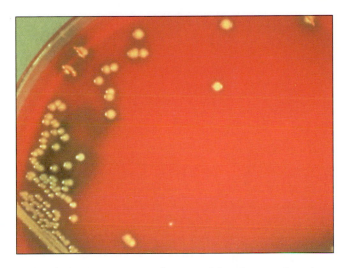

4. *Staphylococcus aures* culture on blood agar showing beta hemolysis and gold color

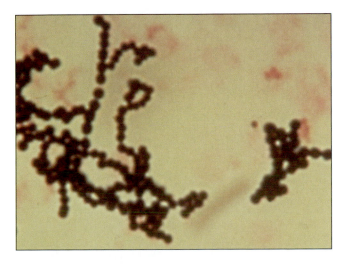

5. Gram positive *Streptococcus pyogenes*, chains

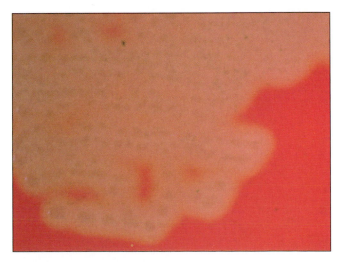

6. Bacitracin disc inhibition of group A *Streptococcus pyogenes*—showing beta hemolysis

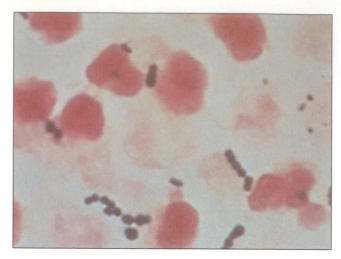

7. Gram stain of throat swab from patient with strep throat—short chains, white blood

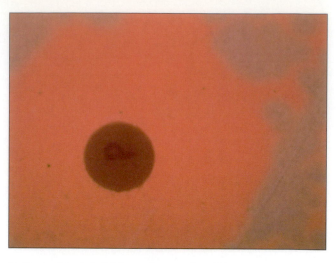

8. Optochin inhibition of growth of *Streptococcus pneumoniae* showing alpha hemolysis

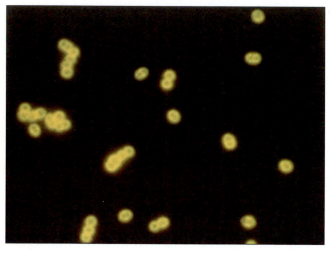

9. Fluorescent antibody stained group B streptococcus

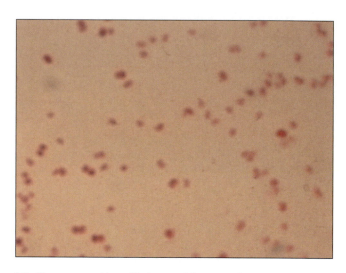

10. Gram negative diplococci from culture

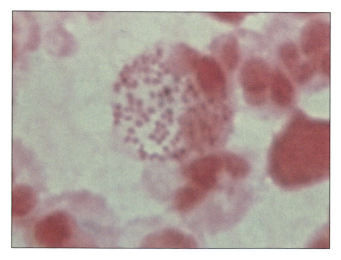

11. Intracellular *Neisseria gonorrhoeae* from patient with gonorrhea showing large numers of WBCs

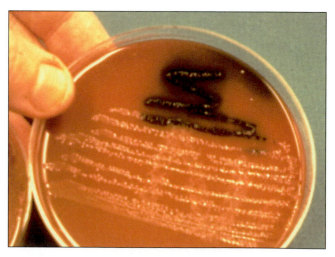

12. *Neisseria gonorrhoeae* cultured on chocolate agar

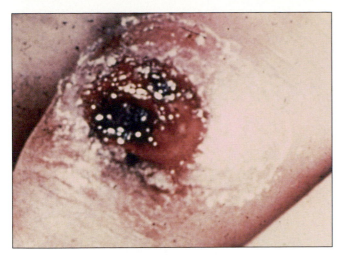

13. Primary syphilis—chancre on finger

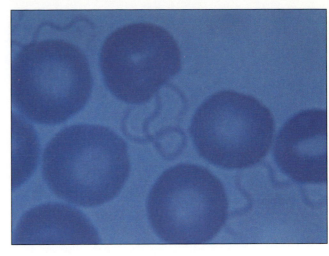

14. Blood smear showing *Borrelia recurrentis* spirochete

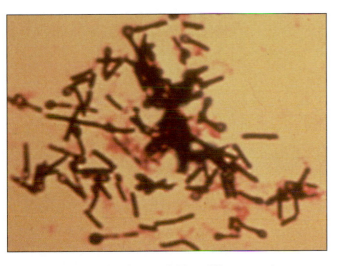

15. Gram stain of culture of *Clostridium tetani*

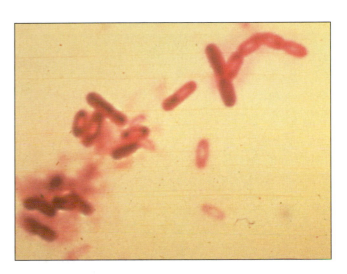

16. Gram stain of *Bacillus* with large endospores

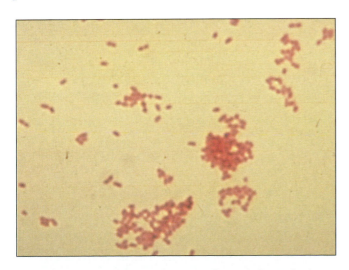

17. Gram stain of gram negative *Bacteriodes melanino-genicus*, an anaerobic bacterium

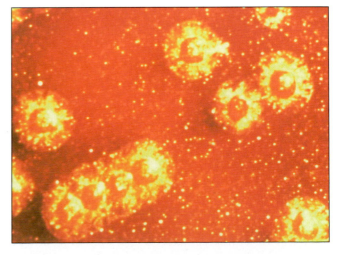

18. Colonies of *Clostridium botulinum*; colonies are umbonate

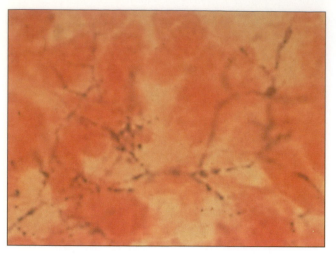

19. Gram stain of sputum from patient with pulmonary infection due to *Nocardia*

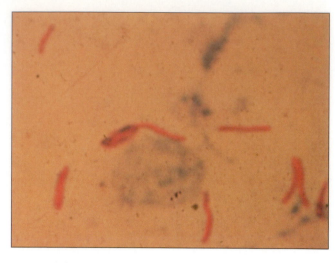

20. Acid fast stain of *Mycobacterium tuberculosis* from sputum specimen

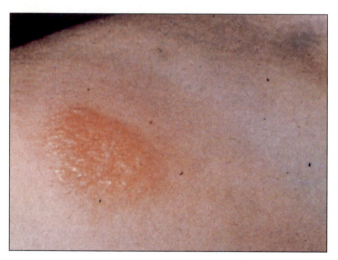

21. Positive skin test on person infected with *Mycobacterium tuberculosis*

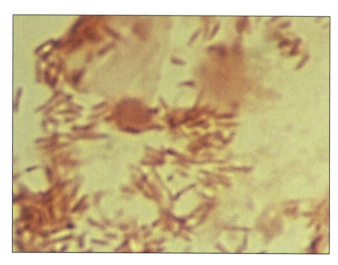

22. Gram negative *Escherichia coli* from urine of patient with cystitis

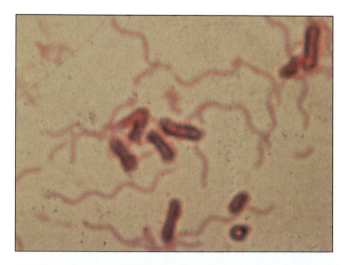

23. Flagella stain of *Salmonella typhi*

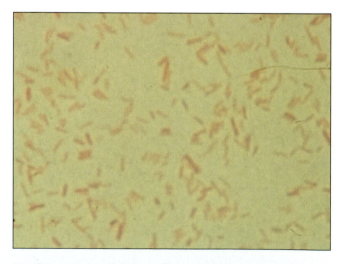

24. Gram stain of *Salmonella typhi* from culture

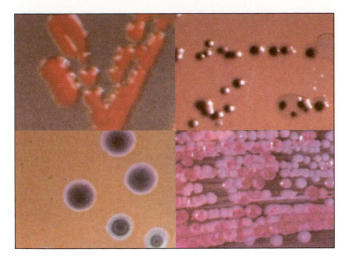

25. Colonies of several species of *Enterobacteriaceae* grown on differential agars

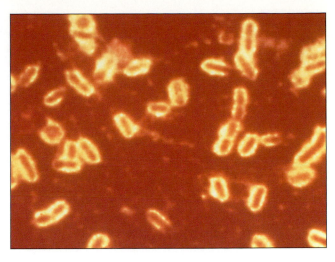

26. Fluorescent antibody stain of *Yersinea pestis*

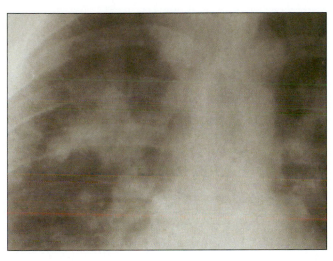

27. Chest X-ray of patient infected with *Yersinea pestis*

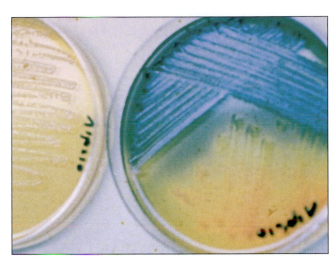

28. *Vibrio choloerae* grown on TCBS and Mac Conkey agar

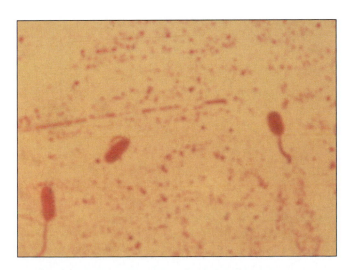

29. Intestinal biopsy showing *Vibrio cholerae*

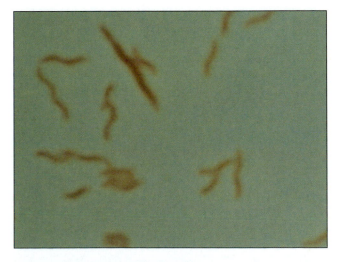

30. Gram stain of cholera vibrio

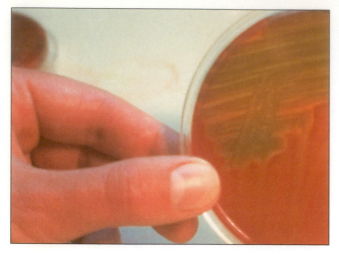

31. Culture of *Pseudomonas aeruginosa* on blood agar showing pyocyanin pigment and haemolysis

32. Pearl drops colonies of *Bordetella pertussis* on blood agar

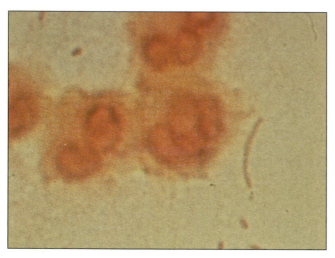

33. Gram stain of cerebral spinal fluid from patient with meningitis due to *Haemophilus influenzae*

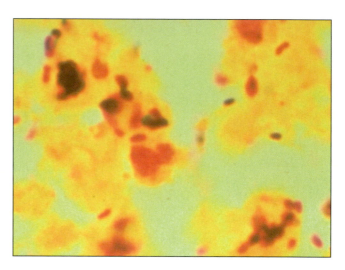

34. Silver stain of lung impression smear from patient with pneumonia due to *Legionella pneumophila*

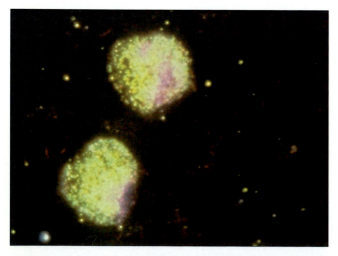

35. Immunofluorescence stain of cells infected by *C. trachomatis*

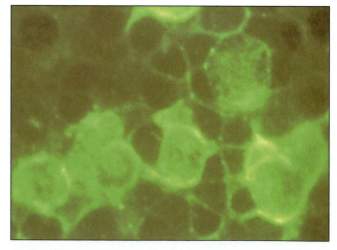

36. Immunofluorescense of cells infected by HSV-1

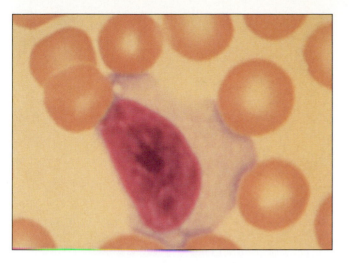

37. Blood smear from patient with infectious mononucleosis showing abnormal monocyte

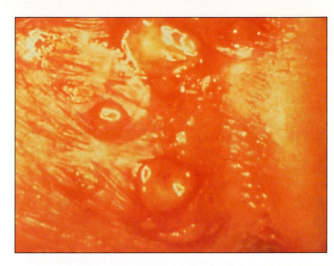

38. Genital herpes lesions

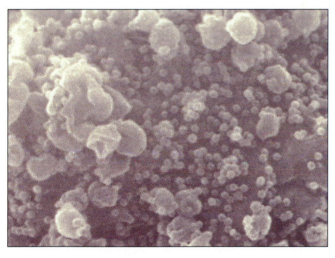

39. HIV-infected lymphocyte

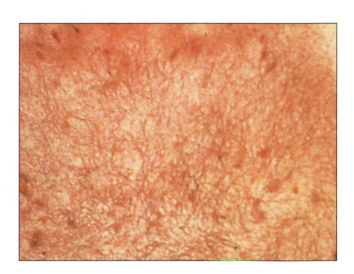

40. Kaposi's sarcoma in AIDS patient

41. Colonies of molds on agar

42. *Microsporium canis*—dermatophyte culture

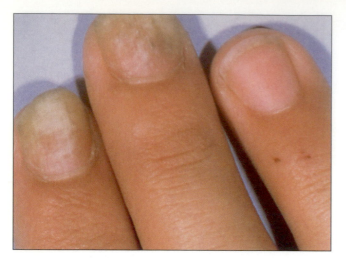
43. Fungus infected fingernails

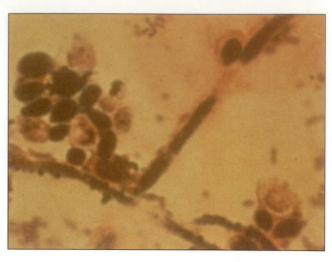
44. Sputum from patient with *Candida* pneumonia showing yeast cells and pseudohyphae

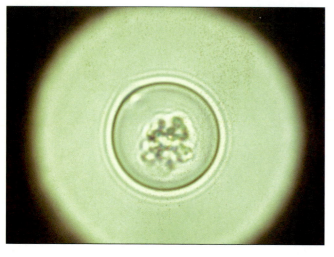
45. Single cell of *Cryptococcus neoformans* showing large capsule

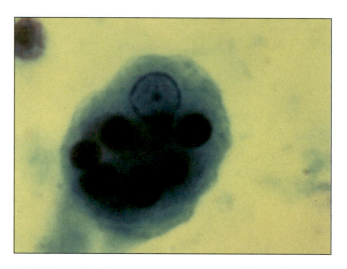
46. *Endamoeba histolytica* containing red blood cells

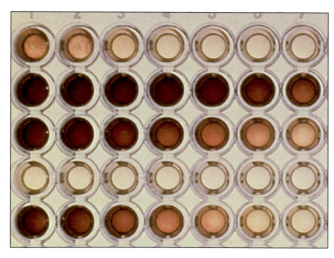
47. Enzyme linked immunoassay for detection of antibody to *E. histolytica* in patients serum

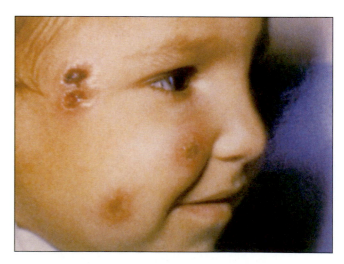
48. Leishmaniasis—cutaneous, face

Index

Panum, Peter, 4
Parainfluenza viruses, 425–26, 434
Paramyxoviruses, 425–34
Parasite, 188
Parenteral administration, 128
Parvoviruses, 477
Passive immunity, 173–74
Pasteur, Louis, 7, 8, 16, 22, 118, 449
Pasteurella multocida, 16, 337–38
Pasteurization, 7, 22, 112
Pasteur treatment, 449
Pathogen, 188
Pathogenic bacteria, 143
Pathogenicity, 188, 194–97
Pelvic inflammatory disease (PID), 231, 262, 356
Penems, 134
Penicillin, 10, 127–28, 220
Pentamers, 160
Pentose phosphate pathway, 55
Peptide bond, 60
Peptide fragment, 155
Peptidoglycan, 38
Perforin, 164
Periplasmic space, 53
Pertussis, 328–31
Petroff-Hausser chamber, 100–101
Phage, 77
  typing, 206
Phagocytosis, 19, 146–47
Phagolysosome, 46, 157
Phagosome, 147
Pharyngitis, 211–12, 269
Pharyngoconjunctival fever, 387, 389
Phase-contrast microscopy, 28–29
Phenetic/phenotypic classification, 14
Phenol/phenolics, 118
Phenol coefficient method, 121
Phenotype, 74
Phili, 41
Phospholipids, 37
Phosphorylation, 54
Photoautotrops, 51, 88
Photosynthesis, 17, 51
Phylogenetic classification, 14
Phytoplankton, 18
Picornaviruses 437–43
Pigbel, 260
Pinocytosis, 374
Pinta, 244
Plague, 2, 300
  bacteria, 339
  clinical summary table, 343
  diagnosis, 341
  epidemic, 338–39
  human, 339, 342
  pathogenesis and clinical diseases, 339–40

pneumonic, 340
prevention and control, 342
septicemic, 339
sylvatic, 339
transmission and epidemiology, 340
treatment, 341
urban/domestic, 339
Plankton, 18
Plasma cells, 159
Plasma membrane, 36
Plasmids, 42, 75, 76
Plasmodium, 502–5
Plate counts, 101–2
Platelets, defense mechanism role, 148
Pleurisy, 218
Pleuropneumonia, 347
Pneumocystis carinii, 508–9
Pneumonia, 204, 427
  AIDS and, 455
  Chlamydia, 359
  Klebsiella, 301
  Mycoplasma, 345, 347–48, 352
  Staphylococcus, 203
  Streptococcus, 216–21
Point mutations, 74–75
Polioviruses, 439–42, 443
Polyenes, 134
Polymers, 53, 58, 59
Polymicrobic, 262
Polymorphic, 155
Polymorphonuclear leukocytes (PMNs), 146
Polymyxins, 132–33
Polypeptides, 60–61
Polyribosomes, 42, 74
Polysaccharides, 38s, 58
Pontiac fever, 349
Porins, 39
Postpolio syndrome, 439
Poxviruses, 388–90, 391
Precipitation tests, 177–79
Prevost, Isaac, 2
Primary response, 160–61
Prions, 21
Prokaryotes, 19–20
Prokaryotic/procaryotic cells
  cell envelope structures, 36–40
  components of, 36
  cytoplasm, 41–43
  difference between eukaryotic cells and, 13–14, 44
  endospores, 43–44
  external appendages, 40–41
  sizes and shapes, 35–36
Promoter site, 71
Prontosil, 11
Prophage, 77
Prophylactic treatment, 220

Prophylaxis, 132
Protein(s)
  membrane transport, 37
  molecular structure of, 59–61
  receptors, 155
  regulatory, 157
  structure and functions of, 58–62
  synthesis, 69–70
Proteus, 301
Proton motive force, 58
Protozoa, 19
  diseases, 496–509
  general characteristics, 493–96
Providencia, 301
Prowazek, S. von, 361
Pseudomembranous colitis, 131, 260
Pseudomonas
  aeruginosa, 121, 307–8, 310
  cepacia, 307, 309
  mallei, 307
  maltophelia, 307
  pseudomallei, 307
Psittacosis, 358
Psychrophiles, 98
Puerperal fever, 4, 213
Pure cultures, 91–92
Pyoderma, 214
Pyruvic acid, 55–56

Q
Q-fever, 363–64, 365
Quinolones, 133–34

R
Rabies, 7, 445
  clinical summary table, 451
  diagnosis, 448–49
  pathogenesis and clinical diseases, 446–47
  prevention and control, 449–51
  transmission and epidemiology, 447–48
  treatment, 449
Radiation, 113–14
Radioimmunoassay (RIA), 181
Reagin test, 243
Recombination/recombinant DNA technology, 24, 76, 81
Redi, Francesco, 6
Reduced state, 51
Relapsing fever, 245–47, 250
Reoviruses, 472–74, 478
Replication, 67–68
Resistance plasmids/R factors, 136
Resolving power, 26
Respiration
  aerobic, 55–58
  anaerobic, 56–57
  defined, 56

DISEASES AND THE ORGANISMS THAT CAUSE THEM (CONTINUED)

PROTOZOAL DISEASES

Disease	Organism	Disease	Organism
Acanthamoeba keratitis	Acanthamoeba sp.	cryptosporidiosis	Cryptosporidium sp.
African sleeping sickness (trypanosomiasis)	Trypanosoma brucei gambiense and T. brucei rhodesiense	giardiasis	Giardia intestinalis
		leishmaniasis	Leishmania sp.
amoebic dysentery	Entamoeba histolytica	malaria	Plasmodium sp.
babesiosis	Babesia microti	Pneumocystis pneumonia	Pneumocystis carinii
balantidiasis	Balantidium coli	toxoplasmosis	Toxoplasma gondii
Chagas' disease	Trypanosoma cruzi	trichomoniasis	Trichomonas vaginalis

NOTIFIABLE DISEASES

Disease	1994	1993	1992	1991	1990	1989	1988	1987	1986	1985
AIDS	78,279	103,533	45,472	43,672	41,595	33,722	31,001	21,070	12,932	8,249
Amebiasis	2,983	2,970	2,942	2,989	3,328	3,217	2,860	3,123	3,532	4,433
Anthrax	----	----	1	----	----	----	2	1	----	----
Aseptic meningitis	8,932	12,848	12,223	14,526	11,852	10,274	7,234	1,487	11,374	10,619
Botulism, total (including wound and unsp.)	143	97	91	114	92	89	84	82	109	122
Foodborne	50	27	21	27	23	23	28	17	23	49
Infant	85		65	66	81	65	60	50	59	79 70
Brucellosis	119	120	105	104	85	95	96	129	106	153
Chancroid	773	1,399	1,886	3,476	4,212	4,692	5,001	4,998	3,756	2,067
Cholera	39	18	103	26	6	---	8	6	23	4
Diphtheria	2	---	4	5	4	3	2	3	---	3
Encephalitis, primary 717	919	774	1,021	1,341	981	882	1,418	1,302	1,376	
post-infectious	143	170	129	82	105	88	121	121	124	161
Escherichia coli O157:H7	1,420	---	---	---	---	*	---	---	---	---
Gonorrhea	418,068	439,673	501,409	620,478	690,169	733,151	719,536	780,905	900,868	911,419
Granuloma Inguinale	3	19	6	29	97	7	11	22	61	44
Haemophilus influenza invasive	1,174	1,419	1,412	2,764	---	---	*	---	---	---
Hansen disease (leprosy)	136	187	172	154	198	163	184	238	270	361
Hepatitis A	29,796	24,238	23,112	24,378	31,441	35,821	28,507	25,280	23,430+	23,210+
Hepatitis B	12,517	13,361	16,126	18,003	21,102	23,419	23,177	25,916	26,107+	26,611+
Hepatitis, C/non-A, non-Bss	4,470	4,786	6,010	3,582	2,553	2,529	2,619	2,999	3,634+	4,184+
Hepatitis, unspecified	444	627	884	1,260	1,671	2,306	2,470	3,102	3,940+	5,517+
Legionellosis	1,615	1,280	1,339	1,317	1,370	1,190	1,085	1,038	948	830
Leptospirosis	38	51	54	58	77	93	54	43	41	57
Lyme disease	13,043	8,257	9,895	9,465	---	---	*	---	---	---
Lymphogranuloma venereum	235	285	302	471	277	189	185	303	396	226
Malaria	1,229	1,411	1,087	1,278	1,292	1,277	1,099	944	1,123	1,049
Measles (rubeola)	963	312	2,237	9,643	27,786	18,193	3,396	3,655	6,282	2,822
Meningococcal disease	2,886	2,637	2,134	2,130	2,451	2,727	2,964	2,930	2,594	2,479